The Viscera of the Domestic Mammals

R. Nickel · A. Schummer · E. Seiferle

The Viscera of the Domestic Mammals

Second revised edition

by August Schummer, Richard Nickel
and Wolfgang Otto Sack

With 559 illustrations, some in color, in the text and on 13 plates

1979

Springer-Verlag
Berlin Heidelberg GmbH 1979

This work is an authorized translation and revision of R. Nickel, A. Schummer, E. Seiferle, (Ed.), *Lehrbuch der Anatomie der Haustiere (Textbook of the Anatomy of Domestic Animals)*, Volume II: *Eingeweide (The Viscera of the Domestic Mammals)* by A. Schummer and R. Nickel, 4th Edition, © 1979. Verlag Paul Parey, Berlin und Hamburg, Germany.

Richard Nickel †, Dr. med. vet., Professor and Head of the Department of Anatomy, Tieraerztliche Hochschule Hannover, D-3000 Hannover, Germany

August Schummer †, Dr. med. vet., Professor and Head of the Department of Veterinary Anatomy, Justus-Liebig-Universitaet Giessen, D-6300 Giessen, Germany

Eugen Seiferle, Dr. med. vet., Dr. med. vet. h.c., Professor and Head of the Department of Veterinary Anatomy, Universitaet Zurich, CH-8057 Zurich, Switzerland

Wolfgang Otto Sack, D.V.M., Ph. D., Dr. med. vet., Professor of Anatomy, New York State Veterinary College, Cornell University, Ithaca, N.Y. 14850 USA.

Synopsis of the English edition: *Textbook of the Anatomy of the Domestic Animals*

Volume I: *Locomotor System of the Domestic Mammals.* By R. Nickel, A. Schummer, E. Seiferle, J. Frewein and K.-H. Wille. Translation from the German. Approx. 560 pages, with about 517 illustrations in the text and on 11 colour plates. In preparation.

Volume II: *The Viscera of the Domestic Mammals.* By A. Schummer, R. Nickel and W. O. Sack. 2nd revised edition. Translated and revised from the 4th German edition. 1979. 446 pages, with a total of 559 illustrations in the text and on 13 colour plates.

Volume III: *Circulatory System, Skin and Skin Organs of the Domestic Mammals.* By A. Schummer, H. Wilkens, B. Vollmerhaus and K.-H. Habermehl. Translated from the German by W. G. Siller and P. L. A. Wight. 1980. Approx. 662 pages, with a total of about 439 illustrations, about 172 in colour. In preparation.

Volume IV: *Nervous System, Sensory Organs, Endocrine Glands of the Domestic Mammals.* By E. Seiferle. Translation from the German. Approx. 442 pages, with a total of about 250 illustrations, about 95 in colour, in the text and on 10 colour plates. In preparation.

Volume V: *Anatomy of the Domestic Birds.* By A. Schummer. Translated from the German by W. G. Siller and P. A. L. Wight. 1977. 214 pages, with 141 illustrations in the text and on 7 colour plates.

Synopsis of the German edition: *Lehrbuch der Anatomie der Haustiere*

Volume I: *Bewegungsapparat.* By R. Nickel, A. Schummer, E. Seiferle, J. Frewein and K.-H. Wille. 4th revised edition. 1977. 560 pages, with a total of 517 illustrations in the text and on 11 colour plates.

Volume II: *Eingeweide.* By A. Schummer and R. Nickel. 4th edition. 1979. 446 pages, with a total of 559 illustrations in the text and on 13 colour plates.

Volume III: *Kreislaufsystem, Haut und Hautorgane.* By A. Schummer, H. Wilkens, B. Vollmerhaus and K.-H. Habermehl. 1976. 662 pages, with a total of 439 illustrations, 172 in colour.

Volume IV: *Nervensystem, Sinnesorgane, Endokrine Drüsen.* By E. Seiferle. 1975. 442 pages, with a total of 250 illustrations, 95 in colour in the text and on 10 colour plates.

Volume V: *Anatomie der Hausvögel.* By A. Schummer. 1973. 215 pages, with a total of 141 illustrations in the text and on 7 colour plates.

CIP-Kurztitelaufnahme der Deutschen Bibliothek

Nickel, Richard:
The viscera of the domestic mammals / R. Nickel ; A. Schummer ; E. Seiferle. Transl. and revision by Wolfgang Otto Sack. – 2., rev. ed. / by August Schummer and Richard Nickel.

Dt. Ausg. u.d.T.: Nickel, Richard: Lehrbuch der Anatomie der Haustiere. Bd. 2. Eingeweide.
ISBN 978-1-4757-6816-9 ISBN 978-1-4757-6814-5 (eBook)
DOI 10.1007/978-1-4757-6814-5

NE: Schummer, August:; Seiferle, Eugen:

1st. Ed. 1973
ISBN 978-1-4757-6816-9

Originally published by Springer-Verlag in 1979
Softcover reprint of the hardcover 2nd edition 1979

ISBN 978-1-4757-6816-9

Dedicated with Admiration and Gratitude to

PAUL MARTIN

PAUL MARTIN (1861—1937), Professor, Dr. med. vet., Dr. phil. h.c., Dr. med. vet. h.c.
Veterinary Anatomist from 1886 to 1901 in Zurich, and from 1901 to 1928 in Giessen

Preface to the Second English Edition

Since the publication of this book in 1973, its principal author, Professor AUGUST SCHUMMER, former Head of the Veterinary Anatomy Department, Justus Liebig-Universität, Giessen, West Germany, has passed away. Over the years an amiable understanding as between colleagues dedicated to the same scientific discipline developed between us. This was especially true during the years when I was translating this volume, which was perhaps his most cherished work. This understanding seems to have been flavored with paternal feelings on his part. The loss of this man, whose advice was often sought and whose help was freely given, and of this kind of friendship is deeply regretted by the writer, as it is by many others who knew him well.

During the last six years the remaining two volumes of the originally planned five-volume set of the *Lehrbuch der Anatomie der Haustiere* (Textbook of the Anatomy of the Domestic Animals) was published. Volume V (Anatomy of the Domestic Birds) — also principally SCHUMMER'S contribution — has since 1977 been available in English. An English edition of Volume III (Circulatory System and Integument) will be published next year. Volumes I and IV (Musculoskeletal and Nervous Systems) will follow in due course.

The second edition of this translation, in addition to numerous editorial, typographical and some nomenclatorial corrections, includes two substantive changes. A section on the erection of the ruminant penis has been added, and the pathways by which the spermatozoa leave the testis for the ducts in the epididymis have been clarified.

The writer is grateful, particularly to Professor ROBERT HABEL, former Head of the Veterinary Anatomy Department at the New York State College of Veterinary Medicine, Cornell University; to his colleagues in the same department; and to the many reviewers of the first edition for their valuable suggestions, all of which have been considered in the revision of this book for the second edition.

In keeping with the goal of providing as complete and as modern a gross-anatomical bibliography as possible, about fifty recent references to the veterinary anatomical and appropriate clinical literature have been added.

Ithaca, New York, January 1979 W. O. SACK

Preface to the First English Edition

Soon after the first two volumes of the *Lehrbuch der Anatomie der Haustiere* by R. NICKEL, A. SCHUMMER, and E. SEIFERLE were published in German, inquiries were made by persons in various countries about the possibility of having this textbook translated into other languages. Therefore, the publisher decided to produce an English edition limited at first to Volume II.

The concept and plan of the original German work by AUGUST SCHUMMER and the late RICHARD NICKEL has been preserved in this translation, and what was said in the prefaces to the first and second German editions about the purpose and scope of the book applies equally to this first English edition. The work deals with the body cavities, digestive system and teeth, spleen, and with the respiratory and urogenital systems of the dog, cat, pig, ox, sheep, goat, and horse. Each organ system is described in a general and comparative chapter, which is followed by shorter special chapters for the carnivores, pig, ruminants, and horse.

In agreement with the original authors, substantive changes were made in several instances to take into account the results of recent research and to eliminate conflicts between views commonly held by German anatomists and those outside of Europe, but foremost to profit by the advances in *Nomina anatomica veterinaria** (NAV), a uniform international nomenclature, which came into existence while this translation was in progress. This nomenclature lists a single, usually descriptive term for homologous structures in all domestic mammals, and wherever possible for the same structure in man; and thus has the potential of simplifying student instruction and promoting interdisciplinary understanding. The work of the International Committee on Veterinary Anatomical Nomenclature in many instances included re-evaluations of existing anatomical concepts; and it was these that necessitated most of the changes in the present work.

The nomenclature conforms, with very few exceptions, to the second edition of the NAV. In keeping with the textbook character of the present work, most of the official Latin terms have been translated to accepted English equivalents. Only where the Latin differed greatly from the English, or where it enhanced understanding, has the official term been added parenthetically.

In this edition, a bibliography has been added at the end of each of the major sections, and an attempt has been made to compile citations of the more recent studies dealing with the gross anatomy of the viscera of the domestic mammals. Dr. K. H. Wille of the Department of Veterinary Anatomy, Justus-Liebig-Universitaet Giessen has gathered most of the predominantly German references.

The work on the English edition was begun at the Ontario Veterinary College, Guelph, Ontario, Canada, and completed at the New York State Veterinary College, Cornell University, Ithaca, New York. At the latter institution generous financial support was received for the project through Dr. G. C. Poppensiek, Dean of the Veterinary Faculty, from the General Research Support Grant of the National Institutes of Health, U.S. Department of Health, Education and Welfare. This made it possible to defray typing and editorial expenses and to obtain the valuable assistance of Mrs. Antoinette M. Wilkinson, Ph.D., who greatly enhanced the idiomatic quality of the manuscript.

It is a pleasure to acknowledge the interest and continuous support given this project by Prof. Robert E. Habel, Head of the Department of Anatomy, New York State Veterinary College, and the friendly help received from the other members of his department. Thanks are due to Prof. Stephen J. Roberts, former Chairman of the Department of Large Animal Medicine, Obstetrics and Surgery, New York State Veterinary College, for his advice, often sought, on clinical questions; to Prof. C. J. G. Wensing, Department of Anatomy, Faculty of Veterinary Medicine at the University in Utrecht, Holland, for reading and commenting on the section on the Descent of the Testis and for contributing a figure; and to the many persons who in the past several years helped with typing, editing, and proof reading.

The Paul Parey Publishing Company, particularly its co-owner Dr. Friedrich Georgi, is to be commended for undertaking to publish this English edition, and for its traditionally excellent and careful production.

It is hoped that this work will be as well received as the original German edition, and that it will contribute to filling the need for a modern, comprehensive textbook of Veterinary Anatomy.

Ithaca, New York, January 1973 W. O. Sack

* Schaller, O., R. E. Habel and J. Frewein, Editors. *Nomina anatomica veterinaria,*, 2nd. ed. Vienna, International Committee on Veterinary Anatomical Nomenclature, 1972.

Preface to the Second German Edition

The favorable reception given again the second volume of our Textbook of the Anatomy of Domestic Animals and the need for a second edition of this volume after the relatively short interval of six years, strengthens our conviction that we have produced a book useful especially to students of veterinary medicine.

The second edition differs from the first only in that minor corrections were made in the text and in some references to figures. In this connection we thank Dr. W. O. Sack, Associate Professor, New York State Veterinary College at Cornell University, for his valuable suggestions. We also thank Dr. K. H. Wille, Assistant at the Department of Veterinary Anatomy, Justus-Liebig-Universität Giessen, for his suggestions and technical assistance during the preparation of the manuscript for this edition.

No changes have been made in nomenclature*. It was decided to delay such a revision until the work of the International Commission on Veterinary Anatomical Nomenclature is completed and full agreement has been reached on the terms to be used in the future.

Unfortunately, our friend and colleague, Richard Nickel, did not live to see the new edition of the second volume which he helped to write with much enthusiasm and great skill.

Again, sincere thanks are due to the Paul Parey Publishing Company for their excellent production of this edition of the second volume. We hope this book serves the purpose intended by the authors and meets the needs of the readers.

Zürich and Giessen, April 1967

Eugen Seiferle August Schummer

* In the present first English edition the nomenclature conforms to the NAV (See also footnote to *Preface to the First English Edition*).

Preface to the First German Edition

The second volume of our Textbook of the Anatomy of Domestic Animals follows the first volume after a longer interval than anticipated, because of the extremely time-consuming preparation, especially in connection with the illustrations, and the ever-increasing administrative burdens of the authors.

As announced in the preface to the first volume, this volume, which is written by A. SCHUMMER and R. NICKEL, deals with the viscera of the digestive, respiratory and urogenital systems. In keeping with the basic plan of the work, each organ system and its principal functions are described first in a general and comparative chapter, followed by more detailed descriptions in shorter special chapters for the carnivores, pig, ruminants, and horse. The viscera are given especially extensive treatment, both in text and illustrations, because an accurate knowledge of the viscera of domestic mammals is fundamental to all branches of Veterinary Medicine.

For good reason, an anatomy textbook is judged also by the quality and instructiveness of its illustrations. The 480 new illustrations presented here, many of which were carefully selected to show topographical relationships, are testimony to the skill and perception of our medical artists Miss VALERIE GUBE of the Department of Veterinary Anatomy, Justus Liebig Universitaet Giessen, and WALTER HEINEMANN and GERHARD KAPITZKE of the Department of Anatomy, Tieraerztliche Hochschule Hannover. We are also grateful to Dr. DIETMAR HEGNER for drawing several illustrations.

We thank especially our faithful co-workers Dr. BERND VOLLMERHAUS, assistant at the Department of Veterinary Anatomy, Justus-Liebig-Universität Giessen, and Dr. HELMUT WILKENS, prosector at the Department of Veterinary Anatomy, Tierärztliche Hochschule Hannover, who with great skill saw to the labeling of the illustrations, prepared the legends, and helped in many ways with the preparation of the manuscript. We are indebted to Dr. habil. KARLHEINZ HABERMEHL, Dr. KLAUS LOEFFLER, Dr. RUDOLF SCHWARZ, and HEINZ KOLBE for their assistance in reading the proofs.

Sincere thanks are due to the PAUL PAREY Publishing Company, especially to Mr. FRIEDRICH GEOGRI, co-owner of the company, for his interest in the work, his understanding of the authors' intentions, and for the personal attention to this volume also, giving it their traditionally excellent and careful production.

We hope that the second volume of the Textbook of the Anatomy of Domestic Animals will find as wide acceptance as the first.

Zürich, Hannover, and Giessen, in the Fall of 1959.

EUGEN SEIFERLE RICHARD NICKEL AUGUST SCHUMMER

Contents

Spleen

List of Abbreviations

(The last letter of the abbreviation is duplicated to indicate the plural.)

a.	= arteria, artery	l.n.	= lymphonodus, lymph node
caud.	= caudalis, caudal	longitud.	= longitudinalis, longitudinal
com.	= communis, common	m.	= musculus, muscle
cran.	= cranialis, cranial	maj.	= major
dext.	= dexter, right	med.	= medialis, medial
dors.	= dorsalis, dorsal	min.	= minor
duct.	= ductus, duct	n.	= nervus, nerve
ext.	= externus, external	proc.	= processus, process
for.	= foramen	prof.	= profundus, deep
gl.	= glandula, gland	sin.	= sinister, left
int.	= internus, interna	supf.	= superficialis, superficial
lam.	= lamina	transv.	= transversus, transverse
lat.	= lateralis, lateral	v.	= vena, vein
lig.	= ligamentum, ligament	ventr.	= ventralis, ventral

References to Figures

These appear in parentheses in the text, mostly in this form: (36/*a*). The number preceding the oblique dash refers to the figure; that which follows the oblique dash in italics refers to the part so labeled in that figure. Thus the notation (36/*a*) refers to part *a* in Figure 36. However, the notation (36, 37, 38/*a*) refers to part *a* in Figures 36, 37, and 38; whereas the notation (39; 40; 41/*a*) refers to Figures 39, 40, and part *a* in Figure 41, that is, part *a* refers only to Figure 41. Numerals and letters in italics but not preceded by a number, such as (*a*, *2*), refer to parts in the figure whose number was last quoted.

Color Plates

Figures printed in color are combined on Plates and, with a few exceptions, are numbered consecutively with the text figures. They are distributed as follows:

Fig. 132	= Plate I	next to p. 96
Fig. 144—147	= Plate II and III	next to p. 104
Fig. 162 and 163	= Plate IV	next to p. 120
Fig. 292—303	= Plate V—IX	next to p. 224
Fig. 351—353	= Plate X—XII	next to p. 240
Fig. 505—512	= Plate XIII	next to p. 352

Introduction

The viscera* of the body include the digestive, respiratory, urinary, and genital organs.

1. The **DIGESTIVE ORGANS** are concerned with the nutrition of the animal. This function includes the prehension of food, its mastication, digestion, and absorption, and the initial storage of the nutrients released during digestion. The digestive organs also provide for the expulsion of the unabsorbed portion of the food, and of those substances that are added to the digestive tract by its large accessory glands.

2. The **RESPIRATORY ORGANS** provide for the exchange of gases between the blood and the atmosphere, and produce the voice.

3. The **URINARY ORGANS**, notably the kidneys, eliminate fluid wastes and foreign substances from the blood, and regulate the water and salt metabolism of the body.

4. The **GENITAL ORGANS** are concerned with reproduction. Except for the production of the germ cells, the male and female organs have different functions to perform and consequently differ markedly in their morphology.

These four organ systems are closely related functionally to the **BLOOD VASCULAR** and **LYMPHATIC SYSTEMS,** to the **NERVOUS SYSTEM** which controls their functioning, and to the **SYSTEM OF ENDOCRINE GLANDS.**

Most of the viscera are contained in the large body cavities of the trunk. Some of them, however, are embedded in the tissues of the head, neck, and in the caudal part of the pelvis, where special cavities for them do not develop. The viscera occupying the body cavities are covered with the same serous membrane that lines the cavities, and are separated from one another and from the walls of the cavities by narrow capillary spaces filled with serous fluid. Their loose attachment to the walls of the cavities allows them a certain amount of mobility.

All viscera have either a lumen or an internal duct system with which they communicate either directly or indirectly with the outside, through the **mouth, nose, anus,** or the **urogenital openings,** as the case may be.

---

* *Viscera* (L.) is the plural of *viscus*, organ.

Body Cavities

(1, 5—7, 15—17)

With the evolution of the **diaphragm** (1/*7—10*) in premammalian forms, the general body cavity has come to be divided into a smaller, cranial **THORACIC CAVITY** (*a*) and a larger, caudal **ABDOMINAL CAVITY** (*b*), which is continuous caudally with the **PELVIC CAVITY** (*c*). The body wall surrounding these cavities consists of the integument or skin, followed by a double layer of fascia, a musculoskeletal layer, and an internal layer of fascia.

The body cavities are lined with a serous membrane, known as the parietal* pleura in the thoracic cavity and as the parietal peritoneum (2/*b*) in the abdominal cavity. The serous membrane encloses two large serous cavities. That of the thoracic cavity is the **PLEURAL CAVITY,** which is divided into right and left pleural sacs. Between the pleural sacs is the heart surrounded by its own serous **PERICARDIAL CAVITY** (6). The serous membrane lining the abdominal and pelvic cavities forms the large undivided peritoneal sac, which encloses the **PERITONEAL CAVITY.**

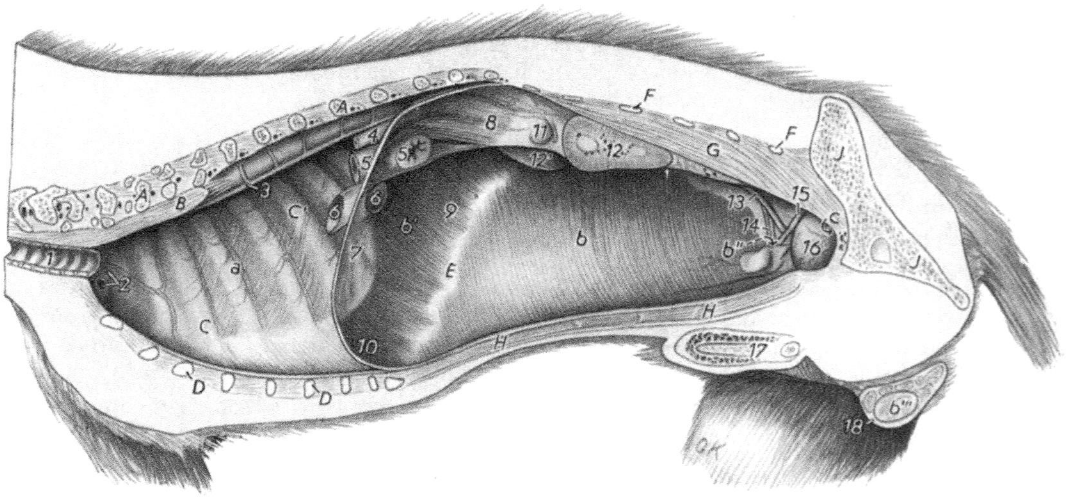

Fig. 1. Sagittal section of thoracic and abdominal cavities of a dog with the thoracic and digestive organs removed. Left aspect.

A Thoracic vertebrae and ribs; *B* Thoracic part of longus colli; *C* Right third rib; *C'* Right sixth rib; *D* Costal cartilages near sternum; *E* Right costal arch; *F* Lumbar transverse processes; *G* Psoas musculature; *H* Abdominal muscles; *J* Pelvis

a Right pleural cavity; *b* Right half of peritoneal cavity; *b'* Intrathoracic part of peritoneal cavity; *b''* Vaginal ring, entrance into tunica vaginalis; *b'''* Testis; *c* Entrance to pelvic cavity

1 Trachea; *2* Brachiocephalic trunk and brachiocephalic vein, passing through thoracic inlet; *3* Right azygous vein; *4* Aorta; *5* Esophagus, in esophageal hiatus; *6* Caudal vena cava, passing through its foramen in diaphragm; *7* Tendinous center of diaphragm; *8, 9, 10* Lumbar, costal, and sternal parts of diaphragm; *11* Lumbar aortic lymph nodes; *12* Left kidney in sagittal section; *12'* Right kidney; *13* Right ureter; *14* Testicular artery and vein; *15* Ductus deferens; *16* Urinary bladder; *17* Penis in longitudinal section; *18* Scrotum

---

* From *paries* (L.) wall, pertaining to the wall; in contrast to visceral, pertaining to the viscera.

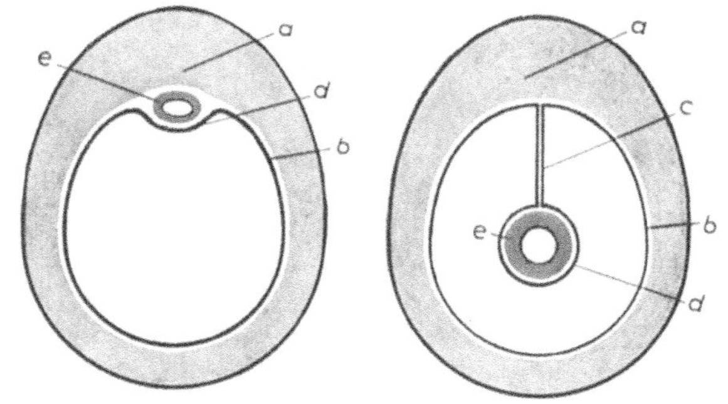

Fig. 2

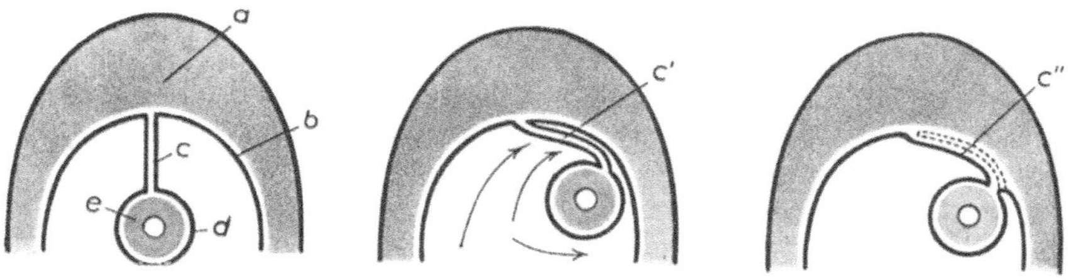

Fig. 3

Figs. 2 and 3. Development of the peritoneal coverings and mesenteries in the abdominal cavity.

a Abdominal wall; *b* Parietal peritoneum; *c* Mesentery; *c'* Mesentery, lying against parietal peritoneum; *c''* Union of mesentery with parietal peritoneum: the original parietal peritoneum in this area and one of the layers of the mesentery have disappeared, and the other layer of the mesentery has become parietal peritoneum; *d* Visceral peritoneum; *e* Wall of intestine

The organs contained in the body cavities are invested by the same serous membrane, the visceral pleura or peritoneum (3/*d*), which is continuous with that lining the walls by double-layered serosal folds known in the wider sense, but particularly with regard to the intestines, as **mesenteries** (*c*). The mesenteries contain the blood vessels, lymphatics, and nerve supply (4), and attach the organs to the body wall. The attachment, however, does not prevent the organs from moving freely in their functioning. The length and thickness of these mesenteric folds determine the degree of mobility afforded to the suspended organs, and whether they are designated as mesenteries, serosal ligaments, or serosal folds.

Somewhat simplified, the mesenteries can be thought of as having developed in the following way. An organ originates retroperitoneally, close to the body wall (2/*e left*). As it sinks into the cavity it obtains its peritoneal covering and is followed by two sheets of peritoneum which unite back to back and form its mesentery (*c*). Some organs, for example the kidneys, never leave the body wall and are covered with peritoneum only on the exposed surface. Others, after first having drawn out a mesentery, rejoin the body wall by adherence of the mesentery to the parietal peritoneum and loss of some of the serosal layers (3).

The **SEROUS MEMBRANES** (pleura or peritoneum) are thin and translucent and have a smooth, moist, and shiny surface. Histologically they consist of serous and subserous layers. The **serous layer** (tunica serosa) consists of simple squamous epithelium of mesodermal origin perforated by microscopic **stomata*** at the cell boundaries, and a thin connective tissue lamina propria. The **subserous layer** (tela subserosa) is a layer of looser connective tissue which, depending on species and nutritional state, contains varying quantities of fat.

The serous membranes release minute amounts of serumlike **serous fluid** known as pleural, pericardial, or peritoneal fluid, depending on the cavity. The fluid fills the capillary spaces between the organs and reduces friction between them (e.g., in respiratory

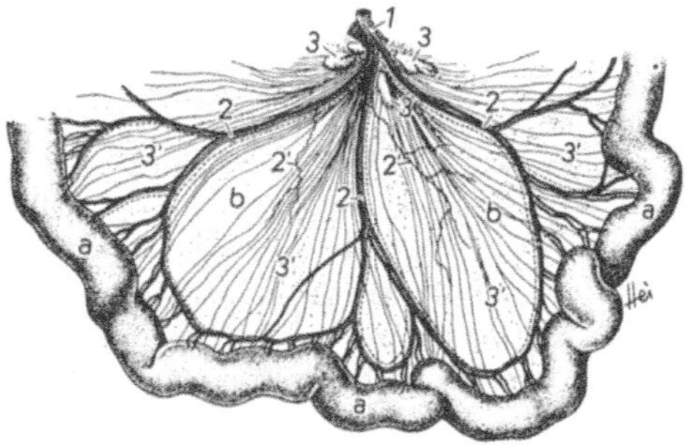

Fig. 4. Loop of jejunum with mesentery, vessels, and nerves of the horse. Semischematic.

a Jejunum; *b* Mesentery

1 Cranial mesenteric artery and vein in the root of the mesentery; *2* Jejunal arteries and veins, and nerve branches (broken lines) of the cranial mesenteric plexus; *2'* Blood vessels supplying the mesentery; *3* Cranial mesenteric lymph nodes; *3'* Lymphatics

* Plural of *stoma* (Gr.) mouth, opening.

movements of the lungs). The thoracic and abdominal organs completely fill the aforementioned cavities so that only capillary spaces are left between them. The serous membranes can resorb this fluid, and absorb fluids introduced into the body cavities from the outside, and can take up small particles that may be suspended in these fluids. When injured mechanically or by chemical or other toxic substances, the serous membranes become inflamed more readily and severely than other tissues.

Thoracic Cavity and Pleura

(1, 5—7)

The rib cage, or thorax, is a part of the skeleton and consists of the thoracic vertebrae, the ribs and their cartilages, and the sternum. It has the shape of a laterally flattened cone open at both ends; at the apex (cranially) is the small thoracic inlet, and at the base (caudally) the very wide thoracic outlet. The inlet is formed by the first thoracic vertebra, the first pair of ribs, and the manubrium sterni. The outlet is formed by the last thoracic vertebra, the last pair of ribs, the costal arch (consisting of the costal cartilages not attaching to the sternum), and the last sternebra and xiphoid process.

When the bony thorax is in situ, i.e., when the remaining components of the thoracic wall (skin, fasciae, and muscles) are present, and when it is closed caudally by the diaphragm, a cavity known as the **THORACIC CAVITY** (1/*a*) results. The thoracic cavity occupies only the cranial portion of the bony thorax. The caudal portion, the intrathoracic part of the abdominal cavity (*b'*), contains abdominal organs. The thoracic cavity, therefore, is smaller than the thorax and varies in size constantly with the respiratory movements of the ribs and diaphragm.

The **thoracic inlet** (apertura thoracis cranialis) is an important passageway for organs and vessels passing between the neck and the thoracic cavity. It is marked externally by the palpable cranial end of the sternum, and, in roughly dorsoventral sequence, contains the longus colli (*B*); esophagus; trachea (*1*); the arteries and veins supplying head, neck, forelimbs, and lateral thoracic wall (*2*); lymphatics and nerves; and in young animals the thymus. These structures are ambedded in loose connective and adipose tissue.

The **endothoracic fascia** (5—7/*d*), the internal layer of trunk fascia that lines the thoracic cavity, is a sheet of fibrous and elastic tissue attached to the deep surfaces of the ribs, intercostal muscles, sternum, and the transversus thoracis. It is reflected caudally onto the cranial surface of the diaphragm and blends with its tendinous center. The sternopericardiac and phrenicopericardiac ligaments (*g'*) detach themselves from the endothoracic fascia at the sternum and diaphragm respectively and unite with the fibrous pericardium surrounding the heart.

The **PLEURA** (5—7/*red lines*) covers the endothoracic fascia and the organs in the thoracic cavity. It is a serous membrane like the peritoneum and forms two laterally flattened semicones, the pleural sacs, each enclosing a **pleural cavity** (5/*7'*), of which the right is larger than the left. The pleura forming the lateral walls of the pleural cavities, the **costal pleura** (*e*), is applied against the ribs. Caudally, the pleura covering the diaphragm, the **diaphragmatic pleura** (*e'*), forms the bases of the cone-shaped pleural cavities. Medially, where the walls of the two pleural cavities lie back to back forming the mediastinum, the pleura is called the **mediastinal pleura** (*f*). The mediastinum is thus a sagittally placed partition consisting of two serous membranes extending from the thoracic inlet in front to the diaphragm behind, and attaching dorsally to the thoracic vertebrae and ventrally to the sternum. Between right and left mediastinal pleura is a supporting layer of connective tissue.

Inserted at about the middle of the mediastinum, and spreading the right and left mediastinal pleura far apart, is the heart with its fibrous and serous pericardial coverings (*g, h, i*). The mediastinum is thus divided into a cranial mediastinum lying cranial to the heart, a middle mediastinum which contains the heart, and a caudal mediastinum caudal to the heart. In the **cranial mediastinum** are found the thoracic part of the longus colli; part of the trachea (5/*1*); part of the esophagus (*2*); the large vessels (*3', 4*) supplying the lateral thoracic wall, forelimbs, neck, and head; the sympathetic trunks, vagi, phrenic, and recurrent nerves (*13, 14*); the cranial mediastinal lymph nodes; the end of the thoracic duct (*15*); and the

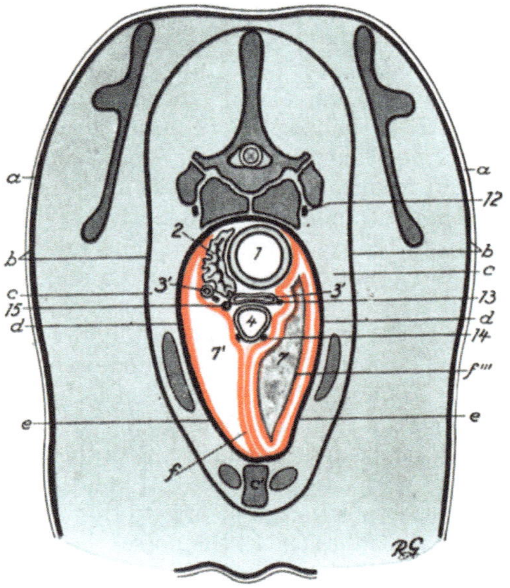

Fig. 5

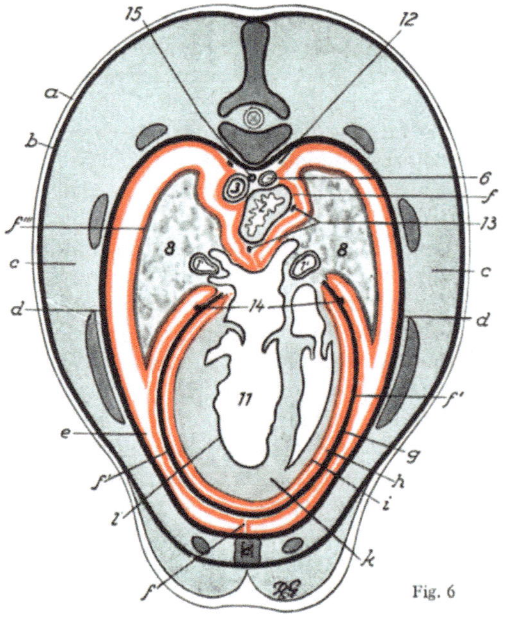

Fig. 6

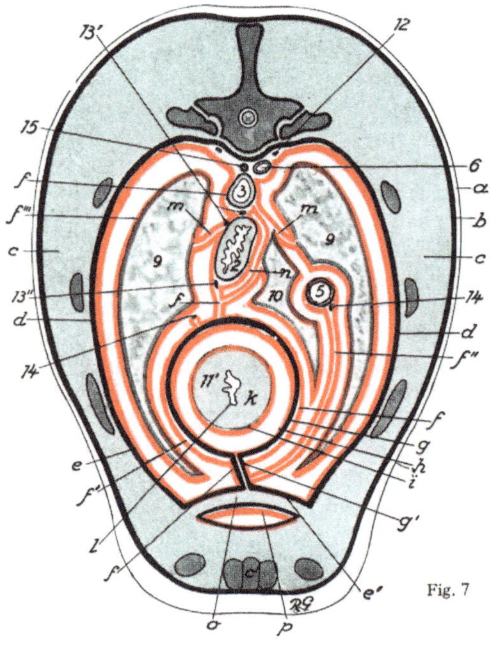

Fig. 7

Figs. 5, 6 and 7. Transverse sections of the thoracic cavity of the dog, cranial to the heart (5), through the heart (6), and caudal to the heart (7). Caudal aspect, semischematic, the serosal membranes in red. (After Zietzschmann, unpublished).

a Supf. fascia; *b* Double-layered deep fascia; *c* Musculoskeletal part of thoracic wall; *c'* Sternum; *d* Endothoracic fascia; *e* Costal pleura; *e'* Diaphragmatic pleura; *f* Mediastinal pleura; *f'* Pericardiac pleura; *f''* Plica venae cavae; *f'''* Pulmonary pleura; *g* Fibrous pericardium; *g'* Phrenicopericardiac ligament; *h* Parietal pericardium; *i* Visceral pericardium (epicardium); *k* Myocardium; *l* Endocardium; *m* Pulmonary ligament; *n* Cavum mediastini serosum; *o* Diaphragm; *p* Peritoneum

1 Trachea; *1'* Main bronchi; *2* Esophagus; *3* Descending aorta; *3'* (left) Subclavian artery; *3'* (right) Brachiocephalic trunk; *4* Cranial vena cava; *5* Caudal vena cava; *6* Right azygous vein; *7* Cranial lobe of right lung; *7'* Left pleural cavity; *8* (left) Cranial lobe of lung; *8* (right) Middle lobe of lung; *9* Caudal lobe of lung; *10* Accessory lobe of right lung; *11* Left ventricle; *11'* Apex of heart; *12* Sympathetic trunk; *13* Vagi; *13'* Radicles of dorsal vagal trunk; *13''* Ventral vagal trunk; *14* Phrenic nerve; *15* Thoracic duct

thymus in young animals. In the **middle mediastinum** are found the heart and pericardium (6/*g*, *h*, *i*), the large blood vessels at the base of the heart, parts of the trachea and esophagus (*2*), the vagi (*13*), and the phrenic nerves (*14*). In the **caudal mediastinum** are found the aorta (7/*3*), part of the esophagus (*2*), dorsal and ventral vagal trunks (*13′*, *13″*), caudal mediastinal lymph nodes, and the left phrenic nerve (*14*) in its separate serosal fold. Ventral to the aorta and to the right of the esophagus is a small, closed serosal cavity (cavum mediastini serosum, *n*), which was cut off from the omental bursa in the abdominal cavity by the developing diaphragm. It is small in the ruminants and horse, but in the dog (8/*h*) and pig, it extends forward from the diaphragm to the root of the lung, and may extend caudally through the esophageal hiatus of the diaphragm into the space between the two layers of the gastrophrenic ligament.

The lungs develop as buds of the trachea and grow laterally into the pleural cavities. They push the pleura ahead of them, and thus become invested with a serous covering, the visceral or pulmonary pleura (5—7/*f″*). Caudal to the root of each lung there is a horizontal fold of pleura, the **pulmonary ligament** (*m*), which connects the mediastinal surface of the lung with the mediastinum or, when it extends farther caudally, with the diaphragm, as in the carnivores and pig. In the ruminants, the mediastinal surface of the lungs caudal to the root adheres to the mediastinum without the interposition of pleura, so that there is only a short pulmonary ligament at the caudal end of the adhesion. In the horse, the union between lung and mediastinum is even more extensive, so that the short pulmonary ligament is at the diaphragm.

At birth the mediastinum is a complete sagittal partition between right and left pleural cavities. In the carnivores and horse, however, openings appear postnatally in the ventral part of the caudal mediastinum through which the two pleural cavities can communicate. Such openings are absent in the ox and goat, and are rare in sheep, but have been observed in the middle mediastinum of carnivores, and in the cranial mediastinum of lean sheep. Ciliga et al. (1966) found no mediastinal openings in asses but state that mules, like horses, have them. It seems from observations in the dog (v. Recum, 1977) that, although fenestrated, the mediastinum provides an effective barrier to fluids, air and infection.

The laterally flattened apices of the pleural sacs, the **cupulae pleurae,** are at the thoracic inlet; the right one, in carnivores and ruminants, projects beyond the cranial border of the first rib (by 6—7 cm. in the ox), while the left one projects beyond the cranial border of the first rib only in the carnivores.

Because of the convexity of the diaphragm, the costal pleura adjacent to the diaphragm lies against the diaphragmatic pleura, with only a narrow capillary space intervening. This space is the **costodiaphragmatic recess,** and is in full communication craniodorsally with the pleural cavity. It is opened by the caudoventral movement of the lungs during inspiration.

In the caudoventral part of the right pleural cavity is a **mediastinal recess** produced by the caudal vena cava and the serosal fold (plica venae cavae) that encloses it. The caudal vena cava (7/*5*) passes through the right pleural cavity from the foramen venae cavae in the diaphragm to the right atrium of the heart. The **plica venae cavae** (*f″*) extends from the ventral border of the vena cava to the floor of the pleural cavity and is attached cranially to the heart and caudally to the diaphragm, thus separating the mediastinal recess from the rest of the right pleural cavity. The walls of the recess are as follows: left, the caudal mediastinum proper; cranially, the pericardium; right, the plica venae cavae; and caudally, the diaphragm. The recess is open dorsally, and through the opening hangs the accessory lobe of the right lung (*10*) which fills the recess.

For the structure and function of the pleura see the section on the serous membranes on page 3.

Abdominal Cavity, Pelvic Cavity, and Peritoneum
(1, 9, 10, 15—17)

The abdomen is the part of the trunk that extends from the costal arch (1/*E*) and last rib to the linea terminalis which surrounds the entrance to the pelvic cavity. This segment of the trunk contains the **ABDOMINAL CAVITY** (*b*). The wall of the abdominal cavity is formed cranially by the diaphragm, which, because of its cranial convexity, extends a considerable distance into the thorax. Dorsally, the wall of the abdominal cavity is formed by the lumbar vertebrae and associated musculature (*F, G*); laterally and ventrally, it consists of the abdominal muscles (*H*). Caudally, it is continuous through the pelvic inlet with the pelvic cavity. The dorsal abdominal wall, or roof of the abdominal cavity, consists of: skin, superficial and deep (thoracolumbar) fasciae, the epaxial muscles (iliocostalis, longissimus, and multifidus), the lumbar vertebrae with their long transverse processes, ventral to these the hypaxial muscles (quadratus lumborum, iliopsoas, and psoas minor), and the iliac fascia. The lateral and ventral abdominal wall is attached cranially to the ribs and sternum, dorsally to the lumbar transverse processes, caudally to the pelvis; and consists of: skin, superficial and deep fasciae, a layer of several abdominal muscles, and an internal layer of fascia, known as the transverse fascia. The superficial fascia encloses the cutaneus trunci. The prominent deep fascia (tunica flava*) of the herbivores contains many yellow elastic fibers and helps to support the heavy abdominal viscera. The external and internal oblique muscles have wide aponeuroses, which unite to form the external lamina of the sheath surrounding the rectus abdominis in the ventral

Right **Left**

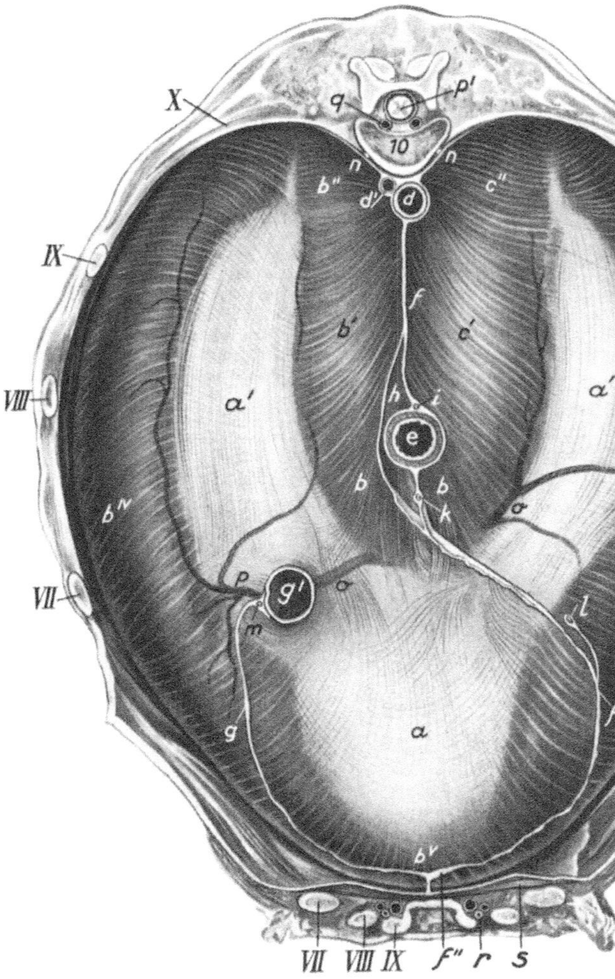

Fig. 8. Cranial surface of diaphragm of the dog. (After Zietzschmann 1936)

a, a' Tendinous center; *b, b', b''* Medial, intermediate, and lateral parts of right crus; *c', c''* Intermediate and lateral parts of left crus; *b*IV Costal part; *b*V Sternal part; *d* Aorta; *d'* Right azygous vein; *e* Esophagus, in esophageal hiatus; *f, f'* Mediastinum; *f''* Phrenicopericardiac ligament; *g* Plica venae cavae; *g'* Caudal vena cava, in foramen venae cavae; *h* Cavum mediastini serosum; *i* Dorsal vagal trunk; *k* Ventral vagal trunk; *l* Left phrenic nerve; *m* Right phrenic nerve; *n* Sympathetic trunk; *o, p* Phrenic veins; *p'* Spinal cord; *q* Int. vertebral venous plexus; *r* Int. thoracic artery and vein; *s* Transversus thoracis; *10* Tenth thoracic vertebra; *VII-X* Sections of ribs of like number

---

* L., yellow tunic.

abdominal wall. The aponeurosis of the transversus abdominis, the deepest of the abdominal muscles, forms the internal lamina of the rectus sheath.

The **PERITONEUM,** the serous membrane lining the abdominal cavity, forms a large peritoneal sac enclosing the **peritoneal cavity.** The peritoneal sac extends caudally into the pelvic cavity for distances varying with the species.

The wall of the abdominal cavity has a number of openings through which vessels and other tubular organs enter or leave. Three of them are in the diaphragm. Through the **aortic hiatus,** which is in the dorsal part of the diaphragm and flanked by the crura, passes the aorta (8/*d*). Ventral to the aortic hiatus is the **esophageal hiatus,** through which the esophagus (*e*) enters the abdominal cavity. The pleura and peritoneum on their respective sides of the diaphragm form the seal around the aorta and esophagus, which are attached loosely to the diaphragm so that a certain amount of movement is possible. The third opening, the **foramen venae cavae** (*g'*) is in the summit or vertex of the tendinous center of the diaphragm. It transmits the caudal vena cava, which is firmly anchored to the diaphragm at this point.

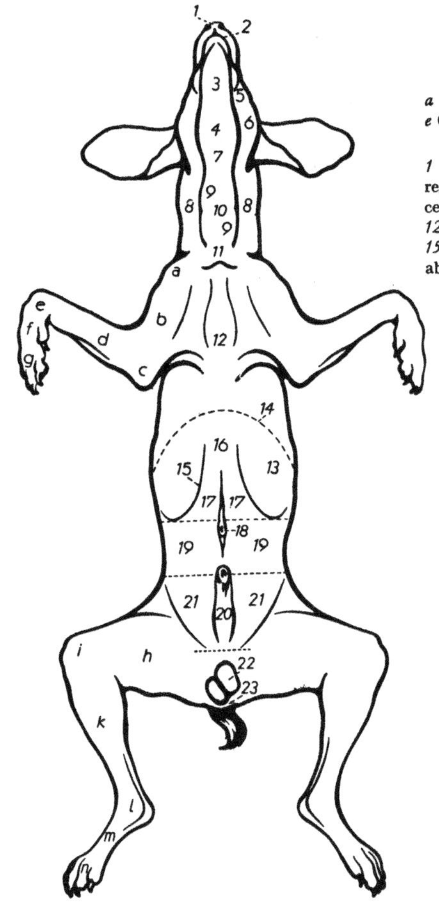

Fig. 9. Regions of the body, illustrated on the dog. Ventral aspect.

a Shoulder joint; *b* Brachial region; *c* Cubital region; *d* Antebrachial region; *e* Carpal region; *f* Metacarpal region; *g* Digits; *h* Femoral region; *i* Stifle region; *k* Crural region; *l* Tarsal region; *m* Metatarsal region; *n* Digits

1 Region of the nostrils; *2* Oral region; *3* Intermandibular region; *4* Subhyoid region; *5* Buccal region; *6* Masseteric region; *7* Laryngeal region; *8* Lateral cervical region; *9* Ventral cervical region; *10* Tracheal region; *11* Presternal region; *12* Sternal region; *13* Hypochondriac region; *14* Most cranial extent of diaphragm; *15* Costal arch; *16, 17* Xiphoid region; *18* Umbilical region; *19* Lateral abdominal region (flank); *20* Pubic region with penis; *21* Inguinal region; *22* Scrotum; *23* Perineal region

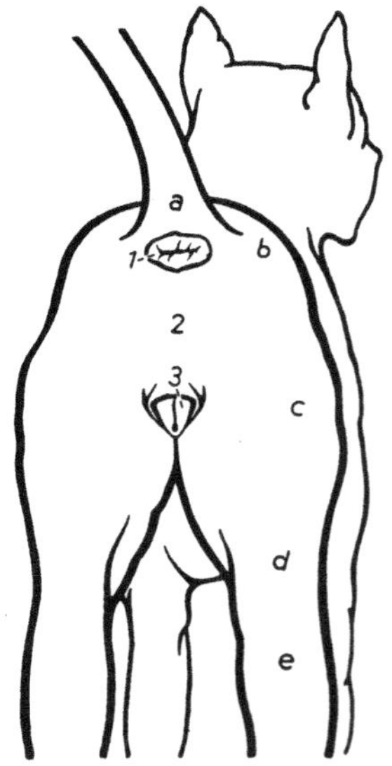

Fig. 10. Regions of the body, illustrated on the dog. Caudal aspect.

a Root of tail; *b* Gluteal region; *c* Femoral region; *d* Popliteal region; *e* Crural region

1 Anus; *2* Perineal region, between, but also surrounding, 1 and 3; *3* Vulva

In the ventral abdominal wall of the fetus and newborn is the **umbilical opening** for the umbilical vessels and the stalks of the allantois and yolk sac. The umbilical opening closes during the first few days of life, leaving a scar known as the umbilicus (9/*18*). In the inguinal area of the ventral abdominal wall (*21*) are two intermuscular spaces known as the inguinal canals. Each **inguinal canal** in the male permits a peritoneal evagination (tunica vaginalis, 493/*a*) to reach the scrotum. These evaginations are not present in females, except in the bitch.

In male animals the peritoneal cavity is closed; in female animals, it is open to the outside through the genital tract. The internal orifices of the tract are the small openings of the uterine tubes through which ova leave the peritoneal cavity and spermatozoa, and possibly also bacteria, enter it.

By far the greatest part of the abdominal cavity is occupied by the gastrointestinal tract, including its large accessory glands, the liver and pancreas (11/*f, g*); the spleen and parts of the urogenital tract occupy the rest of the cavity. The terminal portion of the intestinal tract (*p*) passes through the pelvic cavity and pierces the caudal body wall to end at the anus.

To permit accurate description of the position of an organ, or of pain, swelling, or other lesions in the abdomen, the abdominal cavity may be divided into three transverse segments, represented externally by three bandlike regions on the abdominal wall. The most cranial segment, externally the **cranial abdominal region** (9/*13, 16, 17*), extends from the diaphragm (*14*) to a transverse plane at the most caudal point on the costal arch (*15*). The diaphragm, it must be remembered, does not bulge uniformly into the thorax, but slopes cranioventrally from about the first lumbar vertebra to the caudal end of the sternum (1). Its summit, which usually contains the foramen venae cavae, is fixed at the transverse level of the sixth to seventh intercostal space, and dorsoventrally at the junction of the dorsal and middle thirds of the dorsoventral diameter of the thoracic cavity. The size of the cranial segment of the abdominal cavity depends therefore on the shape and position of the diaphragm, which in turn is related to the number and position of the asternal ribs. As a result of the dome-shaped configuration of the diaphragm, the organs in the cranial portion of this segment are covered by the lungs and thoracic wall and are therefore difficult to examine by external palpation. The middle segment, externally the **middle abdominal region** (9/*18, 19*), extends from the plane through the caudalmost point on the costal arch to the plane through the most cranial point on the tuber coxae. The caudal segment, externally the **caudal abdominal region** (*20, 21*), begins here and extends to the pelvic inlet. Each of these three segments may be divided by the median plane and by a dorsal plane at the middle of the dorsoventral abdominal diameter into right and left dorsal and right and left ventral subsegments or quadrants. The three external abdominal regions may also be subdivided into smaller fields by boundaries that are not always natural and that do not correspond to the planes that subdivide the internal segments. The cranial abdominal region is subdivided into hypochondriac and xiphoid regions. The **hypochondriac region** (*13*) is cranial to the costal arch, and as the name indicates, is over the costal cartilages. The triangular **xiphoid region** (*16, 17*) lies between the costal arches. The middle abdominal region consists of the **flank** (regio abdominis lateralis, *19*) and the **umbilical region** (*18*) on the ventral midline. In the dorsal part of the flank is the **paralumbar fossa,** which is bounded dorsally by the tips of the lumbar transverse processes, ventrally by the part of the internal abdominal oblique muscle passing from the tuber coxae to the last rib, and cranially by the last rib. The caudal abdominal region consists of a median **pubic region** (*20*), the area in front of the pubic bones; and the **inguinal region** (*21*), which extends laterally to the fold of the flank and thigh.

Because of the slope of the diaphragm, the long axis of the abdominal cavity is oblique from cranioventral to caudodorsal. The longest diameter in this direction extends from the caudal end of the sternum to the cranial end of the pelvic symphysis. The greatest dorsoventral diameter of the abdominal cavity is at about the level of the first lumbar vertebra, while the greatest transverse diameter lies between the second-last or third-last pair of ribs.

The **PELVIC CAVITY** (15—17) is enclosed dorsally by the sacrum and the first three or four caudal vertebrae, and laterally and ventrally by the ilium, ischium, and pubis, of which the latter two meet in the median pelvic symphysis. In the ungulates, the lateral wall of the pelvic cavity is formed, in addition, by the wide sacroischiatic ligament (15, 16/*D*). This ligament is represented in the dog only by a narrow strand (sacrotuberal ligament) extending from the last sacral transverse process to the tuber ischiadicum; in the cat even this is absent. Surrounding the pelvis is the gluteal, thigh, and tail musculature, which completes the wall of the pelvic cavity.

The entrance to the pelvic cavity, the **pelvic inlet** (apertura pelvis cranialis), is an osseous oval ring known as the **linea terminalis,** which consists dorsally of the base of the sacrum with its median promontorium and lateral wings, laterally of the body of the ilium, and ventrally of the pecten of the pubis, and is palpable rectally in the large domestic species. The **pelvic outlet** (apertura pelvis caudalis) is formed dorsally by the third or fourth caudal vertebra; laterally by the sacrotuberal part of the sacroischiatic ligament in the ungulates and by the sacrotuberal ligament in the dog; and ventrally by the tubera ischiadica and the arch that connects them. Except in the carnivores, the pelvic outlet is smaller than the pelvic inlet. The pelvic outlet can enlarge slightly (e.g., during parturition) which the inlet, being entirely osseous, cannot.

The pelvic cavity contains the rectum and anal canal and varying portions of the urinary bladder, the pelvic part of the urethra and the accessory genital glands in the male, and the caudal parts of the genital organs in the female (15—17). It is lined with the **pelvic fascia,** which is continuous cranially with the iliac and transverse fascia lining the abdominal cavity. The abdominal peritoneum extends into the pelvic cavity and lines its cranial part, but also invests the pelvic organs and forms the ligaments associated with them. The part of the pelvic cavity so lined is the peritoneal part; caudal to it is the retroperitoneal part, which is essentially the body wall that closes the pelvic outlet and is known as the perineum. The boundary between the two parts of the pelvic cavity is not at the same level in all domestic mammals. In the dog it is at the second caudal vertebra, in the cat at second to third, in the pig at first to second, in the ox at the first, and in the small ruminants at first to second; in the horse, however, it is at the third to fourth sacral segment.

The **PERINEUM,** when compared to the body wall of the thorax or abdomen, is complicated by the terminal part of the digestive tube and the urogenital tract that pass through it. It thus includes many muscles and fibrous structures associated with these organs that are not present in the rest of the body wall. Its principal component as regards the containment of the pelvic viscera, however, is the **pelvic diaphragm,** consisting of the levator ani and coccygeus, and a layer of internal and external fascia on each side of these muscles. In man, with his erect posture, the pelvic diaphragm supports the weight of the pelvic and to some extent also the abdominal organs, and is thus well developed. It forms a concave muscular plate, which spans the outlet of the pelvis. At its summit or most caudal part is the opening for the anal canal, and ventral to that is a space for the urogenital tract. Except when a person is reclining, the pelvic diaphragm in man carries the viscera and prevents prolapse of the anus, or of the vagina and uterus.

In quadrupeds, the pelvic organs are supported principally by the bony floor of the pelvis, with the result that the pelvic diaphragm is not as well developed as in man. Nevertheless, it functions to contain the pelvic viscera during abdominal press, i.e., during defecation, urination, copulation, during the latter part of gestation, during parturition, when abdominal viscera are abnormally full, during labored breathing, and (in draft animals) when pulling heavy loads. If it functions inadequately during such stresses, prolapse of the rectum or of the vagina and uterus may occur.

In the female there is a cutaneous bridge (526—529/*b*) between the anus and the vulva, which often tears during difficult births (perineal laceration). Deep to this cutaneous bridge, and present also in the male between anal canal and bulb of the penis, is an accumulation of fibrous and muscular tissue without lateral boundaries on which several of the perineal muscles converge. This is the **perineal body** which is part of the perineum and often tears with the skin in severe perineal lacerations.

The deep boundaries of the perineum are the structures that form the pelvic outlet. The superficial boundaries of the perineum coincide with those of the **perineal region** on the surface of the body. Dorsally and laterally, the deep boundaries are by and large the same as the superficial boundaries. Ventrally, however, the superficial boundaries extend to the base of the udder or scrotum. In the cat and pig where the scrotum is just ventral to the anus, the scrotum is included in the perineal region.

Omenta and Mesenteries

The omenta and the mesenteries are serosal sheets associated with the gastrointestinal tract in the abdominal and to a lesser extent in the pelvic cavities. The parts of the gastrointestinal tract (11) occupying these cavities are:

STOMACH	LARGE INTESTINE	LIVER
	Cecum	
SMALL INTESTINE	Colon	PANCREAS
Duodenum	Ascending colon	
Jejunum	Transverse colon	
Ileum	Descending colon	
	Rectum	
	Anal canal	

Early in its development, the gastrointestinal tract is a straight tube that extends longitudinally through the body cavity of the embryo, and is suspended from the dorsal abdominal wall by the **primitive dorsal mesentery.** A **primitive ventral mesentery** is present only cranially and extends from the lesser curvature of the gastric primordium and the first part of the duodenal primordium to the ventral abdominal wall (14*A*). With differentiation of the primitive digestive tube into the above segments, the dorsal mesentery is divided into approximately the same number of segments (right column below) as the gastrointestinal tract:

Stomach:	Dorsal mesogastrium, known as the greater omentum in the adult (the lesser omentum is derived from the ventral mesogastrium)	
Duodenum:	Mesoduodenum	
Jejunum:	Mesojejunum	Mesentery
Ileum:	Mesoileum	
Cecum:	———	
Colon:	Mesocolon	
Ascending colon:	Ascending mesocolon	
Transverse colon:	Transverse mesocolon	
Descending colon:	Descending mesocolon	
Rectum:	Mesorectum	
Anal canal:	———	

The intestinal segments suspended by the descending mesocolon and mesorectum are supplied by the caudal mesenteric artery, while all intestinal segments proximal to the descending mesocolon are supplied by the cranial mesenteric artery, which descends from the roof of the abdominal cavity in the root of the mesentery.

Because of the rotation of the developing stomach and intestines, and the uneven elongation of the various intestinal segments, changes of varying degrees of complexity take place in the initially sheetlike primitive dorsal mesentery. Some segments of the mesentery become extremely long, allowing the intestine considerable range of movement. Others fail to lengthen with fetal growth, or become shorter, with the result that the mesentery disappears and the organ becomes applied directly to the dorsal abdominal or pelvic wall. Sheets or folds of mesentery coming to lie against each other in this rearrangement of organs may adhere, with a loss of serosal surfaces, and may displace the original line of mesenteric attachment on the body wall (3/*c*, *c′*, *c″*).

The changes taking place in the dorsal and ventral mesogastrium, the forerunners of the **OMENTA,** during rotation and enlargement of the stomach primordium can only be understood when the **rotation of the stomach primordium** is understood. The stomach primordium (14*A*/*a*) is a spindle-shaped enlargement of the primitive digestive tube and, before rotation, is oriented with its long axis parallel to the long axis of the embryo. It has a convex

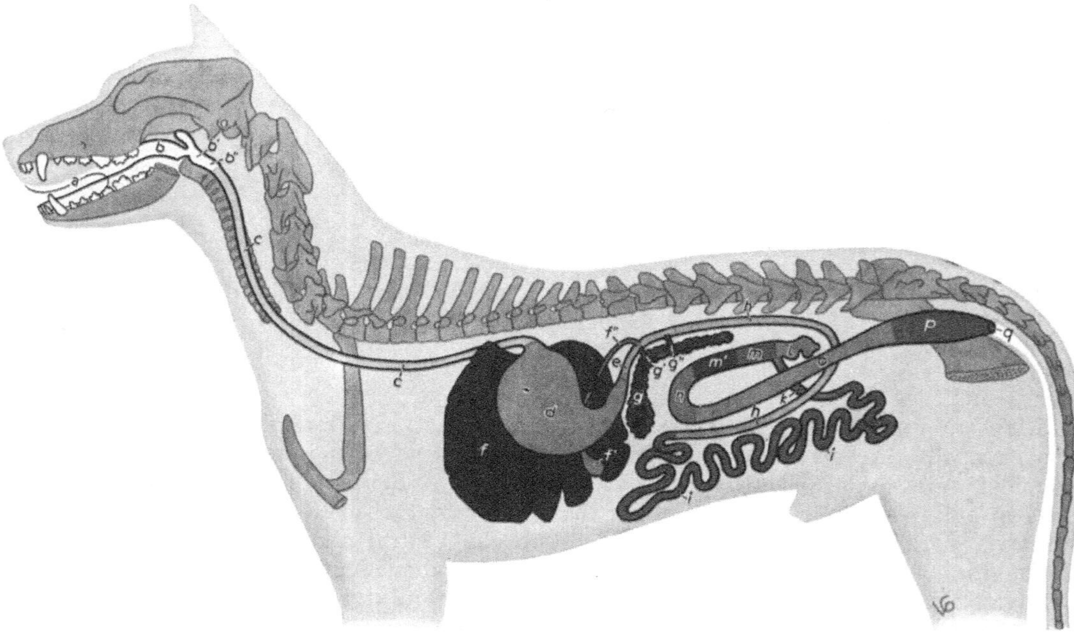

Fig. 11. Digestive system of the dog. Schematic. The salivary glands are not shown.

a Oral cavity; *b* Oropharynx; *b'*, *b''* Laryngopharynx; *c* Esophagus; *d* Stomach; *e* Cranial part of duodenum; *f* Liver; *f'* Gall bladder; *f''* Bile duct; *g* Pancreas; *g'* Major pancreatic duct; *g''* Minor pancreatic duct; *h*, *h* Descending and ascending parts of duodenum; *i* Jejunum; *k* Ileum; *l* Cecum; *m*, *m'* Ascending colon; *n* Transverse colon; *o* Descending colon; *p* Rectum; *q* Anus

greater curvature, the greater curvature of the adult stomach, which is directed dorsally, and a slightly concave lesser curvature, the lesser curvature of the adult stomach, which is directed ventrally. The dorsal mesogastrium (12*I*/*a*) is attached along the greater, and the ventral mesogastrium (*b*, *c*) along the lesser curvature of the primordium. From this original position, the simple stomach, by rotation and displacement, moves into the position seen in the adult. This rotation may be divided into two phases. The rotation around the longitudinal or craniocaudal axis is counterclockwise when viewed from a caudal position (12*I*/*B*), and brings the greater curvature around to the left, so that it is directed lateroventrally. The rotation around the dorsoventral axis is counterclockwise when viewed from a dorsal position, and brings the caudal end of the primordium over to the right, so that the greater curvature becomes directed more caudally. Thus, after rotation, the greater curvature faces to the left, caudally and ventrally, and the lesser curvature faces to the right, cranially and dorsally. During gastric rotation, the dorsal mesogastrium follows the greater curvature to the left and ventral, greatly increasing in length in the process, and forms the **greater omentum**, which lies against the visceral surface of the stomach (12*II*/*a*). The ventral mesogastrium is pulled up and to the right. The liver (*C*) develops in the ventral mesogastrium and divides it into a distal part (*c*) connecting the liver with the diaphragm, and a proximal part (*b*) connecting it with the lesser curvature of the stomach and the cranial part of the duodenum. The proximal part is the **lesser omentum** in the adult.

Fig. 12. Gastric rotation and development of the omenta. Schematic. (After Zietzschmann 1955)

I Transverse section, early stage of stomach development; *II* Sagittal section, after gastric rotation has taken place

A Spleen in dorsal mesogastrium; *B* Stomach with (*1*) greater curvature and (*2*) lesser curvature; *C* Liver in ventral mesogastrium

a Dorsal mesogastrium, greater omentum of adult; *b*, *c* Ventral mesogastrium; *b* Lesser omentum of adult, between lesser curvature of stomach and liver; *c* Hepatic ligaments, between liver and abdominal wall; *d* Caudal recess of omental bursa; *e* Vestibule of omental bursa

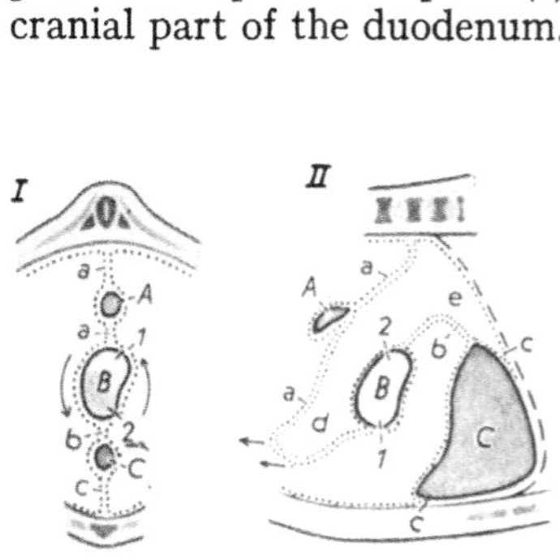

The **GREATER OMENTUM** (omentum majus)* is a serosal fold of considerable size, usually folded on itself to form a superficial wall (paries superficialis, 13/*2*) and a deep wall (paries profundus, *3*). Its lacy, netlike appearance is due to the many blood vessels and lymphatics, which course through it embedded in strands of fat. Between the finer strands of fat, the omentum is very thin and translucent (179). Small opaque patches, so-called milky spots, found in the omentum are temporary aggregations of lymphocytes, histiocytes, and other migratory cells, and are thought to be sites for the production of lymphocytes and antibodies.

Before rotation of the stomach, the line of attachment of the greater omentum (dorsal mesogastrium) along the dorsal abdominal wall begins at the esophageal hiatus ventral to the vertebral column and extends caudally in a straight line to become continuous with that of the mesoduodenum. During the rotation of the stomach and the rearrangement of the intestines, this linear arrangement becomes distorted and changes with the type of stomach, so that the greater omentum does not originate from the same area on the dorsal abdominal wall in all species (see pp. 126 [carnivores], 138 [pig], 166 [ruminants], and 183 [horse] for details).

In domestic mammals having simple stomachs, the deep wall of the greater omentum (13/*3*) extends ventrally from the region of the pancreas (*g*) on the dorsal abdominal wall, passes the visceral surface of the stomach (*b*) and turns caudally toward the pelvic inlet. Here it folds on itself ventrally and becomes the superficial wall (*2*) which runs cranially in contact with the deep wall and ends at the greater curvature of the stomach (*1*), where it blends with the visceral peritoneum of that organ. The line of attachment of the deep wall on the roof of the abdominal cavity and that of the superficial wall on the stomach, come together in the vicinity of the spleen (*h*) on the left and in the vicinity of the duodenum (*c*) on the right, forming a circle. The deep and superficial walls of the greater omentum thus enclose a potential space, the **caudal recess of the omental bursa,** the entrance of which is formed by this circular line of attachment. Access from the general peritoneal cavity (greater peritoneal sac) to the **omental bursa** (lesser peritoneal sac) is through the epiploic foramen (*11*) close to the visceral surface of the liver. The foramen opens into the **vestibule of the omental bursa** which communicates over the lesser curvature of the stomach with the caudal recess (*12*). The vestibule of the omental bursa is formed in part by the lesser omentum and is described more fully in the section on the lesser omentum below.

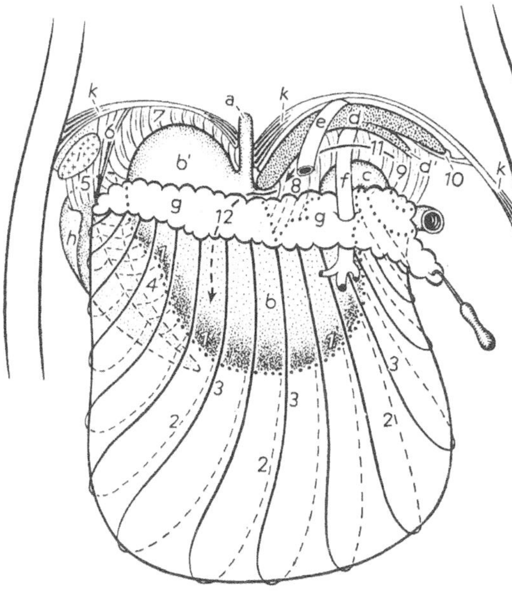

Fig. 13. Omenta and ligaments associated with the stomach of the horse. Caudodorsal aspect. Schematic. (After Zietzschmann, unpublished)

a Esophagus; *b* Body of stomach; *b'* Fundus (saccus cecus) of stomach; *c* Cranial part of duodenum; *d* Liver; *d'* Caudate process of liver; *e* Caudal vena cava; *f* Portal vein; *g* Pancreas; *h* Spleen; *i* Left kidney; *k* Diaphragm

1—7 Greater omentum; *1* Attachment of greater omentum to greater curvature of stomach; *2* Supf. wall of greater omentum; *3* Deep wall of greater omentum; *4* Gastrosplenic ligament; *5* Renosplenic ligament; *6* Phrenicosplenic ligament; *7* Gastrophrenic ligament, drawn out here by pulling stomach away from diaphragm; *8*, *9* Lesser omentum; *8* Hepatogastric ligament; *9* Hepatoduodenal ligament; *10* Right triangular ligament of liver; *11* Arrow passing through epiploic foramen into vestibule of omental bursa; *12* Arrow from vestibule to caudal recess of omental bursa

* Also *epiploon* (Gr.); hence epiploic.

In the carnivores, the greater omentum lies between the intestines and the ventral abdominal wall, covering the intestinal coils ventrally and to a certain extent laterally (179). Its free edge lies just cranial to the pelvic inlet. Superficial and deep walls are separable back to the line of reflection, so that the omental bursa in these species extends the full length of the omentum. The greater omentum of the pig is similar to that of the carnivores, but does not extend as far caudally. In the horse, the caudal parts of the superficial and deep walls adhere to each other, partly eliminating the bursa. The equine omentum is distributed among the coils of the jejunum; it may reach the inguinal area, and has been reported to enter the tunica vaginalis and be visible during castration.

The greater omentum, in which are embedded the spleen and part (dorsal primordium) of the pancreas, has undergone such a fundamental transformation from the simple sheet of mesogastrium, that it can no longer suspend the stomach from the dorsal abdominal wall as it did in the embryo. It does, like the other mesenteries, carry blood vessels (celiac artery, and branches of the portal vein), lymphatics, and nerves to the organs, but its functional significance is thought to be broader than this. Because a single sheet of omentum consists of two serosal layers applied to each other back to back, the omentum represents a considerable enlargement of the serosal surfaces in the abdomen and therefore an increase in their capacity to produce and absorb abdominal fluids. Further, experimental and clinical observations indicate that cellular reactions of serous membranes are most intense in the greater omentum. Except in the horse, the greater omentum can store large amounts of fat, which may insulate the abdominal organs and prevent loss of body heat, or, in the form of fatfilled appendages (appendices epiploicae) fill spaces between organs. It is also thought that the large number of blood vessels in the greater omentum may play a role in regulating blood pressure in the abdominal cavity. Furthermore, the omentum is capable of closing breaks in the abdominal wall, such as may occur in the diaphragm for instance, first by plugging the break and then initiating closure by adhering to the edges of the wound.

The **LESSER OMENTUM** (omentum minus, 13/*8*, *9*) is a serosal sheet passing from the lesser curvature of the stomach and cranial part of the duodenum to the visceral surface of the liver. It originates from the proximal part of the ventral mesogastrium that connects the same organs in the embryo (12/*b*). (The distal part of the embryonic mesogastrium (*c*) between liver and body wall gives rise to the falciform, coronary, and triangular ligaments associated with the liver.) According to its attachment on stomach and duodenum, the lesser omentum consists of **hepatogastric** and **hepatoduodenal ligaments** (13), and takes part in the formation of the **vestibule of the omental bursa.** The vestibule is bounded cranially by the liver (*d*) and caudally by the stomach (*b'*), cranial part of duodenum (*c*), and pancreas (*g*). The left wall of the vestibule is formed by the gastrophrenic ligament (*7*) which is connected to the lesser omentum, and to the right the vestibule is bounded by the lesser omentum, pancreas, caudal vena cava (*e*) and portal vein (*f*). The vestibule is accessible from the general peritoneal cavity (greater peritoneal sac) through the **epiploic foramen** (*11*), which is a slitlike opening to the right of the median plane ventral to the base of the caudate process of the liver. The craniodorsal boundary of the foramen is formed by the caudate process and caudal vena cava; the caudoventral boundary is formed by the portal vein, pancreas, and the free border of the hepatoduodenal ligament.

The **omental bursa** (lesser peritoneal sac) consists then of a vestibule and a number of recesses (dorsal, caudal, and splenic) of which the caudal—the one enclosed by the greater omentum—is the most prominent. Vestibule and caudal recess communicate freely over the lesser curvature of the stomach.

This description of the omenta applies in general to animals with a simple stomach, such as the carnivores, pig, and horse. In the ruminants, which have a stomach consisting of several compartments, the omenta, although similar to the general pattern, are arranged differently (see p. 166).

MESENTERIES. The mesenteries are the serosal folds that suspend the intestinal tract from the roof of the abdominal cavity. Their complex arrangement and differences between species will best be understood by briefly considering the **rotation of the embryonic gut,** which, like the rotation of the stomach, is an important phase in intestinal development.

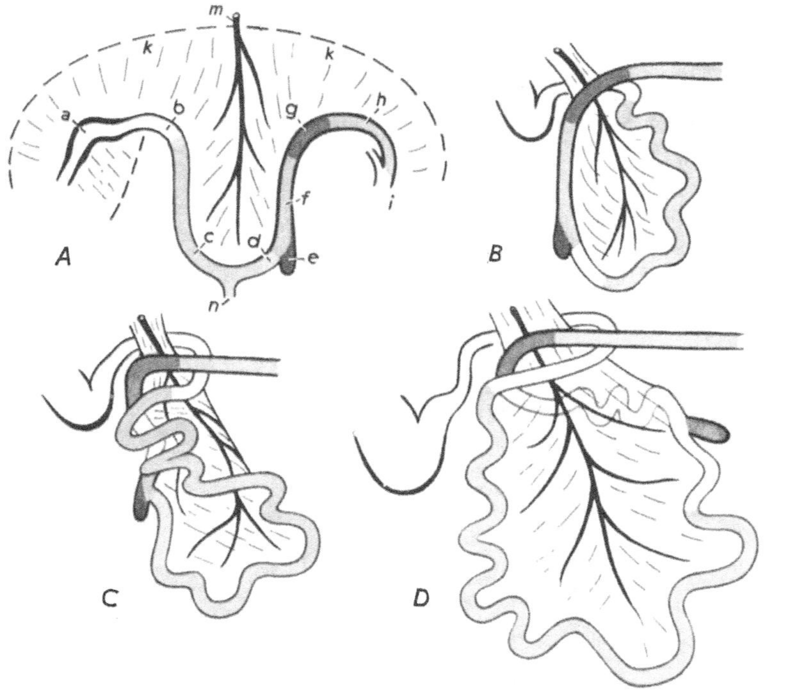

Fig. 14. Rotation of the intestinal tract during development of the mammalian embryo. Schematic. Left lateral aspect. (After Zietzschmann 1955)

A Stage of primitive intestinal loop; *B* Stage after rotation through 180 degrees; *C* Stage after rotation through about 270 degrees; *D* Stage after completed rotation

a Stomach; *b* Duodenum; *c* Jejunum; *d* Ileum; *e* Cecum; *f* Ascending colon; *g* Transverse colon; *h* Descending colon and rectum; *i* Cloaca; *k* Primitive dorsal mesentery; *l* Primitive ventral mesentery; *m* Cranial mesenteric artery; *n* Yolk stalk

As was mentioned previously, the primordium of the intestinal tract is a straight tube passing through the abdominal cavity, suspended by the **primitive dorsal mesentery.** Longitudinal growth of the intestinal tube exceeds that of the embryo so that a loop results with a ventrally directed flexure, from which the yolk stalk continues into the umbilical cord (14*A*). Beginning at the gastric primordium, the intestinal loop consists of a short longitudinal part, a descending limb, a flexure, an ascending limb, and a terminal longitudinal part which ends at the cloaca. The primordium of the cecum (*e*) on the ascending limb marks the division between small and large intestines.

At this stage the intestinal loop is placed more or less sagittally with the descending limb cranial to the ascending. With continued intestinal elongation, rotation begins around a dorsoventral axis which coincides with the cranial mesenteric artery (*m*). In the initial 180 degrees, the ascending limb passes from its caudal position along the left side of the body to a cranial position, while the descending limb passes from cranial to caudal on the right. The rotation is clockwise when viewed from a dorsal position. The ascending limb is now cranial to the descending and the large intestine (*f*, *g*, *h*) crosses, and lies to the left of, the duodenum (14*B*). In the following roughly 150 degree rotation (14*C*, *D*), the ascending limb with its cecal primordium moves dorsocaudally and to the right, while the descending limb, consisting mainly of jejunum, passes cranially on the left side toward its original cranial position. The intestinal rotation has thus gone almost full circle, and the originally flat mesentery suspending the loop has been gathered around the cranial mesenteric artery to form the **root of the mesentery** (radix mesenterii). The duodenum passes caudally on the right of the root of the mesentery, hooks around its caudal aspect, and passes forward on the left of the mesentery continuing as the jejunum. The rotation, thus, has bent the duodenum and divided it into descending and ascending parts. The colon, similarly, begins on the right side of the mesentery as the ascending colon, passes from right to left cranial to the root of the mesentery as the transverse colon, and descends on the left as the descending colon (14*D*). In the carnivores, the ascending colon is a simple short tube (144/*F*, *F*′). In the other domestic mammals, which are all ungulates, the ascending colon (and in the horse the descending colon also) varies greatly in length and position from the simple pattern.

The primitive dorsal mesentery of the embryo is continuous cranially with the dorsal mesogastrium and extends as a simple sheet from the pylorus into the pelvic cavity. The parts into which it is divided when differentiation of the gut into the postnatally recognized segments has taken place have already been given on page 11. Its line of attachment on the roof of the abdomen is relatively short and straight, more or less following the course of the abdominal aorta and including the cranial and caudal mesenteric arteries, the principal vessels supplying the intestinal tract. In contrast to this, the attachment of the mesentery along the intestine with its tremendous elongation particularly of the jejunum (but in the ungulates also the colon), is many times as long as the dorsal attachment. Because of this, any section of mesentery, when spread out, is fan-shaped (4). The blood vessels (*2*) supplying the gut converge toward their parent artery in the center, which in the case of the cranial mesenteric artery (*1*) lies embedded in the root of the mesentery. Extending from the root, the undisturbed mesentery is arranged in deep folds on the ends of which are the intestinal "coils" (230/*2*). The simplest arrangement of the mesenteries is found in the carnivores, since they have the simplest intestinal tract. In the pig, ruminants, and horse, principally because of the extraordinary elongation and repositioning of the ascending colon, but also because of extensive adhesions of parts of the intestine and their mesenteries with one another, with

Figs. 15 and 16. Transverse sections through the pelvis of stallion and mare at the level of the hip joint. Semischematic. Peritoneum in red.

A Sacrum; *B* Acetabulum; *B'* Ischiatic spine; *C* Pubis; *D* Sacroischiatic ligament; *E* Femur

a Rectum; *b* Ureter; *c* Ampulla of ductus deferens (stallion), vagina (mare); *c'* Seminal vesicle; *d* Bladder; *e* Pelvic flexure of great colon; *f* Loops of small colon

1 Rectogenital pouch; *2* Vesicogenital pouch; *3* Right and left parts of pubovesical pouch; *4* Mesorectum; *5* Genital fold (stallion), broad ligament (mare); *6* Lateral ligament of bladder; *7* Median ligament of bladder

Fig. 15 (Stallion)

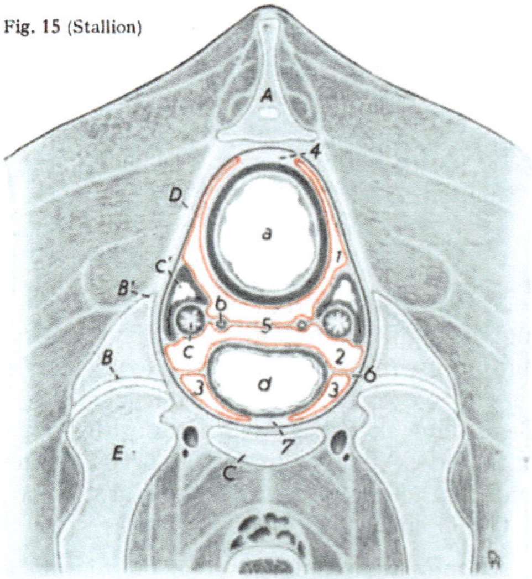

Fig. 16 (Mare)

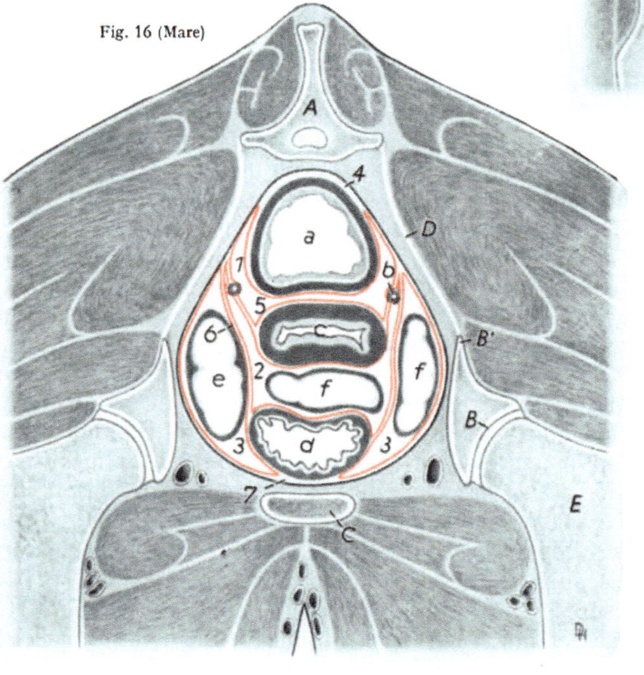

neighboring organs, and with the dorsal abdominal wall, the mesenteries differ markedly from the simple pattern. They are described with the digestive organs of these species. Fat is embedded in some parts of the mesenteries along blood and lymphatic vessels, as it is in the greater omentum. Fatfilled **appendices epiploicae** are also found, especially on the descending mesocolon of the horse.

Peritoneal Folds Associated with the Urogenital Organs

(15—17)

The urogenital organs are located partly in the abdominal and partly in the pelvic cavity, and on leaving the body wall in development, draw out peritoneal folds similar to those of the intestines. The kidneys in general are retroperitoneal; only their ventral exposed surface is covered with peritoneum. In ruminants and occasionally in lean dogs, the left kidney is suspended from the dorsal abdominal wall by a peritoneal fold (554). The ureter is also largely retroperitoneal, but it leaves the wall in the pelvic cavity and enters the genital fold in the male and the braod ligament in the female.

The **BROAD LIGAMENTS** (ligamenta lata uteri, 554/*11*) arise from the laterodorsal wall of the pelvic and caudal part of the abdominal cavity and contain varying amounts of smooth muscle fibers. The cranial part suspends the ovaries, and is known therefore as the **mesovarium** (547B/*1*). From the lateral surface of the mesovarium a fold enveloping the uterine tube is detached; this is the **mesosalpinx*** (*3*). The largest part of the broad ligaments, however, attaches to, and supports, the horns and body of the uterus and is known therefore as the **mesometrium**** (*2*). Figure 16 shows the mesometrium (*5*) in the pelvic cavity and its relation to the parietal peritoneum and adjacent ligaments of the bladder (*6*) in the mare.

The **GENITAL FOLD** of the male (15/*5*) is much smaller than the homologous paired broad ligaments. It contains the ampullae of the deferent ducts (*c*), ureters (*b*), the vesicular glands (seminal vesicles in the horse, *c'*) except in the carnivores, and occasionally the uterus masculinus, which is the remnant of the paramesonephric ducts***.

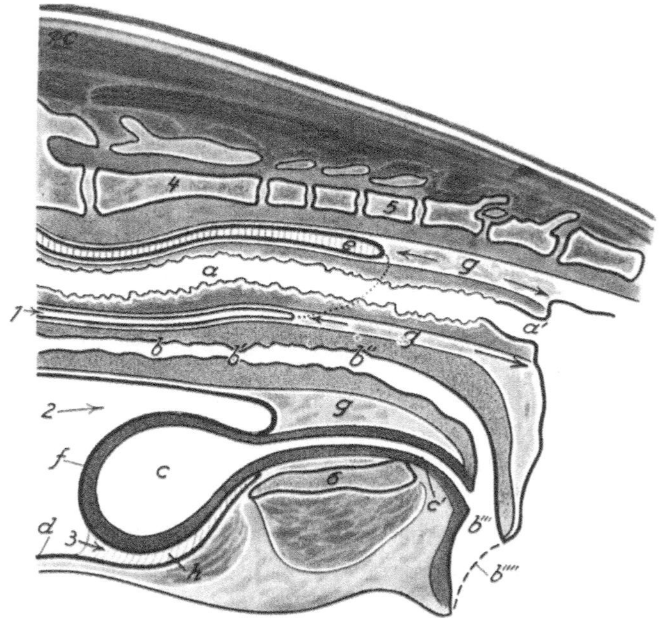

Fig. 17. Median section through the pelvis of the bitch. Semischematic. (After Zietzschmann, unpublished)

a Rectum; *a'* Anus; *b* Body of uterus; *b'* Cervix; *b''* Vagina; *b'''* Vestibule; *b''''* Vulva; *c* Bladder; *c'* Urethra; *d* Parietal peritoneum; *e* Mesorectum; *f* Visceral peritoneum; *g* Retroperitoneal part of pelvic cavity; *h* Median ligament of bladder

1 Rectogenital pouch; *2* Vesicogenital pouch; *3* Pubovesical pouch; *4* Sacrum; *5* Third caudal vertebra; *6* Symphysis pelvis

* *Salpinx* (Gr., trumpet), an earlier name for uterine tube, because of its trumpet-shaped abdominal end; hence mesosalpinx and salpingitis.

** Gr. *metra*, uterus.

*** Formerly Müllerian ducts.

Ventral to the genital fold or the uterus in the female is the urinary bladder, supported by the two **lateral ligaments of the bladder** (15, 16/*6*) in the free cranial edges of which are rudimentary umbilical arteries, known as the **round ligaments of the bladder.** Connecting the bladder to the floor of pelvic and abdominal cavities is the **median ligament of the bladder** (*7*). This ligament supported the urachus in the fetus and may still be traced occasionally to the umbilicus.

Dorsal to the genital fold or the uterus is the rectum, which is attached to the dorsal wall by the short **mesorectum** (*4*).

The genital fold, the broad ligaments, and the lateral ligaments of the bladder project into the peritoneal part of the pelvic cavity from the lateral and caudal walls, and divide it into three recesses or pouches, which are open cranially. The most dorsal of these is the **rectogenital pouch** (excavatio rectogenitalis, *1*), which is bounded dorsally by the rectum and ventrally by the uterus and broad ligaments in the female, or the genital fold in the male. The parts of the rectogenital pouch on each side of the mesorectum are the **pararectal fossae.** Ventral to the rectogenital pouch is the **vesicogenital pouch** (excavatio vesicogenitalis, *2*) formed dorsally by the uterus and broad ligaments or the genital fold and ventrally by the bladder and its lateral ligaments. The most ventral is the **pubovesical pouch** (excavatio pubovesicalis, *3*); it is between the floor of the pelvis and the bladder and its lateral ligaments, and is bisected in the median plane by the median ligament of the bladder.

BIBLIOGRAPHY

Body Cavities, Omenta and Mesenteries

Ackerknecht, E.: Über Höhlen und Spalten des Säugetierkörpers. Schweiz. Arch. Tierheilk. **62,** 1920.

Ackerknecht, E.: Über den Begriff und das Vorkommen der Spatien im Säugetierkörper. Anat. Anz. **54,** 1921.

Agduhr, E.: Kommen intravitale Kommunikationen zwischen den Pleurahöhlen der Pleurasäckchen bei den Haustieren vor? Svensk Veterinärtidskrift, H. 10, 1922.

Badoux, D. M., and C. J. G. Wensing: De bursa omentalis en adnexa bij herkauwers en vleeseters. Tijdschr. Dierg. **90,** 1965.

Böcker, H.: „Omentum lienale". Anat. Anz. **78** (Erg. H.), 1934; Verh. anat. Ges. **42,** 1934.

Broman, I.: Die Entwicklungsgeschichte der Bursa omentalis und ähnlicher Rezeßbildungen bei den Wirbeltieren. Wiesbaden, 1904.

Broman, I.: Warum wird die Entwicklung der Bursa omentalis in Lehrbüchern fortwährend unrichtig beschrieben? Anat. Anz. **86,** 1938.

Bucher, H.: Topographische Anatomie der Brusthöhlenorgane des Hundes mit besonderer Berücksichtigung der tierärztlichen Praxis. Leipzig, Diss. med. vet., 1909.

Ciliga, T., S. Rapic, U. Bego, and I. Huber: A contribution to the anatomy of the mediastinum in domestic equines. Vet. Arhiv **36,** 1966.

Dumont, H. A.: Contribution iconographique à la connaissance de la topographie viscerale des ovins. Diss. med. vet. Toulouse, 1972.

Eichbaum, F.: Die Brusthöhle des Pferdes vom topographisch-anatomischen Standpunkte und mit besonderer Berücksichtigung der physikalischen Diagnostik. Vorträge für Thierärzte, II. Serie, H. 1., 1879.

Fransen, J. W. P.: Form und funktionelle Bedeutung des großen Netzes. Zschr. ges. Anat. II. **1,** 1914.

Goldschmidt u. Schloss: Großes Netz und Bauchfell. Arch. klin. Chir. **160,** 1930.

Gouffé, D., J. Michel: Contribution iconographique à la connaissance de la topographie viscerale des bovins. Diss. École Nat. Vét. Toulouse, Fac. Méd. Pharmac. 1968.

Gräper, L.: Zwerchfell, Lunge und Pleurahöhlen in der Tierreihe. Anat. Anz. **66,** (Erg. H.) 1928.

Gräper, L.: Lungen, Pleurahöhlen und Zwerchfell bei den Amphibien und Warmblütern. Morph. Jb. **60,** 1928.

Heller, O.: Über Appendices epiploicae und sonstige Fettanhängsel in der Bauchhöhle bei Pferd und Hund. Zschr. Anat. Entw. gesch. **98,** 1932.

Iwanoff, St.: Die Topographie der Brustkorbwände und der Brustorgane beim Schafe. Zschr. Anat. Entw. **109**, 1939.

Iwanoff, St.: Über die Lage der Cupula pleurale beim Schaf und beim Schwein. (Russian with German summary) Jb. Univ. Sofia, Vet.-med. Fak. **16**, 1939/40.

Iwanoff, St.: Über die Topographie der Brustkorbwände und Brustorgane beim Schaf. (Russian with German summary) Jb. Univ. Sofia, Vet.-med. Fak. **14**, 1937/38.

Iwanoff, St.: Die Stellung des Zwerchfells und die kaudalen Grenzen der Pleurahöhlen bei einigen Haussäugern (Wiederkäuer und Schwein). (Russian with German summary) Jb. Univ. Sofia, Vet.-med. Fak. **15**, 1938/39.

Jacobi, W.: Zur Topographie der Brusthöhlenorgane des Hausschweines (Sus scrofa domestica). Berlin (Humboldt-Univ.), Diss. med. vet., 1962.

Koch, T., und H. Sajonski: Beitrag zur vergleichend-topographischen Anatomie des Mediastinum und der Serosabekleidung des Oesophagus bei Mensch und Hund. Zbl. Vet.-med. **1**, 1954.

Krüger, W.: Allgemeines zur Frage der Homologisierung der Darmgekröseabschnitte bei den Säugetieren. Dtsch. Tierärztl. Wschr., 1928.

Krüger, W.: Zur vergleichenden Anatomie des Darmgekröses bei den Säugetieren. Anat. Anz. **67** (Erg. H.), 1929.

Krüger, W.: Die vergleichende Entwicklungsgeschichte im Dienste der Lösung des Homologisierungsproblems an den Darm- und Gekröseabschnitten des Menschen und einiger Haussäugetiere (Hund, Katze, Pferd, Schwein und Wiederkäuer). Zschr. Anat. Entw. gesch. **90**, 1929.

Law, M. E.: Histology of omental reactivity. Nature **194**, 1962.

Martin, P.: Die Gekröseverhältnisse und Lageveränderungen des Hüftblind-Grimmdarmgebietes bei Pferdeembryonen. Festschrift Zschokke, Zürich, 39—57, 1925.

Martin, P.: Zur Entwicklung der Bursa omentalis und der Mägen beim Rinde. Österr. Mschr. Tierheilk., 1890.

Martin, P.: Zur Entwicklung des Netzbeutels der Wiederkäuer. Österr. Mschr. Tierheilk., 1895.

Mierswa, K.: Beziehungen zwischen äußeren und inneren Massen des Brustkorbes und seinen Organen beim Rinde. Arch. Tierernährung u. Tierzucht **7**, 1932.

Palmgren, A.: Zur Anatomie und Entwicklungsgeschichte des Mittelfelles (Mediastinum) der Haussäugetiere. Zschr. Anat. Entw. gesch. **87**, 1928.

Preuss, F., G. Fabian und E. Henschel: Zur Nomenklatur des Brustfells und seiner angewandten Anatomie beim Hund. Berl. Münch. Tierärztl. Wschr. **77**, 1964.

Rawe, B.: Verhalten und Entwicklung des Netzes (Omentum maius) beim Wiederkäuer. Hannover, Diss., 1921.

Recum, A. F. von: The mediastinum and hemothorax, pyothorax and pneumothorax in the dog. J. Am. Vet. Med. Assoc. **171**, 1977.

Richter, H.: Einiges über die Entstehung und Bedeutung der serösen Räume im Säugetierkörper und über die Lappenbildung an gewissen Organen, mit einem Beitrage zur Erklärung der eigenartigen Pleuraverhältnisse beim Elefanten. Festschr. Baum, Schaper, Hannover, 1929.

Schmaltz, R.: Topographische Anatomie der Körperhöhlen des Rindes. Berlin, Enslin (Schoetz), 1890.

Schmaltz, R.: Über die Plica gastropancreatica oder das Lig. gastro-duodenale und das Foramen epiploicum beim Pferde. Berl. Tierärztl. Wschr. H. 33, 1897.

Schrauth, O.: Beiträge zur Entwicklung des Netzbeutels, der Milz und des Pankreas beim Wiederkäuer und beim Schwein. Gießen, Diss. med. vet., 1909.

Simon, Ph.: Die Appendices epiploicae am Colon des Menschen und der Säugetiere. Berlin, Diss. med. vet., 1922.

Sussdorf, M. v.: Gibt es ein wirkliches Cavum mediastini? Ein Beitrag zur Anatomie des Mittelfelles der Fleischfresser. Dtsch. Zschr. Tiermed. vergl. Path. **18**, 1892.

Sussdorf, M. v.: Das Netz in seinem Verhältnis zum Bauchfell und zu den Baucheingeweiden bei den Haussäugetieren. Arch. wiss. prakt. Tierhkd. **63**, 1931.

Tenschert, H.: Zur Anatomie und Physiologie der Zwerchfellkuppel bei Hund und Katze. München, Diss. med. vet., 1952.

Verine, H.: La relation droite-gauche de quelques organes pairs du chien. Bull. Sic. Sci. vét. et Med. comp. Lyon **74**, 237—244 (1972).

Walker, F. C., and A. W. Rogers: The greater omentum as a site of antibody synthesis. Brit. J. Exp. Path. **42**, 1961.

Wiethölter, G.: Topographische Anatomie der Bauch- und Beckenorgane von Hund und Katze im Röntgenbild. Diss. med. vet. Leipzig, 1964.

Wilkens, H.: Mesogastrium dorsale der Katze. Hannover, Diss. med. vet., 1951.

Zietzschmann, O.: In Baum, H., und O. Zietzschmann: Handbuch der Anatomie des Hundes; Vol. 1, 2nd ed. Berlin, Paul Parey, 1936.

Zietzschmann, O.: Das Mesogastrium dorsale des Hundes mit einer schematischen Darstellung seiner Blätter. Morph. Jb. **83**, 1939.

Zietzschmann, O.: In Zietzschmann, O., u. O. Krölling: Lehrbuch der Entwicklungsgeschichte der Haustiere; 2nd ed. Berlin, Hamburg, Paul Parey, 1955.

Zimmermann, G.: Das Netz des Schafes. Dtsch. tierärztl. Wschr. **47**, 1939.

Zimmermann, G.: Die Ausbildung der kaudalen Grenze des Peritonaeum in der Beckenhöhle. Acta. Vet. Acad. Sci. Hung. **7**, 1957.
Zimmermann, G.: Eine Revision der Beschreibung der Bauchfellduplikaturen des Beckens. Acta. Vet. Acad. Sci. Hung. **13**, 1963.
Zschokke, M.: Cavum mediastini serosum s. bursa infracardiaca. (Kritisches über das kaudale Mittelfell). Anat. Anz. **53**, 1920.

Digestive System

The organs of the digestive system are concerned with the nutrition of the body. This function includes the prehension of food, its mastication, digestion, and absorption, and the initial storage of the nutrients released during digestion. The digestive organs also take care of the expulsion of the unabsorbed portion of the food and of some of the substances that are discharged into the digestive tube by its large accessory glands.

The digestive system includes the mouth, the pharynx, the alimentary canal, and several accessory glands. The **ALIMENTARY CANAL** is a muscular tube which begins with the esophagus at the caudal end of the pharynx and ends at the anus. Its wall consists of three layers. On the inside is a **mucous membrane** (tunica mucosa) which contains many glands. This is followed by a **muscular coat** (tunica muscularis), and an external layer of connective tissue. In the large body cavities, most of the connective tissue layer is covered with mesothelium and is termed a **serous coat** (tunica serosa); it forms part of the visceral pleura in the thoracic cavity and most of the visceral peritoneum in the abdominal cavity. Outside of the body cavities, the connective tissue layer blends with the surrounding tissues and is known as the **tunica adventitia.** The accessory glands of the digestive system are the large **SALIVARY GLANDS,** which are located in the head, the **LIVER,** and the **PANCREAS.** They are slightly removed from the digestive tube, but retain a connection with it through their excretory ducts.

The digestive tube consists of the following consecutive segments: the **MOUTH, PHARYNX, ESOPHAGUS, STOMACH, SMALL INTESTINE, LARGE INTESTINE,** and **ANAL CANAL** (11). The gross and histological appearance of these segments differs with the functions each segment performs. There are also marked species-specific differences reflecting the type and nutritive value of the food that is normally consumed.

Mouth and Pharynx

General and Comparative

Oral Cavity

(25, 57—62, 79, 81, 83, 85)

The oral cavity and its accessory organs, the tongue, teeth, and salivary glands, are concerned with the prehension, selection, mastication, and insalivation of food, in short, with the conversion of the food to a bolus that can be swallowed. In addition, taste buds in its mucosa scrutinize the food for palatability. The oral cavity extends from the lips to the entrance into the pharynx. Its osseous support is provided by the incisive bone, the palatine and alveolar processes of the maxilla, the horizontal lamina of the palatine bone, and by the mandible. The oral cavity is bounded rostrally by the **lips** and laterally by the **cheeks.** Its dorsal limit, or roof, is the **hard palate;** ventrally there is the **tongue** and under its apex and lateral margins there is a crescent-shaped space which is the actual **floor of the oral cavity.** Caudally, the oral cavity communicates with the **oropharynx** (57/*d*), which is a narrow isthmus (isthmus faucium) formed by the root of the tongue (*1''*) and the soft palate (*k*) and is usually closed. When the jaws are closed, the oral cavity is divided by the teeth

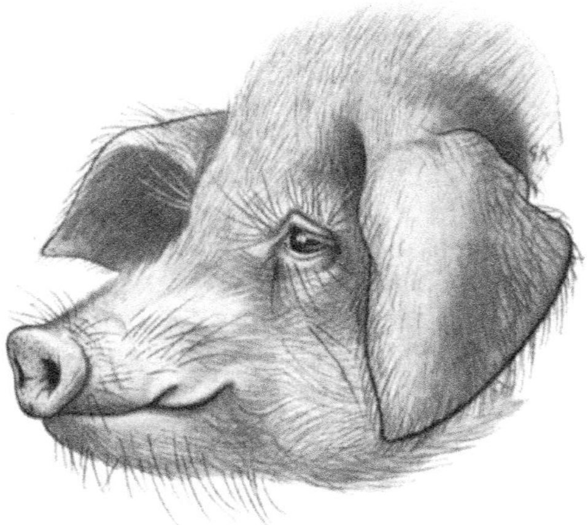

Fig. 20

Fig. 21

Fig. 24

Fig. 23

Figs. 18—24. Heads of the domestic mammals, showing the nostrils, the lips, and the distribution of tactile hairs

and the alveolar processes into the **VESTIBULE** (*b'*) and the **ORAL CAVITY PROPER** (*b*). The two cavities communicate via the interdental spaces, especially the large space (diastema) between the incisors and the cheek teeth, and the space behind the last molars. Rostrally, two narrow **incisive ducts** connect the oral cavity with the nasal cavity. The ducts open on the **incisive papilla** (26, 28/*1*) caudal to the upper incisors, and are described more fully on page 219. The **mucous membrane** of the oral cavity is usually pink, but may be pigmented (black) in places. It is well supplied with blood vessels and in its submucosa contains serous or mucous glands known as **labial, buccal,** and **lingual glands,** depending on the location. In addition, there are the large **salivary glands** which shed their secretions into the oral cavity through special ducts.

Lips

(18—24, 47—62, 269—276)

The entrance to the oral cavity, the **oral cleft** (rima oris), is bounded by the edges of the **upper** and **lower lips,** which unite on each side at the **angle of the mouth.** The lips of the domestic mammals assist in sucking and in the prehension of food; or they may act as tactile organs. They differ, therefore, in shape and mobility from species to species — the upper lip more than the lower. Thus, the lips of the horse, sheep, goat, and carnivores are quite mobile; while those of the ox and pig do not have much freedom of motion.

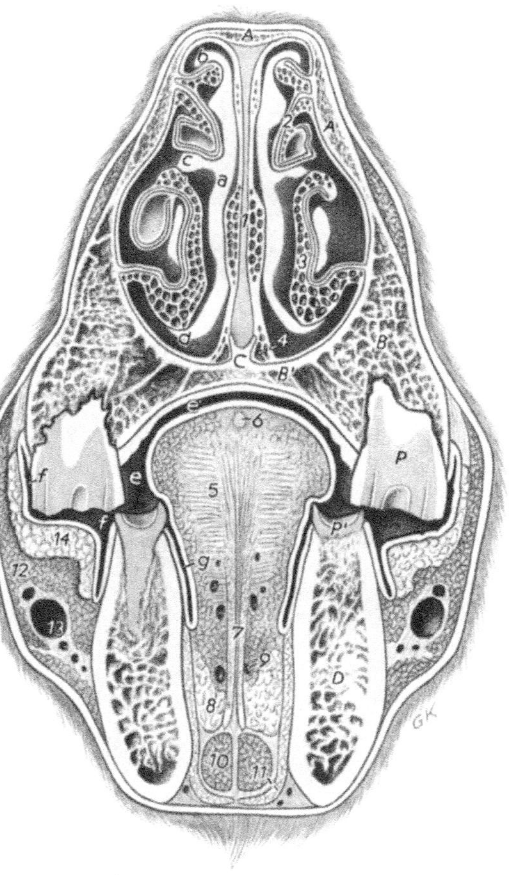

Fig. 25. Transverse section of the head of a horse at the level of the third premolar.

A Nasal bone; *B* Maxilla, *B'* its palatine process; *C* Vomer; *D* Mandible; *P*, *P'* Third premolar

a Common nasal meatus; *b* Dorsal nasal meatus; *c* Middle nasal meatus; *d* Ventral nasal meatus; ***e* Oral cavity proper; *f* Vestibule; *g* Lateral sublingual recess**

1 Nasal septum with venous plexus; *2* Dorsal nasal concha with venous plexus; *3* Ventral nasal concha with venous plexus; *4* Vomeronasal organ; *5* Tongue; *6* Lingual cartilage; *7* Genioglossus; *8* Polystomatic sublingual gland; *9* Mandibular duct; *10* Geniohyoideus; *11* Rostral part of mylohyoideus; *12* Buccal musculature; *13* Buccal venous plexus; *14* Buccal glands

The lips are attached to the incisive bone and the incisive part of the mandible, and consist of three layers. The middle layer, consisting of muscle, tendons, connective and adipose tissue, forms the bulk of the lip. It is covered externally by the skin, and internally by the labial mucosa which meet in a sharp line at the edge of the lip. The labial mucosa is continuous at the base of the lips with the gums (89/*f*, *h*). The **labial glands,** found in the submucosa or in the muscular layer, are especially well developed near the angle of the mouth. They are most numerous in the horse, and decrease in number in the following sequence: horse, ox, goat, sheep, pig, dog, cat. In the carnivores, small ruminants, and horse, the skin covering the lips has by and large the same appearance as the skin elsewhere, but usually bears tactile hairs. In the pig and ox there is an area of modified skin, known as the **rostral** and **nasolabial plate** respectively, that incorporates the central portion of the upper lip (29/*R;* 271; 274). The upper lip of the carnivores and small ruminants is divided by a distinct median cleft, **philtrum,** which is shallow or absent in the other species (269—276).

The lower lip is noticeably smaller than the upper lip in the carnivores and pig. In the ox and horse it presents the **chin** (mentum*), a protuberance formed by muscular and adipose tissue. (The osseous mental protuberance, which forms the base fo the chin in man, is not present on the mandible of domestic mammals.)

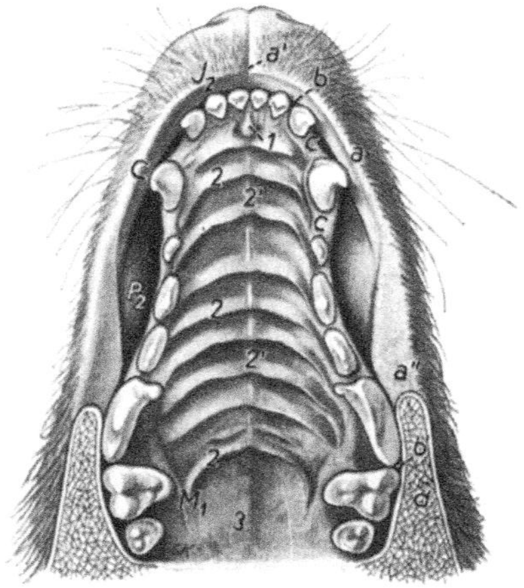

Fig. 26

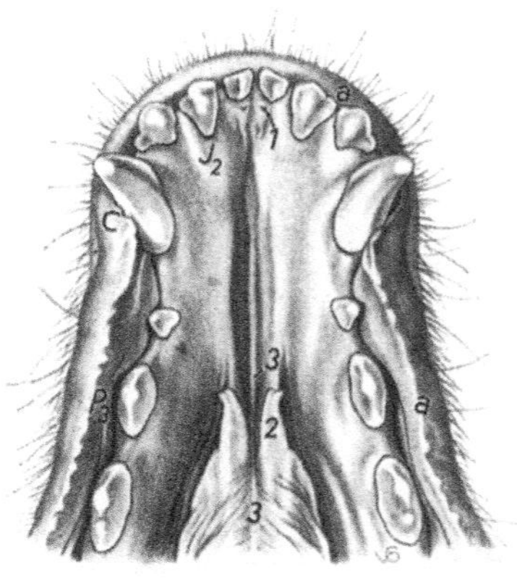

Fig. 27

Fig. 26. Roof of the oral cavity of a dog. *J2* Second incisor; *C* Canine; *P2* Second premolar; *M1* First molar. *a* Upper lip; *a'* Philtrum; *a''* Angle of the mouth; *b* Labial vestibule; *b'* Buccal vestibule; *c, c* Diastema; *d* Cheek with buccal musculature; *1* Incisive papilla with openings of the incisive ducts; *2* Hard palate with palatine ridges; *2'* Palatine raphe; *3* Soft palate

Fig. 27. Sublingual floor of the oral cavity of the dog. *J2* Second incisor; *C* Canine; *P3* Third premolar; *a* Lower lip; *1* Orobasal organ; *2* Sublingual caruncle with openings of the major sublingual and mandibular ducts; *3, 3* Frenulum linguae

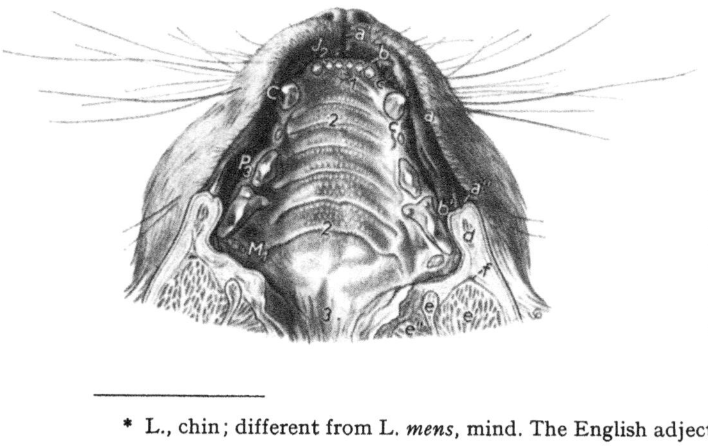

Fig. 28. Roof of the oral cavity of a cat.

J2 Second incisor; *C* Canine; *P3* Third premolar; *M1* First molar

a Upper lip; *a'* Philtrum; *a''* Angle of the mouth; *b* Labial vestibule; *b'* Buccal vestibule; *c, c* Diastema; *d* Cheek; *e* Ramus of the mandible; *e'* Masseter muscle; *e''* Pterygoid muscle; *f* Facial vein

1 Incisive papilla with openings of the incisive ducts; *2* Hard palate with palatine ridges and rows of cornified papillae; *3* Soft palate

* L., chin; different from L. *mens,* mind. The English adjective derived from both, however, is "mental".

Cheeks

(25, 26, 28, 29, 31, 35)

The lateral walls of the buccal vestibules are formed by the cheeks (buccae). The cheeks are attached to the alveolar margins of the maxilla and the mandible in the region of the cheek teeth, and extend from the angle of the mouth to the pterygomandibular fold, which connects the palate with the mandible behind the last cheek teeth. The caudal portion of the cheek contains the powerful masseter muscle. The cheeks, like the lips, consist of three layers: skin, buccal mucosa, and an intermediate layer of glands and muscles. The buccal mucosa is continuous with the gums of the cheek teeth. In the ruminants, it forms cone-shaped cornified **papillae,** which are also present on the lips and are directed caudally (31). The **buccal glands** are located either between the mucosa and the musculature or between the buccal muscles. Dorsal, ventral, and in the ox also middle buccal glands (49, 50/*8*, *9*, *10*) can be distinguished. Because the dorsal buccal gland in the carnivores is located medial to the zygomatic arch, it is known as the **zygomatic gland** (52/*e*). The ducts of the buccal glands open into the buccal vestibule.

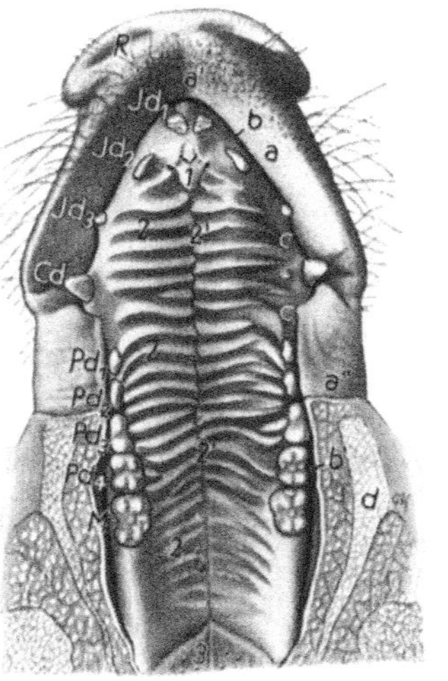

Fig. 29

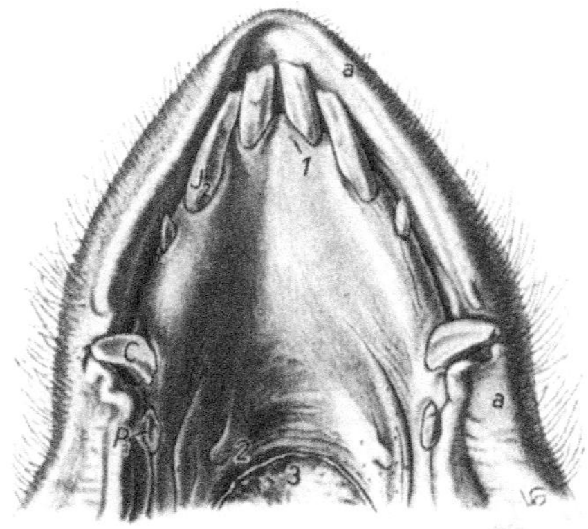

Fig. 30

Fig. 29. Roof of the oral cavity of a pig. *Jd 1—3* Deciduous incisors; *Cd* Deciduous canine; *Pd 1—4* Deciduous premolars; *M* First molar; *R* Rostral plate; *a* Upper lip; *a'* Philtrum; *a''* Angle of the mouth; *b* Labial vestibule; *b'* Buccal vestibule; *c*, *c* Diastema; *d* Cheek with buccal musculature and dorsal buccal glands; *1* Incisive papilla with openings of the incisive ducts; *2* Hard palate with palatine ridges; *2'* Palatine raphe; *3* Soft palate with tonsillar fossules

Fig. 30. Sublingual floor of the oral cavity of the pig. Note the unusually distinct sublingual caruncles. *J2* Second incisor; *C* Canine; *P1* First premolar; *a* Lower lip; *1* Orobasal organ; *2* Sublingual caruncle, covering the openings of the mandibular and major sublingual salivary ducts; *3* Ventral attachment of the frenulum linguae

Gums

(25, 89)

The parts of the oral mucosa that are intimately united to the periosteum of the alveolar processes of the jaws are known as the gums (gingivae). The gums closely encircle the necks of the teeth and exchange fibers with the alveolar periosteum. In the ruminants, the gums are modified to form the **dental pad** (pulvinus dentalis, 31/*D*) which takes the place of the upper incisors.

Hard Palate

(25, 26, 28, 29, 31, 35, 57—62, 79, 81, 83, 85)

The osseous palate and the mucosa that covers its oral surface is kown as the hard palate (palatum durum, 29/*2*). The hard palate is bounded laterally and rostrally by the upper dental arch. Its tough mucosa, which in the horse contains rich venous plexuses, is continuous laterally with the gums and caudally with the mucous membrane of the soft palate (*3*). The hard palate is divided into two symmetrical halves by a median **palatine raphe** (*2'*) which usually takes the form of a shallow groove; in the dog it is an indistinct median crest. On either side of the palatine raphe are the transversely directed **palatine ridges** (rugae palatinae). The ridges are cornified, are usually slightly concave caudally, and have a gradual rostral slope and a steeper caudal slope. In the ox, their crests are studded with caudally-directed cornified papillae (31). The number of palatine ridges differs with the species: the dog has 6—10 pairs, the cat 7, the pig 20—23, the ox 15—20, the sheep about 14, the goat 12, and the horse 16—18.

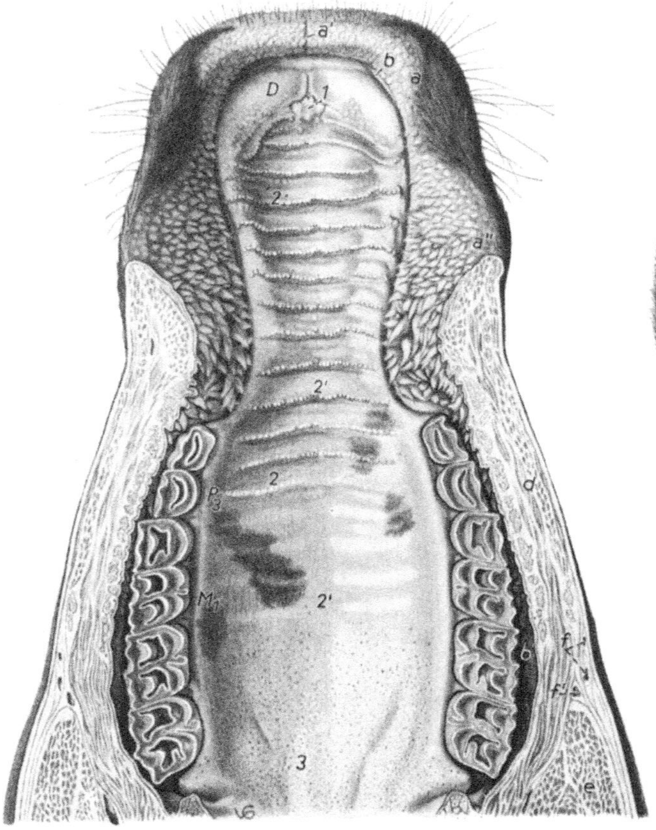

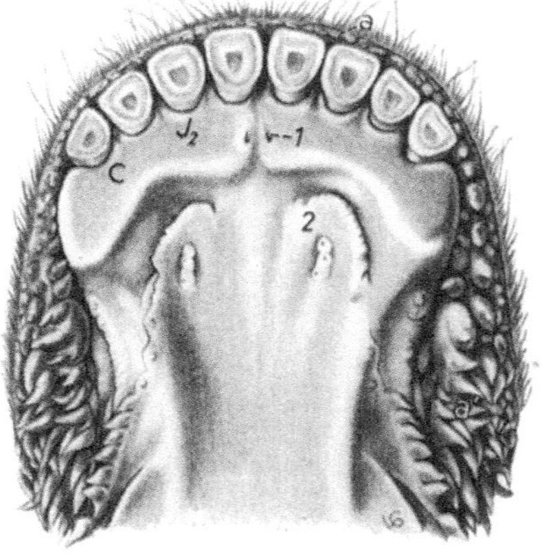

Fig. 32. Sublingual floor of the oral cavity of the cow.
J 2 **Second incisor;** ***C*** **Corner incisor (I 4)** ***a*** **Lower lip;** ***a'*** **Labial papillae** ***1*** **Orobasal organ;** ***2*** **Sublingual caruncle with openings of the mandibular and major sublingual ducts**

Fig. 31. Roof of the oral cavity of a cow. *D* Dental pad; *P3* Third premolar; *M1* First molar; *a* Upper lip; *a'* Philtrum; *a''* Angle of the mouth; *b, b'* Vestibule; *d* Cheek with buccal musculature and dorsal buccal glands; *e* Masseter muscle; *f* Facial artery and vein; *f'* Parotid duct; *1* Incisive papilla with openings of the incisive ducts; *2* Hard palate with palatine ridges, the latter with cornified papillae; *2'* Palatine raphe; *3* Soft palate

In the pig and horse, the palatine ridges extend to the soft palate; in the other species, the caudal portion of the hard palate is smooth. Generally, the mucosa of the hard palate is without glands. There are glands, however, in the caudal portion of the hard palate in the dog and ruminants, and in the rostral portion in the pig. The palatine mucosa may be more or less pigmented (black) in all domestic mammals (31). The central prominence just behind the upper incisors, or just behind the dental pad in the ruminants, is the **incisive papilla** (*1*). Found on it, except in the horse, are the two minute openings of the incisive ducts. The hard palate lies opposite the dorsum of the tongue and with its ridges assists the tongue during prehension and mastication and in moving the bolus into the oropharynx.

Tongue

(25, 36—41, 57—62)

The tongue (lingua*) fills the oral cavity when the upper and lower teeth are in contact. It is supported caudally by the hyoid bone, and consists of striated musculature, connective and adipose tissue, some glands, and externally of a thick mucous membrane. Its shape differs slightly from species to species. The tongue is a very mobile and versatile organ. It is essential for the prehension and sorting of solid food in some animals, for the intake of liquids in others (e.g., the lapping of liquids by carnivores), and plays an important role in the sucking of the newborn. The tongue is also an important tactile organ and is capable of either mechanical or,

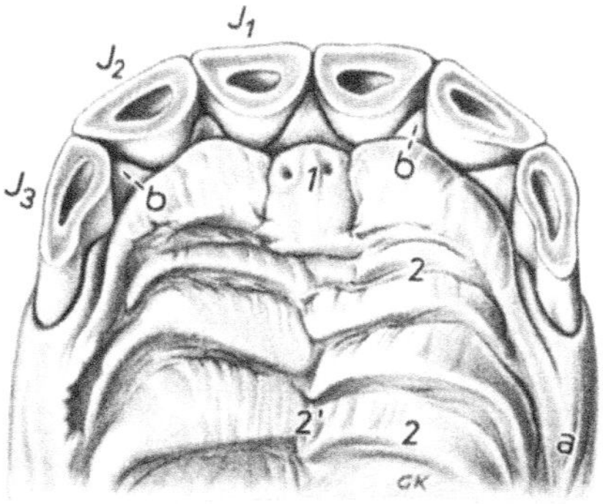

Fig. 33

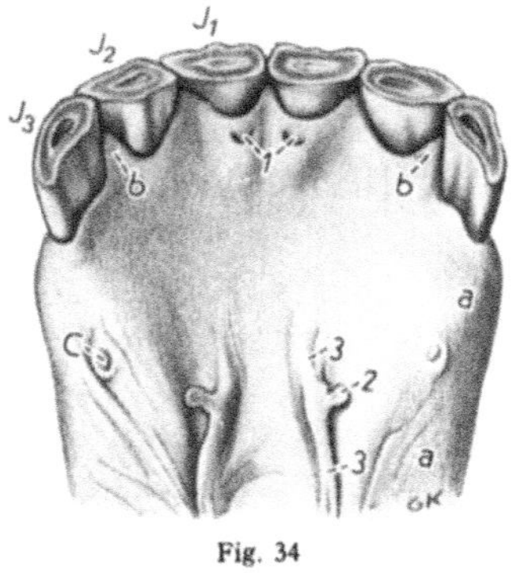

Fig. 34

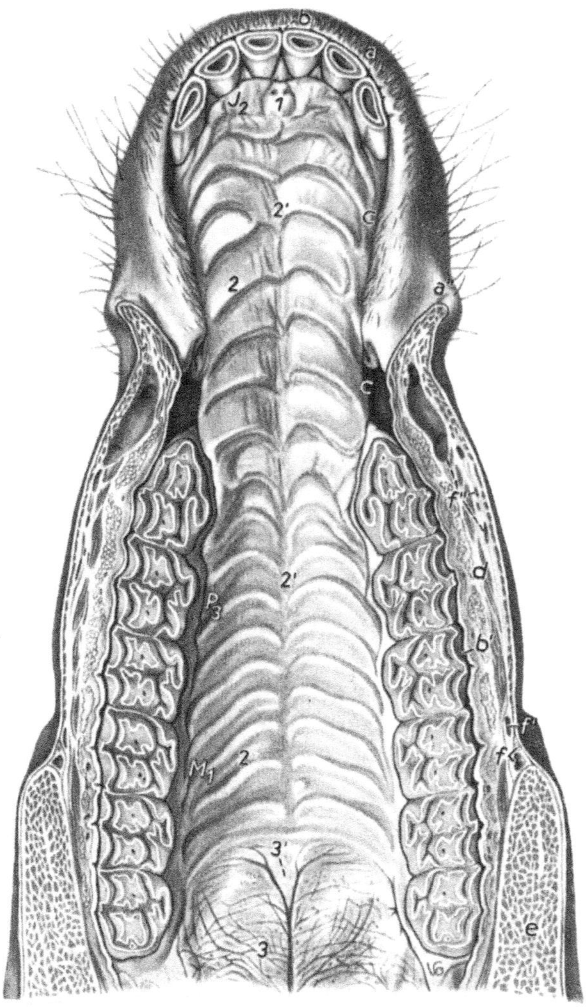

Fig. 35

Fig. 33. Rostral portion of the hard palate of a mare about six years old. *J 1—3* Incisors with cups; *a* Diastema; *b* Gums; *1* Incisive papilla with two small depressions (obliterated openings of the incisive ducts); *2* Hard palate with palatine ridges; *2'* Palatine raphe

Fig. 34. Sublingual floor of the oral cavity of a mare about six years old. *J 1—3* Incisors with cups; *C* Rudimentary canine; *a, a* Diastema; *b* Gums; *1* Orobasal organ; *2* Sublingual caruncle, the opening of the mandibular duct is on its lateral surface; *3* Caruncular fold

Fig. 35. Roof of the oral cavity of the horse. *J2* Second incisor; *P3* Third premolar; *M1* First molar

a Upper lip; a'' Angle of the mouth; b, b' Vestibule; c, c Diastema; d Cheek with buccal musculature and dorsal buccal glands; e Masseter muscle; f Facial artery and veins; f' Parotid duct; f'' Part of the buccal venous plexus

1 Incisive papilla with two small depressions (obliterated openings of the incisive ducts); *2* Hard palate with palatine ridges; *2'* Palatine raphe; *3* Soft palate; *3'* Tonsil of the soft palate with tonsillar fossules

* L., also Gr. *glossa;* hence, lingual and glossal.

by means of taste buds, chemical selection of food. The tongue is necessary to both mastication and deglutition and, like a plunger, it delivers the insalivated bolus into the pharynx. Some species use the tongue to rid themselves of insects or to scratch themselves; others use it to clean their skins and groom their hair coats.

The **INNERVATION** of the tongue is derived from five pairs of cranial nerves. They are the mandibular branch of the trigeminal nerve (V), the facial nerve (VII), the glossopharyngeal nerve (IX), the vagus (X), and the hypoglossal nerve (XII). The hypoglossal is the only motor nerve to the tongue; the other four are sensory and mediate gustatory, tactile, pain, and temperature stimuli.

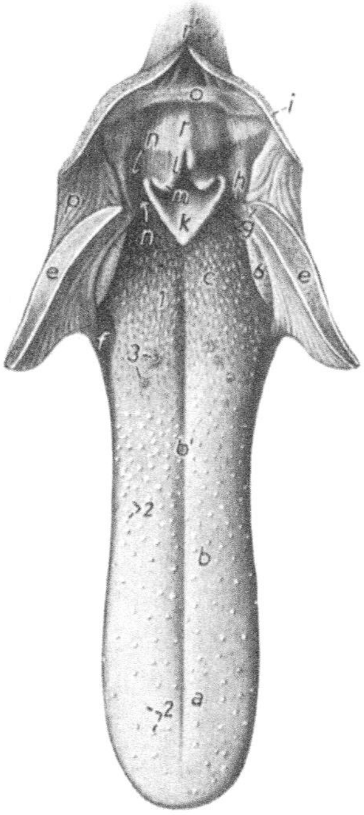

Fig. 36 (Dog)

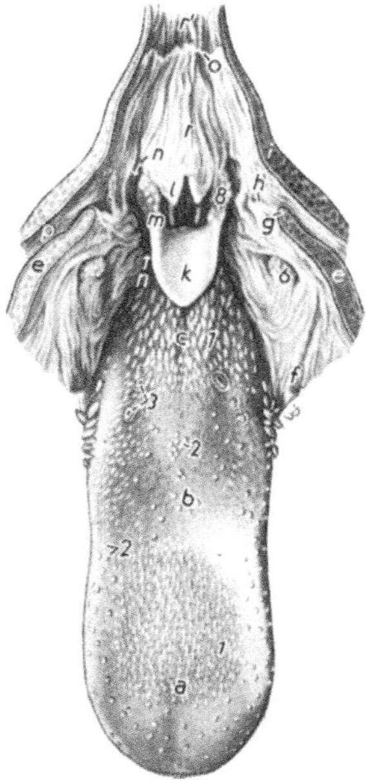

Fig. 37 (Cat)

Fig. 38 (Pig)

FORM AND STRUCTURE. The surface of the tongue opposite the palate is the **dorsum*** **linguae;** the free rostral portion is the **apex.** The apex presents dorsal and ventral surfaces which meet in either a sharp or rounded border, depending on the species the ventral surface is connected to the floor of the oral cavity by a median fold (frenulum linguae, 43/*e*). Caudal to the apex and representing the bulk of the tongue is the **body** (corpus linguae) followed by the **root** (radix linguae), which slopes ventrally toward the base of the epiglottis. Both body and root of the tongue lie in the intermandibular space, which is quite narrow in the herbivores. The tongue is anchored to the mandible and hyoid apparatus by the extrinsic lingual muscles (25) that enter the body of the tongue from behind and below. In the herbivores and pig, the body of the tongue has extensive lateral surfaces. Farther caudally, only the dorsal surface is exposed and is covered with mucous membrane. In the ruminants, the caudal portion of the dorsum is raised to form an elliptical prominence, **torus linguae** (39/*d*), rostral to which in the ox is the funnel-shaped **fossa linguae.** (This fossa is not homologous with the foramen cecum of the human tongue.) The horse's tongue usually contains a slender bar of cartilage (cartilago dorsi linguae, 25/*6*) in the median plane, just below the mucous mem-

* L. the back.

brane of its dorsal surface. In the dog, the dorsum of the tongue is divided by a **median groove** (sulcus medianus, 36/*b'*) into two equal halves; the apex contains the characteristic **lyssa** (42/*a*; 43/*h*), a median filiform structure embedded in the musculature along the ventral surface of the apex.

The **MUCOUS MEMBRANE** of the tongue has a dense submucosa by which it adheres to the subjacent tissues. Along the ventral and lateral surfaces of the tongue the mucous membrane is thin and delicate. On the dorsum, however, where the wear of the tongue is greaest, the mucosa is thick and tough, because of the cornification of its epithelium, especially in the ruminants and the cat.

The mucous membrane of the tongue presents numerous papillae, which are named according to their shape. The filiform and conical papillae have mechanical functions, while the fungiform, vallate, and foliate papillae serve mainly a gustatory function.

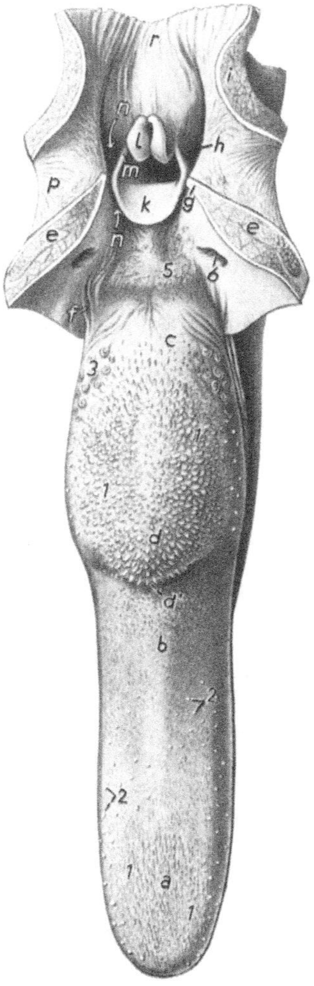

Fig. 39 (Ox)

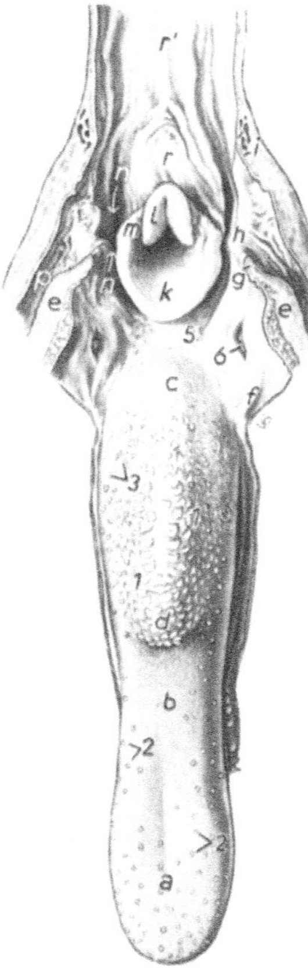

Fig. 40 (Goat)

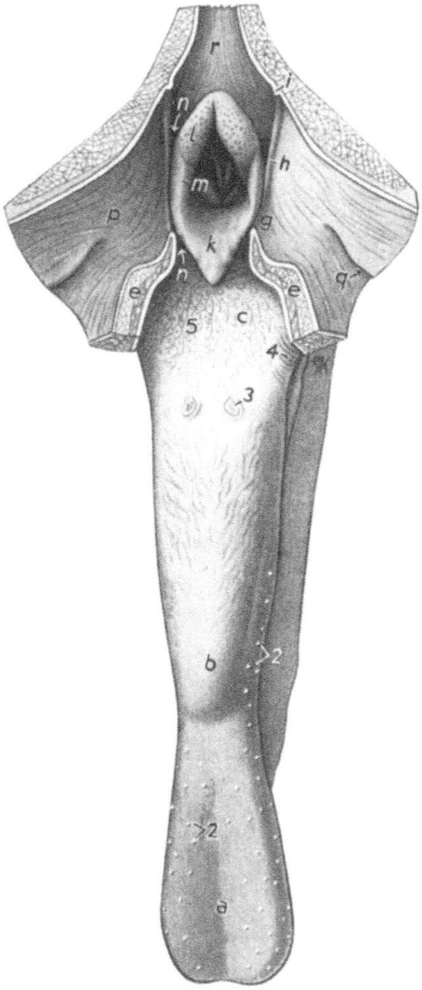

Fig. 41 (Horse)

Figs. 36—41. Tongue and opened pharynx. Dorsal aspect.

a Apex; *b* Body; *b'* Median groove (dog); *c* Root; *d* Torus linguae (ruminants); *d'* Fossa linguae (ox); *e* Soft palate (roof of oropharynx), cut in the median plane to expose the root of the tongue; *f* Palatoglossal arch; *g* Free edge of soft palate; *h*, *i* Palatopharyngeal arch; *k* Epiglottis; *l* Corniculate process of arytenoid; *m* Aryepiglottic fold; *n* Piriform recess; *o* Pharyngo-esophageal junction (dog, cat); *p* Nasopharynx; *q* Pharyngeal opening of auditory tube (horse); *r* Caudal part of laryngopharynx; *r'* Esophagus (dog, cat, goat)

1 Filiform, conical, and lenticular papillae (filiform papillae are not shown in the horse); *2* Fungiform papillae; *3* Vallate papillae; *4* Foliate papillae (pig, horse); *5* Lingual tonsil (not visible with the naked eye in the dog and cat; contained in the papillae of the root of the tongue in the pig); *6* Palatine tonsil (drawn out of the fossa tonsillaris in the dog and cat; not present in the pig; only the entrance to the tonsillar sinus is visible in the ox); *7* Tonsil of the soft palate (visible only in the pig); *8* Paraepiglottic tonsil (visible only in the cat)

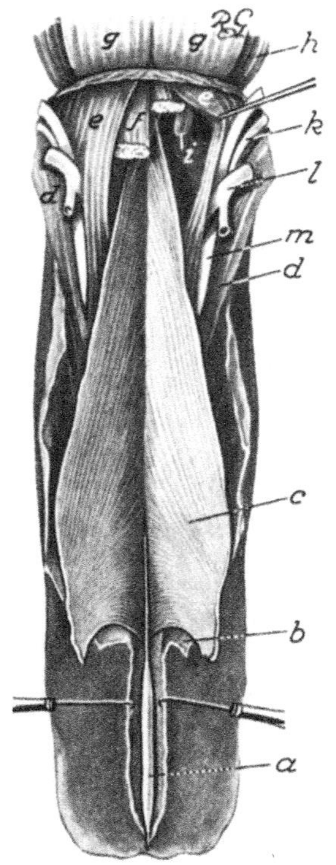

Fig. 42. Ventral surface of the tongue of the dog. (After Zietzschmann, unpublished)

a Lyssa, exposed; *b* Frenulum; *c* Genioglossus; *d* Styloglossus; *e* Hyoglossus; *f* Geniohyoideus; *g* Sternohyoideus; *h* Thyrohyoideus; *i* Epihyoid; *k* Lingual artery; *l* Lingual vein; *m* Hypoglossal nerve

The **filiform papillae** are soft, horny threads which cover the dorsum of the tongue in the pig, goat, and horse, and give the mucosa a velvety appearance. In the ox, sheep, cat, and to some extent also in the dog, the filiform papillae are small, have a core of connective tissue, and are directed caudally. They are heavily cornified in the cat and the ox. Scattered among the pointed filiform papillae of the ox, goat, and sheep, especially on the torus linguae, are the **conical papillae** and the flat **lenticular papillae** (39,40/*1*). At the root of the tongue of the carnivores and pig the filiform papillae are long and soft. In the other species the root of the tongue is free of papillae.

The **fungiform papillae** (36–41/*2*) are less numerous, but larger, than those just described. Because they contain **taste buds,** they are usually classified as gustatory papillae. Taste bud-free fungiform papillae are occasionally encountered in the ox and horse. The fungiform papillae are distributed over the dorsum of the tongue, especially along the borders. They are also found on the lateral, and even on the ventral surface of the tongue.

The **vallate papillae** (41/*3*) are located on the dorsum, just rostral to the root of the tongue. They have a core of connective tissue and are larger than the fungiform papillae. They are surrounded by a circular cleft and do not project above the surface of the tongue. There are many taste buds in the epithelium of the papilla facing the cleft, but only a few on the opposite wall (vallum) of the cleft. **Serous glands** (gll. gustatoriae) are found in the neighborhood of these papillae; they are either submucosal or intermuscular in position and shed their secretions into the depth of the cleft. The serous glands are believed to flush the cleft and to provide a neutral environment for the taste buds. The number of the

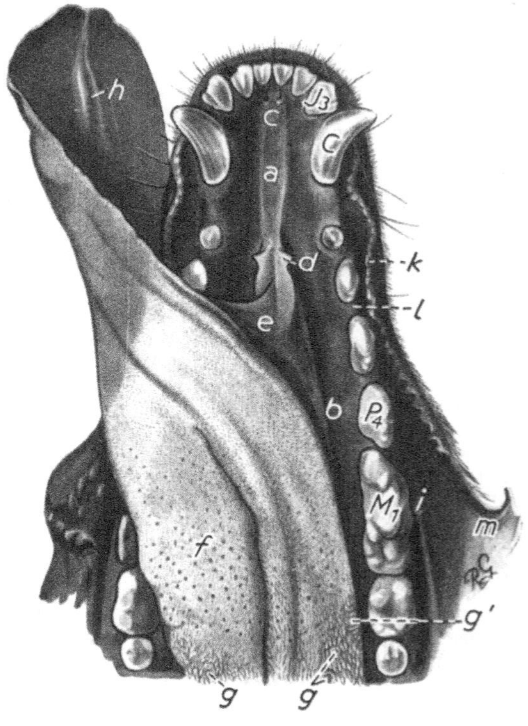

Fig. 43. Floor of the oral cavity of a dog. (After Zietzschmann, unpublished)

J3 Third incisor; *C* Canine; *P4* Fourth premolar; *M1* First molar (sectorial)

a Sublingual floor of oral cavity; *b* Lateral sublingual recess; *c* Orobasal organ; *d* Sublingual caruncle with openings of mandibular and major sublingual ducts; *e* Frenulum; *f* Dorsal surface of tongue; *g* Vallate papillae; **g' Foliate papilla; h Lyssa, exposed; i Vestibule; k Lower lip; l Gums; m Angle of the mouth**

vallate papillae varies with the species. The pig and horse have only one pair of very large papillae. The horse occasionally has an additional unpaired papilla. The carnivores have 2—3 papillae on each side, the ox 8—17, the sheep 18—24, and the goat 12—18.

The **foliate papillae** (41/*4*) are found on the border of the tongue immediately rostral to the palatoglossal arch, which is the thick mucosal fold connecting the root of the tongue with the ventral surface of the soft palate. These papillae consist of a series of parallel leaves of connective tissue separated from one another by small furrows. They are about 20 mm. long in the horse, 7—8 mm. long in the pig, extremely small in the dog, and rudimentary in the cat. The foliate papillae are altogether absent in the ruminants, except the ox, which occasionally has rudimentary ones. In the depth of the furrows are the taste buds. Serous glands are also present deep to the foliate papillae; they have a function similar to that of the glands associated with the vallate papillae.

In addition to the gustatory glands, the tongue contains other glandular accumulations of the serous, mucous, or seromucous type. They lie in the submucosa or between the muscle bundles of the root of the tongue, along the border and on the lateral surfaces, and in some species, also in the neighborhood of the frenulum. The glossoepiglottic fold and the palatoglossal arches (84/*2, 2'*) also contain numerous glands. The mucosa over the root of the tongue is characterized by the presence of lymphatic tissue. This takes the form of diffuse lymphocytic accumulations, solitary and aggregate lymph nodules, or tonsillar follicles, and is called, collectively, the **lingual tonsil** (*1*).

Lingual Muscles

The lingual muscles may be divided into an intrinsic **lingual muscle proper,** which forms the bulk of the tongue, and a group of extrinsic muscles. The intrinsic muscle consists of a system of fibers which are not attached to the skeleton and which run in longitudinal, perpendicular, and transverse directions. The **extrinsic muscles** originate from the skeleton and enter the tongue from behind and below. They blend with the fibers of the lingual muscle proper inside the tongue. A thin, median **lingual septum** divides the tongue into symmetrical halves.

The **LINGUAL MUSCLE PROPER** (25/*5*) consists of **deep** and **superficial longitudinal,** of **transverse,** and of **perpendicular fiber bundles.** The longitudinal bundles extend in both superficial and deep layers from the apex to the root of the tongue, the transverse fibers run from side to side, and the perpendicular fibers have a dorsoventral orientation. These three groups of muscle fibers are more or less diffusely distributed and, acting either singly or in unison, can alter the shape of the tongue. For instance, simultaneous contraction of the perpendicular and transverse fibers results in the lengthening of the tongue, if the longitudinal fibers remain relaxed. Contraction of the longitudinal fibers causes shortening of the tongue, if the transverse and perpendicular fibers remain inactive. Simultaneous contraction of all three groups produces rigidity of the tongue. Through the influence of the extrinsic muscles, the range of lingual movements and shapes is increased even further.

The **EXTRINSIC MUSCLES** of the tongue are as follows.

GENIOGLOSSUS (25/*7;* 42/*c*). This semipennate, flat, fan-shaped muscle lies next to the median plane and is separated from its fellow on the opposite side by the lingual septum. Its ventral margin is tendinous and extends from the incisive part of the mandible to the hyoid bone. From the tendinous margin the muscle bundles enter the tongue from below in fan-shaped fashion. The muscle draws the tongue rostrally and ventrally and may produce a median groove on the dorsum.

HYOGLOSSUS (45/*d*). This muscle lies ventrolateral to the root of the tongue. It is roughly rectangular, and is inserted between the genioglosses medially and the styloglossus laterally. It originates from the basihyoid and its lingual process and from the thyrohyoid. Its fibers enter the root of the tongue from behind and extend into the apex. Its contraction causes the tongue to be drawn caudally; therefore, it is an antagonist of the genioglossus. With the hyoid bone fixed, simultaneous action of the hyoglossus and genioglossus depresses the tongue.

STYLOGLOSSUS (45/*c*). This long, slender muscle originates with a flat tendon from the ventral end of the stylohyoid. It enters the tongue from behind, runs forward in the lateral part, and ends at the apex. Acting together, right and left styloglossi shorten the tongue and elevate its apex; acting singly, they pull the tongue laterally.

Hyoid Muscles

(44—46, 53—56)

Because of their close functional relationship to the muscles of the tongue, the hyoid muscles will be described at this point. A dorsal and a ventral group of hyoid muscles can be distinguished. The dorsal hyoid muscles are as follows.

MYLOHYOIDEUS (25/*11*). This is a flat muscle with fibers that run transversely. Together with its mate, it bridges the intermandibular space and forms a sling for the support of the tongue. In the pig, ruminants, and horse, the muscle consists of rostral and caudal parts. The rostral part in the ruminants and horse partly covers the caudal part (46/*15*, *15'*). In the pig, the two parts lie one behind the other and do not overlap. In the carnivores, the muscle is uniform. The mylohyoideus arises from the mylohyoid line on the lingual surface of the body of the mandible. It extends first ventrally, and then medially, and meets its fellow in a median fibrous raphe. The most caudal fibers of the muscle are attached on the basihyoid in the carnivores and pig and on the lingual process of that bone in the ruminants and horse. The action of the mylohyoideus serves to support and to elevate the tongue. This slinglike muscle is relatively weak in animals with a narrow intermandibular space (herbivores), but is well developed in the carnivores and the pig, which have a wide intermandibular space.

GENIOHYOIDEUS (25/*10*). This long, fusiform muscle lies in the intermandibular space under cover of the mylohyoideus. It extends from the incisive part of the mandible to the basihyoid (the lingual process in the ruminants and horse) and it moves the hyoid bone and, with it, the tongue rostrally (57, 58/*m*).

STYLOHYOIDEUS (46/*17*). This muscle originates from the dorsal end of the stylohyoid (stylohyoid angle (*c'*), where present) and ends on the thyrohyoid. It moves the hyoid bone and the larynx caudally and dorsally. In the horse, the tendon of insertion of the stylohyoid is perforated for the passage of the intermediary tendon of the digastricus (*S*, *S'*).

OCCIPITOHYOIDEUS (46/*17'*). This is a small muscle which originates from the paracondylar process and is inserted on the dorsal end of the stylohyoid. Its contraction helps to move the ventral end of the stylohyoid and, with it, the root of the tongue and larynx caudoventrally.

CERATOHYOIDEUS (369, 370/*1*). A thin muscular plate which, under cover of the hyoglossus, occupies the angle between the ceratohyoid and thyrohyoid. It arises from the rostral border of the thyrohyoid and ends on the caudal border of the ceratohyoid and the adjacent part of the stylohyoid. By elevating the thyrohyoid, it moves the larynx rostrodorsally.

HYOIDEUS TRANSVERSUS (84/*16*). This small muscle connects the two ceratohyoids. It is divided by an indistinct median tendon and is absent in the carnivores and pig.

The following three pairs of ventral hyoid muscles are thought to form the cranial continuation of the rectus abdominis muscles that connect the sternum with the pelvis along the floor of the abdominal cavity. They are straplike muscles running along the ventral surface of the neck and surrounding the trachea and larynx ventrally and laterally. The sternohyoidei lie next to the midline, followed laterally by the sternothyroidei and omohyoidei. Their names indicate their origin and insertion.

STERNOHYOIDEUS (46/*18*; 132/*h*). This is a flat band which extends from the manubrium sterni to the basihyoid. It lies next to the median plane and is separated from its fellow by a narrow raphe of connective tissue. By incising this raphe and separating the muscles, the surgeon gains access to the trachea and larynx. In the horse, the muscle is intersected by a tendon in the middle of the neck. The sternohyoideus pulls the hyoid bone, the root of the tongue, and the larynx caudally.

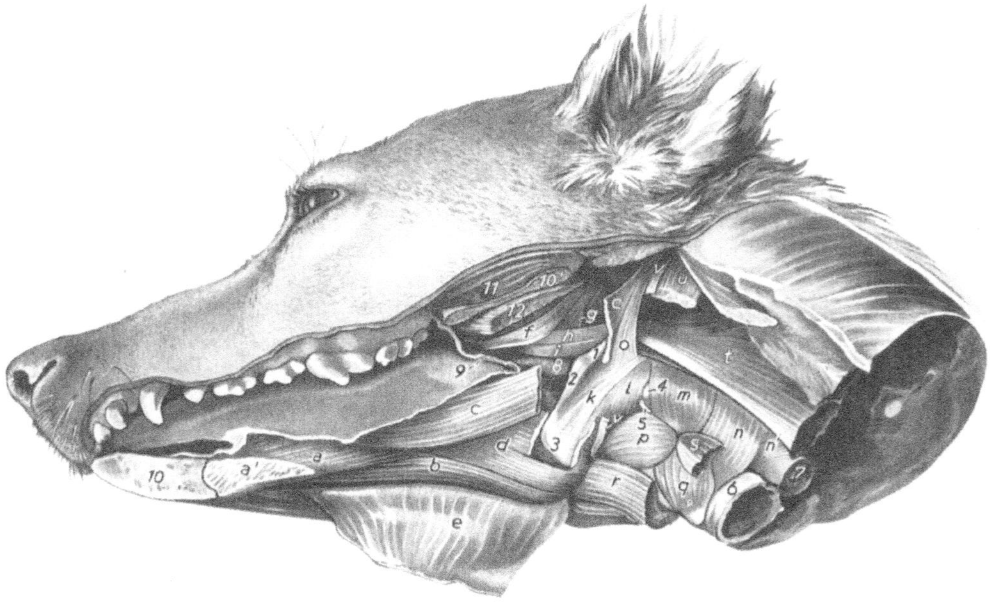

Fig. 44. Lingual and pharyngeal muscles of a dog. (After Zietzschmann, unpublished)

a Genioglossus, transected at *a'*; *b* Geniohyoideus; *c* Styloglossus, partly removed; *d* Hyoglossus; *e* Mylohyoideus; *f* Tensor veli palatini; *g* Levator veli palatini; *h* Pterygopharyngeus; *i* Palatopharyngeus; *k, l* Hyopharyngeus; *m* Thyropharyngeus; *n* Cricopharyngeus; *n'* Crico-esophageus; *o* Stylopharyngeus caudalis; *p* Thyrohyoideus; *q* Cricothyroideus; *r* Sternohyoideus; *s* Sternothyroideus; *t* Longus capitis; *u* Digastricus; *v* Stylohyoideus, partly removed

1 Stylohyoid; *2* Epihyoid; *3* Ceratohyoid; *4* Thyrohyoid; *5* Cranial laryngeal nerve; *6* Trachea; *7* Esophagus; *8* Glands of soft palate; *9* Oropharynx; *10* Mandibular symphysis; *10'* Ramus of mandible; *11* Masseter; *12* Medial pterygoideus

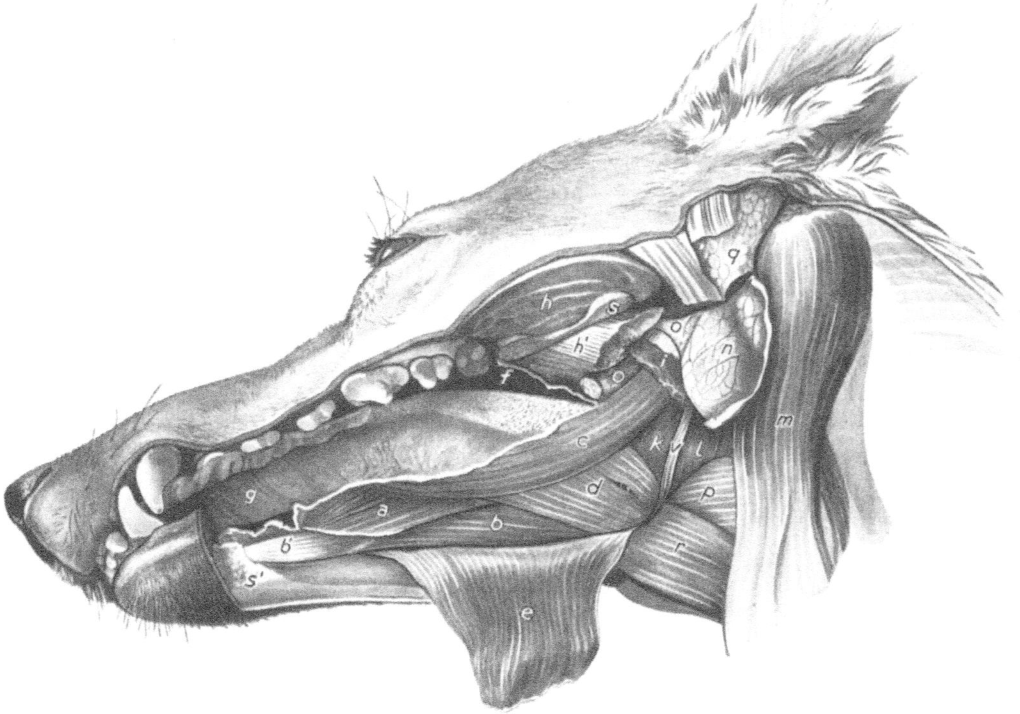

Fig. 45. Lingual and pharyngeal muscles of a dog. (After Zietzschmann, unpublished)

a Genioglossus; *b, b'* Geniohyoideus; *c* Styloglossus; *d* Hyoglossus; *e* Mylohyoideus; *f* Oropharynx; *g* Apex of the tongue; *h* Masseter; *h'* Medial pterygoideus; *i* Stump of digastricus; *k, l* Hyopharyngeus; *m* Sternocephalicus; *n* Mandibular gland; *o, o* Caudal portion of monostomatic sublingual gland; *p* Thyrohyoideus; *q* Parotid gland; *r* Sternohyoideus; *s, s'* Left mandible, partly removed; *v* Stylohyoideus

STERNOTHYROIDEUS (46/*20;* 132/*i*). This muscle originates with the sternohyoideus on the manubrium and covers first the ventral surface and then more cranially the lateral surface of the trachea. It ends with a flat tendon on the lateral surface of the thyroid lamina and draws the larynx caudally. Rostrally, it is continued by the thyrohyoideus (46/*20'*) which originates on the thyroid lamina immediately rostral to the insertion of the sternothyroideus and passes to the caudal border of the thyrohyoid. When the two muscles act together, the hyoid bone and the tongue are pulled caudally; and when the hyoid bone is fixed, the thyrohyoideus pulls the larynx closer to the hyoid bone.

OMOHYOIDEUS. This is a thin muscular sheet originating in the pig and horse from the subscapular fascia (in the horse also from the transverse processes of the second to fourth cervical vertebrae). In the ruminants (132/*g*), it arises from the deep fascia of the neck at the level of the third cervical vertebra. It is absent in the carnivores. The muscle is related laterally

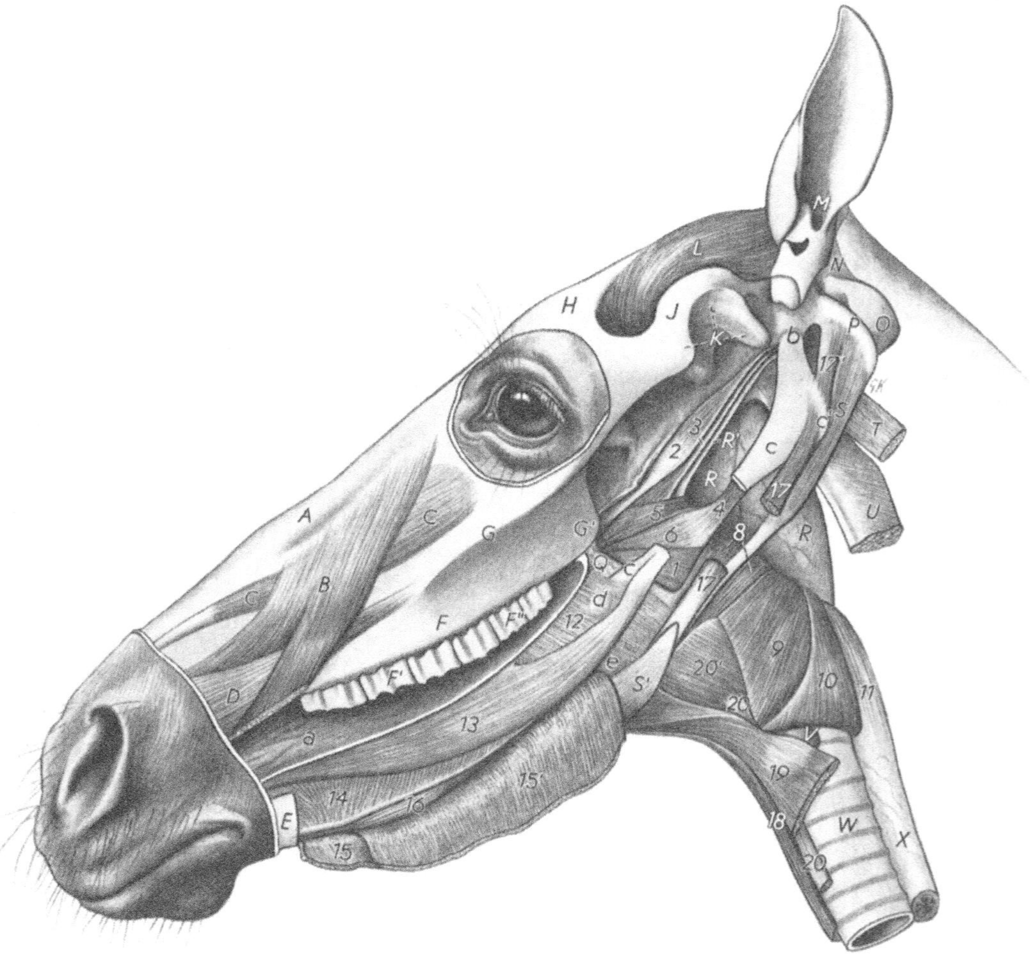

Fig. 46. Lingual and pharyngeal muscles of the horse.

A **Nasal bone;** ***B*** **Levator nasolabialis;** ***C*** **Levator labii superioris;** ***D*** **Caninus;** ***E*** **Stump of left mandible;** ***F*** **Alveolar process of maxilla;** ***F'*** **Third upper premolar;** ***F''*** **Second upper molar;** ***G*** **Facial crest;** ***G'*** **Maxillary tuberosity;** ***H*** **Frontal bone;** ***J*** **Zygomatic arch;** ***K*** **Articular** tubercle, mandibular fossa, and retroarticular process of temporal bone; *L* Temporalis; *M* Auricular cartilage; *N* Squamous part of occipital bone; *O* Occipital condyle; *P* Paracondylar process; *Q* Glands of the soft palate; *R, R* Guttural pouch; *R'* Auditory tube; *S, S'* Digastricus; *T* Rectus capitis ventralis; *U* Longus capitis; *V* Thyroid gland; *W* Trachea; *X* Esophagus

a Tongue; *b* Position of tympanohyoid; *c* Stylohyoid, partly removed; *c'* Stylohyoid angle; *d* Position of epihyoid; *e* Position of the junction between ceratohyoid and basihyoid (*d* and *e* are hidden by the hyoglossus)

1 Palatopharyngeus, ventral portion; *2* Tensor veli palatini; *3* Levator veli palatini; *4* Stylopharyngeus caudalis; *5* Pterygopharyngeus; *6* Palatopharyngeus, dorsal portion; *8* Hyopharyngeus; *9* Thyropharyngeus; *10* Cricopharyngeus; *11* Cricoesophageus; *12* Hyoglossus; *13* Styloglossus; *14* Genioglossus; *15, 15'* Mylohyoideus; *16* Geniohyoideus; *17* Stylohyoideus, partly removed; *17'* Occipitohyoideus; *18* Sternohyoideus; *19* Omohyoideus; *20* Sternothyroideus; *20'* Thyrohyoideus

to the brachiocephalicus with which it fuses in places. In the cranial third of the neck it crosses the trachea and passes between the external jugular vein laterally and the common carotid artery medially (131/*16*). Because of the protection this muscle gives to the common carotid artery in this region, the cranial third of the neck is the preferred site for venipuncture in the large domestic mammals. The omohyoideus is inserted on the basihyoid adjacent to the sternohyoideus, except in the pig where it is inserted on the thyrohyoid. It draws the hyoid bone and the larynx caudally.

The combined action of the ventral hyoid muscles pulls the hyoid apparatus and the larynx caudally and, by doing so, helps to dilate the pharynx. Raising and extending the animal's head has the same result, because the ventral hyoid muscles and the trachea are stretched, and they pull the hyolaryngeal complex out of the intermandibular space. If the animal is drenched with its head in this position, it may, therefore, experience difficulty in swallowing. The reason for this is that the caudal fixation of the hyoid bone and larynx prevents these two structures from attaining the elevated position that is essential prior to the act of swallowing (see also p. 56).

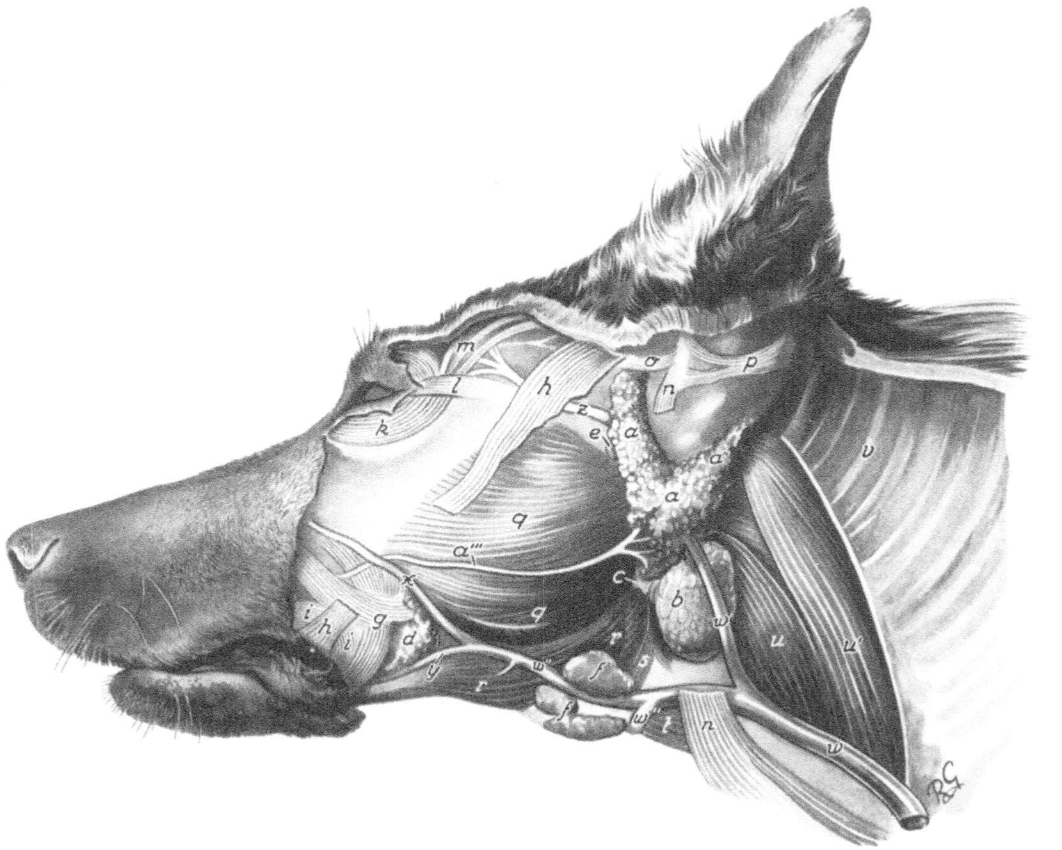

Fig. 47. Superficial salivary glands of a dog. (After Zietzschmann, unpublished)

a Parotid gland, *a'* its preauricular angle, *a''* its retroauricular angle; *a'''* Parotid duct; *b* Mandibular gland; *c* Caudal portion of monostomatic sublingual gland; *d* Buccal glands; *e* Parotid lymph node; *f, f* Mandibular lymph nodes; *g* Buccinator; *h* Zygomaticus, partly removed; *i* Orbicularis oris; *k* Orbicularis oculi; *l* Retractor anguli oculi; *m* Frontoscutularis; *n* Parotidoauricularis, partly removed; *o* Zygomaticoauricularis; *p* Cervicoauricularis profundus; *q* Masseter; *r* Digastricus; *s* Stylohyoideus; *t* Sternohyoideus; *u* Sternomastoideus; *u'* Sterno-occipitalis; *v* Platysma; *w* Ext. jugular vein; *w'* Maxillary vein; *w''*, *x* Facial vein; *w'''* Lingual vein; *y* Inferior labial vein; *z* Auriculopalpebral nerve

Sublingual Floor of Oral Cavity

(25, 27, 30, 32, 34, 43)

As mentioned previously, this part of the floor of the oral cavity is a crescent-shaped space that becomes visible when the tongue is raised. It consists of a rostral, prefrenular part and two lateral sublingual recesses which extend caudally between the mandible and the tongue.

The **prefrenular part** (43) lies inside the arch of the incisor teeth and is supported ventrally by the incisive part of the mandible. Its mucous membrane forms the median **frenulum linguae** (*e*) for connection to the ventral surface of the tongue. On either side of the frenulum but slightly more rostral are the **sublingual caruncles** (*d*). They vary with the species in shape, size, and position and are often absent in the pig. The mandibular and major sublingual salivary ducts (in the horse only the mandibular) open on the sublingual caruncles (32/*2*). In the horse and goat, small **paracaruncular glands** are found near the sublingual caruncles. Just caudal to the lower central incisors is the small **orobasal organ** (*1*). This consists of two

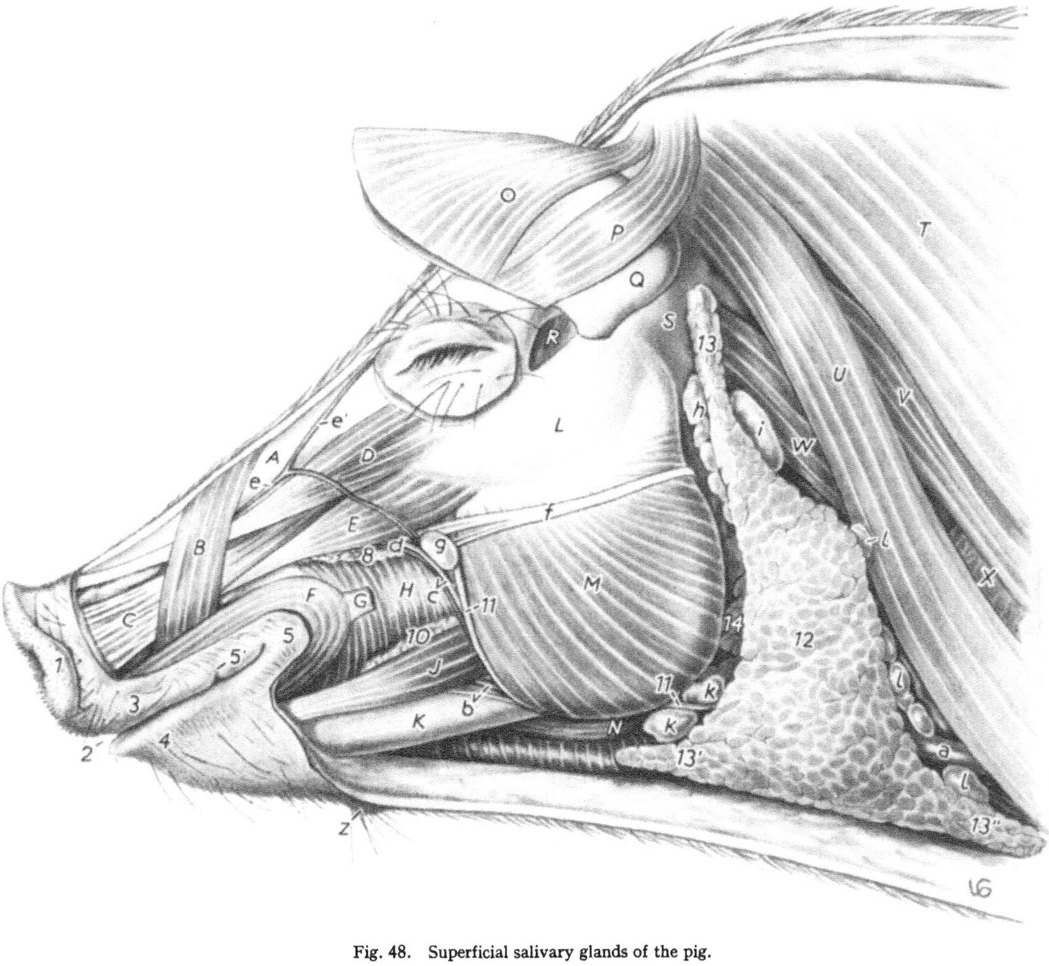

Fig. 48. Superficial salivary glands of the pig.

A Nasal bone; *B* Levator nasolabialis; *C* Caninus; *D* Levator labii superioris; *E* Depressor labii superioris; *F* Orbicularis oris; *G* Zygomaticus; *H* Buccinator; *J* Depressor labii inferioris; *K* Left mandible; *L* Zygomatic arch; *M* Masseter; *N* Digastricus; *O* Cervicoauricularis supf.; *P* Cervicoauricularis profundus; *Q* Auricular cartilage; *R* Temporalis; *S* Position of temporomandibular joint; *T* Cleidooccipitalis; *U* Cleidomastoideus; *V* Omotransversarius; *W* Sternomastoideus; *X* Omohyoideus; *Z* Mental hairs and gland

a Ext. jugular vein; *b* Inferior labial artery and vein; *c* Facial artery and vein; *d* Superior labial artery and vein; *e* Dorsal nasal vein; *e'* Angularis oculi vein; *f* Dorsal buccal branch of facial nerve; *g* Adipose tissue; *h* Parotid lymph nodes; *i* Lateral retropharyngeal lymph nodes; *k, k* Mandibular lymph nodes; *l, l, l* Ventral supf. cervical lymph nodes

1 Rostral plate; *2* Oral cleft; *3* Upper lip; *4* Lower lip; *5* Angle of the mouth; *5'* Notch for the canine tooth; *8, 10* Dorsal and ventral buccal glands; *11, 11* Parotid duct; *12* Parotid gland, *13* its auricular angle, *13'* its mandibular angle, *13''* its cervical angle; *14* Mandibular gland

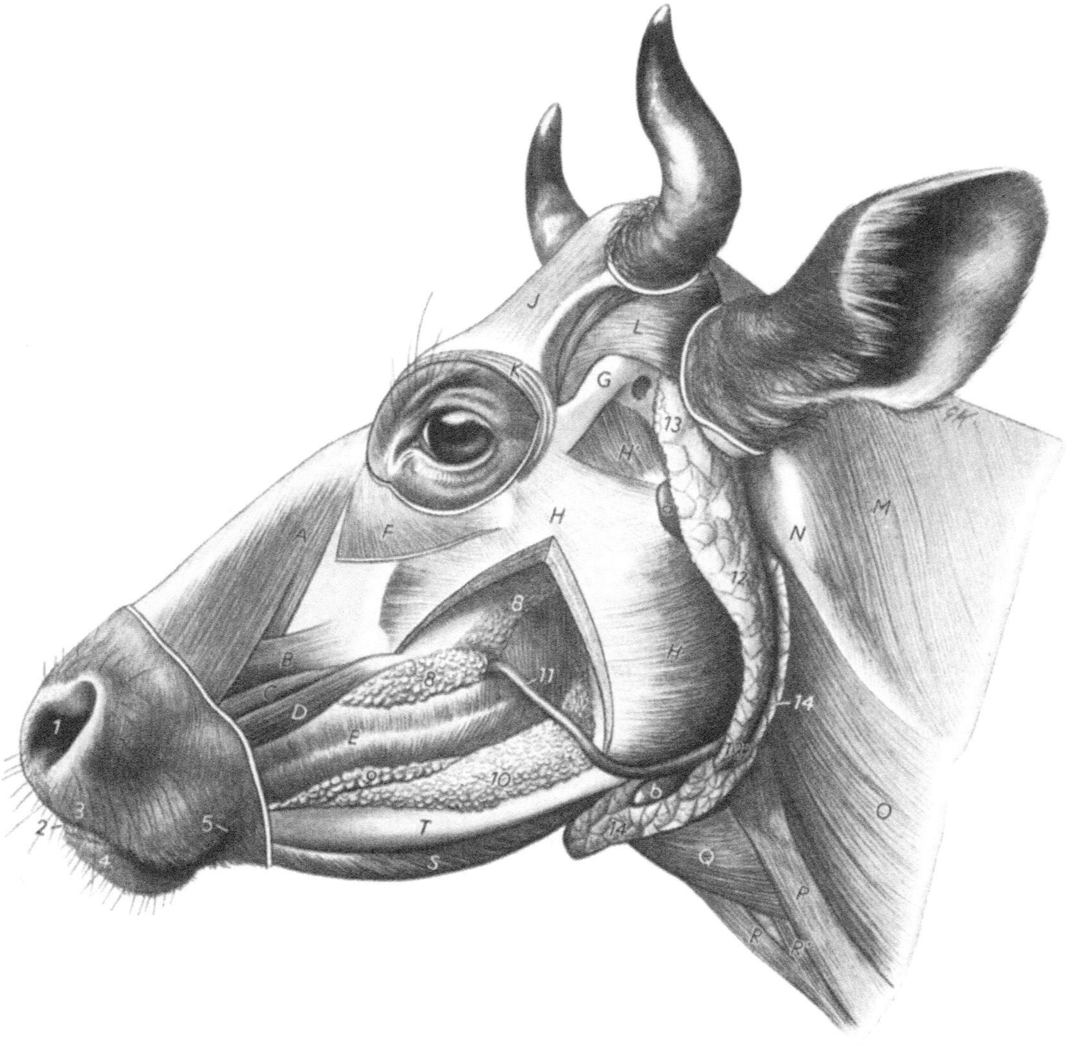

Fig. 49. Superficial salivary glands of the ox.

A Levator nasolabialis; *B* Levator labii superioris; *C* Caninus; *D* Depressor labii superioris; *E* Buccinator; *F* Malaris; *G* Zygomatic arch; *H* Masseter; *H'* Deep part of masseter; *J* Frontalis; *K* Orbicularis oculi; *L* Temporalis; *M* Cleidooccipitalis; *N* Wing of atlas; *O* Cleidomastoideus; *P* Sternocephalicus; *Q* Omohyoideus; *R* Sternohyoideus; *R'* Sternothyroideus; *S* Mylohyoideus; *T* Left mandible

a Parotid lymph node; *b* Mandibular lymph node

1 Left nostril; *2* Oral cleft; *3* Upper lip and nasolabial plate; *4* Lower lip and chin; *5* Angle of the mouth; *8, 9, 10* Dorsal, middle, and ventral buccal glands; *11* Parotid duct; *12, 13, 13''* Parotid gland; *14, 14* Mandibular gland

narrow epithelial strands or ducts which extend from the two external openings into the tunica propria. They are thought to be rudiments of the rostral sublingual glands found in reptiles. The small amount of lymphatic tissue often present in the vicinity of the sublingual caruncles is the **sublingual tonsil.**

The **lateral sublingual recesses** (25/*g*; 43/*b*) are bounded laterally by the lower cheek teeth and gums, and medially by the lateral surface of the tongue. Along the floor of each recess lies the polystomatic sublingual salivary gland inside a fold of mucous membrane (plica sublingualis). Along the edge of the fold are the openings of the many excretory ducts of this gland (53/*10'*).

Fig. 50. Superficial salivary glands of the goat.

A Levator nasolabialis; *B* Malaris; *C* Buccinator; *D* Depressor labii inferioris; *E* Masseter; *F* Left mandible; *G* Parotidoauricularis; *H* Cleidocephalicus; *J* Sternocephalicus; *J'* Tendon of insertion of sternocephalicus; *K* Sternohyoideus; *L* Wing of atlas; *M* Laryngeal prominence

a Ext. jugular vein; *b* Linguofacial vein; *c* Maxillary vein; *d* Caudal auricular vein; *e* Facial vein; *f* Inferior labial vein; *g* Superior labial vein; *h* Dorsal nasal vein; *i* Transverse facial vein and buccal branch of facial nerve; *k* Lateral retropharyngeal lymph node; *l* Mandibular lymph node

1 Left nostril; *2* Oral cleft; *3* Upper lip; *3'* Dental pad; *4* Lower lip; *4'* Third and fourth incisors; *5* Angle of the mouth; *8, 9, 10* Dorsal, middle, and ventral buccal glands; *11* Parotid duct; *12* Parotid gland, *13* its preauricular angle, *13'* its retroauricular angle, *13''* its mandibular angle; *14, 14* Mandibular gland; *15* Thyroid gland

Fig. 51 (Legend on facing page)

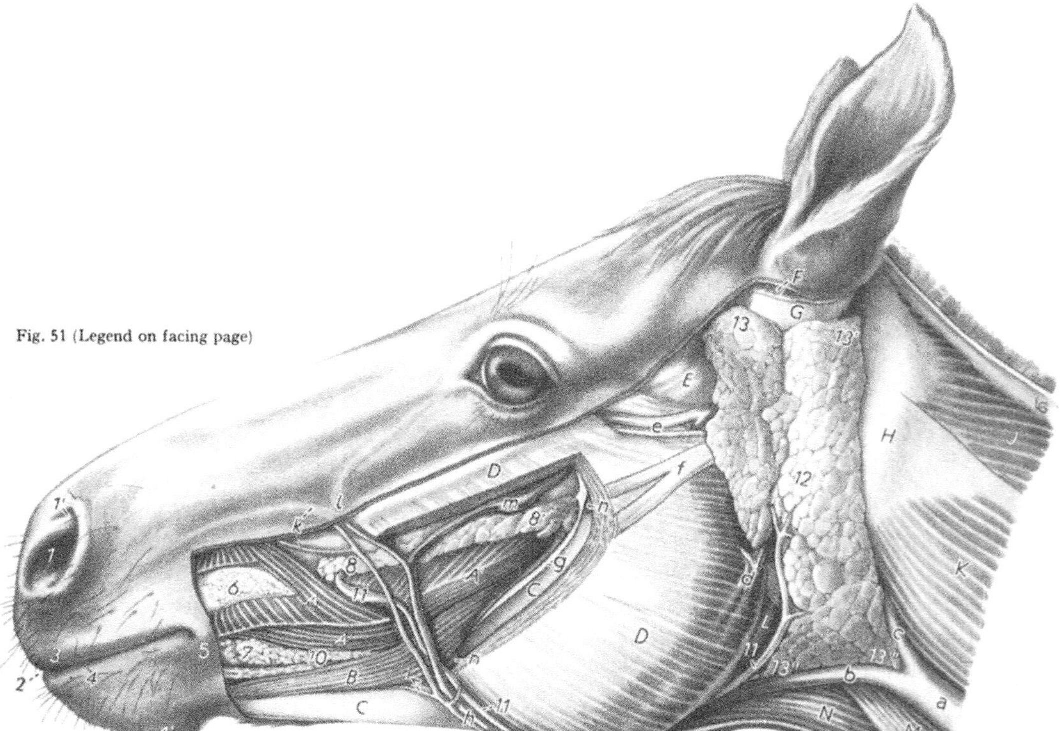

Salivary Glands

(47—56)

The small glands of the oral cavity, i.e., the labial, buccal, and lingual glands, as well as those on the hard palate and near the sublingual caruncles, have been described. They are only of local importance and provide the necessary moisture for the area in which they are found.

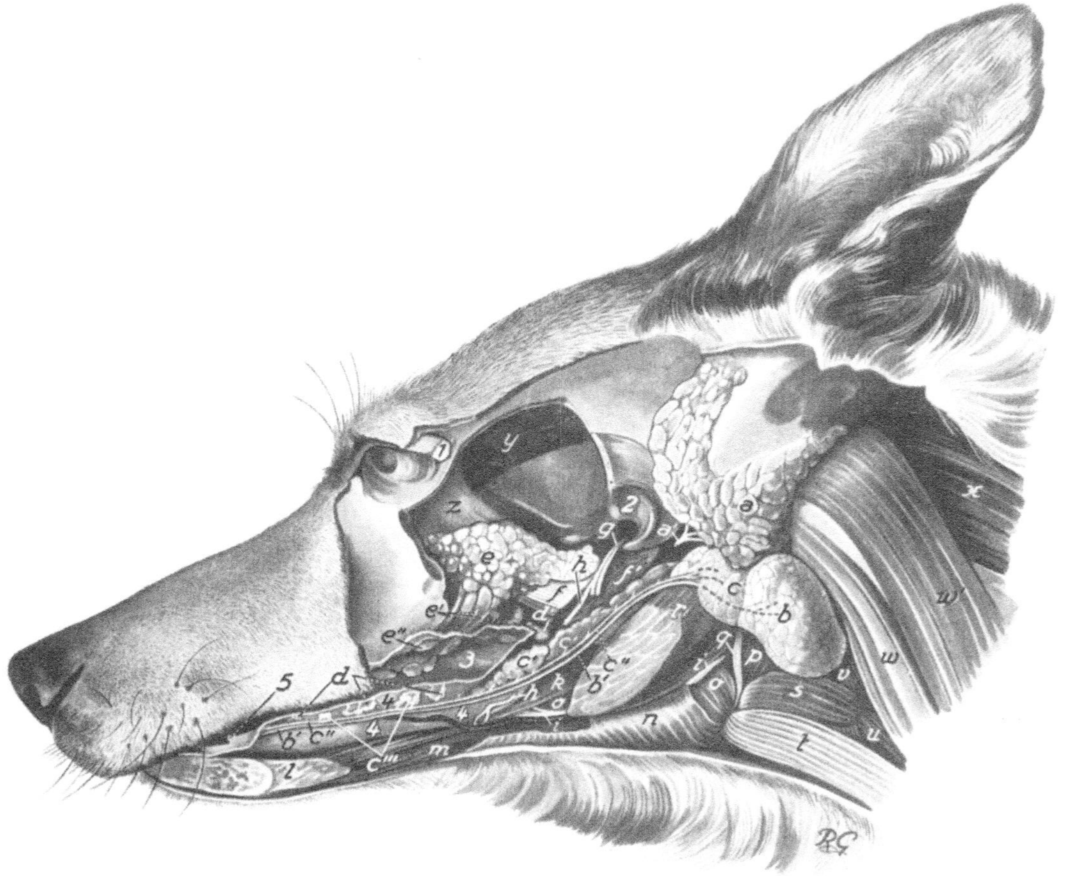

Fig. 52. Deep salivary glands, lingual and pharyngeal muscles of a dog. (After Scheuerer 1933)

a Parotid gland; *a'* Parotid duct; *b* Mandibular gland; *b', b'* Mandibular duct; *c* Caudal portion of monostomatic sublingual gland; *c'* Narrow rostral portion of monostomatic sublingual gland; *c'', c''* Major sublingual duct; *c'''* Solitary lobules of monostomatic sublingual gland next to lobules of the polystomatic sublingual gland (*d*); *d, d'* Polystomatic sublingual gland; *e* Zygomatic gland; *e'* Zygomatic ducts, *e''* their openings in the buccal vestibule; *f* Pterygoideus, *f'* its cut surface; *g* Masseteric nerve; *h* Lingual nerve, partly removed; *i* Hypoglossal nerve; *k* Styloglossus; *l* Cut surface of genioglossus; *m* Geniohyoideus; *n* Mylohyoideus; *o* Hyoglossus; *p* Hyopharyngeus; *q* Stylohyoideus; *r* Digastricus; *s* Thyrohyoideus; *t* Sternohyoideus; *u* Sternothyroideus; *v* Thyropharyngeus; *w* Sternomastoideus; *w'* Sterno-occipitalis; *x* Splenius; *y* Temporalis; *z* Periorbita

1 Orbital ligament, transected; *2* Mandibular fossa of temporal bone; *3* Tongue; *4* Oral mucosa on lingual surface of mandible; *5* Sublingual caruncle with openings of mandibular and major sublingual ducts

Fig. 51. Superficial salivary glands of the horse.

A, A Buccal and molar parts of buccinator; *B* Depressor labii inferioris; *C* Left mandible; *D* Masseter; *E* Position of temporomandibular joint; *F* Stump of parotidoauricularis; *G* Base of ear; *H* Wing of atlas; *J* Splenius; *K* Cleidomastoideus; *L* Occipitomandibular part of digastricus; *M* Sternomandibularis; *N* Sternohyoideus and omohyoideus

a Ext. jugular vein; *b* Linguofacial vein; *c* Maxillary vein; *d* Masseteric artery and vein; *e* Transverse facial artery and vein; *f* Buccal branches of facial nerve; *g* Buccal nerve; *h* Facial artery and vein; *i* Inferior labial artery and vein; *k* Superior labial artery and vein; *l* Dorsal nasal artery and vein; *m* Deep facial vein; *n, n* Stumps of buccal vein

1 Left nostril; *1'* Nasal diverticulum; *2* Oral cleft; *3* Upper lip; *4* Lower lip; *4'* Chin; *5* Angle of the mouth; *6* Upper labial glands; *7* Lower labial glands; *8, 8'* Dorsal buccal glands; *10* Ventral buccal glands, continuous rostrally with the lower labial glands; *11* Parotid duct; *12* Parotid gland, *13* its preauricular angle, *13'* its retroauricular angle, *13''* its mandibular angle, *13'''* its cervical angle

The three large salivary glands, the **parotid, mandibular,** and **sublingual glands,** are of more general importance. Their secretion, the saliva, is serous or mucous in character, and is produced in large quantities, e.g., 40—50 liters a day in the horse and the ox. During mastication the saliva is mixed with the food and aids in the formation of the bolus. It also acts as a lubricant during swallowing. In addition, the saliva, especially that of the pig, contains an enzyme, ptyalin, which initiates the hydrolysis of starch in the mouth. In general, the large salivary glands of the herbivores are better developed than those of the carnivores.

The salivary glands may be clasisfied on the basis of their secretions as serous, mucous, or seromucous (mixed) glands. The distribution of these types varies from species to species. Histology textbooks should be consulted for details.

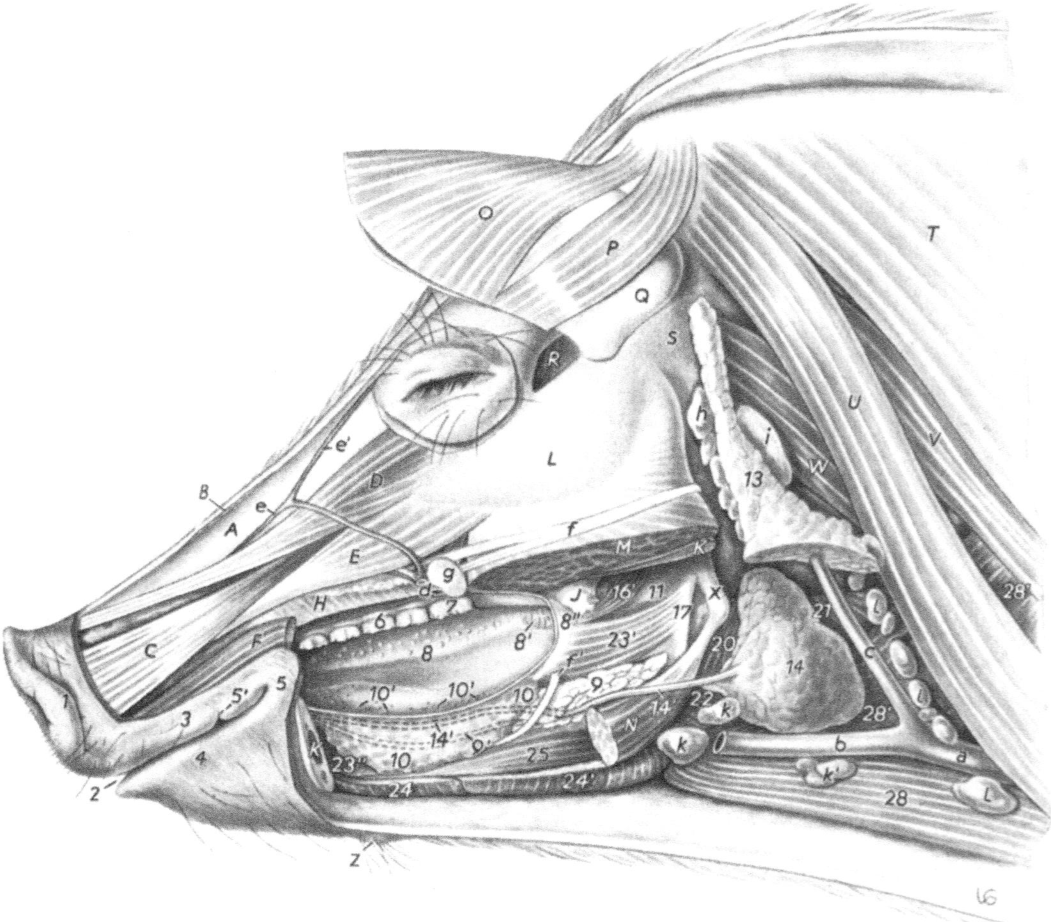

Fig. 53. Deep salivary glands, lingual and pharyngeal muscles of the pig.

A Nasal bone; *B* Levator nasolabialis, transected; *C* Caninus; *D* Levator labii superioris; *E* Depressor labii superioris; *F* Orbicularis oris, partly removed; *H* Buccinator; *J* Sphenoid and pterygoid processes of the palatine and sphenoid bones, respectively; *K, K* Left mandible, partly removed; *L* Zygomatic arch; *M* Masseter; *N* Digastricus; *O* Cervicoauricularis supf.; *P* Cervicoauricularis profundus; *Q* Auricular cartilage; *R* Temporalis; *S* Position of temporomandibular joint; *T* Cleido-occipitalis; *U* Cleidomastoideus; *V* Omotransversarius; *W* Sternomastoideus; *X* Paracondylar process; *Z* Mental hairs and gland

a Ext. jugular vein; *b* Linguofacial vein; *c* Maxillary vein; *d* Facial artery and vein at the point of division into superior labial artery and vein, deep facial vein, and dorsal nasal vein; *e* Dorsal nasal vein; *e'* Angularis oculi vein; *f* Dorsal buccal branch of facial nerve; *f'* Lingual nerve; *g* Adipose tissue; *h* Parotid lymph nodes; *i* Lateral retropharyngeal lymph nodes; *k, k'* Mandibular lymph nodes; *l, l, l* Ventral supf. cervical lymph nodes

1 Rostral plate; *2* Oral cleft; *3* Upper lip; *4* Lower lip; *5* Angle of the mouth; *5'* Notch for canine tooth; *6* Fourth upper premolar; *7* Second upper molar (the third molar has not yet erupted); *8* Body of tongue with fungiform papillae; *8'* Foliate papilla; *8''* Pterygomandibular fold; *9* Monostomatic sublingual gland; *9'* Major sublingual duct; *10, 10* Polystomatic sublingual gland; *10', 10'* Openings of the minor sublingual ducts; *11* Glands of the soft palate; *13* Auricular angle of parotid gland; *14* Mandibular gland; *14'* Mandibular duct; *16'* Tensor veli palatini; *17* Stylohyoid; *20* Hyopharyngeus; *21* Caudal pharyngeal constrictors; *22* Thyrohyoideus; *23'* Styloglossus; *23''* Genioglossus; *24, 24'* Mylohyoideus; *25* Geniohyoideus; *28* Sternohyoideus; *28'* Omohyoideus

Parotid Gland

(47—52)

The lobules of the parotid gland are visible to the naked eye. The color of the gland depends on its functional state and changes with the amount of blood present in the gland. It is, however, always of a lighter red than the adjacent skeletal muscles. The gland more or less fills the **retromandibular fossa,** which is the depression caudal to the ramus of the mandible and ventral to the wing of the atlas. The gland is in contact dorsally with the base of the ear;

Fig. 54. Deep salivary glands, lingual and pharyngeal muscles of the ox.

A Levator nasolabialis; *B* Levator labii superioris; *C* Caninus; *D* Depressor labii superioris; *E* Alveolar process of maxilla; *F* Malaris; *G* Zygomatic arch; *G'* Mandibular fossa of temporal bone; *G''* Base of ear; *H* Masseter; *J* Frontalis; *K* Orbicularis oculi; *L* Temporalis; *M* Funicular part of lig. nuchae; *M'* Multifidus; *M''* Rectus capitis dorsalis; *N* Atlas; *N'* Paracondylar process; *O* Axis; *P* Third cervical vertebra; *Q* Vertebral canal with spinal cord; *R* Longus colli; *S* Longus capitis; *T* Digastricus; *U* Stump of left mandible

1 Left nostril; *2* Oral cleft; *3* Upper lip and nasolabial plate; *4* Lower lip and chin; *5* Angle of the mouth; *6* Third upper premolar; *7* Second upper molar; *8* Body of tongue; *8'*, *8'* Buccal mucosa with papillae; *8''* Vallate papillae; *9* Monostomatic sublingual gland; *10* Polystomatic sublingual gland; *11* Palatine glands; *12* Palatine tonsil; *14'* Mandibular duct; *15* Thyroid gland; *16'* Tensor veli palatini; *16''* Levator veli palatini; *17*, *17* Stylohyoid, partly removed; *18* Stylopharyngeus caudalis; *19* Rostral pharyngeal constrictors; *20* Hyopharyngeus; *21* Thyropharyngeus; *21'* Cricopharyngeus; *22* Thyrohyoideus; *22'* Cricothyroideus; *23*, *23* The two parts of the hyoglossus; *23'* Styloglossus; *24*, *24'* Mylohyoideus; *25* Geniohyoideus; *26* Stylohyoideus; *26'* Occipitohyoideus; *28* Sternohyoideus; *28'* Omohyoideus; *28''* Sternothyroideus; *29* Esophagus; *30* Trachea

a Medial retropharyngeal lymph node

ventrally, it extends into the neck or into the intermandibular space for varying distances. Its lateral surface is covered by the parotid fascia and the parotidoauricularis. Medially, it is intimately related to the branches of the common carotid artery and external jugular vein, to the hyoid bone and its muscles, to branches of the facial and trigeminal nerves, and to lymph nodes. In the horse, it also makes contact with the guttural pouch, the occipitomandibular part of the digastricus, and the sternomandibularis. Because of the clinical importance of the parotid region in the horse, a section on parotid topography is found on page 71 in the chapter dealing with the mouth and pharynx of that species.

Fig. 55. Deep salivary glands, lingual and pharyngeal muscles of the goat.

A Levator nasolabialis; *B* Malaris; *C* Alveolar process of maxilla; *D* Pterygoideus; *E* Masseter; *E'* Deep part of masseter; *F, F* Left mandible, partly removed; *G* Stump of parotidoauricularis; *G'* Base of ear; *H* Cleidocephalicus; *J* Sternocephalicus; *J'* Tendon of insertion of sternocephalicus; *K* Digastricus; *L* Wing of atlas; *M* Laryngeal prominence

a Ext. jugular vein; *b* Supf. temporal vein; *c* Parotid lymph node; *d* Medial retropharyngeal lymph node; *e* Mandibular lymph node; *f* Lateral retropharyngeal lymph node

1 Left nostril; *2* Oral cleft; *3* Upper lip; *3'* Dental pad; *4* Lower lip; *4'* Third and fourth incisors; *5* Angle of the mouth; *6* Third upper premolar; *7* Second upper molar; *8* Body of tongue with fungiform papillae; *8'* Buccal mucosa with papillae; *8''* Pterygomandibular fold; *9* Monostomatic sublingual gland; *9'* Major sublingual duct; *10* Polystomatic sublingual gland; *11* Palatine glands; *12* Palatine tonsil; *14* Mandibular salivary gland; *14'* Mandibular duct; *15* Thyroid gland; *17* Stylohyoid; *19* Pterygopharyngeus; *20* Stylopharyngeus rostralis; *21* Hyopharyngeus; *23* Hyoglossus; *23'* Styloglossus; *24, 24'* Mylohyoideus; *25* Geniohyoideus; *26* Stylohyoideus; *26'* Occipitohyoideus; *27* Ceratohyoideus; *28* Sternohyoideus

SHAPE, SIZE, AND POSITION. The parotid gland of the carnivores is small and roughly triangular (47/*a*). Its apex is directed ventrally and covers only a small portion of the mandibular gland. Its concave base embraces the base of the auricular cartilage from below. In the pig, the parotid gland is large and triangular in shape (48/*12*). The auricular angle of the gland (*13*) does not quite reach the base of the ear; the mandibular angle (*13'*) passes rostrally beyond the level of the cranial border of the masseter; and the cervical angle (*13''*) extends caudally almost the full length of the neck. The parotid gland of the ox is club-shaped (49/*12, 13, 13'*). The thicker end of the gland is directed toward the ear and makes extensive contact rostrally with the masseter muscle. Its narrower ventral end extends roughly to the angle of the mandible and partly covers the mandibular gland. The parotid gland of the small ruminants (50) is similar to that of the ox. The parotid gland of the horse is large in comparison to that of the ox, and fills the retromandibular fossa completely (51). Its dorsal end is at the base of the ear. The ventral end of the gland is wider than the dorsal and occupies the angle between the linguofacial and external jugular veins (*b, c*). It has short mandibular and cervical angles (*13''*, 13''').

The **parotid duct** (47/*a‴*) is formed by the union of numerous radicles within the gland. In the cranivores, and usually also in the small ruminants, it crosses the lateral surface of the masseter muscle. In the pig, ox, and horse (48—51/*11*) it runs first on the medial surface of the mandible after leaving the gland, and it then winds around the ventral border of the mandible to gain the lateral surface of the face. It passes rostrodorsally in the cheek to the buccal vestibule. The parotid duct ends on the lateral wall of the buccal vestibule on a small elevation (papilla parotidea) which is located in the cat opposite the second upper cheek tooth; in dog, goat, and horse opposite the third; in the sheep opposite the fourth; in the ox opposite the fifth; and in the pig opposite the third or fourth upper cheek tooth.

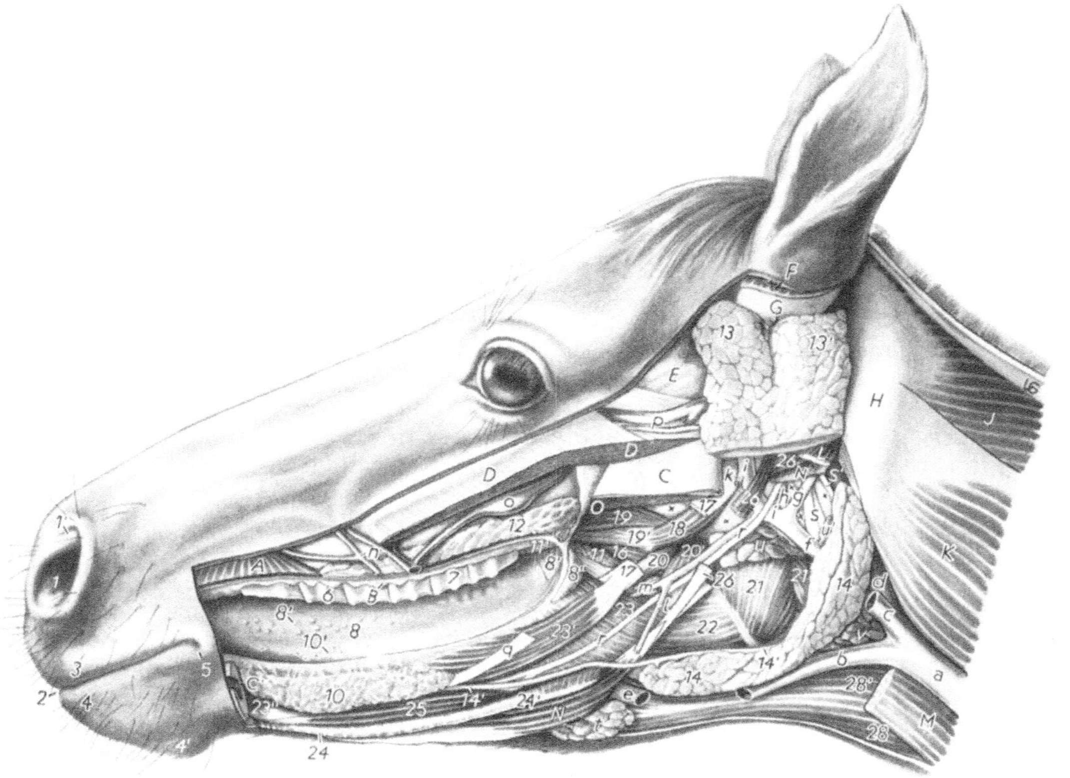

Fig. 56. Deep salivary glands, lingual and pharyngeal muscles of the horse.

A Stump of buccinator; *B* Alveolar process of maxilla; *C, C'* Left mandible, partly removed; *D* Stump of masseter; *E* Position of temporomandibular joint; *F* Stump of parotidoauricularis; *G* Base of ear; *H* Wing of atlas; *J* Splenius; *K* Cleidomastoideus; *L* Occipitomandibular part of digastricus; *M* Sternomandibularis; *N, N'* Rostral and caudal bellies of digastricus; *O* Pterygoideus

a Ext. jugular vein; *b* Linguofacial vein; *c* Maxillary vein; *d* Occipital vein; *e* Lingual vein; *f* Common carotid artery; *g* Occipital artery; *h* Int. carotid artery; *i, i* Ext. carotid artery; *k* Masseteric artery; *l* Facial artery medial to mandible; *m* Lingual artery; *n* Facial artery and vein; *o* Deep facial vein; *p* Transverse facial artery and vein; *q* Lingual nerve; *r* Hypoglossal nerve; *s* Vagus and sympathetic nerves; *s'* Accessory nerve; *t* Mandibular lymph node; *u* Medial retropharyngeal lymph node; *u'* Lateral retropharyngeal lymph node; *v* Cranial cervical lymph node; *x, x, x, x* Guttural pouch

1 Left nostril; *1'* Nasal diverticulum; *2* Oral cleft; *3* Upper lip; *4* Lower lip; *4'* Chin; *5* Angle of the mouth; *6* Third upper premolar; *7* Second upper molar; *8* Body of tongue; *8'* Fungiform papillae; *8‴, 8″* Foliate papilla with associated glands; *10* Polystomatic sublingual gland; *10'* Openings of minor sublingual ducts; *11* Palatine glands; *11'* Palatine tonsil; *12* Dorsal buccal glands; *13* Preauricular angle of parotid gland; *13'* Retroauricular angle of parotid gland; *14* Mandibular gland; *14'* Mandibular duct; *16* Palatopharyngeus; *17* Stylohyoid, partly removed; *18* Stylopharyngeus caudalis; *19* Pterygopharyngeus; *19'* Palatopharyngeus; *20* Stylopharyngeus rostralis; *20'* Hyopharyngeus; *21* Thyropharyngeus; *21'* Cricopharyngeus; *22* Thyrohyoideus; *23* Hyoglossus; *23'* Styloglossus; *23″* Genioglossus; *24, 24'* Mylohyoideus; *25* Geniohyoideus; *26* Stylohyoideus, partly removed; *26'* Occipitohyoideus; *28* Sternohyoideus; *28'* Omohyoideus

Mandibular Gland

(47—50, 52, 53, 55, 56)

The mandibular gland occupies the space between the basihyoid and the wing of the atlas, and is partly covered by the parotid gland. In the carnivores (52/*b*), it is oval und usually larger than the parotid gland. The mandibular gland of the pig (53/*14*) is similar in shape to that of the carnivores; it is smaller than the parotid gland and has a small rostral angle. In the ruminants (55/*14*), the mandibular gland is large and extends from the wing of the atlas well into the intermandibular space where it is markedly thickened. The mandibular gland of the horse (56/*14*) is much smaller than the parotid gland; it is long and narrow and reaches to the basihyoid rostrally.

The **mandibular duct** (*14'*) passes rostrally between the mylohyoid and hyoglossus muscles, medial to the sublingual salivary glands. It opens on the sublingual caruncle on the floor of the oral cavity.

Sublingual Glands

(52—56)

The sublingual salivary glands, two in number, lie under the mucosa of the lateral sublingual recess and of the lateral surface of the tongue. As its name implies, the **monostomatic* sublingual gland** (55/*9*), which is absent in the horse, has only one excretory duct, the **major sublingual duct** (*9'*). In the ruminants, carnivores, and pig, this duct opens on the sublingual caruncle, which is not well developed in the carnivores and pig. The **polystomatic** sublingual gland** (53/*10*) consists of a loose chain of small lobules, each of which secretes through its own short **minor sublingual duct** (*10'*) into the lateral sublingual recess. The polystomatic gland is rostral to the monostomatic gland in the carnivores and pig, while in the ruminants it is somewhat dorsocaudal to the monostomatic gland. The two glands extend together from the palatoglossal arch to the symphysis of the mandible.

Pharynx

(44, 46, 56, 57—66, 79—88)

The pharynx (61/*b*, *c*, *d*, *e*) is a funnel-shaped, musculo-membranous passage that connects the oral cavity with the esophagus, and the nasal cavity with the larynx. The concave roof of the pharynx is related to the base of the cranium (vomer and the bodies of the sphenoids) and to the rectus capitis ventralis and longus capitis muscles. In the dog it extends caudally to the second, and in the cat to the third, cervical vertebra. The pharynx of the horse is pushed away from the sphenoids and the above-mentioned muscles by the guttural pouches (62/*14'*). The lateral walls of the pharynx are related to the stylohyoids and the pterygoid muscles, and in the horse also to the guttural pouches. The floor of the pharynx extends from the root of the tongue over and around the laryngeal entrance to about the level of the cricoid cartilage of the larynx. The rostral portion of the pharyngeal cavity is divided by the **soft palate** (61/*36*) into dorsal and ventral channels, the **NASOPHARYNX** (*b*) and **OROPHARYNX** (*c*), respectively. The narrower caudal portion of the pharyngeal cavity is known as the **LARYNGOPHARYNX** (*d*, *d*, *e*). The free border of the soft palate (*38*) and the paired palatopharyngeal arches (*39*) surround the **intrapharyngeal opening** (ostium intrapharyngeum), which is located above the entrance (aditus) to the larynx. Through the ostium the nasopharynx communicates with the laryngopharynx.

The pharyngeal cavity has the following openings:

1. The paired **choanae** (*13*) rostrodorsally; they connect the nasopharynx with the nasal cavity.

2. The paired **pharyngeal openings of the auditory tubes** (*14*) dorsolaterally; they connect the nasopharynx with the auditory tubes and, thus, with the middle ears.

---

* From Gr., having one opening.
** From Gr., having many openings.

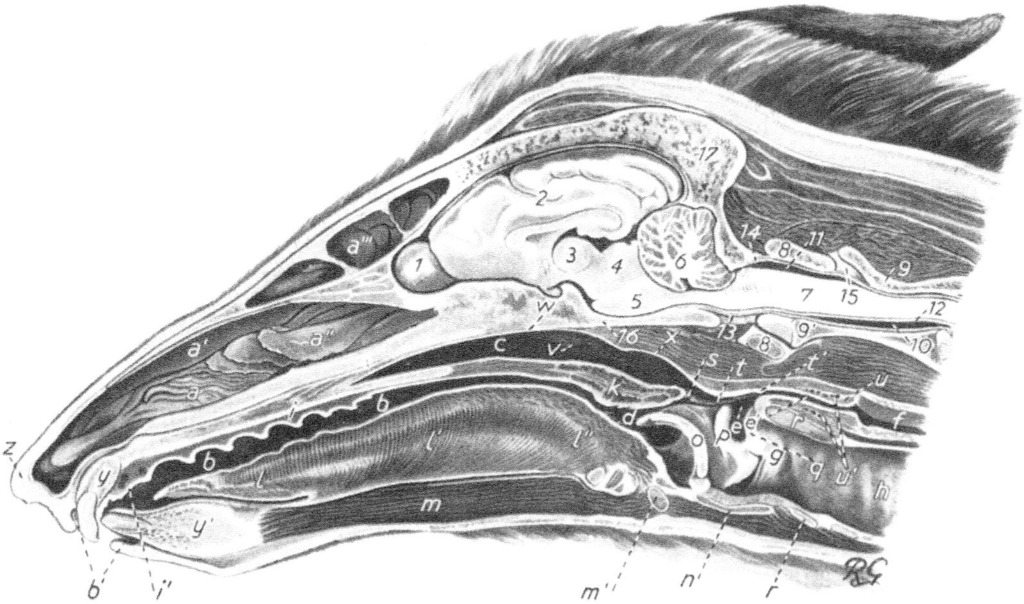

Figs. 57 and 58. Sagittal section through the head of a dog and a cat. (Fig. 57 after Zietzschmann, unpublished)

***a—a''* Right nasal cavity, exposed by the removal of the nasal septum; *a* Ventral nasal concha; *a'* Dorsal nasal concha; *a''* Ethmoid conchae; *a'''* Frontal sinus, with septum of the frontal sinus in the cat;** *b* Oral cavity proper; *b'* Vestibule; *c* Nasopharynx; *d* Oropharynx; *e, e'* Laryngopharynx; *f* Esophagus; *g* Larynx; *h* Trachea; *i* Hard palate; *i'* Incisive papilla; *k* Soft palate; *k'* Palatine tonsil (cat); *l* Apex, *l'* body, and *l''* root of tongue; *m* Geniohyoideus; *m'* Basihyoid; *n* Thyroid cartilage; *o* Epiglottis; *o'* Aryepiglottic fold with paraepiglottic tonsil (cat); *p* Cuneiform process (dog); *q* Arytenoid cartilage; *r, r* Cricoid cartilage; *s—t'* Intrapharyngeal opening; *s* Free border of soft palate; *t, t'* Palatopharyngeal arch (only visible in the dog); *u* Cricopharyngeus; *u'* Venous plexus; *v* Pharyngeal opening of auditory tube; ***w* Pharyngeal fornix (dog); *w'* Pharyngeal tonsil (cat); *x* Pharyngeal raphe; *y* Alveolar process of incisive bone; *y'* Incisive part of mandible; *z* Nasal plate**

1 Olfactory bulb; *2* Cerebrum; *3* Interthalamic adhesion; *4* Area of corpora quadrigemina; *4'* Hypophysis (cat); *5* Pons; *6* Cerebellum; *7* Spinal cord; *8, 8* Atlas; *9* Axis; *9'* Dens of axis. Illustrated only in the dog: *10* Epidural space; *11* Subarachnoid space, cerebello-medullary cistern; *12* Dura mater; *13* Ventral atlanto-occipital membrane; *14* Dorsal atlanto-occipital membrane; *15* Interarcual space between atlas and axis; *16* Base of cranium; *17* Roof of cranium

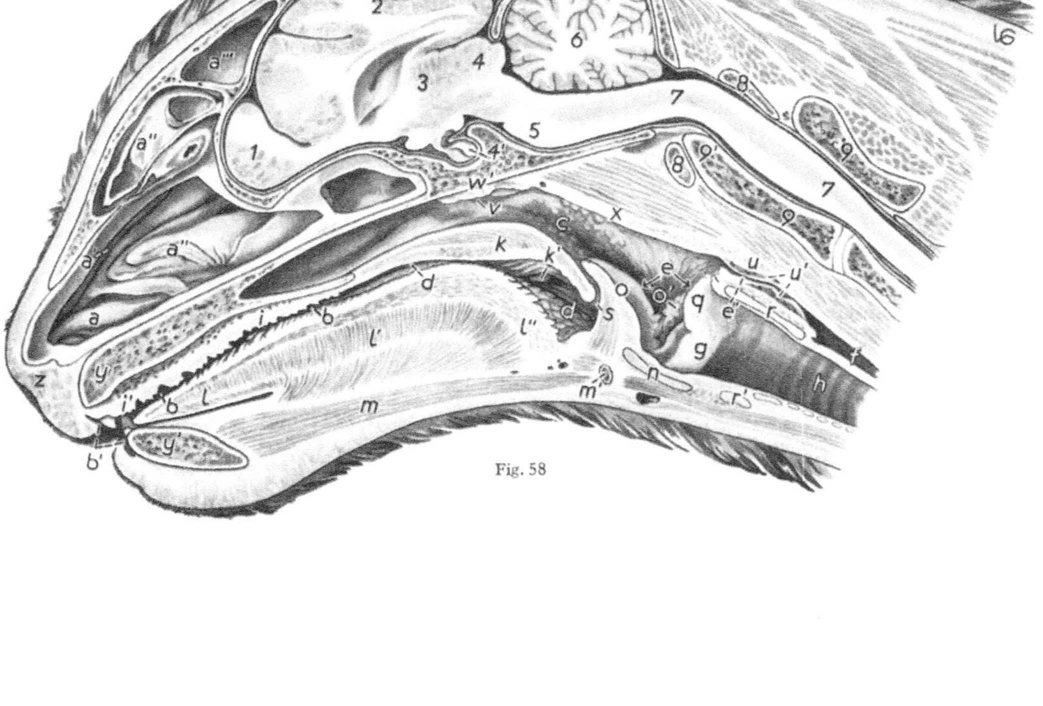

Fig. 58

3. The slitlike **isthmus faucium** (59/between *37* and *30*) leading from the oral cavity into the oropharynx and bounded laterally by the palatoglossal arches, dorsally by the soft palate, and ventrally by the root of the tongue.

4. The **aditus laryngis** caudoventrally. This opening is surrounded by the rostral laryngeal cartilages (61/*16*, *17*), which project upward from the floor of the laryngopharynx. When the animal swallows, the aditus is closed by the epiglottis (*16*).

5. The **entrance into the esophagus** (*e*) at the caudal end of the laryngopharynx.

Both respiratory air and food are channeled through the pharynx. The air passes from the nasal cavity to the larynx—that is, from rostrodorsal to caudoventral—during inspiration and in the opposite direction during expiration. Food passes from the oral cavity to the esophagus during deglutition and, in the ruminants, also in the reversed direction during regurgitation, and crosses the respiratory passageway in the pharyngeal cavity, specifically, in the laryngopharynx. It is the function of the pharynx and related structures to direct the air or the food, as the case may be, toward its proper destination. The mechanism of swallowing is described on page 56.

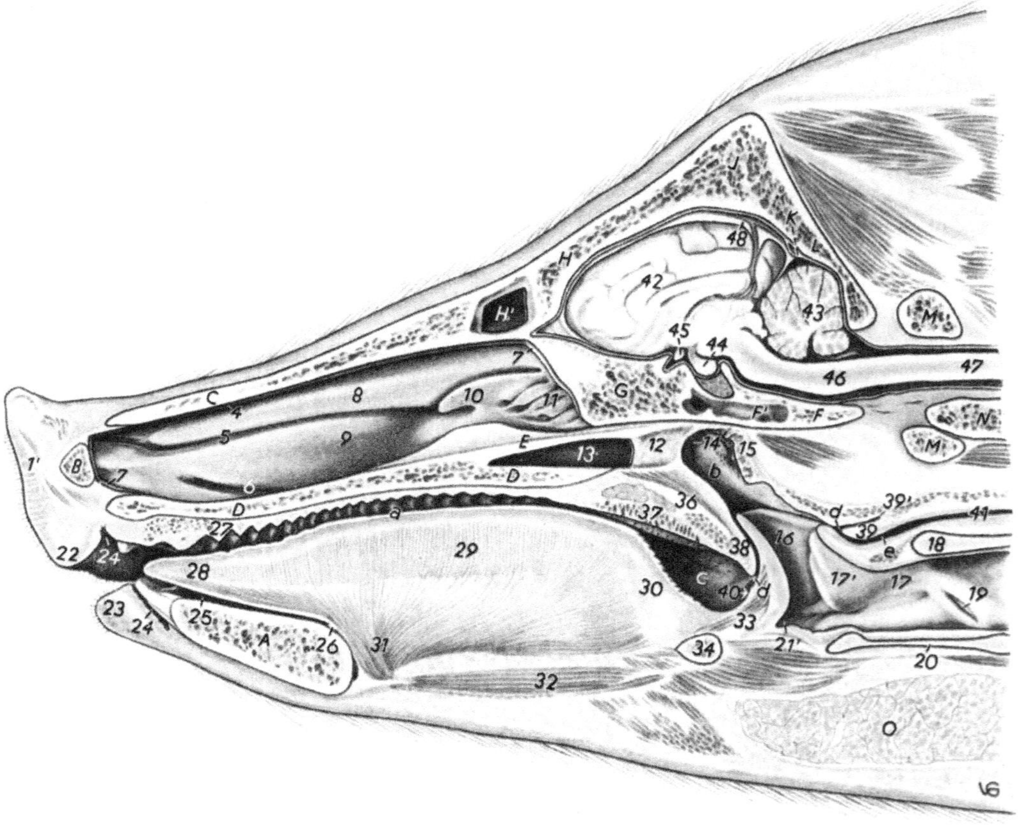

Fig. 59. Sagittal section of the head of a pig. The right nasa lcavity has been opened by removing the nasal septum.

a Oral cavity; *b* Nasopharynx; *c* Oropharynx; *d*, *d*, *e* Laryngopharynx

A Incisive part of mandible with first incisor; *B* Rostral bone; *C* Nasal bone; *D* Osseous palate; *E* Vomer; *F* Sphenoid bone; *F'* Sphenoid sinus; *G* Ethmoid bone; *H* Frontal bone; *H'* Frontal sinus; *J* Parietal bone; *K* Int. occipital protuberance; *L* Occipital bone; *M*, *M* Atlas; *N* Dens of axis; *O* Mandibular angle of parotid gland

1' Rostrum with rostral plate; *4* Dorsal, *5* middle, and *6* ventral nasal meatuses; *7* Nasal septum, almost entirely removed; *8* Dorsal, *9* ventral, and *10* middle nasal conchae; *11* Ethmoid conchae; *12* Pharyngeal septum, fenestrated; *13* Choana; *14* Pharyngeal opening of auditory tube with tubal tonsil; *15* Pharyngeal tonsil; *16* Epiglottis; *17* Arytenoid cartilage; *17'* Corniculate process; *18* Lamina of cricoid cartilage; *19* Entrance to lateral laryngeal ventricle; *20* Thyroid cartilage; *21'* Middle laryngeal ventricle; *22* Upper lip; *23* Lower lip; *24* Vestibule; *25* Sublingual floor of oral cavity; *26* Frenulum linguae; *27* Hard palate with venous plexus and palatine ridges; *28* Apex, *29* body, and *30* root of tongue; *31* Genioglossus; *32* Geniohyoideus; *33* Hyoepiglotticus in glossoepiglottic fold; *34* Basihyoid; *36* Soft palate with glands and muscles; *37* Tonsil of the soft palate; *38*, *39* Rostral and caudal boundaries of the intrapharyngeal opening; *38* Free border of the soft palate; *39* Caudal end of palatopharyngeal arch; *39'* Pharyngeal diverticulum; *40* Piriform recess; *41* Esophagus; *42* Cerebrum; *43* Cerebellum; *44* Hypophysis; *45* Optic chiasma; *46* Brain stem; *47* Spinal cord; *48* Falx cerebri with sagittal sinus

The **NASOPHARYNX** (pars nasalis pharyngis, 61/*b*) is part of the respiratory channel (see also p. 221). It lies dorsal to the soft palate and extends from the choanae to the intrapharyngeal opening. The roof of the nasopharynx (fornix pharyngis) is concave, both from front to back and from side to side, and in the ruminants and pig is divided by the median **pharyngeal septum** (*12*). The **pharyngeal tonsil** (*15*) is located high in the roof of the nasopharynx, either in or next to the median line, and forms the end of the pharyngeal septum in the ruminants. The slitlike **pharyngeal openings of the auditory tubes** (*14*) are in the lateral walls of the nasopharynx.

The **OROPHARYNX** (pars oralis pharyngis, *c*) is part of the digestive tract, although air passes through it when the animal coughs or breathes orally. The oropharynx extends from the palatoglossal arches to the base of the epiglottis. Its roof is formed by the soft palate and its floor is the root of the tongue. In its lateral walls are the palatine tonsils and the structures associated with them*. When the animal breathes through its nose (which it does most of the time) the ventral surface of the soft palate is in contact with the root of the tongue,

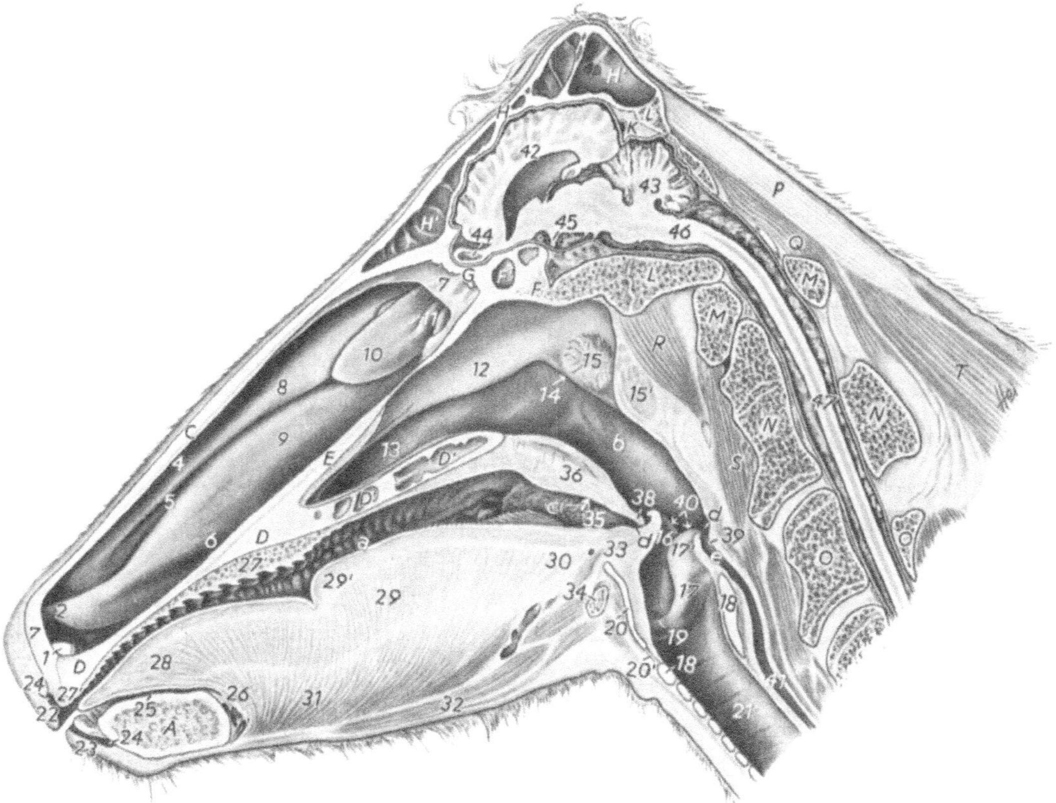

Fig. 60. Sagittal section of the head of a cow. The right nasal cavity has been opened by removing the nasal septum.

a Oral cavity; *b* Nasopharynx; *c* Oropharynx; *d, d, e* Laryngopharynx

A Incisive part of mandible with first incisor; *C* Nasal bone; *D* Osseous palate; *D'* Palatine sinus; *E* Vomer; *F* Sphenoid bone; *F'* Sphenoid sinus; *G* Ethmoid bone; *H* Frontal bone; *H', H'* Frontal sinus; *K* Int. occipital protuberance; *L, L* Occipital bone; *M, M* Atlas; *N, N* Axis; *O, O* Third cervical vertebra; *P* Funicular part of lig. nuchae; *Q* Rectus capitis dorsalis; *R* Longus capitis; *S* Longus colli; *T* Semispinalis capitis

1 Right nostril; *2* Alar fold; *4* Dorsal, *5* middle, and *6* ventral nasal meatuses; *7* Nasal septum, almost entirely removed; *8* Dorsal, *9* ventral, and *10* middle nasal conchae; *10, 11* Ethmoid conchae; *12* Pharyngeal septum; *13* Choana; *14* Pharyngeal opening of auditory tube, hidden by the pharyngeal septum; *15* Pharyngeal tonsil; *15'* Medial retropharyngeal lymph node; *16* Epiglottis; *17* Arytenoid cartilage; *17'* Corniculate process; *18, 18* Cricoid cartilage; *19* Vocal fold; *20* Thyroid cartilage; *20'* Cricothyroid ligament; *21* Trachea; *22* Upper lip; *23* Lower lip; *24* Vestibule; *25* Sublingual floor of oral cavity; *26* Frenulum linguae; *27* Hard palate with venous plexus and palatine ridges; *27'* Dental pad; *28* Apex of tongue; *29* Body of tongue; *29'* Torus linguae, in front of it the fossa linguae; *30* Root of tongue; *31* Genioglossus; *32* Geniohyoideus; *33* Hyoepiglotticus in glossoepiglottic fold; *34* Basihyoid; *35* Entrance to the sinus of the palatine tonsil; *36* Soft palate with glands and muscles; *38, 39* Rostral and caudal boundary of intrapharyngeal opening; *38* Free border of soft palate; *39* Caudal end of palatopharyngeal arch; *40* Piriform recess; *41* Esophagus; *42* Cerebrum; *43* Cerebellum; *44* Olfactory bulb; *45* Optic tract; *46* Brain stem; *47* Spinal cord

* The part of the oropharynx bounded by the palatine tonsils may be called fauces.

and the lumen of the oropharynx is obliterated (65A). During swallowing, however, the palate is lifted away from the tongue and the bolus can pass through the oropharynx (65B).

The **LARYNGOPHARYNX** (pars laryngea pharyngis, 63/*e, e'*) is common to the respiratory and the digestive channels. It is the caudal continuation of the oropharynx (*d*) and extends from the base of the epiglottis (*o*) to about the level of the cricoid cartilage (*r*). It contains the rostral parts of the larynx (*o, p*) which project rostrodorsally from the floor. On each side of the base of the epiglottis are the **piriform recesses** (88/7) which continue the floor of the oropharynx around the entrance of the larynx. With the soft palate close to the tongue in the normal breathing position, the rostral parts of the larynx protrude through the intrapharyngeal opening, and the laryngopharynx is largely obliterated (65B). Liquids (milk in the newborn, saliva) may, however, pass from the mouth to the esophagus through the piriform recesses without swallowing movements. The caudal part of the laryngopharynx that is covered by the caudal pharyngeal constrictors is also known as the vestibulum esophagi (63/*e'*). In the carnivores, the caudal extent of the laryngopharynx is marked by an annular fold of mucous membrane (limen pharyngoesophageum, *u*).

STRUCTURE OF THE PHARYNGEAL WALL. The pharyngeal wall consists, from the inside out, of mucous membrane, fascia, a layer of pharyngeal muscles, fascia, and a tunica adventitia.

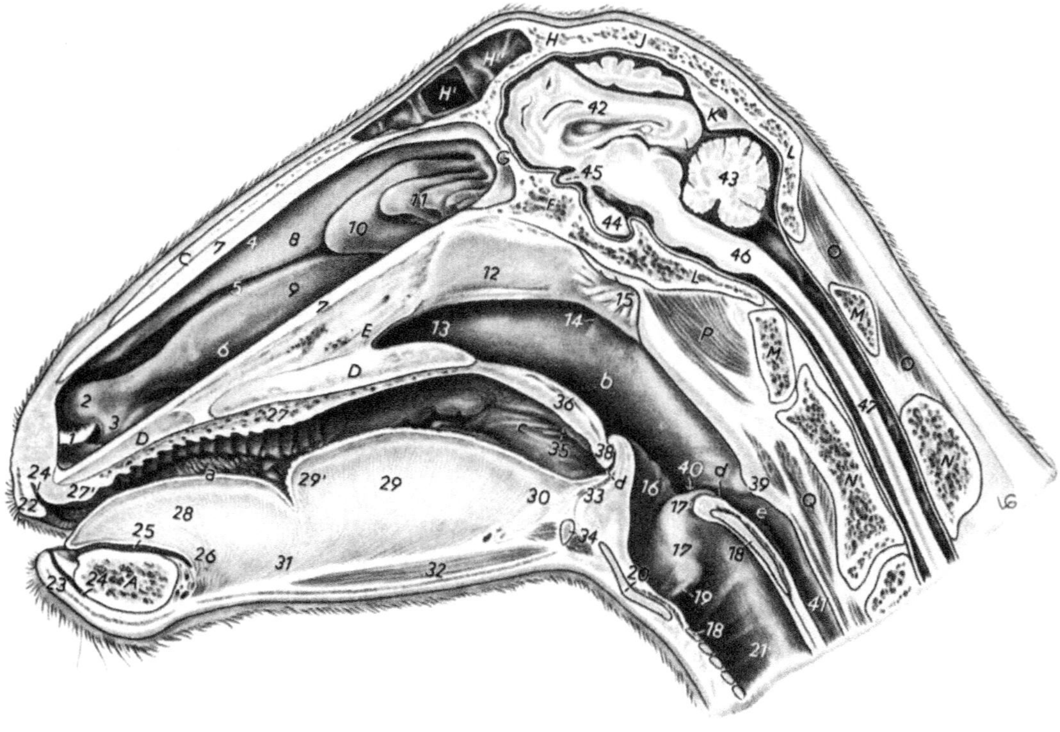

Fig. 61. Sagittal section of the head of a sheep. The right nasal cavity has been opened by removing the nasal septum.

a Oral cavity; *b* Nasopharynx; *c* Oropharynx; *d, d, e* Laryngopharynx

A Incisive part of mandible with first incisor; *C* Nasal bone; *D* Osseous palate; *E* Vomer; *F* Sphenoid bone; *G* Ethmoid bone; *H* Frontal bone; *H'* Frontal sinus; *H''* Sagittal septum of frontal sinuses, fenestrated; *J* Parietal bone; *K* Int. occipital protuberance; *L, L* Occipital bone; *M, M* Atlas; *N, N* Axis; *O* Rectus capitis dorsalis; *P* Longus capitis; *Q* Longus colli

1 Right nostril; *2* Alar fold; *3* Basal fold; *4* Dorsal, *5* middle, and *6* ventral nasal meatuses; *7* Nasal septum, almost entirely removed; *8* Dorsal, *9* ventral, and *10* middle nasal conchae; *10, 11* Ethmoid conchae; *12* Pharyngeal septum; *13* Choana; *14* Pharyngeal opening of auditory tube; *15* Pharyngeal tonsil; *16* Epiglottis; *17* Arytenoid cartilage; *17'* Corniculate process; *18, 18* Cricoid cartilage; *19* Vocal fold; *20* Thyroid cartilage; *21* Trachea; *22* Upper lip; *23* Lower lip; *24* Vestibule; *25* Sublingual floor of oral cavity; *26* Frenulum linguae; *27* Hard palate with venous plexus and palatine ridges; *27'* Dental pad; *28* Apex of tongue; *29* Body of tongue; *29'* Torus linguae, in front of it the fossa linguae; *30* Root of tongue; *31* Genioglossus; *32* Geniohyoideus; *33* Hyoepiglotticus in glossoepiglottic fold; *34* Basihyoid; *35* Palatine tonsil; *36* Soft palate with glands and muscles; *38, 39* Rostral and caudal boundary of intrapharyngeal opening; *38* Free border of soft palate; *39* Caudal end of palatopharyngeal arch; *40* Piriform recess; *41* Esophagus; *42* Cerebrum; *43* Cerebellum; *44* Hypophysis; *45* Optic chiasma; *46* Brain stem; *47* Spinal cord

The **mucous membrane** lining the nasopharynx is similar to that of the respiratory region of the nasal cavity. It is slightly folded, contains **glands** (gll. pharyngeae) and regional accumulations of lymphoid tissue, and it has a ciliated pseudostratified columnar epithelium. The remaining pharyngeal mucosa is like that of the oral cavity; it also contains glands, but is has a stratified squamous epithelium. At the entrance to the esophagus, the mucosa overlies extensive venous plexuses. The **fascia** between the mucosa and the pharyngeal muscles is thin and is attached to the **pharyngeal raphe;** the fascia on the outside of the pharyngeal muscles is thicker.

The **PHARYNGEAL MUSCLES** are striated and consist of a series of bilateral pairs. They function under the swallowing reflex and therefore are not under direct voluntary control. Except for the stylopharyngeus caudalis, which is a dilator, the pharyngeal muscles are constrictors and are all inserted on the pharyngeal raphe. Since they differ little from species to species as far as their origins and insertions are concerned, they are described only once, in the section that follows.

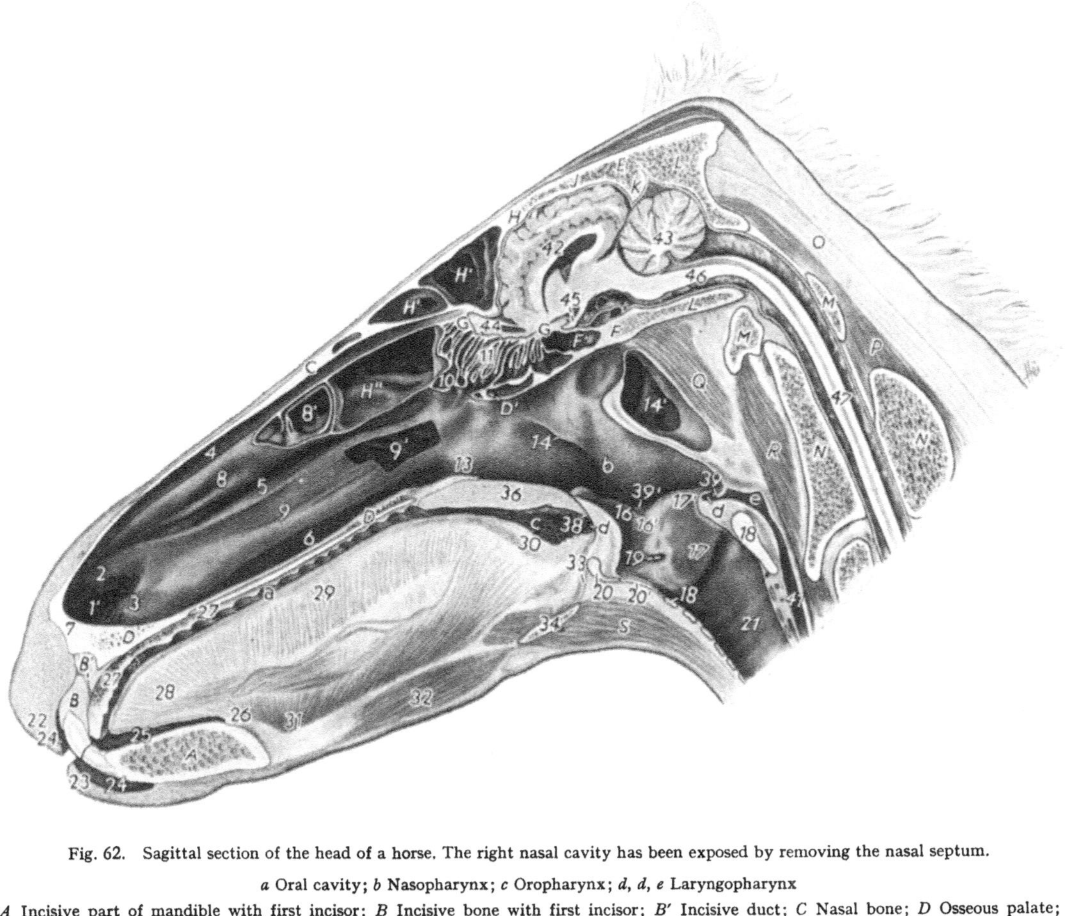

Fig. 62. Sagittal section of the head of a horse. The right nasal cavity has been exposed by removing the nasal septum.

a Oral cavity; *b* Nasopharynx; *c* Oropharynx; *d, d, e* Laryngopharynx

A Incisive part of mandible with first incisor; *B* Incisive bone with first incisor; *B'* Incisive duct; *C* Nasal bone; *D* Osseous palate; *E* Interparietal bone; *F* Sphenoid bone; *F'* Sphenoid sinus; *G* Ethmoid bone; *H* Frontal bone; *H', H''* Conchofrontal sinus; *H'* Frontal sinus; *H''* Sinus of dorsal nasal concha; *J* Parietal bone; *K* Tentorium cerebelli osseum; *L, L* Occipital bone; *M, M* Atlas; *N, N* Axis; *O* Funicular part of lig. nuchae; *P* Rectus capitis dorsalis; *Q* Longus capitis; *R* Longus colli; *S* Sternohyoideus and omohyoideus

1' Nasal vestibule; *2* Alar fold; *3* Basal fold; *4* Dorsal, *5* middle, and *6* ventral nasal meatuses; *7* Nasal septum, almost entirely removed; *8* Dorsal nasal concha; *8'* Conchal cells; *9* Ventral nasal concha; *9'* Sinus of ventral nasal concha; *10* Middle nasal concha, opened; *10, 11* Ethmoid conchae; *13* Choana; *14* Pharyngeal opening of auditory tube; *14'* Right guttural pouch, opened; *16* Epiglottis; *16'* Aryepiglottic fold; *17* Arytenoid cartilage; *17'* Corniculate process; *18, 18* Cricoid cartilage; *19* Entrance to lateral laryngeal ventricle; *20* Thyroid cartilage; *20'* Cricothyroid ligament; *21* Trachea; *22* Upper lip; *23* Lower lip; *24* Vestibule; *25* Sublingual floor of oral cavity; *26* Frenulum linguae; *27* Hard palate with venous plexus and palatine ridges; *28* Apex, *29* body, and *30* root of tongue; *31* Genioglossus; *32* Geniohyoideus; *33* Hyoepiglotticus in glossoepiglottic fold; *34* Basihyoid with lingual process; *36* Soft palate with glands and muscles; *38, 39* Rostral and caudal boundaries of intrapharyngeal opening; *38* Free border of soft palate; *39* Caudal end of the palatopharyngeal arch; *41* Esophagus; *42* Cerebrum; *43* Cerebellum; *44* Olfactory bulb; *45* Optic chiasma; *46* Brain stem; *47* Spinal cord

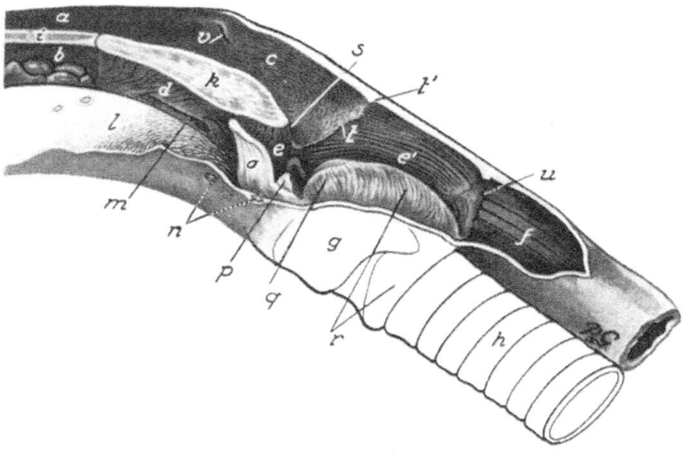

Fig. 63. Pharynx of a dog, opened from the left side. The soft palate is in mouth breathing position. (After Zietzschmann, 1939.)

a Choana; *b* Oral cavity; *c* Nasopharynx; *d* Oropharynx; *e, e′* Laryngopharynx; *f* Esophagus; *g* Thyroid cartilage; *h* Trachea; *i* Palatine bone; *k* Soft palate; *l* Root of tongue; *m* Fossa tonsillaris; *n* Hyoid bone; *o* Epiglottis; *p* Cuneiform process; *q* Arytenoid cartilage; *r* Cricoid cartilage; *s, t, t′* Intrapharyngeal opening; *s* Free border of soft palate; *t, t′* Palatopharyngeal arch; *u* Limen pharyngoesophageum; *v* Pharyngeal opening of auditory tube

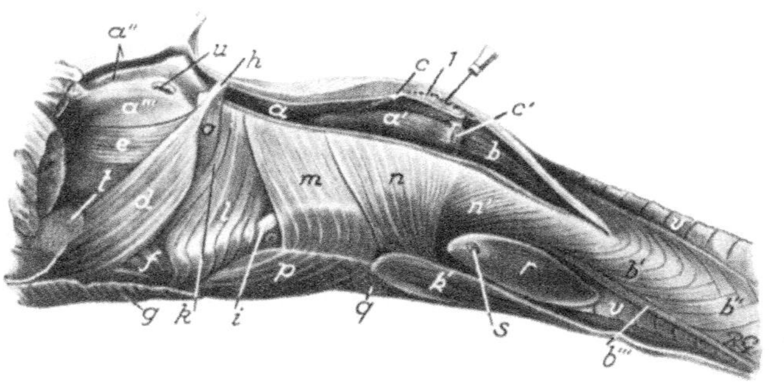

Fig. 64. Pharynx of the dog. Lateral aspect. The pharyngeal roof has been split. (After Zietzschmann, unpublished)

a Nasopharynx; *a′* Caudal part of laryngopharynx; *a″* Edge of nasopharyngeal mucosa; *a‴* Pharyngeal fornix; *b* Esophagus with *b′* elliptical, *b″* decussating, and *b‴* longitudinal muscle fibres; *c* Caudal end of palatopharyngeal arch; *c′* Limen pharyngoesophageum; *d* Styloglossus; *e* Palatopharyngeus and pterygopharyngeus; *f* Hyoglossus; *g* Mylohyoideus; *h* Stylohyoid; *i* Thyrohyoid; *k, l* Hyopharyngeus; *m* Thyropharyngeus; *n* Cricopharyngeus, *n′* its fibers to the esophagus; *o* Stylopharyngeus caudalis; *p* Thyrohyoideus; *q* Cricothyroideus; *r* Thyroid gland; s Parathyroid III; *t* Root of tongue; *u* Pharyngeal opening of auditory tube; *v, v* Trachea

1 Venous plexus in the roof of the laryngopharynx

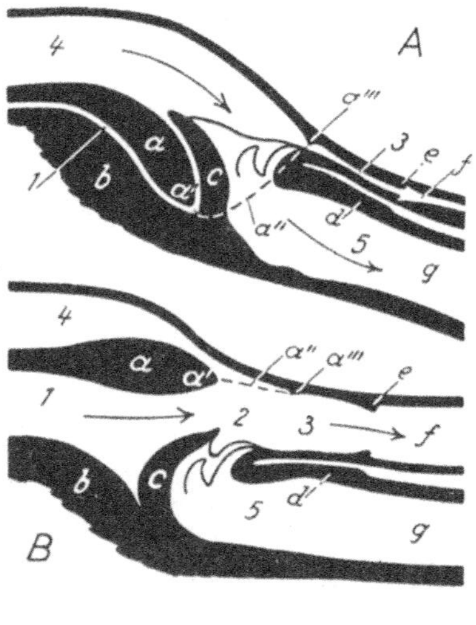

Fig. 65. Pharyngeal cavity of the dog; schematic. A. Nose-breathing position. B. Swallowing position. (After Zietzschmann, 1939.)

1 Oropharynx; *2, 3* Laryngopharynx; *4* Nasopharynx; *5* Larynx

a Soft palate; *a′, a″, a‴* Boundary of intrapharyngeal opening; *a′* Free border of soft palate; *a″, a‴* Palatopharyngeal arch; *b* Root of tongue; *c* Epiglottis; *d* Lamina of cricoid cartilage; *e* Limen pharyngoesophageum; *f* Esophagus; *g* Trachea

ROSTRAL PHARYNGEAL CONSTRICTORS:

Pterygopharyngeus (44/*h;* 46/*5;* 54—56/*19*), O. (origin): pterygoid bone, I. (insertion): pharyngeal raphe.

Palatopharyngeus (44/*i;* 46/*6;* 54/*19;* 56/*19'*), O.: together with the palatinus from the edge of the palatine and pterygoid bones, I.: the rostral border of the thyroid cartilage and the pharyngeal raphe.

Stylopharyngeus rostralis (often absent; 55, 56/*20*), O.: medial surface of the rostral end of the stylohyoid, I.: pharyngeal raphe.

Hyopharyngeus (also known as the middle pharyngeal constrictor; 44, 45/*k, l;* 46/*8;* 53, 54/*20;* 55/*21;* 56/*20'*), O.: caudal end of thyrohyoid, I.: pharyngeal raphe.

CAUDAL PHARYNGEAL CONSTRICTORS:

Thyropharyngeus (44/*m;* 46/*9;* 54, 56/*21*), O.: oblique line of the thyroid lamina, I.: pharyngeal raphe.

Cricopharyngeus (44/*n;* 46/*10;* 54, 56/*21'*), O.: lateral surface of cricoid cartilage, I.: pharyngeal raphe.

The DILATOR OF THE PHARYNX is the **stylopharyngeus caudalis** (44/*o;* 46/*4;* 54, 56/*18*). It arises from the medial surface of the dorsal third of the stylohyoid, passes rostroventrally between the hyopharyngeus and the caudal pharyngeal constrictors, and ends in the lateral wall of the pharynx.

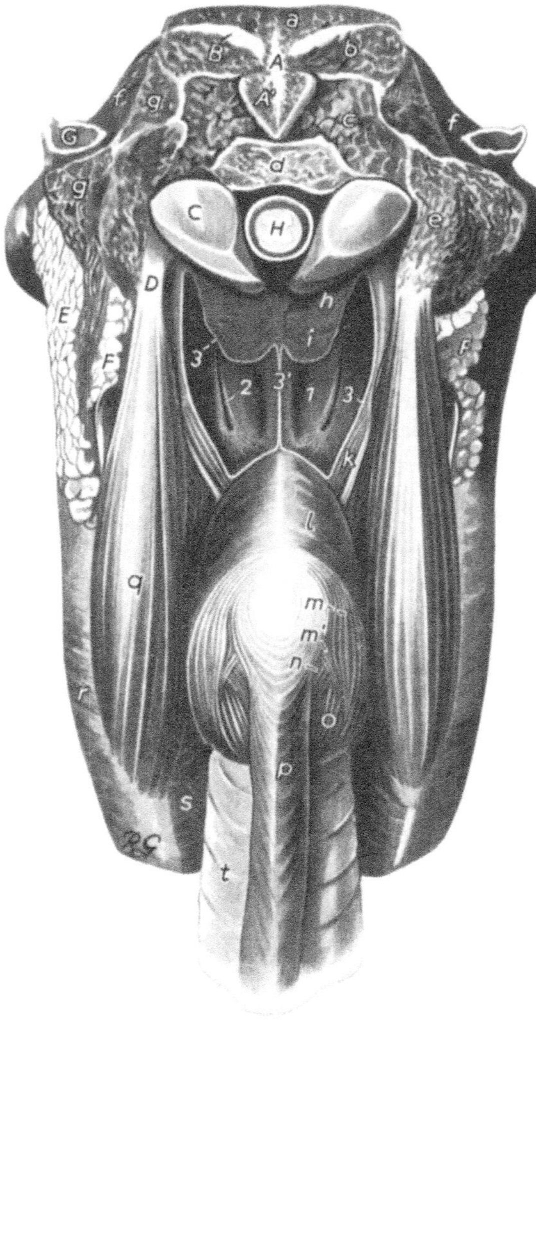

Fig. 66. Pharynx and guttural pouches of a horse. Caudal aspect. (After Zietzschmann, unpublished.) The guttural pouches have been opened and the esophagus and trachea have been drawn ventrally.

A Ext. occipital protuberance; *A'* Insertion of the paired funicular part of the lig. nuchae; *B* Nuchal crest; *C* Occipital condyle; *D* Paracondylar process; *E* Parotid gland; *F* Mandibular gland; *G* Ext. acoustic meatus; *H* Spinal cord

a Auricular muscles; *b* Insertion of semispinalis capitis; *c* Rectus capitis dorsalis major; *d* Rectus capitis dorsalis minor and dorsal atlanto-occipital membrane; *e* Rectus capitis lateralis; *f, g* Obliquus capitis cranialis, *f* its lateral aspect, *g* its cut surface; *h* Rectus capitis ventralis; *i* Longus capitis; *k* Stylopharyngeus caudalis; *l* Hyopharyngeus; *m* Thyropharyngeus; *m'* Cricopharyngeus; *n* Esophageus longitud. lateralis; *o* Cricoarytenoideus dorsalis; *p* Esophagus; *q* Occipitomandibular part of digastricus; *r* Masseter; *s* Pterygoideus medialis; *t* Trachea

1 Medial recess of right guttural pouch; *2* Left pharyngeal opening of auditory tube; *3* Cut edge of guttural pouch mucosa; *3'* Medial walls of guttural pouches in contact with each other

Soft Palate

(57—63, 65, 79—88)

The soft palate (velum palatinum, palatum molle) is a substantial musculomucosal shelf (62/*36*) which forms the caudal continuation of the hard palate (*D*). It extends into the pharyngeal cavity from the free border of the palatine bones and divides the rostral portion of this cavity, as already mentioned, into nasopharynx and oropharynx. The free caudal border of the soft palate (*38*) normally lies near the base of the epiglottis so that the rostral parts of the larynx protude through the intrapharyngeal opening. In the pig the caudal free border presents a rudimentary median **uvula** (81/*8*). The soft palate of the horse is exceptionally long. Except during deglutition, its free border is wedged against the base of the epiglottis causing the rostral surface of the epiglottis to lie against the dorsal surface of the soft palate (86/*3*, *7*). The horse seems unable to raise the soft palate from this position. It is, therefore, very difficult, if not impossible, for the horse to breathe through its mouth. The ventral surface of the soft palate forms the roof of the oropharynx and is in contact with the root of the tongue in normal breathing; it is covered with stratified squamous epithelium. The dorsal surface of the soft palate is the floor of the nasopharynx and is covered with pseudostratified ciliated epithelium. The **palatoglossal arches** (84/*2'*) are two mucosal pillars that connect the soft palate with the root of the tongue and form the lateral boundaries of the isthmus faucium. The **palatopharyngeal arches** (63/*t*, *t'*) proceed caudally from the free border of the soft palate along the lateral walls of the pharynx, and anchor the soft palate to the pharyngeal wall. The mucosa on the ventral surface of the soft palate contains varying amounts of lymphoid tissue. Deep to the mucosa is a thick layer of **glands** (gll. palatini, 84/*3*). The respiratory mucosa on the opposite side contains little lymphoid tissue and few glands.

Three paired muscles are responsible for the movement of the soft palate:

The median **PALATINUS** originates from the free border of the palatine bones and extends into the caudal border of the soft palate. It is closely related laterally to the palatopharyngeus, which is described above with the pharyngeal muscles. The palatinus shortens the soft palate. Künzel et al. (1966) made a detailed study of the palatine muscles of the domestic mammals, and concluded that the palatinus is homologous to the m. uvulae of man.

The **TENSOR VELI PALATINI** (44/*f*; 46/*2*; 53, 54/*16'*) arises from the muscular process of the tympanic part of the temporal bone and passes rostroventrally accompanying the auditory tube. Its tendon is reflected around the hamulus of the pterygoid bone and turns medially to expand into the **aponeurosis** of the soft palate. Between tendon and hamulus is a small synovial bursa. The tensor tenses and straightens the rostral part of the soft palate.

The **LEVATOR VELI PALATINI** (44/*g*; 46/*3*; 54/*16''*) originates with the tensor from the temporal bone. It lies medial to the tensor, crosses the medial surface of the pterygopharyngeus muscle in the lateral wall of the nasopharynx, and enters the soft palate to unite with the contralateral muscle in the median plane. The levator raises the soft palate toward the base of the cranium.

Lymphatic Organs of the Pharynx (Tonsils)

More or less organized accumulations of lymphatic tissue are found in several places in the pharyngeal mucosa. These accumulations may take the form of diffuse, unorganized lymphocytic infiltrations, which appear in certain areas and disappear again; or they may take the form of permanent solitary lymph nodules, or of small or large aggregates of such lymph nodules, which may be surrounded by a connective tissue capsule.

Lymph nodules (noduli lymphatici), when observed in histological section, appear to consist of a dark, peripheral zone of tightly packed lymphocytes and a lighter, central zone containing fewer cells. The lighter zone is known as a **germinal center** because in it lymphocytes are produced by active proliferation. These centers also play a role in the defense against toxic substances or microorganisms that have penetrated the epithelium, and are also referred to as **reaction centers.** In addition to mature lymphocytes and reticulocytes, the lymph nodules contain lymphoblasts, macrophages, other types of leukocytes, plasma cells, and cel-

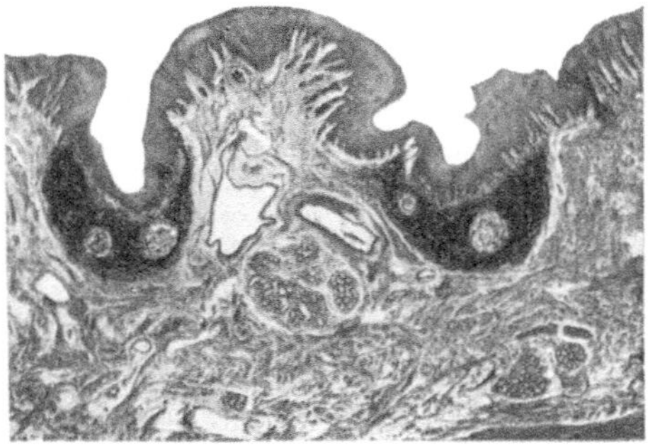

Fig. 67. Lingual tonsil of the horse, consisting of isolated tonsillar follicles. Microphotograph. The tonsillar fossulae are surrounded by lymphoid tissue which contains lymph nodules

Fig. 68. Lingual tonsil of the pig. The lymphoid tissue is inside the lingual papillae (papillae tonsillares). Microphotograph. Glandular lobules are between the muscle bundles

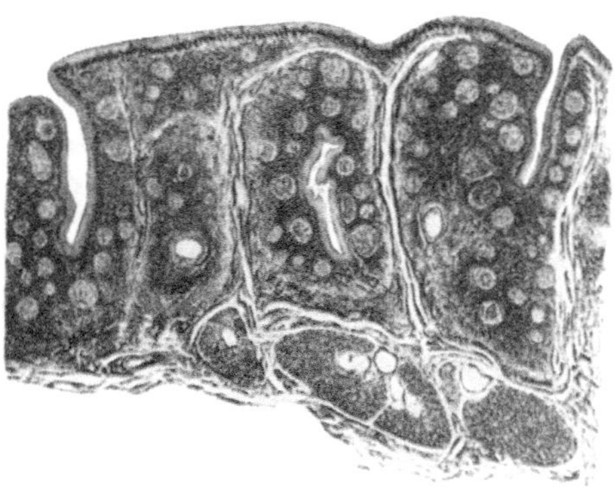

Fig. 69. Palatine tonsil of the horse. Microphotograph. The tonsillar follicles are close together and are surrounded by connective tissue. Glandular lobules are present in the vicinity

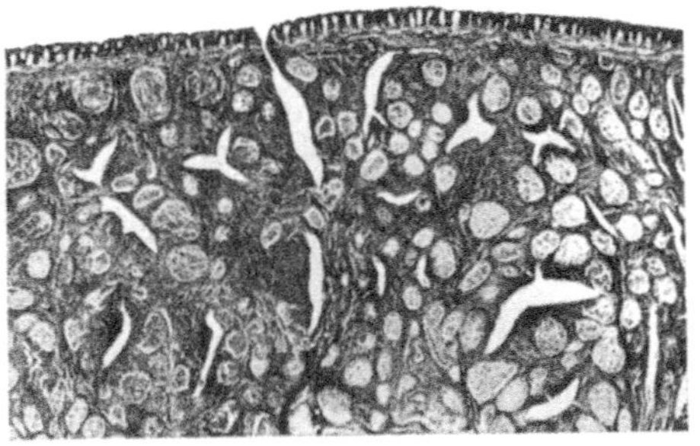

Fig. 70. The tonsil of the soft palate of the pig. Microphotograph

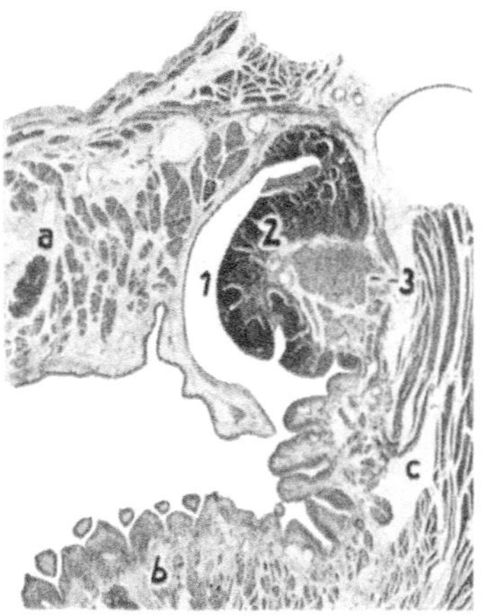

Fig. 71. Palatine tonsil of the dog. Microphotograph.
a Soft palate; *b* Root of tongue; *c* Lateral wall of oropharynx; *1* Fossa tonsillaris; *2* Palatine tonsil; *3* Glandular lobules

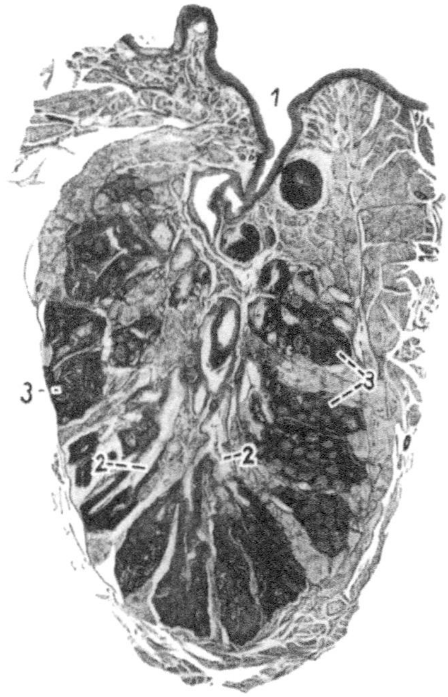

Fig. 72. Palatine tonsil of the ox. Microphotograph. *1* Sinus tonsillaris; *2, 2* Fossulae tonsillares; *3, 3* Tonsillar follicles, between the follicles are glandular lobules

lular debris. The lymph nodules are supplied by numerous blood capillaries and are surrounded by a network of lymphatic capillaries, which are responsible for the removal of cells and fluids. Afferent lymphatics are absent. When irritated, the lymph nodules enlarge and become visible in the mucous membrane as minute reddish tubercles.

When large numbers of lymph nodules combine with diffuse lymphatic tissue, they form independent lymphatic organs known as **TONSILS.** There are two types of tonsils: those composed of tonsillar follicles (follicular tonsils) and those that are not (nonfollicular tonsils).

The surface of the **follicular tonsil** facing the lumen of the pharynx presents numerous invaginations (fossulae tonsillares, 67) by which the exposable surface epithelium is increased. Each fossula is surrounded by lymphatic tissue with lymph nodules, forming a **tonsillar**

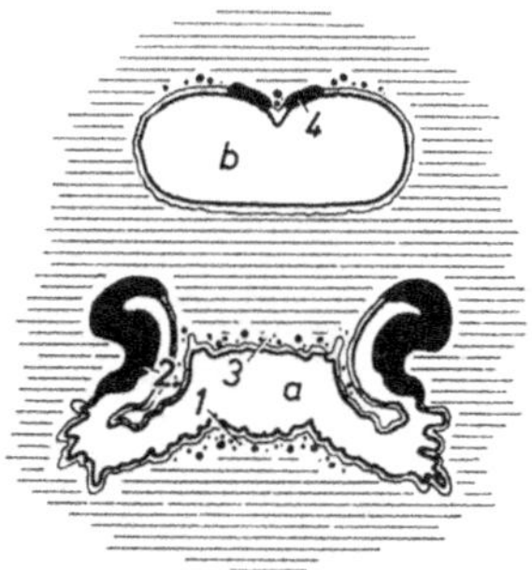

Fig. 73 (Dog)

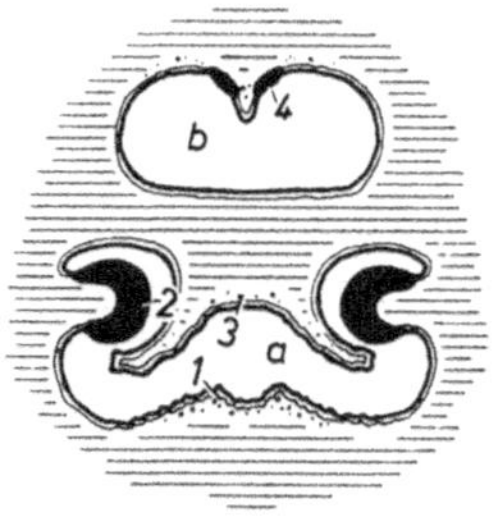

Fig. 74 (Cat)

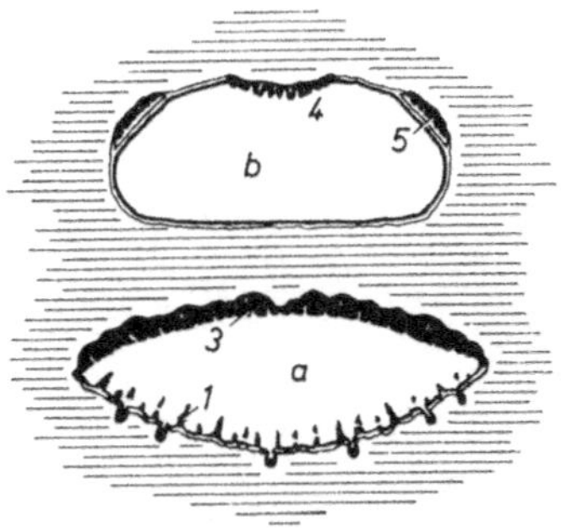

Fig. 75 (Pig)

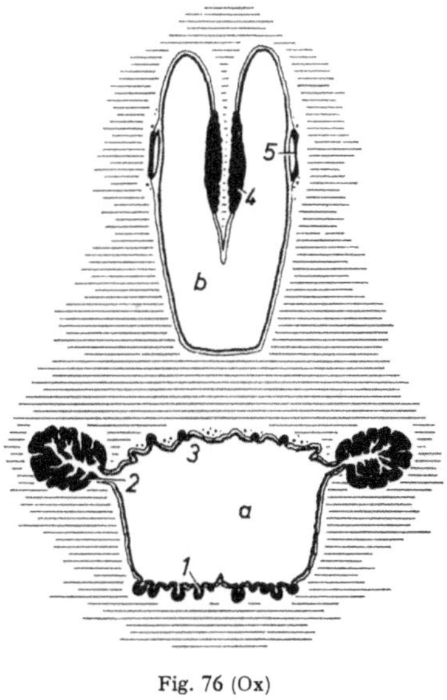

Fig. 76 (Ox)

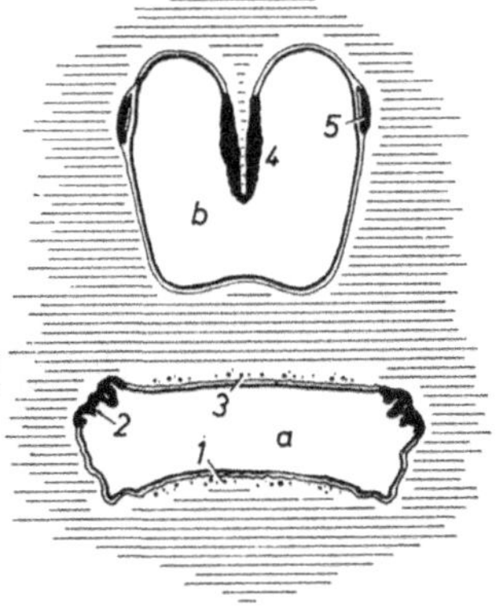

Fig. 77 (Goat)

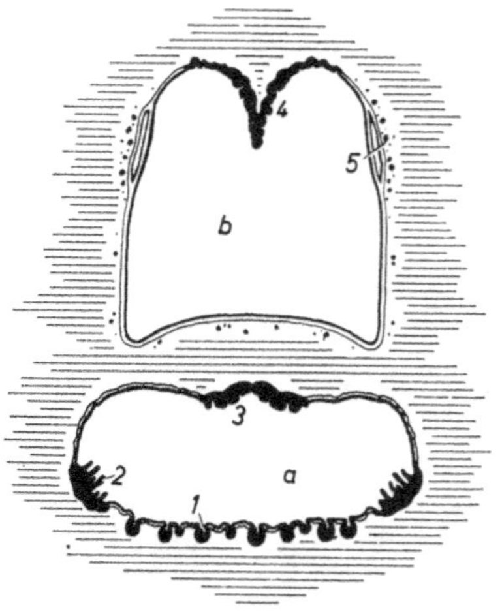

Fig. 78 (Horse)

Figs. 73—78. Ring of lymphatic tissue in the pharynx of domestic mammals. Schematic. (After Vollmerhaus 1959.) *a* Oropharynx; *b* Nasopharynx; *1* Lingual tonsil; *2* Palatine tonsil; *3* Tonsil of the soft palate; *4* Pharyngeal tonsil; *5* Tubal tonsil

follicle (folliculus tonsillaris). Tonsillar follicles may be present singly, i.e., separated from one another by nonlymphatic tissue, as for example in the lingual tonsil of the horse (67). More often they are closely packed into a solid mass of lymphatic tissue (69). Generally, the surface epithelium and the fossulae of the follicular tonsils are freely exposed on the pharyngeal wall. An exception is the palatine tonsil of the ox (72), the fossulae of which open into a deep **sinus tonsillaris** (*1;* 76/*2*). Thus only the opening of the sinus is visible (39/*6*) in the pharyngeal wall.

In the **nonfollicular tonsils,** the lymphatic tissue lies under a flat sheet of epithelium. The smooth surface of this type of tonsil may either be exposed directly to the pharyngeal lumen or it may be hidden from view inside a **fossa tonsillaris** (71/*1*).

Both follicular and nonfollicular tonsils have a connective tissue capsule and are well supplied with blood vessels. They have efferent lymphatics, which carry lymph toward more centrally located lymph nodes. Afferent lymphatics again are absent. Mucous or seromucous glands are usually found in the vicinity of the tonsils. The epithelium covering the tonsils is invaded in many places by masses of lymphocytes which, thus, are exposed directly to the pharyngeal lumen. Many of them are flushed away from the surface by saliva. The tonsillar fossulae are often filled with debris consisting of lymphocytes, desquamated epithelial cells, and food particles; saprophytic and pathogenic microorganisms may also be present. The tonsils, more than the solitary lymph nodules, are an important defense against toxic substances and microorganisms. They produce lymphocytes, take part in the production of antibodies, and are said to "secrete" protective substances into the pharyngeal cavity.

The tonsils are named according to their position in the pharynx. The following are present in the domestic mammals: the lingual tonsil, the palatine tonsils, the tonsil of the soft palate, the paraepiglottic tonsils, the pharyngeal tonsil, and the tubal tonsils. Together with the solitary lymph nodules and the diffuse lymphatic tissue present in the pharyngeal wall, they form a ring of lymphatic tissue around the nasopharynx and oropharynx (73—78).

The tonsils of the carnivores are described on page 60, those of the pig on page 63, those of the ruminants on page 69, and those of the horse on page 73. A comparative summary of their principal features is given below.

LINGUAL TONSIL: in the mucosa of the root of the tongue (73—78/*1*):

Carnivores. No tonsillar follicles, only areas of diffuse lymphoid tissue and solitary nodules.

Pig. A few isolated tonsillar follicles; however, the connective tissue cores of the papillae on the root of the tongue contain large amounts of lymphoid tissue with many lymph nodules (68).

Ruminants. Numerous large tonsillar follicles with distinct fossulae in the ox. The small ruminants have only small amounts of diffuse lymphoid tissue (39, 40/*5*).

Horse. Many tonsillar follicles with distinct fossulae (67; 86, 88/*1*).

PALATINE TONSIL: located either on or in the lateral wall of the oropharynx (73—78/*2*):

Carnivores. Nonfollicular, cylindrical tonsil on the lateral wall of the oropharynx in the fossa tonsillaris, covered medially by the semilunar fold. The cat's palatine tonsil, otherwise similar, is relatively shorter and more spherical (36, 37/*6;* 71).

Pig. Absent.

Ox. Follicular, spherical tonsil, about 3 cm. in diameter, embedded in the lateral wall of the oropharynx in the underlying musculature and connective tissue; the fossulae open into a branching tonsillar sinus (72; 76/*2*).

Small Ruminants. Tonsil with three to six deep fossulae on the surface of the pharyngeal wall (40/*6;* 77/*2*).

Horse. Follicular tonsil, 10—12 cm. long and 2 cm. wide, lying on the surface of the pharyngeal wall between the base of the epiglottis and the palatoglossal arch (69; 78/*2;* 88/*2'*).

TONSIL OF THE SOFT PALATE (tonsilla veli palatini): on the ventral surface of the soft palate (73—78/*3*):

Carnivores. Only small amounts of diffuse lymphoid tissue or isolated lymph nodules.

Pig. Large flat follicular tonsil with numerous distinct fossulae (59/*37;* 70; 81/*5*).

Ruminants. Only small amounts of lymphoid tissue, but some tonsillar follicles in the ox.

Horse. Flat follicular tonsil, median in position, about 4 cm. long and 2.5 cm. wide (35/*3';* 78/*3*).

PARAEPIGLOTTIC TONSIL:

Carnivores. Small flat tonsil on the base of the epiglottis (37/*8*); absent in the dog and inconstant in the cat.

Pig. Small group of tonsillar follicles in the floor of a small depression (5—8 mm. long and 3—4 mm. wide) on each side of the base of the epiglottis.

Ruminants. Small group of tonsillar follicles in the sheep and goat; absent in the ox.
Horse. Absent.

PHARYNGEAL TONSIL: in the roof of the nasopharynx (73—78/*4*):
Carnivores. Flat nonfollicular tonsil between the openings of the auditory tubes.
Pig. Irregularly raised follicular tonsil on the caudal end of the pharyngeal septum (59/*15*).
Ruminants. Irregularly raised nonfollicular tonsil on the caudal end of the pharyngeal septum (60, 61/*15*).
Horse. Follicular tonsil between the openings of the auditory tubes (78/*4*).

TUBAL TONSIL (tonsilla tubaria): in the lateral wall of the pharyngeal opening of the auditory tube (75—78/*5*):
Carnivores. Absent.
Pig. Large flat follicular tonsil (75/*5*).
Ruminants. Flat nonfollicular tonsil (76, 77/*5*).
Horse. Small amounts of diffuse lymphoid tissue and solitary nodules (78/*5*).

Deglutition

(65)

Deglutition, or swallowing, is a complicated neuromuscular act which causes a bolus of food to be transferred from the oral cavity through the pharynx into the esophagus and finally into the stomach. This sequence of events may be divided into two stages, the first being voluntary and the second involuntary, or reflex.

During the **FIRST, VOLUNTARY STAGE OF DEGLUTITION** the insalivated bolus or fluid is pressed into the oropharynx by a wavelike motion of the tongue similar to the action of a plunger in a syringe. This motion consists of caudally-progressing pressure against the palate. The successful accomplishment of this action requires that the mouth be tightly closed, which is brought about by closing the jaws and pressing the lips and cheeks against the teeth. The passage of the bolus is facilitated by the caudally-directed filiform papillae and by the palatine ridges.

The **SECOND, INVOLUNTARY STAGE OF DEGLUTITION** begins as the bolus enters the oropharynx. In order to convey the bolus toward the esophageal entrance, the pharynx must change its shape in such a way as to prevent the escape of the bolus through any of the other pharyngeal openings. That is to say, it should be impossible for the bolus to enter either the nasopharynx or the larynx, or to re-enter the oral cavity. The intrapharyngeal opening is closed by the elevation of the soft palate against the roof of the nasopharynx and by the contraction of muscle bundles inside the palatopharyngeal arches. The oropharynx is tightly closed by the raised root of the tongue pressing against the ventral surface of the soft palate. Closure of the larynx is largely passive and comes about in the following way. At the end of the wavelike raising of the tongue comes the elevation of its root against the soft palate. For this to occur the ventral end of the hyoid bone is moved rostrodorsally by the action of the geniohyoid, stylohyoid, mylohyoid, digastric, and ceratohyoid muscles; and this also draws the larynx forward. As the larynx is pulled forward with the hyoid bone, the rostral parts of the larynx are tucked under the elevated root of the tongue and the epiglottis is pressed, like a lid, on the laryngeal entrance. With the rostral movement of the larynx, the esophageal entrance is dilated and pulled toward the oncoming bolus. If the bolus is fairly liquid, it is divided by the epiglottis and much of it passes along the piriform recesses. If the bolus is coarse or insufficiently insalivated, it passes directly over the epiglottis and is pressed into the esophagus by the action of the pharyngeal constrictors. Liquids or semifluid boluses may be propelled by the plungerlike action of the tongue with such force that they are hurled through the pharynx and even through the esophagus, making active constriction of these structures superfluous. After swallowing, the soft palate, the hyoid bone, and the larynx return to their normal breathing positions, and air can once again pass freely through the intrapharyngeal opening into the larynx.

Mouth and Pharynx of the Carnivores

Oral Cavity

The shape and size of the oral cavity of the dog varies with the breed. In the dolichocephalic breeds, the oral cavity is long and relatively narrow, while in the brachycephalic breeds it is short and wide. The oral cavity of the cat is always short and wide. The oral mucosa, including occasionally that of the tongue, may be partly or completely pigmented (black).

The **LIPS** present many long tactile hairs in addition to their ordinary hair. The tactile hairs (whiskers) on the upper lip of the cat are especially well developed and noticeable (19). The upper lip of the dog is long and mobile, especially at the angle of the mouth. A small area in the middle of the upper lip is devoid of hair and is confluent with the **nasal plate** (270/*a*). The **philtrum** is usually shallow, but it may be a deep cleft in some breeds. The small **frenulum** connecting the mucosa of the upper lip with the gums is occasionally double. The lower lip is shorter than the upper lip, and thus is overhung by the upper lip in front and to some extent laterally. The lips of the cat contain numerous sebaceous **circumoral glands;** the lower lip more than the upper. The secretion of these glands is distributed over the hair coat when the cat cleans itself, and is thought to keep the coat in condition. The **angle of the mouth** is fairly far caudal on the face and is at the level of the third or fourth cheek tooth. This causes the **oral cleft** and the **labial vestibule** to be large, and allows the carnivore to open its mouth wide. This is of advantage to the veterinarian when examining mouth and pharynx or when performing oral surgery. Small groups of **labial glands** are present in both lips.

The **BUCCAL VESTIBULE** is short. In the dog, the **parotid duct** opens into it opposite the third or fourth upper cheek tooth. The parotid duct of the cat ends opposite the second upper cheek tooth. The **buccal glands** are well developed (47/*d*). The dorsal buccal gland (52/*e*) is known as the **zygomatic gland** in the carnivores. It is a globular, seromucous gland which lies medial to the rostral end of the zygomatic arch and is in contact with the periorbita dorsally and with the pterygoideus medially. Its major and its 3 or 4 minor ducts (*e' e''*) open into the buccal vestibule at the level of the last cheek tooth. The **ventral buccal glands** extend from the canine tooth to the level of the third lower cheek tooth.

The **HARD PALATE** of the carnivores is narrow rostrally, but widens at the level of the molars (79). Its epithelium is heavily cornified in the cat, but only slightly cornified in the dog. The **incisive papilla** (*1*) located just caudal to the central incisors, may be either roundish or triangular; lateral to it are the openings of the **incisive ducts.** There are 6—10 **palatine ridges** in the dog and 7—9 in the cat. They are concave caudally, not always equally spaced, and some of them are present only on one side. There are rows of papillae between the palatine ridges in the cat (28). The slightly raised **palatine raphe** (79/*4*) is indistinct and may be absent. The portion of the hard palate which is caudal to the level of the last cheek teeth is smooth and contains glands. Glands are also present in the neighborhood of the incisive papilla.

The **TONGUE** of the carnivores (36, 37) is a very mobile organ and important for the intake of liquids. In the dog, the apex is wide and flat with sharp borders and can be made to assume the shape of a shallow ladle for drinking. During forced breathing or panting the dog's tongue protrudes from the mouth. It has a shallow **median sulcus** (36/*b'*) which divides the dorsum into two symmetrical halves. The median **glossoepiglottic fold** (80/*2*) connects the root of the tongue with the epiglottis.

The dorsal surface of the tongue is covered with **filiform papillae** (36, 37/*1*) which are arranged in diagonal rows. Filiform papillae are absent on the ventral and lateral surfaces. In the dog, the filiform papillae are soft; they increase in size gradually toward the root of the tongue where they are longest but reduced in number. In the cat, they are caudally directed, hooklike, and firm; the animal uses them to advantage in eating and grooming. The **fungiform papillae** (*2*) are distributed over approximately the rostral two-thirds of the dorsum of the tongue. They are red, about 1 mm. in diameter, and have especially large taste buds. Just rostral to the root of the tongue are 2—3 pairs of **vallate papillae** (*3*). At the same level, but on the edges of the tongue, are the **foliate papillae** (43/*g'*). The latter are not very distinct in the dog, less so in the cat, and contain taste buds only in the dog.

Diffuse **lymphatic tissue** and lymph nodules are regularly found in the mucosa of the root of the tongue and of the glossoepiglottic fold, and are referred to as the **lingual tonsil** (73, 74/*1*). Seromucous glands are also present in this area.

A peculiar structure in the tongue of the carnivores is the **lyssa.** This is a median, spindle-shaped spicule, which lies along the undersurface of the tongue (42/*a*; 43/*h*). It begins a few millimeters caudal to the tip and is embedded between the genioglossus muscles (42/*c*).

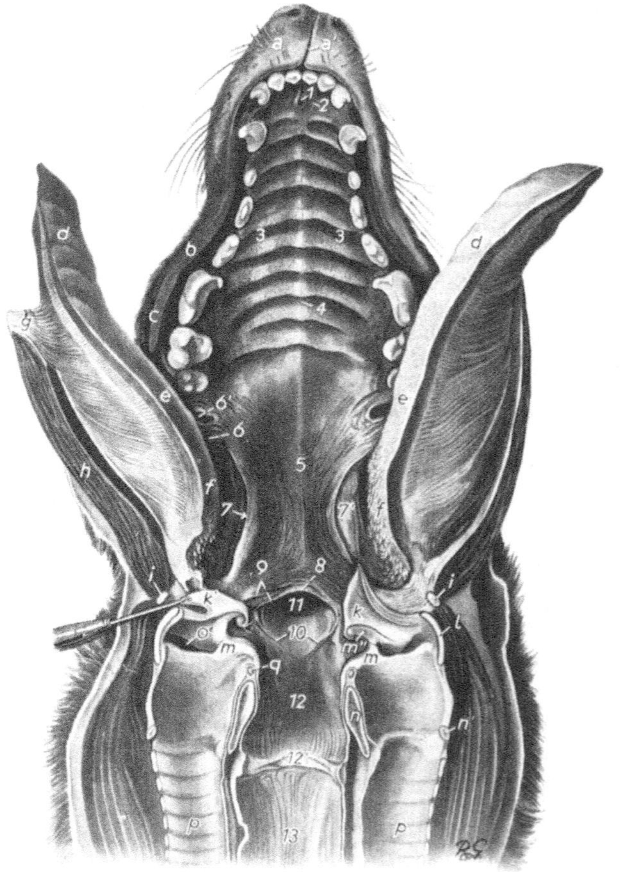

Fig. 79. Roof of the oral cavity, oropharynx, laryngopharynx, and esophagus of a dog. The tongue, larynx, trachea, and esophagus have been split in the median plane. Ventral aspect. (After Zietzschmann 1939.)

a Upper lip; *a'* Philtrum; *b* Angle of the mouth; *c* Cut surface of cheek; *d* Apex, *e* body, and *f* root of the tongue; *g* Genioglossus; *h* Geniohyoideus; *i* Basihyoid; *k* Epiglottis, split and drawn laterally at k'; *l* Thyroid cartilage; *m* Arytenoid cartilage; *m''* Cuneiform process; *n, n'* Cricoid cartilage; *o* Entrance to lateral laryngeal ventricle, caudal to it is the vocal fold; *p* Trachea; *q* Interarytenoid cartilage

1 Incisive papilla; *2* Opening of incisive duct; *3* Hard palate with palatine ridges; *4* Palatine raphe; *5* Soft palate; 6 Palatoglossal arch; *6'* Pterygomandibular fold; *7* Fossa tonsillaris of right palatine tonsil; *7'* Left palatine tonsil, withdrawn from fossa; *8* Free border of soft palate; *9, 10* Palatopharyngeal arch; *11* Intrapharyngeal opening; *12* Caudal part of laryngopharynx; *12'* Limen pharyngoesophageum; *13* Esophagus

Caudally, it is tapered to a slender thread which blends with the septum of the tongue, but does not reach the hyoid bone. In the dog, it raises the overlying mucosa slightly. The lyssa is a tube of connective tissue that is filled with adipose tissue and in its middle segment contains striated muscle fibers and small amounts of cartilage. In the cat, muscle is only occasionally present, and cartilage has never been demonstrated.

The **sublingual caruncles** (27/*2*) of the carnivores are small mucosal elevations at the base of the lingual frenulum. The **orobasal organ** (*1*) consists of two shallow grooves; in the cat, it is often only a small paired depression.

Salivary Glands

The **PAROTID GLAND** (52/*a*) is small and triangular, and with its preauricular and postauricular angles embraces the base of the ear. Its blunt ventral end covers the dorsal part of the mandibular salivary gland. The parotid gland is related medially to branches of the facial nerve, to the maxillary artery and vein and their branches, and to the parotid lymph node (47/*e*). The flat parotidoauricularis lies on the lateral surface of the gland. The **parotid duct** (*a'''*) emerges from the rostral border of the gland, and on its way to the mouth crosses the lateral surface of the masseter. The duct opens into the buccal vestibule opposite the third or fourth upper cheek tooth in the dog and opposite the second in the cat. Small accessory parotid lobules may be present along the course of the duct.

The **MANDIBULAR GLAND** (47/*b*) is globular and often larger and lighter in color than the parotid gland. Its lobules are less distinct because the connective tissue separating them is scant. Like the parotid gland, it occupies a superficial position caudal to the ramus of the mandible in the angle formed by the linguofacial and maxillary veins. The mandibular lymph nodes (*f*) are rostral and to some extent also ventral to the mandibular salivary gland. The **mandibular duct** passes rostrally between the lingual muscles and the mandible to the indistinct sublingual caruncle (52/*5*).

The **SUBLINGUAL GLANDS** consist of polystomatic and monostomatic glands. The **polystomatic sublingual gland** (52/*d, d'*) is composed of a chain of isolated lobules situated between the styloglossus and the mandible and extending from the first to the last cheek tooth. Its numerous **minor sublingual ducts** open into the lateral sublingual recess. The **monostomatic sublingual gland** (*c, c'*) is more compact and lies between the digastricus medially and the mandible and pterygoideus medius laterally. It extends far enough caudally to make contact with the mandibular salivary gland, with which it forms a unit and shares a common connective tissue capsule (*c, b*). The **major sublingual duct** (*c''*) accompanies the mandibular duct and with it opens on the sublingual caruncle. Small salivary lobules (*c'''*) of the monostomatic gland are usually found along the course of the major sublingual duct, lying among the lobules of the polystomatic gland.

Pharynx

(57, 58, 63, 64, 79, 80)

The pharynx of the carnivores is long and extends into the neck to the level of the second cervical vertebra in the dog and to the level of the third in the cat (57, 58). In the dog, the **palatopharyngeal arches** (63/*t, t'*) are distinct, while in the cat they are not very prominent. The **pharyngeal opening of the auditory tube** (*v*) is a small slit in the laterodorsal wall of the nasopharynx; it is 1 cm. long in the dog and about 4 mm. long in the cat. Caudal to it is a slight mucosal elevation (torus tubarius). The roof of the pharynx between the openings of the auditory tubes is occupied by the flat **pharyngeal tonsil** (58/*w'*). During normal breathing, the rostral portion of the larynx extends through the **intrapharyngeal opening** into the nasopharynx, with the epiglottis assuming a position caudal to the soft palate. During forced breathing, the carnivores elevate the soft palate so that much of the air passes through the mouth (63).

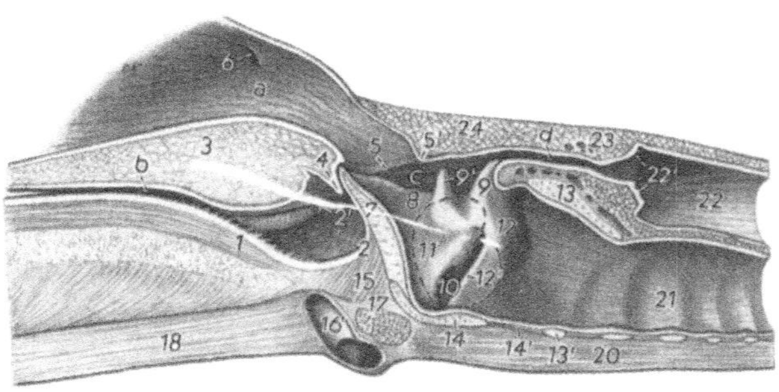

Fig. 80. Sagittal section of the pharynx and larynx of the dog. Breathing position.

a Nasopharynx; *b* Oropharynx; *c, d* Laryngopharynx

1 Root of tongue; *2* Glossoepiglottic fold; *2'* Right palatine tonsil; *3* Soft palate; *4, 5, 5'* Boundaries of intrapharyngeal opening; *4* Free border of soft palate; *5, 5'* Palatopharyngeal arch; *6* Pharyngeal opening of auditory tube; *7* Epiglottis; *8* Aryepiglottic fold; *9* Arytenoid cartilage; *9'* Cuneiform process; *10* Entrance to lateral laryngeal ventricle, the broken lines mark the extent of the ventricle; *11* Laryngeal vestibule; *12* Vocal process of arytenoid cartilage; *12'* Vocal fold; *13, 13'* Cricoid cartilage; *14* Thyroid cartilage; *14'* Cricothyroid ligament; *15* Hyoepiglotticus; *16'* Lingual vein; *17* Basihyoid; *18* Geniohyoideus; *20* Sternohyoideus; *21* Trachea; *22* Esophagus; *22'* Limen pharyngoesophageum; *23* Caudal pharyngeal constrictors; *24* Hyopharyngeus

The **SOFT PALATE** of the carnivores is long, but does not reach to the base of the epiglottis. There are small papillae on the ventral surface of the soft palate in the cat, and there is a heavy layer of **glands** and **lymphoid tissue** in both dog and cat. The dorsal surface of the soft palate is covered with pseudostratified ciliated epithelium; it contains few glands and little lymphoid tissue. The **palatine tonsil** (79/*7'*) is caudal to the palatoglossal arch.

The **LARYNGOPHARYNX** is exceptionally long and much of it overlies the larynx. The **piriform recesses** are moderately deep (36, 37/*n*). The caudal extent of the laryngopharynx is marked internally by an annular fold of mucous membrane (limen pharyngoesophageum, 80/*22'*). Immediately rostral to this fold the mucosa is pale, devoid of glands, and lightly folded. The esophageal mucosa caudal to the fold is darker, contains glands, and presents larger folds. In the cat, the limen is a transitory mucosal fold devoid of glands.

Tonsils

LINGUAL TONSIL (73, 74/*1*). An accumulation of diffuse lymphoid tissue and solitary and aggregated lymph nodules at the root of the tongue and at the glossoepiglottic fold.

PARAEPIGLOTTIC TONSIL (37/*8*). Found only in the cat; a small platelike tonsil with nodules, laterally on the base of the epiglottis.

PALATINE TONSIL (71; 79/*7'*; 80/*2'*). This is a cylindrical, nonfollicular tonsil located in the **fossa tonsillaris** and partly covered medially by a thin **semilunar fold** (plica semilunaris). In the dorsal part of the fossa are small recesses, the walls of which are invaded by lymphatic tissue. As in man, the palatine tonsil of the carnivores is easily accessible through the mouth. The palatine tonsil of the cat (37/*6*) is somewhat shorter and rounder than that of the dog.

TONSIL OF THE SOFT PALATE (73, 74/*3*). An accumulation of diffuse lymphoid tissue and solitary and aggregated nodules found in the mucosa of the ventral surface of the soft palate.

PHARYNGEAL TONSIL (58/*w'*; 73, 74/*4*). A platelike accumulation of diffuse and nodular lymphatic tissue between the tubal openings in the roof of the nasopharynx.

The tubal tonsil is absent in the carnivores.

Mouth and Pharynx of the Pig

Oral Cavity

The oral cavity is always relatively long in the pig, but its length varies with the breed. It is widest at the level of the canine teeth (tusks).

The **LIPS** are not very mobile. They have hairs and along their margins also **tactile hairs.** The **angle of the mouth** is caudal to the level of the tusks, with the result that the **oral cleft** is very large. The central portion of the upper lip is without hair and takes part in the formation of the **rostral plate** (271/*a*). The **philtrum** (*b*) is only a shallow median groove. The upper lip presents a transverse notch which accommodates the tusk of the upper jaw (29). This notch is present in both sexes and is independent of the development of the tusk. The lower lip is pointed in front (30). Since it is shorter, its margin is usually covered by the upper lip. The **labial glands** are not very numerous in the pig; they are situated between the bundles of the labial muscles.

The **dorsal** and **ventral buccal glands** (48/*8, 10*) form two bands of glandular tissue. They lie on the surface of the buccinator, roughly parallel to the cheek teeth, and extend from the angle of the mouth to the masseter, which covers their caudal ends. The buccal glands have numerous excretory ducts, which open into the buccal vestibule.

The **HARD PALATE** (29/*2*) increases in width up to the level of the tusks; caudal to the tusks, its width remains fairly constant. The **palatine raphe** (*2'*) of the pig is a median groove from which usually 20--23 staggered, smooth **palatine ridges** proceed laterally. The palatine ridges are high rostrally and have sharp edges. They grow gradually lower toward the pharynx and end at the junction of the hard and soft palates. The **incisive papilla** (81/*1*) is elongated and is in a line with the palatine raphe. Two small depressions on either side of it mark the openings of the **incisive ducts** (*2*). Solitary lymph nodules are found in the mucosa over the entire hard palate; glands are present only in the rostral portion.

The apex of the **TONGUE** is long and pointed (38). The body of the tongue has definite lateral surfaces and presents an elongated median elevation on its dorsum. The epithelium covering the tongue is only moderately cornified. **Filiform papillae** are present on the dorsal surface of the tongue in the form of soft horny threads. On the root of the tongue, the filiform

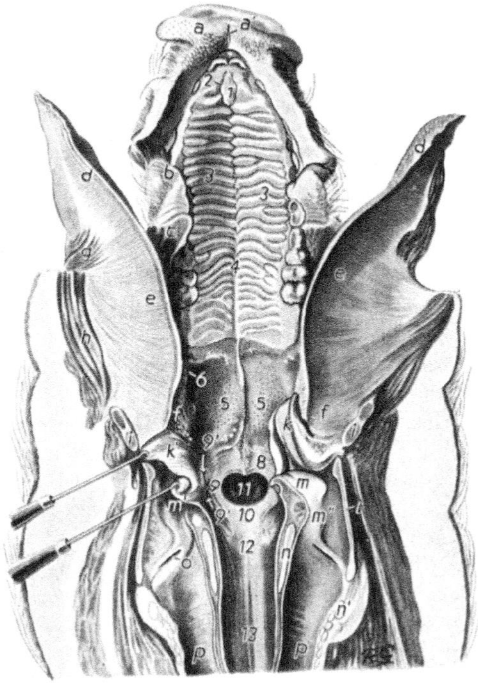

Fig. 81. Roof of the oral cavity, oropharynx, laryngopharynx, and esophagus of a pig. The tongue, larynx, trachea, and esophagus have been split in the median plane. Ventral aspect. (After Zietzschmann, unpublished.)

a Upper lip, rostral plate; *a'* Philtrum; *b* Angle of the mouth; *c* Cut surface of the cheek; *d* Apex, *e* body, *f* root of tongue; *g* Genioglossus; *h* Geniohyoideus; *i* Basihyoid; *k* Epiglottis, split and drawn laterally at k'; *l* Thyroid cartilage; *m* Arytenoid cartilage, drawn laterally at *m'*, *m''* its vocal process; *n*, *n'* Cricoid cartilage; *o* Entrance to lateral laryngeal ventricle; *p* Trachea

1 Incisive papilla; *2* Opening of incisive duct; *3* Hard palate with palatine ridges; *4* Palatine raphe; *5*, *5* Soft palate with glandular openings, solitary lymph nodules and fossulae tonsillares; *6* Palatoglossal arch; *8* Free border of soft palate; *9*, *10* Palatopharyngeal arch; *9'* Piriform recess; *11* Intrapharyngeal opening; *12* Caudal part of laryngopharynx; *13* Esophagus

papillae (*1*) are less numerous, but are larger and more flexible because they have a connective tissue core. The median **glossoepiglottic fold** (82/*2*) connects the root of the tongue with the rostral surface of the epiglottis. Numerous **fungiform papillae** (38/*2*) are distributed over the dorsal and lateral surfaces of the tongue. Only one pair of **vallate papillae** (*3*) is present at the junction of the body and the root of the tongue. The **foliate papilla** (*4*) is about 7—8 mm. long, consists of 5—6 minute leaves, and is situated just rostral to the palatoglossal arch. The gustatory glands associated with the large vallate papillae are usually of the serous type. The remaining **lingual glands** of the area are mucous; their lobules are arranged in a U-shaped band, which is open rostrally. This band extends in an arc from the lateral surface of the tongue to the base of the epiglottis on one side and back to the lateral surface of the tongue on the other. The excretory ducts of these glands open mainly on the lateral surfaces of the tongue. Glands are also present in the glossoepiglottic fold and on each side of it. There is a thin median strand of connective tissue close to the ventral surface of the apex of the tongue. Some authorities consider it to be homologous to the lyssa of carnivores.

The **lingual tonsil** (75/*1*) at the root of the tongue consists of diffuse lymphatic tissue with lymph nodules, and of tonsillar follicles. The latter cause small tubercles on the mucous membrane between the bases of the large papillae, while the former invades the connective tissue cores of the papillae (papillae tonsillares, 68).

The **orobasal organ** (30/*1*) is just caudal to the central incisors on the **SUBLINGUAL FLOOR OF THE MOUTH.** It consists of two solid epithelial sprouts, which extend into the lamina propria of the oral mucosa; two minute depressions on the surface of the oral mucosa indicate their position. The **lingual frenulum** is a double fold in the pig. Lateral to its base are the rather indistinct **sublingual caruncles** (*2*). The ancestor of the domestic pig, the European wild boar, has a paracaruncular gland accompanied by solitary lymph nodules. Whether this gland is also present in the domestic pig has not been established.

Salivary Glands

The yellowish **PAROTID GLAND** (48/*12*) is large, entirely serous, and triangular in shape. Its rostral margin lies against the caudal border of the mandible; its ventral margin runs along the dorsal border of the sternohyoideus; and its caudal margin lies on the sternomastoideus. The auricular angle (*13*) of the parotid gland does not quite reach the base of the ear; the mandibular angle (*13'*) projects into the intermandibular space to the level of the rostral border of the masseter or even beyond it; and the cervical angle (*13''*) extends toward the thorax about two-thirds the length of the neck. The parotid gland of the pig contains large amounts of interlobular adipose tissue and is partly covered laterally by the parotidoauricularis. The medial surface of the gland is uneven and is related to the parotid, lateral retropharyngeal, and mandibular lymph nodes (*h, i, k*); to the mandibular salivary gland (*14*); to the branches of the common carotid artery and jugular vein (*a*); to the sternothyroideus and sternohyoideus; and even to the larynx. The **parotid duct** (*11*) originates at the confluence of several radicles on the medial surface of the gland. It crosses the lateral surface of the mandibular gland and follows the digastricus (*N*) into the intermandibular space. It then accompanies the facial vein around the ventral border of the mandible, ascends on the lateral surface of the face in front of the masseter muscle, and opens into the buccal vestibule on a prominent **papilla** opposite the third or fourth upper cheek tooth.

The **MANDIBULAR GLAND** (53/*14*) is reddish in color and is much smaller than the parotid gland. It is more or less globular in shape, but presents a short rostromedial angle. The rostral part of the gland is medial to the mandible and is inserted between the pterygoideus and the pharyngeal muscles; its caudal part is covered by the parotid gland. The **mandibular duct** (*14'*) arises from the lateral surface of the gland near the base of the rostromedial angle, crosses the lingual nerve (*f'*) medially, and passes rostrally along the medial surface of the polystomatic sublingual gland to the sublingual caruncle.

The yellowish-red **MONOSTOMATIC SUBLINGUAL GLAND** (53/*9*) is a flat band about 4—6 cm. long. It extends rostrally from the intermediate tendon of the digastricus to the point at which the mandibular duct crosses the lingual nerve. Numerous small excretory ducts emerging on the lateral surface of the gland combine to form the **major sublingual duct** (*9'*) which opens rostrally on the sublingual caruncle with the mandibular duct.

The reddish **POLYSTOMATIC SUBLINGUAL GLAND** (*10*) is 7—9 cm. long and continues the more yellowish monostomatic gland rostrally. It lies against the lateral surface of the tongue, extending to the incisive part of the mandible. This gland underlies the **sublingual fold,** on the lateral surface of which the numerous **minor sublingual ducts** (*10'*) open into the lateral sublingual recess.

Pharynx

(59, 81, 82)

The pharynx of the pig is long and narrow and extends caudally to the level of the axis. The **intrapharyngeal opening** (81/*11*), through which the nasopharynx communicates with the laryngopharynx, is small, about 1.5—2 cm. in diameter.

The **SOFT PALATE** is short and thick; it is only a little longer than it is wide. In the carcass, it is usually found lying on the epiglottis (82), and it is likely that this also is its normal position in the live animal. It may, however, lie with its free edge against the base of the epiglottis, allowing the latter to project through the intrapharyngeal opening. There are up to four small conical projections on the free border of the soft palate, which are thought to be homologous to the **uvula** of man. The mucosa on the ventral surface of the soft palate contains the extensive, follicular **tonsil of the soft palate** (75/*3*; 81/*5*), which is surrounded by mucous glands.

The **oropharynx** (59/*c*), like the soft palate, is short. Its floor is the root of the tongue, which slopes steeply caudoventrally toward the base of the epiglottis. Its roof is formed by the soft palate. In almost all specimens the epiglottis is ventral to the soft palate and projects into the oropharynx from behind. The **piriform recesses** are deep (81/*9'*). It is thought that food can pass through the recesses while the laryngeal entrance is open, and that the pig is, thus, capable of breathing and swallowing simultaneously. The **laryngopharynx** extends to about the middle of the cricoid lamina. The pharyngoesophageal junction is not marked internally as it is in the carnivores.

The mucous membrane covering the nasal septum is continued into the nasopharynx where it forms the **pharyngeal septum** (59/*12*), which extends to about the level of the tubal openings. A characteristic feature of the nasopharynx of the pig is the median **pharyngeal diverticulum** (59/*39'*; 82/*5'*). This is a blind mucosal pouch, about 3—4 cm. deep, situated in the roof of the nasopharynx immediately dorsal to the caudal boundary of the intrapharyngeal opening. The pouch is directed caudally and lies just ventral to the pharyngeal raphe under cover of the cricopharyngeus muscles. Contraction of these muscles closes the pharyngeal diverticulum. The **pharyngeal openings of the auditory tubes** (59/*14*) are situated in shallow depressions on the lateral walls of the nasopharynx at the level at which the roof of the nasopharynx meets the base of the cranium. The mucosa of the roof of the nasopharynx contains much lymphatic tissue and tonsillar follicles. The lymphatic tissue is concentrated between the pharyngeal openings of the auditory tubes and also near these openings, forming, respectively, the **pharyngeal** (*15*) and **tubal tonsils.**

Tonsils

LINGUAL TONSIL (68; 75/*1*). Scattered tonsillar follicles and large amounts of lymphatic tissue which contains lymph nodules and invades the connective tissue cores of the filiform papillae at the root of the tongue.

PARAEPIGLOTTIC TONSIL. A group of tonsillar follicles at the bottom of the two depressions (sulci tonsillares) on either side of the base of the epiglottis.

TONSIL OF THE SOFT PALATE (59/*37*; 75/*3*; 81/*5*). An extensive, platelike, follicular tonsil on the ventral surface of the soft palate.

PHARYNGEAL TONSIL (59/*15*; 75/*4*). An unevenly raised follicular tonsil in the roof of the nasopharynx.

TUBAL TONSIL (75/*5*). A flat follicular tonsil in the mucosa of the pharyngeal opening of the auditory tube.

The palatine tonsil is absent in the pig.

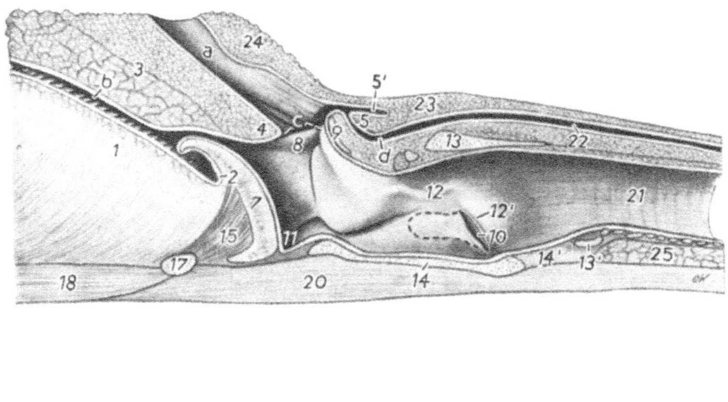

Fig. 82. Sagittal section of the pharynx and larynx of the pig.

a Nasopharynx; *b* Oropharynx; *c, d* Laryngopharynx

1 Root of tongue; *2* Glossoepiglottic fold; *3* Soft palate; *4, 5* Boundaries of intrapharyngeal opening; *4* Free border of soft palate; *5* Palatopharyngeal arch; *5'* Pharyngeal diverticulum; *7* Epiglottis; *8* Aryepiglottic fold; *9* Arytenoid cartilage; *10* Entrance to lateral laryngeal ventricle, the broken line marks extent of ventricle; *11* Median laryngeal ventricle; *12* Vocal process of arytenoid cartilage; *12'* Vocal fold; *13, 13'* Cricoid cartilage; *14* Thyroid cartilage; *14'* Cricothyroid ligament; *15* Hyoepiglotticus; *17* Basihyoid; *18* Geniohyoideus; *20* Sternohyoideus; *21* Trachea; *22* Esophagus; *23* Caudal pharyngeal constrictors; *24* Hyopharyngeus; *25* Thyroid gland

Mouth and Pharynx of the Ruminants

Oral Cavity

The oral cavity of the ruminants is adapted to the efficient handling of the great volumes of plant roughage consumed by these animals. The frequently very bulky food is grasped by the lips or by the highly mobile, protrusile tongue, or, on pasture, may also be seized between the incisors of the lower jaw and the dental pad and torn off. The food is only superficially chewed and insalivated before it is swallowed and deposited in the rumen. This circumstance enables the ruminant to consume large amounts of food in a short period of time. While in the rumen, the ingesta are thoroughly mixed, macerated, and prepared for further digestion by the bacteria and infusoria that are present. At intervals, portions of the ingesta are regurgitated into the mouth for remastication and reinsalivation.

LIPS. The upper lip (274) of the ox is thick, firm, and not very mobile. Its central portion is without hair, and takes part in the formation of the **nasolabial plate** (*a*). Laterally, the upper lip is covered with hair and, in addition, has tactile hairs. In relation to the large oral cavity the lips are quite short. As a consequence, the mouth opening is small and the **angle of the mouth** is far forward on the face (21). The lower lip, being shorter than the upper, is slightly overhung by the margin of the upper lip. The external surface of the lower lip is covered with hair and also has tactile hairs. At the free margin, the hair-covered skin gives way to a narrow hairless zone with small rounded papillae along the edge of the lip. The **chin** of the ox is well developed and consists mainly of adipose tissue. Numerous cornified **labial papillae** are present on the inside of the lips near the angle of the mouth. They are caudally-directed, conical outgrowths of the mucous membrane and are continued on the inside of the cheeks (31, 32). The **labial glands** are concentrated in the vicinity of the angle of the mouth, and are often imbedded between bundles of the labial muscles. The short lips and oral cleft make both inspection of the caudal parts of the oral cavity and oral surgery difficult, even when the animal's mouth is fully opened.

The **LIPS OF THE SHEEP AND GOAT** are very mobile and are important for the prehension of food. The upper lip, which in addition to the fine ordinary hair carries tactile hairs, is divided by a deep **philtrum** (272, 273/*b*). The narrow, hairless **nasal plate** (*a*) does not involve the upper lip as in the ox. The free edges of both lips are set with rows of short, blunt papillae which become more pointed toward the angle of the mouth where they merge with the buccal papillae (85). The **labial glands** of the upper lip are quite numerous, especially near the angle of the mouth; those of the lower lip are only moderately developed.

The **BUCCAL VESTIBULE** of the ruminants is capacious and distensible. Its mucosa carries the large conical **buccal papillae** (31), which are directed toward the pharynx and are covered with a cornified epithelium. These are up to 1 cm. long in the ox and 4—5 mm. long in the small ruminants. They are longest and most numerous around the angle of the mouth, and decrease in size and number caudally.

The buccal glands are well developed and may be divided into dorsal, middle, and ventral glands. The **dorsal buccal glands** (49, 50/*8*) are opposite the alveolar border of the maxilla, and extend from the angle of the mouth to the maxillary tuberosity a short distance under the masseter muscle. The **ventral buccal glands** (*10*) are opposite the lower cheek teeth; they form a wider and more compact chain that stretches from the angle of the mouth to the rostral border of the masseter. The **middle buccal glands** (*9*) consist of more loosely arranged lobules, found along the dorsal border of the ventral glands.

In ox and sheep, the **HARD PALATE** is wide rostrally and caudally with a narrow part in the middle (83, 85). In the goat, the hard palate is of uniform width, but widens abruptly caudal to the level of the first cheek teeth.

Upper incisors are absent in the ruminants and are replaced by a **dental pad** (pulvinus dentalis, 31/*D*). This is a thick semilunar plate of dense connective tissue that is attached to the underlying body of the incisive bone. The tough mucous membrane covering the pad has a thick, heavily cornified epithelium. A narrow caudal zone of the dental pad carries short conical papillae and, centrally, the **incisive papilla** (*1*). The latter is surrounded by a circular groove, into the bottom of which the **incisive ducts** open on each side of the papilla. The mucous membrane of the hard palate overlies a layer of venous plexuses and may either be partly or completely pigmented (black). The **palatine ridges** (*2*) number 15—19 in the ox, 14 in the sheep, and 12 in the goat. In the ox, the ridges are set with caudally-directed, horny papillae and gradually diminish in size, so that caudal to the level of the second cheek teeth the hard palate is smooth. In the sheep and goat, the palatine ridges stop at the level of the third and second cheek teeth respectively. The **palatine raphe** is prominent (85/*4*). A few scattered **palatine glands** are found rostrally near the incisive papilla. There are many glands in the smooth caudal portion of the hard palate, and none in the area of the palatine ridges.

The **TONGUE** of the ruminants is highly mobile and protrusile and has an important function in the prehension of food. In the ox, the tongue is firm and rather plump. The rounded borders of its pointed apex blend caudally with the high lateral surfaces of the body. The **lingual frenulum** is wide. The caudal part of the dorsum of the tongue is raised and forms the characteristic, oval **torus linguae** (39, 40/*d*), rostral to which is the transverse **fossa linguae** (*d'*) of varying depth. Occasionally, barbed awns or husks of grain lodge in the fossa of the ox and cause infection of the tongue. The mucous membrane of the dorsum of the tongue is thick, firm, often pigmented, and adheres intimately to the subjacent tissues. The

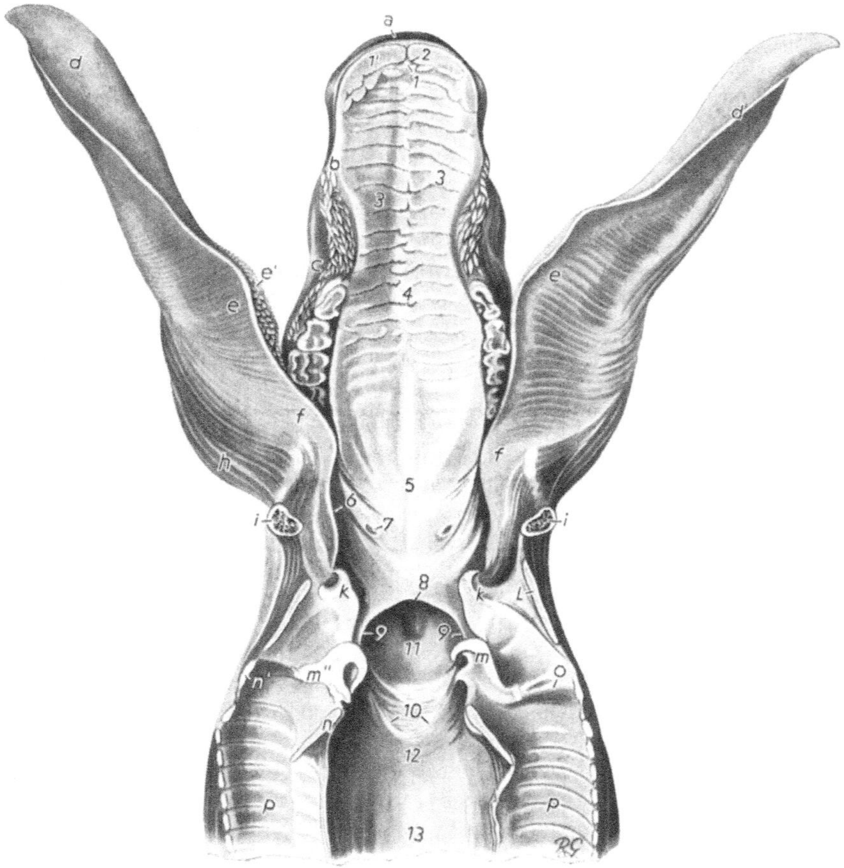

Fig. 83. Roof of the oral cavity, oropharynx, laryngopharynx, and esophagus of a cow. The tongue, larynx, trachea and esophagus have been split in the median plane. Ventral aspect. (After Zietzschmann, unpublished.)

a Upper lip; *b* Angle of the mouth; *c* Cut surface of the cheek; *d* Apex of tongue; *e* Body of tongue; *e'* Torus linguae; *f* Root of tongue; *h* Geniohyoideus; *i* Basihyoid; *k* Epiglottis; *l* Thyroid cartilage; *m* Arytenoid cartilage, *m''* its vocal process; *n, n'* Cricoid cartilage; *o* Vocal fold; *p* Trachea

1 Incisive papilla; *1'* Dental pad; *2* Opening of incisive duct; *3* Hard palate with palatine ridges; *4* Palatine raphe; *5* Soft palate; *6* Palatoglossal arch; *7* Sinus of palatine tonsil; *8* Free border of soft palate; *9, 9, 10* Palatopharyngeal arches; *11* Intrapharyngeal opening; *12* Caudal part of laryngopharynx; *13* Esophagus

filiform papillae (*1*) are directed caudally, pointed, and threadlike, and are found on the dorsum rostral to the fossa. The papillae on the torus are either blunt and conical **(conical papillae)** or round and flat **(lenticular* papillae),** and decrease in number toward the root of the tongue. In the small ruminants, the filiform papillae extend also onto the ventral surface of the apex; and on the torus the papillae are larger and more flattened than in the ox. The **fungiform papillae** (*2*) are numerous especially along the edges of the apex, but are also found in smaller numbers on the dorsum and along the lateral surfaces of the tongue. The **vallate papillae** (*3*) number 8—17 on each side in the ox, 18—24 in the sheep, and 12—18 in the goat. They vary in size, and form irregular rows on each side of the caudal part of the torus. **Foliate papillae** are usually absent in the ruminants; rudimentary ones may be present, however, in the ox.

The rasplike roughness of the tongue, the papillae on the palatine ridges, and the large caudally-directed papillae on the lips and cheeks facilitate the prehension and swallowing of coarse food. For the same reasons it is difficult for the ruminants to discharge undesirable food or foreign bodies from the mouth. This, together with the fact that they consume their food quickly and with little mastication, explains why the domestic ruminants, particularly the ox, swallow many foreign bodies, such as nails and pieces of wire, which usually settle in the reticulum of the stomach and give rise to traumatic gastritis or, in the language of the layman, "hardware" disease.

The **lingual tonsil** (39/*5*; 76/*1*) of the ox consists of many tonsillar follicles, and extends to the base of the epiglottis on either side of the glossoepiglottic fold. In the sheep and goat, the lingual tonsil consists only of diffuse lymphoid tissue and a few solitary lymph nodules (77/*1*). In addition to the mucous, serous, or seromucous glands at the root of the tongue and in the vicinity of the vallate papillae, there is a further group of **lingual glands** along the lateral and ventral surfaces of the tongue. The excretory ducts of this chain of glands open along a row of papillae, which corresponds to the plica fimbriata of man, and curve from the lateral surface of the tongue to the frenulum. A second row of papillae, nearly parallel to the first, but more ventrally situated and extending farther caudally, marks the position of the excretory ducts of the **polystomatic sublingual gland** which is described on page 67. In the sheep and sometimes also in the ox there is an apical lingual gland embedded in the musculature in the ventral surface of the tongue just rostral to the frenulum.

SUBLINGUAL FLOOR OF ORAL CAVITY. The **sublingual caruncle** (32/*2*) is flat and firm and has a slightly serrated edge. The goat has a small paracaruncular gland in its vicinity. In addition, a small mucous gland (Gl. lingualis apicalis) is present on the ventral surface of the tongue on each side of the frenulum in the small ruminants and occasionally in the ox. In the vicinity of this gland in the goat are a few tonsillar follicles (tonsilla sublingualis).

The **orobasal organ** of the ox is found on a low median ridge just caudal to the central incisors; it consists of two slightly diverging epithelial grooves which end in minute depressions (*1*). In the sheep, the orobasal organ consists of two minute grooves, and in the goat of two equally small depressions.

Salivary Glands

The **PAROTID GLAND** of the ox (49/*12*) is club-shaped and lies along the caudal border of the masseter. Its thicker dorsal end reaches the region of the temporomandibular joint rostral to the base of the ear, and with its rostral border lies on the masseter. Its narrow ventral end (*13″*) follows the caudal border of the mandible and is related deeply to the large mandibular gland. In the sheep and goat, the parotid gland is more rectangular in shape (50/*12*). It extends from the base of the ear to the angle formed by the linguofacial and maxillary veins, where it is in contact with the mandibular gland. The parotid gland of the ox covers only part of the parotid lymph node (49/*a*), while in the sheep and goat the node is completely covered. Proceeding from the deep face of the gland in the ox, the **parotid duct** (*11*), together with the facial artery and vein, follows the ventral and rostral borders of the

* Lens or lentil-shaped.

masseter and ascends on the lateral surface of the face. It opens on the **parotid papilla** in the buccal vestibule opposite the fifth upper cheek tooth. In the small ruminants, the parotid duct crosses the lateral surface of the masseter muscle at about the middle, and opens into the buccal vestibule opposite the third or fourth upper cheek tooth. Occasionally, the parotid duct of the goat is found to follow a course similar to that of the ox (50/*11*).

The **MANDIBULAR GLAND** of the ox (49/*14*) is much larger than the parotid gland; it is 18—20 cm. long, 8—10 cm. wide, and 2—4 cm. thick. The gland extends in a curve from the atlantal fossa to the basihyoid, and is so situated that its middle part is related to the parotid gland. The ventral end of the gland, which is palpable in the live animal, is rounded and lies near the midline, almost touching the ventral end of the gland on the other side. Lateral to it is the mandibular lymph node (*b*). The mandibular gland is related laterally to the branches of the external jugular vein and of the facial nerve. Medially, it is related to the common carotid artery and its branches, to the vagus and sympathetic nerves, to branches of the trigeminal nerve, to the pharynx and larynx, and to the sternocephalicus and brachiocephalicus muscles. The **mandibular duct** (54/*14'*) leaves the gland at about the middle of the rostral border and, crossing the lateral surface of the digastricus, passes forward on the deep surface of the mylohyoideus to the sublingual caruncle. It ends just lateral to the caruncle so that the opening of the duct is concealed by it. The mandibular gland in the small ruminants is similar to that of the ox. It is relatively large in the goat (55/*14*).

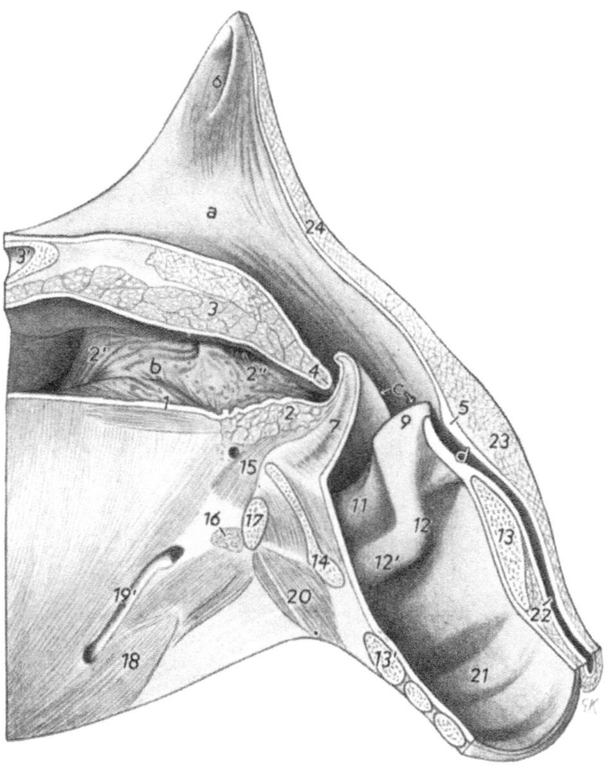

Fig. 84. Sagittal section of the pharynx and larynx of the ox. Breathing position.

a Nasopharynx; *b* Oropharynx; *c, d* Laryngopharynx

1 Root of tongue with fossulae of lingual tonsil; *2* Glossoepiglottic fold; *2'* Palatoglossal arch; *2''* Sinus of palatine tonsil; *3* Soft palate with thick layer of palatine glands; *3'* Palatine bone and palatine sinus; *4, 5* Boundaries of intrapharyngeal opening; *4* Free border of soft palate; *5* Palatopharyngeal arch; *6* Pharyngeal opening of auditory tube; *7* Epiglottis; *9* Arytenoid cartilage; *11* Laryngeal vestibule; *12* Vocal process of arytenoid cartilage; *12'* Vocal fold; *13, 13'* Cricoid cartilage; *14* Thyroid cartilage; *15* Hyoepiglotticus; *16* Hyoideus transversus; *17* Basihyoid; *18* Geniohyoideus; *19'* Lingual vein; *20* Sternohyoideus; *21* Trachea; *22* Esophagus; *23* Caudal pharyngeal constrictors; *24* Hyopharyngeus

The seromucous **SUBLINGUAL GLAND** of the ruminants also consists of two parts, the more ventral monostomatic and the more dorsal polystomatic gland. They are deep to the mylohyoideus and are related medially to the styloglossus, genioglossus, and geniohyoideus. In the ox, the **monostomatic sublingual gland** (54/*9*) is 10—12 cm. long and 2—3 cm. wide, and extends from the incisive part of the mandible to about the middle of the polystomatic gland, which lies along its dorsal border. Its excretory duct, the **major sublingual duct,** arises from the confluence of many radicles, passes forward medial to the gland, and accompanies the mandibular duct to the sublingual caruncle, lateral to which it opens with the mandibular duct. The **polystomatic sublingual gland** (*10*) lies dorsal to the monostomatic gland and consists of a chain of lobules, about 15—18 cm. in length, that extends from the incisive part of the mandible to the palatoglossal arch. The minute **minor sublingual ducts** open along the row of long papillae found on the floor of the lateral sublingual recess. The sublingual glands of the small ruminants are similar to those of the ox.

Pharynx

(60, 61, 83—85)

The pharynx of the ruminants is relatively short and does not extend caudally beyond the base of the skull.

The **SOFT PALATE,** which in the ox has a length of 8.5—12 cm., reaches the base of the epiglottis and is held in this position during normal breathing (60, 61). It may be raised, however, to permit forced mouth breathing. Elevation of the soft palate occurs also during regurgitation. This dilates the oropharynx, and the regurgitated bolus is directed into the mouth where it can be remasticated. Eructated* gas from the rumen may also pass through the oral cavity. Most of the gas, however, is forced into the lungs (Dougherty et al., 1962). This is accomplished by closing the intrapharyngeal opening and the mouth, and by opening the glottis. The intrapharyngeal opening is closed by the elevation of the soft palate and the action of the rostral pharyngeal constrictors. If eructation is inhibited for some reason, gas accumulates rapidly in the rumen (bloat) and may kill the animal in a short time.

The submucosa of the ventral surface of the soft palate contains a thick layer of **glands** (gll. palatini, 84/*3*), which produce mucus and make up approximately one-third to one-half of the thickness of the soft palate. These glands extend dorsally between the palatine muscles and laterally into the wall of the pharynx (54, 55/*11*). Diffuse lymphoid tissue and lymph nodules are also present on the ventral surface of the soft palate; the ox, in addition, has sporadic tonsillar follicles (76, 77/*3*). The delicate mucous membrane on the dorsal surface of the soft palate is covered with ciliated pseudostratified epithelium. In its lamina propia and submucosa are scattered glands as well as diffuse lymphoid tissue and solitary lymph nodules. The **palatopharyngeal arches** (83/*9*), small folds on the lateral wall of the pharynx, are not very prominent.

The **oropharynx** (84/*b*) is wide and dilatable. The rostral part of the larynx projects upward from the floor of the **laryngopharynx** through the **intrapharyngeal opening.** The **piriform recesses** are deep in the ox. The laryngohparynx (*c*, *d*) is relatively short and extends to about the rostral third of the cricoid lamina. No internal demarcation is present at the pharyngoesophageal junction.

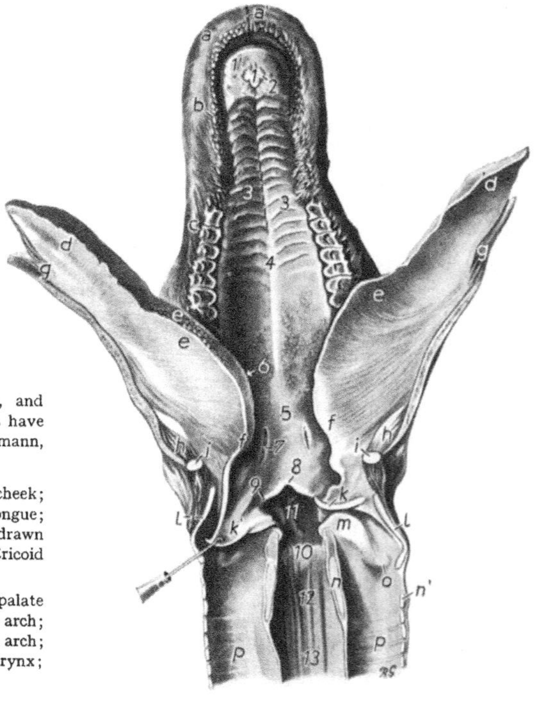

Fig. 85. Roof of the oral cavity, oropharynx, laryngopharynx, and esophagus of a sheep. The tongue, larynx, trachea, and esophagus have been split in the median plane. Ventral aspect. (After Zietzschmann, unpublished.)

a Upper lip; *a'* Philtrum; *b* Angle of the mouth; *c* Cut surface of cheek; *d* Apex of tongue; *e* Body of tongue; *e'* Torus linguae; *f* Root of tongue; *g* Genioglossus; *h* Geniohyoideus; *i* Basihyoid; *k* Epiglottis, split and drawn laterally at *k'*; *l* Thyroid cartilage; *m* Arytenoid cartilage; *n*, *n'* Cricoid cartilage; *o* Vocal fold; *p* Trachea

1 Incisive papilla; *1'* Dental pad; *2* Opening of incisive duct; *3* Hard palate with palatine ridges; *4* Palatine raphe; *5* Soft palate; *6* Palatoglossal arch; *7* Palatine tonsil; *8* Free border of soft palate; *9*, *10* Palatopharyngeal arch; *11* Intrapharyngeal opening; *12* Caudal part of laryngopharynx; *13* Esophagus

---

* From (L.) *ructus,* a belch.

The nasal septum is continued into the **nasopharynx** as the **pharyngeal septum** (60, 61/*12*), which divides the dorsal part of the nasopharynx into right and left recesses. The pharyngeal septum is high in the calf and somewhat lower in the adult ox. The **pharyngeal openings of the auditory tubes** (*14*) are small slits located in the caudal part of the nasopharynx.

Tonsils

LINGUAL TONSIL (39, 40/*5*; 76, 77/*1*). Consists in the ox of numerous large tonsillar follicles; is present in the small ruminants only as a small amount of diffuse lymphoid tissue.

PARAEPIGLOTTIC TONSIL. Present only in the small ruminants as an aggregation of tonsillar follicles at the base of the epiglottis.

TONSIL OF THE SOFT PALATE (76, 77/*3*). Consists of small amounts of lymphoid tissue on the ventral surface of the soft palate. The ox has also some scattered tonsillar follicles.

PALATINE TONSIL. In the ox (39/*6*; 72; 76/*2*; 83/*7*) about 3 cm. in diameter and concealed in the connective tissue and musculature of the wall of the oropharynx. It consists of numerous follicles which surround a branching **sinus tonsillaris.** The fossulae of the follicles open into the branches of the sinus, and the sinus in turn communicates with the oropharynx. In the small ruminants (40/*6*; 77/*2*; 85/*7*), the palatine tonsil is in the same location, but it consists of only three to six follicles with cleftlike fossulae.

PHARYNGEAL TONSIL (60, 61/*15*; 76, 77/*4*). An irregular elevation on the caudal end of the pharyngeal septum.

TUBAL TONSIL (76, 77/*5*). An accumulation of lymphatic tissue in the mucosa of the pharyngeal opening of the auditory tube.

Mouth and Pharynx of the Horse

Oral Cavity

The oral cavity of the horse is unusually long and relatively narrow. Its length is due to the remarkable development of the facial skeleton, particularly the structures concerned with mastication, while its width is determined by the narrow intermandibular space.

LIPS. The large upper lip and the smaller lower lip of the horse (275, 276) are highly mobile musculo-membranous folds and are very sensitive tactile and prehensile organs. Both upper and lower lip are covered externally with fine hair; on the free borders of the lips the hair is short, stiff, and bristly. Numerous tactile hairs surround the oral cleft. The hairs on each side of the middle of the upper lip are often quite long. Below the lower lip is the **chin** (56/*4'*) consisting of the poorly developed mentalis muscle and of adipose and connective tissue. The **oral cleft** extends to the level of the first cheek teeth, and is relatively small compared to the length of the oral cavity. This makes examination of the caudal parts of the oral cavity and oral surgery difficult.

The **labial glands** (51/*6*, *7*) are more numerous and better developed in the upper lip than in the lower lip. They increase in number toward the angles of the mouth, where they form compact masses, and empty into the labial vestibule through numerous visible openings. The buccal mucosa is smooth and it presents, opposite the third cheek tooth, the distinct **parotid papilla** with the relatively wide opening of the parotid duct.

The **dorsal buccal glands** (51/*8*, *8'*) lie in the submucosa or between bundles of the buccal muscles opposite the upper row of cheek teeth, and may be divided into rostral and caudal chains. The rostral chain consists of a loose collection of lobules which extends from the angle of the mouth to the rostral border of the masseter muscle. The caudal chain, more compact than the rostral, is about 6—8 cm. long and lies on the maxilla between the masseter and the molar part of the buccinator. The **ventral buccal glands** (*10*) lie along the ventral border of the molar part of the buccinator, extending from the angle of the mouth to the rostral border of the masseter.

The **HARD PALATE** (33, 35) extends from the incisors to the level of the last cheek teeth and is of almost equal width throughout. Immediately caudal to the incisors, the hard palate bulges ventrally and may, particularly in the foal, be level with the occlusal surface of the upper incisors. This bulge is sometimes thought by the layman to be a pathological swelling. The oval **incisive papilla** (*1*) is found in the center of this thick portion. On the papilla are two small depressions which do not, however, lead to the incisive ducts. In the horse, the **incisive ducts** end blindly just under the palatine epithelium of this region (282/*b*). There are 16—18 well-developed **palatine ridges** (35) on either side of the deep **palatine raphe.** They are closely spaced rostrally and caudally, but are farther apart at the diastema (*c*). Deep to the firm, nonglandular palatine mucosa are several layers of venous plexuses.

The **TONGUE** (41), like the intermandibular space it occupies, is long and narrow and has tall lateral surfaces. Its **apex**, long, spatular in front, and highly mobile, has rounded borders and is connected with the floor of the mouth by a well-developed **frenulum.** The **filiform papillae** are soft and thin and give a velvety texture to the dorsum of the tongue. The **fungiform papillae** (*2*) are scattered over the dorsal surface of the apex and the lateral surfaces of the body. There is usually only one pair of large **vallate papillae** (*3*) at the junction of the body and root of the tongue. The vallate papillae are about 7 mm. in diameter and have an uneven surface (88/*1'*). Occasionally a second pair, and very rarely a third, may be present caudal to the regular papillae. The **foliate papillae** (*1''*) are located on the lateral borders of the tongue just rostral to the palatoglossal arches, forming rounded eminences about 20—25 mm. long. Because of the presence of numerous tonsillar follicles

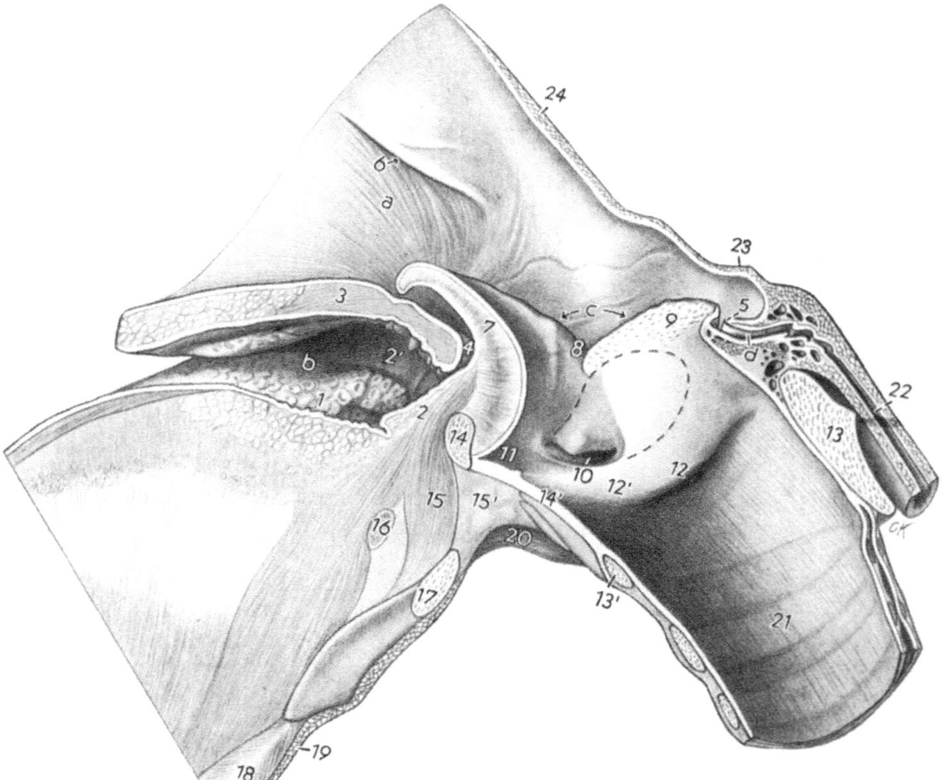

Fig. 86. Sagittal section of the pharynx and larynx of the horse. Breathing position.

a Nasopharynx; *b* Oropharynx; *c, d* Laryngopharynx

1 Root of tongue with tonsillar follicles (lingual tonsil); *2* Glossoepiglottic fold; *2'* Palatoglossal arch; *3* Soft palate; *4, 5* Boundaries of intrapharyngeal opening; *4* Free border of soft palate; *5* Palatopharyngeal arch; *6* Pharyngeal opening of auditory tube; *7* Epiglottis; *8* Aryepiglottic fold; *9* Corniculate process of arytenoid cartilage; *10* Entrance to lateral laryngeal ventricle, the broken line marks the extent of the ventricle; *11* Median laryngeal ventricle; *12* Vocal process of arytenoid cartilage; *12'* Vocal fold; *13, 13'* Cricoid cartilage; *14* Thyroid cartilage; *14'* Cricothyroid ligament; *15* Hyoepiglotticus; *15'* Fat; *16* Hyoideus transversus; *17* Basihyoid with lingual process; *18* Geniohyoideus; *19* Mylohyoideus; *20* Sternohyoideus; *21* Trachea; *22* Esophagus; *23* Caudal pharyngeal constrictors; *24* Hyopharyngeus

(lingual tonsil, *1*), the mucosa at the root of the tongue is very uneven. Under the mucosa of the body of the tongue is often a median, fibrous cord (cartilago dorsi linguae, 25/*6*) about 11—17 cm. long and 4—6 mm. in diameter. The cord is cylindrical rostrally, more filiform caudally, and consists predominantly of dense elastic fibers, interspersed with numerous adipose cells, and single or groups of cartilage cells. Mucous, serous, or seromucous **lingual glands** are regularly encountered in the vicinity of the vallate papillae and at the root of the tongue. A further chain of glands of variable width is found beneath the mucosa on the lateral surface. It extends from the root to about the middle of the body of the tongue, its many small excretory ducts opening directly over the chain.

The **sublingual caruncles** (34/*2*) are flat and project laterally from two narrow caruncular folds, which are lateral to the frenulum on the **SUBLINGUAL FLOOR OF THE ORAL CAVITY.** The **mandibular duct** opens on the lateral border of the caruncles. Lymphoid tissue that is present in and around the caruncular folds constitutes the sublingual tonsil. The horse has also a **paracaruncular gland;** its excretory ducts open on and in front of the sublingual caruncle. The **orobasal organ** (*1*) of the horse, two slitlike openings leading into minute epithelial canals about 9 mm. long, is located a short distance caudal to the lower central incisors.

Salivary Glands

The salivary glands of the horse, like those of the other herbivores, are relatively large. The slightly tuberculate, yellowish-red **PAROTID GLAND** (51/*12*) is the largest. It is 20—26 cm. long, 5—10 cm. wide, about 2 cm. thick, and weighs 200—225 gm. It occupies the retromandibular fossa, which is the space caudal to the ramus of the mandible and ventral to the wing of the atlas. The rostral border of the gland reaches the temporo-mandibular joint dorsally and is intimately related to the caudal border of the mandible, overlapping the masseter to some extent. The caudal border follows the slope of the wing of the atlas. Dorsally, the gland is notched to receive the base of the ear between the preauricular and retroauricular angles of the gland (*13, 13'*). The ventral end of the gland has a mandibular angle (*13''*) of variable length that extends into the intermandibular space along the medial surface of the occipitomandibular part of the digastricus, and a cervical angle (*13'''*) in the fork formed by the linguofacial and maxillary veins. The maxillary vein passes obliquely through the middle portion of the gland. Many large radicles unite in the region of the mandibular angle of the gland to form the **parotid duct** (*11*). The duct runs forward along the medial surface of the mandible and winds around its ventral border with the facial artery and vein (*h*), lying caudal to the vein at this point. It then ascends along the rostral border of the masseter, crosses under the facial vessels, and, becoming wider at the end, opens into the buccal vestibule on the **parotid papilla** opposite the third upper cheek tooth.

Because of the close relations between the parotid gland, the guttural pouch (56/*x*), and the retropharyngeal and cranial cervical lymph nodes, and because of the clinical importance of the retromandibular fossa in the horse, the **TOPOGRAPHY OF THE PAROTID REGION** is briefly described here. The lateral surface of the parotid gland is covered by the parotidoauricularis. Its ventral border follows the linguofacial vein (51/*b*), while the maxillary vein (*c*) passes obliquely through the middle of the gland. The medial surface of the parotid gland is very uneven and is related to the following structures: to the dorsal end of the mandibular gland (56/*14*) and the cranial cervical lymph nodes (*v*) through the space between the linguofacial and maxillary veins, to the tendon of insertion of the sternomandibularis (51, 56/*M*), to the occipitomandibular part of the digastricus (51/*L*), to the tendon of insertion of the cleidomastoideus (51, 56/*K*), to the branches of the maxillary vein and the external and internal carotid arteries (56/*h, i*), and to the branches of the facial nerve which are partly embedded in the gland. The hypoglossal nerve (*r*) and the glossopharyngeal nerve pass rostroventrally toward the tongue on each side of the external carotid artery, while the vagus and sympathetic nerves (*s*) pass into the neck on the deep surface of the common carotid artery. Directly ventral to the temporo-mandibular joint on the caudal border of the mandible and under cover of the parotid gland are the parotid lymph nodes. The medial retropharyngeal nodes (*u*) are dorsal to the pharynx, while the lateral retropharyngeal nodes (*u'*) are on the caudoventral border of the occipitomandibular part of the digastricus in the depth of the atlantal fossa.

The parotid gland is in direct contact with the guttural pouch in two places. First, between the caudal border of the mandible and the rostral border of the occipitomandibular part of the digastricus. Second, between the caudal border of the digastricus and the atlantal fossa. The guttural pouch is accessible surgically by two routes, both involving the parotid gland. First, through Viborg's triangle in a rostrodorsal direction toward the stylohyoid. (Viborg's triangle is formed by the caudal border of the mandible cranially, by the linguofacial vein ventrally, and by the tendon of the sternomandibularis caudodorsally.) Second, from a point about 1 cm. cranial to the border of the wing of the atlas around the caudal borders of both the parotid gland and the digastricus or through the digastricus. With this second route care must be taken not to injure the caudal auricular vein and nerve.

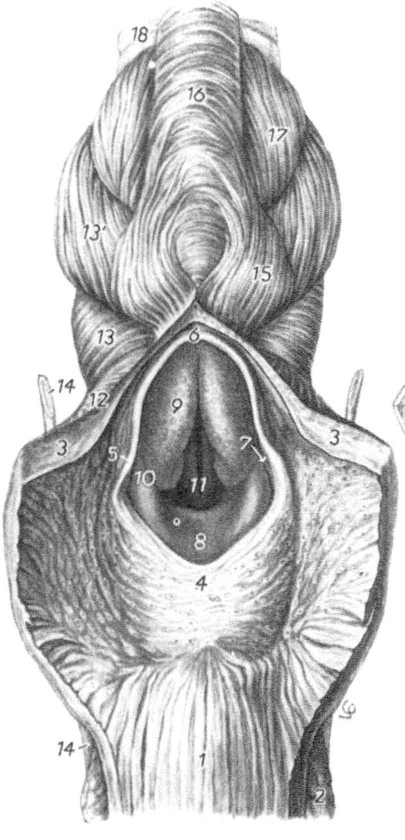

Fig. 87

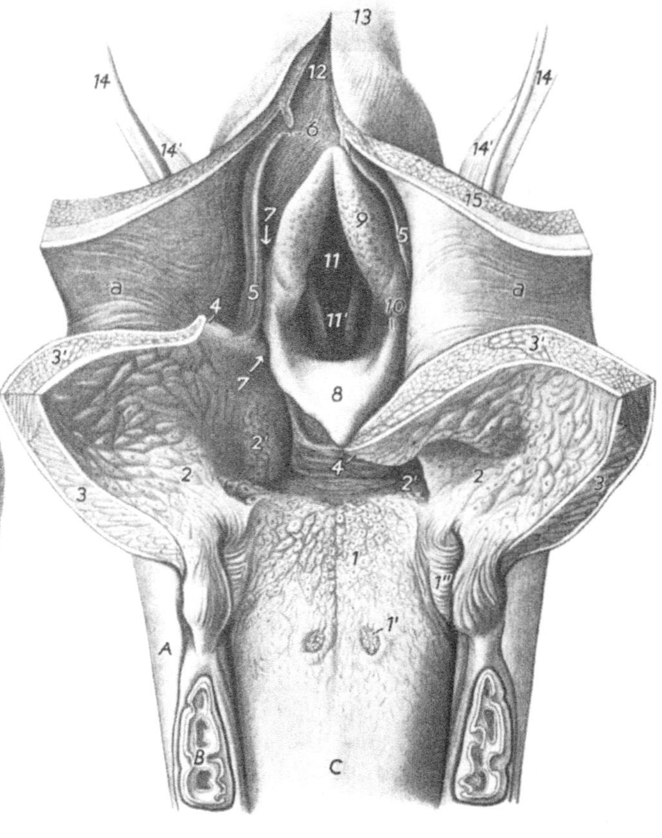

Fig. 88

Fig. 87. Pharynx of the horse. Dorsal aspect. The roof of the nasopharynx has been split in the median plane.

1 Dorsal surface of the soft palate with glandular openings and fossulae tonsillares; *2* Palatine glands; *3* Roof of nasopharynx, split; *4, 5, 6* Boundaries of intrapharyngeal opening; *4* Free border of soft palate; *5, 6* Palatopharyngeal arch; *7* Piriform recess; *8* Epigottis, its apex is covered by the soft palate; *9* Corniculate process, covered with solitary lymph nodules; *10* Aryepiglottic fold; *11* Laryngeall entrance, visible through the laryngeal entrance is the glottis; *12* Hyopharyngeus; *13* Thyropharyngeus; *13'* Cricopharyngeus; *14* Stylohyoid; *15* Lateral longitud. esophageal muscles; *16* Esophagus; *17* Cricoarytenoideus dorsalis; *18* Trachea

Fig. 88. Pharynx of the horse. Dorsal aspect. The roof of the pharynx and the soft palate have been split in the median plane.

A Mandible; *B* Third molar tooth; *C* Dorsum of tongue; *a* Nasopharynx

1 Root of tongue with tonsillar follicles (lingual tonsil); *1'* Vallate papilla; *1''* Foliate papilla; *2* Palatoglossal arch; *2'* Palatine tonsil; *3, 3'* Soft palate with glandular opening and tonsillar follicles; *4, 5, 6* Boundaries of intrapharyngeal opening; *4* Free border of soft palate; *5, 6* Palatopharyngeal arch; *7* Piriform recess; *8* Epiglottis; *9* Corniculate process, covered with solitary lymph nodules; *10* Aryepiglottic fold; *11* Laryngeal entrance; *11'* Intermembranous part of glottic cleft; *12* Caudal part of laryngopharynx; *13* Esophagus; *14* Stylohyoid; *14'* Caudal stylopharyngeus; *15* Pharyngeal constrictors

The **MANDIBULAR GLAND** of the horse (56/*14*), which is smaller than the parotid, is 20—23 cm. long, 2—3.5 cm. wide, up to 1 cm. thick, and weighs 45—60 gm. It extends from the atlantal fossa to the vicinity of the basihyoid. Its dorsal end is covered laterally by the parotid gland, the tendon of the sternomandibularis, and the maxillary vein. Farther ventrally, its lateral relations are the digastricus and the pterygoideus muscles. Its medial surface is related to the flexors of the head, guttural pouch (*x*), common carotid artery (*f*), vagus and its branches, sympathetic nerve (*s*), pharynx, and larynx. The **mandibular duct** (*14'*) emerges from the concave border of the gland. It follows this border rostrally, crosses the intermediate tendon of the digastricus (*N*), gains the medial surface of the polystomatic sublingual gland, and opens on the lateral surface of the sublingual caruncle.

Of the two sublingual salivary glands found in the domestic mammals, only the **POLYSTOMATIC SUBLINGUAL GLAND** is present in the horse (56/*10*). It lies under the mucous membrane of the lateral sublingual recess and extends from the incisive part of the mandible to about the level of the third lower cheek tooth. Its thin dorsal border lies inside the **sublingual fold.** The polystomatic sublingual gland is 12—15 cm. long, 1.5—3 cm. wide, 4—6 mm. thick, and weighs about 15—16 gm. Its medial surface is related to the styloglossus and genioglossus and to the mandibular duct (25/*7*, *9*). Its many small excretory ducts, the **minor sublingual ducts** open on visible pores along the sublingual fold (56/*10'*).

Pharynx

(62, 86—88)

The pharynx of the horse has a length of 19—20 cm., its caudal limit, however, does not extend beyond the caudal limit of the skull. Only the rostral third of the pharyngeal roof is attached to the base of the cranium (62/*F*). The caudal two-thirds are related to the guttural pouches (*14'*) and so are the dorsolateral walls.

The **SOFT PALATE** is long, and from its free border to its attachment on the osseous palate measures 10—13 cm. The mucous membrane on the ventral surface of the soft palate is wrinkled and presents numerous small openings of the large **palatine glands** (35/*3*). Rostrally, there is the median **tonsil of the soft palate** (*3'*). Diffuse lymphatic tissue and lymph nodules are also present in the mucosa of the dorsal surface. During normal breathing, the free border of the soft palate lies against the base of the epiglottis, and the epiglottis and part of the arytenoid cartilages protrude through the intrapharyngeal opening into the **nasopharynx**. The horse apparently is unable to elevate the soft palate sufficiently for mouth breathing. The same is true when the horse vomits; the soft palate directs the vomitus into the nasopharynx and nasal cavities so that it is ejected through the nostrils. The **palatopharyngeal arches** (88/*5*, *6*) continue the free border of the soft palate along the lateral walls of the pharynx. They measure as much as 1 cm. in height and meet dorsal to the arytenoid cartilages (87). Together with the free border of the soft palate, they enclose the slightly oval **intrapharyngeal opening,** which is about 5.5 cm. long and 5 cm. wide.

The **oropharynx** (86/*b*), like the soft palate, is relatively long in the horse. Except during swallowing, it is no more than a narrow elongated cleft between the root of the tongue and the soft palate, being slightly wider caudally than rostrally. The **laryngopharynx** (*c*, *d*) is relatively short, and extends from the base of the epiglottis to the front of the cricoid lamina. The **piriform recesses** (88/*7*) on each side of the entrance to the larynx are 3 cm. deep, measured from the edge of the aryepiglottic folds.

Tonsils

LINGUAL TONSIL (67; 78/*1*; 86, 88/*1*). Follicles at the root of the tongue and in the vicinity of the glossoepiglottic fold.

TONSIL OF THE SOFT PALATE (35/*3'*; 78/*3*). An oval, slightly elevated follicular tonsil located rostrally on the ventral surface of the soft palate.

PALATINE TONSIL (78/*2*; 88/*2'*). An elongated, flat follicular tonsil, 10—12 cm. long and 2 cm. wide, located on the floor of the oropharynx, lateral to the glossoepiglottic fold, and extending caudally to the base of the epiglottis.

PHARYNGEAL TONSIL (78/*4*). An accumulation of tonsillar follicles at the caudal end of the nasal septum and in the vicinity of the choanae.

TUBAL TONSIL (78/*5*). A triangular area of lymphoid tissue found either on or between the two laminae of the pharyngeal opening of the auditory tube; also diffuse lymphoid tissue or solitary nodules in the lateral walls of the nasopharynx and the dorsal surface of the soft palate.

BIBLIOGRAPHY

Mouth and Pharynx

Ackerknecht, E.: Ein eigenartiges Organ im Mundhöhlenboden der Säugetiere. Anat. Anz. **41,** 1912.

Ackerknecht, E.: Zur Topographie des präfrenularen Mundhöhlenbodens vom Pferde; zugleich Feststellungen über das regelrechte Vorkommen parakarunkulären Tonsillengewebes (Tonsilla sublingualis) und einer Glandula paracaruncularis beim Pferde. Arch. Anat. Physiol., 1912.

Ackermann, O.: Neues über das Vorkommen des Ackerknecht'schen Organs in der Säugetierreihe. Anat. Anz. **57,** 1923.

Bärner, M.: Über die Backendrüsen der Haussäugetiere. Arch. wiss. prakt. Tierheilk. **19,** 1893.

Behrendt, E.: Beitrag zur topographischen Anatomie des Schweinekopfes. Diss. med. vet. Berlin, 1966.

Bock, E., u. A. Trautmann: Die Glandula parotis bei Ovis aries. Anat. Anz. **47,** 1914.

Bosma, J. F.: Myology of the pharynx (dog, cat, monkey). Ann. Otol. Rhin. Laryng. **65,** 1956.

Bräter, H.: Funktionelles vom Zungenbein des Pferdes. Leipzig, Diss. med. vet., 1940.

Ciliga, T.: Prilog pozuavanju mišica mekog nepca (Zur Anatomie der Gaumensegelmuskeln des Pferdes) (Jugoslav. with German summary) Veterinarski arkiv, Zagreb, **12,** 1942.

Dougherty, R. W., K. J. Hill, F. L. Campeti, R. C. McClure, and R. E. Habel: Studies of pharyngeal and laryngeal activity during eructation in ruminants. Am. J. Vet. Res. **23,** 1962.

Dyce, K. M.: The muscles of the pharynx and palate of the dog. Anat. Rec. **127,** 1957.

Eberle, W.: Zur Entwicklung des Ackerknecht'schen Organs. Untersuchungen bei Katze, Hund und Mensch Anat. Anz. **60,** 1925.

Fernandes Filho, A. A. D'Errico, V. Borelli: Topographie der Austrittsstelle des Ductus parotideus beim Büffel (Bubalus, bubalis Linnaeus, 1758). Rev. Fac. Med. Vet. São Paulo **8,** 389—393 (1970).

Freund, L.: Zur Morphologie des harten Gaumens der Säugetiere. Zschr. Morph. Anthrop. **13,** 1911.

Frewein, J.: Die Ursprünge der Mm. tensor und levator veli palatini bei Haussäugetieren. Anat. Anz. **112,** Erg. H. 313—318 (1963).

Ghetie, V.: La musculature de la base de la langue chez le cheval. Anat. Anz. **87,** 1938/39.

Glen, J. B.: Salivary cysts in the dog: Identification of sublingual duct defects by saliography. Vet. Rec. **78,** 1966.

Haller, B.: Die phyletische Entfaltung der Sinnesorgane der Säugetierzunge. Arch. mikrosk. Anat. **74,** 1910.

Hamecher, H.: Vergleichende Untersuchungen über die kleinen Mundhöhlendrüsen unserer Haussäugetiere. Leipzig, Diss. phil., 1905.

Hamon, M. A.: Atlas de la Tete du Chien: Coupes series—Radio-Anatomie—Tomographies. Thesis, Université Paul Sabatier, Toulouse, 1977.

Hauser, H.: Über Bau und Funktion der Wiederkäuerparotis. Zschr. mikrosk.-anat. Forsch. **41,** 1937.

Helber, K.: Die motorische Innervation der Gaumensegelmuskeln des Hundes. Diss. Berlin, 1967.

Herre, W., u. H. Metzdorff: Über das Ackerknecht'sche Organ einiger Primaten. Zool. Anz. **124,** 1938.

Himmelreich, H.: Zur vergleichenden Anatomie der Schlundmuskeln der Haussäugetiere. I. Zur Anatomie der Schlundwandmuskeln des Pferdes. Anat. Anz. **81,** 1935.

Himmelreich, H. A.: Der M. tensor veli palatini der Säugetiere unter Berücksichtigung seines Aufbaus, seiner Funktion und seiner Entstehungsgeschichte. Anat. Anz. **115,** 1—26 (1964).

Hotescheck, H. J.: Die topographische Anatomie des Übergangsgebietes Kopf-Hals und die des Halses als vergleichende Literaturstudie bei Pferd, Wiederkäuer, Schwein und Hund. Diss. med. vet. Berlin, 1968.

Illing, G.: Vergleichende makroskopische und mikroskopische Untersuchungen über die submaxillären Speicheldrüsen der Haustiere. Zürich, Diss. phil., 1904.

Immisch, K. B.: Untersuchungen über die mechanisch wirkenden Papillen der Mundhöhle der Haussäugetiere. Gießen, Diss. med. vet., 1908.

Iwanoff, St.: Das Relief des harten Gaumens beim Schwein mit Berücksichtigung der Variabilität der Plicae palatinae transversae (German summary). Jb. Univ. Sofia, Vet. med. Fak. **16,** 1939/40.

Iwanoff, St.: Das Relief des harten Gaumens beim Rind unter Berücksichtigung der Variabilität der Gaumenstaffeln. Jb. Univ. Sofia, Vet. med. Fak. **17,** 1940/41.

Jaenicke, H.: Vergleichende anatomische und histologische Untersuchungen über den Gaumen der Haussäugetiere. Zürich, Diss. med. vet., 1908.

Keller, E.: Über ein rudimentäres Epithelialorgan im präfrenularen Mundbogen der Säugetiere. Anat. Anz. **55,** 1922.

Kraft, H.: Vergleichende Betrachtungen über den harten Gaumen der Haussäugetiere. Tierärztl. Umsch. **11,** 1956.
Künzel, E., G. Luckhaus, und P. Scholz: Vergleichend-anatomische Untersuchungen der Gaumensegelmuskulatur. Zschr. Anat. Entw. gesch. **125,** 1966.
Michel, G.: Beitrag zur Topographie der Ausführungsgänge der Gl. mandibularis und der Gl. sublingualis major des Hundes. Berl. Münch. Tierärztl. Wschr. **69,** 1956.
Nagy, F.: Kopf- und Vorderdarm der Katze (Felis domestica). (German summary). Budapest, Diss. med. vet., 1932.
Nikolov, D.: Über den Bau des organinneren Blutkreislaufes in den großen Speicheldrüsen des Hundes. Anat. Anz. **125** Erg. H. 705—711 (1969).
Peters, J.: Untersuchungen über die Kopfspeicheldrüsen bei Pferd, Rind und Schwein. Gießen, Diss. med. vet., 1904.
Pinto e Silva, P., A. F. Filho, A. A. D'Errico: Topographie der Austrittsstelle des Ductus parotideus bei Vollblutpferden. Rev. Fac. Med. Vet., São Paulo **8,** 403—409 (1970).
Risberg, A. R.: Ein Beitrag zur Frage des Baues der Lyssa bei Säugetieren. Zürich, Diss. med. vet., 1918.
Scheuerer, E.: Die Unterzungendrüsen des Hundes. Anat. Anz. **77,** 1933.
Schröder, D.: Ein Beitrag zur topographischen Anatomie des Schafkopfes. Diss. med. vet. Berlin, 1970.
Tehver, J.: Über die vordere Zungendrüse der Hauswiederkäuer. Anat. Anz. **90,** 1940.
Trautmann, A.: Der Zungenrückenknorpel von Equus caballus. Morph. Jb. **51,** 1921.
Vitums, A.: Über den Schlingrachen bei Haussäugetieren. Veröff. Univ. Riga, 1940; ref. Jber. Vet. Med. **70,** 1942.
Vollmerhaus, B.: Zur vergleichenden Nomenklatur des lymphoepithelialen Rachenringes der Haussäugetiere und des Menschen. Zbl. Vet. Med. **6,** 1959.
Wehner, G.: Zur Anatomie der Backen-, Masseter- und Parotisgegend des Hausschafes (Ovis aries L.). Leipzig, Diss. med. vet., 1936.
Williams, D. M. and A. C. Rowland: The palatine tonsils of the pig—an afferent route to the lymphoid tissue. J. Anat. **113,** 1972.
Ziegler, H.: Beiträge zum Bau der Unterkieferdrüse der Hauswiederkäuer: Rind, Ziege und Schaf. Zschr. Anat. **82,** 1927.
Ziegler, H.: Lassen sich die Unterkieferdrüsen unserer Hauswiederkäuer morphologisch voneinander unterscheiden? Zschr. Anat. **85,** 1928.
Ziegler, H.: Zur Morphologie gemischter Hauptstücke in sublingualen Speicheldrüsen von Haustieren. Zschr. mikr. anat. Forsch. **39,** 1936.
Zietzschmann, O.: Betrachtungen über den Schlundkopf. Dtsch. Tierärztl. Wschr. **47,** 1939.
Zimmermann, G.: Über den Waldeyer'schen lymphatischen Rachenring. Arch. wiss. prakt. Tierheilk. **67,** 1933.

Teeth

General and Comparative

The dentition of the domestic mammals consists of two dental arches, but the shape, arrangement, and number of teeth vary from species to species. Among mammals in general, the dentition of particular groups is highly characteristic and is, therefore, an important criterion for identification and classification. Also, because of their permanence, teeth are among the most important paleontological finds.

The **TEETH** (dentes) are the principal organs of mastication and function together with the jaws, masticatory muscles, lips, and tongue in the prehension and mastication of food. The dentition of an animal is always intimately related to its mode of nutrition; consequently, such terms as **carnivorous, omnivorous,** and **herbivorous dentition** have come into use. In some species the teeth have developed into formidable weapons.

The relatively simple, unspecialized **brachydont* tooth** (98) consists of a recognizable **crown** (corona dentis), the free distal portion of the tooth projecting into the mouth; a **root** (radix dentis), the embedded portion; and a slightly constricted **neck** (collum dentis) between

* From Gr., low-crowned; as the teeth of man, dog, or pig.

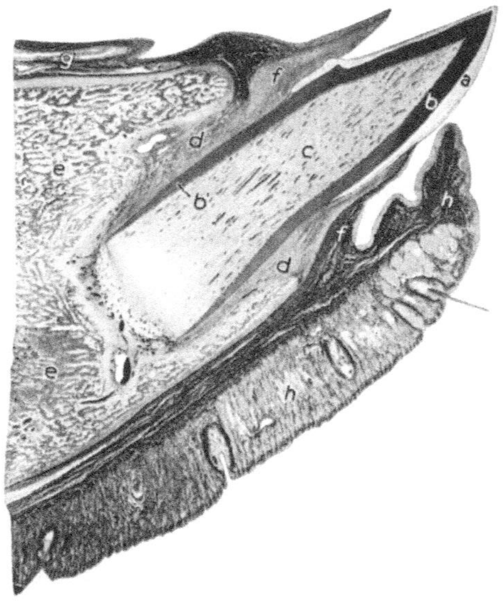

Fig. 89. Sagittal section of lower lip, incisor, and mandible of a newborn calf. Microphotograph.

a Enamel, covering the crown of the tooth; *b* Dentine of the crown; *b'* Dentine of the root; *c* Dental cavity filled with dental pulp; *d* Periodontium; *e* Incisive part of mandible; *f* Gums; *g* Sublingual floor of oral cavity; *h* Lower lip with tactile hairs and labial glands

crown and root at the gum line. In **hypsodont* teeth** (106), which are more specialized than the brachydont teeth, crown and neck are not easily distinguished, and there is only a body and a root. The **body** (corpus dentis) has a free portion (corona clinica), which is surrounded at its base by the gums, and an embedded portion (radix clinica), which is usually long in the young animal. The **root** (radix dentis) is usually short. Teeth such as the tusks of the boar (94) also consist of a long body, partly exposed and partly embedded, but have no root (radix dentis) in the strict sense.

The roots (radix dentis and radix clinica) of the teeth are firmly implanted in the dental alveoli of the mandible, incisive bone, and the alveolar process of the maxilla. They are anchored to the alveolar wall by means of the **periodontium** (89/*d*), which consists of strong collagenous bundles, traversing the space between the alveolar wall and the root and, by their orientation, preventing the tooth from being pressed into its socket during mastication.

Mammalian teeth are composed of three substances: dentine, enamel, and cement. The **DENTINE** (98/*gray*), hard, yellowish-white osseous tissue produced by the odontoblasts, usually constitutes the bulk of the tooth and contains the **dental cavity** in its center. In this cavity is the **dental pulp** (pulpa dentis), which is a mass of delicate connective tissue, fine blood vessels, and nerves. The **ENAMEL** (98/*white*; 89/*a*), brilliantly white, the hardest part of the tooth, and the hardest substance in the body, is a product of the ameloblasts, which are modified ectodermal cells of the embryonic oral epithelium. In brachydont teeth, the enamel, which is covered by a thin **cuticle** (cuticula dentis), envelops only the short, exposed crown. In hypsodont teeth, it also covers the embedded portion of the tooth, but not the short root (radix dentis). The enamel, covering hypsodont cheek teeth, usually forms prominent longitudinal folds (plicae enameli). The **CEMENT**, a product of the cementoblasts and very similar to bone, invests the root of the tooth (98/*black*) and in the hypsodont type also the tall body of the tooth (104/*black*).

The surface of the tooth which faces its antagonist in the opposite jaw is the **occlusal surface.** The surface which faces the adjacent teeth is the **contact surface,** the one toward the first incisor (the tooth next to the median plane) being the mesial and the one on the opposite side the distal contact surface. The **vestibular surface** faces the lips and

Fig. 90

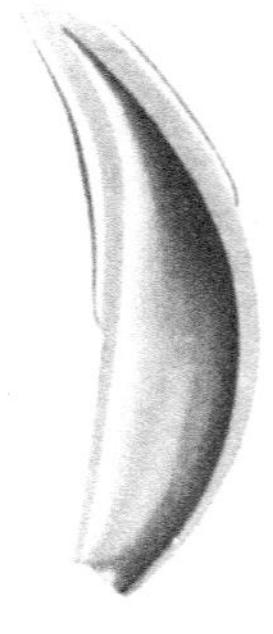

Fig. 91

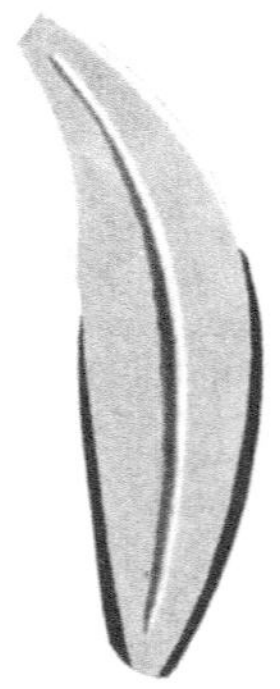

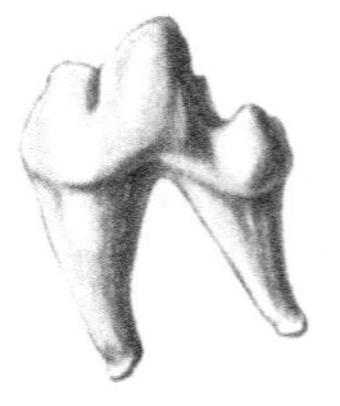

Fig. 92

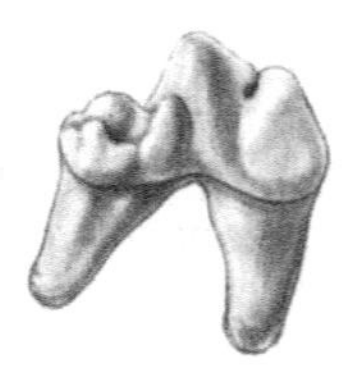

Fig. 93

Fig. 90. Section of deciduous canine (Dc) of a dog. Gray: dentine; White: enamel. The dental cavity is large

Fig. 91. Section of canine (C) of an older dog. Gray: dentine; White: enamel, worn at the distal end; Black: cement. The dental cavity is small

Figs. 92 and 93. Lower sectorial tooth (M1) of an older dog. (92) Vestibular aspect (93) Lingual aspect

* From Gr., high-crowned; as the teeth of the horse

the cheeks, while the **lingual surface** is in contact with the tongue. The opening on the apex of the root, the **apical foramen,** leads through the **root canal** into the **dental cavity.**

Fig. 94. Section of lower tusk (C) of an older pig.

Typical example of a permanently growing tooth. The dental cavity remains wide open proximally. The distal end of the tooth is in wear. Gray: dentine; White: enamel; Black: cement

Teeth with well-developed roots and short crowns (brachydont) complete their growth shortly after they erupt. During development, they have a large dental cavity and a wide root canal, both filled with a pulp rich in vessels and nerves (97). After the tooth ceases to grow, deposition of **secondary dentine** narrows first the apical foramen and later, gradually, the entire dental cavity from the crown toward the root. In contrast to these teeth are those that have no root in the strict sense (radix dentis), e. g., the tusks of boars (94) or the incisors of rodents. They have a large dental cavity, which remains wide open proximally, and they continue to grow throughout the animal's lifetime. Growth usually keeps pace with the wear of the occlusal surface of the teeth. If there is no wear at the distal end, these teeth will grow to great lengths. The cheek teeth of the ruminants and the horse lie between these two extremes in shape and growth pattern.

Fig. 95

Fig. 96

Figs. 95 and 96. Upper M1 (95) and lower M1 (96) of a pig about 20 months old. Vestibular aspect

Replacement of Teeth

In most mammals—and in all the domestic mammals—two sets of teeth develop (diphyodont* animals). The first set, consisting of fewer teeth than the second, appears early in life and is called the deciduous dentition. It is gradually replaced by the permanent dentition during the animal's growth period. The deciduous dentition provides the young mammal with a fully functional, though smaller, set of teeth that can be accommodated by its small jaws. As the jaws grow longer, new (permanent) teeth are added, and the **deciduous teeth** are gradually replaced by **permanent teeth.** Incisors, canines, and premolars, with the exception of the first premolar, are replaced. The molars, which erupt caudal to the premolars, are not present in the deciduous dentition. Replacement is gradual and follows a definite sequence, so that some deciduous and some permanent teeth are in use at the same time. As the developing permanent tooth pushes to the surface, it presses on the roots of the worn-out deciduous tooth and gradually cuts off its nutrition. The deciduous tooth becomes loose, eventually dies, and is displaced. The osseous alveolar walls adjust to these changes either by bone production or by resorption, and thereby provide a new socket for the embedded portion of the new tooth.

Types of Teeth

In mammals, the function of a tooth is related to its location in the mouth, and the morphology or type of tooth is related to its function. Mammals have several types of teeth and are, therefore, said to be heterodont animals. The **incisors** (112/*J1*—*J3*) are embedded in the incisive bone and the incisive part of the mandible, and are usually simple morphologically. They are followed caudally by the **canines** (*C*), and these by the more complex **cheek teeth,** which consist of a rostral group of **premolars** (*P1*—*P4*) and a caudal group of **molars** (*M1*—*M2*).

---

* From Gr., to grow teeth twice.

Dental Formula

The dental formula is an abbreviated statement of the number of teeth of each dental type in an animal, which facilitates the comparison of dentitions among mammalian species. The formula uses the symbols **I** for incisors, **C** for canines, **P** for premolars, and **M** for molars. The symbol is followed by a number that indicates how many teeth of that type are present. Beginning at the median plane, the dental formula lists the permanent teeth of one side: **I** 1, **I** 2, **I** 3 for the **incisors***, **C** for the **canines, P** 1, **P** 2, **etc.** for the **premolars**, and **M** 1, **M** 2, **etc.** for the **molars,** depending on how many teeth are present. For the deciduous teeth, the lower-case letter is used for the symbol preceded by a D (e.g., Di 1, the deciduous incisor next to the median plane), or the lower-case letter alone may be used (e.g., i 1). Because the dentition is the same on both sides, the formula lists only one side, and is enclosed in parentheses and multiplied by 2 to arrive at the total number of teeth. The numbers above the lines are for the teeth of the upper jaw and those below are for the teeth of the lower jaw. Thus, the dental formulas of the pig, which among domestic mammals has the most complete dentition, are as follows:

Deciduous dentition

$$2\ (\mathrm{Di}\ \tfrac{3}{3}\ \ \mathrm{Dc}\ \tfrac{1}{1}\ \ \mathrm{Dp}\ \tfrac{3}{3}) = 28$$

Permanent dentition (117—119)

$$2\ (\mathrm{I}\ \tfrac{3}{3}\ \ \mathrm{C}\ \tfrac{1}{1}\ \ \mathrm{P}\ \tfrac{4}{4}\ \ \mathrm{M}\ \tfrac{3}{3}) = 44$$

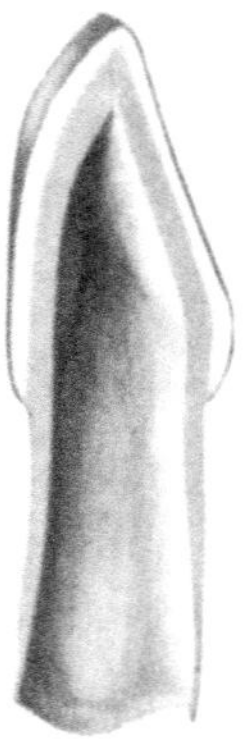

Fig. 97

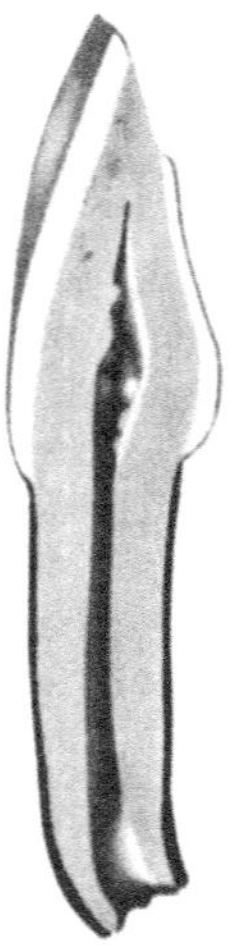

Fig. 98

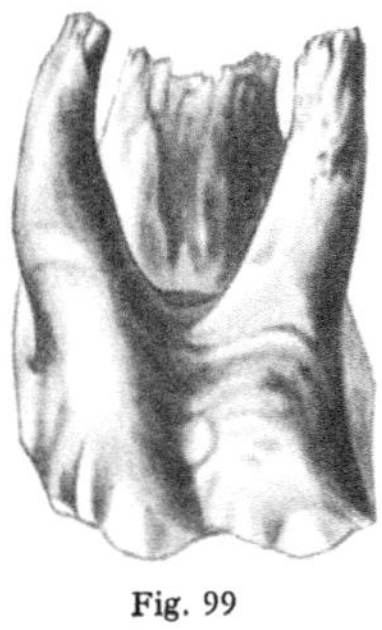

Fig. 99

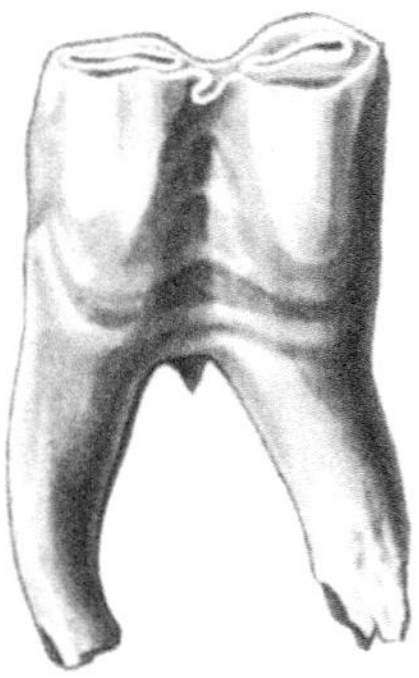

Fig. 100

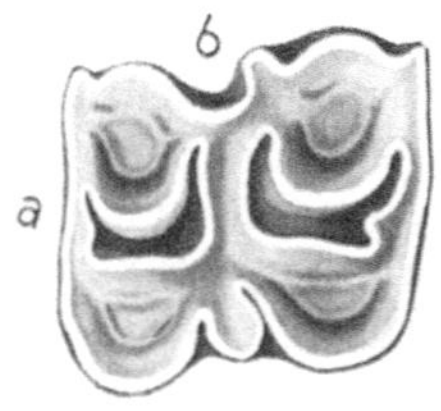

Fig. 101

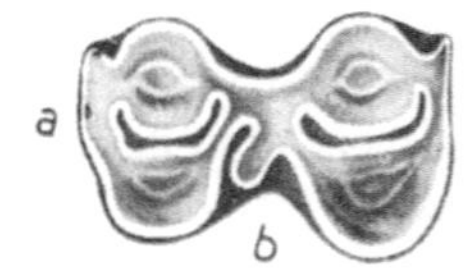

Fig. 102

Fig. 97. Section of deciduous incisor Di1 of a ten-day-old calf. Gray: dentine; White: enamel. Note the large dental cavity

Fig. 98. Section of permanent incisor I3 of an ox about seven years old. Gray: dentine; White: enamel, partly worn; Black: cement. The dental cavity and root canal are still patent

Figs. 99 and 100. Upper M1 (99) and lower M1 (100) of an ox about ten years old. Vestibular aspect.

Figs. 101 and 102. Occlusal surfaces of upper M1 (101) and lower M1 (102) of an ox about ten years old. Dark gray: cement surounding the tooth and in the infundibula; White: external enamel at the periphery, and central enamel (infundibula) in the center; Light gray: dentine *a* Mesial contact surface; *b* Vestibular surface

* Also known as central, intermediate, and corner incisors.

The teeth of the upper jaw form the **upper dental arch** (109) and those of the lower jaw form the **lower dental arch** (110). Each arch is interrupted between the incisors and the cheek teeth by a wide interdental space, the **diastema** (*d*). Animals are said to be **isognathous*** when their upper and lower jaws are of the same width and when, on centric occlusion, the whole occlusal surface of the upper teeth makes contact with the whole occlusal surface of the lower teeth. They are said to be **anisognathous**** when the lower jaw, and with it the dental arch, is narrower than the upper (25).

Morphology of Teeth

The simple, unspecialized tooth, the haplodont*** type, has a conical crown, a slightly constricted neck, and a simple conical root (91). In domestic mammals except the horse, this type, with such slight modifications as the chisel-shaped or shovel-shaped crown (121/*J1*), is found in the incisors and canines. In shape and growth pattern they are brachydont, that

Fig. 103

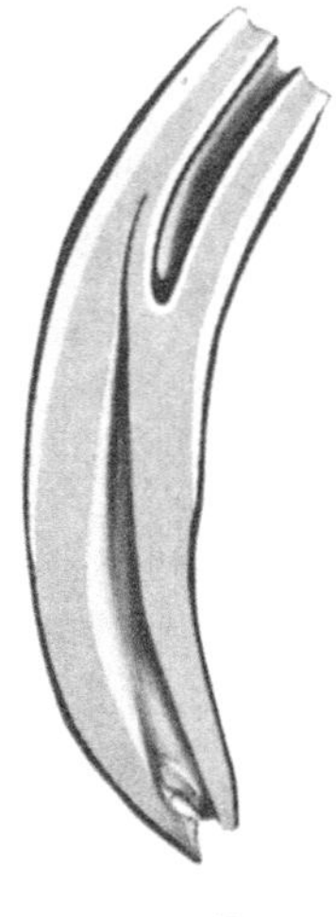

Fig. 104

Fig. 105

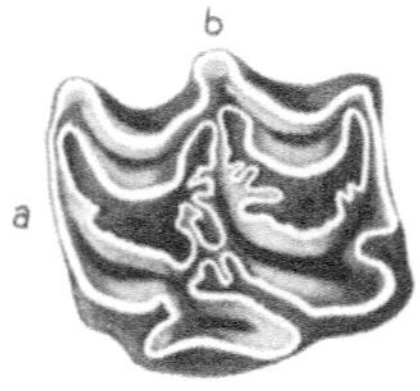

Fig. 107

Fig. 107. Occlusal surface of upper M1 of a horse about seven years old.
The raised enamel crests of the external enamel surround the dentine (light gray). The enamel crests of the infundibula surround the cups which are partly filled with cement (dark gray). Cement also surrounds the entire tooth.
a Mesial contact surface; *b* Vestibular surface

Fig. 103. Section of the lower deciduous incisor Di2 of a foal.
Gray: dentine; White: enamel; Black: cement. Note the large dental cavity, the distinct neck of the tooth, and the cup-shaped invagination of the enamel (infundibulum) on the occlusal surface. This tooth has not been in wear

Fig. 104. Section of permanent incisor I2 of a horse about seven years old.
Gray: dentine; White: enamel; Black: cement. The enamel covers only the long body of this tooth. Note the narrow dental cavity. This tooth has been in wear: the enamel folds on the occlusal surface have become double enamel crests, i. e., the external enamel has been separated from the central enamel (compare with Fig. 103)

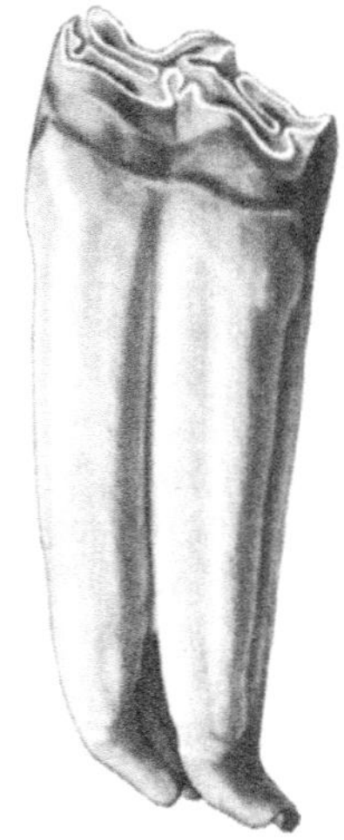

Fig. 106

Figs. 105 and 106. Upper M1 (105) and lower M1 (106) of a horse about seven years old. Vestibular aspect

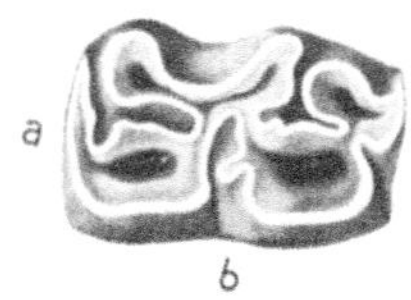

Fig. 108

Fig. 108. Occlusal surface of lower M1 of a horse about seven years old.
The external enamel is extensively folded and surrounds the dentine (light gray). Cement (dark gray) surrounds the external enamel. True infundibula are absent.
a Mesial contact surface; *b* Vestibular surface

* Gr. *isos*, equal; *gnathos*, jaw.
** Gr. *anisos*, unequal.
*** Gr. *haplos*, single, simple.

is to say they are fully developed and have ceased growing at the time of eruption. A much more complex type of tooth in shape and in some species also in growth pattern is found in the cheek teeth of the domestic mammals, which are of the following types. The **tuberculosectorial type,** characteristic of carnivores (109, 110/*P*, *M*), is a multitubercular tooth, which is capable, with its antagonist, of shearing or cutting the food (secodont* dentition). The **bunodont** type,** characteristic of omnivores such as the pig (117, 118/*M*), is also multituberculate, but it has a flatter occlusal surface, which is suited for crushing the food. These two brachydont types of teeth have low, enamel-covered crowns and well-developed roots (92, 93, 95, 96); their growth is completed at the time of eruption.

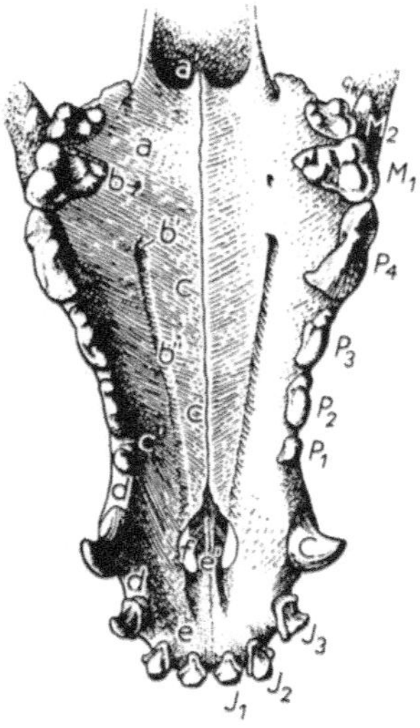

Fig. 109

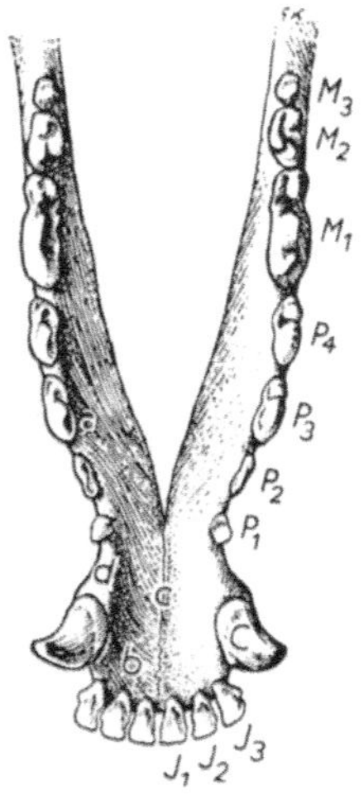

Fig. 110

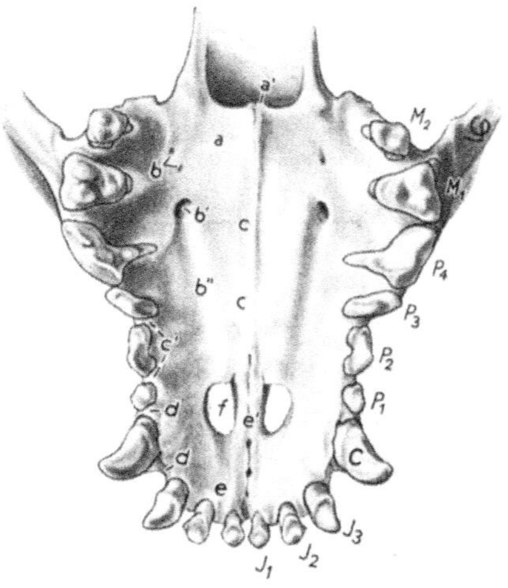

Fig. 111

Fig. 109. Upper dental arch of a dog about one year old. *J1—3* Incisors; *C* Canine; *P1—4* Premolars; *M1 and 2* Molars. *a* Horizontal plate of palatine bone; *a'* Nasal spine; *b* Minor palatine foramen; *b'* Major palatine foramen; *b''* Palatine groove; *c* Palatine process of maxilla; *c'* Alveolar process of maxilla; *d, d* Diastema; *e* Body of incisive bone; *e'* Palatine process of incisive bone; *f* Palatine fissure

Fig. 110. Lower dental arch of a dog about one year old. *J1—3* Incisors; *C* Canine; *P1—4* Premolars; *M1—3* Molars. *a* Molar part of mandible; *b* Incisive part of mandible; *c* Mandibular symphysis; *d* Diastema

Fig. 111. Upper dental arch of a brachycephalic dog (Boxer). Note the placement of P3 and P4. For legend see Fig. 109

The highly specialized, hypsodont cheek teeth of the herbivores result essentially from an invagination of the enamel on the occlusal surface of the tooth. These invaginations, known as **infundibula,** project vertically from the occlusal surface into the dentine of the tall bodies of these teeth (104/*white*, shown in an equine incisor). The enamel on the outside of the tooth is the external enamel; the enamel of the infundibulum is the central enamel. In domestic ruminants, the infundibula and their surrounding dentine are semilunar on cross section (101, 102); therefore, their cheek teeth are known as the selenodont*** type. In solipeds, the enamel on the occlusal surface forms pronounced folds and ridges (107/*white*); therefore, their cheek teeth are known as the lophodont**** type. In both selenodont and lophodont types, the external enamel forms more or less pronounced longitudinal folds (plicae enameli, 105) on the side of the tooth. The external enamel is covered by a layer of cement, which also partly fills

* Having cutting teeth.
** Gr. *bounos*, mound.
*** Gr. *selene*, moon.
**** Gr. *lophos*, ridge.

the infundibulum. Between the external enamel and the internal enamel is the dentine. These cheek teeth continue to grow in height for several years after they erupt, resulting in a hypsodont type with the rather high tooth body. When growth of the tooth body is complete, a relatively short root (radix dentis) develops (106). After a tooth such as this comes into wear, the enamel folds on the occlusal surface (103/*white*) wear off and are succeeded by two **enamel crests** (cristae enameli, 104/*white*) separated by dentine. (Figures 103 and 104 are incisors; however, the changes on the occlusal surface they illustrate are the same in the cheek teeth.) Since these enamel crests are much harder than the adjacent dentine and cement, they stand out over the recessed softer substances and provide the coarse grinding surface (107) that is essential for the mastication of hard, fibrous plant food.

Several theories have been advanced to explain the extraordinary evolutionary development from large numbers of simple teeth to fewer, larger, and more complex teeth that are specially adapted to the particular requirements of each species. The two leading theories are the tritubercular and concrescent. The **tritubercular** or **differentiation theory** suggests that the complex tooth is derived from a single primitive haplodont tooth, which, in the course of evolution developed accessory tubercles and an annular enamel ridge (cingulum) at the base of the crown. The **concrescence theory** proposes that several primitive tooth primordia fused, thus accounting for the multiplication of roots and tubercles. Such fusion may have occurred between teeth of the same generation of teeth or may have involved teeth of successive generations. This process of tooth specialization was accompanied by a gradual reduction in the number of teeth and also by a decrease in tooth succession from the polyphyodont* dentitions of the nonmammalian classes to the diphyodont dentition of mammals.

The Teeth of the Carnivores

(90—93, 109—116, also 26—28, 43, 79)

The dentition of the **DOG** (109—113) consists of tuberculate teeth which have well-developed roots and are of the brachydont type; their growth is completed when they are fully erupted and functional. The upper dental arch is wider than the lower (anisognathism) and as a result of this the lingual surface of the upper teeth slides over the vestibular surface of the lower teeth when the jaws are being closed (shearing action). Accordingly, the structure of the temporomandibular joint permits only dorsoventral movement of the mandible, although slight lateral movement of the mandible is possible for the shearing action of the teeth to come into play on one side or the other.

The formula for the **PERMANENT DENTITION** of the dog is:

$$2\left(I\tfrac{3}{3}\ \ C\tfrac{1}{1}\ \ P\tfrac{4}{4}\ \ M\tfrac{2}{3}\right) = 42.$$

The development of the diverse canine breeds is related to changes in skull conformation, which in turn have resulted in changes in the shape, the placement, and occasionally the number of teeth. The dentition of the dolichocephalic** breeds, such as the collie and the Russian wolfhound, on the one hand, and that of the brachycephalic*** breeds, such as the Boston terrier and Pekingese, on the other, represent the two extremes. Seiferle and Meyer (1942) regard the dentition of the German shepherd as the most normal intermediary type, because it resembles most closely the dentition of the ancestral forms of the canine species. The following description, therefore, is based on the dentition of the German shepherd.

The **INCISORS** have a distinct crown which is well set off by the neck from the long and thick root. The incisors of the upper jaw are larger than those of the lower (112). The roots of the upper incisors are laterally compressed and are implanted in the separate alveoli of the incisive bone, converging slightly toward the median plane. The roots of the lower incisors are smaller, and the interalveolar septa between their alveoli may be missing. The incisors increase in size from medial to lateral in both the upper and the lower jaw. I1 and I2 of the

* From Gr., to grow teeth many times.

** From Gr., long-headed.

*** From Gr., short-headed.

upper jaw have trilobed crowns, while that of upper I3 is conical. The crowns of the lower incisors are bilobed. The vestibular surface of the incisors is convex, and their lingual surface is concave with a central depression. When not worn down very much, they have quite sharp borders. When the jaws are closed, the upper incisors overhang the lower upper and lower I1 and I2 lie opposite each other, and the upper I3 lies between lower I3 and C.

The **CANINES** of the dog are massive, the upper being somewhat larger than the lower. The crowns of the canines are conical and curved somewhat caudally. The massive roots are compressed laterally, curve caudally in the jaws, and lie proximal to the roots of P1 and P2 (112/*C*). There is a longitudinal enamel ridge on the lingual surface, but otherwise the teeth are smooth. When the jaws are closed, the crown of the lower canine occupies the space between upper I3 and C, while the upper canine is between C and P1 of the lower jaw.

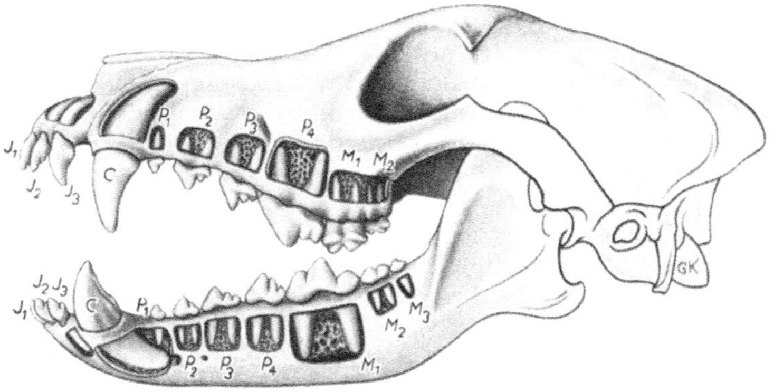

Fig. 112. Permanent dentition of a German shepherd about eight months old. The roots have been exposed by removing the externa alveolar walls. Upper *P4* and lower *M1* are the sectorial teeth; *J1—3* Incisors; *C* Canines; *P1—4* Premolars; *M1—3* Molars

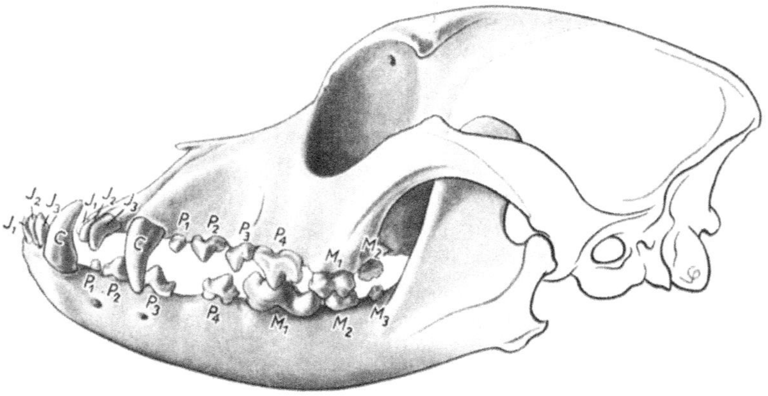

Fig. 113. Dentition of a brachycephalic dog (Boxer). Note the upper brachygnathia of this breed resulting in malocclusion of practically all teeth. *J1—3* Incisors; *C* Canines; *P1—4* Premolars; *M1—3* Molars

The four **PREMOLARS** of the upper and the lower jaws (112/*P*) are sectorial in character. They have laterally compressed, pointed crowns with a distinct **cingulum** and well developed roots. Upper and lower P1 have a single root and a short cone-shaped crown. Upper and lower P2 and P3 and lower P4 have two roots; they are laterally compressed and have tricuspid* crowns. Upper P4, known also as the **sectorial** or carnassial **tooth** (dens sectorius), has one lingual and two vestibular roots; two of its cusps are especially large. When the jaws are closed, there is no contact between the first three upper premolars and the four lower premolars; each crown points into an interdental space of the opposite jaw. The fourth upper premolar, however, meets its antagonist, the equally prominent **sectorial tooth** of the lower

* From (L.) *cuspis*, cusp, point, or tubercle.

jaw (M1). The lingual surface of the upper sectorial tooth slides closely over the vestibular surface of the lower sectorial, producing shearing action. The sectorials are opposite the angle of the mouth and, thus, under the direct influence of the powerful masseter muscle (43/*M1*).

The **MOLARS** are tuberculate, i.e., they have rounded projections or cusps on the occlusal surface. Lower M1 has two roots, and in addition to making contact with its main antagonist, upper P4, it also meets the massive M1 of the upper jaw. Lower M2 is much smaller than M1, has two roots and low cusps. Lower M3 is even smaller, has one root, a short conical crown, and no antagonist. M1 of the upper jaw has three roots and is very large. The tubercles of this tooth are at two levels, the lower level projecting medially into the hard palate (109/*M1*). Upper M2, which is similar to M1 but smaller, also has three roots; with its more flattened crown it contacts the smaller lower M2 and occasionally also lower M3. The morphological and functional diversity of the teeth in the canine dentition enables the members of this species under natural conditions to grasp and hold their prey, to tear pieces from it, and by superficial crushing to prepare large portions for swallowing.

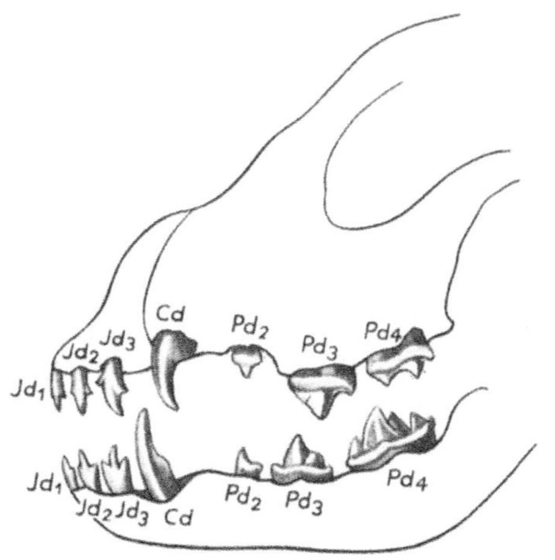

Fig. 114. Deciduous dentition of a pup about six weeks old. (After Weber 1927)
Jd1—3 Deciduous incisors; *Cd* Deciduous canines; *Pd 2—4* Deciduous premolars

For the first three weeks of its life the dog is without teeth. The **DECIDUOUS DENTITION,** which begins to erupt after that time, has the following formula:

$$2\,(\mathrm{Di}\,\tfrac{3}{3}\quad \mathrm{Dc}\,\tfrac{1}{1}\quad \mathrm{Dp}\,\tfrac{3}{3}) = 28.$$

The deciduous dentition (114) of the dog is complete within one and a half months, with the exception of upper and lower P1, which appear between three and a half to six months of age. The first premolars are not replaced, but remain as persisting deciduous teeth to head

Table 1. Eruption and Replacement of Teeth in the Dog

Teeth	Time of Eruption	Teeth	Time of Replacement
Di $\frac{1}{1}$ Di $\frac{2}{2}$ Di $\frac{3}{3}$	4—6 weeks	I $\frac{1}{1}$ I $\frac{2}{2}$ I $\frac{3}{3}$	3—5 months
Dc $\frac{1}{1}$	3—5 weeks	C $\frac{1}{1}$	5—7 months
Dp $\frac{1}{1}$	4—5 months	(is not replaced)	
Dp $\frac{2}{2}$ Dp $\frac{3}{3}$ Dp $\frac{4}{4}$	5—6 weeks	P $\frac{2}{2}$ P $\frac{3}{3}$ P $\frac{4}{4}$	5—6 months
M $\frac{1}{1}$ M $\frac{2}{2}$ M $\frac{}{3}$	4—5 months 5—6 months 6—7 months		

the series of permanent premolars that erupt shortly thereafter. The deciduous teeth are usually smaller than the permanent teeth. They have sharp, pointed crowns and slender roots. Because of the steady enlargement of the jaws, the spaces between the deciduous teeth grow larger as the time approaches for their replacement. By their shape and function, Dp3 of the upper jaw and Dp4 of the lower serve as the **sectorial teeth** of the deciduous dentition. The times of eruption and replacement of the dog's teeth is given in the table on page 83.

Tooth eruption and replacement thus occur remarkably early in the dog, and a fully functional dentition exists in this species at a relatively early age. The times of eruption and of replacement, however, vary from animal to animal and from breed to breed, with the result that they are not very reliable criteria for determining the age of a dog. Similarly, the wear shown by the incisors, as well as by the canines and cheek teeth in older animals, and the loss of teeth in senility are dependent on breed and husbandry and on the use which has been made of the animal. Therefore, these criteria are also not very useful for determining age. According to the investigtaions of Meyer (1942) the chances of judging a dog's age correctly by its teeth are about 41 per cent at best. For further detail, consult the literature on the ageing of the dog.

The general description of the canine dentition also applies to the **CAT.** The dental formulas for the cat are:

Deciduous dentition	Permanent dentition
$2\,(\mathrm{Di}\,\frac{3}{3}\quad \mathrm{Dc}\,\frac{1}{1}\quad \mathrm{Dp}\,\frac{3}{2}) = 26.$	$2\,(\mathrm{I}\,\frac{3}{3}\quad \mathrm{C}\,\frac{1}{1}\quad \mathrm{P}\,\frac{3}{2}\quad \mathrm{M}\,\frac{1}{1}) = 30.$

Because of its shorter jaws, the dentition of the cat (115, 116) is reduced on each side by one premolar and one molar in the upper jaw, and by two premolars and two molars in the lower. The loss of upper M2 and lower M2 and M3, which in the dog were rather flat, results in an entirely secondont dentition. The incisors of the cat are very small, and, as in the dog, they increase in size from medial to lateral. The canines are shaped like sabers; when the jaws are closed, the lower canine occupies the space between upper I3 and C, while the upper, being slightly larger, is lateral to the lower, and with its tip projects well below the alveolar margin of the lower jaw. The canine tooth of the upper jaw is followed by a very small P2, a P3 with pointed cusps, and a massive P4. The cat also is anisognathous, and the lingual surface of upper P3 and P4 slide over the vestibular surface of lower P4 and M1 (shearing) when the jaws are brought together. Upper P4 and lower M1 are the **sectorial teeth** in the cat also. Upper M1 is a small rostrocaudally-flattened tooth; it is turned inward, is caudomedial to the last premolar, and barely makes contact with the crown of lower M1. The lower cheek teeth of the cat all have two roots. In the upper jaw, P2 has one root, P3 has two, P4 has three and M1 has two.

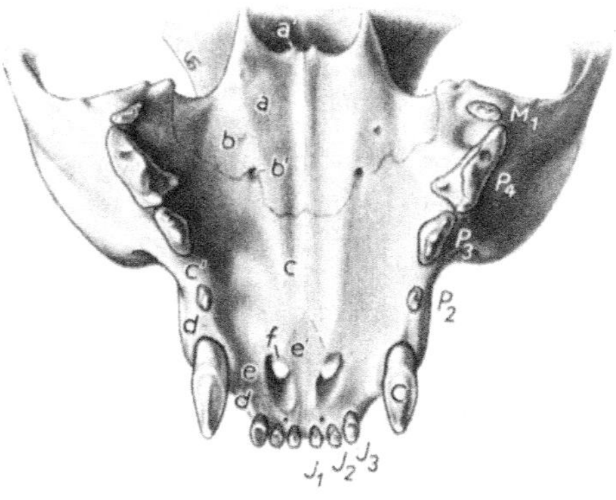

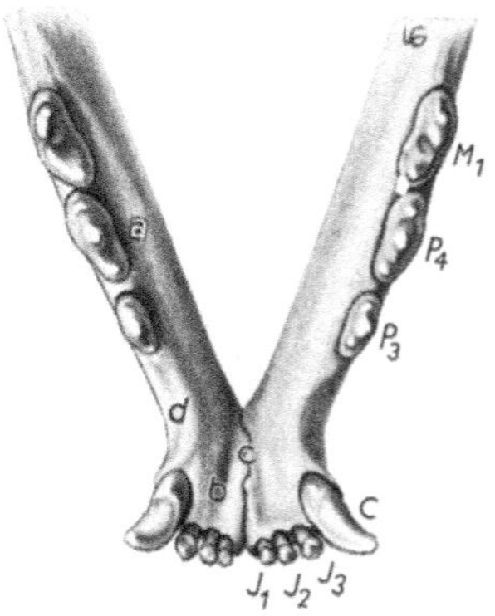

Fig. 115. Upper dental arch of a cat about seven months old. *J1—3* Incisors; *C* Canine; *P2—4* Premolars; *M1* First molar. *a* Horizontal plate of palatine bone; *a'* Nasal spine; *b* Minor palatine foramen; *b'* Major palatine foramen; *c* Palatine process of maxilla; *c'* Alveolar process of maxilla; *d, d* Diastema; *e* Body of incisive bone; *e'* Palatine process of incisive bone; *f* Palatine fissure

Fig. 116. Lower dental arch of a cat about seven months old. *J1—3* Incisors; *C* Canine; *P3 and 4* Premolars; *M1* First molar. *a* Molar part of mandible; *b* Incisive part of mandible; *c* Mandibular symphysis; *d* Diastema

Table 2. Eruption and Replacement of Teeth in the Cat

Teeth	Time of Eruption	Teeth	Time of Replacement
Di $\frac{1}{1}$ Di $\frac{2}{2}$ Di $\frac{3}{3}$	3—4 weeks	I $\frac{1}{1}$ I $\frac{2}{2}$ I $\frac{3}{3}$	$3\frac{1}{2}$—$5\frac{1}{2}$ months
Dc $\frac{1}{1}$	3—4 weeks	C $\frac{1}{1}$	$5\frac{1}{2}$—$6\frac{1}{2}$ months
Dp $\frac{2}{}$ Dp $\frac{3}{3}$ Dp $\frac{4}{4}$	5—6 weeks	P $\frac{2}{}$ P $\frac{3}{3}$ P $\frac{4}{4}$	4—5 months
M $\frac{1}{1}$	5—6 months		

The Teeth of the Pig

(94—96, 117—119, also 29, 30)

The formula for the **PERMANENT DENTITION** of the pig is:

$$2\left(I\tfrac{3}{3}\quad C\tfrac{1}{1}\quad P\tfrac{4}{4}\quad M\tfrac{3}{3}\right) = 44.$$

The pig, being omnivorous, has simple (haplodont) incisors and tuberculate (bunodont) cheek teeth. The cheek teeth have low crowns (brachydont) and, like the incisors, well developed roots (95, 96). With the exception of the canines, or tusks, the growth of the pig's teeth is completed at the time of eruption.

INCISORS. Lower I1 and I2 are long, straight, chisellike rods. They are close together and with their mates on the other side of the median plane they form a shovel-shaped rostral extension of the mandible (119). Their lingual surface bears a longitudinal enamel ridge that is flanked by two grooves. Their roots are nearly square in cross section and are deeply embedded in their sockets. Lower I3 is remarkably short and has a narrow crown and a distinct

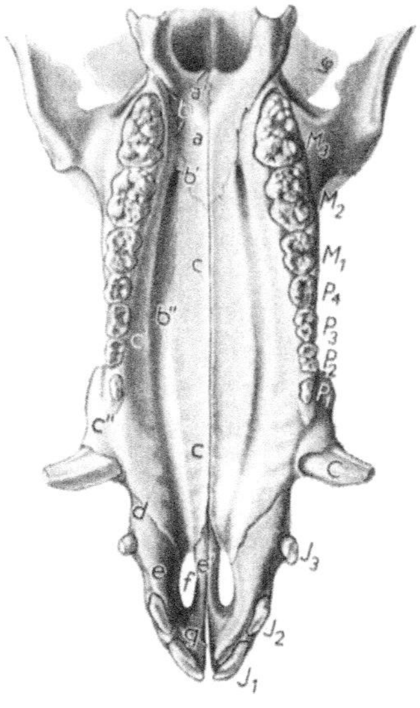

Fig. 117

Fig. 117. Upper dental arch of a boar about two years old.

J1—3 Incisors; *C* Canine; *P1—4* Premolars; *M1—3* Molars

a Horizontal plate of palatine bone; *a'* Nasal spine; *b* Minor palatine foramina; *b'* Major palatine foramen; *b''* Palatine groove; *c* Palatine process of maxilla; *c'* Alveolar process of maxilla; *c''* Eminentia canina; *d* Diastema; *e* Body of incisive bone; *e'* Palatine process of incisive bone; *f* Palatine fissure; *g* Interincisive fissure

Fig. 118. Lower dental arch of a boar about two years old.

J1—3 Incisors; *C* Canine; *P1—4* Premolars; *M1—3* Molars

a Molar part of mandible; *b* Incisive part of mandible; *d* Diastema; *e* One of the mental foramina

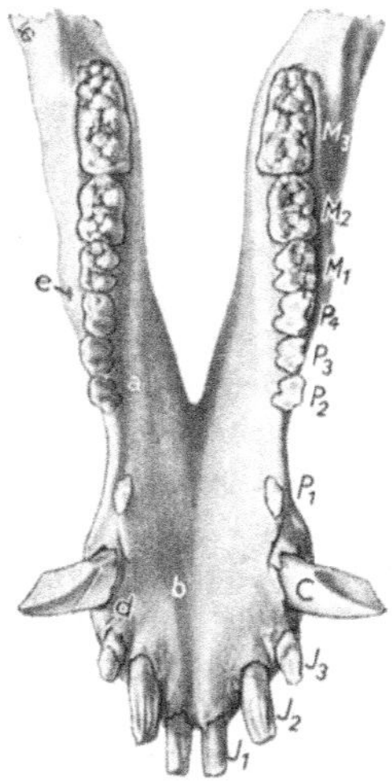

Fig. 118

neck (118). Upper I1 is the largest of the upper incisors and is deeply implanted in the body of the incisive bone. Its crown curves medially, and partly overhangs the interincisive fissure (117/*g*). It is the antagonist of I1 and I2 of the lower jaw. Upper I2 is a smaller tooth than I1 and has a laterally compressed crown and a distinct neck. It is the antagonist of lower I3, but does not make contact with it when the jaws are closed. Upper I3 is the smallest of the pig's incisors. It has a trilobed crown and, not having an antagonist, points downward into the space between lower I3 and C (119).

The **CANINES,** or **TUSKS,** of the pig are of remarkable size and never stop growing. The tusk has not root (radix dentis) in the strict sense and has a large dental cavity which remains wide open proximally throughout the life of the animal (94). The **lower tusk** is only a short

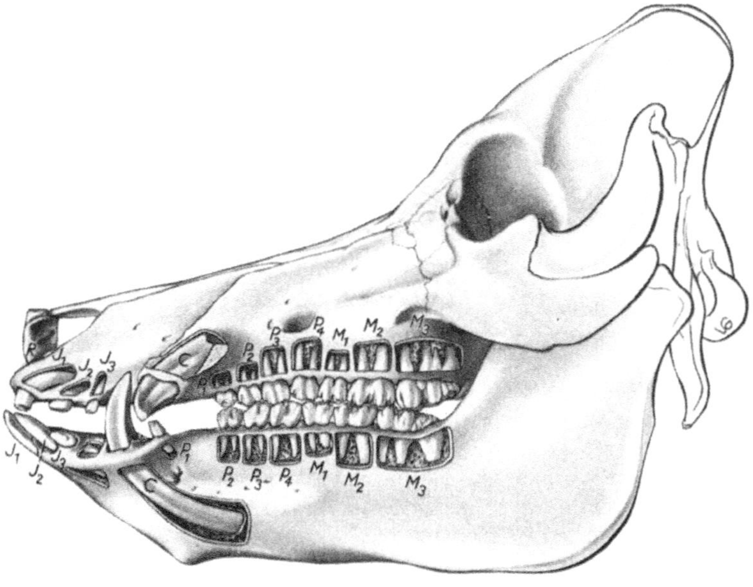

Fig. 119. Permanent dentition of a boar about two years old. (After Habermehl 1957.) The roots have been exposed by removing the external alveolar walls. *J1—3* Incisors; *C* Tusks; *P1—4* Premolars; *M1—3* Molars; *R* Rostral bone

distance behind lower I3 and is larger than its antagonist, the upper tusk. In the boar, it may reach a length of 15—18 cm. and most of it is firmly implanted in a long, curved alveolus (119). The lower tusk is shaped like a curved prism. Its exposed part curves caudolaterally and ends in a sharp point. This point and the lateral and medial edges of the exposed part are kept sharp by the contact between the concave caudal surface of this tusk and the rostral surface of the upper tusk. The **upper tusk** of the boar is only 6—10 cm. long. It is separated from upper I3 by a considerable space and is deeply embedded in the maxilla. Lateral to the embedded part of the tooth is an elongated **eminence** (eminentia canina, 117/*c''*) which supports the tooth. The exposed part of the upper tusk is conical and curves slightly caudolaterally. Its rostral surface is worn from contact with its antagonist. The sow's tusks are much smaller than those of the boar.

Both upper and lower **PREMOLARS** (117—119) increase in size from rostral to caudal. In the lower jaw, P1 and P2 each have two roots, P3 has two or three roots, and P4 has three roots. In the upper jaw, P1 has two roots, P2 and P3 each have three, and P4 has four or five. The crowns of the lower premolars are laterally compressed and shaped like sharp chisels. P1 is not replaced in either the upper or the lower jaw. Lower P1 erupts close to the tusk, but there is a considerable space between P1 and P2. P1 is a small tooth with a laterally compressed, sharp crown, and is occasionally absent. The crowns of the remaining lower premolars each have three cusps, which, however, gradually disappear through wear. Upper P1 erupts close to P2. The space between it and the tusk is large. Its crown is laterally compressed and consists of two indistinct cusps. Upper P3 is trituberculate when it erupts, but after wear shows a wide, triangular occlusal surface. P4 is the largest upper premolar. It also is trituberculate in the beginning, but with wear develops a wide, roundish occlusal surface.

Table 3. Eruption and Replacement of Teeth in the Pig

Teeth	Time of Eruption*	Teeth	Time of Replacement*
Di $\frac{1}{1}$	1—3 weeks 4—14 days	I $\frac{1}{1}$	12—17 months
Di $\frac{2}{2}$	10—14 weeks 8—12 weeks	I $\frac{2}{2}$	17—18 months
Di $\frac{3}{3}$	Before birth	I $\frac{3}{3}$	8—12 months
Dc $\frac{1}{1}$	Before birth	C $\frac{1}{1}$	8—12 months
(D)P $\frac{1}{1}$	$3\frac{1}{2}$—$6\frac{1}{2}$ months		
Dp $\frac{2}{2}$	7—10 weeks	P $\frac{2}{2}$	12—16 months
Dp $\frac{3}{3}$	1—3 weeks 1—5 weeks	P $\frac{3}{3}$	12—16 months
Dp $\frac{4}{4}$	1—4 weeks 2—7 weeks	P $\frac{4}{4}$	12—16 months
M $\frac{1}{1}$	4—6 months		
M $\frac{2}{2}$	7—13 months		
M $\frac{3}{3}$	17—22 months		

* The lower figures (ranges) are for early-maturing breeds, the higher figures for late-maturing breeds.

The **MOLARS** also increase in size from rostral to caudal in both jaws. Lower M1 (96) and M2 each have four roots and lower M3 has six. The three upper molars have six roots each. The molars of the lower jaw are rectangular on cross section. The crowns of the first two have four primary tubercles and a number of smaller secondary tubercles, while M3 has six primary tubercles. Upper M1 (95) and M2 are similar in shape to the three molars of the lower jaw. The crown of the large M3 presents two lingual and two vestibular tubercles and a caudally-placed single primary tubercle as well as numerous secondary ones.

The pig has an isognathous dentition. When the jaws are closed in centric occlusion, the cheek teeth make the following contacts. Upper P1 is removed from lower P1 by the length of the diastema and the two, therefore, do not make contact. Before they are worn, upper P1, P2, and P3 slide over the vestibular surface of their antagonists and, thus, are sectorial in function. Because it is close to the tusk, lower P1 is without an antagonist, and projects into the diastema of the upper jaw. Thus, the four upper premolars have only three of the lower premolars for antagonists. The first and second upper molars make contact with the caudal half of the corresponding lower molars and with the cranial half of the lower molars one higher in number, while upper M3 makes contact only with the corresponding lower third molar. The tubercles on the wide upper molars fit into depressions on the lower molars and vice versa. In this way, unbroken contact is made between the wide occlusal surfaces of these teeth.

In accordance with the rooting and burrowing habits of the pig, its omnivorous dentition falls into various functional categories. The lower incisors form a shovel-like extension of the lower jaw for exposing edibles in the ground. The curved upper incisors are used for grasping the food. The sectorial premolars are for superficially dividing it; and the molars, with their wide tuberculate occlusal surfaces, are used to crush it. The mandibles move mainly in a dorsoventral direction, but some slight lateral movement is also possible.

DECIDUOUS DENTITION. Di3 and Dc are present at birth and often cause injury to the sow's mammary gland during suckling. The formula for the deciduous dentition of the pig is:

$$2\,(\mathrm{Di}\,\tfrac{3}{3}\quad \mathrm{Dc}\,\tfrac{1}{1}\quad \mathrm{Dp}\,\tfrac{3}{3}) = 28.$$

Di1 and Di2 of both upper and lower jaws are similar in form and placement to their permanent counterparts. Di3 and Dc of both jaws are thin pegs. The antagonist of upper Di3 is the lower deciduous canine. The upper deciduous canine has no antagonist and projects into the lower diastema. The first deciduous premolars (Dp1) are absent or, if present, do not erupt. Lower Dp2 and Dp3 are similar in shape and placement to the corresponding permanent teeth. Upper Dp3 and Dp4 and the especially large lower Dp4 are similar in shape and function to the molars of the permanent dentition.

Selective breeding, particularly with the object of producing early-maturing breeds, has resulted in more or less marked changes in skull conformation. These changes, which must often be considered pathological, have greatly influenced the development, eruption time, shape, and placement of the pig's teeth. Eruption and replacement times, therefore, are usually stated in rather wide ranges, and any determination of the age of a pig by its teeth is only approximate.

The Teeth of the Ruminants

(97—102, 120—125, also 31, 32)

The formula for the **PERMANENT DENTITION** of the ruminants is:

$$2\,(\mathrm{I}\,\tfrac{0}{4}\quad \mathrm{C}\,\tfrac{0}{0}\quad \mathrm{P}\,\tfrac{3}{3}\quad \mathrm{M}\,\tfrac{3}{3}) = 32.$$

The herbivorous dentition of ruminants is composed of teeth of the haplodont and selenodont types.

Both **INCISORS** and **CANINES** are absent from the upper jaw (120). Primordia of these teeth have been demonstrated in both bovine and ovine embryos, but disappear before birth. The **dental pad** (pulvinus dentalis, 31/*D*), a tough connective tissue elevation with a thick, cornified epithelial covering, takes the place of the missing teeth and acts as antagonist to the lower incisors. The lower canines (121/*C*) have been incorporated into the row of incisors and, functionally, have become the fourth incisors, and are so designated (I4). In addition to their numerical designation, I1—I4, the incisors of the ruminants are also known, respectively, as **central, first intermediate, second intermediate,** and **corner incisors.** Each of them has a shovel-shaped, asymmetrical **crown,** which is slightly more drawn out laterally. The crown is covered with enamel that forms two longitudinal ridges on the slightly concave lingual surface. The vestibular surface is slightly convex and in young animals meets the lingual surface in a sharp edge (89, 97, 98). In the young animal, the incisors are arranged in a semicircle, are somewhat crowded, and their crowns overlap onto the lingual surfaces of their medial neighbors. As the incisive part of the mandible increases in size, however, and more room is gained, the crowns line up next to one another and form a more flattened arch. Each incisor has a distinct **neck.** The **roots** are either roundish or somewhat square on cross section and surrounded by very high, thick gums at the neck of the tooth. The incisors of the ruminants are not very firmly implanted, allowing limited movement, especially in older animals; they often fall out in aged animals. As the incisors begin to wear, an **occlusal surface** appears (121). This surface consists of a ring of enamel surrounding a core of yellowish dentine. The lingual border of the occlusal surface is notched because of the enamel ridges on the lingual surface of the tooth. When wear has reduced the tooth beyond the proximal end of these ridges, and the notches have disappeared, the tooth is said to be **level** (Habel 1970).

As the occlusal surface approaches the distal end of the dental cavity, secondary dentine is produced, appears as a small "dental star" in the center of the occlusal surface (32), and obliterates the dental cavity in distoproximal progression. In old animals, the crowns of the incisors are often completely worn off, and only the short stumps of the roots are left protruding from the gums.

With their relatively high bodies, the **CHEEK TEETH** of the ruminants are of the hypsodont type, and continue to grow for some time after eruption. As they begin to wear, the loss of substance on the occlusal surface is compensated for, first, by true longitudinal growth at the proximal end of the tooth and, later, as growth ceases, by gradual advancement from their sockets. Thus, in older animals, the cheek teeth are very short and are held in the jaws

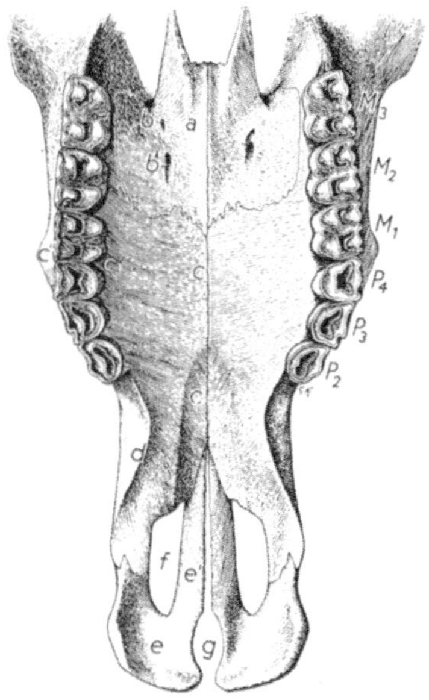

Fig. 120

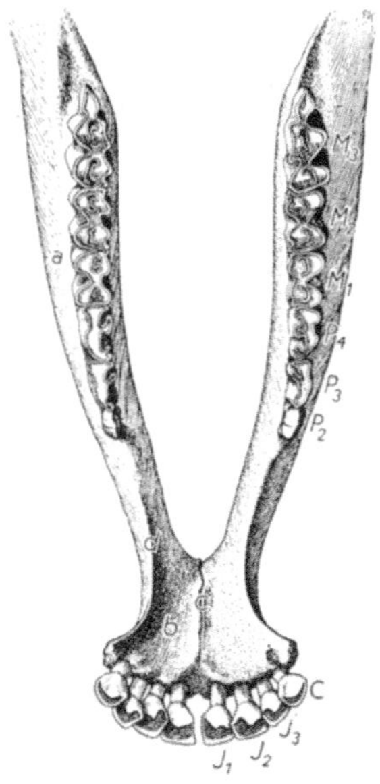

Fig. 121

Fig. 120. Upper dental arch of an ox about six years old.
P2—4 Premolars; *M1—3* Molars
a Horizontal plate of palatine bone; *b* Minor palatine foramen; *b'* Major palatine foramen; *c* Palatine process of maxilla; *c'* Alveolar process of maxilla; *c''* Facial tuberosity; *d* Diastema; *e* Body of incisive bone; *e'* Palatine process of incisive bone; *f* Palatine fissure; *g* Interincisive fissure

Fig. 121. Lower dental arch of an ox about six years old.
J1—3 Incisors; *C* Corner incisor (I4); *P2—4* Premolars; *M1—3* Molars
a Molar part of mandible; *b* Incisive part of mandible; *c* Mandibular symphysis; *d* Diastema

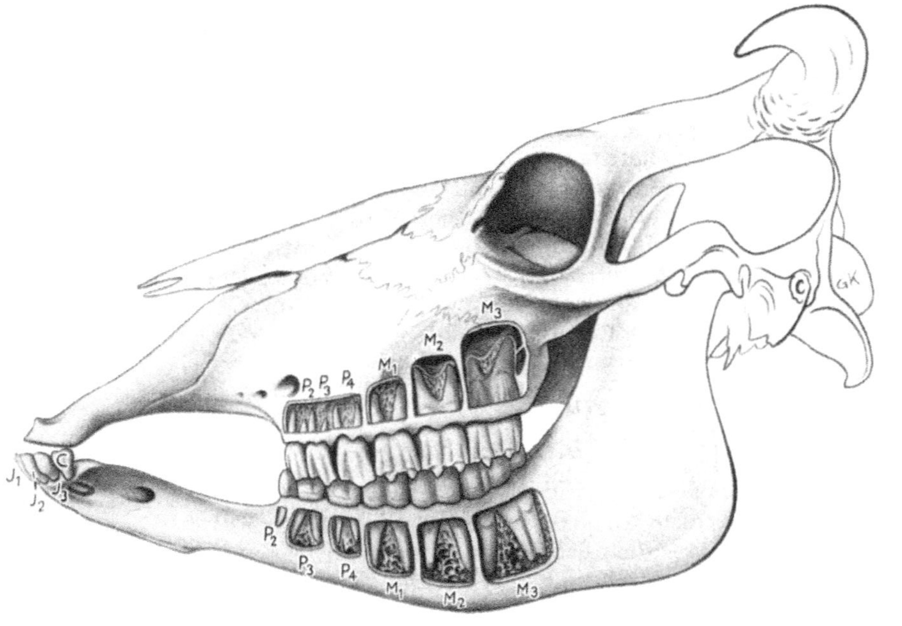

Fig. 122. Permanent dentition of an ox about six years old. The roots of the teeth have been exposed by removing the external alveolar walls. *J1—3* Incisors; *C* Corner incisor (I4); *P2—4* Premolars; *M1—3* Molars

by short, stubby roots. The upper cheek teeth form a solid row, implanted in the massive alveolar process of the maxilla (120, 122). They increase in size from rostral to caudal. Each of the three **UPPER PREMOLARS** has one infundibulum in the center of its wide occlusal surface. The infundibula are semilunar in outline with the convexity toward the tongue. The bodies of the upper premolars consist of a single column, or component, (120) which ends in two vestibular and one wide lingual root. The **UPPER MOLARS** (99, 101) have three roots each. Their bodies are composed of two columns and give the impression of having resulted from the union of two premolars. There are two infundibula per tooth, one for each column, with their convexities directed toward the lingual surface. The six lower cheek teeth (121) also increase in size from front to back and form a solid row caudal to the long diastema. The

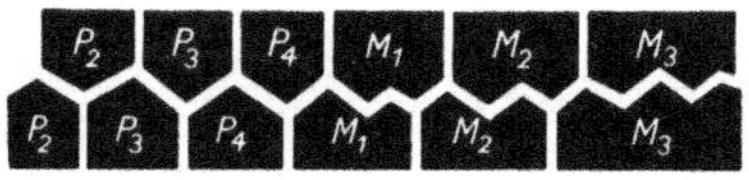

Fig. 123. Occlusal pattern of the cheek teeth of the ox. Schematic. Lateral aspect. *P2—4* Premolars; *M1—3* Molars

LOWER PREMOLARS each have two roots. P2 is small and has a conical crown without marked longitudinal enamel folds. It does not come into wear until the animal is quite old. Lower P3 and P4 are laterally compressed and present deep enamel folds on the lingual surface; true infundibula are usually not formed. The **LOWER MOLARS** (100, 102) have two roots each and are relatively narrow from side to side. Both M1 and M2 are composed of two columns with each column ending in a root, while M3 consists of two large rostral columns and one small caudal column. Each lower molar has two semilunar infundibula, which face the vestibule with their convex surfaces.

OCCLUSION. The occlusal pattern of the upper and lower cheek teeth of the ox is shown in Figure 123. The lower cheek teeth are slightly rostral to the corresponding teeth in the upper jaw, with the result that each upper tooth makes contact not only with its corresponding lower tooth, but also with the latter's caudal neighbor. Upper M3 is an exception and has only one antagonist: M3 of the lower jaw. The occlusal surface of the upper cheek teeth is wide and slopes toward the hard palate from a high vestibular to a low lingual surface. In contrast, the occlusal surface of the lower cheek teeth is narrow and slopes toward the cheek from a high lingual to a low vestibular surface. The occlusal surfaces of the upper and lower rows of cheek teeth are traversed by a number of transverse ridges and alternating depressions, giving the teeth a coarsely serrated appearance on lateral inspection (123). When the jaws are closed, the transverse ridges of the upper teeth interdigitate with those of the lower; and during mastication they glide in lateral motion along the opposing transverse depressions. Domestic ruminants are highly anisognathous, the distance between the right and left lower cheek teeth being considerably less than the distance between the right and left upper cheek teeth. At centric occlusion, only a narrow vestibular strip of the occlusal surface of the lower cheek teeth is in contact with an equally narrow strip along the lingual surface of the broader upper teeth. When the animal chews, the occlusal surfaces on the right side, for instance, are in full contact as they pass transversely across one another, while on the other side contact is lost. If chewing is shifted to the left side, the occlusal surfaces on the right side loose contact with each other. The structure of the cheek teeth, particularly their permanently rough and serrated occlusal surfaces, and the habit of chewing all of their food twice, enables the ruminants to convert the very coarse food they eat into a fine pulp, and makes it possible for them to extract from it all of the few nutrients it contains.

The formula for the **DECIDUOUS DENTITION** is:

$$2\ (\mathrm{Di}\,\tfrac{0}{4}\quad \mathrm{Dc}\,\tfrac{0}{0}\quad \mathrm{Dp}\,\tfrac{3}{3}) = 20.$$

The deciduous incisors are very similar to the permanent incisors in shape, placement, and wear. They are either present at birth or erupt a few days thereafter. For some time after eruption the high gums cover the proximal parts of the crowns (89), but by the 27th postnatal day the crowns protrude suffiently to be fully visible. As the deciduous incisors are being replaced, the permanent incisors undergo a peculiar rotation around their long axes for proper placement after eruption. Abnormal placement of the incisors is occasionally encountered in the ruminants.

The bodies of lower Dp2 and Dp3 consist of a single column, or component, each. They have no infundibulum, but end in two roots. Dp4 of the lower jaw is composed of three columns and has three infundibula and three roots. Dp2 in the upper jaw consists of a single column which ends in two roots and has one infundibulum. Upper Dp3 and Dp4 each consist of two columns and have two infundibula, Dp3 ending in two roots and Dp4 in three. The deciduous premolars consisting of more than one column are quite similar in shape and function to the permanent molars.

Table 4. Eruption and Replacement of Teeth in the Ox

Teeth	Time of Eruption*	Teeth	Time of Replacement*
Di $\frac{}{1}$	Before birth	I $\frac{}{1}$	14—25 months
Di $\frac{}{2}$	Before birth	I $\frac{}{2}$	17—33 months
Di $\frac{}{3}$	Before birth— up to 2—6 days	I $\frac{}{3}$	22—40 months
Di $\frac{}{4}$	Before birth— up to 2—14 days	I $\frac{}{4}$	32—42 months
Dp $\frac{2}{2}$	Before birth— up to 14—21 days	P $\frac{2}{2}$	24—28 months
Dp $\frac{3}{3}$	Before birth— up to 14—21 days	P $\frac{3}{3}$	24—30 months
Dp $\frac{4}{4}$	Before birth— up to 14—21 days	P $\frac{4}{4}$	28—34 months
M $\frac{1}{1}$	5—6 months		
M $\frac{2}{2}$	15—18 months		
M $\frac{3}{3}$	24—28 months		

* The lower figures are for early-maturing breeds, the higher figures for late-maturing breeds.

The **TEETH OF THE SMALL RUMINANTS** (124, 125) by and large resemble those of the ox in number, structure, placement, and function. Only the few differences will be mentioned here. The eight **incisors** are long and narrow, and each tooth has a semilunar outline when viewed from the side. They are arranged in a strongly curved arch and meet the dental pad at a less acute angle than in the ox. The **cheek teeth** are simpler in form. Because the difference in width between the upper and lower jaws is more marked than in the ox, the occlusal surfaces of the cheek teeth slope more steeply, so that the lingual border of the lower occlusal surface and the vestibular border of the upper coclusal surface are quite sharp.

Eruption and replacement of teeth, the size and shape of the occlusal surfaces of the incisors, and the loss of teeth in advanced age may be used in determining the age of domestic ruminants. In addition, the number of rings on the horns of the female and, in the calf, the height of the gums in relation to the incisor crowns and changes in horn buds and hooves, as well as in the umbilical cord after birth, may be useful. It should be remembered, however, that also in the ruminants, management, nutrition, and other factors, by influencing not only the rate of dental development but the other criteria mentioned as well, determine the reliability of these criteria in estimating the age of an animal.

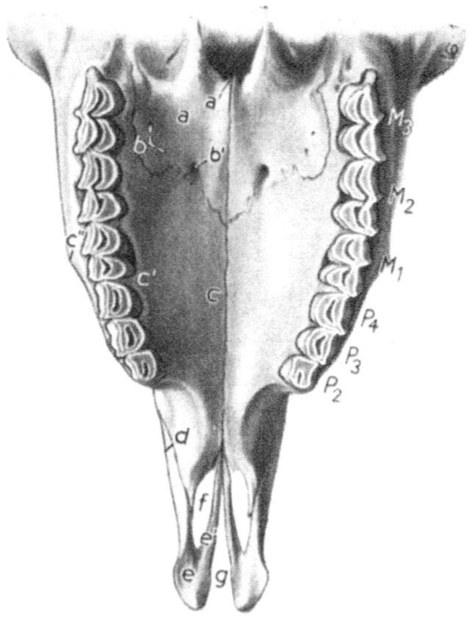

Fig. 124

Fig. 124. Upper dental arch of a goat about three years old.

P2—4 Premolars; *M1—3* Molars

a Horizontal plate of palatine bone; *a'* Nasal spine; *b* Minor palatine foramina; *b'* Major palatine foramen; *c* Palatine process of maxilla; *c'* Alveolar process of maxilla; *c''* Facial tuberosity; *d* Diastema; *e* Body of incisive bone; *e'* Palatine process of incisive bone; *f* Palatine fissure; *g* Interincisive fissure

Fig. 125. Lower dental arch of a goat about three years old.

J1—3 Incisors; *C* Corner incisor (I4); *P2—4* Premolars; *M1—3* Molars

a Molar part of mandible; *b* Incisive part of mandible; *c* Mandibular symphysis; *d* Diastema; *e* Mental foramina

Fig. 125

Table 5. Eruption and Replacement of Teeth in the Sheep

Teeth	Time of Eruption*	Teeth	Time of Replacement*
$\mathrm{Di}_{\bar{1}}$	Before birth—up to 8 days	$\mathrm{I}_{\bar{1}}$	12—18 months
$\mathrm{Di}_{\bar{2}}$	Before birth	$\mathrm{I}_{\bar{2}}$	21—24 months
$\mathrm{Di}_{\bar{3}}$	Before birth	$\mathrm{I}_{\bar{3}}$	27—36 months
$\mathrm{Di}_{\bar{4}}$	Birth—up to 8 days	$\mathrm{I}_{\bar{4}}$	36—48 months
$\mathrm{Dp}\frac{2}{2}$	Before birth—up to 4 weeks	$\mathrm{P}\frac{2}{2}$	21—24 months
$\mathrm{Dp}\frac{3}{3}$	Before birth—up to 4 weeks	$\mathrm{P}\frac{3}{3}$	21—24 months
$\mathrm{Dp}\frac{4}{4}$	Before birth—up to 4 weeks	$\mathrm{P}\frac{4}{4}$	21—24 months
$\mathrm{M}\frac{1}{1}$	3 months		
$\mathrm{M}\frac{2}{2}$	9 months		
$\mathrm{M}\frac{3}{3}$	18 months		

* The lower figures are for early-maturing breeds, the higher figures for late-maturing breeds.

Table 6. Eruption and Replacement of Teeth in the Goat

Teeth	Time of Eruption	Teeth	Time of Replacement
$Di_{\overline{1}}$	At birth	$I_{\overline{1}}$	15 months
$Di_{\overline{2}}$	At birth	$I_{\overline{2}}$	21 months
$Di_{\overline{3}}$	At birth	$I_{\overline{3}}$	27 months
$Di_{\overline{4}}$	1—3 weeks	$I_{\overline{4}}$	36 months
$Dp\frac{2}{2}$	3 months	$P\frac{2}{2}$	17—20 months
$Dp\frac{3}{3}$	3 months	$P\frac{3}{3}$	17—20 months
$Dp\frac{4}{4}$	3 months	$P\frac{4}{4}$	17—20 months
$M\frac{1}{1}$	5—6 months		
$M\frac{2}{2}$	8—10 months		
$M\frac{3}{3}$	18—24 months		

The Teeth of the Horse

(103—108, 126—130, also 25, 33—35)

The formula for the **PERMANENT DENTITION** of the horse is:

$$2\left(I\frac{3}{3}\ C\frac{1}{1}\ P\frac{3(4)}{3}\ M\frac{3}{3}\right) = 40\ (42).$$

The dentition of the horse presents all the features typical of the dentition of a herbivorous animal.

In addition to their numerical designations, I1, I2, and I3, the **INCISORS** of the horse are also known, respectively, as the **central, intermediate,** and **corner incisors.** Their length ranges from 5.5—7 cm., increasing slightly from lateral to medial. They are embedded in the incisive bone and the incisive part of the mandible, their embedded portions (radices clinicae) converging slightly toward the median plane. Each tooth is curved (104), the concavity being toward the tongue; and the curvature of the upper incisors is generally more marked than that of the lower. In young animals, the exposed portions, the crowns (coronae clinicae), of the incisors stand close together, and when the jaws are in apposition, they form a semicircle on both dorsoventral and lateral inspection. The basic curvature of an unworn incisor is not even; the crown is curved more strongly than the embedded part. Because of this the incisor crowns of the young horse meet their antagonists almost vertically, that is, the angle between the upper and the lower incisors when viewed in profile is nearly 180 degrees. As the horse grows older and the incisors wear, the less curved and originally embedded part is pushed out of the socket and the angle becomes more and more acute. At the same time the arch formed by the occlusal surfaces as seen on dorsoventral inspection becomes progressively flatter. The general shape of an equine incisor is that of a three-sided, slender, curved pyramid. The base of the pyramid, the exposed portion of the tooth in the young animal, is flattened rostrocaudally, while the apex, the proximal part, is flattened slightly from side to side. The body of the pyramid is roundish in cross section distally and more triangular proximally. Consequently, as the incisor is slowly pushed from its socket, the **occlusal surface,** being in full contact with the occlusal surface of its antagonist, is at first (in the young animal) oval transversely, then becomes roundish, then somewhat triangular, and finally (in the old horse) it is oval longitudinally. Knowledge of this is of value in estimating the age of a horse. Each incisor has a centrally placed **infundibulum** (104). The infundibula of the upper incisors are about 12 mm. deep, while those of the lower are about 6 mm. deep. Before the tooth is in wear, the enamel of the infundibulum is continuous with the external enamel, and an annular fold is present on the occlusal surface (103). As the tooth is abraded through wear, this connection is lost and two annular enamel crests result (104); each one is independent of the other, and the two are separated by a circular layer of yellowish dentine. When the two enamel crests are fully visible, the horse's incisor is said to be **level** (Habel 1965). The

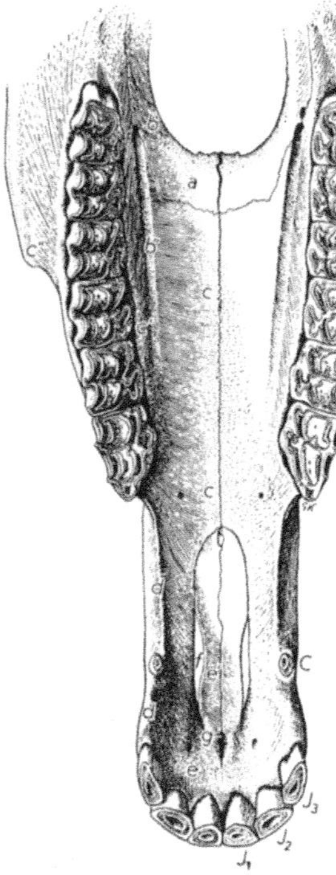

Fig. 126

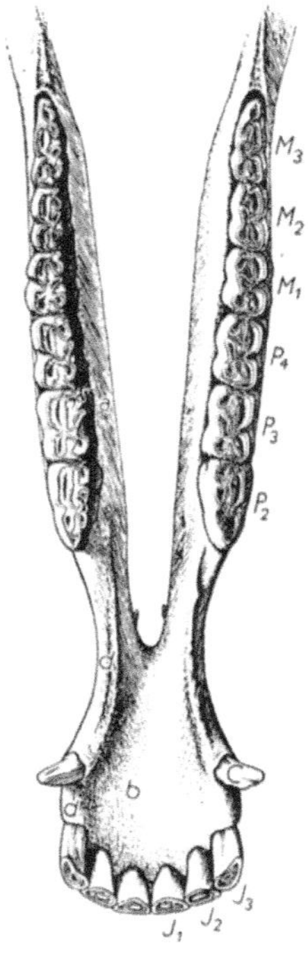

Fig. 127

Fig. 126. Upper dental arch of a gelding about eight years old. *J1—3* Incisors; *C* Canine; *P2—4* Premolars; *M1—3* Molars; *a* Horizontal plate of palatine bone; *b'* Major palatine foramen; *b''* Palatine groove; *c* Palatine process of maxilla; *c'* Alveolar process of maxilla; *c''* Facial crest; *d, d* Diastema; *e* Body of incisive bone; *e'* Palatine process of incisive bone; *f* Palatine fissure; *g* Interincisive canal

Fig. 127. Lower dental arch of a gelding about eight years old. *J1—3* Incisors; *C* Canine; *P2—4* Premolars; *M1—3* Molars; *a* Molar part of mandible; *b* Incisive part of mandible; *d, d* Diastema

infundibulum is lined with a layer of cement; the remaining lumen is filled with decomposing food particles, giving it a black appearance. The black cavity of the indundibulum is known as the **cup** (33/*J1*—*J3*). As the tooth wears (at a rate of about 2 mm. per year), the cup gradually narrows and disappears. The bottom of the infundibulum, however, remains as a small, raised **enamel spot** (34/*J1*) close to the lingual surface of the tooth for some time. As the cup disappears, the **dental star** appears on the occlusal surface between the infundibulum and the vestibular surface of the tooth. The dental star is the darker, secondary dentine that fills the dental cavity as the occlusal surface approaches it.

The **CANINES** (126—128/*C*) develop only in the male, but vestigial canines are encountered occasionally in the mare (34/*C*). When fully developed, the canines have a length of 4—5 cm., of which usually the most distal centimeter is visible as a cone-shaped crown with a simple covering of enamel. The roots are strongly curved caudally. The canines are closer to the corner incisors than to the first cheek teeth, the upper being located at the junction of the incisive bone and the maxilla. Between the canines and the cheek teeth is the extensive diastema (126, 127/*d*). Signs of wear on the canines are present only in old subjects.

The horse has six upper and six lower cheek teeth on each side, three **PREMOLARS** and three **MOLARS** (128). They are placed very close together and form two slightly curved rows, which extend from the diastema to the maxillary tuberosity in the upper jaw and to the ramus of the mandible in the lower (126, 127). The external enamel of these teeth is extensively folded, and the nature of their occlusal surfaces classifies them as lophodont. They complete true longitudinal growth when the horse is 6 to 7 years old, and at that time they are tall columns, about 8—10.5 cm. long. Their exposed parts project about 1.5—2 cm. above the gums. The upper cheek teeth are nearly square on cross section (107), while those of the lower jaw are rectangular (108). The **upper cheek teeth** present two **infundibula** each. Before these teeth are in wear, the infundibular walls are continuous with the external enamel, forming enamel folds on the occlusal surface. With wear, this connection is lost and the typical lophodont occlusal surface with deeply folded, raised enamel ridges and depressions of dentine or cement results. As longitudinal growth ceases, three short, stubby **roots** develop. On the vestibular surface of the upper cheek teeth are three longitudinal ridges of varying heigth which are separated by two longitudinal grooves (105). There is a centrally placed longitudinal ridge on their lingual surface which is accompanied on either side by two longitudinal grooves (107). The enamel of the **lower cheek teeth** is arranged in deep longitudinal folds, but true infundibula are absent (108). The vestibular surface (106) has a distinct longitudinal groove, as has the lingual surface. On the lingual surface, however, the principal groove is accompanied by several shallower ones. The deep, longitudinal folds of the enamel are best observed on the worn occlusal surface. Two deep infoldings are present on the lingual side and one on the vestib-

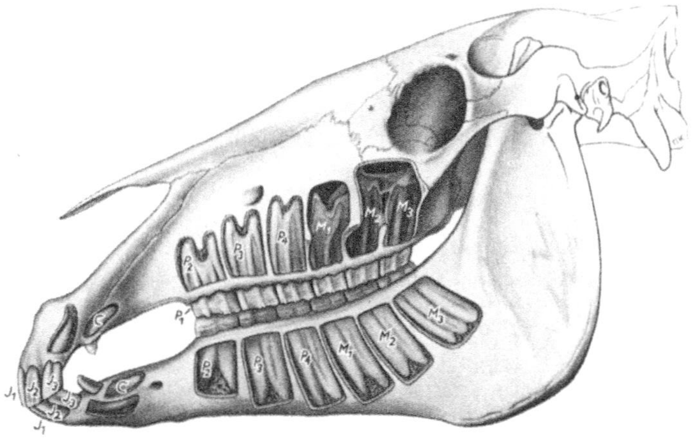

Fig. 128. Permanent dentition of an eight year old gelding.The roots of the teeth have been exposed by removing the external alveolar walls *J1—3* Incisors; *C* Canines; *P1—4* Premolars; *M1—3* Molars. Note the presence of a wolf tooth (P1) in the upper jaw

ular side. The lower cheek teeth have two short **roots.** Occasionally, a rudimentary upper P1 is present. It is known as the **wolf tooth** (dens lupinus, 128/*P1*), and falls out when the animal is still young, and is not replaced. It is also present in the lower jaw, but does not erupt. The embedded parts of the cheek teeth diverge as shown in Figure 128. In the upper jaw, P2 and P3 incline rostrally, P3 less than P2. P4 and M1 are more or less vertical to the masticatory surface, while the two remaining cheek teeth incline caudally, M3 more than M2. In the lower jaw, P2 inclines slightly rostrally, P3 is vertical, and P4 and the remaining teeth have an increasingly caudal direction.

OCCLUSION. Because the horse is anisognathous, only about the lingual third of the occlusal surface of the upper cheek teeth is in contact with the vestibular half of the lower cheek teeth at centric occlusion (129*A*). As the mandible moves to one side during mastication, all contact between the upper and the lower cheek teeth on the opposite side is lost (129*B*).

The lingual surface of the lower cheek teeth is higher than the vestibular surface. Conversely, the lingual surface of the upper cheek teeth is lower than the vestibular (25). As a result, the occlusal surface of the lower cheek teeth slopes toward the vestibule, while the occlusal surface of the upper cheek teeth slopes toward the hard palate. Occasionally, due to incomplete lateral

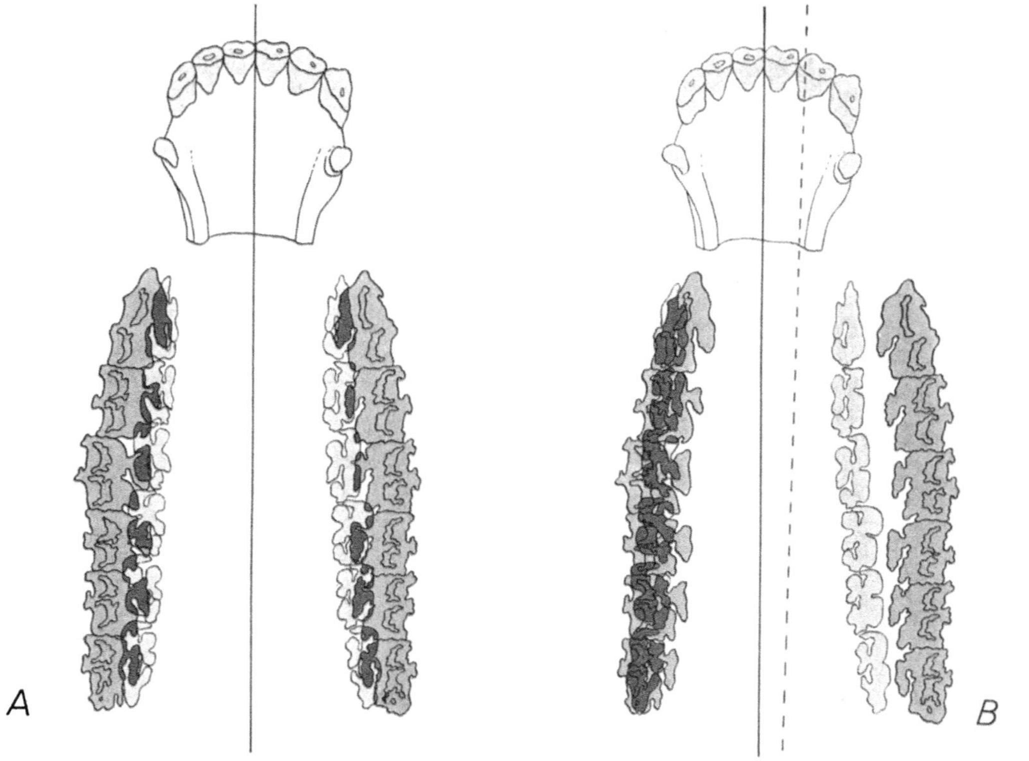

Fig. 129. Occlusal pattern of the cheek teeth of the horse, (A) in centric and (B) in left-sided occlusion. (After Küpfer 1936, redrawn.) *Light gray:* incisors, canines, and occlusal surfaces of the lower cheek teeth; ***Medium gray:*** occlusal surfaces of the upper cheek teeth; *Dark gray:* the part of both upper and lower occlusal surfaces that are in contact. *A* Centric occlusion; *B* Left-sided occlusion

movement of the mandible during mastication, only parts of the occlusal surfaces are worn, and the lingual edge of the lower cheek teeth and the vestibular edge of the upper become very sharp and damage the buccal and lingual mucosa. This is known as "sharp teeth" or, in its extreme form, "shear mouth."

The occlusal surfaces of the upper and lower rows of cheek teeth are divided by a series of alternating transverse ridges and depressions, which give the teeth a coarsely serrated appearance on lateral inspection (130). When the jaws are closed, the ridges of the upper teeth interdigitate with those of the lower, and during mastication glide in lateral motion through the transverse depressions in the opposing teeth. As illustrated schematically in Figure 130, the lower cheek teeth are placed slightly in advance of the upper, with the result that most of them have both primary and secondary antagonists, each one making contact with about three-quarters of the occlusal surface of the primary antagonist and about one-quarter of the occlusal surface of the secondary antagonist. Only upper M3 and lower P2 have a single antagonist: the corresponding tooth in the opposing jaw.

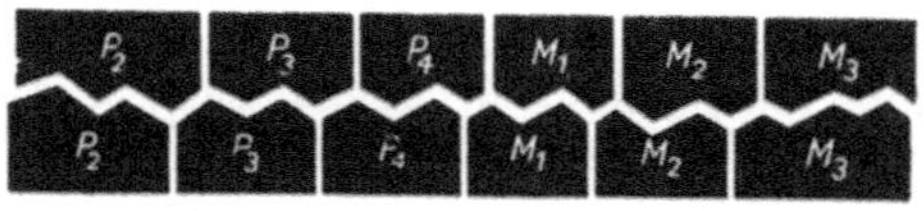

Fig. 130. Occlusal pattern of the cheek teeth of the horse. Schematic. Lateral aspect. *P2—4* Premolars; *M1—3* Molars

The formula for the **DECIDUOUS DENTITION** of the horse is:

$$2\ (\mathrm{Di}\,\tfrac{3}{3}\ \ \mathrm{Dc}\,\tfrac{1}{1}\ \ \mathrm{Dp}\,\tfrac{3}{3}) = 28.$$

The **deciduous incisors** are smaller and whiter than the permanent incisors. Their crowns are shovel-shaped, and there is a distinct neck and a relatively short root (103). An infundibulum is present, but it is only about 4 mm. deep. Short vestiges of the **deciduous canines**

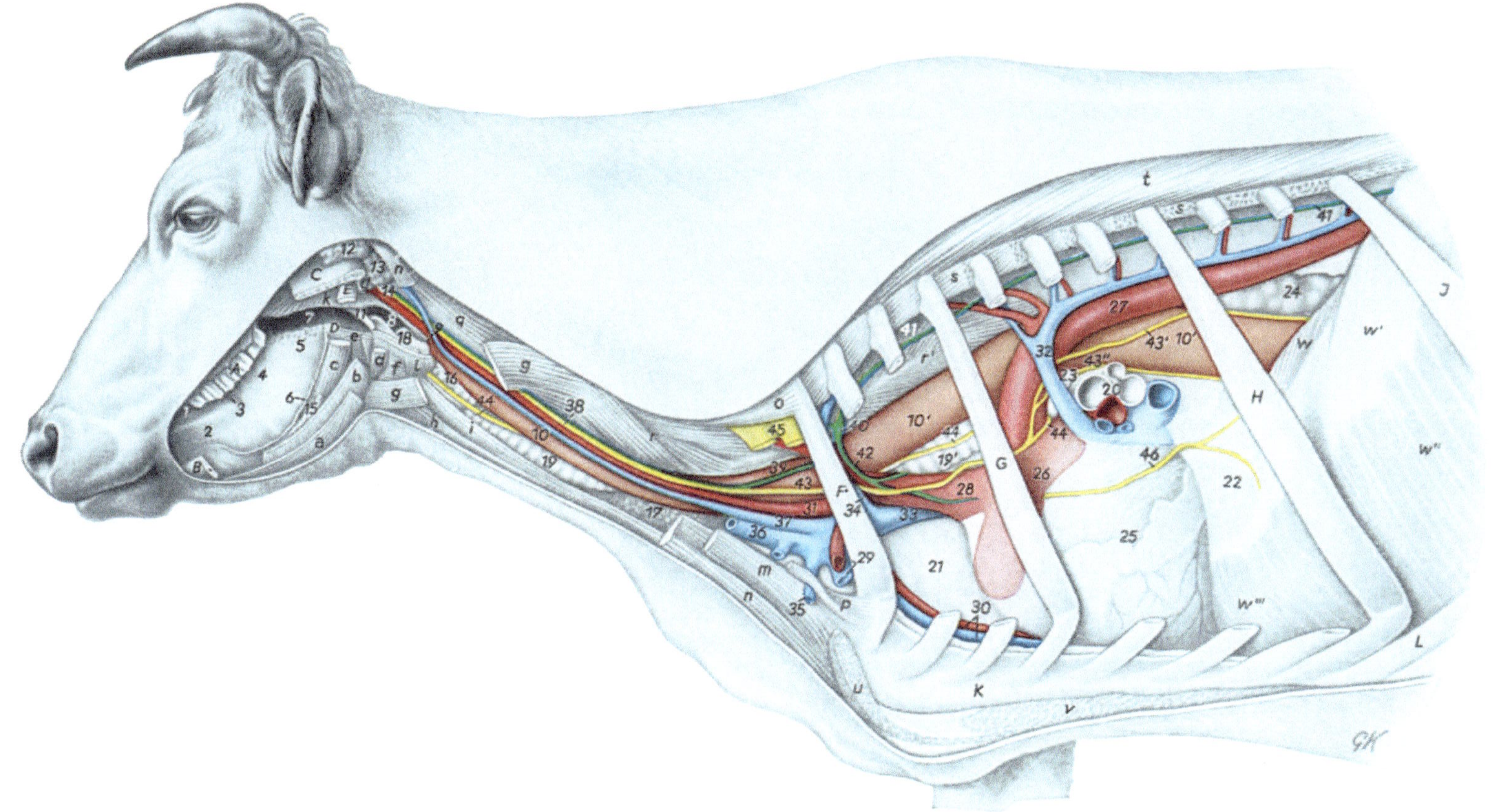

Fig. 132. Topography of the neck and the thoracic cavity of the cow. (After Wilkens and Rosenberger 1957.) (*Light brown:* esophagus; *Red:* arteries; *Blue:* veins; *Yellow:* nerves; *Green:* sympathetic trunk).

A Upper cheek teeth; *B* Body of mandible; *C* Ramus of mandible; *D* Ceratohyoid; *E* Stump of stylohyoid; *F* First rib; *G* Fourth rib; *H* Eighth rib; *J* Eleventh rib; *K* Sternum; *L* Cartilage of ninth rib

a Digastricus; *b* Mylohyoideus; *c* Styloglossus; *d, d* Stylohyoideus; *e* Ceratohyoideus; *f* Thyrohyoideus; *g, g* Omohyoideus; *h* Sternohyoideus; *i* Sternothyroideus; *k* Palatopharyngeus; *l* Caudal pharyngeal constrictors; *m* Sternomandibularis; *n* Sternomastoideus; *o* Dorsal part of scalenus medius; *p* Scalenus ventralis; *q* Longus capitis; *r, r'* Longus colli; *s* Intercostal muscles; *t* Longissimus; *u* Supf. pectoral muscles; *v* Deep pectoral muscle; *w* Diaphragm, part of its right crus, *w'* its tendinous center, *w''* its costal part, *w'''* its sternal part

1 Oral cavity; *2* Apex of tongue; *3* Fossa linguae; *4* Torus linguae; *5* Root of tongue with vallate papillae; *6* Cut edge of oral mucosa; *7* Oropharynx; *8, 9* Laryngopharynx; *10, 10'* Esophagus; *11* Soft palate; *12* Parotid salivary gland; *13* Mandibular salivary gland; *14* Lateral retropharyngeal lymph node; *15* Polystomatic sublingual gland; *16* Thyroid gland; *17* Thymus; *18* Rostral part of larynx; *19, 19'* Trachea; *20* Bronchi; *21* Cranial lobe of right lung, and *22* accessory lobe of right lung, both covered by the mediastinum; *23* Left tracheobronchial lymph nodes; *24* Caudal mediastinal lymph node; *25* Heart inside pericardium; *26* Pulmonary artery; *27* Aorta; *28* Brachiocephalic trunk; *29* Axillary artery and vein; *30* Int. thoracic artery and vein; *31* Common carotid artery; *32* Left azygous vein; *33* Cranial vena cava; *34* Venous costocervical trunk; *35* Cephalic vein; *36* Left ext. jugular vein, giving off supf. cervical vein above *m*; *37* Left int. jugular vein; *38* Vagosympathetic trunk; *39* Sympathetic trunk; *40* Cervicothoracic ganglion; *41* Sympathetic trunk; *42* Cardiac branch of sympathetic trunk; *43, 43', 43''* Left vagus and its terminal branches, which unite with similar branches of the right vagus to form the dorsal and ventral vagal trunks; *44, 44, 44* Caudal laryngeal nerve; *45* Stumps of ventral branches of 7th and 8th cervical nerves (brachial plexus); *46* Left phrenic nerve, caudal stump

develop in both sexes, but never erupt. The three **deciduous premolars** of both upper and lower jaws are smaller than the permanent premolars, but resemble them in shape and structure.

Estimating the age of a horse by its teeth has been practiced for many centuries, and with a normal set of teeth and normal wear, a fair degree of accuracy is possible. The more useful criteria for determining the age of a horse are as follows: eruption and wear of the deciduous incisors, eruption and wear of the permanent incisors, the shape of the occlusal surfaces of the permanent incisors, and the profile angle between the upper and lower incisors. The eruption of the cheek teeth is rarely considered. For details, publications on the ageing of domestic animals should be consulted.

Table 7. Eruption and Replacement of Teeth in the Horse

Teeth	Time of Eruption	Teeth	Time of Replacement
Di $\frac{1}{1}$	Before or shortly after birth	I $\frac{1}{1}$	$2\frac{1}{2}$—3 years
Di $\frac{2}{2}$	3—4 weeks and, rarely, up to 8 weeks	I $\frac{2}{2}$	$3\frac{1}{2}$—4 years
Di $\frac{3}{3}$	5—9 months	I $\frac{3}{3}$	$4\frac{1}{2}$—5 years
Dc $\frac{1}{1}$	Rarely erupt	C $\frac{1}{1}$	4—5 years
Dp $\frac{2}{2}$	Before birth or during first week after birth	P $\frac{2}{2}$	$2\frac{1}{2}$ years
Dp $\frac{3}{3}$		P $\frac{3}{3}$	$2\frac{1}{2}$ years
Dp $\frac{4}{4}$		P $\frac{4}{4}$	$3\frac{1}{2}$ years
M $\frac{1}{1}$	6—9 months, and, rarely, up to 14 months		
M $\frac{2}{2}$	2—$2\frac{1}{2}$ years		
M $\frac{3}{3}$	$3\frac{1}{2}$—$4\frac{1}{2}$ years		

BIBLIOGRAPHY

Teeth

Adloff, P.: Zur Frage nach der Entstehung der heutigen Säugetierzahnformeln. Zschr. Morph. Antr. **5,** 1902.

Adloff, P.: Zur Entwicklung des Säugetiergebisses. Anat. Anz. **26,** 1905.

Andres, J.: Hat die Hauskatze im Unterkiefer Molaren? Anat. Anz. **61,** 1926.

Andrews, A. H.: The relationship between age and development of the anterior teeth in cattle as determined by the oral examination of 2900 animals between the ages of 12 and 60 months. Brit. Vet. J. **131** (1975): 152—159.

Barnicoat, C. R., D. M. Hall: Attrition of incisors of grazing sheep. Nature London **185,** 179 (1960).

Baum, H.: Anatomische Betrachtungen über die Zähne der Säugetiere. Anat. Anz. **53,** (Erg. H.) 1920.

Benzie, D., and E. Cresswell: Studies of the dentition of sheep. II. Radiographic illustrations of stages in the development and shedding of the permanent dentition of Scottish Blackface sheep. Res. Vet. Sci. **3,** 1962.

Bodurov, N., K. Binev: Altersveränderungen bei der Entwicklung der Unterkieferzähne beim Rind. Naučni trudove. Visš. vet. med. inst., Sofija, **19,** 229—240 (1968).

Boue, C. P.: Anatomie fonctionelle des dents labiales des ruminants. Thesis, Univ. Paris, VI; Rev. Méd. Vet. **123,** 1141 (Abstr.) 1972.

Brown, W. A. B., P. V. Christofferson, M. Massler and M. B. Weiss: Postnatal tooth development in cattle. Am. J. Vet. Res. **21,** 1960.

Butler, P.: Studies of the mammalian dentition. 2. Differentiation of the poscanine dentition. Proc. Roy. Soc. London **109**, 1939.

Dreyhaupt, R.: Das Lageverhältnis der Oberkieferbackzähne zu der Kieferhöhle beim Pferd. Leipzig, Diss. med. vet., 1934.

Duckworth, J., R. Hill, D. Benzie and A. C. Dalgarno: Studies of the dentition of sheep. I. Clinical observations from investigations into the shedding of permanent incisor teeth by hill sheep. Res. Vet. Sci. **3**, 1962.

Finger, H.: Beitrag zur Kenntnis der postembryonalen Entwicklung der Backzähne des Pferdes. Leipzig, Diss. med. vet., 1920.

Garlick, N. L.: The teeth of the ox in clinical diagnosis. II. Gross anatomy and physiology. Am. J. Vet. Res. **15**, 1954.

Gredig, M.: Der Prämolarenverlust beim deutschen Schäferhund. Schweizer Hundepost, **71**, 1955.

Habel, R. E.: Applied veterinary anatomy. 2nd. ed. Ithaca, New York: author, 1978.

Habel, R. E.: Guide to the dissection of domestic ruminants. 3rd ed. Ithaca, New York: author, 1977.

Habermehl, K.-H.: Der Einfluß der Oberkieferspalten auf den Milchzahndurchbruch beim Schwein. Anat. Anz. **103**, 1956.

Habermehl, K.-H.: Über das Gebiß des Hausschweines (Sus scrofa dom. L.) mit besonderer Berücksichtigung der Backzahnwurzeln. Zbl. Vet. Med. **4**, 1957.

Habermehl, K.-H.: Die Altersbestimmung bei Haus- und Labortieren. 2nd. ed. Berlin und Hamburg, Paul Parey, 1975.

Habermehl, K.-H.: Besitzt das weibliche Hausschwein permanent wachsende Hakenzähne? Berl. Münch. Tierärztl. Wschr. **75**, 1962.

Hauck, E.: Untersuchungen über die Form und Abänderungsbreite des Hundegebisses. Wien. Tierärztl. Mschr. **29**, 249 u. 273 (1942).

Helmcke, H.-J.: Ergebnisse und Probleme aus der vergleichenden Anatomie der Wirbeltierzähne. Studium Generale **18** (1965).

Hilzheimer, M.: Variationen des Canidengebisses mit Berücksichtigung des Haushundes. Z. Morph. Anthropol. **9**, 1906.

Hirsch, M.: Der Lückenzahn von Sus domesticus, ein Beitrag zur Entwicklungsgeschichte des Gebisses von Sus domesticus und zur Kenntnis des Wesens der Dentitionen. Anat. Anz. **54**, 1921.

Hornickel, E.: Über die Lageverhältnisse der Unterkieferbackzähne beim Pferd. Leipzig, Diss. med. vet., 1934.

Joest, E.: Odontologische Notizen. Berl. Tierärztl. Wschr. **31**, 1915.

Joest, E.: Studien über das Backenzahngebiß des Pferdes mit besonderer Berücksichtigung seiner postembryonalen Entwicklung und seines Einflusses auf den Gesichtsschädel und die Kieferhöhle. Berlin, Schoetz, 1922.

Keil, A.: Grundzüge der Odontologie. Allgemeine und vergleichende Zahnkunde als Organwissenschaft. Berlin, Borntraeger, 1966.

Kretzer, H.: Prämolarverlust bei Caniden. Gießen, Diss. med. vet., 1951.

Kroon, H. M.: Die Lehre der Altersbestimmung bei den Haustieren. 3rd ed. Hannover, Schaper, 1929.

Kükenthal, W.: Über den Ursprung und die Entwicklung der Säugetierzähne. Jen. Zschr. Naturwiss. **27**, 1892.

Küpfer, M.: Über die Bildung der Backenzähne am Kiefer des großen und kleinen Wiederkäuers, bei Rind und Schaf. Schweiz. Landw. Mhefte **13**, 1935a.

Küpfer, M.: Beiträge zur Erforschung der baulichen Struktur der Backenzähne des Hausrindes (Bos taurus L.). Die Prämolar- und Molarentwicklung auf Grund röntgenologischer, histogenetischer und morphologischer Untersuchungen. Die gegenseitigen Beziehungen der einzelnen Gebißkonstituenten und ihre Heranziehung zur physiologischen Leistung. Denkschr. Schweiz. Naturforsch. Ges. LXX, Abh. **1**, 1935b.

Küpfer, M.: Backenzahnstruktur und Molarentwicklung bei Esel und Pferd. Schweiz. Landw. Mhefte **10**, 1936.

Lawson, D. D., G. S. Nixon, H. W. Noble, W. L. Weipers: Development and eruption of the canine dentition. Brit. Vet. J. **123**, 26—30 (1967).

Meyer, L.: Das Gebiß des deutschen Schäferhundes mit besonderer Berücksichtigung der Zahnaltersbestimmung und der Zahnanomalien. Zürich, Diss. med. vet., 1942.

Mohr, E.: Normalgebiß und richtige Benennung der Zähne des Haushundes. Der Terrier, **47**, 8. 1954.

Mohr, E.: Hundegebiß und -biß. Schweiz. Hundesport, **71**, 16, 1955.

Mohr, E.: Der Zahnschluß im Gebiß der Wildraubtiere und des Haushundes. Z. Säugetierk. **26**, 50—56 (1961).

Peyer, B.: Comparative Odontology. Translated and edited by R. Zangerl. Chicago: University of Chicago Press, 1968.

Pirilä, H.: Untersuchungen an 16 Pferdeschädeln über die Formveränderungen der Zähne und ihre Lage im Kiefer in den verschiedenen Altersstadien. Z. Anat. Entw. **102**, 1933.

Ripke, E.: Beitrag zur Kenntnis des Schweinegebisses. Anat. Anz. **114**, 181—211 (1964).

St. Clair, L. E., and N. D. Jones: Observations on the cheek teeth of the dog. J. A. V. M. A. **130,** 1957.
Schlaak, W.: Beitrag zur Mechanik der Backzähne des Unterkiefers des Pferdes. Leipzig, Diss. med. vet., 1938.
Schlosser, M.: Das Milchgebiß der Säugetiere. Biol. Zbl. **10,** 1890—91.
Schwalbe, G.: Über Theorien der Dentition. Anat. Anz. **9,** (Erg. H.) 1894.
Seiferle, E.: Zum Gebißproblem des Hundes. Schweiz. Hundesport Nr. 11, 1956.
Seiferle, E., u. L. Meyer: Gebiß des deutschen Schäferhundes in Altersstufen. Vjschr. naturforsch. Ges. Zürich, 1942.
Simon, Ch.: Untersuchungen über den Bau der Zähne beim Rind und Altersbestimmungen unter besonderer Berücksichtigung der Gebißanomalien. Kühn-Arch. **22,** 1929.
Steenkamp, J. D. G.: Wear in bovine teeth. Proc. Symp. Animal Prod., Salisbury **2,** 11—23 (1969).
Terra, P. de: Vergleichende Anatomie des menschlichen Gebisses und der Zähne der Vertebraten. Jena, 1911.
Tonge, C. H. and R. A. McCance: Normal development of the jaws and teeth in pigs, and the delay and malocclusion produced by calorie deficiencies. J. Anat. **115,** (1973): 1—22.
Virchow, H.: Über das Schweinegebiß mit Ausblicken auf die Gebisse anderer Säugetiere. Sitzgsber. Ges. naturforsch. Freunde, Berlin, 1937.
Weaver, M. E., E. B. Jump, and C. F. McKean: The eruption pattern of deciduous teeth in miniature swine. Anat. Rec. **154,** 1966.
Weber, M.: Die Säugetiere, Vol. 2, 2nd ed. Jena, Fischer, 1927.
Weinreb, M. M., and Y. Sharav: Tooth development in sheep. Am. J. Vet. Res. **25,** 1964.
Weiss, H.: Vergleichende Untersuchungen über die Zähne der Haussäugetiere. Zürich, Diss. med. vet., 1911.
Weller, J. M.: Evolution of mammalian teeth. J. Paleontol. **42,** 1968.
Westin, R.: Zahndurchbruch und Zahnwechsel. Z. mikrosk.-anat. Forsch. **52,** 1942.
Wolf, F.: Untersuchungen an der Pulpahöhle der maxillaren Backzähne des Pferdes. Leipzig, Diss. med. vet., 1939.

The Alimentary Canal

General and Comparative

The alimentary canal (11) consists of the **esophagus, stomach, small intestine, large intestine,** and the **anal canal.** Two large glands, the **liver** and the **pancreas,** are associated with the alimentary canal and release their secretions into its lumen.

Esophagus

(131—133)

The esophagus is a musculo-membranous tube which connects the pharyngeal cavity with the stomach. It is the direct continuation of the laryngopharynx (132/*8, 9*) and is divided into a cervical, a thoracic, and a short abdominal part. The **cervical part** lies for the most part between the longus colli, the muscle running along the ventral surface of the cervical vertebrae, and the trachea, but as it approaches the thoracic inlet it shifts from its median position and passes to the left side of the trachea. Because the esophagus lies relatively close to the skin in this area, boluses of food or water, or bubbles of air, can be seen passing through it; and the tip of a stomach tube can be seen or palpated here as it is passed through the esophagus. In this region the esophagus is also accessible for surgery. The more important relations of the cervical part of the esophagus are: the common carotid artery, the internal jugular vein, the tracheal duct, the cervical lymph nodes, the vagosympathetic trunk, the caudal (recurrent) laryngeal nerve, and, in young animals, the thymus (131). The **thoracic part** of the esophagus begins at the thoracic inlet, and after passing through the inlet on the left side of the trachea returns to its position dorsal to the trachea (132). It runs caudally in the dorsal mediastinum, passes dorsal to the tracheal bifurcation, and crosses the right side of the aortic arch (*27*). Caudal to the base of the heart, the esophagus lies between the lungs ventral to the thoracic aorta. It is accompanied by the dorsal and ventral branches of the vagus nerves and farther caudally by the dorsal and ventral vagal trunks, and before reaching the diaphragm it is related on the right side to the cavum mediastinum serosum (7/*n*). Finally it passes through the esophageal hiatus of the diaphragm and terminates with a very short **abdominal part** at the cardia of the stomach.

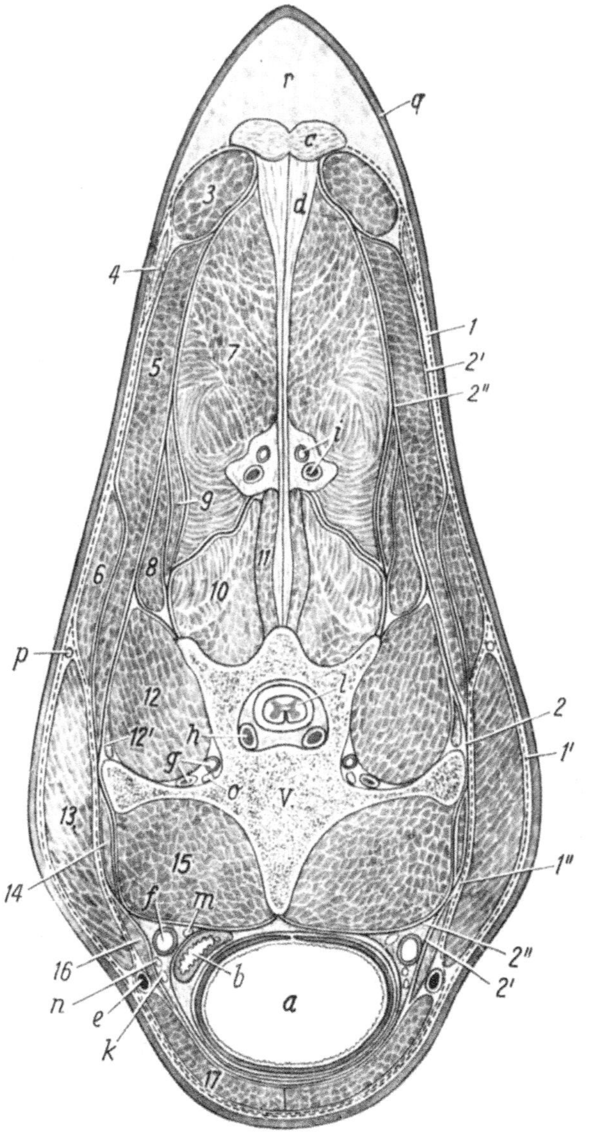

Fig. 131. Cross section of the neck of the horse at the level of the fifth cervical vertebra. Caudal aspect. Semischematic.

a Trachea; *b* Esophagus; *c* Funicular part of lig. nuchae; *d* Laminar part of lig. nuchae; *e* Ext. jugular vein in jugular groove; *f* Common carotid artery; *g* Vertebral artery and vein; *h* Vertebral venous sinus; *i* Deep cervical artery and vein; *k* Tracheal lymphatic trunk; *l* Spinal cord; *m* Vagosymphathetic trunk; *n* Caudal (recurrent) laryngeal nerve; *o* Vertebral nerve; *p* Dorsal branch of accessory nerve; *q* Skin; *r* Nuchal fat

1 Supf. cervical fascia, *1'* its superficial, *1''* its deep lamina; *2* Deep cervical fascia, *2'* its superficial, *2''* its deep lamina (in the region of the longus colli muscle (*15*) known as prevertebral fascia); *3* Rhomboideus cervicis; *4* Cervical part of trapezius; *5* Splenius cervicis; *6* Serratus ventralis cervicis; *7* Semispinalis capitis; *8* Longissimus atlantis; *9* Longissimus capitis; *10* Multifidus; *11* Spinalis cervicis; *12* Intertransversarius dorsalis; *12'* Intertransversarius intermedius; *13* Brachiocephalicus; *14* Longus capitis; *15* Longus colli; *16* Omohyoideus; *17* Sternomandibularis, sternohyoideus, and sternothyroideus; *V* Fifth cervical vertebra

STRUCTURE. The esophagus is a tube of varying diameter. Where its diameter is narrow, its muscular wall is thick; where the diameter is wide, the muscular wall is thin. Except in the ruminants, the muscular wall of the esophagus increases gradually in thickness from cranial to caudal; this is especially noticeable in the horse.

The wall of the esophagus consists of three layers: a connective tissue adventitia externally, a muscular coat, and a mucous membrane internally. The **tunica adventitia** (133/*h*) loosely connects the esophagus to the neighboring structures and allows it freedom to move during swallowing and when the animal bends its neck. The adventitia of the thoracic part of the esophagus is covered to some extent with pleura. The **muscular coat** (tunica muscularis) by a wavelike contraction moves the bolus toward the stomach or, during regurgitation in ruminants, toward the mouth. In the dog and ruminants, the muscular coat consists en-

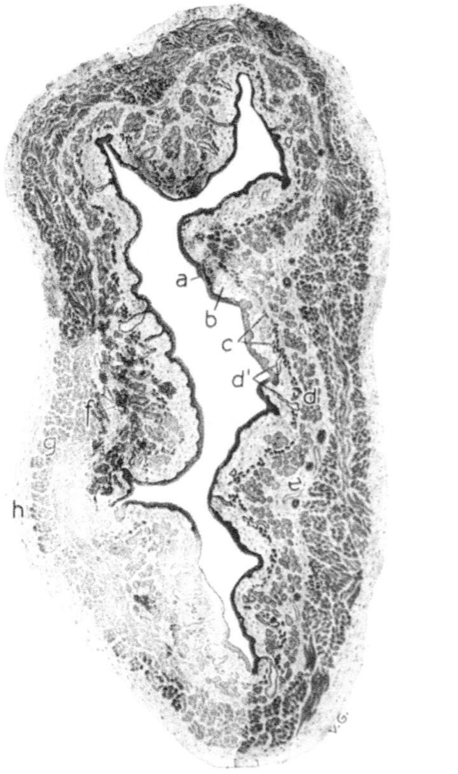

Fig. 133. Cross section of the esophagus of the dog.

a—d' Mucous membrane; *a* Epithelium; *b* Lamina propria; *c* Muscularis mucosae; *d* Esophageal glands, *d'* their excretory ducts; *e* Submucosa; *f*, *g* Muscular coat; *f* Inner circular muscle layer; *g* Outer longitudinal muscle layer; *h* Adventitia

tirely of striated muscle; in the pig, there is a short part with smooth muscle near the cardia; in the cat and horse, the cranial two thirds of the esophagus consists of striated muscle and the caudal third of smooth muscle. The muscular coat of the esophagus (*f*, *g*) consists of inner and outer layers, which are continuous cranially with the pharyngeal muscles. Here, the esophageal musculature can be separated into independent muscle bands that surround the mucosa in long elliptical loops. In the middle portion of the esophagus, the two muscle layers are arranged in intercrossing spirals. In the caudal portion, the fibers of the outer layer become more and more longitudinal and those of the inner layer more and more circular and thicker. The **mucous membrane** (tunica mucosa) is covered with stratified squamous epithelium which is cornified, particularly in the herbivores. The lamina propria (*b*) is without glands. The lamina muscularis mucosae (*c*) is incomplete and forms a more or less continuous sheet only in the caudal part of the esophagus. Where present, it lies on a well-developed tela submucosa (*e*). In the dog, the submucosa contains mucous **glands** (*d*) over the entire length of the esophagus. In the pig, glands are present only in the cranial half of the esophagus, while in the horse, ruminants, and cat, they are present only at the pharyngo-esophageal junction. The mucosa (because of the abundant submucosa) and the muscular coat are capable of great expansion when the lumen of the tube is dilated by the passing bolus. In the collapsed state the mucosa is arranged in deep longitudinal folds which give the lumen a rosette-like appearance when viewed in cross section.

BLOOD VESSELS, LYMPHATICS, AND INNERVATION. The cervical part of the esophagus is supplied by branches of the right and left common carotid arteries; the thoracic part by the bronchoesophageal artery. The veins of the cervical part of the esophagus enter the external jugular veins, while the thoracic part is drained by the esophageal vein. The carnivores have a pair of esophageal veins that empty into the azygous veins via the bronchoesophageal veins; whereas in the small ruminants and the horse, the esophageal vein enters the azygous vein directly. **Lymphatic drainage** of the cervical part of the esophagus is into the cranial, middle, and caudal deep cervical nodes. In the thorax, lymph from the esophagus enters the cranial and caudal mediastinal lymph nodes. The **innervation** of the esophagus is derived from the vagus and the sympathetic trunk. Intramural ganglia of the myenteric plexus are found between the internal and external muscle layers.

Stomach

(134—143, 169—171, 191, 245, 246)

The stomach receives the insalivated boluses of food from the esophagus and temporarily stores them. The gastric juice secreted by glands in the stomach wall consists chiefly of pepsin, rennin, and hydrochloric acid, and initiate enzymatic and chemical digestion. The ingesta are mixed with gastric juice by the muscular contractions of the stomach and are gradually moved into the duodenum.

The stomachs of the domestic mammals differ with the nutritional habits that have evolved; that is, the kind of food that is usually consumed. These are differences not only in the external shape and size of the organ, but also in the composition of its lining. The carnivores, pig, and horse have what is known as a simple stomach (134—137) consisting of a single compartment. The stomach of the ruminants is much larger and more complex and has four compartments (203). As regards the lining there are two general tpyes: a nonglandular mucosa, which is covered with stratified squamous epithelium; and a glandular mucosa covered with simple columnar epithelium. The former is similar to the mucosa of the alimentary canal proximal (cranial) to the stomach, while the latter resembles the mucosa distal to the stomach. Stomachs lined exclusively with glandular mucosa are called glandular stomachs; those lined with both types are called composite stomachs, the part lined with the nonglandular mucosa being **the nonglandular part,** and the part lined with the glandular mucosa being the **glandular part.**

Because of the similarity of the proventricular lining to the mucosa of the esophagus, it had been assumed that the proventricular part of the stomach was a dilatation of the embryonic esophagus that had become part of the stomach. Studies of the embryonic development

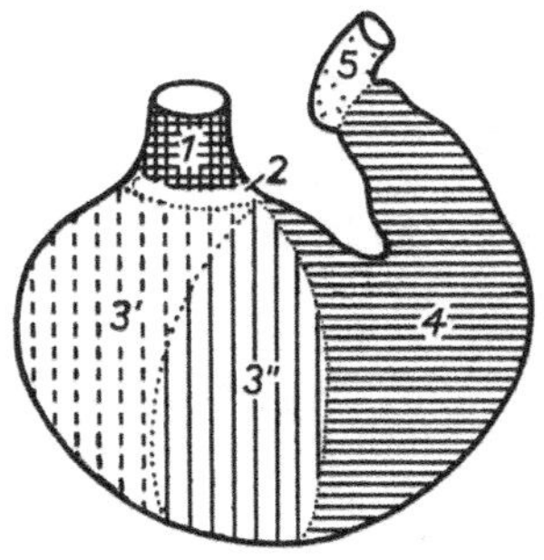

Fig. 134 (Dog)

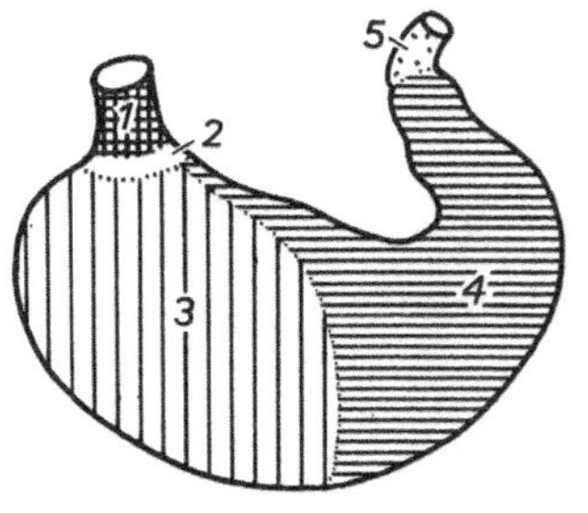

Fig. 135 (Cat)

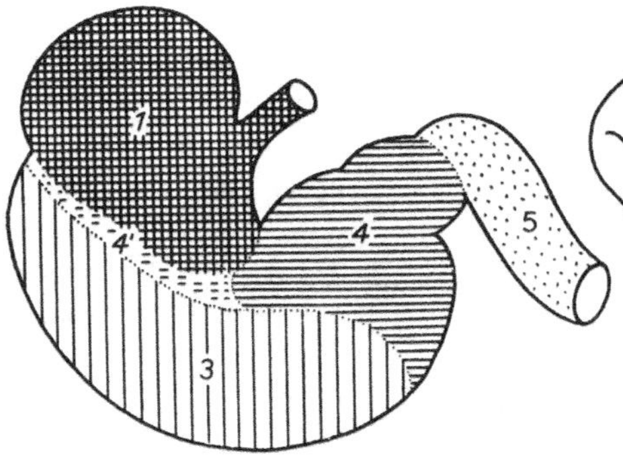

Fig. 136 (Horse)

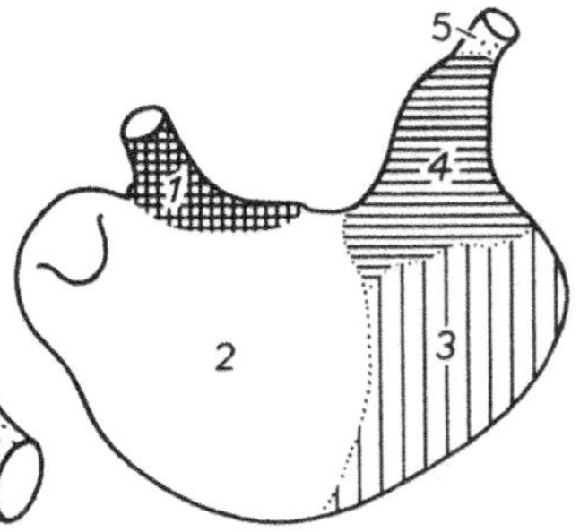

Fig. 137 (Pig)

Figs. 134—137. Mucosal regions of the stomachs of domestic mammals. Schematic. (For the ruminant stomach see fig. 203.)

1 Nonglandular mucosa of the esophagus and the proventricular part of the stomach (cross-hatched); *2* Cardiac gland region (white); *3* Region of proper gastric glands (vertical lines); *3'* Light (broken vertical lines) and *3''* dark (vertical lines) regions of proper gastric glands of the dog; *4* Pyloric gland region (horizontal lines); *4'* Region of mixed cardiac and pyloric glands of the horse (broken horizontal lines); *5* Duodenal mucosa (stippled)

of these two organs, however, have shown that even the complex ruminant stomach (having a very large proventriculus) originates from the same spindle-shaped primordium that gives rise to the simple stomach, and that the embryonic esophagus is well set off from the stomach in all stages of gastric development (Pernkopf 1931; Warner 1958). It is possible, therefore, to correlate parts of the simple stomach with the various compartments of the complex ruminant type; for instance, fundus and part of the body of the simple stomach is homologous to the rumen and reticulum of the ruminant stomach.

The carnivores have a simple stomach lined entirely with glandular mucous membrane. The stomach of the pig and horse is also simple in its external shape, but has a composite lining, consisting of both glandular and nonglandular parts; the nonglandular part is small in the pig but quite extensive in the horse (136/*1*). The ruminants have a complex stomach consisting of four compartments. The proximal three—the rumen, reticulun, and omasum—having a nonglandular lining, comprise the **forestomach** (proventriculus), while the most distal compartment—the abomasum—because of its glandular mucosa, is the glandular stomach (203).

SHAPE. The stomach is a saclike enlargement of the alimentary canal between the esophagus and the duodenum (11). The opening at the esophageal end, through which the food enters, is the **cardia,** or **cardiac opening** (ostium cardiacum), and the opening at the duodenal end, through which the ingesta leave, is the **pyloric opening** (ostium pyloricum). The part of the stomach that surrounds the cardiac opening is the **pars cardiaca** (169/*a*), while the very muscular part surrounding the pyloric opening is the **pylorus** (*f*).

The shape of the stomach is not constant and depends on the amount of ingesta it contains, and on the presence or absence of muscular contractions at the time of inspection. Nevertheless, it does have a basic form, and this is best demonstrated by a moderately filled organ exposed shortly after death. Only the simple stomach is described here; the ruminant stomach is described on page 148.

The basic form of the simple stomach is that of a J-shaped, curved sac (143) flattened craniocaudally with greater and lesser curvatures extending from the cardia to the pylorus. The surface that is directed cranially and that is in contract with the diaphragm is the **parietal surface,** while the surface that is directed caudally and in contact with the other abdominal viscera is the **visceral surface.** The **greater curvature** (curvatura major, 170/*i*) is convex and directed toward the left and ventrally. It is much longer than the opposite **lesser curvature** (curvatura minor, *k*) which is concave and directed to the right and dorsally. The lesser curvature of the simple stomach is marked by an incisure (incisura angularis) at the level of which the stomach is usually sharply bent. Proximal to, or in situ to the left of this flexure, is the **body** (corpus) of the stomach (143/*c*), which reaches to the level of the cardia. The part of the stomach bulging above the level of the cardia is the **fundus** (*b*), which is separated from the cardia by an indentation (incisura cardiaca). In the live animal, the fundus usually contains a sizable bubble of air. Distal to the body of the stomach is the **pyloric part,** consisting of a wider proximal portion, the **pyloric antrum** (*d*), and a narrower distal portion, the **pyloric canal** (*e*), which terminates at the pylorus (*f*). Along the inside of the lesser curvature and reaching from the cardia almost to the pylorus is the area of the **gastric groove** (sulcus ventriculi, 191/*6*). The stomach of the horse and pig deviate slightly from the basic shape. In the horse, the fundus is very extensive and rises considerably above the level of the cardia; because of this and the presence of an additional circular muscle layer, the fundus is called the **blind sac** (saccus cecus, 246/*b*). In the pig, the fundus has a small **diverticulum** (191/*1*).

STRUCTURE. The wall of the stomach consists from inside out of a mucous membrane, a muscular, and a serous coat. The **mucous membrane** (tunica mucosa) can be subdivided into the usual layers: surface epithelium (138/*a*), lamina propria mucosae (*b*), lamina muscularis mucosae (*c*), and a tela submucosa (*d*).

The **nonglandular mucosa** of the stomach is often slightly folded. It is whitish, smooth and firm and is covered with a thick, cornified stratified squamous epithelium (139/*a*). The junction with the glandular mucosa is sharp. In the horse, the junction is marked by an

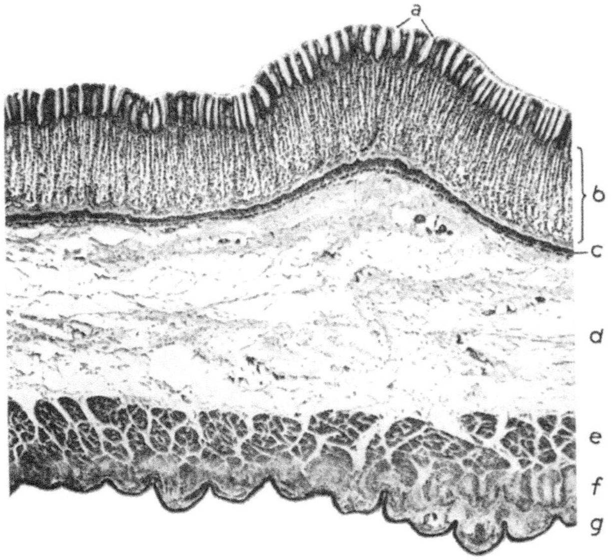

Fig. 138. Section of the stomach wall of the cat taken from the region of the proper gastric glands. Microphotograph.

a—c Mucous membrane; *a* Foveolae lined with surface epithelium; *b* Lamina propria containing proper gastric glands; *c* Muscularis mucosae internal to it the stratum compactum; *d* Loose submucosa; *e, f* Double-layered muscular coat; *g* Tunica serosa

irregular, raised ridge, **margo plicatus** (246/*3*), and the nonglandular mucosa, particularly near the cardia, is usually dotted with craterlike perforations caused by the larvae of the bot fly *(Gastrophilus intestinalis)*. These parasites are often found still attached to the interior of the stomach in the embalmed horse.

The **glandular mucosa** (pars glandularis) of the stomach often forms high transient folds (plicae gastricae), especially in the pyloric part, and when examined with a lens, the surface appears uneven and is divided by shallow grooves (228) into raised areas (areae gastricae) a few millimeters in diameter. The gastric pits (foveolae gastricae, 140—142/*a*) open on the

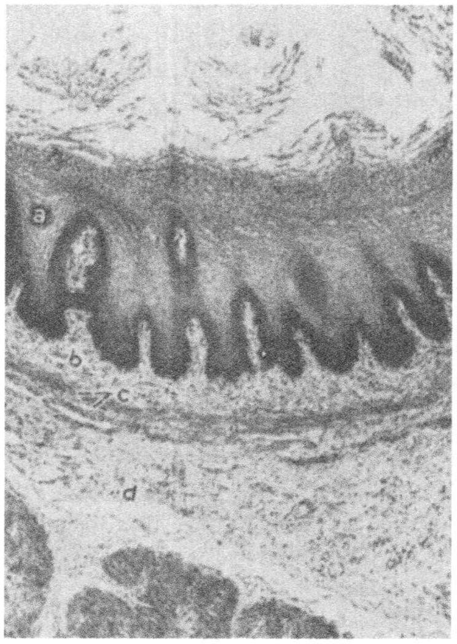

Fig. 139. Nonglandular mucous membrane of the proventricular part of the pig's stomach. Microphotograph.
a Stratified squamous epithelium partly cornified; *b* Lamina propria with papillary layer; *c* Muscularis mucosae; *d* Submucosa

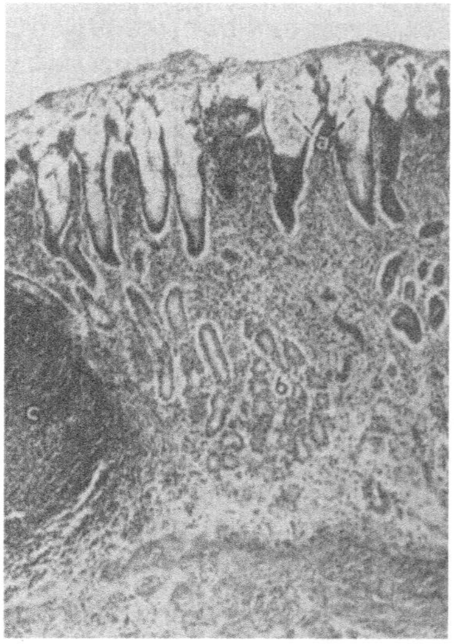

Fig. 140. Mucous membrane of the cardiac gland region of the pig's stomach. Microphotograph.
a Foveolae gastricae; *b* Cardiac glands in lamina propria, which is rich in lymphocytes; *c* Lymph nodule

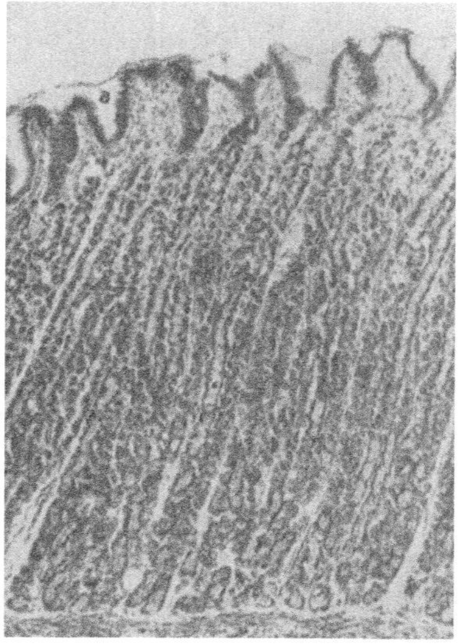

Fig. 141. Mucous membrane of the dog's stomach in the region of the proper gastric glands. Microphotograph.
a Foveola gastrica; the long, tubular proper gastric glands are in the lamina propria

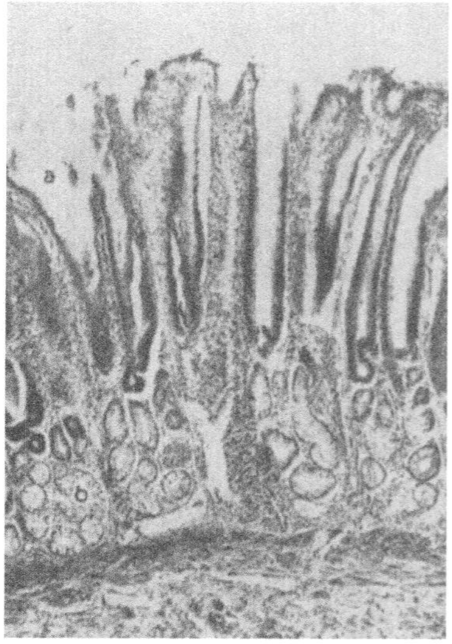

Fig. 142. Mucous membrane of the pyloric gland region of the dog's stomach. Microphotograph.
a Foveola gastrica; *b* Pyloric glands in lamina propria

PLATE II

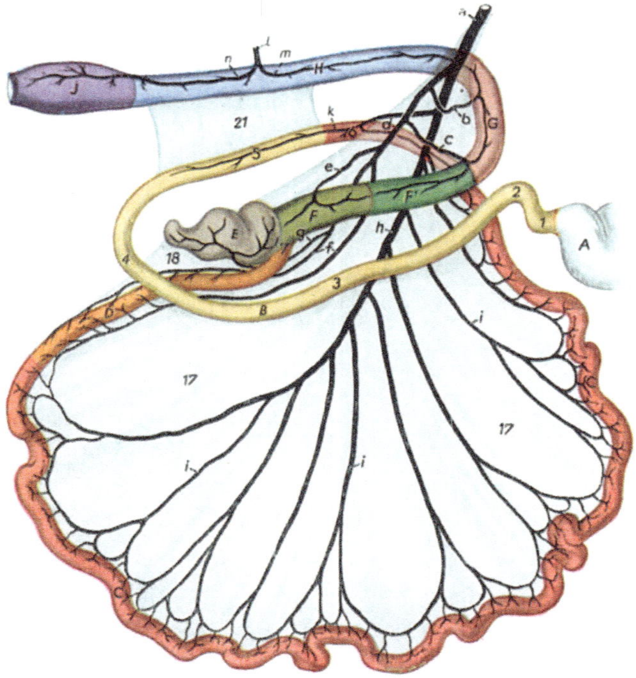

Fig. 144 (Dog)

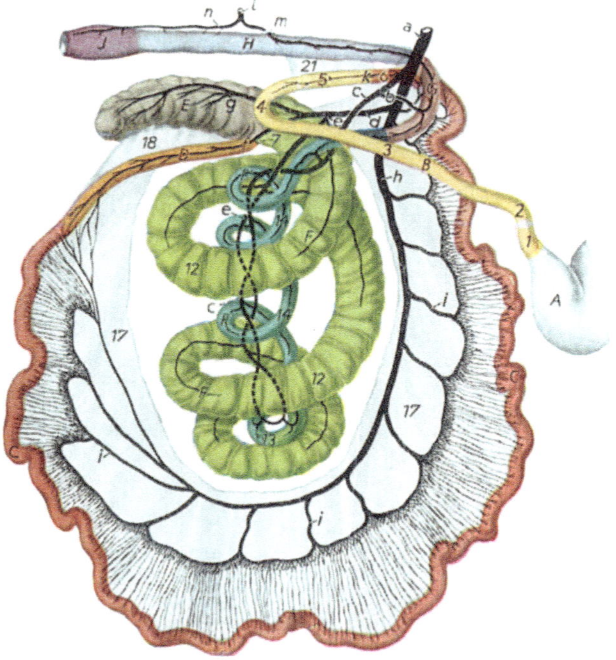

Fig. 145 (Pig)

Figs. 144—147. The intestines of the domestic mammals. Schematic.

A Pyloric part of stomach or abomasum; *B* Duodenum; *C* Jejunum; *D* Ileum; *E* Cecum; *F, F'* Ascending colon; *G* Transverse colon; *H* Descending colon; *J* Rectum; *K* Anus

a, h Cranial mesenteric artery; *b* Middle colic artery; *c* Right colic artery (formerly known as the dorsal colic artery in the horse); *d* Ileocolic artery; *d', g* Cecal artery; *e* Colic branch (formerly known as the ventral colic artery in the horse); *f* Ileal branch of cecal artery; *g* Cecal artery, in the horse medial and lateral cecal arteries; *h* Continuation of cranial mesenteric artery; *h'* Collateral branch (ruminants); *i* Jejunal arteries; *k* Caudal pancreatico-duodenal artery; *l* Caudal mesenteric artery; *m* Left colic artery; *n* Cranial rectal artery

1—6 Duodenum: *1* cranial part, *1'* sigmoid loop, *2* cranial flexure, *3* descending duodenum, *4* caudal flexure, *5* ascending duodenum, *6* duodenojejunal flexure; *7* Ileal opening; *7'* Cecocolic opening (horse); *8* Base, *9* body, and *10* apex of cecum (horse); *11—15* Ascending colon: *11, 11, 11* proximal loop (ruminants), *12* centripetal turns (pig, ruminants), ventral colon (horse), *12'* sternal flexure (horse), *13* central flexure (pig, ruminants), pelvic flexure (horse), *14* centrifugal turns (pig, ruminants), dorsal colon (horse), *14'* diaphragmatic flexure (horse), *15, 15* distal loop (ruminants); *16* Sigmoid colon (ruminants); *17* Mesentery; *18* Ileocecal fold; *19* Cecocolic fold (horse); *20* Ascending mesocolon (horse); *21* Duodenocolic fold; *22* Descending mesocolon (shown only in the horse)

Fig. 146 (Ox)

Fig. 147 (Horse)

surface of these areas and into the grooves, and are separated by microscopic annular folds (plicae villosae). The single layer of tall columnar cells that makes up the surface epithelium is continued into the foveolae and secretes mucus, which forms a thin layer on the surface of the mucosa.

The **lamina propria mucosae** contains three types of glands: the proper gastric glands, the pyloric glands, and the cardiac glands. The glandular lining of the stomach is divided into three regions based on the type of gland present: the region of the proper gastric glands, the pyloric gland region, and the cardiac gland region. These regions differ in extent from species to species.

In the **region of the proper gastric glands** (134—137/*3*, *3'*, *3''*) the mucous membrane is thick, reddish brown, and irregularly mottled. In the carnivores it lines about two-thirds of the stomach, i.e., the fundus and almost the entire body; in the pig it lines the distal portion of the body near the greater curvature; and in the horse it lines most of the body of the stomach except the fundus. The **proper gastric glands** (141) are closely packed, tubular glands. They have a narrow neck portion and open in groups into the foveolae. The body of the proper gastric gland consists of chief cells and parietal cells; closer to the neck of the gland are the mucous neck cells.

In the **pyloric gland region** (134—137/*4*) the mucous membrane is thinner than in the region of the proper gastric glands. It is grayish yellow or light gray, and is arranged in mostly high, coarse folds. The mucus-producing **pyloric glands** (142/*b*) are more branched and tortuous than the proper gastric glands and open into deeper foveolae. The pyloric gland region in the dog takes up about one-third of the stomach and is found mainly in the pyloric part, but extends also into the body, reaching the cardia in the vicinity of the lesser curvature. In the the pig's stomach it lines most of the pyloric part, except for an area near the greater curvature. In the horse, the pyloric gland region is coextensive with the pyloric part of the stomach.

The mucosa of the **cardiac gland region** has highly branched, coiled seromucous glands, the **cardiac glands,** and is rich in both diffuse and nodular lymphoid tissue (140/*c*). In the carnivores, the cardiac gland region (134/*2*) is a narrow annular zone at the cardia, while in the pig it is extensive and occupies the fundus, including the diverticulum, and most of the body (137/*2*). In the horse, the cardiac glands are mixed with pyloric glands along a narrow zone between the margo plicatus and the region of the proper gastric glands (136/*4'*).

The **lamina muscularis mucosae** (138/*c*) is present in all regions of the stomach and consists mostly of two layers. It separates the lamina propria from the well-developed **submucosa** (*d*). The latter is composed of loose connective tissue which permits the mucosa to slide on the subjacent muscular coat. The loose submucosa further permits the mucosa to be raised into transient folds by the action of the muscularis mucosae when the stomach as a whole contracts.

The **muscular coat** (tunica muscularis) of the stomach consists, as in the rest of the alimentary canal, of an outer longitudinal and an inner circular layer. Peculiar to the stomach, however, is a third layer, the **internal oblique fibers,** which is restricted to the fundus, and to the body in the vicinity of the greater curvature. It is in these areas that the greatest expansion takes place during the development of the stomach, and leads to a rearrangement of the outer longitudinal and inner circular fibers and to the addition of the third muscle layer for the expanded parts. The direction of the muscle fibers in the various layers follows a similar pattern in all domestic mammals with simple stomachs. And it will be shown later (page 156) that the ruminant stomach, despite its marked external dissimilarity, also follows this basic musculatur pattern, again underscoring its common origin with the simple stomach.

The arrangement of the three muscle layers of the simple stomach is shown schematically in Figure 143. The **outer longitudinal layer** (143/solid lines) coming from the esophagus, spreads out as it passes over the stomach and forms two longitudinal bands of muscle fibers along each curvature, both bands becoming less distinct as they approach the pyloric part. On the parietal and visceral surfaces of the stomach, the longitudinal fibers run obliquely to the longitudinal axis of the stomach and are known as the **external oblique fibers.** These spread out from the cardia over the fundus, then curve and run parallel with the greater curvature over the body, and gradually come together again as they approach the pyloric part of the stomach. The external oblique fibers are especially prominent in the stomachs of the horse and pig, but do not follow the general pattern shown in Figure 143. From the cardia they pass over the fundus in a fan-shaped manner, and have a circular arrangement over the body of the stomach. True longitudinal bundles are present only on the curvatures, where they can be demonstrated as discrete muscular bands to about the pyloric part.

In all animals, the wall of the pyloric canal has a heavy longitudinal muscle coat that is separate from the longitudinal muscle bundles that follow the curvatures.

The **circular muscle layer** (143/dotted lines), the innermost layer, is present only in the body and pyloric part of the stomach, and does not encircle the fundus. The bundles of this layer are closely packed over the lesser curvature and fan out from there toward the greater curvature. The circular layer is thickest at the pyloric canal where it forms the **pyloric sphincter.** In the pyloric part, both circular and longitudinal muscle layers are thick, and perform the strong peristaltic contractions that move the ingesta into the duodenum.

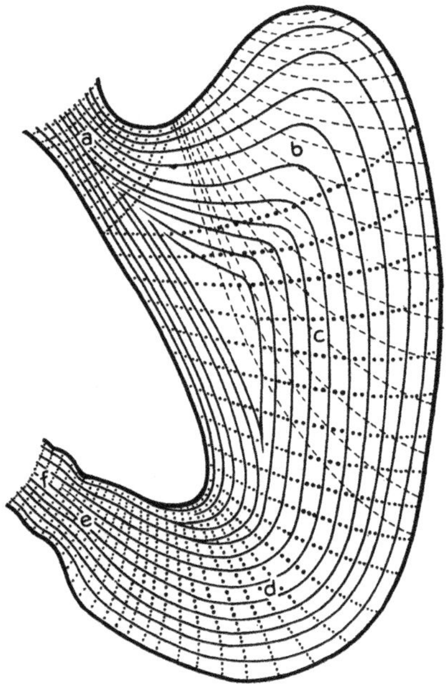

Fig. 143. The muscular coat of the simple stomach, based on the carnivore. Schematic. (Redrawn from Pernkopf 1930.)

a Cardia; *b* Fundus; *c* Body; *d, e* Pyloric part; *d* Pyloric antrum; *e* Pyloric canal; *f* Pylorus

Solid lines: *longitudinal layer.* Coming from the esophagus, the fibers of the longitudinal layer pass distally along the greater and lesser curvatures to be continued as the longitudinal layer of the duodenum. On the fundus of the stomach the longitudinal layer forms the *external oblique fibers*

Dotted lines: *circular layer.* The circular fibers form part of the *cardiac sphincter*; on the body and the pyloric part they form the circular muscle layer, and at the pylorus the *pyloric sphincter*

Broken lines: *internal oblique fibers.* The internal oblique fibers surround the fundus and part of the body of the stomach. The fibers closest to the lesser curvature form the boundaries of the gastric groove and at the cardia lie in the incisura cardiaca as the *cardiac loop*, which, together with the circular fibers, forms the *cardiac sphincter*

The third layer, the **internal oblique fibers** (143/broken lines), is present only on the fundus and body of the stomach. These fibers encircle the fundus and at the cardia, where they are thicker and more closely packed, they form the **cardiac loop** (ansa cardiaca). The crest of the cardiac loop lies in the incisura cardiaca, and the limbs of the loop descend parallel to the lesser curvature toward the pyloric part of the stomach. The internal oblique fibers diverge from the loop, and pass over the body of the stomach in the direction of the greater curvature. The longest bundles are those closest to the lesser curvature, and may reach to the level of the incisura angularis. They are particularly thick (247/*f'*) and form the boundaries of the **gastric groove** (sulcus ventriculi) which extends from the cardia to the pylorus. The fibers of the circular muscle layer (*g, g'*) cross the bundles of the cardiac loop (*f, f'*) and form the floor of the gastric groove. In the pig, the muscular boundaries of the gastric groove are prominent enough that they raise two longitudinal folds of mucosa (191/*6*), which correspond to the lips of the much more conspicuous gastric groove of ruminants (207/*1*).

The cardiac loop and the heavy bundles of the circular muscle layer near the cardia combine to form the **cardiac sphincter,** which is especially well developed in the horse. The spiral fold at the base of the gastric diverticulum in the pig is also formed by bundles of the internal oblique fibers.

The **serous coat** (tunica serosa) of the stomach invests the greater part of the organ and is continuous at the greater curvature with the **greater omentum,** toward the diaphragm with the **gastrophrenic ligament,** and at the lesser curvature with the **lesser omentum.** Subserous fat and connective tissue present along the lesser curvature, and in lesser amounts also along the greater curvature, leave bandlike areas of stomach wall that are not covered with serous membrane, as shown in Figure 205 for the ruminant stomach. In view of the tensile strength of serous membranes, the areas devoid of this covering, particularly the one along the greater curvature, cannot withstand abnormally high gastric pressures as well as the stomach wall directly invested with serosa.

BLOOD VESSELS, LYMPHATICS, AND INNERVATION. The stomach is supplied largely by the left gastric artery, but also by the other two branches of the celiac artery, the splenic and hepatic arteries. The gastric veins are branches of the portal vein. The **lymphatics** of the stomach go to the gastric, splenic, celiac, and pancreatico-duodenal lymph nodes. The **innervation** is from the vagus and sympathetic nerves via the celiac and cranial mesenteric plexuses, and the intramural myenteric and subserosal plexuses.

Intestines

(11, 144—152, 192, 230, 249)

The intestines extend from the pylorus of the stomach to the anus. Their proximal part, because of its relatively small lumen, is called the **small intestine;** the wider distal part is known as the **large intestine.** The small intestine is subdivided into **duodenum** (11/*h*), **jejunum** (*i*), and **ileum** (*k*), with the jejunum by far the longest of the three. The large intestine consists of the **cecum** (*l*), the **colon** (*m—o*), and the **rectum** (*p*). The **anal canal** is the short end portion of the digestive tract.

The total length of the intestinal tract and the length of its segments differ with the species and breed, but may vary also from animal to animal of the same species. It is difficult to determine the exact length of the intestines in the live animal. Although measurements are easier to obtain on dead material, they are comparable only if taken under similar conditions. It is necessary therefore to keep this limitation in mind for the measurements given in the Table below.

Table 8. Length of Intestines in Meters

	Dog	Cat	Pig	Ox	Sheep Goat	Horse
Small intestine	1.8 – 4.8	.8 – 1.3	16.0 – 21.0	27.0 – 49.0	18.0 – 35.0	19.0 – 30.0
Duodenum	.2 – .6	.1 – .12	.7 – .95	.9 – 1.2	.6 – 1.2	1.0 – 1.5
Jejunum } Ileum	1.6 – 4.2	.7 – 1.2	15.0 – 20.0	26.0 – 48.0	17.5 – 34.0	17.0 – 28.0 (Jejunum); .7 – .8 (Ileum)
Large intestine	.28 – .9	.2 – .45	3.5 – 6.0	6.5 – 14.0	4.0 – 8.0	6.0 – 9.0
Cecum	.08 – .3	.02 – .04	.3 – .4	.5 – .7	.25 – .42	.8 – 1.3
Colon and rectum	.2 – .6	.2 – .4	3.0 – 5.8	6.0 – 13.0	3.5 – 7.5	5.5 – 8.0 *
Total	2.0 – 5.7 (7.0) **	1.0 – 1.8	20.0 – 27.0	33.0 – 63.0	22.0 – 43.0	25.0 – 39.0

* The length of the ascending (great) colon of the horse is considered to be 3–4 m., while that of the descending (small) colon is 2.5–4 m.

** Maximum length

Generally, the length of the intestinal tract is considered to be 5 times the body length in the dog, 15 times in the pig, 20 times in the ox, 25 times in the small ruminants, and 10 times in the horse. These figures show that the carnivores have much shorter intestines than the herbivores or the omnivorous pig, except for the horse, which although a herbivore has a relatively short intestinal tract, shorter in fact than that of the pig.

Diameter alone cannot be used to distinguish the small intestine from the large intestine with certainty, because in some domestic mammals, portions of the large intestine are no wider than the small intestine. The most reliable criterion to use is the structure of the mucous membrane, which bears villi in the small intestine and none in the large intestine. It is also possible, of course, to distinguish the various intestinal segments by their shape, by the length and the mode of attachment of their mesenteries, and by their blood supply.

The **duodenum, jejunum,** and **ileum,** the three segments of the small intestine, can be distinguished by the course and position of each in the abdominal cavity, by their relations to other organs, and by their peritoneal attachments and blood supply. The **cecum** is also easily recognized. The segments of the colon that follow the cecum are named according to their relatively simple topography in man and the carnivores. In these species the colon first passes cranially on the right side of the abdominal cavity as the **ascending colon** (144/*F*, *F'*), turns to the left and crosses the median plane cranial to the cranial mesenteric artery as the **transverse colon** (*G*), and then descends on the left side as the **descending colon** (*H*). In man follows a short **sigmoid* colon,** which is followed by the rectum. In all domestic mammals the transverse colon is always found passing from right to left cranial to the cranial mesenteric artery (144—147/*a*).

In the pig, ruminants, and horse, despite marked specialization of the colon; ascending, transverse, and descending parts can also be recognized, and developmental studies have shown them to be homologous to the corresponding segments in the less specialized carnivores. The **ascending colon** is much longer than that of the carnivores, however, and is considerably modified. In the pig it forms a cone-shaped coil (145/*F*, *F'*), in the ruminants it forms a flat, disc-shaped coil (146/*F*, *F'*), and in the horse it is doubled on itself twice and forms a large horseshoe-shaped loop (147/*F*, *F'*). Because of these variations, the ascending colon in these species occupies different parts of the abdominal cavity and has different relations. The **transverse colon,** however, retains its typical position cranial to the cranial mesenteric artery. The **descending colon** of the pig, like that of the carnivores, follows a straight course before it joins the rectum (144, 145). In the ruminants it presents a sigmoid flexure (146/*16*) at the pelvic inlet. In the horse, however, the descending colon is greatly elongated and suspended from the roof of the abdominal cavity by a long mesentery. The **rectum** (144—147/*J*) is the straight terminal part of the alimentary canal in the pelvic cavity. It is followed by the short **anal canal** which surrounds the **anus.**

Small Intestine

The small intestine begins at the pylorus and terminates at the cecocolic junction. It consists of duodenum, jejunum, and ileum.

The **DUODENUM**** is the first part of the small intestine and extends from the pylorus to the beginning of the jejunum. Its mesentery, the mesoduodenum, is relatively short, except in the carnivores. Two flexures divide the duodenum into three parts. The **cranial part** (144—147/*1*) passes to the right along the visceral surface of the liver and ends at the **cranial flexure** (*2*). The **descending duodenum** (*3*) runs caudally from the cranial flexure toward the right kidney. Caudal to the right kidney is the **caudal flexure** (*4*) which turns to the left and cranially. This is followed in the vicinity of the left kidney by the **ascending duodenum** (*5*) which passes cranially and, as its mesoduodenum becomes longer, turns ventrally at the **duodenojejunal flexure** (*6*) to be continued by the jejunum. At the duodenojejunal flexure the duodenum is attached to the descending colon by the **duodenocolic fold** (*21*).

* From Gr., S-shaped or, occasionally, C-shaped.

** From L. *duodeni,* twelve each; because in man the duodenum was thought to have the length of twelve fingers placed side by side.

The descending and ascending parts of the duodenum (184/*g*—*g'''*) form a U-shaped loop around the caudal aspect of the root of the mesentery and the cranial mesenteric artery (*f*). The cranial part of the duodenum is closely related to the liver and pancreas (248/*k*, *l*), and forms a **sigmoid loop** (ansa sigmoidea, 146/*1'*) in the horse, ruminants, and pig. It is attached to the liver by means of the **hepatoduodenal ligament** and receives the bile duct from the liver and the pancreatic ducts from the pancreas.

The **JEJUNUM*** (144—147/*C*) is the longest part of the small intestine. When it is opened during dissection, it is usually empty or contains only small amounts of liquid ingesta. The jejunum begins at the duodenojejunal flexure at the cranial end of the duodenocolic fold. Its mesentery is long and allows the jejunum great range, especially in the carnivores and horse (144, 147/*17*). The particularly long jejunum of the ruminants and pig is tightly coiled along the edge of the sheetlike mesentery to the side of which the coiled ascending colon is attached (146).

In the carnivores, the mass of jejunal loops occupies the ventral part of the abdominal cavity and, covered by the greater omentum, lies against the lateral and ventral abdominal walls (179); the rest of the intestines is crowded against the roof of the abdominal cavity. The jejunum of the pig is found mainly in the ventral part of the right half of the abdominal cavity, but may extend along the floor into the left half and lie ventral to the coiled ascending colon and cecum (195, 196). In the ruminants, the large stomach fills the left half of the abdominal cavity and displaces the intestines entirely to the right. The disc-shaped ascending colon is sagittally placed and surrounded on its cranial, ventral, and caudal aspects by the tightly coiled jejunum. In the horse, because of the great length of the mesentery, the large loops of the jejunum are not confined to a particular area of the abdominal cavity as much as in the other species, although they are found mostly in the left dorsal quadrant (257).

The **ILEUM** (144—147/*D*) is the short terminal part of the small intestine and forms the connection to the large intestine. It is suspended by the caudal part of the mesentery (meso-ileum) and is attached, in addition, to the cecum by the **ileocecal fold** (*18*). This fold arises from the antimesenteric surface of the ileum, and its proximal extent is often considered to mark the jejuno-ileal junction. The ileum terminates at the cecocolic junction of the large intestine forming the **ileal orifice** (*7*). The anatomy of the terminal part of the ileum suggests that the junction of ileum and large intestine is not only an anatomical, but also an important functional division of the alimentary canal.

Large Intestine

The large intestine consists of the cecum, the colon with its three subdivisions, and the rectum.

The **CECUM**** (144—147/*E*) is the initial blind part of the large intestine, and is joined to the colon at the ileal orifice. It is shortest in the cat, and is increasingly longer in the dog, pig, ruminants, and horse, where it is a large, elongated pouch (254). The vermiform appendix of man is absent on the ceca of the domestic species considered here; although the hare and rabbit, for instance, have an appendix. In the carnivores (184/*k*; 185/*g*), ruminants (234/*r*; 236/*o*; 238/*k*), and horse (258/*t*, *t'*, *t''*), the cecum lies on the right side of the abdominal cavity; in the pig (195/*m*), it is on the left. Except in the pig, the attached part (base) of the cecum is usually found high in the right flank; the position of the free part, however, varies with the species and will be described in the special chapters.

The differences in the **COLON** (144—147/*F*, *G*, *H*) found in the various species have already been pointed out. Its simple arrangement in man gave rise to dividing it into an ascending colon passing cranially on the right, a transverse colon passing from right to left in front of the cranial mesenteric artery, and a descending colon passing caudally on the left. The course

---

* L. *jejunus*, hungry, empty.

** From L. *cecus*, blind.

and topography of the colon in the carnivores (184/*l, l', l''*) resemble the simple human arrangement. The position and direction of the transverse colon is the same in all domestic mammals.

The **ascending colon** of the pig, ruminants, and horse is greatly elongated and modified. In the pig and ruminants it is coiled on itself and forms the **spiral loop** (ansa spiralis) of the colon, which consists of **centripetal turns** (gyri centripetales, 145, 146/*12*) that spiral toward the center of the coil, a **central flexure** (*13*), and **centrifugal turns** (gyri centrifugales, *14*) that spiral away from the center of the coil. In the pig, the spiral loop is between the cecum and the transverse colon, and its turns are piled up to form a thick cone (145/*F, F'*; 192/*e, f, g*). In the ruminants, however, the spiral loop is preceded by a **proximal loop** (ansa proximalis, 146/*11*) and followed by a **distal loop** (ansa distalis, *15*), and its turns are arranged more or less in a single plane, forming a disc. The **transverse** and **descending colons** of the ruminants and pig resemble the simple arrangement found in man and the carnivores.

The **ascending colon** of the horse (147/*F, F'*) is not only elongated, but also for the most part has a greatly increased diameter, and is therefore also called the **great colon.** This large segment of gut doubles on itself twice to form two large U-shaped loops which occupy the ventral half of the abdominal cavity. These loops are open caudally and lie in dorsal planes one on top of the other. With two sections of the great colon, then, on the right side and two on the left, its parts can be designated as follows. The first part, the **right ventral colon,** runs cranially from the cecum and ends at the **sternal flexure** (*12'*). The second part, the **left ventral colon,** passes caudally from the sternal flexure on the left side of the abdomen and doubles back on itself in front of the pelvic inlet, forming the **pelvic flexure** (*13*). The third part, the **left dorsal colon,** passes cranially dorsal to the left ventral colon to the **diaphragmatic flexure** (*14'*) continuing on the right as the short but very wide fourth part, the **right dorsal colon.** This is followed by the **transverse colon** (*G*), which passes from right to left in front of the cranial mesenteric artery. This continues on the left side as the **descending colon** (*H*), also known as the **small colon,** which is very long and is suspended by the long descending mesocolon. The large coils of the descending colon are found in the left dorsal quadrant of the abdominal cavity.

The cecum and colon of the horse and pig have a peculiar sacculated appearance. In these organs, the longitudinal muscle coat is concentrated into flat longitudinal **bands** (teniae, 193, 249), between which the intestinal wall is gathered into rows of **sacculations** (haustra). Between adjacent sacculations, **semilunar folds** project into the interior of the gut and increase the internal surface area (255).

The **RECTUM*** (144—147/*J*) is a straight piece of gut which continues from the descending colon into the pelvic cavity. Before ending at the short anal canal, it becomes enlarged, forming the **ampulla recti**. This is very prominent in the horse (559/*24*), but absent in the cat, sheep, and goat.

Anal Canal

The anal canal is the short terminal portion of the digestive tract; the orifice it surrounds is the **anus** (526—529/*a*). The mucous membrane of the anal canal is covered with stratified squamous epithelium and meets the rectal mucosa at the **anorectal line,** and the pigmented skin, which surrounds the anus, at the **anocutaneous line.** The structure of the anal wall differs from species to species and will be considered in the special chapters.

External and internal sphincters surround the anus and keep the opening closed. The **internal anal sphincter** is a thickened continuation of the circular smooth muscle coat of the rectum. The striated **external anal sphincter** (559/*26*) arises from the caudal vertebrae and lies superficial to the internal sphincter. Most of it encircles the anus, but some of its fibers pass the anus laterally and join urogenital muscles ventral to the anus, a portion of these decussating below the anus before reaching the urogenital muscles. Immediately cranial to the

* L. *rectus*, straight.

external anal sphincter is the **retractor penis** in the male, or the **retractor clitoridis** (*26'*) in the female except in the bitch. This muscle arises bilaterally from the caudal vertebrae, passes the rectum on each side, and continues ventrally to the genital organs (490). Its rectal part* surrounds the terminal portion of the rectum ventrally and thus suspends it and the anal canal from the caudal vertebrae. The **levator ani** (554/*25*) originates from the sacroischiatic ligament or the ischiatic spine, passes caudally, and blends into the external anal sphincter.

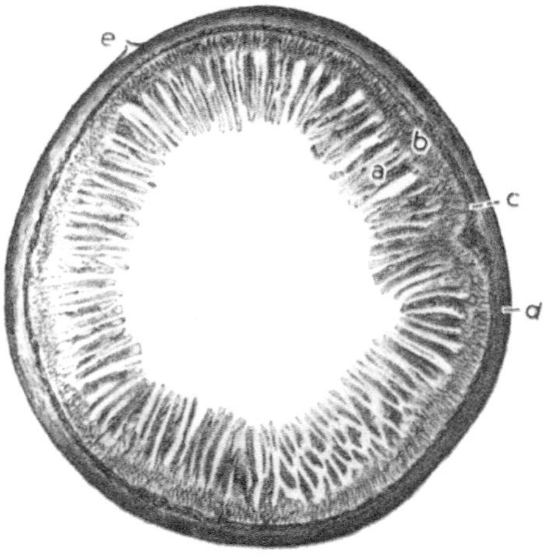

Fig. 148

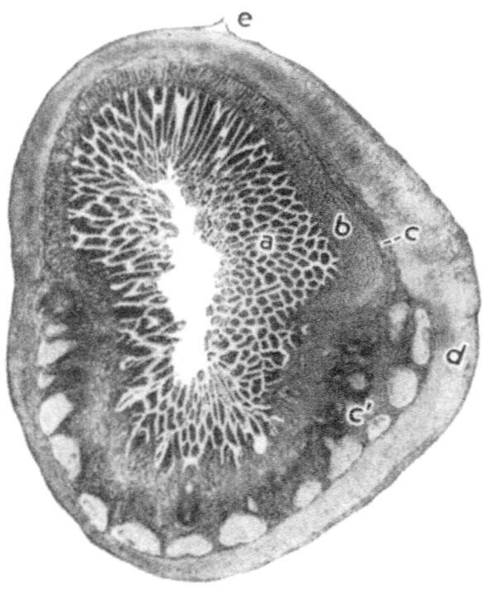

Fig. 149

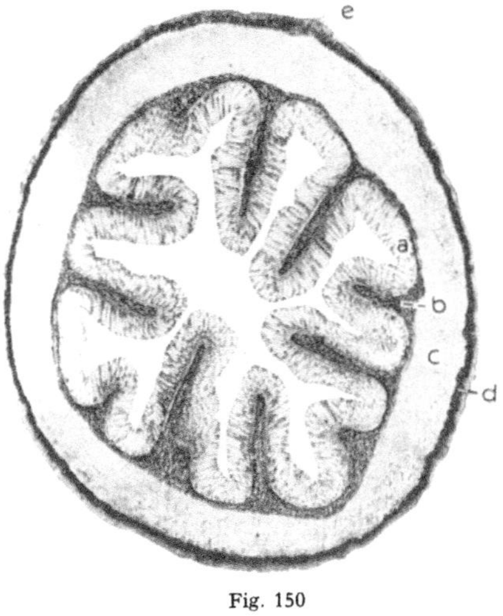

Fig. 150

Fig. 148. Cross section of the jejunum of the cat. Microphotograph. (For details see fig. 151.)
a, b, c Mucous membrane; *a* Intestinal villi; *b* Intestinal glands; *c* Muscularis mucosae and submucosa; *d* Muscular coat covered externally with serosa; *e* Attachment of the mesentery

Fig. 149. Cross section of the ileum of the cat. Microphotograph.
a, b Mucous membrane; *a* Intestinal villi; *b* Intestinal glands; *c* Submucosa; *c'* Aggregate lymph nodules in the propria and submucosa; *d* Muscular coat, covered externally with serosa; *e* Attachment of the mesentery

Fig. 150. Cross section of the colon of the dog. Microphotograph. (For details see fig. 152.)
a Mucous membrane forming high transient folds; *b* Submucosa; *c* Circular, *d* longitudinal muscle layers, the latter covered with serosa; *e* Attachment of the mesocolon

STRUCTURE OF THE INTESTINAL WALL. The function of the intestines is to break down the ingested food by chemical and enzymatic action and to absorb the released nutrients into the body. The enzymes required for these assimilative changes are produced by the pancreas and liver, and by glands present in the wall of the intestine itself. A flora of bacteria and protozoa in the stomach of ruminants and the large intestine of the horse, plays an important role in the initial breakdown and assimilation of the coarse, fibrous plant food of the herbivores.

The peristaltic action of the muscular intestinal wall mixes the ingesta with the secretions of the digestive glands, propels the ingesta distally, and eliminates the unabsorbable residue as the feces. In general, the digestion and absorption of nutrients takes place in the small intestine, while the residue of wastes is collected, thickened, and stored in the large intestine

---

* Formerly known as the suspensory ligament of the anus.

prior to elimination. Digestion and absorption may, however, also occur in the large intestine of the horse; and the large intestines of the other domestic mammals have also been credited with some capacity for absorption of nutrients.

The intestinal wall consists of a mucous membrane, a muscular coat, and a serous coat. The structural variations encountered especially in the mucous membrane reflect the functions of the different parts of the gut. The **mucous membrane** (148, 149/*a*, *b*, *c*; 150/*a*, *b*) forms transient folds of varying height and number as it adjusts to the continual changes of the intestinal lumen. Permanent mucosal folds are also present, for instance in the large intestine of the horse and pig. The surface epithelium throughout the entire intestinal

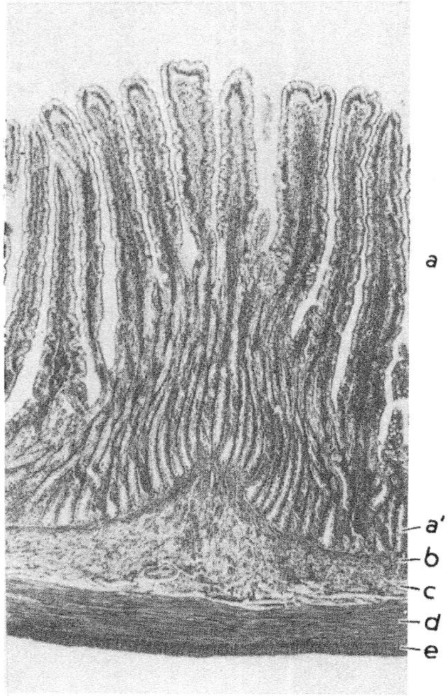

Fig. 151. Section of the jejunum of the cat. Microphotograph. (Higher magnification of fig. 148.)
a, *a'* Lamina propria; *a* Intestinal villi; *a'* Intestinal glands; *b* Muscularis mucosae; *c* Submucosa; *d* Circular, *e* longitudinal muscle layers, the latter covered with serosa

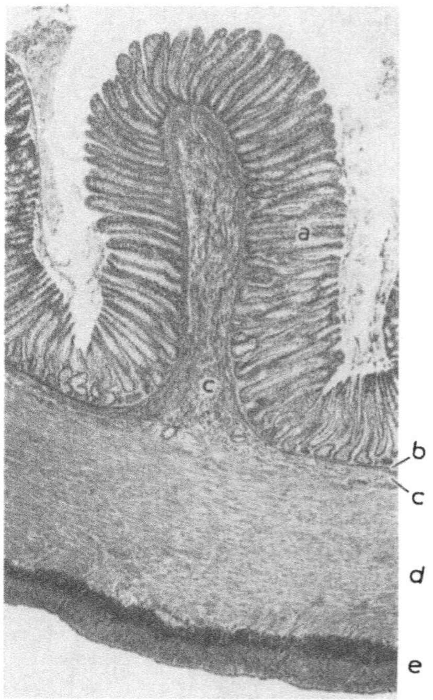

Fig. 152. Section of the colon of the dog. Microphotograph. (Higher magnification of fig. 150.)
a Lamina propria with intestinal glands containing many goblet cells; *b* Muscularis mucosae; *c*, *c* Submucosa; *d* Circular, *e* longitudinal muscle layers, the latter covered with serosa

tract consists of a single layer of tall columnar cells with a distinct cuticular border (151). These cells absorb the released nutrients and mediate their passage from the intestinal lumen to the blood vessels and lymphatics of the gut wall. Scattered among the columnar cells are mucus-secreting goblet cells, which are especially numerous in the large intestine (152). The mucus secreted by these cells forms a protective and lubricating layer on the surface of the epithelium, and is a normal, though minor, constituent of the feces; however, in certain diseases of the bowel, mucus may be eliminated in great quantities. Tubular **intestinal glands** (151/*a'*; 152/*a*) are present in the lamina propria throughout the length of the intestines and constitute an enormous extension of the secretory surface of the mucosa. Their openings on the surface of the epithelium are visible with a magnifying glass.

In addition to the intestinal glands, **duodenal glands** are found in the submucosa of the proximal part of the small intestine, from the pylorus distally for 1.5—2 cm. in the carnivores, 20—25 cm. in the goat, 60—70 cm. in the sheep, 3—5 m. in the pig, 4—5 m. in the ox, and 5—6 m. in the horse, although there are individual variations in the extent of this zone.

The small intestine has **intestinal villi** (151/*a*), minute (.5—1 mm.), roughly conical projections from the mucosal surface into the lumen that give the mucosa of the small intestine its velvety appearance. They increase the absorptive surface enormously; and it has been estimated that the dog, for instance, has about four million villi. They vary somewhat from species to species, but in general, each villus consists of a core that is covered by the intestinal epithelium. The core, or stroma, is made up of reticular connective tissue that contains smooth muscle cells, and blood and lymph capillaries. The intestinal villi function like minute pumps: by intermittent contraction they move blood and lymph, containing absorbed nutrients, from the villi toward larger vessels in the intestinal wall.

The **lamina muscularis mucosae** (*b*), which by its contraction produces the transient folds of the mucous membrane, consists of two thin layers. Internal to the muscularis mucosae of the carnivores is a **stratum compactum** which is peculiar to them. The **submucosa**(*c*) contains blood vascular and lymphatic networks and nerves. It is loosely arranged and thus facilitates the formation of mucosal folds.

The **lymphatic tissue** of the intestinal wall is of particular importance in the body defense mechanism. Scattered lymphocytes are present in large numbers in the lamina propria and between the epithelial cells, and are the first line of defense against microorganisms entering the intestinal wall. They may, like other leucocytes, pass into the intestinal lumen. Accumulations of lymphocytes, **solitary lymph nodules** (lymphonoduli solitarii), visible on the surface as minute tubercles, are embedded in the propria but may extend through the muscularis mucosae into the submucosa. **Patches of aggregate nodules*** (lymphonoduli aggregati, 149/*c'*; 233/*2*) occur primarily on the antimesenteric wall of the gut and are visible on the mucosal surface as irregularly raised plaques or bands ranging in length from a few centimeters to several meters. In the pig, for example, patches up to 3.5 m. long have been measured. The number, size, and shape of the two types of nodules vary with the species, age, intestinal region, and diet. Solitary nodules are more common in the large intestine, while aggregate nodules are more common in the small intestine.

The **muscular coat** (151, 152/*d*, *e*) consists of a thick, inner circular layer and a thin, outer longitudinal layer, with a thin lamina of connective tissue separating the two. In the pig and horse most of the longitudinal fibers combine to form the **bands** (teniae) of the large intestine (see p. 110).

BLOOD VESSELS, LYMPHATICS, AND INNERVATION. The **arteries** of the small intestine come mainly from the cranial mesenteric artery (144—147/*a*); those of the proximal part of the duodenum come also from the celiac artery. The blood for the large intestine comes from both cranial and caudal mesenteric (*l*) arteries, the rectum receiving its blood also from the internal iliac arteries. The **veins** of the entire intestinal tract go to the portal vein, but blood from the rectum is returned to the caudal vena cava.

The intestinal **lymphatics** pass to the following **lymph nodes:** those of the duodenum to the hepatic, pancreatico-duodenal, cranial mesenteric, and cecal; those of the jejunum go to the jejunal lymph nodes; those of the ileum to the jejunal, cecal, and colic lymph nodes; and those of the cecum, colon, rectum, and anal canal go to the jejunal, cecal, colic, and anorectal lymph nodes (230).

The muscular coat and the glands of the intestines receive their **innervation** from the sympathetic and parasympathetic parts of the autonomic nervous system. In general, the sympathetic part retards, while the parasympathetic part accelerates the activity of the gut. Sympathetic fibers to the various segments of intestine pass through the celiac and cranial mesenteric ganglia and the cranial mesenteric plexus, through the caudal mesenteric ganglion and plexus, and through the lumbar ganglia and cranial rectal plexus. The parasympathetic fibers for the intestine originate in both cranial and sacral regions. Those from the cranial region pass along the vagus to the abdominal ganglia, and those of sacral origin pass to the pelvic ganglia. After synapsing, their postganglionic fibers pass along the arteries to the intestines.

The **intramural nervous system** of the intestines (plexus entericus) consists of subserous, myenteric, and submucosal plexuses, which, as their names imply, lie below the serosa, between the two muscle layers, and in the submucosa, respectively. Ganglion cells are associated with these plexuses. Although they are under the overriding influence of the parasympathetic and sympathetic systems, the intramural nerve plexuses are thought to be responsible for the largely independent muscular and secretory activity of the gut.

---

* Formerly Peyer's patches.

Liver

(153—163, 188, 189)

The duct system of the liver (hepar) arises in the embryo as an outgrowth of the glandular mucosa of the primitive duodenum. Thus the enormous amount of glandular tissue that constitutes this organ, although physically removed from the gut, remains associated with it by a duct. The secretion of the liver, the bile, passes through this duct into the lumen of the duodenum.

The liver is the largest gland of the body and its size reflects the multiplicity of its functions. Only the more important are listed here. The most apparent function, the secretion of bile, has already been mentioned. In embryonic life, the liver develops hemopoietic* centers and is an important blood-forming organ of the fetus, but it does not continue this function postnatally. In the newborn animal it still occupies a considerable portion of the abdominal cavity, but becomes relatively smaller after birth. In young dogs, for instance, the liver represents 40—50 gm. of one kilogram of body weight; in older dogs, the ratio is only 20 gm. per kilogram. Despite the reduction in relative weight, it remains the largest gland of the body.

An important function of the liver is the storage of glycogen, which it synthesizes from the carbohydrates it receives from the portal blood. It can also store fats and small amounts of protein. It converts end products of protein catabolism to urea and uric acid, which are discharged into the blood stream and then removed by the kidneys. While active as a hemopoietic organ before birth, postnatally it removes waste products resulting from the breakdown of red blood cells in the spleen from the blood. These wastes are the end products of the hemoglobin catabolism and are discharged in the bile as bile pigments. The liver is also capable of extracting harmful substances from the blood and detoxifying them.

The **COLOR** of the liver depends on factors such as the amount of blood it contains, the species, the age of the animal, but primarily on the nutritional state of the animal and the type of food eaten prior to death. In the pig, ox, sheep, and horse, the liver is reddish brown but becomes brown when bled out. In the dog and cat it is reddish brown. Suckling and pregnant animals, and those on a fattening diet have a yellowish brown liver because of the presence of fats, whereas emaciated or starving animals have a dark reddish brown liver.

SIZE and **WEIGHT** of the liver vary greatly. Because it stores fats and glycogen, it weighs more in well-fed animals than in emaciated ones. The weight of the liver always decreases with age. The following data may serve as a guide when evaluating liver weight of domestic mammals.

Table 9. Absolute and relative liver weights of domestic mammals.

Animal	*Weight of Liver*	*Per cent of Body Weight*
Dog	127 gm.—1.35 kg.	1.33—5.95
Cat	average of 68.5 gm.	2.46
Pig	1—2.5 kg.	1.7
Ox	3—10 kg.	1.03—1.54
Calves (up to 3 mos. old)	500 gm.—2 kg.	1.9
Sheep	500 gm.—1.26 kg.	1.45
Horse	2.5—7 kg. (average 5 kg.)	1.2—1.5
Old horses	2.5—3 kg.	

* From Gr., blood-forming.

The serosal covering gives the liver a smooth and glossy appearance. The liver itself consists of a great number of small **lobules,** each not larger than 1.5—2 mm. The lobules are visible to the naked eye only when surrounded by a noticeable amount of interlobular connective tissue, as they are in the pig (201, 202). The granular appearance of the freshly ruptured surface is further indication that the liver is generally composed of lobules.

SHAPE. The liver is firm, yet somewhat elastic to the touch and quite friable. When in situ, it molds itself readily to the neighboring structures, but flattens out when removed from the carcass in the fresh state. The liver lies in the intrathoracic portion of the abdominal cavity. Its **diaphragmatic surface** (189) is convex and lies against the concavity of the diaphragm. Its **visceral surface** (188) faces mostly caudally and is related to the stomach, duodenum, colon, jejunum, and to the right kidney. These structures indent the liver and, according to the organ involved, produce such impressions as the **esophageal notch** (impressio esophagea, 260/*f*), the **gastric impression** (*11*), the **duodenal impression**, the **colic impression** (*12'*), the **pancreatic impression** (*14*), and the **renal impression** (*13*). These impressions disappear when the liver is removed in the fresh state, but remain if the liver was embalmed (hardened) in situ.

The border of the liver between the esophageal notch and the renal impression is thick and rounded, while along the remainder of the periphery it is sharp and thin, an important characteristic in distinguishing a swollen from a normal liver. The rounded border is crossed by the esophagus and the caudal vena cava. By orienting the livers of the domestic mammals as shown in Figures 153—160; i. e., with the rounded border up (or dorsal), directional names can be assigned to the borders. The rounded border can then be called the **dorsal border** (margo dorsalis), and the sharp border can be divided into **left, ventral,** and **right borders** (margo sinister, ventralis, and dexter). Only in the carnivores and the pig is the liver oriented in this way in the body. In the ruminants, the liver is displaced entirely to the right of the abdominal cavity (243). It is oriented in such a way that the rounded border lies in the median plane, and that the sharp border is directed mainly to the right, but ventral to the esophageal notch (*m*) also to the left, and in the region of the right triangular ligament (*g*) also dorsally. In the horse, the liver is placed obliquely (262). Its rounded border crosses the median plane and faces dorsolaterally and to the left, while its sharp border is directed chiefly ventrolaterally and to the right, and ventral to the esophageal notch also to the left. Because of these differences in orientation, the directional designations of the liver borders do not pertain to the same border throughout the domestic species considered here. For instance, the ventral border of the dog's or pig's liver is not formed by the same lobes as the border that is directed ventrally in the ox.

Because of the differences in shape and position of the liver in domestic mammals, it is important to recognize the homology of its parts and to adhere to a meaningful nomenclature of its lobes. The landmarks established for the division of the human liver are suitable for this purpose.

An imaginary line on the visceral surface of the liver from the esophageal notch (153—160/*6*) to the notch for the round ligament (*5*) on the opposite border separates the **left lobe** (*a, a'*) from the rest of the liver. Similarly, a line from the point where the caudal vena cava (*4*) crosses the rounded border to the fossa for the gall bladder (*e*) on the opposite border separates the **right lobe** (*b, b'*) from the rest of the liver. Between these lines, which in the human liver are represented by two sagittal grooves, is the **quadrate lobe** (*c*) ventral to the hepatic porta, and the **caudate lobe** (*d, d'*) dorsal to the porta. (The porta is the depression on the visceral surface of the liver through which the portal vein, hepatic artery, and hepatic ducts pass.)

The left lobe of the liver may be subdivided into **left medial** (*a'*) and **left lateral** (*a*) **lobes,** the right lobe into **right medial** (*b'*) and **right lateral** (*b*) **lobes.** The caudate lobe dorsal to the porta has **caudate** (*d*) and **papillary** (*d'*) **processes.**

The liver of the carnivores (153, 157, 188) consists of a left lateral, a left medial, and a right lateral and a right medial lobe, a quadrate lobe (*c*), and a caudate lobe with a caudate process (*d*) on the right and a papillary process (*d'*) on the left. The pig's liver (154, 158, 201) is similarly lobated, but there is no papillary process, and its quadrate lobe (*c*) is small and short and does not reach the ventral border. The liver of the ruminants (155, 159, 239—242) is a compact organ and is not divided by interlobar notches. It consists of undivided right and left lobes (*a*, *b*), a quadrate lobe (*c*) between the notch for the round ligament and the gall bladder, and a caudate lobe with a small papillary process (*d'*) and an exceptionally large caudate process (*d*). In the horse (156, 160, 260), the left lobe (*a*, *a'*) is subdivided into left medial and lateral lobes, while the right lobe (*b*) remains undivided; the quadrate lobe (*c*) presents several notches on its border and is separated from the right lobe by a deep interlobar notch; and the caudate lobe has a caudate process (*d*) but no papillary process.

BLOOD VESSELS, LYMPHATICS, AND INNERVATION. The **portal vein** (157—160/*2*) enters the liver at the **hepatic porta** carrying venous blood, rich in freshly absorbed nutrients, from the intestines, and blood from the stomach, pancreas, and spleen. This blood is the **functional blood supply** of the liver and constitutes the raw material for

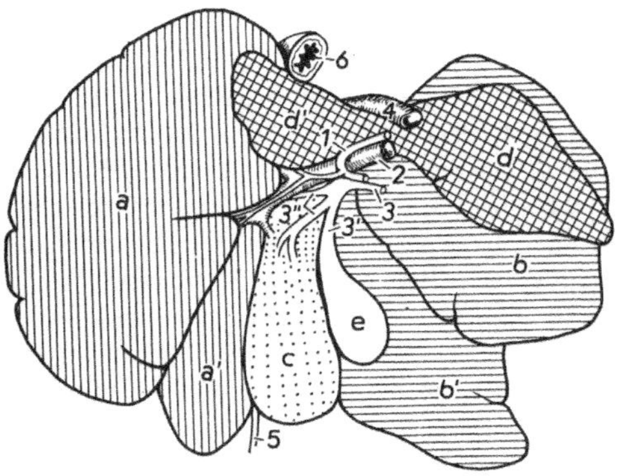

Fig. 153 (Dog)

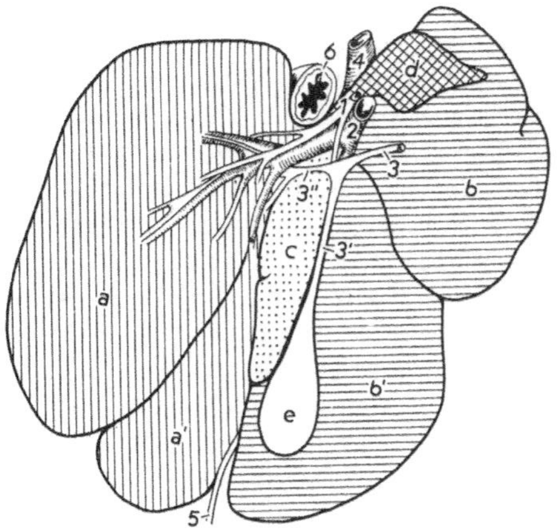

Fig. 154 (Pig)

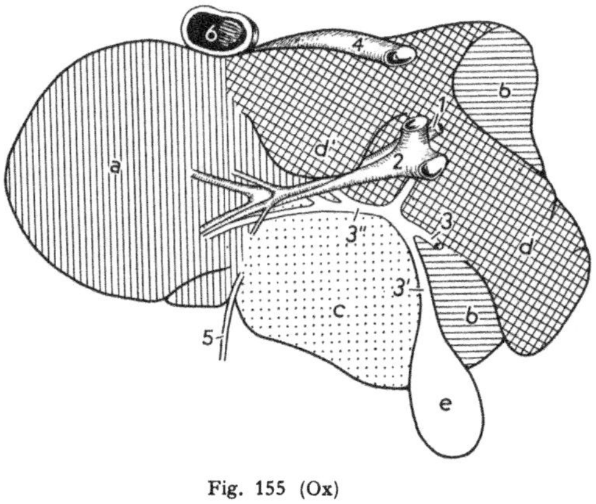

Fig. 155 (Ox)

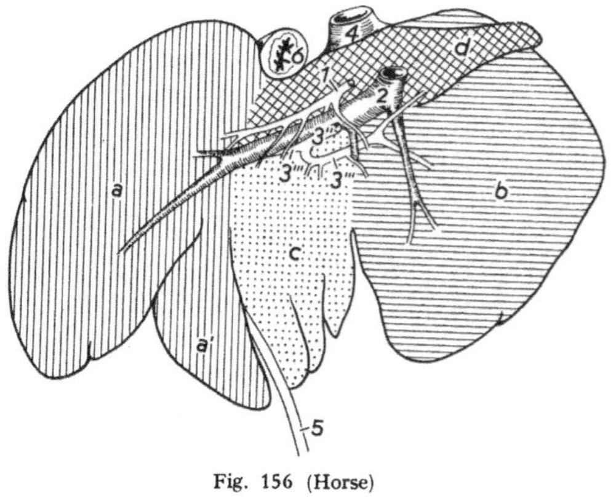

Fig. 156 (Horse)

its metabolic functions. The **nutritional blood supply** also enters at the porta. This is arterial blood brought into the liver by the hepatic branches (*1*) of the hepatic artery and is intended solely for the nourishment of the liver itself. The bile-carrying **hepatic ducts** and the deep **lymphatics** leave at the porta. The deep lymphatics pass to the hepatic lymph nodes near the porta. The superficial lymphatics from the serous coat of the liver pass also to the caudal mediastinal, phrenic, and sternal lymph nodes. The **innervation** of the liver is by branches of the vagus and sympathetic nerves, which reach it from the celiac ganglion and also enter at the porta. The **hepatic veins** (261/*3*) open directly into the caudal vena cava where it is embedded on the diaphragmatic surface of the liver.

BILE PASSAGES. Branches of the **right and left hepatic ducts** carry bile from all parts of the liver to the porta. Here right and left ducts unite and form the **common hepatic duct**

(157—160/*3''*), which, after receiving the **cystic duct** (*3'*) from the gall bladder, becomes the **bile duct** (ductus choledochus, *3*) and passes in the hepatoduodenal ligament to the duodenum which it enters at the **major duodenal papilla** (164, 167/*1'*). Despite the absence of a gall bladder and cystic duct in the horse, the wider, terminal portion of the common hepatic duct is usually called the bile duct. Species differences in the arrangement of the extra-hepatic bile passages and the termination of the bile duct will be described in the special chapters. The wall of the hepatic, cystic, and bile ducts consists of a tunica serosa, a muscular coat, and a mucous membrane. The **mucous membrane** is of the glandular type and is covered with a high, simple columnar epithelium which in the ruminants contains goblet cells. The **glands,** mostly mucus-producing, are especially numerous in the ox but parse in the pig and horse.

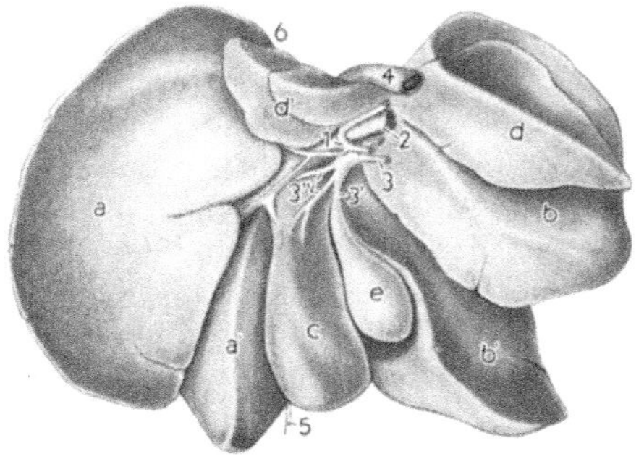

Fig. 157 (Dog)

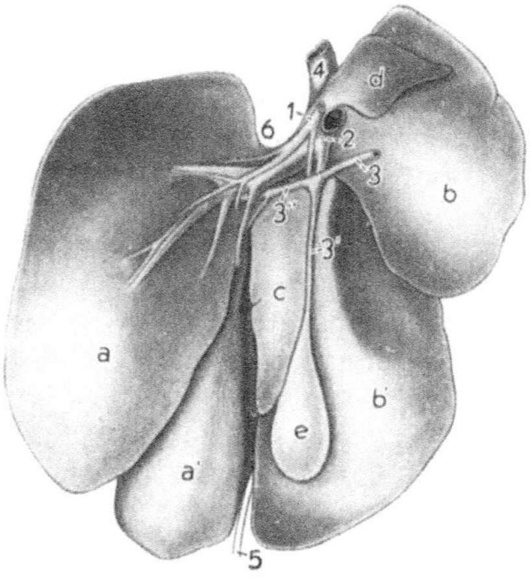

Fig. 158 (Pig)

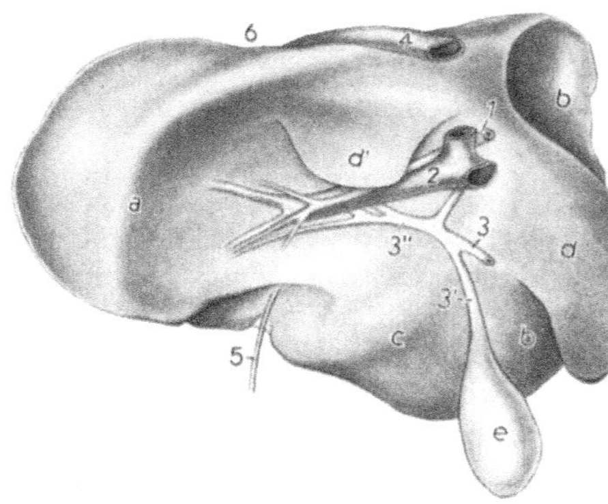

Fig. 159 (Ox)

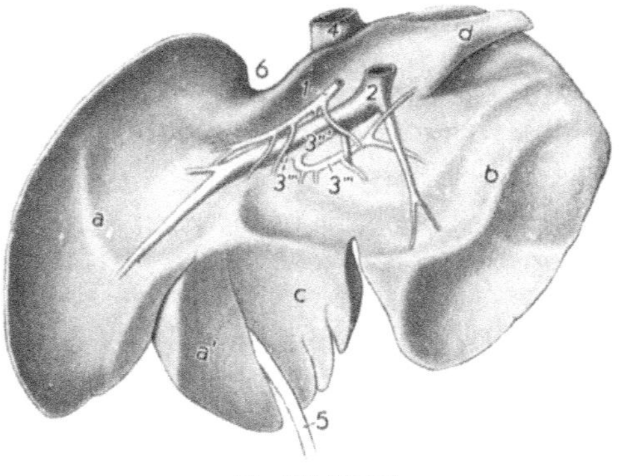

Fig. 160 (Horse)

Figs. 153—160. The liver of domestic mammals. Visceral surface. (Fig. 153—156 schematic.)

a, a' Left lobe (dog, pig, horse); *a* Left lateral lobe, left lobe (ox); *a'* Left medial lobe; *b, b'* Right lobe (dog, pig); *b* Right lateral lobe, right lobe (ox, horse); *b'* Right medial lobe; *c* Quadrate lobe; *d, d'* Caudate lobe; *d* Caudate process; *d'* Papillary process; *e* Gall bladder (except in the horse)

1—3 Blood vessels and bile passages at the hepatic porta; *1* Hepatic branch of hepatic artery; *2* Portal vein; *3* Bile duct (except in the horse); *3'* Cystic duct (except in the horse); *3''* Common hepatic duct (pig, horse), left hepatic duct(s) (dog, ox); *3'''* Left and right hepatic ducts (horse); *4* Caudal vena cava; *5* Round ligament (vestigial umbilical vein); *6* Esophagus or esophageal notch

The pear-shaped **GALL BLADDER** (vesica fellea, 157—159/*e*) lies in a **fossa** on the visceral surface of the liver with which it is firmly united. The gall bladder stores the bile temporarily and discharges it into the duodenum when food enters it from the stomach. It protrudes from the border of the liver only in the ruminants. The unattached portion of the gall bladder is covered with serosa. Deep to the serosa is a thin muscular coat followed by a folded

mucous membrane. The latter bears a high, simple columnar epithelium which contains goblet cells in the ox. Mucous and serous **glands** are abundant in the gall bladder mucosa of ruminants but sparse or absent in the carnivores and pig. Bile enters and leaves the gall bladder through the cystic duct. In the carnivores, ox, and sheep, part of the bile enters the gall bladder directly through small **hepatocystic ducts,** which penetrate the wall of the gall bladder where it is united with the liver.

LIGAMENTS OF THE LIVER. When in situ, the liver is encased between the diaphragm cranially, to which it is closely applied, and the viscera caudally, and is prevented from sliding on the surface of the diaphragm by the hepatic ligaments.

Since in its embryonic development the liver expands inside the ventral mesogastrium (12), it becomes connected to the stomach by the **lesser omentum** (*b*), which is the chief derivative of the ventral mesogastrium. The lesser omentum consists of the **hepatogastric** and **hepatoduodenal ligaments** (239/*8*, *9*) and extends from the area of the hepatic porta to the lesser curvature of the stomach and the proximal portion of the duodenum. In addition, the visceral surface of the liver is attached to the root of the mesentery by the portal vein. Cranially, the liver is attached to the diaphragm by the caudal vena cava, the **coronary*** **ligament** (189/*8*), which is ventral and lateral to the caudal vena cava at the caval foramen, and by the crescent-shaped **falciform ligament** (*7*), which originates ventrally on the liver in the region of the notch for the round ligament and extends to the sternal part of the diaphragm. In its free dorsocaudal edge, the falciform ligament contains the **round ligament of the liver** (lig. teres hepatis), which is the vestige of the obliterated umbilical vein. In the suckling animal, the umbilical vein is a nonfunctional, thick-walled vessel, which poses a threat (especially in calves) as a route for infection as long as the stump of the umbilical cord has not dried up. The round ligament leaves the liver at the notch for the round ligament and passes caudoventrally toward the abdominal floor, where it can be followed retroperitoneally beyond the caudal extent of the falciform ligament until it reaches the umbilicus. The right and left lobes of the liver are attached to the diaphragm by the **right and left triangular ligaments** (*6*, *5*) respectively, which are continuous medially with the coronary ligament.

Owing to its intrathoracic position, the examination of the liver in the live animal is difficult. In the ox, the liver can be percussed in a narrow zone along the basal border of the lung high in the right eleventh or twelfth intercostal space. Extrathoracic portions of the liver may be palpated in the carnivores and the pig at the costal arch. Habel (1965) states that this is possible in the dog only when the liver is enlarged.

STRUCTURE OF THE LIVER. The liver cells receive nutritive and metabolic substances mainly from the portal blood carried to them by the **blood capillaries.** The bile they secrete is gathered by the **bile canaliculi** and is conducted to the porta by numerous ducts of various orders. Other metabolic products of the hepatic cells or substances temporarily stored by them are returned directly to the blood in the manner of an endocrine gland. There is therefore an intimate relationship between the hepatic cells and the vascular channels of the liver, and it is necessary to understand the closeness of this relationship to understand the microscopic structure of this organ.

The structural elements or units of the liver are the **liver lobules** (161, 162), which are irregularly shaped prisms roughly 1.5—2 mm. long and about 1 mm. in diameter. The **interlobular connective tissue** separating them represents the finest ramifications of the connective tissue stroma of the liver, which consists of a thin external **fibrous coat** (tunica fibrosa) and internal **perivascular fibrous tissue** (capsula fibrosa perivascularis**). The fibrous coat, lying deep to the serosa, invests the organ and at the porta is continuous with the perivascular fibrous tissue which, as its name indicates, follows the vessels into the interior. In the pig, the interlobular connective tissue completely surrounds each lobule and makes it visible to the naked eye on both the external and cut surfaces (161). In the other species, the interlobular tissue is reduced and is evident only where several lobules come together. Inside the lobule is a delicate network of reticular fibers.

---

* L. *corona*, crown; in man this ligament looks like a crown.

** Formerly Glisson's capsule.

The fine terminal branches of the portal vein and of the hepatic artery ramify on the surfaces of the liver lobules. Here the veins give rise to **sinusoidal blood capillaries,** which enter the lobule and form a centrally oriented, delicate network (163/*blue*) ending at the **central vein** (162), which opens into collecting veins (*e*) at the base of the lobule. The **collecting veins,** after uniting with many others of their kind, form the large **hepatic veins,** which empty into the caudal vena cava.

The **hepatic cells,** arranged in plates and cords, lie in the meshes of the intralobular blood capillary network (163). This arrangement allows for maximum contact with the blood capillaries. Arterial blood is brought to the hepatic lobules by the fine terminal branches of the hepatic artery (162/*a*), and is channeled into the sinusoidal blood capillaries at the periphery of the lobules.

Fig. 161. Section of pig liver. Microphotograph. The lobules are surrounded by interlobular connective tissue. In the center of each lobule is the central vein; afferent blood vessels and interlobular ductules are in the interlobular connective tissue

The **bile canaliculi** (163/*green*) have no walls, but are tubelike spaces between adjacent hepatic cells running toward the periphery of the lobule in the center of the plates and cords of liver cells. Near the periphery of the lobule the bile canaliculi join to form short bile ductules which carry the bile into the interlobular space to **interlobular ductules.** The latter are lined with simple cuboidal cells on a basement membrane and unite to form the many branches of the right and left hepatic ducts (162).

The wall of the sinusoidal blood capillaries consists of a syncytial endothelium with few nuclei which is covered externally by a delicate network of reticular fibers. Present in the wall are also **stellate* endothelial cells** (Kupffer cells) which often project into the lumen of the sinusoid. They are part of the **reticulo-endothelial system** and play an important role in the defense against infectious diseases; they are also said to have a mediating function between the blood and the hepatic cells.

The **liver cells** themselves are polygonal in structure and often have two or more nuclei. Glycogen, fat, and protein deposits can be demonstrated in the cytoplasm depending on the functional state of the liver when the sample was taken.

Pancreas

(164—168, 190, 210, 248, 439)

The pancreas develops from the embryonic duodenum by dorsal and ventral budlike primordia. Like the liver, which arises with the ventral pancreatic primordium, the pancreas represents an extension of the glandular mucosa of the duodenum, and remains connected to it by secretory ducts. Either the dorsal or ventral primordium may involute again during development, so that in some species the pancreas has only one duct and is a development of either the dorsal or the ventral primordium, whereas in other species, in which involution did not occur, the pancreas has a double origin and two ducts.

In man, the part of the pancreas that lies in the curve of the proximal half of the duodenum is the head of the pancreas (caput pancreatis). The small projection from the head of the pancreas toward the ascending duodenum is the uncinate process. Continuing from the head of the pancreas to the left is the body (corpus) and then the tail

---

* From L., star-shaped.

(cauda) of the pancreas. Because of the considerable difference between the human pancreas and that of domestic mammals, it is not possible to apply the nomenclature of the human pancreas meaningfully to the pancreas in domestic mammals.

In the domestic mammals, that part of the pancreas that lies against the cranial part of the duodenum is the **body** (corpus, 164—167/*a*); the part that continues from the body to the left is the **left lobe** (lobus sinister, *b*); and the part that continues from the body to the right along the descending duodenum (except in the horse) is the **right lobe** (lobus dexter, *c*).

The body of the pancreas in the carnivores and ruminants is notched (incisura pancreatis) and that of the pig and horse perforated (anulus pancreatis) by the portal vein, which crosses the dorsal border of the pancreas at this point (*3*).

The pancreas of the carnivores (164, 190) forms a U-shaped loop consisting of a centrally placed body and right and left lobes. In the ruminants (166, 210) the body of the pancreas is relatively small and is continued on the left by a wide left lobe and on the right by a long, caudally directed right lobe. The pancreas of the pig (165) consists of an extensive body,

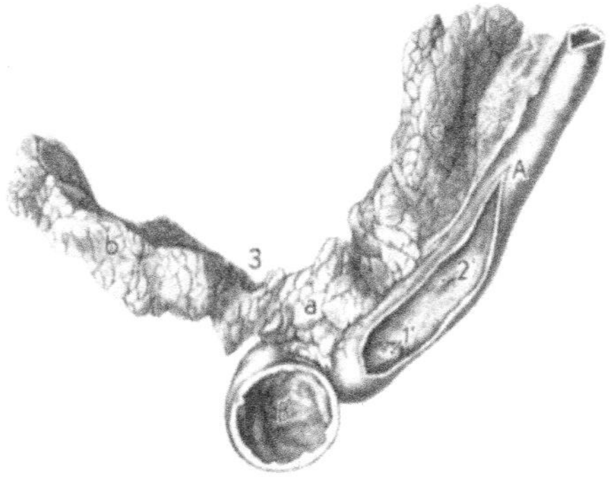

Fig. 164. Pancreas of the dog. Caudoventral aspect. *A* Descending duodenum; *B* Pylorus; *a* Body of pancreas; *b* Left lobe; *c* Right lobe; *1'* Major duodenal papilla with openings of bile and pancreatic ducts; *2'* Minor duodenal papilla with opening of accessory pancreatic duct; *3* Incisura pancreatis

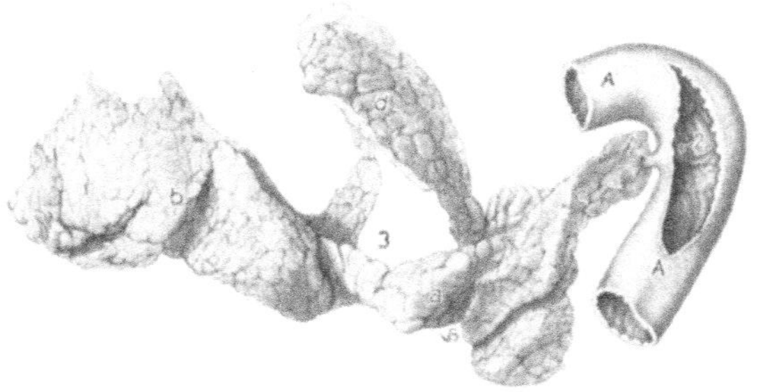

Fig. 165. Pancreas of the pig. Caudoventral aspect. *A* Cranial part of duodenum; *A'* Beginning of descending duodenum; *a, d* Body, forming anulus pancreatis; *b* Left lobe; *c* Right lobe; *2* Accessory pancreatic duct (the only duct in the pig); *2'* Minor duodenal papilla with opening of accessory pancreatic duct; *3* Anulus pancreatis

which surrounds the portal vein with a caudally directed process, and a large left and a small right lobe. The pancreas of the horse (167, 248, 439) is a compact organ, consisting of a large body to which a long left lobe and a short right lobe are attached like processes.

Because of the regression of one or the other pancreatic primordium in some species, the arrangement of the pancreatic ducts varies. In the horse and dog, both primordia persist and fuse, and the two original ducts are retained. The duct of the ventral primordium is the **pancreatic duct*** and opens with the bile duct on the **major duodenal papilla** (167/*1, 1'*). The duct of the dorsal primordium is the **accessory pancreatic duct**** and opens on the **minor duodenal papilla** (*2, 2'*). In the horse, the minor duodenal papilla is opposite the major duodenal papilla, while in the dog it is a few centimeters distal to the major duodenal papilla (164). In the pig and ox, only the duct of the dorsal primordium, the accessory pancreatic

* Formerly Wirsung's duct.
** Formerly Santorini's duct.

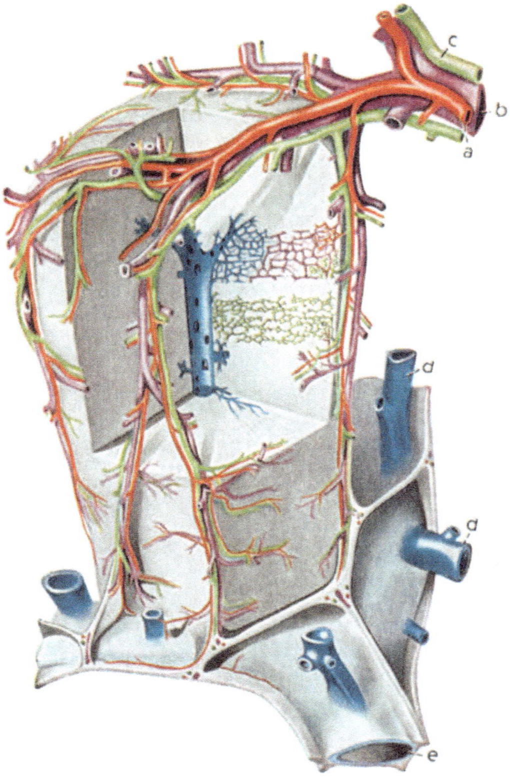

Fig. 162. Liver lobule of the pig (After Vierling, taken from Braus 1924.)

a (red) Branches of hepatic artery; *b* (purple) Branches of portal vein; *c* Branch of hepatic duct, (green) interlobular bile ductules and, in the lobule, bile canaliculi; *d* Central veins of adjacent lobules; *e* Collecting vein. A segment of the liver lobule has been removed to show blood capillaries and bile canaliculi

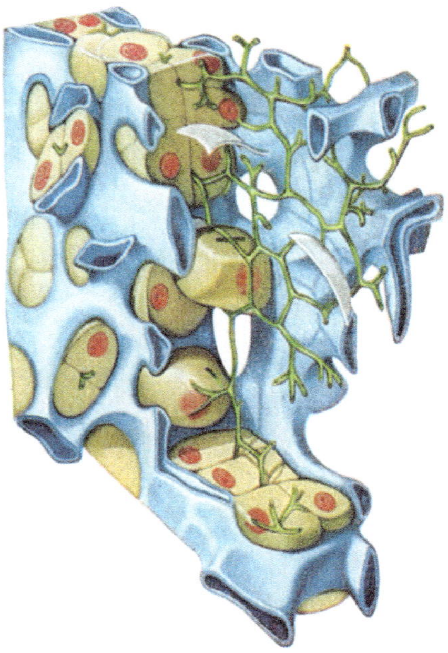

Fig. 163. Wax-plate model of a small segment of a liver lobule, illustrating the arrangement of the bile canaliculi and the intimate relationship between liver cells and sinusoidal blood capillaries. (After Vierling, taken from Braus 1924.)

Blue: Three-dimensional network of sinusoidal blood capillaries; Yellow: Liver cells arranged in cords or plates filling the meshes of the blood capillary network; Green: Intercellular bile canaliculi; White: Isolated connective tissue septa

duct, persists, and although this is the only duct in these animals, comparative considerations dictate adherence to the name "accessory duct." In the small ruminants and the cat, the duct of the ventral primordium, the pancreatic duct, persists, and opens with the bile duct on the major duodenal papilla.

STRUCTURE. The pancreas consists of lobules loosely united by small amounts of interlobular connective tissue (168). When fresh, it is pale red. After death it rapidly decomposes because of autolysis and bacterial invasion from the intestine. The pancreas is both an exocrine and an endocrine gland. Its exocrine secretion is the pancreatic juice, which contains various enzymes and precursors of others. The pancreatic juice passes through the pancreatic and accessory pancreatic ducts to the duodenum where its three principal enzymes effect the

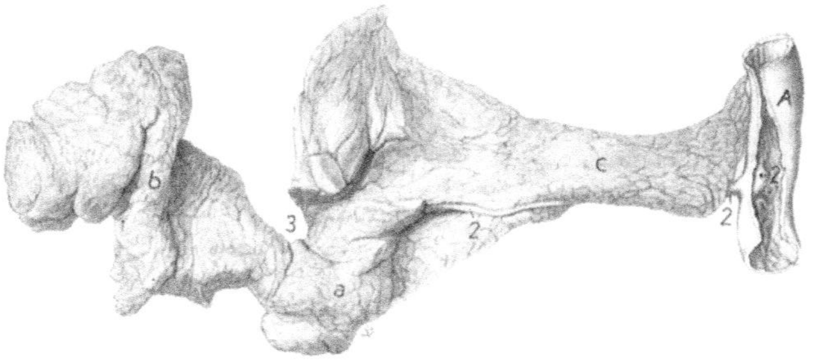

Fig. 166. Pancreas of the ox. Caudoventral aspect. *A* Descending duodenum; *a* Body; *b* Left lobe; *c* Right lobe; *2* Accessory pancreatic duct (the only duct in the ox); *2'* Minor duodenal papilla with opening of accessory pancreatic duct; *3* Incisura pancreatis

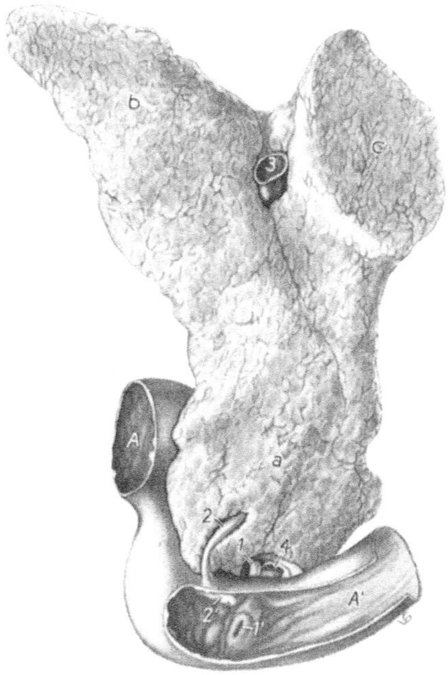

Fig. 167. Pancreas of the horse. Caudoventral aspect. *A*, *A'* Cranial part of duodenum, *A* its proximal pear-shaped enlargement; *a* Body; *b* Left lobe; *c* Right lobe; *1* Pancreatic duct; *1'* Major duodenal papilla with openings of pancreatic and bile ducts inside the ampulla hepatopancreatica; *2* Accessory pancreatic duct; *2'* Minor duodenal papilla with opening of accessory pancreatic duct; *3* Portal vein passing through anulus pancreatis; *4* Bile duct

digestion of proteins, fats, and carbohydrates. The endocrine secretion of the pancreas consists of the hormones insulin and glucagon, which are produced by the cells of the **pancreatic islets** (*c*) and are important in carbohydrate and particularly sugar metabolism. The pancreatic islets are scattered throughout the pancreas and consist of clusters of epithelial cells that are not connected to the excretory duct system of the gland. Each islet is richly supplied with capillaries into which the hormones diffuse. Diabetes, a fatal disease, sets in if the cells of the pancreatic islets fail to produce insulin. The pancreas of animals commonly brought to slaughter is used by the pharmaceutical industry as the raw material for the commercial production of insulin. The principal enzymes of this gland are also extracted and utilized.

BLOOD VESSELS, LYMPHATICS, AND INNERVATION. The pancreas is supplied by the celiac and cranial mesenteric arteries. Its veins enter the portal vein. The **lymphatics** draining the pancreas pass mainly to the pancreatico-duodenal and hepatic nodes. In addition, there are lymphatics to the splenic and gastric nodes in the pig, to the splenic and jejunal nodes in the dog, and to the celiac, cranial mesenteric or colic nodes in the horse. The pancreas is **innervated** by parasympathetic fibers of the vagus, which activate the secretory cells, and by sympathetic fibers, which reduce secretory activity.

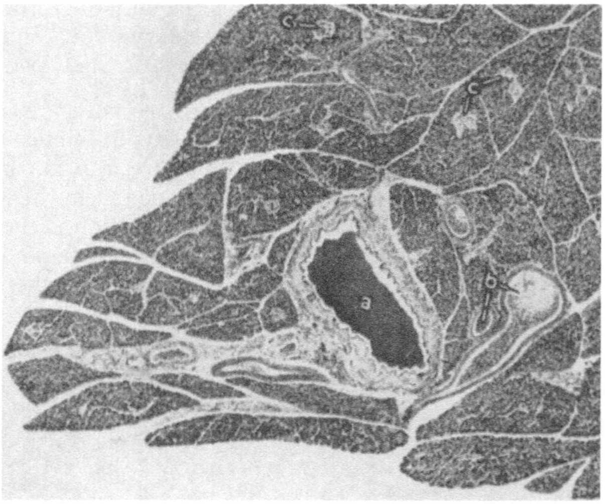

Fig. 168. Section of the pancreas of the cat. Microphotograph. Note the lobulation of this gland; *a* Secretory duct; *b* Interlobular blood vessels; *c* Pancreatic islets

The Alimentary Canal of the Carnivores

Esophagus

The esophagus of the carnivores begins at the caudal border of the cricoid cartilage opposite the middle of the axis in the dog and the third cervical vertebra in the cat (57, 58). The pharyngoesophageal junction is marked on the inside by an annular fold (limen pharyngoesophageum, 79/*12'*) which contains glands in the dog. The wall of the esophagus in the dog is thin cranially and increases gradually in thickness toward the stomach; it is thickest a few centimeters cranial to the cardia where there is a constriction of the lumen. The lumen is widest cranially and narrowest caudally. The esophagus joins the stomach at the funnel-shaped cardia to the right of the fundus (169—171). In the cat, the esophageal lumen is reduced a few centimeters away from the cranial end and again just cranial to the stomach, at which points the wall is relatively thick. **Esophageal glands** are present in the submucosa along the entire esophagus in the dog; in the cat they occur only in the first few centimeters. Carnivores have no difficulty swallowing or vomiting large boluses of food even if they contain bone fragments.

Stomach

(169—177, 180, 186)

The shape, size, and position of the stomach of the carnivores depend to a large extent on its degree of fullness. When the stomach is moderately full (169) it presents the basic form of a simple stomach that has been described on page 102. Fundus and body (*c, b*) are on the left side of the animal, and the pyloric part (*d, e*) is ventral and to the right, turning dorsally and slightly cranially. The body and the pyloric antrum (*d*) are of about equal diameter; the pyloric canal (*e*), which follows distally, decreases in width toward the pylorus. The empty stomach (170) is a U-shaped, narrow tube with the rather wide fundus (*b*) at the end of its left limb. In the greatly distended stomach (171), the U-shaped appearance is lost because of the expansion of the body and the pyloric antrum in the center of the stomach to form a uniformly distended sack; the pyloric canal remains relatively narrow.

POSITION AND RELATIONS OF THE DOG'S STOMACH. As the stomach increases in size with the intake of food it changes its position and displaces the abdominal organs, particularly the spleen and the intestines. The **empty stomach** (172, 175/*c, c'*) cannot be palpated because it lies entirely in the intrathoracic part of the abdominal cavity and fails to make contact with the ventral abdominal wall, extending ventrally only to about a level of the tenth costochondral junction. The fundus is in contact with the diaphragm below the

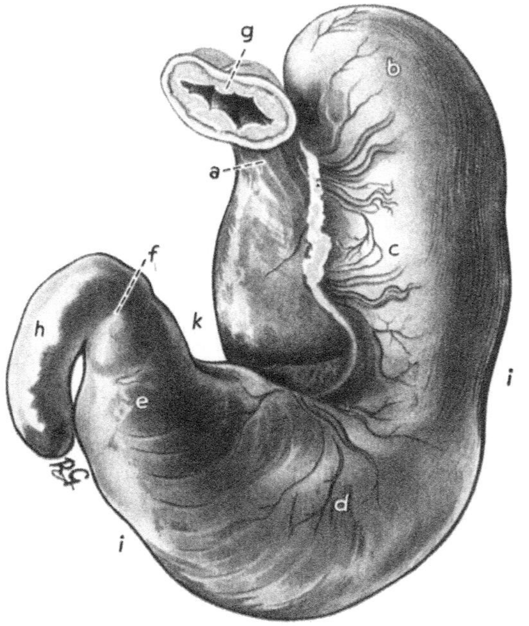

Fig. 169

Figs. 169—171. Stomach of the dog filled to various degrees. (After Zietzschmann 1938.)

Fig. 169. Moderately full stomach (basic form). Parietal surface. Fig. 170. Almost empty stomach. Parietal surface. Fig. 171. Very full stomach. Visceral surface.

a Cardiac part; *b* Fundus; *c* Body; *d, e* Pyloric part; *d* Pyloric antrum; *e* Pyloric canal; *f* Pylorus; *g* Esophagus; *h* Cranial part of duodenum; *i, i* Greater curvature; *k* Lesser curvature

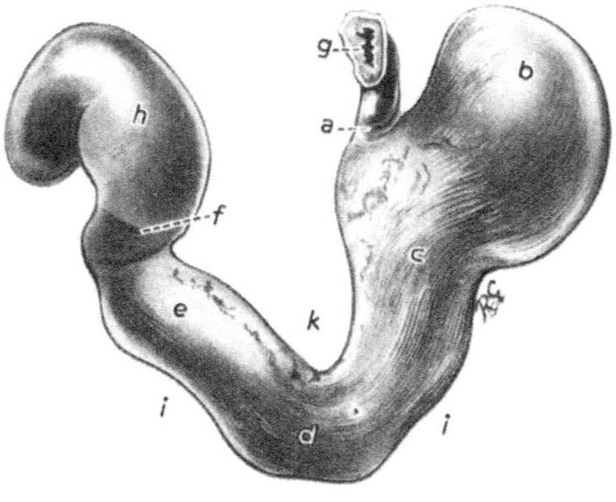

Fig. 170

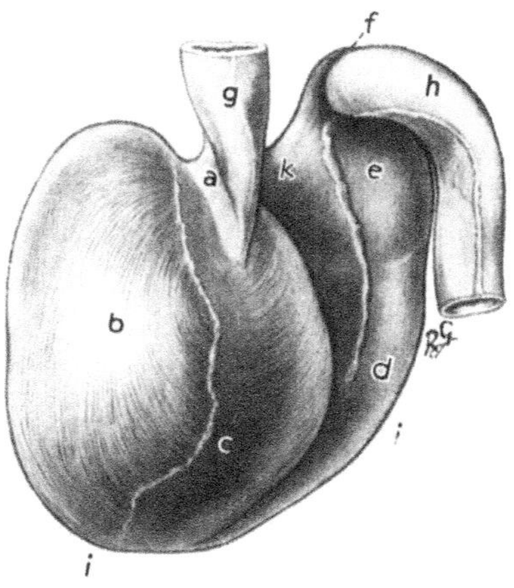

Fig. 171

angles of the left tenth, eleventh, and twelfth ribs. The pyloric antrum rests on the fatfilled falciform ligament and on loops of the jejunum, while on the right the pylorus and adjacent pyloric canal lie against the hepatic porta opposite the middle of the tenth rib. The liver is cranial to the stomach and only allows the fundus to make contact with the diaphragm. The spleen (*d*) lies against the greater curvature of the body of the stomach and with its caudal border more or less follows the left costal arch.

The body of the **moderately full stomach** (173, 176/*c, c'*) enlarges both cranially and caudally and, pushing the diaphragm slightly cranially, extends from the level of the ninth rib to the level of the twelfth rib, the caudal limit being equal to the level of the first or second lumbar vertebra. The ventral part of the body and parts of the pyloric antrum leave the

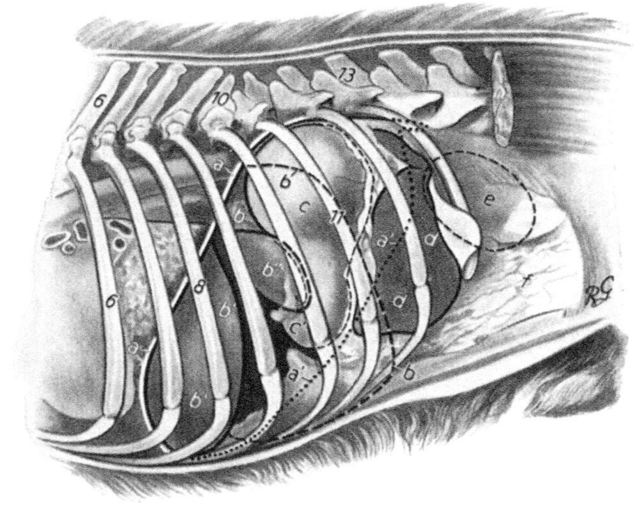

Fig. 172 (Stomach almost empty)

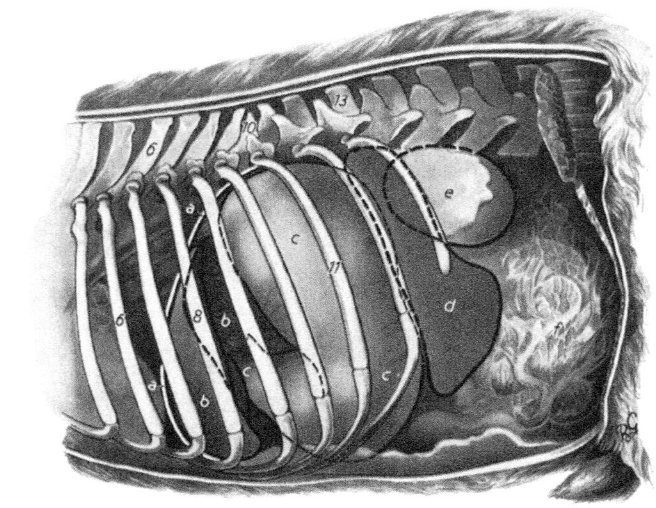

Fig. 173 (Stomach moderately full)

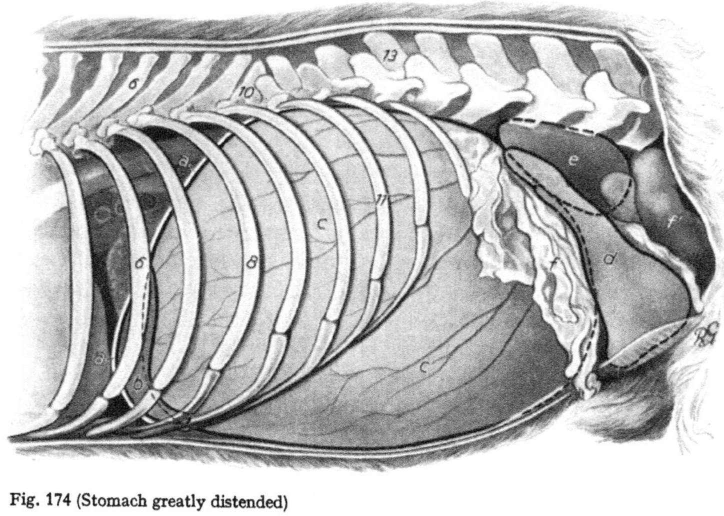

Fig. 174 (Stomach greatly distended)

Figs. 172—174. Topography of the canine stomach filled to various degrees. Left lateral aspect. (After Zietzschmann 1938.)

a Diaphragm; *a'* Line of diaphragmatic attachment (fig. 172); *b—b''* Liver, some of its outlines in heavy broken lines: *b* left lateral lobe (removed in figs. 172 and 173), *b'* left medial lobe (not visible in fig. 174), *b''* papillary process (visible only in fig. 172); *c* Stomach, some of its outlines in thin broken lines; *c'* Pylorus (not visible in fig. 174); *d* Spleen, some of its outlines in broken lines; *e* Left kidney, some of its outlines in broken lines; *f* Greater omentum; *f'* Jejunum (fig. 174)
6, 10, 13 Thoracic vertebrae of like number; *6, 8, 11* Left ribs of like number

intrathoracic part of the abdominal cavity and make contact with the abdominal floor. There is little change in the position of the pylorus, but portions of the spleen (*d*) protrude over the left costal arch.

The body and pyloric antrum of the **greatly distended stomach** (174, 177/*c, c'*) form a uniform sack, which extends into the middle or even caudal abdominal region, reaching a level of the second or third lumbar vertebra (third or fourth if the stomach is filled to capacity). The stomach makes extensive contact with the abdominal floor and is easily palpated. The spleen (*d*) and the left kidney (*e*) are displaced caudally to or beyond the same transverse level reached by the stomach, and the small intestine may be completely crowded away from the left abdominal wall. The liver is also slightly displaced to the right, and the pylorus and adjacent, now slightly distended, pyloric canal move with it from the level of the ninth through the level of the twelfth rib toward the right costal arch. The diaphragm may bulge cranially by one costal segment to the level of the fifth intercostal space. Many

Figs. 175—177. Topography of the canine stomach filled to various degrees. Right lateral aspect. (After Zietzschmann 1938.)

a Diaphragm; *a′* Line of diaphragmatic attachment (fig. 175); b—b^{V} Liver, some of its outlines in heavy broken lines; *b* caudate process (removed in fig. 176), b' right lateral lobe (removed in fig. 175), b'' right medial lobe (removed in figs. 175 and 176), b''' quadrate lobe, b^{IV} left medial lobe, b^{V} left lateral lobe (visible only in fig. 176); b^{VI} Gall bladder; *c* Stomach, some of its outlines in thin broken lines (figs. 176 and 177); *c′* Pylorus; *d* Descending duodenum; *e* Right kidney, some of its outlines in broken lines; *f* Greater omentum; *f′* Jejunum (fig. 177); *g* Right lobe of pancreas

6, 10, 13 Thoracic vertebrae of like number; *6, 8, 11* Right ribs of like number

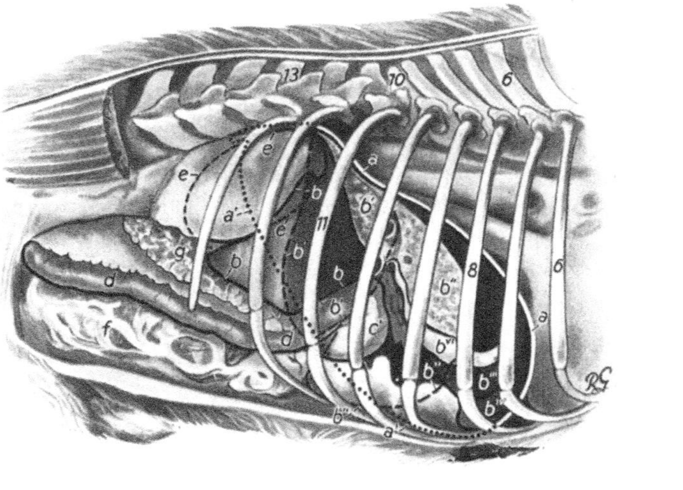

Fig. 175 (Stomach almost empty)

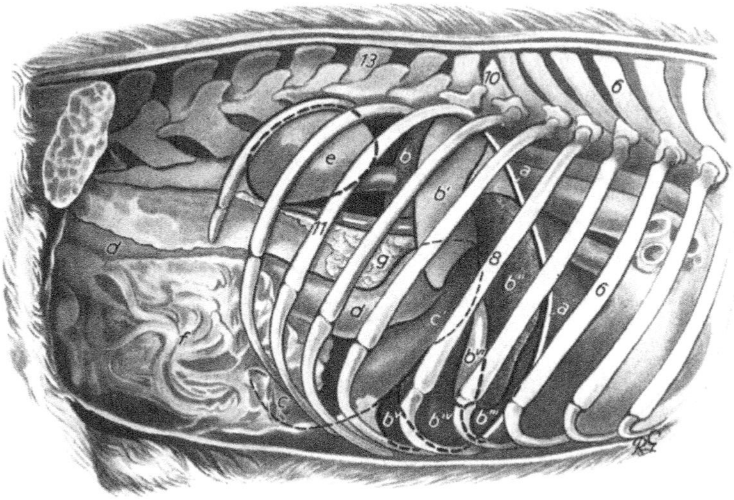

Fig. 176 (Stomach moderately full)

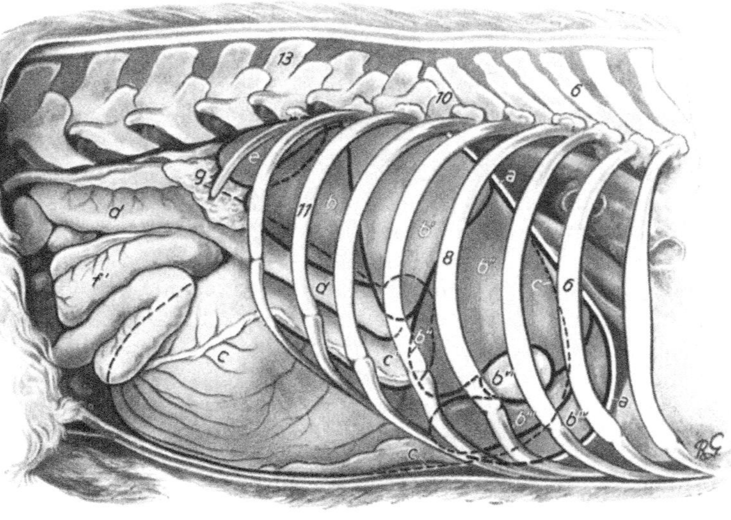

Fig. 177 (Stomach greatly distended)

gradations of course exist between the stages of stomach enlargement that have been described.

The simple stomach of the carnivores is relatively large. In the dog its capacity varies from 1—9 liters, and when empty it weighs from 65—270 gm., which is .621—1.385 per cent of the body weight.

STRUCTURE. The arrangement of the **muscular coat** corresponds to that shown for the simple stomach in Figure 143. The outer **longitudinal layer** of the esophagus is continued as narrow muscular bands along the greater and lesser curvatures and as **external oblique fibers** on the two surfaces of the body of the stomach. Distal to the body, the longitudinal fibers regain their true longitudinal course and uniformly surround the pyloric part to be continued as the longitudinal muscle coat of the duodenum. The fibers of the **inner circular**

layer diverge from the lesser curvature and encircle the entire stomach except the fundus. They are especially numerous at the pylorus where they form the **pyloric sphincter,** and a short distance proximal to the pylorus where they cause a slight constriction at which the pyloric part is divided into pyloric antrum and pyloric canal. The **internal oblique fibers** that make up the third muscle layer, are concentrated in the incisura cardiaca where they form the crest of the **cardiac loop.** The limbs of the cardiac loop pass distally on each side of the lesser curvature to about the level of the incisura angularis and form the lateral boundaries of the gastric groove. The floor of the groove is furnished by the fibers of the inner circular layer which cross the loop. From the cardiac loop thin internal oblique fibers diverge to encircle the fundus and pass obliquely over the body of the stomach to the greater curvature.

The stomach of the carnivores is entirely lined with a glandular mucous membrane. The **cardiac gland region** is a narrow annular zone at the cardia (134, 135). The **region of the proper gastric glands** is relatively large and lines two-thirds to three-fourths of the stomach. It is uniform in the cat (*3*), but can be divided in the dog into a lighter proximal zone (*3'*) which has a thinner mucous membrane and distinct foveolae, and a reddish brown distal zone (*3''*) with a thicker mucous membrane and less distinct foveolae. The **pyloric gland region** (*4*) lines the remaining distal part of the stomach and is pale red to yellowish in color. Bile sometimes enters the pyloric part of the stomach after death and stains much of the pyloric gland region green. In keeping with the stomach's ability to expand greatly, the submucosa is well developed and, depending on the state of contraction of the organ, allows the mucous membrane to form numerous transient folds.

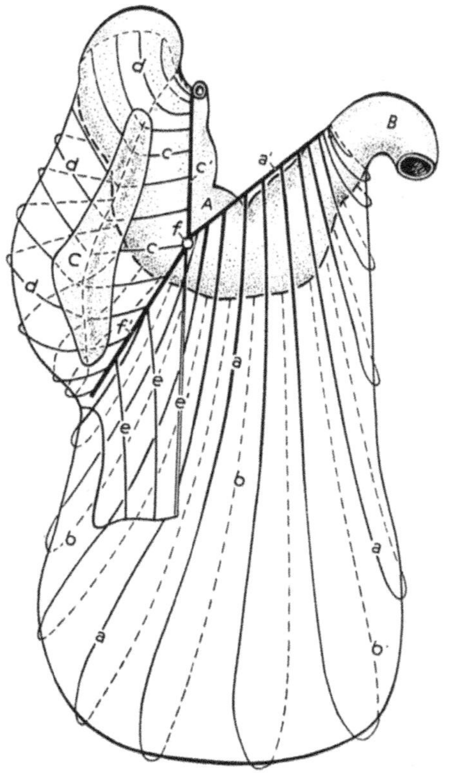

Fig. 178. The greater omentum of the dog. Schematic. Dorsal aspect. (After Zietzschmann 1939.) *A* Stomach; *B* Cranial part of duodenum; *C* Spleen; *a, c, d* Deep wall of greater omentum; *a', c'* Line of attachment of deep wall; *b* Supf. wall of greater omentum; *c* Phrenicosplenic ligament; *d* Gastrosplenic ligament; *e* Omental veil, *e'* its attachment to the descending mesocolon; *f* Celiac artery; *f'* Splenic artery

The **GREATER OMENTUM** is similar in dog and cat; its description is based on the dog and illustrated schematically in Figure 178. The **deep wall** of the greater omentum (*a, c, d*) originates from a line (*c', a'*) that extends from the esophageal hiatus along the left crus of the diaphragm to the celiac artery (*f*) and then to the right along the left lobe of the pancreas to the ventral border of the epiploic foramen. From this line the deep wall descends along the visceral surface of the stomach and, after first attaching to the hilus of the spleen (*C*) on the left, passes caudally ventral to the intestinal mass to the pelvic inlet. The part between the line of origin and the hilus of the spleen is the **phrenicosplenic ligament** (*c*), the part that continues from the spleen to the greater curvature of the stomach is the **gastrosplenic ligament** (*d*), and the cranial part extending directly from the line of origin (*c'*) to the fundus of the stomach is the **gastrophrenic ligament.** At its periphery, the deep wall of the greater omentum folds on itself ventrally and retraces its course to the stomach as the **superficial wall** (*b*). The latter lies between the deep wall and the abdominal floor and ends at the greater curvature of the stomach.

Deep and superficial walls of the greater omentum enclose the **caudal recess of the omental bursa** and cover the ventral and lateral aspects of the abdominal organs except the spleen and descending colon on the left, the duodenum on the right, the liver cranially, and the urinary bladder caudally (179).

Attached to the greater omentum but not taking part in the formation of the omental bursa is a sheetlike appendage, the **omental veil** (Miller, Christensen, Evans 1964). This is a rectangular fold (178/*e*) with cranial and dorsal attached borders and caudal and ventral free borders. The cranial border arises from the splenic artery (*f'*), and the dorsal border (*e'*), blending with the descending mesocolon, arises from a sagittal line extending medial to the left kidney to the fourth lumbar vertebra (184/c^{IV}). From this attachment high in the abdominal cavity the veil passes ventrolaterally to about the lateral border of the caudal recess of the omental bursa. The omental veil is thought to function as a suspensory apparatus for the spleen and the greater omentum.

The **LESSER OMENTUM** of the dog (180/*o*) and cat corresponds to the general description on page 14. It consists of **hepatogastric** and **hepatoduodenal ligaments** and forms the ventral boundary of the **vestibule of the omental bursa,** which communicates through the **epiploic foramen** (180/arrow between *a* and *b*) with the peritoneal cavity proper.

Intestines

(11, 144, 179, 184—187)

The intestines of the dog and cat, unlike those of the other domestic species, have the simple arrangement found in man. There is little difference between the diameter of the small and large intestines.

The **SMALL INTESTINE,** consisting of duodenum, jejunum, and ileum, has a length in the dog of 1.80—4.80 m. and in the cat of about 1.30 m. By far the longest part is jejunum.

The **DUODENUM** begins at the pylorus to the right of the median plane. Its **cranial part** passes dorsally and to the right at the level of the ninth intercostal space, being closely related to the liver, to which it is attached by the **hepatoduodenal ligament,** and to the pancreas. At the **cranial flexure** (184/*g*) the duodenum turns caudally as the **descending duodenum,** which has a relatively wide mesoduodenum that encloses the right lobe of the pancreas (185/*c'*). The descending duodenum is not covered by the greater omentum and lies directly against the right dorsolateral abdominal wall (179/*k*). It passes the caudal pole of the right kidney, and at the level of the fifth or sixth lumbar vertebra forms the **caudal flexure** (184/*g''*), a wide arc, open cranially, by means of which the duodenum passes from right to left around the cecum and the root of the mesentery. Continuing from the caudal flexure is the **ascending duodenum** (*g'''*) which is suspended by a short mesoduodenum and lies between the cecum, ascending colon, and root of the mesentery on the right, and the descending colon and left kidney on the left. The ascending duodenum is connected with the descending mesocolon and rectum by the **duodenocolic fold** (l^{IV}) and, cranially, forms the **duodenojejunal flexure** (*h*) to the left of the root of the mesentery, turns ventrally, and with the mesentery becoming longer is continued by the jejunum.

The **JEJUNUM,** by far the longest section of the small intestine, is suspended by the long mesentery, which is gathered at the roof of the abdominal cavity to form the **root of the mesentery** (*f*). The jejunum consists of about six to eight large loops which constitute the large intestinal mass between the stomach and the pelvic inlet (179/*d*). The jejunal mass is covered ventrally and laterally by the greater omentum and in general is nearly equally distributed on each side of the median plane. However, when the stomach becomes greatly distended and migrates caudally, the jejunum is crowded away from the left abdominal wall and is displaced dorsally and to the right (174, 177).

The **ILEUM** (184/*i*; 185/*f*) is the short terminal segment of the small intestine. Since there is no macroscopic demarcation at the jejunoileal junction, the length of the ileum is usually determined by the proximal extent of the **ileocecal fold** (144/*18*) and ileal arteries. The ileum arises caudally from the jejunal mass and passes cranially to open into the proximal end of the ascending colon. The **ileal orifice** is located at the level of the first or second lumbar vertebra, and is surrounded by a ringlike mucosal fold (181/*1*).

The **LARGE INTESTINE** of the carnivores consisting of cecum, colon, rectum, and a short anal canal is only slightly larger in diameter than the small intestine, and as a whole is relatively short. The cecum of the dog is 8—30 cm. long and that of the cat is 2—4 cm. long; the colon and rectum of the dog are 20—60 cm. long and those of the cat are about 30 cm. long.

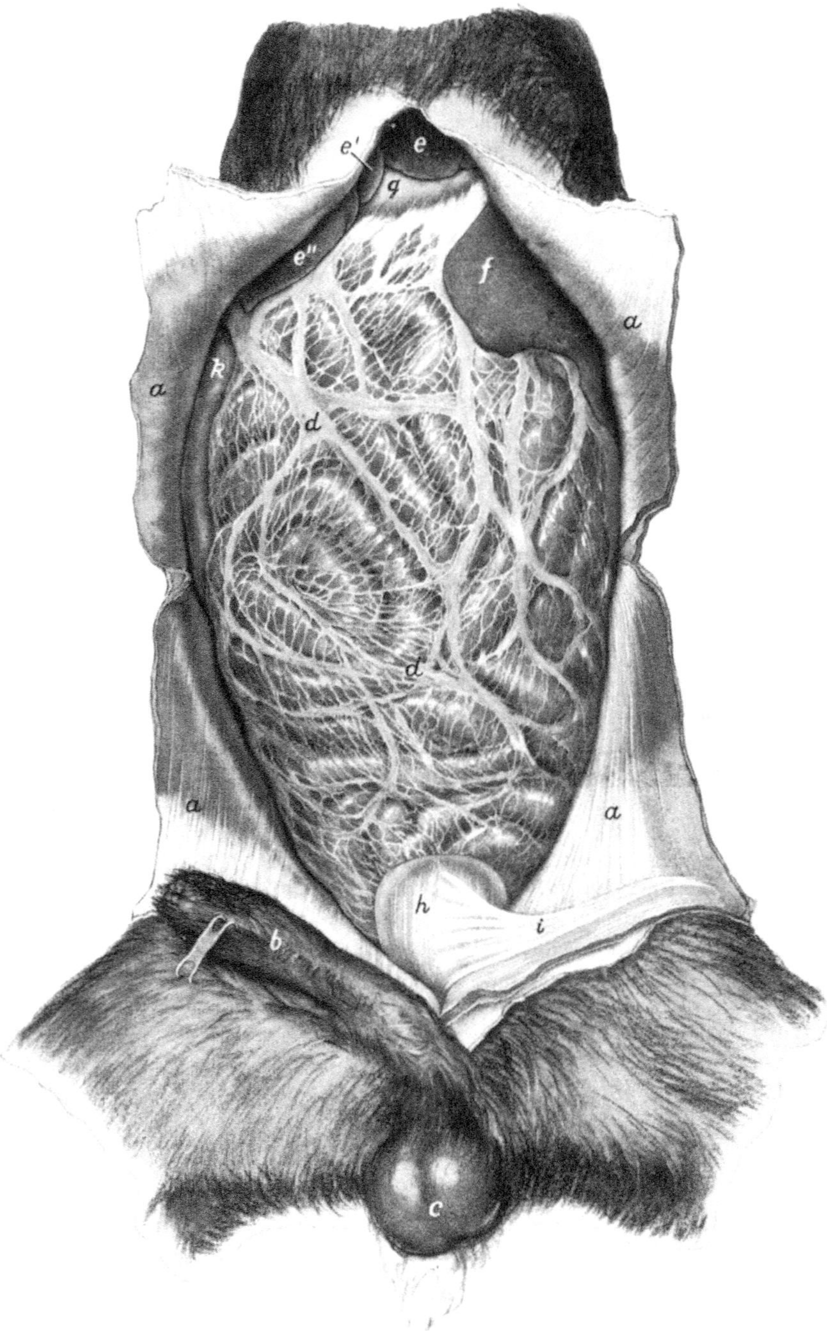

Fig. 179. Abdominal cavity of the dog, opened ventrally. (After Zietzschmann 1943.)

a Abdominal wall folded back; *b* Penis; *c* Scrotum; *d* Greater omentum, covering the jejunum; *e* Left medial lobe, *e'* quadrate lobe, *e''* right medial lobe of liver; *f* Spleen, ventral end; *g* Greater curvature of stomach; *h* Bladder; *i* Median ligament of bladder; *k* Descending duodenum

The **CECUM** of the dog (181, 182/*b*) is an irregularly twisted tube which is attached to the ileum and ascending colon by short peritoneal folds. It is located on the right a short distance ventral to the transverse processes of the second to the fourth lumbar vertebrae. The general direction of the cecum is caudal, but its blind end is bent and often points cranially. The cecum

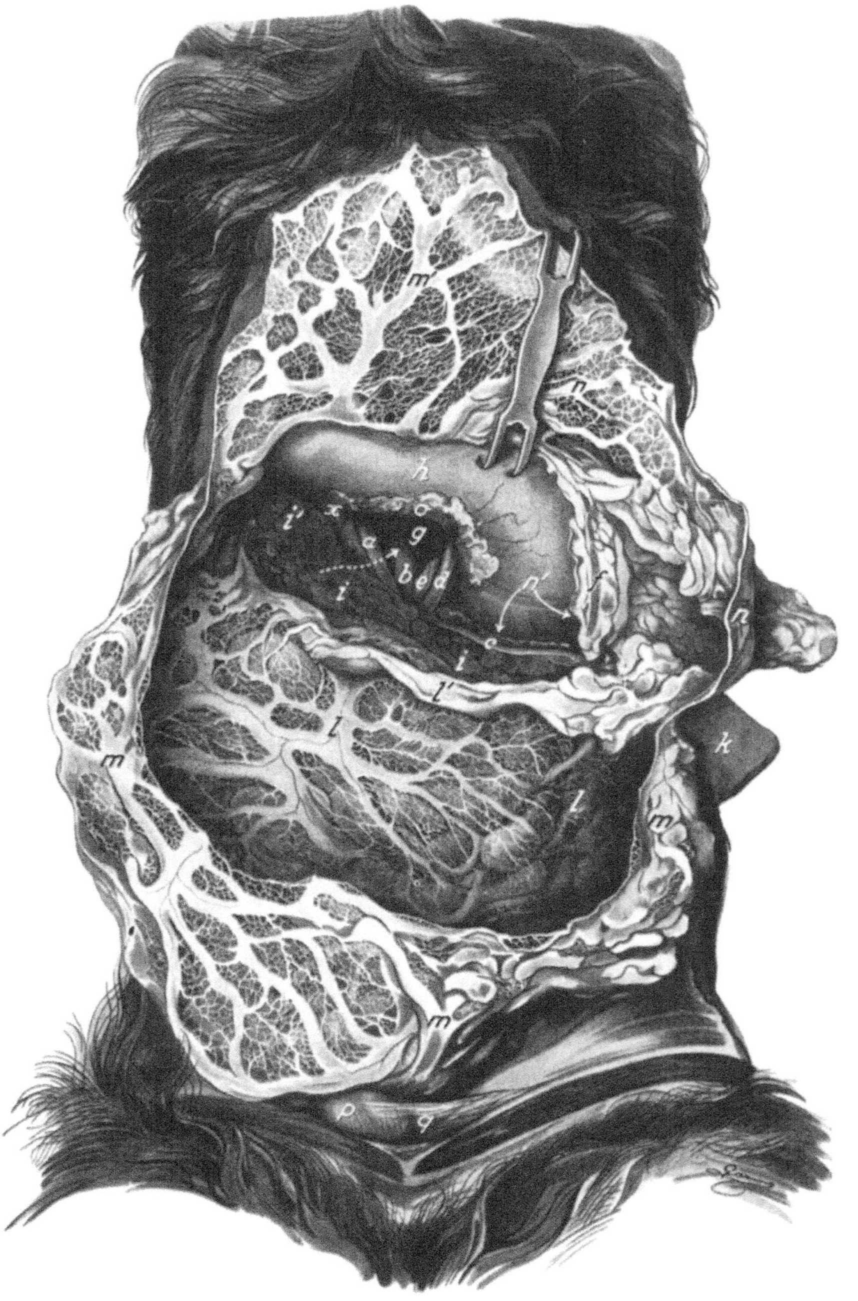

Fig. 180. The omental bursa of a dog, opened ventrally. Ventral aspect. (After Zietzschmann 1939.)

a Portal vein, ventral to epiploic foramen; *b* Caudal vena cava, dorsal to epiploic foramen; *c* Splenic vein; *d* Left gastric vein; *e* Left crus of diaphragm; *f* Left gastroepiploic vein, following the greater curvature of the stomach in the gastrosplenic ligament; *g* Papillary process of liver; *h* Visceral surface of stomach; *i* Left lobe of pancreas in deep wall of greater omentum; *i'* Body of pancreas, lying against cranial part of duodenum; *k* Ventral end of spleen; *l, l'* Deep wall of greater omentum; *m, m', n* Supf. wall of greater omentum, cut and folded cranially; *n'* Entrance to the splenic recess; *o* Lesser omentum; *p* Empty bladder; *q* Median lig. of bladder; *x* Point at which the greater and lesser omenta come together. The arrow nearest *x* passes between the portal vein ventrally and the caudal vena cava dorsally through the epiploic foramen into the vestibule of the omental bursa

(184/*k*) is related ventrally to the ileum and jejunum, dorsally to the right kidney, laterally to the descending duodenum (*g'*) and right lobe of the pancreas, caudally to the caudal flexure of the duodenum (*g''*), and medially to the root of the mesentery (*f*).

The cecum of the cat (183/*b*; 185/*g*) is an unusually short, comma-shaped diverticulum of the large intestine.

The **COLON** of the carnivores is short, and its three segments, the ascending, transverse, and descending colons, are arranged as their names indicate. The **ascending colon** (184/*l*; 185/*h*), the shortest of the three, begins at the cecum at about the level of the second lumbar vertebra and has a short cranial course. Owing to its narrow mesocolon, it is closely applied to the roof of the abdominal cavity and the right kidney. Laterally, it is related to the descending duodenum and its mesoduodenum and to the right lobe of the pancreas; and medially, it lies against the root of the mesentery. The **transverse colon** (184/*l'*; 186/*f*) passes to the left between the stomach and the cranial mesenteric artery, crossing the median plane at the level of the twelfth thoracic vertebra. The **descending colon** (184/*l''*; 186/*g*) which follows, extends to the pelvic inlet where it is continued by the rectum. It is the longest of the three colic segments and has a slightly wider mesocolon. The descending colon lies dorsally in the left half of the abdominal cavity, inclining slightly toward the median plane. It is related medially to the ascending duodenum (184/*g'''*) to which it is connected by the **duodenocolic fold** (l^{IV}). The colon as a whole (*l*, *l'*, *l''*) describes a loop, open caudally, that passes around the cranial aspect of the root of the mesentery (*f*). The duodenum (*g'*, *g''*, *g'''*) forms a similar loop, open cranially however, which passes around the caudal aspect of the root of the mesentery.

Fig. 181. Ileocolic junction of the dog.

a Ileum; *a'* Ileocecal fold; *b* Cecum; *c* Proximal part of ascending colon, opened; *1* Ileal orifice surrounded by annular fold

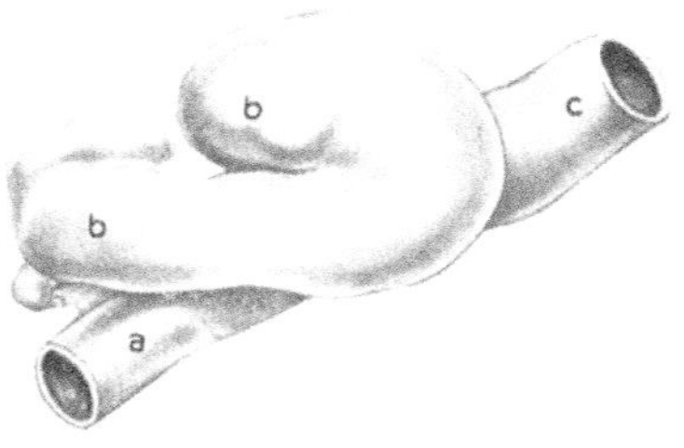

Fig. 182 (Dog)

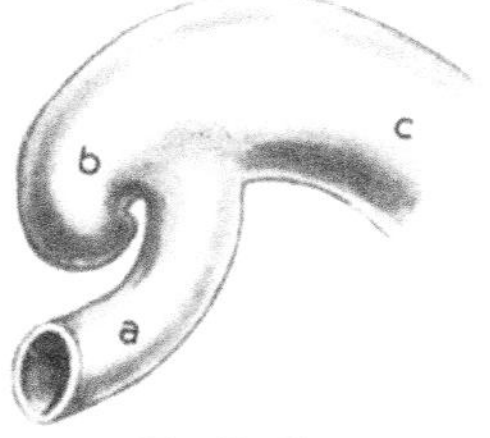

Fig. 183 (Cat)

Fig. 182 and 183. Cecum of the dog and of the cat.

a Ileum; *b* Cecum; *c* Proximal part of ascending colon

The **RECTUM** of the carnivores is short (471/*23*; 539, 541/*24*, *24'*). It begins at about the level of the seventh lumbar vertebra and enters the pelvic cavity ventral to the short sacrum to which it is attached by the mesorectum. In the retroperitoneal part of the pelvic cavity the rectum widens slightly (ampulla recti) and ends at the level of the fourth caudal vertebra to be continued by the short anal canal. The outer longitudinal muscle layer of the rectum is relatively well developed; its fibers are condensed dorsally and form the distinct **rectococcygeus** (471/*24*) which passes to the ventral surface of the first few caudal vertebrae. Ventral to the rectum is the urinary bladder (184/*r*) and in the male also the small genital fold (*q*), the deferent ducts (*p*), the pelvic urethra, the prostate gland, and in the tomcat also the bulbourethral glands (471/*15*). In the female, the rectum is related ventrally to the cervix of the uterus and the vagina (17; 539; 541).

The **ANAL CANAL** (17/*a'*; 187) is the very short terminal portion of the large intestine. Its mucous membrane is divided into three consecutive annular zones: from cranial to caudal, a columnar, an intermediate, and a cutaneous zone. The **columnar zone** in the dog is 5—12 mm. wide, but is rather indistinct in the cat. Its mucosa is darker than the rectal mucosa, is covered by stratified squamous epithelium, and is arranged in longitudinal folds (columnae anales) with grooves (sinus anales) between them. It contains lymph nodules and the **anal glands** which are peculiar to the dog. The **intermediate zone,** also covered by stratified squamous epithelium, is only .5—1.5 mm. wide and more appropriately referred to as the **ano-**

Right **Left**

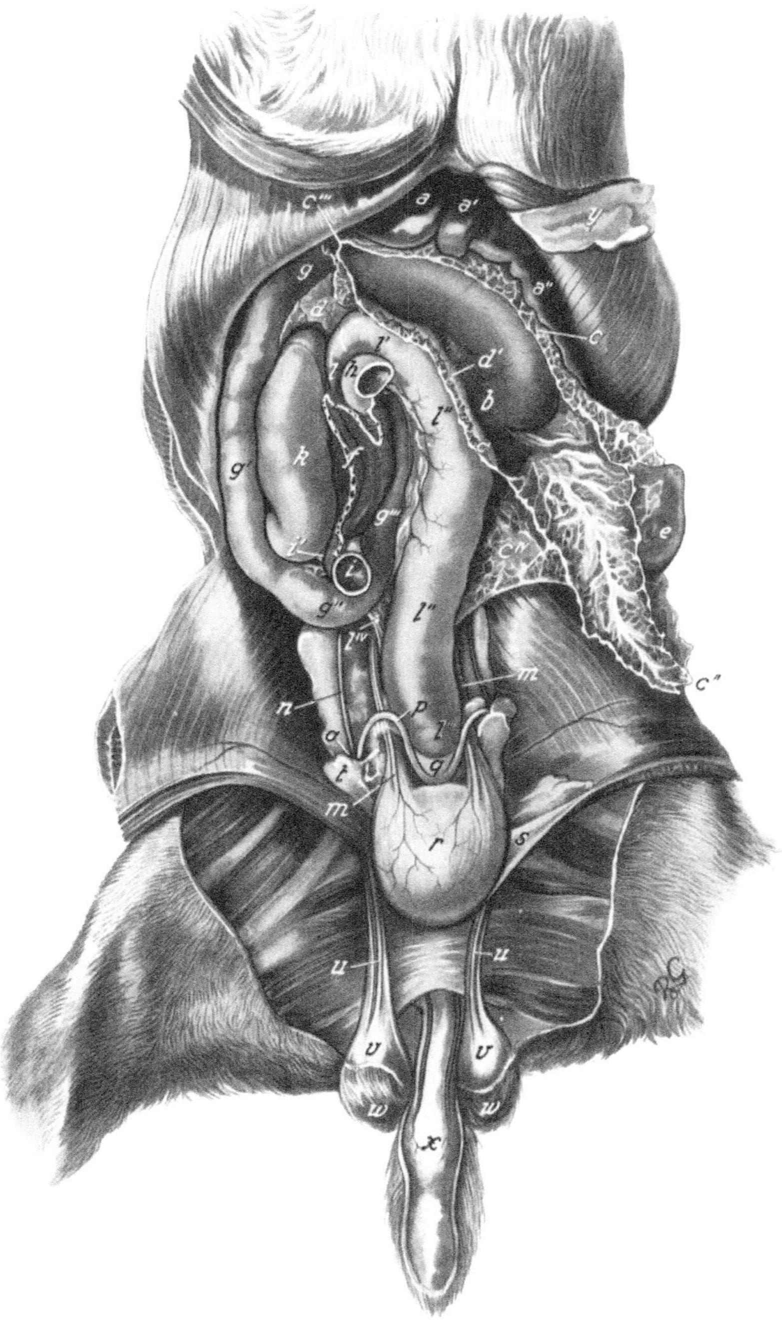

Fig. 184. The abdominal viscera of a male dog after removal of the jejunum, ileum, and most of the greater omentum. Ventral aspect. (After Zietzschmann, unpublished.)

a, a', a'' Ventral border of liver; *a* Right medial lobe, *a'* quadrate lobe, *a''* left medial lobe of liver; *b* Stomach, moderately full; *c*—c^{IV} Greater omentum, cut just caudal to the stomach so as to open omental bursa: *c* supf. wall, *c'* deep wall, *c''*, *c'''* union of deep and supf. walls, c^{IV} omental veil; *d* Body of pancreas; *d'* Left lobe of pancreas in deep wall of greater omentum; *e* Ventral end of spleen displaced laterally; *f* Mesentery cut just ventral to its root, with branches of cranial mesenteric artery; *g* Cranial part and cranial flexure of duodenum; *g'* Descending duodenum; *g''*, *g'''* Caudal flexure and ascending duodenum; *h* Duodenojejunal flexure; *i* Ileum; *i'* Ileocecal fold; *k* Cecum; *l* Ascending colon; *l'* Transverse colon; *l''* Descending colon; *l'''* Rectum; l^{IV} Duodenocolic fold; *m, m* Lateral lig. of bladder; *n* Testicular artery; *o* Vaginal ring; *p* Deferent duct; *q* Genital fold; *r* Urinary bladder; *s* Median lig. of bladder; *t* Fat associated with right lateral lig. of bladder; *u, v* Tunica vaginalis, at *v* enclosing the testicle; *w* Scrotum; *x* Penis reflected caudally; *y* Fatfilled falciform ligament

cutaneous line. The **cutaneous zone** forms the transition with the skin. It is bluish red and up to 4 cm. wide in the dog, is covered with cornified stratified squamous epithelium and fine hairs, and contains the **circumanal glands** (187/*3*). In the dog and cat the ducts (*2*) of the two laterally placed **anal sacs** (sinus paranales) open on this zone. The anal sacs are small reservoirs for a foul-smelling secretion of glands (gll. sinus paranalis) that surround them.

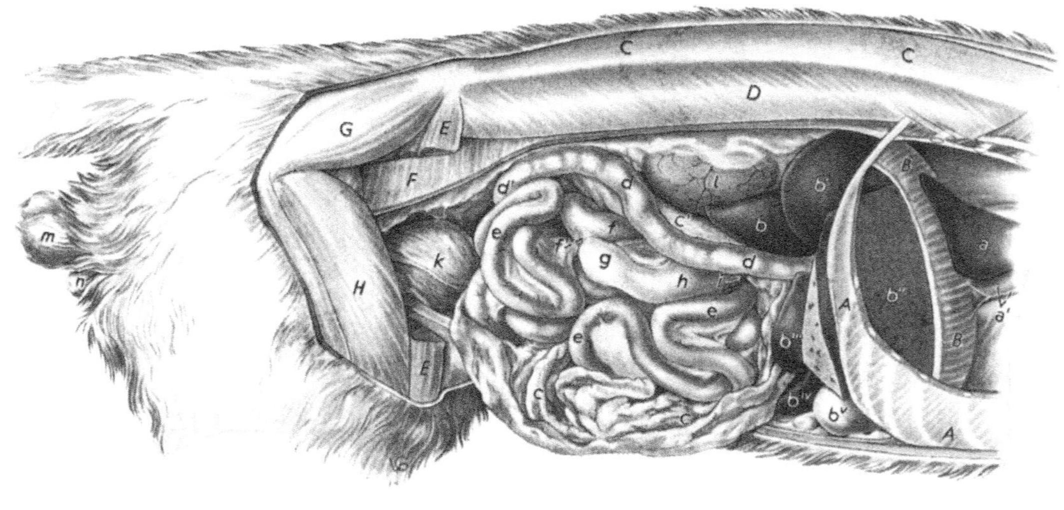

Fig. 185. Abdominal organs of a tomcat. Right lateral aspect.

A Costal arch and intercostal muscles; *B* Diaphragm, partly removed; *C* Longissimus; *D* Iliocostalis; *E*, *E* Sartorius, partly removed; *F* Transversus abdominis; *G* Gluteus medius; *H* Quadriceps femoris

a Accessory lobe of right lung; *a'* Caudal vena cava and right phrenic nerve; *b*—*b*IV Liver: *b* caudate process, *b'* right lateral lobe, *b''* right medial lobe, *b'''* quadrate lobe, *b*IV left medial lobe; *b*V Gall Bladder; *c* Greater omentum folded ventrally; *c'* Right lobe of pancreas; *d* Descending duodenum; *d'* Caudal flexure of duodenum; *e* Jejunum; *f* Ileum; *f'* Ileocecal fold; *g* Cecum; *h* Ascending colon; *i* Transverse colon; *k* Urinary bladder; *l* Right kidney; *m* Scrotum; *n* Prepuce

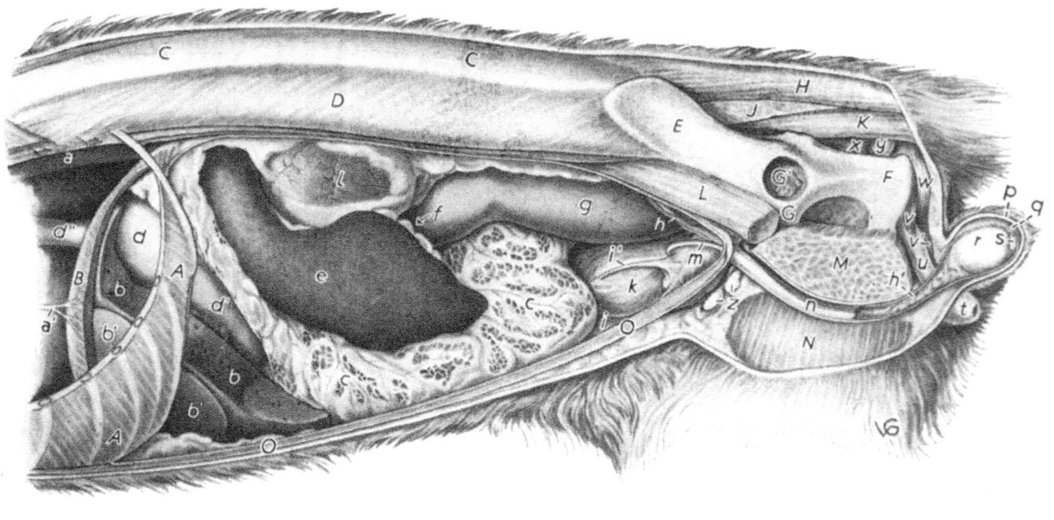

Fig. 186. Abdominal organs of a tomcat. Left lateral aspect.

A Costal arch and intercostal muscles; *B* Diaphragm, partly removed; *C* Longissimus; *D* Iliocostalis; *E* Ilium; *F* Ischium; *G* Pubis; *G'* Acetabulum; *H* Sacrocaudalis dorsalis lateralis; *J* Stumps of intertransversarii caudae; *K* Sacrocaudalis ventralis lateralis; *L* Iliopsoas; *M* Stump of left adductor; *N* Right gracilis; *O* Rectus abdominis

a Thoracic aorta; *a'* Left phrenic nerve; *b* Left lateral lobe of liver; *b'* Left medial lobe of liver; *c* Greater omentum, enveloping the jejunal mass; *d* Fundus of stomach; *d'* Body of stomach; *d''* Esophagus; *e* Spleen; *f* Transverse colon; *g* Descending colon; *h* Testicular artery and vein entering vaginal ring; *h'* Testicular artery and pampiniform plexus in tunica vaginalis; *i* Median lig. of bladder; *i'* Left lateral lig. of bladder; *k* Urinary bladder; *l* Left kidney; *m* Deferent duct; *n*, *q* Tunica vaginalis, fenestrated; *p* Scrotum; *r* Left testis; *s* Tail of epididymis; *t* Prepuce; *u* Penis; *v* Left ischiocavernosus; *v'* Retractor penis; *w* Ext. anal sphincter; *x* Rectum; *y* Anal sac; *z* Supf. inguinal lymph nodes

ANAL MUSCLES. The **internal anal sphincter** is the terminal thickened portion of the circular muscle layer of the rectum. The **external anal sphincter** (187/*n*, *n'*; 471/*22*), which is striated, consists of three successive rings in the male dog and of two successive rings in the bitch. These rings are difficult to separate; the most caudal (pars cutanea) is the best developed. The external anal sphincter surrounds the anus and is associated laterally in the male dog with the paired **rectrator penis** of which the penile part (pars penina, 187/*r*) passes caudoventrally, unites below the anus, and gains the ventral surface of the penis. In the bitch, the muscle that is associated similarly with the external anal sphincter is the constrictor vulvae (539/*19*) which enters the labia. A retractor clitoridis is apparently present in the cat but not in the bitch (see also p. 365).

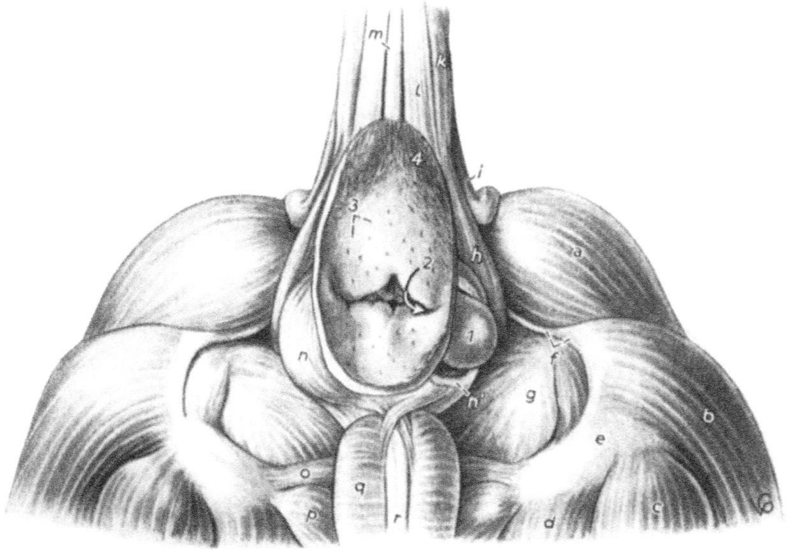

Fig. 187. Topography of the anus in the dog.

a Gluteus medius; *b* Biceps femoris; *c* Semitendinosus; *d* Semimembranosus; *e* Tuber ischiadicum; *f* Sacrotuberous ligament; *g* Int. obturator; *h* Levator ani; *i* Sacrocaudalis dorsalis lateralis; *k* Intertransversarii caudae; *l* Sacrocaudalis ventralis lateralis; *m* Sacrocaudalis ventralis medialis; *n* Ext. anal sphincter, fenestrated on the right at *n'*; *o* Ischiourethralis; *p* Ischiocavernosus; *q* Bulbospongiosus; *r* Retractor penis, penile part

1 Right anal sac, exposed; *2* Opening of right anal sac on the cutaneous zone of the anal canal; *3* Excretory pores of the circumanal glands; *4* Skin

The **levator ani** (187/*h*; 539/*S*) is a very extensive muscle in the carnivores. It is a thin triangular plate with its base on the pelvis and its apex on the caudal vertebrae. The levator can be separated into a larger medial part (m. pubocaudalis) which arises from the floor of the pelvis just caudal to the pecten of the pubis and along the pelvic symphysis, and a smaller lateral part (m. iliocaudalis) which comes from the medial surface of the shaft of the ilium near the iliopubic eminence. Both parts pass dorsocaudally and are inserted on the fourth to the seventh caudal vertebrae with only a few fibers attaching on the external anal sphincter. Despite this weak attachment on the wall of the anus, comparative anatomical reasons dictate adherence to the name levator ani.

In addition to what has been said in the general chapter (pp. 111—113) about the histological structure of the intestinal tract, the **LYMPHORETICULAR TISSUE** in the intestinal wall of the carnivores needs special mention. **Solitary lymph nodules** are present throughout the small intestine, but are rarely noticed with the naked eye because they are mostly submucosal. In the colon they are visible as small whitish gray nodules. **Patches of aggregate lymph nodules** (about 20—25 in the dog) are present throughout the small intestine in the wall opposite the mesenteric attachment. They are round, oval, or elongated, .7—8.2 cm. long and 3—11 mm. wide, and, with a pitted or nodular surface, are raised slightly above the mucosa. The most distal patch is just proximal to the ileal opening; there are none in the colon.

The appearance and distribution of the solitary lymph nodules in the cat is similar to that of the dog. The cat has only 4—6 patches of aggregate nodules .4—3 cm. long, many fewer than in the other domestic mammals. The patch closest to the ileal opening is longer and measures from 4—10 cm. The apex of the short feline cecum contains an unusually large accumulation of lymph nodules.

Liver

(153, 157, 172—177, 179, 184—186, 188, 189)

The reddish brown liver of the carnivores lies almost entirely within the intrathoracic portion of the abdominal cavity. Its **diaphragmatic surface** (189) is strongly convex in adaptation to the dome-shaped concavity of the diaphragm; its **visceral surface** is deeply concave. Depending on the size of the animal, the liver of the dog weighs 127—1,350 gm. and that of the cat 75—80 gm.

The liver of the carnivores is well separated into lobes by deep interlobar notches. To the left of the notch for the round ligament are the **left lateral** and **left medial lobes** (188, 189/*a*, *a'*) and to the right of the gall bladder are the **right lateral** and **right medial lobes** (*b*, *b'*). Between the two medial lobes and ventral to the hepatic porta is the **quadrate lobe** (*c*); and dorsal to the porta is the **caudate lobe** with a large **caudate process** (*d*) projecting to the right, and a distinct **papillary process** (*d'*). Apart from the prominent interlobar notches, there are smaller secondary notches on the left, ventral, and right borders of the liver. The esophagus and the caudal vena cava lie in their respective impressions on the dorsal border. The vein crosses the base of the caudate process and, occasionally surrounded by liver tissue, passes directly to the foramen venae cavae of the diaphragm without lying between the liver and the diaphragm for a short distance as in the other species. Therefore, the openings of the hepatic veins into the caudal vena cava are close to the dorsal border of the liver (189). The deep **renal impression** (188/*13*) for the cranial pole of the right kidney is formed by the caudate process and the dorsal part of the right lateral lobe.

LIGAMENTS OF THE LIVER. The **left triangular ligament** (189/*5*) is well developed and attaches the dorsal part of the left lateral lobe to the left crus and adjacent tendinous center of the diaphragm. The **right triangular ligament** (*6*) is short and attaches the dorsal part of the right lateral lobe to the right crus of the diaphragm. The middle part of the **falciform ligament** has disappeared. Its cranial part (*7*) leaves the liver between the left medial lobe and the quadrate lobe and extends to the tendinous center of the diaphragm ventral to the foramen venae cavae. It is a delicate membrane and contains in its free border the **round ligament of the liver** (157/*5*), the vestige of the umbilical vein. The caudal part is an irregular, fatfilled fold cranial to the umbilicus that may weigh several pounds in fat dogs (Miller, Christensen, Evans 1964). The **coronary ligament** (189/*8*) is a narrow peritoneal band which also attaches the liver to the diaphragm. It continues the two triangular ligaments medially and ventrally, and forms an arc around the ventral surface of the caudal vena cava. The **hepatorenal ligament** connects the liver and the caudal vena cava with the cranial pole of the right kidney. The **hepatogastric** and **hepatoduodenal ligaments,** in the latter of which the bile and pancreatic ducts pass to the duodenum, have been described on page 14. They extend as the **lesser omentum** from the porta to the lesser curvature of the stomach and to the cranial part of the duodenum.

The **GALL BLADDER** (189/*e*) is embedded in its fossa between the quadrate and the right medial lobe at the level of the eighth intercostal space, and does not reach the ventral border of the liver. The gall bladder is visible on the visceral as well as on the diaphragmatic surface of the liver and thus is in contact with the diaphragm. There is considerable variation in the number and arrangement of the **hepatic ducts** in the dog. Usually, three to five ducts leave the liver and empty separately into the cystic duct, which becomes the **bile duct** (157/*3*) after the last hepatic duct has been received. The bile duct opens into the duodenum together with the pancreatic duct about 2.5—6 cm. distal to the pylorus on the **major duodenal papilla** (164/*1'*).

In the cat, either complete or partial duplication of the gall bladder is occasionally encountered. One or more hepatic ducts leave the liver and join the tortuous **cystic duct** to form the **bile duct.** The latter ends together with the pancreatic duct about 3 cm. distal to the pylorus on the **major duodenal papilla,** which is inside the ampulla hepatopancreatica.

POSITION AND RELATIONS. The strongly convex diaphragmatic surface of the liver lies against the concavity of the diaphragm and almost covers it (188). The ventral end of the left lateral lobe (*a*) extends across the median plane to the right and leaves the left dorsal portion of the diaphragm opposite the eleventh and twelfth intercostal spaces uncovered. It is here that the fundus of the stomach contacts the diaphragm. The right lateral lobe and the caudate process (*b, d*) extend dorsocaudally on the right side, and at the level of the thirteenth rib are in contact with the cranial pole of the right kidney (175). At about the eighth intercostal space, the liver (184/*a, a', a''*) crosses the costal arch ventrally and overlies the xiphoid region. It is in contact here with the fatfilled falciform ligament (*y*) and reaches the vicinity of the umbilicus where it may be palpated in thin specimens. Habel (1978) states that this is possible in the dog only when the liver is enlarged. The movements of the diaphragm and positional changes of the animal cause minor displacements of the liver.

Fig. 188. Liver of a dog in situ. Visceral surface (After Baum, unpublished.)

A Thirteenth thoracic vertebra
a Left lateral lobe, covering the left medial lobe; *b, b* Right lateral lobe; *b'* Right medial lobe; *c* Quadrate lobe; *d, d'* Caudate lobe; *d* Caudate process; *d'* Papillary process; *e* Gall Bladder; *f* Esophagus in esophageal notch; *g* Medial portion of right crus of diaphragm; *g'* Left crus of diaphragm; *g''* Costal part of diaphragm; *h, h* Transversus abdominis; *i* Psoas musculature; *k* Epaxial musculature; *r* Costal part of diaphragm

1 Hepatic branch of hepatic artery; *1'* Abdominal aorta; *2* Portal vein, entering the liver at the porta; *3* Cystic duct; *4* Caudal vena cava; *7* Falciform ligament; *11* Gastric impression; *13* Renal impression

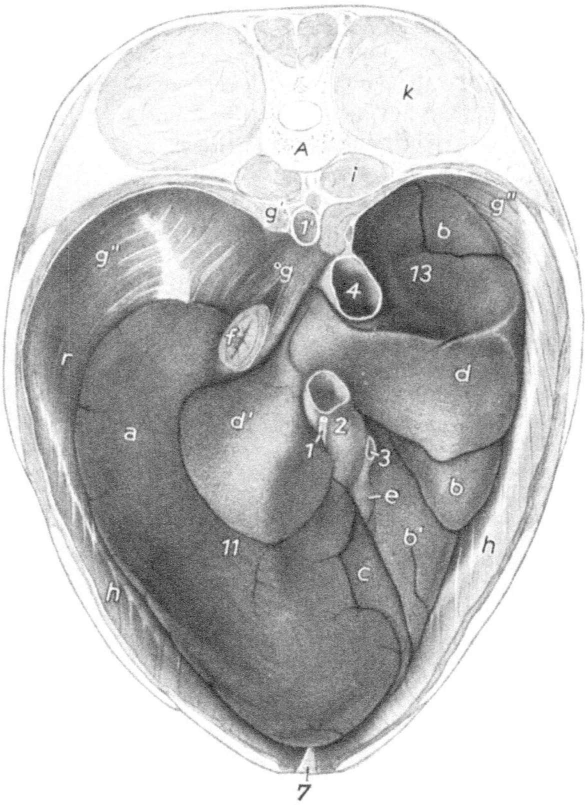

Fig. 188

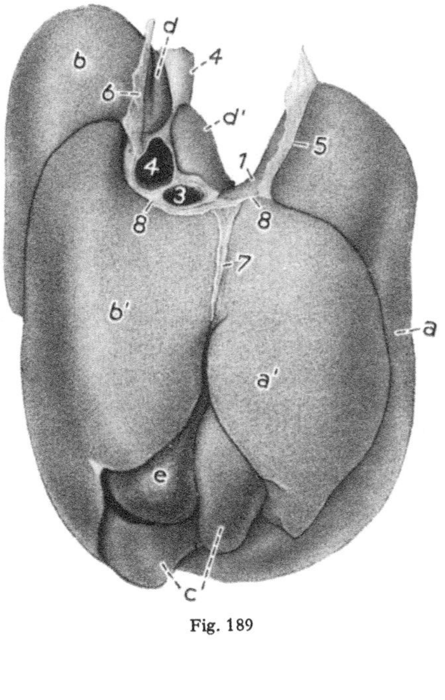

Fig. 189

Fig. 189. Liver of a dog fixed in situ. Diaphragmatic surface. (After Baum, unpublished.)

a Left lateral lobe; *a'* Left medial lobe; *b* Right lateral lobe; *b'* Right medial lobe; *c* Quadrate lobe; *d, d'* Caudate lobe; *d* Caudate process; *d'* Papillary process; *e* Gall bladder

1 Esophageal notch; *3* Trunk of the hepatic veins; *4, 4* Caudal vena cava embedded in the dorsal border; *5* Left triangular ligament; *6* Right triangular ligament; *7* Falciform ligament; *8* Coronary ligament

The dorsal border of the liver is intimately related to the esophagus, to the right crus of the diaphragm, and to the caudal vena cava (188). The concave visceral surface of the liver is related mainly to the stomach, which nestles with its parietal surface in the **gastric impression** (*11*), but also lies against the diaphragm dorsally and to the left. On the right side, the body of the pancreas, the cranial part of the duodenum, and the cranial pole of the right kidney contact the visceral surface of the liver.

Pancreas

(164, 180, 190)

Depending on the amount of blood it contains, the pancreas of the carnivores is lighter or darker red. The absolute and relative weights and the length of the canine pancreas vary greatly; generally, light animals have a higher relative pancreatic weight than heavier animals. The absolute weight of the pancreas of the dog ranges from 13—108 gm., and the relative weight ranges from .135—.356 per cent of the body weight. As in the other domestic mammals, the pancreas of the carnivores (184/*d*, *d'*; 190/*k*, *k'*, *k''*) forms close relations with the stomach, liver, and duodenum, and as a result of its development in the dorsal mesogastrium is associated with the mesoduodenum and the deep wall of the greater omentum. In the dog, the pancreas forms a horizontal loop, open caudally and to the left, and constists of a central **body** (190/*k*), which lies in the bend of the cranial part of the duodenum, and elongated right and left lobes. The portal vein on its way to the liver crosses the body of the pancreas. The long **right lobe** (*k'*) is in the mesoduodenum and accompanies the descending duodenum to about the caudal flexure of the duodenum where it may reach the cecum (*f*). The **left lobe** (*k''*) is slightly thicker than the right and lies in the deep wall of the greater omentum. It accompanies the pyloric part of the stomach to the left and makes contact with the liver, transverse colon, and usually also with the left kidney.

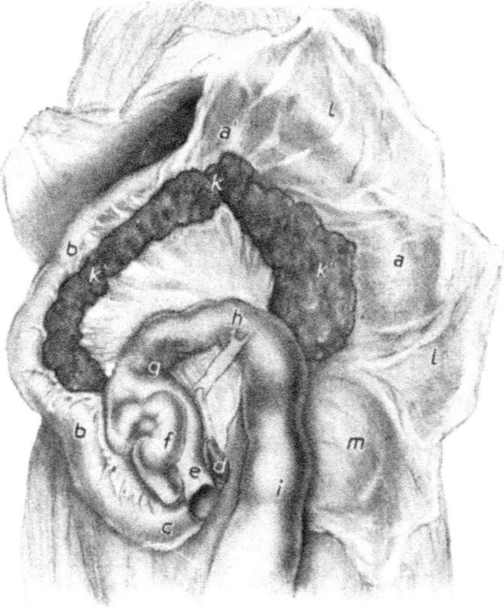

Fig. 190. Pancreas of the dog in situ. Ventral aspect. (After Baum, unpublished.)

a, *a'* Greater curvature of stomach, and pylorus seen through greater omentum; *b* Descending duodenum; *c*, *d* Caudal flexure and ascending duodenum; *e* Ileum; *f* Cecum; *g* Ascending colon; *h* Transverse colon, displaced slightly caudally; *i* Descending colon; *k*—*k''* Pancreas: *k* its body, *k'* its right lobe in mesoduodenum, *k''* its left lobe in the deep wall of greater omentum; *l* Greater omentum, reflected cranially and to the left; *m* Left kidney seen through greater omentum

There are usually two secretory ducts which carry the pancreatic juice to the duodenum. The **pancreatic duct** ends with the bile duct on the **major duodenal papilla** (164/*1'*). The **accessory pancreatic duct,** the larger of the two, opens 23—80 mm. distal to the first on the **minor duodenal papilla** (*2'*). In some dogs only the accessory pancreatic duct is present, and all the pancreatic juice enters the duodenum at the less distinct minor duodenal papilla. Further variations occur.

The pancreas of the cat weighs 8—10 gm., which is .27 per cent of the body weight, and is similar in shape and position to the pancreas of the dog. The **right lobe** (185/*c'*) accompanies the descending duodenum. The thicker **left lobe** follows the lesser curvature of the stomach, is related to the transverse colon, and reaches the spleen. A peritoneal fold connects the left lobe to the mesocolon. The **pancreatic duct,** which opens on the **major duodenal papilla,** is always present. The **accessory pancreatic duct** is present in only 20 per cent of the cats examined. There are many lamellar corpuscles in the feline pancreas.

The Alimentary Canal of the Pig

Esophagus

The esophagus of the pig begins at the caudal border of the caudal pharyngeal constrictors opposite the middle of the cricoid cartilage (59/*41*). It is attached to the larynx by muscle bands of the thyropharyngeus and cricopharyngeus, arising from the caudal border of the thyroid and cricoid cartilages. The beginning and the end of the esophagus have approximately the same diameter; its narrowest portion is in the middle segment. For the topography of the cervical and thoracic parts of the esophagus see page 99.

The **muscular coat** of the esophagus consists almost entirely of striated muscle, except for the short abdominal part, which consists of smooth muscle. **Esophageal glands** are abundant and closely packed proximally, but decrease rapidly toward the cardia. Lympho-recticular tissue is also abundant in the mucosa in the cranial portion of the esophagus, but decreases more gradually toward the stomach. It takes the form of either **solitary nodules** or **tonsillar follicles;** the latter are circular, raised areas about .5 cm. in diameter with a central depression.

Stomach

The average capacity of the stomach of pigs more than three months old is 3.8 liters. In a series of 25 pigs that died of natural causes, the capacity ranged from 1—6 liters. The pig has a simple stomach with a composite lining. When the stomach is moderately full, it resembles a sharply bent, tapering sack (191). The fundus presents a flattened conical **diverticulum** (diverticulum ventriculi, *1*), which distinguishes the porcine stomach from other simple stomachs. The diverticulum points to the right and caudally and is set off from the fundus by an annular groove.

POSITION AND RELATIONS. When distending with food, the stomach of the pig, like that of the carnivores, changes in both shape and position and displaces the other abdominal viscera. The **moderately full stomach** lies almost entirely within the intrathoracic part of the abdominal cavity, mainly to the left of the median plane; only the pyloric part is on the right. The parietal surface of the stomach lies in the gastric impression of the liver and against the left dorsal part of the diaphragm. The greater curvature is directed toward the left and ventrally, the lesser curvature is directed toward the right and dorsally. The cardia is at the level of the eleventh to the thirteenth (mostly twelfth) thoracic vertebra. The most cranial extent of the stomach is at the level of the seventh (sometimes sixth) intercostal space, while the diverticulum may reach caudally to the level of the twelfth or thirteenth thoracic vertebra. The moderately filled stomach is not in contact with the floor of the abdominal cavity; jejunal loops intervene between it and the abdominal wall. Neither does it reach the left ab-

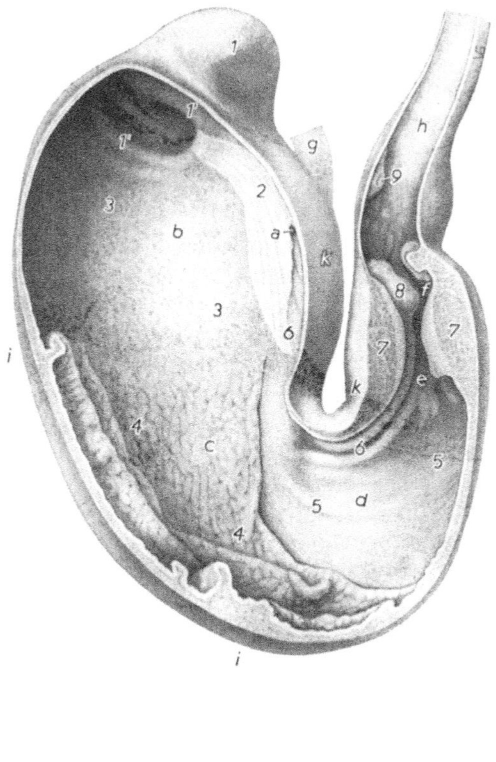

Fig. 191. Stomach of the pig, opened at the visceral surface.

a Cardia; *b* Fundus; *c* Body; *d, e* Pyloric part; *d* Pyloric antrum; Pyloric canal; *f* Pyloric opening; *g* Esophagus; *h* Cranial part of duodenum; *i* Greater curvature; *k* Lesser curvature

1 Gastric diverticulum; *1'* Spiral fold at base of gastric diverticulum; *2* Proventricular part of mucous membrane; *3* Cardiac gland region; *4* Region of proper gastric glands with high folds; *5* Pyloric gland region; *6, 6* Gastric groove; *7* Pyloric sphincter; *8* Torus pyloricus; *9* Major duodenal papilla with opening of bile duct

dominal wall. On the right side, however, it may lie against the abdominal wall opposite the eleventh to twelfth costal cartilages. Its lesser curvature is related to the pancreas and to the medial portion of the right crus of the diaphragm. Its visceral surface is in contact (through the greater omentum) on the left with the spleen and the coiled ascending colon, and on the right with the jejunum. The **empty stomach** (195/*f*; 197/*i*) lies entirely within the intrathoracic part of the abdominal cavity and in general has the same relations as the moderately filled organ. The coiled ascending colon, however, lies more cranial than in the animal with a moderately full stomach, and jejunal coils are found also to the left of the stomach. In the **greatly distended stomach** (199/*g*; 200/*f*) mainly the greater curvature expands caudoventrally and to the left, making extensive contact with the abdominal wall opposite the left ninth to twelfth costal cartilages. The stomach may displace the left lateral lobe of the liver and become related to the entire left costal arch. On the right side it makes more extensive contact with the abdominal wall ventral to the eleventh to thirteenth or twelfth to fourteenth costal cartilages. Ventrally, it makes extensive contact with the abdominal wall between the liver cranially and the intestines caudally, pushing the ascending colon on the left and the jejunum on the right toward the pelvis.

The stomach is attached to the diaphragm by the **gastrophrenic ligament,** which encloses the short abdominal part of the esophagus and passes dorsally and to the left, attaching on the fundus of the stomach, where it is continued by the greater omentum. From here the line of attachment of the **GREATER OMENTUM** (199/*9*; 200/*12*) can be traced along the greater curvature of the stomach to the duodenum, then along the ventral surface of the pancreas to the transverse colon and back along the left lobe of the pancreas to the gastrophrenic ligament. In its superficial wall the greater omentum contains the spleen (195/*g*; 197/*k*; 199/*h*). The greater omentum lies between the stomach and the coiled ascending colon, and on the right side also between the coils of the jejunum. The **epiploic foramen** is at the base of the caudate process of the liver and is bounded ventrally by the portal vein, dorsally by the caudal vena cava, and caudally by the body of the pancreas. It leads into the **vestibule of the omental bursa,** which is continuous caudoventrally with the larger **caudal recess** of the bursa. The **LESSER OMENTUM** (201/*8*, *9*) passes from the visceral surface of the liver to the lesser curvature of the stomach and to the duodenum as in the other animals.

STRUCTURE. The **muscular coat** of the stomach in the pig consists of the typical three layers, which are arranged largely as described in the general chapter on page 105. The **internal oblique fibers** are condensed at the lesser curvature and form the **cardiac loop** and the edges of the gastric groove. At the cardia, the loop is complemented by fibers of the **inner circular muscle layer** to form an efficient **cardiac sphincter.** The spiral fold (191/*1'*) at the base of the gastric diverticulum is also formed by internal oblique fibers. At the pylorus, the **circular muscle layer** is thick and forms an incomplete **pyloric sphincter.** At the break in the sphincter is an elongated **torus pyloricus** (*8*) which protrudes into the lumen of the pyloric canal and consists of adipose tissue and some muscle fibers.

The nonglandular **proventricular part of the gastric mucosa** lines an elongated area at the cardia (137/*1*; 191/*2*). It is covered with stratified squamous epithelium, is almost white, and extends into the gastric diverticulum, which it partly lines. The **cardiac gland region** (137/*2*; 191/*3*), which is unusually extensive in the pig, lines nearly one-third of the stomach, including the fundus, diverticulum, and the proximal portion of the body. It is soft and smooth, light red or light gray in color, and is sharply demarcated from the proventricular part and from the region of the proper gastric glands. At the lesser curvature it borders on the pyloric gland zone. The **region of the proper gastric glands** (137/*3*, 191/*4*) also lines about one-third of the stomach, lying between the cardiac gland region and the pyloric gland region, and mainly lining the distal portion of the body. It is brownish red and presents many folds and easily distinguishable areae and foveolae gastricae. The **pyloric gland region** (137/*4*; 191/*5*) lines most of the pyloric part of the stomach, again about one-third of the stomach wall. It is not as plicated as the preceding region and has a pale pink to yellowish color.

The gastric mucosa is well supplied with **LYMPHORETICULAR TISSUE** especially in the cardiac gland region (140). Both **solitary** and **aggregate lymph nodules** are present.

The aggregate nodules are found mainly at the periphery of the cardiac gland region and are especially numerous at the lesser curvature, where the three glandular regions meet. They are slightly raised above the surface with a craterlike central depression and are large enough to be visible to the naked eye.

Intestines

(145, 192—200)

The intestines of the pig are arranged as in the other domestic mammals, except that the greatly elongated ascending colon is characteristically coiled upon itself to form a cone-shaped mass.

The **SMALL INTESTINE** of fully grown pigs is 16—21 m. long, of which .70—.95 m. is duodenum, 14—19 m. jejunum, and .7—1 m. ileum.

The **DUODENUM** (145/*B*) begins at the pylorus on the right side of the body at the level of the tenth to twelfth (most often at the eleventh) intercostal space. Its **cranial part** ascends caudodorsally along the visceral surface of the liver, forms a horizontal sigmoid loop just cranial to the right kidney, and ends at the cranial flexure of the duodenum (196/*g*). The **descending duodenum** passes caudally ventral to the right kidney. It is suspended by a 6—10 cm. long mesoduodenum and ends at the caudal flexure. After crossing to the left side of the body, the duodenum ascends just to the left of the median plane in a craniodorsal and slightly lateral direction, and is related here to the descending colon (192/*a'*, *i*), with which it is connected by the **duodenocolic fold.** At the level of the cranial mesenteric artery, the ascending duodenum turns sharply to the right in front of the artery, and lies in close relation to the transverse colon (*h*). The bile duct opens into the duodenum on the small **major duodenal papilla,** 2—5 cm. distal to the pylorus; the accessory pancreatic duct, the only duct of the pancreas in the pig, enters the duodenum on the **minor duodenal papilla,** 12—20 cm. distal to the major duodenal papilla (165).

The **JEJUNUM** (145/*C*) consists of a large number of small loops which are suspended in the sublumbar region by the fan-shaped mesentery. Most of the jejunum lies to the right of the median plane, but a few loops may be found ventrally in the left half of the abdomen (195 to 200). The jejunal mass extends from the stomach and liver to the pelvic inlet, making extensive contact with the right abdominal wall. Medially, it is related to the ascending colon

Fig. 192. Intestines of the pig. Seen from the left and slightly dorsally.

a Descending duodenum; *a'* Ascending duodenum; *a''* Duodenojejunal flexure; *b* Proximal and distal jejunal loops, the bulk of the jejunum is behind the ascending colon; *c* Ileum; *d* Cecum; *e* Centripetal (outer) turns of ascending colon; *f* Central flexure; *g*, *g'* Proximal and distal centrifugal (inner) turns of ascending colon; *h* Transverse colon; *i* Descending colon; *k* Rectum; *1* Cranial mesenteric artery

and, at the pelvic inlet, also to the cecum, which are mainly on the left. Dorsally, the jejunum is related to the duodenum, pancreas, right kidney, the caudal portion of the descending colon, the bladder, and in the female also to the uterus.

When the stomach is empty, jejunal loops move to the left, cranial to the coiled ascending colon, making more extensive contact with the visceral surface of the stomach and the left lobes of the liver (197). The distended stomach displaces the jejunal mass slightly caudally (200).

The **ILEUM** (145/*D*) can be identified by its slightly thicker muscular coat and the **ileocecal fold.** It arises from the jejunal mass in the left caudoventral quadrant of the abdomen (195/*l*), where it is related to the urinary bladder; and passes cranially, dorsally, and medially to enter the large intestine obliquely at the cecocolic junction. The **ileal orifice** is at the summit of a 2—3 cm. long projection, the **papilla ilealis** (194/*1*), which is bent slightly toward the cecum. The borders of the papilla are connected to the wall of the colon by mucosal folds (frenula, *3*), and the circular muscle layer here is twice as thick as in the other parts of the ileum and functions as an **ileal sphincter.** The patches of aggregate lymph nodules (*2*) present in the pig's intestines are described in the section on lymphoreticular tissue on page 145.

The **LARGE INTESTINE** of the fully grown pig is on the average 3.5—6 m. long, .3—.4 m. of which is cecum; the rest is colon and rectum.

The **CECUM** (193) has a capacity of 1.5—2.2 liters. It is a cylindrical, slightly tapering blind sac with three longitudinal muscle **bands** alternating with three rows of **sacculations.** The ventral band gives attachment to the ileocecal fold; the lateral and medial bands are free and join at the apex of the cecum. The cecocolic junction lies ventral to the left kidney (195/*i*). From here the cecum extends caudoventrally along the left abdominal wall, so that the apex comes to lie in the inguinal region. When the stomach is empty, the apex (198/*l*) may move over to the right side into space vacated by the jejunum which has moved cranially.

COLON. The **ascending colon** of the pig (145/*F*, *F'*) is greatly elongated and characteristically rolled up on itself, forming a conical mass of spiraling coils, the **spiral loop,** which is suspended by the ascending mesocolon. The **ascending mesocolon** together with the arteries for the ascending colon, enters the base of the cone and attaches it to the left side of the **root of the mesentery.** The central axis of the cone is directed from the root of the mesentery mainly ventrally, and slightly left-laterally and cranially. Arising from the somewhat wider cecum at the level of the third lumbar vertebra, the ascending colon passes around the central axis of this cone in a clockwise direction when viewed from a dorsal position. It makes three and one-half fairly wide **centripetal turns,** which bring it to the apex of the cone. The centripetal turns form the outside of the cone and are visible without dissection (192/*e*). At the apex, the ascending colon reverses direction and describes the **central flexure** (*f*), after which it returns in a counterclockwise direction in tighter and steeper **centrifugal turns** inside the centripetal turns to the base of the cone. Because the centrifugal turns are inside the centripetal turns, they must be dissected in order to be seen, except at the apex (*g*). After emerging from the base of the cone, the last of the centrifugal turns crosses the

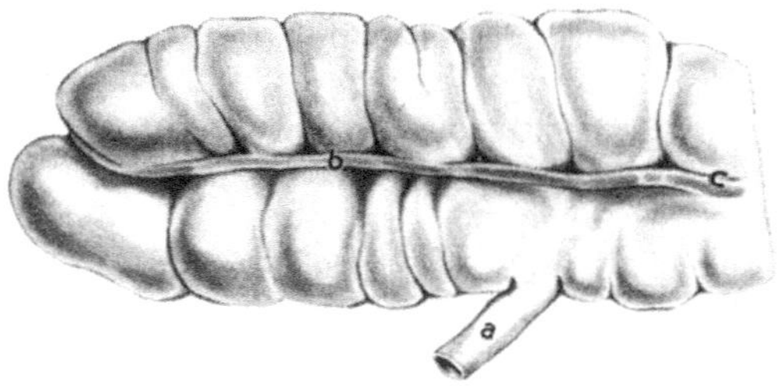

Fig. 193. Cecum of the pig.
a Ileum; *b* Cecum; *c* Proximal part of ascending colon. Note the muscular band and sacculations

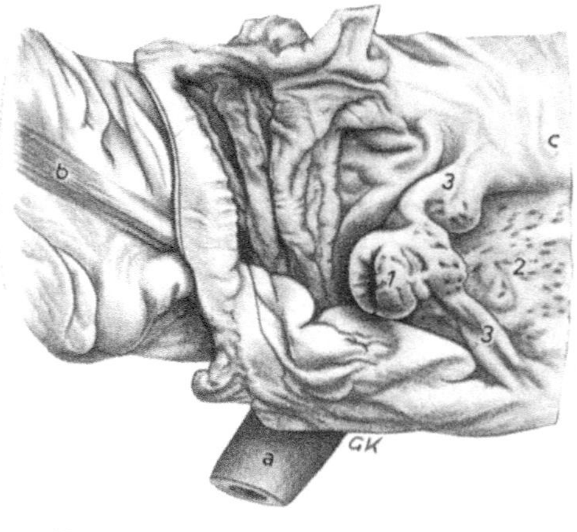

Fig. 194. The termination of the ileum in the pig.
a Ileum; *b* Cecum with muscular band and sacculations; *c* Proximal part of ascending colon, opened; *1* Papilla ilealis with ileal opening; *2* Patch of aggregate lymph nodules; *3* Frenulum papillae ilealis

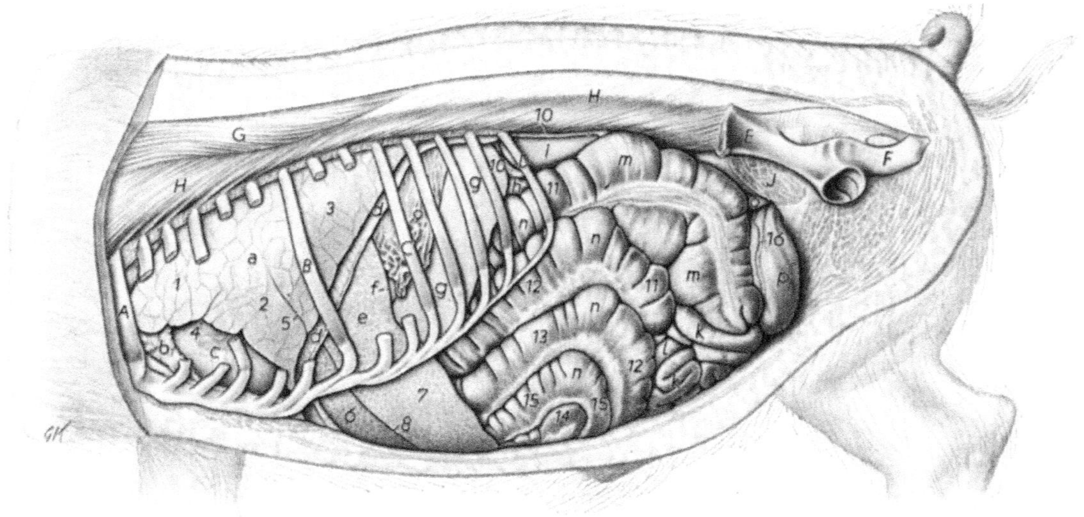

Fig. 195. Topography of the thoracic and abdominal organs of a castrated male pig with an empty stomach. Left aspect. (After Graeger 1957.)

A First rib; *B* Seventh rib; *C* Tenth rib; *D* Fourteenth rib; *E* Ilium; *F* Ischium; *G* Spinalis; *H* Longissimus; *J* Cut surface of iliopsoas; *a* Left lung; *b* Thymus; *c* Heart inside pericardium; *d* Diaphragm, its tendinous center, *d'* its costal part; *e* Liver; *f* Stomach; *g* Spleen; *h* Pancreas; *i* Left kidney (perirenal fat partly removed); *k* Jejunum; *l* Ileum; *m* Cecum; *n* Ascending colon; *o* Descending colon; *p* Urinary bladder

1—5 On the lung: *1*, *2* divided cranial lobe, *3* caudal lobe, *4* cardiac notch, *5* interlobar fissure; *6—8* On the liver: *6* left medial lobe, *7* left lateral lobe, *8* interlobar notch; *9* Greater omentum (gastrosplenic ligament); *10* Cut surface of perirenal fat; *11—15* On the ascending colon: *11* first, *12* second, *13* third, and *14* fourth centripetal turns, *15* first centrifugal turn; *16* Lateral lig. of bladder

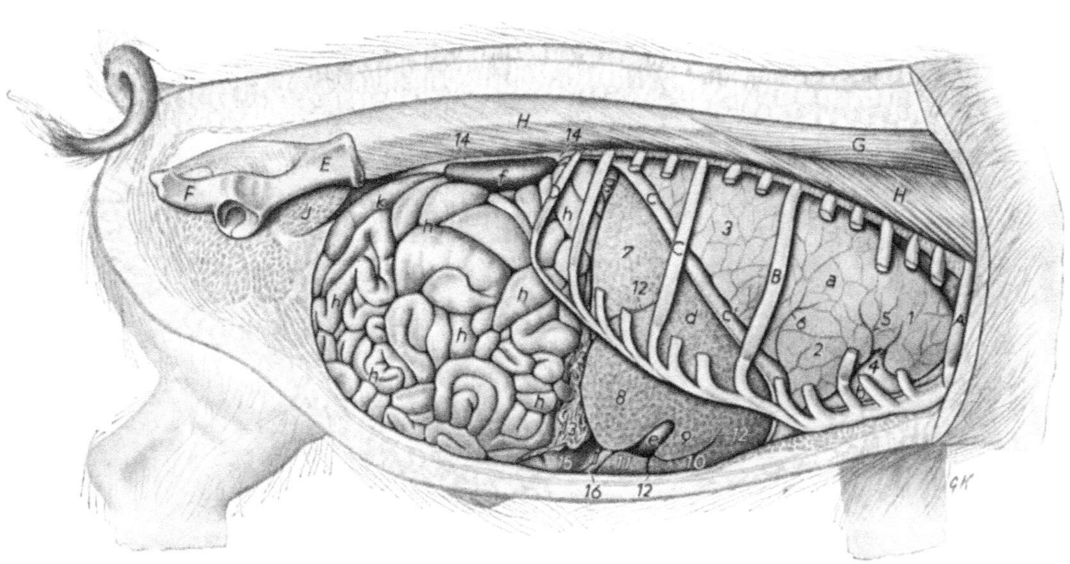

Fig. 196. Topography of the thoracic and abdominal organs of a castrated male pig with an empty stomach. Right aspect. (After Graeger 1957.)

A First rib; *B* Seventh rib; *C* Tenth rib; *D* Fourteenth rib; *E* Ilium; *F* Ischium; *G* Spinalis; *H* Longissimus; *J* Cut surface of iliopsoas; *a* Right lung; *b* Heart inside pericardium; *c* Diaphragm, its tendinous center, *c'* its costal part; *d* Liver; *e* Gall bladder; *f* Right kidney (perirenal fat partly removed); *g* Cranial part of duodenum; *h* Jejunum; *i* Ascending colon; *k* Descending colon

1—6 On the lung: *1* cranial lobe, *2* middle lobe, *3* caudal lobe, *4* cardiac notch, *5* cranial interlobar fissure, *6* caudal interlobar fissure; *7—12* On the liver: *7* right lateral lobe, *8* right medial lobe, *9* quadrate lobe, *10* left medial lobe, *11* left lateral lobe, *12* interlobar notches; *13* Greater omentum; *14* Cut surface of perirenal fat; *15*, *16* Centripetal turns of ascending colon

ventral aspect of the ascending duodenum and passes cranially along the right side of the root of the mesentery, and is followed by the transverse colon. The outer centripetal turns are wider in diameter than the inner centrifugal turns and have two distinct muscle bands. The centrifugal turns lack bands, except for some indistinct ones near the central flexure.

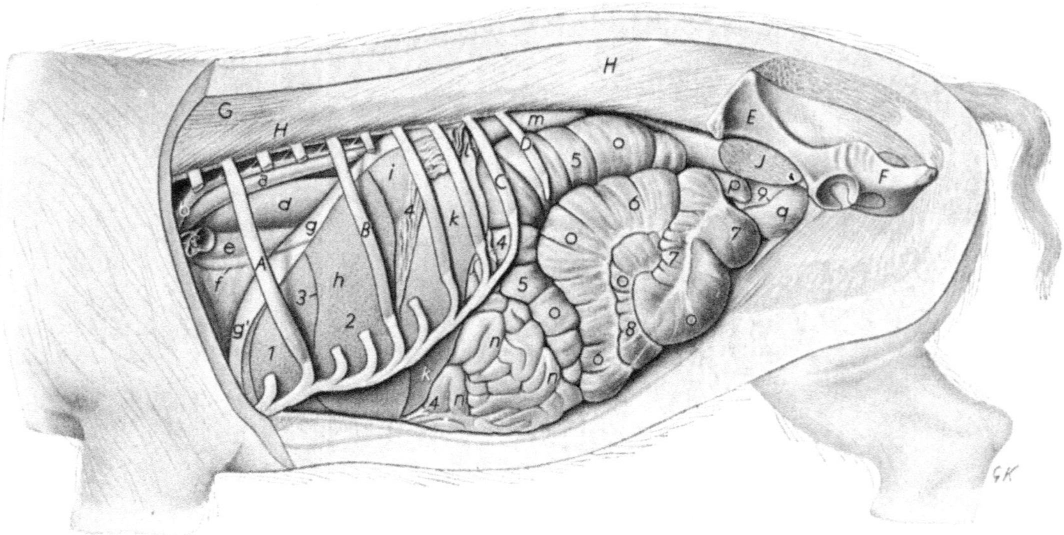

Fig. 197. Topography of the thoracic and abdominal organs of a female pig with an empty stomach. Left aspect. (After Graeger 1957.)

A Seventh rib; *B* Tenth rib; *C* Fourteenth rib; *D* Fifteenth (floating) rib; *E* Ilium; *F* Ischium; *G* Spinalis; *H* Longissimus; *J* Cut surface of iliopsoas

a Aorta; *b* Left azygous vein; *c* Root of left lung; *d* Esophagus; *e* Mediastinum; *f* Left phrenic nerve; *g* Diaphragm, its tendinous center, *g'* its sternal part; *h* Liver; *i* Stomach; *k* Spleen; *l* Pancreas; *m* Left kidney (perirenal fat removed); *n* Jejunum; *o* Ascending colon; *p* Left ovary; *q* Urinary bladder

1—3 On the liver: *1* Left medial lobe, *2* left lateral lobe, *3* interlobar notch; *4* Greater omentum; *5—8* On the ascending colon: *5* first, *6* second, and *7* third centripetal turns, *8* first centrifugal turn; *9* Lateral lig. of bladder

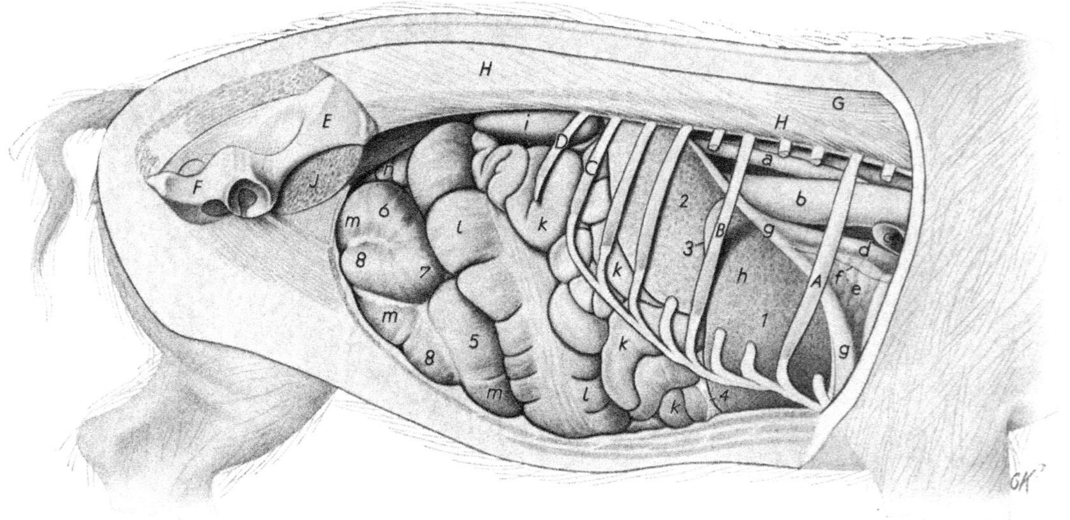

Fig. 198. Topography of the thoracic and abdominal organs of a female pig with an empty stomach. Right aspect. (After Graeger 1957.)

A Seventh rib; *B* Tenth rib; *C* Fourteenth rib; *D* Fifteenth (floating) rib; *E* Ilium; *F* Ischium; *G* Spinalis; *H* Longissimus; *J* Cut surface of iliopsoas

a Aorta; *b* Esophagus; *c* Right principal bronchus; *d* Caudal vena cava; *e* Plica venae cavae, covering accessory lobe of the lung; *f* Right phrenic nerve; *g* Diaphragm, its tendinous center, *g'* its sternal part; *h* Liver; *i* Right kidney (perirenal fat removed); *k* Jejunum; *l* Cecum; *m* Ascending colon; *n* Uterus

1—3 On the liver: *1* right medial lobe, *2* right lateral lobe, *3* interlobar notch; *4* Greater omentum; *5—8* On the ascending colon: *5* second, and *6* third centripetal turns, *7* central flexure, *8* first centrifugal turn

When the stomach is moderately full, the cone-shaped ascending colon occupies the middle, and to some extent also the cranial, third of the left half of the abdomen and makes extensive contact with the left abdominal wall. Its axis is directed ventrally and slightly cranially, and it is related cranially to the stomach and the spleen. The ascending colon is surrounded by the jejunum on the right, caudally, and ventrally. Dorsally, it is related to the pancreas, left

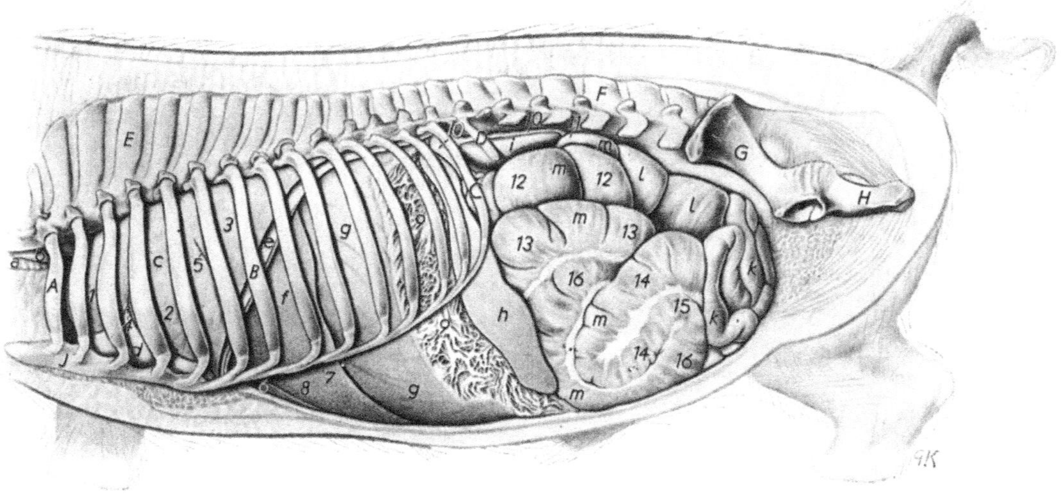

Fig. 199. Topography of the thoracic and abdominal organs of a castrated male pig with a greatly distended stomach. Left aspect. (After Graeger 1957.)

A First rib; *B* Seventh rib; *C* Fourteenth rib; *D* Fifteenth (floating) rib; *E* Spinous process of third thoracic vertebra; *F* Spinous process of third lumbar vertebra; *G* Ilium; *H* Ischium; *J* Sternum

a Trachea; *b* Esophagus; *c* Left lung; *d* Heart inside pericardium; *e* Diaphragm, its costal part; *f* Liver; *g* Stomach; *h* Spleen; *i* Left kidney (perirenal fat partly removed); *k* Jejunum; *l* Cecum; *m* Ascending colon

1—5 On the lung: *1*, *2* divided cranial lobe, *3* caudal lobe, *4* cardiac notch, *5* interlobar fissure; *6—8* On the liver: *6* Left medial lobe, *7* left lateral lobe, *8* interlobar notch; *9* Greater omentum (gastrosplenic ligament); *10* Cut surface of perirenal fat; *11—16* On the ascending colon: *11* first, *12* second, *13* third, and *14* fourth centripetal turns, *15* central flexure, *16* first centrifugal turn

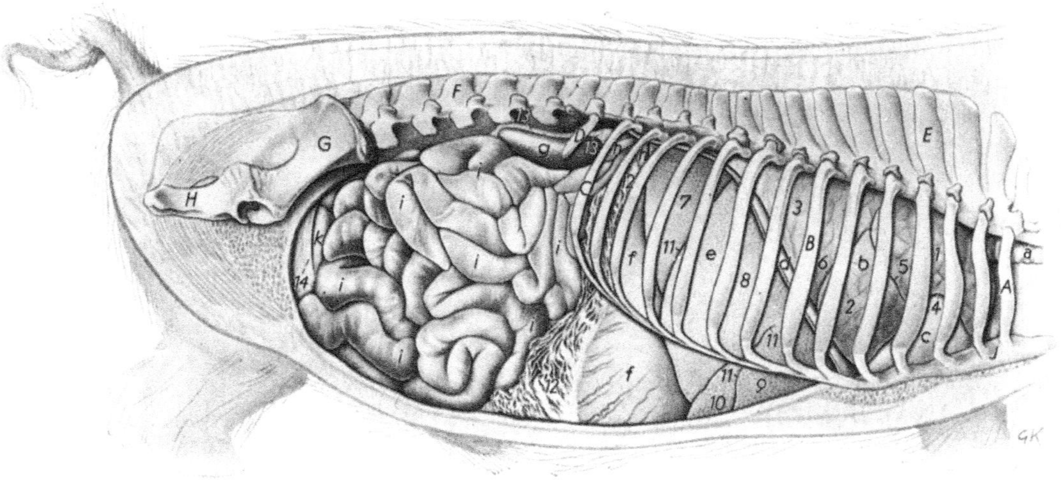

Fig. 200. Topography of the thoracic and abdominal organs of a castrated male pig with a greatly distended stomach. Right aspect. (After Graeger 1957.)

A First rib; *B* Seventh rib; *C* Fourteenth rib; *D* Fifteenth (floating) rib; *E* Spinous process of third thoracic vertebra; *F* Spinous process of third lumbar vertebra; *G* Ilium; *H* Ischium; *J* Sternum

a Trachea; *b* Right lung; *c* Heart inside pericardium; *d* Diaphragm, its costal part; *e* Liver; *f* Stomach; *g* Right kidney (perirenal fat partly removed); *h*, *h* Duodenum, cranial and descending parts; *i* Jejunum; *k* Urinary bladder

1—6 On the lung: *1* cranial lobe, *2* middle lobe, *3* caudal lobe, *4* cardiac notch, *5* cranial interlobar fissure, *6* caudal interlobar fissure; *7—11* On the liver: *7* right lateral lobe, *8* right medial lobe, *9* left medial lobe, *10* left lateral lobe, *11* interlobar notches; *12* Greater omentum; *13* Cut surface of perirenal fat; *14* Right lateral lig. of bladder

kidney, ascending duodenum, and the transverse and descending colons. These relations may be retained when the stomach is empty. However, if the jejunal mass moves cranially and some of its coils come to lie between the ascending colon and the empty stomach, then the ascending colon may be displaced caudally and to the right and may occupy, with its central axis now directed ventrally and slightly caudally, the caudal third of the abdominal cavity, reaching the pelvic inlet and making contact with the right abdominal wall (198).

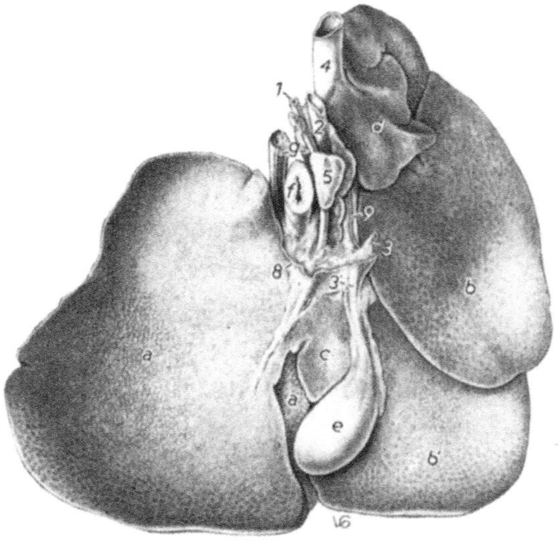

Fig. 201. Liver of the pig, fixed in situ. Visceral surface.

a Left lateral lobe; *a'* Left medial lobe; *b* Right lateral lobe; *b'* Right medial lobe; *c* Quadrate lobe; *d* Caudate process; *e* Gall bladder; *f* Esophagus in esophageal notch; *g* Medial parts of right crus of diaphragm

1 Hepatic branch of hepatic artery; *2* Portal vein; *3* Bile duct; *3'* Cystic duct; *4* Caudal vena cava; *5* Hepatic lymph nodes; *8*, *9* Lesser omentum; *8* Hepatogastric ligament; *9* Hepatoduodenal ligament

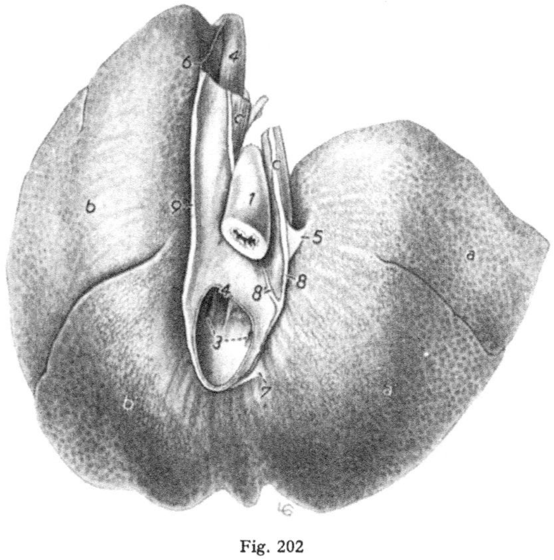

Fig. 202

Fig. 202. **Liver of the pig, fixed in situ. Diaphragmatic surface.**

a Left lateral lobe; *a'* Left medial lobe; *b* Right lateral lobe; *b'* Right medial lobe; *c* Medial parts of right crus of diaphragm

1 Esophagus in esophageal notch; *3* Openings of the hepatic veins; *4*, *4* Caudal vena cava; *5* Left triangular ligament; *6* Right triangular ligament; *7* Falciform ligament; *8*, *8'*, *9* Left, middle, and right laminae of coronary ligament

When the stomach is very distended, the central axis of the cone-shaped ascending colon remains in a ventral and slightly caudal direction, but owing to the general caudal displacement of the jejunum, the ascending colon is prevented from entering the right caudoventral quadrant of the abdominal cavity and remains in full contact with the left abdominal wall (199, 200).

The last centrifugal turn of the ascending colon winds around the caudal and right aspects of the root of the mesentery and is continued by the short **transverse colon** (192/*h*), which passes from right to left cranial to the root of the mesentery. The **descending colon** (196/*k*), suspended by a short, fatfilled descending mesocolon, lies close to the median plane and passes in a straight line to the pelvic inlet.

The **RECTUM** (545/*24*) is embedded in fat. Before it ends at the anal canal it widens to form a distinct **ampulla recti.**

The lining of the short **ANAL CANAL** can be divided, as in the carnivores, into columnar, intermediate, and cutaneous zones. The **columnar zone** is narrow, covered with stratified squamous epithelium, light red in color, and presents longitudinal depressions (sinus anales) associated with accumulations of lymphoreticular tissue. It is set off from the rectal mucosa by the **anorectal line.** The **intermediate zone** is also light red and covered with stratified squamous epithelium. It is separated from the **cutaneous zone** by the **anocutaneous line.**

The following **ANAL MUSCLES** are present in the pig. The **internal anal sphincter,** a smooth muscle, surrounds the anal canal. The **external anal sphincter** is striated and consists of cranial and caudal parts. The cranial part arises from the caudal fascia and surrounds the anal canal. In the boar it joins the bulbospongiosus (474/*i*); in the sow it unites with the constrictor vulvae (545/*19'*). The caudal part (pars cutanea) of the external anal sphincter surrounds the caudal edge of the anus, and in the sow continues into the lateral walls of the vestibule of the vagina. The **rectococcygeus** arises from the dorsal surface of the ampulla recti and passes caudally to attach on the ventral surface of the second and third caudal vertebrae. The **levator ani** arises from the medial surface of the sacroischiatic ligament and is inserted on the lateral wall of the anal canal. The **rectal part** (*26'*) **of the retractor penis** (clitoridis) is thin, passes around the ventral aspect of the rectum, and is independent of the other part of the retractor. The **penile part** of the retractor (474/*h*) consists of two cord-like components originating bilaterally from the second to fourth sacral vertebrae, passing ventrally across the lateral surfaces of the rectum, and attaching on the distal bend of the sigmoid flexure of the penis.

LYMPHORETICULAR TISSUE OF THE INTESTINES. Both solitary lymph nodules and patches of aggregate lymph nodules are present in large numbers in the pig. The **solitary nodules** are whitish and have a diameter of 1—2 mm. They are embedded in the mucosa of the entire intestinal tract, being less numerous in the cecum and more concentrated in the distal parts of the tract. **Patches of aggregate lymph nodules** are found mostly in the small intestine, which has 20—30, averaging about 10 cm. in length. Numerous deep, irregular depressions give their surface an unevenly pitted appearance. The most distal patch has the extraordinary length of 1.15 to 3.20 m. It extends along the entire ileum, including the papilla ilealis, and with increased width is continued up to 10 cm. into the ascending colon (194/*2*). The cecum is free of these patches, and usually so is the colon. Occasionally, small patches are found in the ascending colon in the vicinity of the extensive patch coming from the small intestine.

Liver

(154, 158, 195—202)

Depending on the age and the condition of the animal, the liver of the pig is either light or dark brownish red, and weighs 1—2.5 kg., which is 1.7 per cent of the body weight. The high content of interlobular connective tissue, which is characteristic of the pig's liver, makes the small (1—2 mm.) hepatic lobules readily visible and is a means of identifying it (161, 201, 202). The interlobar notches are deep and divide the liver into several distinct lobes. To the left of an imaginary line connecting the esophageal notch with the notch for the round ligament are the **left medial** and **left lateral lobes** (154, 158, 201/*a, a'*). To the right of a line connecting the caudal vena cava with the fossa for the gall bladder are the **right medial** and **right lateral lobes** (*b, b'*). The **quadrate lobe** (*c*) ventral to the hepatic porta is small and does not reach the ventral border of the liver. The **caudate lobe** above the porta is represented only by a **caudate process** (*d*), which projects dorsally and to the right. There is no papillary process.

The **diaphragmatic surface** of the liver (202) is strongly convex in adaptation to the concavity of the diaphragm. The **visceral surface** (201) is deeply concave, and in the fixed state presents the impressions of the organs that lie against it. There is no renal impression, because the liver of the pig does not make contact with the right kidney.

The caudal vena cava, in crossing the dorsal border of the liver, is usually completely embedded in liver tissue. It receives the hepatic veins (202/*3*) and continues directly to the foramen venae cavae in the diaphragm.

The **coronary ligament** (*8, 8', 9*), a narrow band that connects the liver to the diaphragm, extends from the short **left triangular ligament** (*5*) to the **right triangular ligament** (*6*), passing around the ventral surface of the caudal vena cava. The **round ligament** (158/*5*) arises from the notch for the round ligament and passes to the diaphragm in the caudal edge of the rather small **falciform ligament.** It continues its median course toward the umbilicus, but is usually lost under the peritoneum before reaching it. In older animals, the falciform ligament is represented only by a narrow peritoneal band on the diaphragmatic surface of the liver (202/*7*).

POSITION AND RELATIONS. The liver lies against the diaphragm almost entirely within the intrathoracic part of the abdominal cavity. The greater part of the liver lies to the right of the median plane, allowing the stomach, which is more to the left, to make contact with the left dorsal portion of the diaphragm (195—200). The most cranial point of the liver lies, with the most cranial point of the diaphragm, directly over the sternum and reaches the level of the fifth intercostal space. The caudal extent of the liver is along the eighth and ninth ribs on the left (195, 197, 199), and reaches a caudally convex line on the right, which begins at the proximal end of the thirteenth or fourteenth rib, passes to the costochondral junction of the tenth or eleventh rib, and from there nearly transversely to the ventral midline (196, 198, 200). The two left lobes and often also the right medial lobe make extensive contact with the abdominal wall ventrally between the costal arches. The deep interlobar notches allow the liver lobes considerable freedom of movement. The left lateral lobe is especially mobile and is often displaced to the right, apparently yielding to the pressure of the stomach. Most of the concave visceral surface is related to the stomach (197, 199). Only parts of the right medial and right lateral lobes make contact with the jejunum (196, 198).

The **hepatic ducts** leave the liver at the porta and form the **common hepatic duct,** which unites with the **cystic duct** from the gall bladder to form the unusually long **bile duct** (158, 201/*3*). The bile duct runs to the duodenum in the lesser omentum and ends 2—5 cm. distal to the pylorus on the indistinct major duodenal papilla.

The **GALL BLADDER** is embedded in a deep fossa between the quadrate and right medial lobes. It is long and pear-shaped, but does not reach to the ventral border of the liver. Its long cystic duct passes to the hepatic porta, where it unites with the common hepatic duct to form the bile duct.

Pancreas

The weight of the pancreas seems to depend more on the nutritional state than on the body weight of the animal, and ranges in pigs, weighing more than 100 kg., from 110—150 gm. As in the pancreas of the carnivores and ruminants, that of the pig (165) consists of a small **right** and a large **left lobe** connected by a centrally placed **body,** which forms a ring (anulus pancreatis, *3*) of tissue, through which the portal vein passes to the liver. The cranial part of the ring (*a*) is ventral to the portal vein and connects the two lobes (*b, c*). The caudal part of the ring (*d*) lies dorsal to the portal vein, also connects the lobes, is long, and directed nearly longitudinally.

The body of the pancreas lies against the lesser curvature of the stomach and the cranial part of the duodenum. The left lobe passes to the left and dorsally, and is attached to the dorsal body wall along the line of origin of the deep wall of the greater omentum, extending far enough to the left to make contact with the dorsocaudal border of the spleen, the cranial pole of the left kidney, and the left abdominal wall (195/*h*). The left lobe is also related to the transverse colon and to the base of the cone-shaped ascending colon. The right lobe accompanies the cranial part of the duodenum in the lesser omentum, and the descending duodenum in the mesoduodenum, and ends at the level of the cranial pole of the right kidney. Ventrally, it is in contact with the last section of the ascending colon and the transverse colon. The caudal part of the anulus (165/*d*) extends caudally on the right side of the root of the mesentery to the caudal flexure of the duodenum, and is related cranially to the ventral surface of the caudate process of the liver.

In the pig, only the duct of the dorsal pancreatic primordium, the **accessory pancreatic duct,** remains. The duct leaves the right lobe of the gland and ends at the **minor duodenal papilla** in the descending duodenum 20—25 cm. distal to the pylorus.

The Alimentary Canal of the Ruminants

Esophagus

The esophagus of the ruminants deserves detailed description because of its clinical importance especially in the ox.

The esophagus (60/*e*) begins at the laryngopharynx and lies with its initial portion on the lamina of the cricoid cartilage. Fibers of the **cricopharyngeus** and the **cricoarytenoideus dorsalis** enter the esophageal wall and attach it to the larynx, as does the paired **esophageus longitudinalis lateralis.** In the goat, the latter muscle has ventral and occasionally dorsal parts in addition to the lateral parts; in the ox and sheep, only the lateral parts are present.

The length of the bovine esophagus is 90—95 cm., the **cervical part** being 42—45 cm., and the thoracic part 48—50 cm. In the cranial third of the neck it lies between the longus colli and the trachea (132/*r*, *19*); in the caudal half of the neck it deviates to the left and lies against the lateral surface of the trachea for the remainder of its cervical course. The relations of the cervical part of the esophagus are described in the general chapter on page 99.

The **thoracic part** of the esophagus extends caudally in the mediastinum, and soon after passing through the thoracic inlet it returns to its original position between the longus colli and the trachea, until it reaches the end of the muscle at the level of the sixth thoracic vertebra. The esophagus passes dorsal to the tracheal bifurcation and the base of the heart, crosses the right surface of the aorta (*27*) opposite the fourth to seventh intercostal spaces, and passes through the esophageal hiatus of the diaphragm at the level of the eighth intercostal space. Before reaching the diaphragm it is related dorsally to the long **caudal mediastinal lymph node** (208, 213, 216/*i*). This node may enlarge when diseased and affect the esophagus and the accompanying dorsal vagal trunk (132/*43'*).

The muscular wall of the esophagus consists of striated muscle, which varies in thickness in the different segments of the tube. In the ox, the muscular wall is 4—5 mm. thick in the cervical part, but only 2—3 mm. thick in the thoracic part. Also the lumen of the esophagus varies in the different segments. At the junction of the middle and caudal thirds of the neck, the lumen narrows, but caudal to this constriction steadily widens again. In the cervical part it is rosette-shaped in cross section. Caudal to the heart, the lumen is large and oval in cross section, measuring 7—8 cm. dorsoventrally and 4—5 cm. from side to side. Except for correspondingly smaller dimensions, the esophagus of the small ruminants is similar to that of the ox (213, 214, 216). The ampulliform, thin-walled segment of the esophagus in the caudal mediastinum (207/*n*) is thought to play an important role during eructation* and regurgitation.

Treatment of esophageal obstruction in cattle** necessitates exact knowledge of the topography and the three flexures of the esophagus. The first, at the junction of the head and neck, is formed by the pharynx and the proximal segment of the esophagus, and is convex dorsally (132). The second flexure, at the thoracic inlet, extends along the ventral surface of the last few cervical and the first few thoracic vertebrae, and is concave dorsally. Because of their position at the beginning and the end of the neck, the degree of directional change at the first two flexures depends on the position of the head. If the head is held low, as in grazing, these flexures are nearly eliminated. The third flexure, in the middle of the thoracic cavity where the esophagus passes over the tracheal bifurcation and the base of the heart, is a very slight but permanent bend, and is convex dorsally.

Functional studies of the esophagus, particularly in relation to eructation, indicate that the caudal pharyngeal constrictors (thyropharyngeus and cricopharyngeus) act as a cranial esophageal sphincter, while the circular muscle layer in the wall of the esophagus a short distance cranial to the diaphragm, and the cardiac part and cardiac loop of the stomach combine to act as a caudal esophageal sphincter. At least in the ox and sheep, there seems to be a functional relationship between these two sphincters, since contraction of the cranial is followed by relaxation of the caudal and vice versa. Dougherty (1968) describes the mechanism of eructation and cites further references on the esophageal sphincters.

---

* From *ructus* (L.), the belch.

** Usually caused by unchewed apples or potatoes.

Ruminant Stomach

(203—229)

Because of the marked structural, functional, and topographic differences between the complex stomach of the ruminants and the simple stomach of the other domestic mammals, the ruminant stomach requires separate and detailed description.

The ruminant stomach consists of four compartments: the first three, the **rumen** (205, 206/*A*, *A'*), **reticulum** (*B*), and **omasum** (*C*), comprise the **forestomach** (proventriculus) and are lined with nonglandular mucous membrane; the fourth compartment, the **abomasum*** (*D*), is lined with glandular mucous membrane and is therefore the glandular part of the ruminant stomach.

Because of the distinct compartmentalization of the ruminant stomach and the characteristic functional differences of the compartments, it was long thought that the nonglandular rumen, reticulum, and omasum developed from the esophagus, and that only the glandular abomasum was the homologue of the simple stomach. It has been established, however, that the ruminant stomach, like the simple stomach, develops from a simple spindle-shaped primordium, and that the three parts of the forestomach are outgrowths of that primordium, from areas that correspond to the fundus and body of the simple stomach. In the ruminant as in the species with a simple stomach, the embryonic esophagus ends at the cardia, i.e., proximal to the beginning of the stomach primordium (Pernkopf 1931, and Warner 1958).

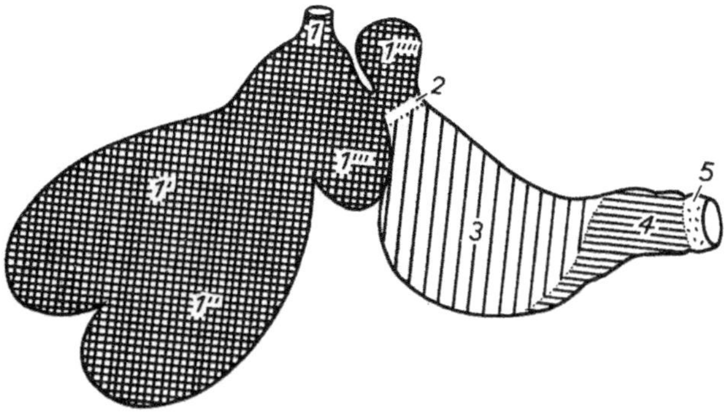

Fig. 203. Mucosal regions of the ruminant stomach. Schematic.

1—1'''' Nonglandular mucosa of the esophagus and proventriculus (cross hatched): *1* esophagus, *1'* dorsal sac of rumen, *1''* ventral sac of rumen, *1'''* reticulum, *1''''* omasum; *2, 3, 4* Abomasum; *2* Cardiac gland region (white); *3* Region of proper gastric glands (vertical lines); *4* Pyloric gland region (horizontal lines); *5* Duodenum (stippled)

The **CAPACITY** of the stomach of the adult ox depends on the size and the breed of the animal and ranges from 110—235 liters. The rumen and reticulum, which together are known as the ruminoreticulum, hold on the average 84 per cent of the total capacity, the rumen alone having a capacity of 102—148 liters. The abomasum, the next in size after the rumen, has a capacity of 10—20 liters, and the omasum 7—18 liters. In the small ruminants, the capacity of the stomach depends largely on the breed of the animal; but in general is 13—23 liters for the rumen, 1—2 liters for the reticulum, .3—.9 liters for the omasum, and 1.75—3.3 liters for the abomasum. The capacity of the four compartments would rank as follows in the ox: rumen, abomasum, omasum, and reticulum, and would differ in the small ruminant only in that the omasum would be smaller than the reticulum. The values given were established post mortem by removing the stomach contents and filling the stomach with measured amounts of water. Because the stomach is rarely completely filled with ingesta, the stated absolute capacities, particularly of the rumen, are never fully utilized in the live animal.

During the growing period of the animal, the stomach compartments change in shape and in their relative capacities. These changes are particularly noticeable in the rumen (204/*a*) and abomasum (*d*) and are related to the gradual change from an initial diet of milk to one consisting exclusively of plant material. At birth, the abomasum of the calf has a capacity of

---

* The rumen, reticulum, omasum, and abomasum are popularly known as the paunch, honeycomb, manyplies, and rennet or true stomach, respectively.

about 2 liters and the ruminoreticulum .75 liters. At eight weeks of age, the capacity of the ruminoreticulum roughly equals that of the abomasum. As the changeover to plant food is gradually completed, the abomasum lags behind more and more until in the fully grown animal the ratio between ruminoreticulum and abomasum is 9 to 1. The omasum (*c*) appears contracted and quiescent in the first few weeks, but increases steadily in size, keeping pace with the absolute increase in size of the entire stomach. In the small ruminants, the stomach goes through comparable phases. Figure 204 illustrates the relative sizes of the various compartments during growth in the ox.

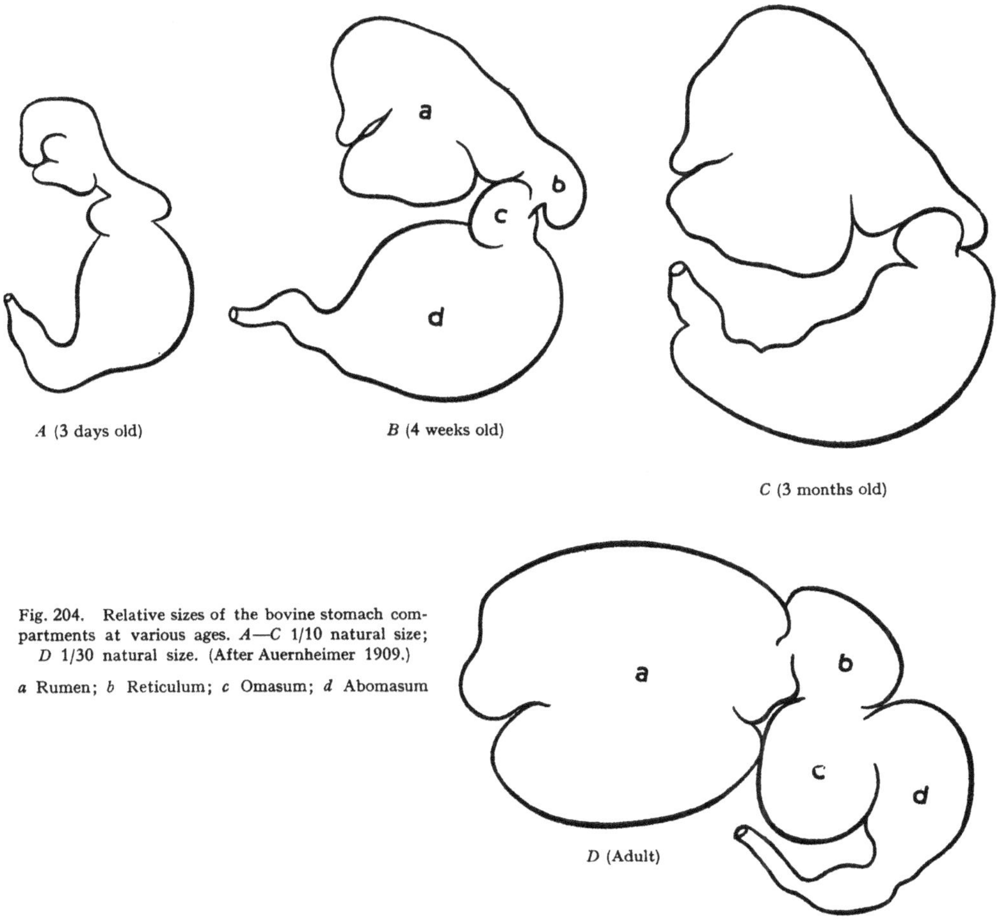

Fig. 204. Relative sizes of the bovine stomach compartments at various ages. *A—C* 1/10 natural size; *D* 1/30 natural size. (After Auernheimer 1909.)

a Rumen; *b* Reticulum; *c* Omasum; *d* Abomasum

SHAPE AND POSITION OF THE RUMINANT STOMACH. The relative position of the four compartments to each other is such that the rumen lies on the left, the reticulum cranially, and the omasum on the right. The abomasum lies ventrally, with its proximal portion below the rumen, reticulum, and omasum (206—208, 214, 216).

The **RUMEN** is a huge, laterally compressed sac which occupies a major portion of the abdominal cavity. It extends from the diaphragm to the pelvic inlet, filling the left half of the abdominal cavity (207); and at times its caudoventral part extends well over the median plane into the right half of the abdominal cavity (229).

The **parietal surface** of the rumen faces mainly to the left and is related to the diaphragm, the left abdominal wall, and the floor of the abdomen. The **visceral surface** faces to the right and is related chiefly to the intestines, the liver, the omasum, and the abomasum. The **dorsal curvature** (205/*a*) lies against the diaphragm and the roof of the abdominal cavity; the **ventral curvature** (*b*) follows the contour of the abdominal floor.

The rumen is divided into several parts by a number of grooves of varying depth. Shallow left and right **longitudinal grooves** (205/*c*; 206/*c*, *c'*) on the parietal and visceral surfaces respectively are connected cranially and caudally by two deep transverse grooves, the **cranial** and **caudal grooves** (*d*, *e*). These four grooves form a nearly horizontal constriction, which divides the rumen into dorsal and ventral sacs. The **dorsal sac** (*A*) lies to the left of the median plane, while the **ventral sac** (*A'*) extends often into the right half of the abdominal cavity (229). The **left longitudinal groove** begins at the cranial groove and, passing at first dorsocaudally, extends along the left side of the rumen to the caudal groove, giving off an accessory groove (205/*c'*) that extends for a short distance along the surface of the dorsal sac.

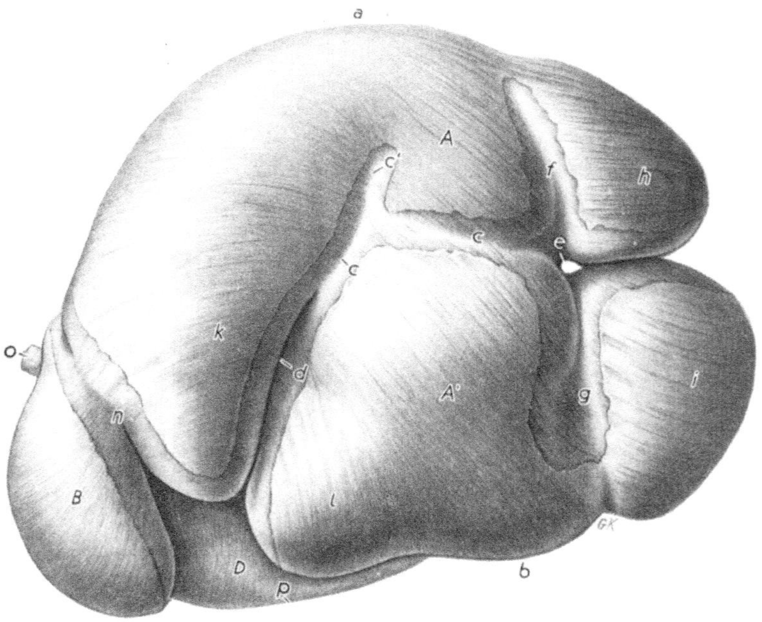

Fig. 205. Bovine stomach. Left lateral aspect. The ruminal grooves appear more prominent in this preparation because the fat, vessels, lymph nodes, and nerves they contain have been removed.
A, *A'* Rumen, its parietal surface, *A* dorsal sac, *A'* ventral sac; *B* Reticulum; *D* Abomasum; *a* Dorsal curvature of rumen; *b* Ventral curvature of rumen; *c* Left longitudinal groove; *c'* Left accessory groove; *d* Cranial groove; *e* Caudal groove; *f* Dorsal coronary groove; *g* Ventral coronary groove; *h* Caudodorsal blind sac; *i* Caudoventral blind sac; *k* Cranial sac of rumen; *l* Recessus ruminis; *n* Ruminoreticular groove; *p* Greater curvature of abomasum

The **right longitudinal groove** splits into two limbs (206/*c*, *c'*) which enclose an elongated area of the wall of the rumen (insula ruminis, *c''*). Although the dorsal limb is more prominent, the deep wall of the greater omentum attaches to the ventral limb. The **dorsal** and **ventral coronary grooves** (*f*, *g*) extend in opposite directions from the caudal end of the longitudinal grooves and mark off the **caudodorsal** and **caudoventral blind sacs** (*h*, *i*). The ventral coronary groove extends completely around the base of the caudoventral blind sac, but the dorsal groove is deficient dorsally. In the ox, the two blind sacs are of about equal length; in the small ruminants, however, the caudoventral blind sac extends farther caudally than the caudodorsal sac.

The two projections (205/*k*, *l*) on the cranial end of the rumen above and below the cranial groove used to be considered cranial blind sacs, but developmental studies have shown that they are not true blind sacs but flexures of the tubelike rumen primordium (Pernkopf, 1931); moreover, they lack the distinct circular muscle layer that characterizes blind sacs. The dorsal projection (*k*), the most cranial part of the rumen, is called the **cranial sac of the rumen** (atrium ruminis); the ventral (*l*) is the **recessus ruminis.** The cranial sac of the rumen is continuous caudally with the dorsal sac. In front it communicates with the reticulum at the wide **ruminoreticular opening** (207/*2*, *2*) through which food from the reticulum passes into the cranial sac of the rumen and from there into the other sacs and vice versa. The cranial sac of the rumen also plays an important role in the regurgitation of food for remastication.

The esophagus (206/*o*) enters the stomach at the junction of rumen and reticulum. Opposite the cardia is the deep **ruminoreticular groove** (205, 206/*n*), which separates the rumen from the reticulum.

The ruminal grooves contain fat and most of the blood vessels, lymphatics, lymph nodes, and nerves of the rumen. They are bridged over by the visceral peritoneum and in some cases also by muscle fibers, and unless they are dissected appear rather shallow. With the exception of the ruminoreticular groove, the ruminal grooves are represented on the inside by muscular pillars, which project to varying degrees into the interior of the rumen. They are described on page 160.

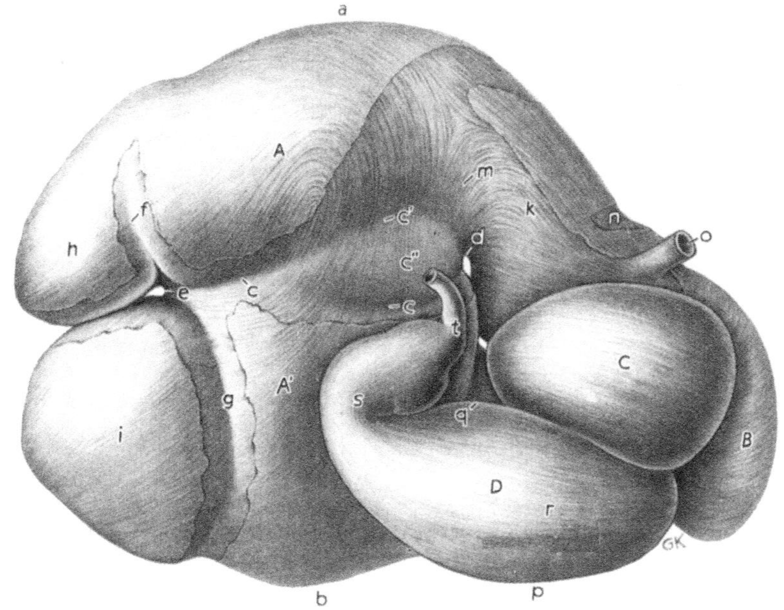

Fig. 206. Bovine stomach. Right lateral aspect.

A, A' Rumen, its visceral surface, *A* dorsal sac, *A'* ventral sac; *B* Reticulum; *C* Omasum; *D* Abomasum; *a* Dorsal curvature of rumen; *b* Ventral curvature of rumen; *c, c'* Right longitudinal groove; *c''* Insula ruminis; *d* Cranial groove; *e* Caudal groove; *f* Dorsal coronary groove; *g* Ventral coronary groove; *h* Caudodorsal blind sac; *i* Caudoventral blind sac; *k* Cranial sac of rumen; *m* Shallow groove between dorsal and cranial sac; *n* Ruminoreticular groove; *o* Esophagus; *p*—*t* On the abomasum: *p* greater curvature, *q* lesser curvature, *r* fundus and body (to the right of *r*: fundus, to the left of *r*: body), *s* pyloric part, *t* pylorus

The **RETICULUM** (205, 206/*B*) is the most cranial compartment of the ruminant stomach. It is spherical but slightly flattened craniocaudally, and lies between the diaphragm and the rumen (207) at the level of the sixth to the ninth intercostal spaces, about equally to the right and left of the median plane. Dorsally, it is continued without demarcation by the cranial sac of the rumen, while ventrally and to the sides it is sharply separated from the rumen by the deep ruminoreticular groove. Its **diaphragmatic surface** is convex, in adaptation to the curvature of the diaphragm; its **visceral surface** is applied against the rumen. On the right, the reticulum is related to the left lobe of the liver, the omasum, and the abomasum; on the left, it lies against the costal part of the diaphragm and occasionally is in contact with the ventral end of the spleen. Its ventral relations are the sternal part of the diaphragm, the caudal end of the sternum, and the xiphoid cartilage (212—216).

The **OMASUM** of the ox (208/*q*) is a spherical organ which is somewhat compressed laterally between its visceral and parietal surfaces. It has a **curvature** facing dorsocaudally and to the right, and opposite the curvature is the flat **base** which faces in the opposite direction. The omasum is clearly set off from the reticulum by a necklike constriction (collum omasi) and from the abomasum by a similar, but wider, constriction (sulcus omasoabomasicus). In the small ruminants, the omasum is oval and smaller than the reticulum (214).

The bovine omasum lies ventrally in the intrathoracic part of the abdominal cavity, to the right of the median plane, between the ventral sac of the rumen on the left and the abdominal wall on the right. Craniodorsally, it is related to the liver. Its **visceral surface** faces mainly

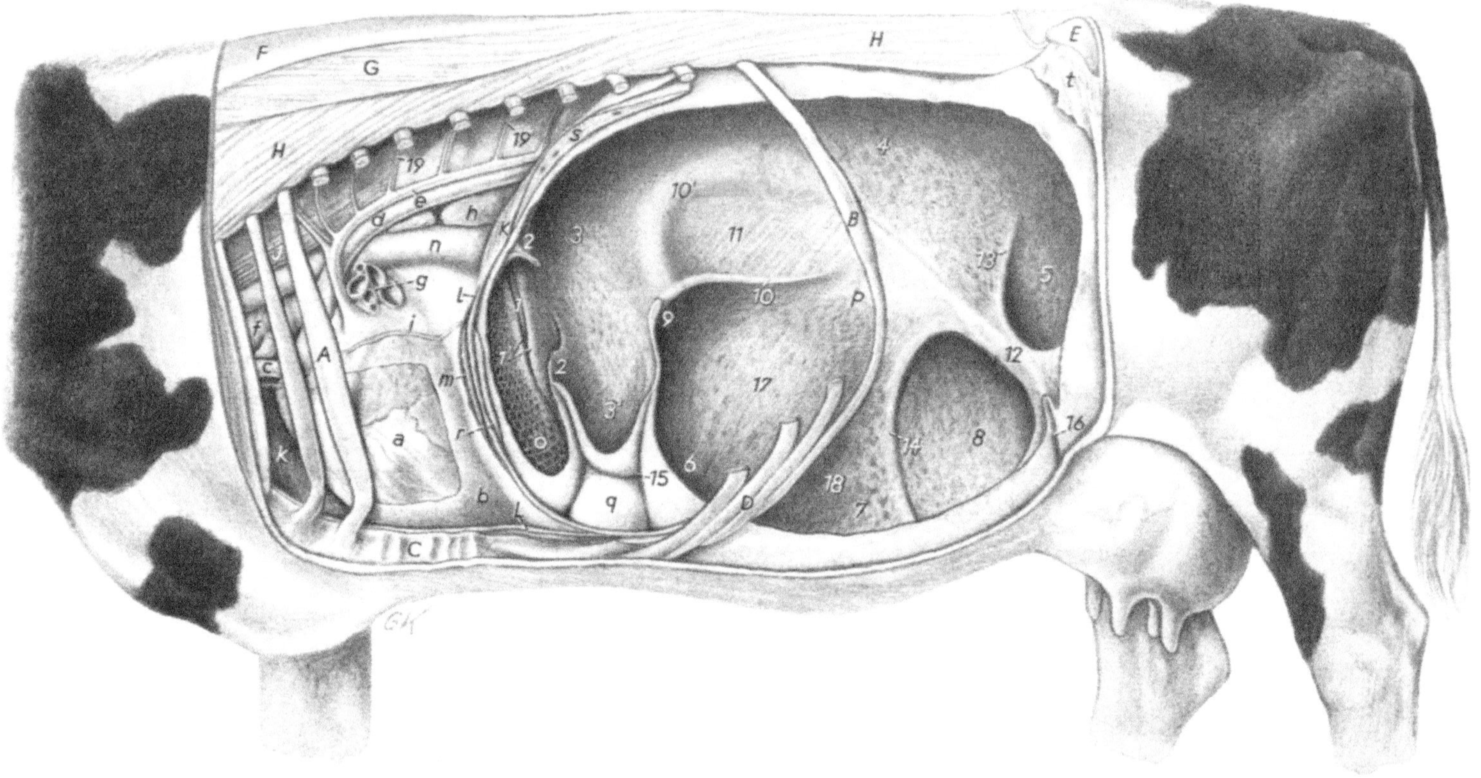

Fig. 207. Topography of the thoracic and abdominal organs of an adult cow. The left lung has been removed and the reticulum and rumen opened to show the interior. Left aspect. (After Nickel and Wilkens 1955.)

A Fourth rib; *B* Thirteenth rib; *C* Sternum; *D* Costal arch; *E* Tuber coxae; *F* Ligamentum nuchae; *G* Spinalis et semispinalis thoracis et cervicis; *H*, *H* Longissimus; *J* Longus colli; *K* Diaphragm, part of its right crus; *L* Transversus thoracis

a Heart; *b* Pericardium, opened; *c* Brachiocephalic trunk; *d* Aorta; *e* Left azygous vein; *f* Trachea; *g* Root of left lung; *h* Caudal mediastinal lymph nodes; *i* Left phrenic nerve; *k* Cranial mediastinum; *l* Caudal mediastinum; *m* Accessory lobe of right lung; *n* Esophagus; *o* Reticulum; *p* Rumen; *q* Abomasum; *r* Liver; *s* Spleen, cut surface; *t* Fat

1 Reticular groove, *1'* its lips; *2*, *2* Ruminoreticular fold, enclosing ruminoreticular opening; *3*—*18* On the rumen: *3*, *3'* cranial sac, *4* dorsal sac, *5* caudodorsal blind sac, *6* recessus ruminis, *7* ventral sac, *8* caudoventral blind sac, *9* cranial pillar, *10*, *10'* right longitudinal pillar, *11* insula ruminis, *12* caudal pillar, *13* dorsal coronary pillar, *14* ventral coronary pillar, *15* cranial groove, *16* caudal groove, *17* omasal bulge (see 209/*6*), *18* abomasal bulge (see 209/*7*); *19* Dorsal intercostal artery and vein

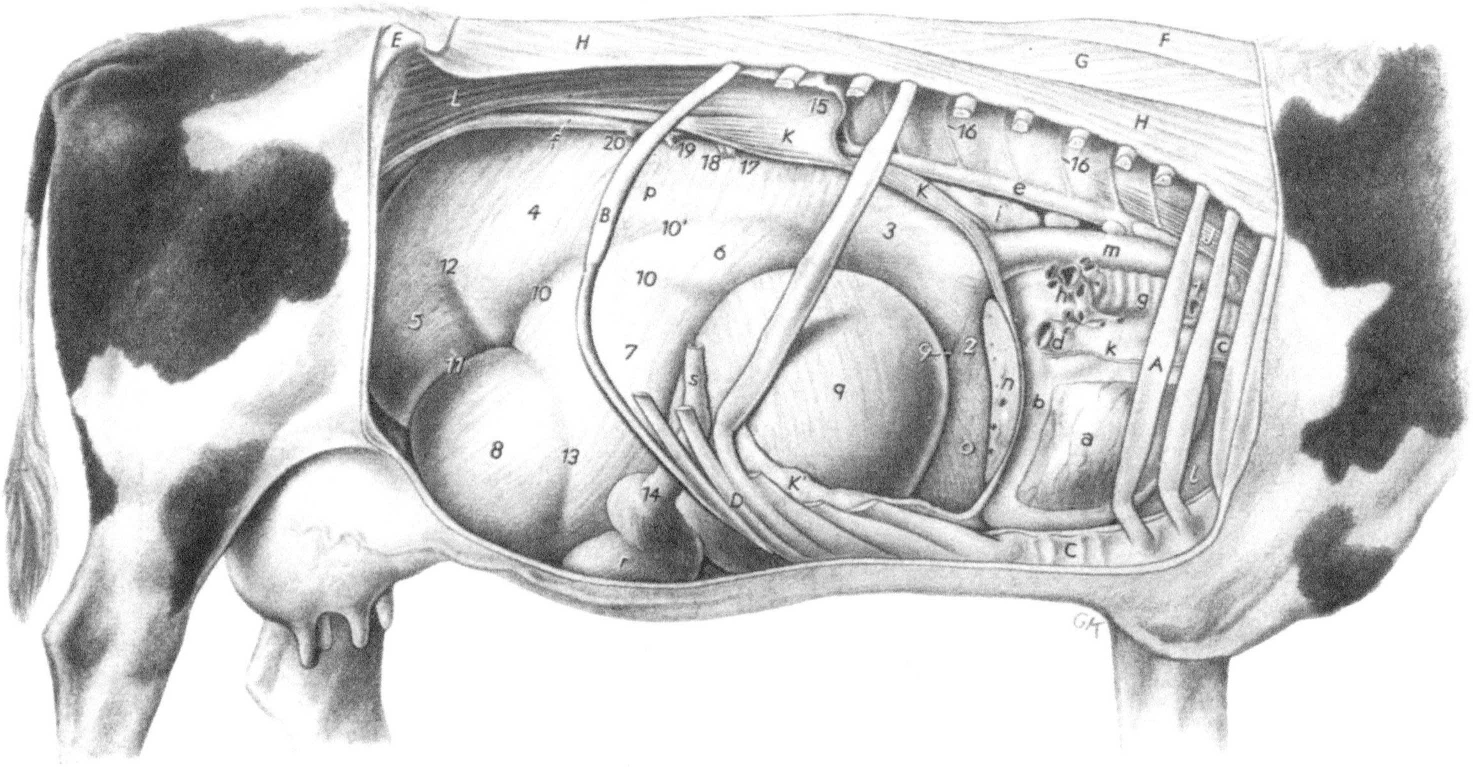

Fig. 208. Topography of the thoracic and abdominal organs of an adult cow. The right lung, omenta, intestines, kidneys, and most of the liver have been removed. Right aspect. (After Nickel and Wilkens 1955.)

A Fourth rib; *b* Thirteenth rib; *C* Sternum; *D* Costal arch; *E* Tuber coxae; *F* Ligamentum nuchae; *G* Spinalis et semispinalis thoracis et cervicis; *H, H* Longissimus; *J* Longus colli; *K, K'* Diaphragm; *L* Psoas musculature

a Heart; *b* Pericardium, opened; *c* Cranial vena cava; *d* Caudal vena cava, cut; *e, f* Aorta; *g* Trachea; *h* Root of right lung; *i* Caudal mediastinal lymph nodes; *k* Right phrenic nerve; *l* Cranial mediastinum; *m* Esophagus; *n* Left lobe of liver, cut surface; *o* Reticulum; *p* Rumen; *q* Omasum; *r* Abomasum; *s* Duodenum

1 Tracheal bronchus; *2* Area of reticular groove; *3—13* On the rumen: *3* cranial sac, *4* dorsal sac, *5* caudodorsal blind sac, *6* insula ruminis, *7* ventral sac, *8* caudoventral blind sac, *9* ruminoreticular groove, *10, 10'* right longitudinal groove, *11* caudal groove, *12* dorsal coronary groove, *13* ventral coronary groove; *14* Pylorus; *15* Recess in the diaphragm for the dorsal part of the liver; *16* Dorsal intercostal artery; *17* Celiac artery; *18* Cranial mesenteric artery; *19* Right renal artery; *20* Left renal artery

to the left, but also slightly caudally and is applied against the ventral sac of the rumen. Its **parietal surface** faces in the opposite direction and is related to the diaphragm, liver, and gall bladder. Between the sixth and eleventh intercostal spaces, the omasum is in contact with the right abdominal wall, protruding ventrally by about 10 cm. from the costal arch. Ventrally, the omasum is related to the reticulum and abomasum, and caudally to the jejunum. The omasum of the sheep and goat lies more medially at about the level of the eighth to tenth ribs and is not in contact with the right abdominal wall, although its other relations are similar to the ones described for the ox.

The **ABOMASUM** (206/*D*), the most distal compartment, follows the three compartments of the forestomach. It is a bent, pear-shaped sac, which is set off from the omasum by a deep annular constriction. The abomasum looks much like a simple stomach and consequently

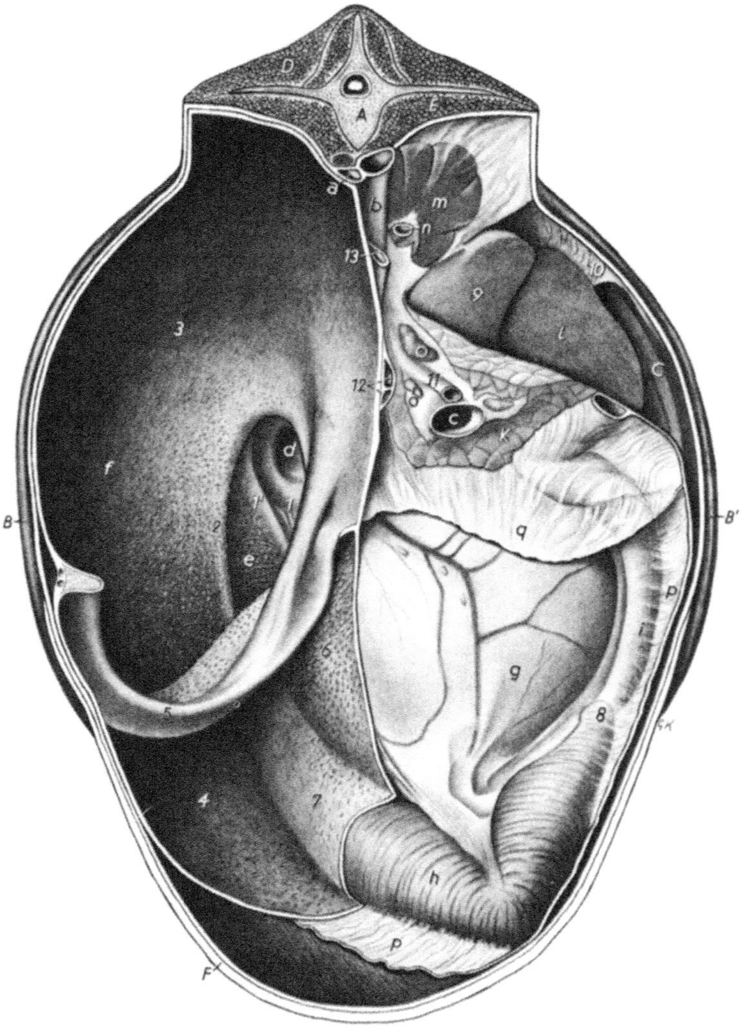

Fig. 209. Cranial abdominal organs of an adult ox. Caudoventral aspect. (After Nickel and Wilkens 1955.)

A Third lumbar vertebra; *B*, *B'* Left and right thirteenth rib; *C* Right twelfth rib; *D* Epaxial musculature in cross section; *E* Psoas musculature in cross section; *F* Ventral abdominal wall

a Aorta; *b* Caudal vena cava; *c* Portal vein; *d* Cardiac opening; *e* Reticulum; *f* Rumen; *g* Omasum; *h* Abomasum; *i* Duodenum; *k* Pancreas; *l* Right lobe of liver; *m* Right kidney; *n* Ureter; *o* Mesenteric lymph nodes; *p* Supf. wall of greater omentum, the portion that descends from the left longitudinal groove of the rumen has been removed; *q* Deep wall of greater omentum, passing to the right longitudinal groove of the rumen

1 Reticular groove, *1'* its left lip; *2* Ruminoreticular fold; *3—7* On the rumen: *3* cranial sac, *4* recessus ruminis, *5* cranial pillar, *6* omasal bulge, *7* abomasal bulge; *8* Pylorus; *9* Caudate process of liver; *10* Right triangular ligament; *11* Cranial mesenteric artery; *12* Right ruminal artery and vein; *13* Left renal vein

has been divided into a fundus, a body, and a pyloric part. It has a **greater curvature** (*p*) facing ventrally and to the left, and a **lesser curvature** (*q*) facing dorsally and to the right. The **fundus** and **body** (*r*) lie on the abdominal floor caudal to the reticulum; the longitudinal axis of this portion crosses the midline somewhat obliquely from left cranial to right caudal, overlying the region caudal to the xiphoid cartilage between the ventral ends of the costal arches (209, 211, 212). The **pyloric part** of the abomasum (206/*s*) is directed dorsolaterally behind the omasum and is followed at the pylorus by the duodenum. The abomasum is

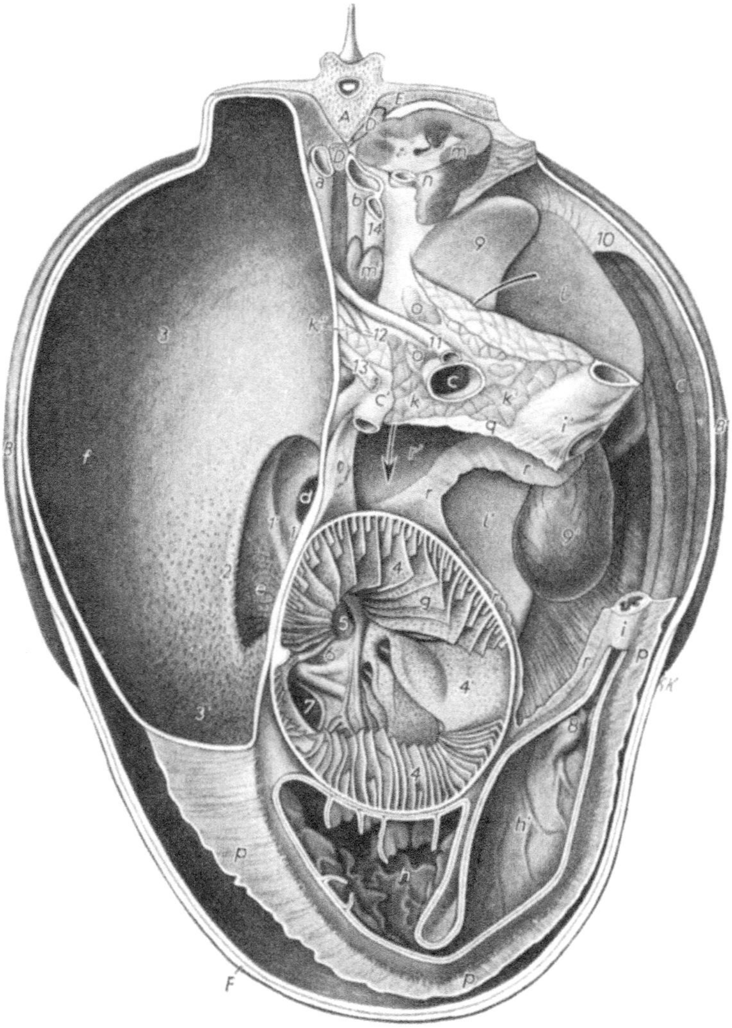

Fig. 210. Cranial abdominal organs of an adult ox. Caudoventral aspect.

A Second lumbar vertebra; *B, B'* Left and right thirteenth rib; *C* Right twelfth rib; *D* Medial part of right crus of diaphragm; *D'* Lateral part of right crus of diaphragm; *E* Psoas musculature; *F* Ventral abdominal wall

a Aorta; *b* Caudal vena cava; *c* Portal vein; *c'* (the venous trunk dorsal to the letter) Splenic vein; *d* Cardiac opening; *e* Reticulum; *f* Rumen; *g* Omasum; *h* Body of abomasum with spiral folds; *h'* Pyloric part of abomasum; *i* Duodenum; *i'* Sigmoid loop of duodenum; *k, k', k''* Body, right lobe, and left lobe of pancreas; *l* Right lobe of liver; *l'* Quadrate lobe of liver; *m* Cut surface of right kidney; *m'* Right adrenal gland; *n* Right ureter; *o* Mesenteric lymph nodes; *p* Supf. wall of greater omentum, attaching to rumen, greater curvature of abomasum, and duodenum; *q* Deep wall of greater omentum, attaching to liver, pancreas, and duodenum; *r* Lesser omentum, attaching to liver, omasum, lesser curvature of abomasum, and duodenum; *r'* Vestibule of omental bursa, the arrow passes through epiploic foramen

1 Reticular groove, *1'* its left lip; *2* Ruminoreticular fold, enclosing ruminoreticular opening; *3, 3'* Cranial sac of rumen; *4* Omasal laminae, the longer ones have been shortened; *4'* Surface of a long omasal lamina; *5* Reticulo-omasal opening; *6* Omasal groove at the base of the omasum; *7* Omasoabomasal opening; *8* Pyloris with torus pyloricus; *9* Caudate process of liver; *9'* Gall bladder; *10* Right triangular ligament; *11* Cranial mesenteric artery; *12* Hepatic artery; *13* Right ruminal artery and vein; *14* Left renal vein

closely related on the left to the recessus ruminis and may, when distended with ingesta, extend to the left and come to lie ventral to the cranial sac of the rumen, making contact with the left abdominal wall (207/*q*; 213/*o*; 216/*n*).

STRUCTURE OF THE RUMINANT STOMACH. The walls of all four compartments are composed of the three typical layers found elsewhere in the alimentary canal: serous coat, muscular coat, and mucous membrane. Except for an area of the dorsal sac that is attached to the roof of the abdominal cavity, the stomach is invested with visceral peritoneum, which bridges the ruminal grooves and most of the deeper constrictions between the compartments, and thus obscures the vessels, lymphatics, and nerves running in them.

The **MUSCULAR COAT** of the ruminant stomach consists of smooth muscle; however, some striated muscle fibers from the longitudinal muscle layer of the esophagus continue onto the stomach and radiate over parts of the rumen and reticulum.

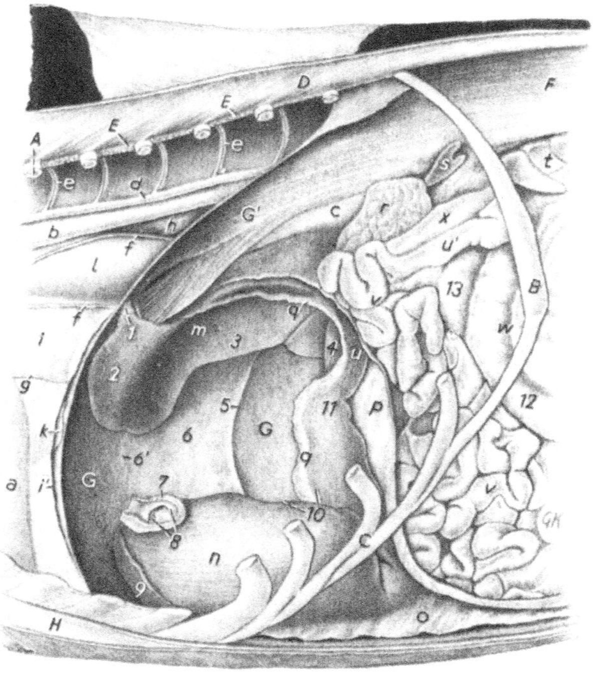

Fig. 211. Topography of some thoracic and abdominal organs of an adult ox. Detail of Fig. 235 after complete removal of the reticulum and omasum. Left aspect. (After Nickel and Wilkens 1955.)

A Stump of seventh rib; *B* Thirteenth rib; *C* Costal arch; *D* Longissimus; *E* Intercostal muscles; *F* Psoas musculature; *G, G'* Diaphragm; *H* Rectus abdominis

a Pericardium; *b* Aorta; *c* Caudal vena cava; *d* Left Azygous vein; *e* Dorsal intercostal artery and vein; *f, f* Dorsal and ventral vagal trunks; *g* Phrenic nerve; *h* Caudal mediastinal lymph node; *i* Caudal mediastinum; *i'* its cut edge; *k* Accessory lobe of right lung; *l* Esophagus; *m* Liver; *n* Abomasum; *o* Supf. wall of greater omentum; *p* Deep wall of greater omentum; *q* Lesser omentum; *r* Pancreas; *s* Left adrenal gland; *t* Left kidney; *u* Cranial part of duodenum; *u'* Ascending duodenum; *v* Jejunum; *w* Ascending colon; *x* Descending colon

1 Left triangular ligament; *2* Reticular impression on liver; *3* Omasal impression on liver; *4* Gall bladder; *5* Round ligament of liver; *6* Falciform ligament; *6'* its line of attachment on the diaphragm; *7* Omasoabomasal opening; *8* Vela abomasica; *9, 10* Greater and lesser curvatures of abomasum; *11* Pylorus; *12, 13* Centripetal and centrifugal turns of ascending colon

The complicated arrangement of the various smooth muscle layers in the wall of the ruminant stomach is best understood in relation to its development and in relation to the homology of its compartments to the corresponding parts of the simple stomach. The part of the spindle-shaped stomach primordium that develops into the fundus and body of the simple stomach enlarges markedly and becomes the ruminoreticulum. The distal part of the stomach primordium, which gives rise to the pyloric part of the simple stomach, becomes the abomasum. The omasum is an outgrowth of the lesser curvature of the primordium and has no homologue in the simple stomach.

The same muscle layers found in the wall of the simple stomach are generally present in the ruminant stomach. Similarly, the rather complicated three-layered arrangement on the fundus and body of the simple stomach is reflected in the muscular arrangement of the homologous ruminoreticulum, while the omasum and abomasum, like the pyloric part of the simple stomach, have only longitudinal and circular layers.

The arrangements of the muscle layers in the ruminant stomach, then, is as follows. Beginning at the cardia, one part of the **longitudinal layer** (217/*solid lines*) passes along the region of the reticular groove to the omasum (*d*) then over the omasum and abomasum to the pylorus. It forms the longitudinal muscle layer of these two compartments, and becomes gradually thicker as it approaches the duodenum. The other part of the longitudinal layer spreads out over the dorsal sac of the rumen (*b*), representing here the external oblique fibers of the simple stomach. The **circular layer** *(dotted lines)* provides the floor of the reticular groove (between *a* and *d*), encircles the reticulum (*c*), and continues along the walls of the ventral sac to the apex of the caudoventral blind sac. It also provides the circular muscle layer of the omasum and abomasum, and of the **pyloric sphincter.** A third layer *(broken lines)* corresponding to the **internal oblique fibers** of the fundus and body of the simple stomach, is present on the ruminoreticulum. As in the simple stomach, the bundles of this layer are concentrated at the cardia and along the lesser curvature, where they form the **cardiac loop.** The cardiac loop passes around the dorsal aspect of the cardiac opening and with its two straight limbs follows the lesser curvature to the reticulo-omasal opening, where the limbs exchange fibers. The concentrated muscle bundles of the cardiac loop and its two limbs provide the muscular base of the lips of the reticular groove (213/*1*), which is described on page 162. The remaining, less concentrated internal oblique fibers radiate from the cardiac loop over the reticulum and surround in nearly circular fashion all parts of the rumen, especially the two blind sacs. They continue into the cranial, caudal, and longitudinal grooves and thus form the muscular pillars of the rumen.

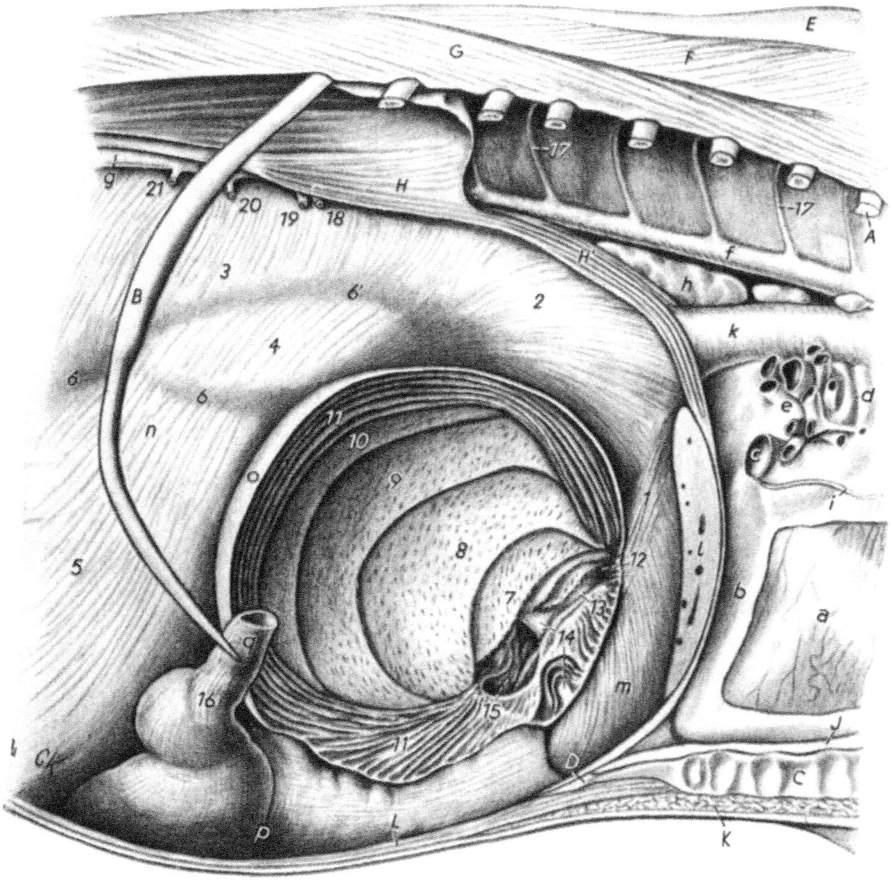

Fig. 212. Topography of some thoracic and abdominal organs of an adult ox. Detail of Fig. 208 after opening the omasum and removing the costal arch. Right aspect. (After Nickel and Wilkens 1955.)

A Stump of sixth rib; *B* Thirteenth rib; *C* Sternum; *D* Xiphoid cartilage; *E* Ligamentum nuchae; *F* Spinalis et semispinalis thoracis et cervicis; *G* Longissimus; *H, H'* Right crus of diaphragm; *J* Transversus thoracis; *K* Deep pectoral muscle; *L* Ventral abdominal wall

a Heart; *b* Pericardium, opened; *c* Caudal vena cava, cut; *d* Trachea; *e* Root of right lung; *f, g* Aorta; *h* Caudal mediastinal lymph nodes; *i* Right phrenic nerve; *k* Esophagus; *l* Left lobe of liver, cut surface; *m* Reticulum; *n* Rumen; *o* Omasum, opened from the right; *p* Abomasum; *q* Duodenum

1 Area of reticular groove; *2* Cranial sac of rumen; *3* Dorsal sac of rumen; *4* Insula ruminis; *5* Ventral sac of rumen; *6, 6'* Right longitudinal groove; *7, 8, 9, 10* Omasal laminae of various sizes; *11* Stumps of removed laminae; *12* Reticulo-omasal opening; *13* Omasal groove at the base of the omasum; *14* Omasal pillar; *15* Omasoabomasal opening, flanked by the vela abomasica; *16* Pylorus; *17* Dorsal intercostal artery; *18* Celiac artery; *19* Cranial mesenteric artery; *20* Right renal artery; *21* Left renal artery

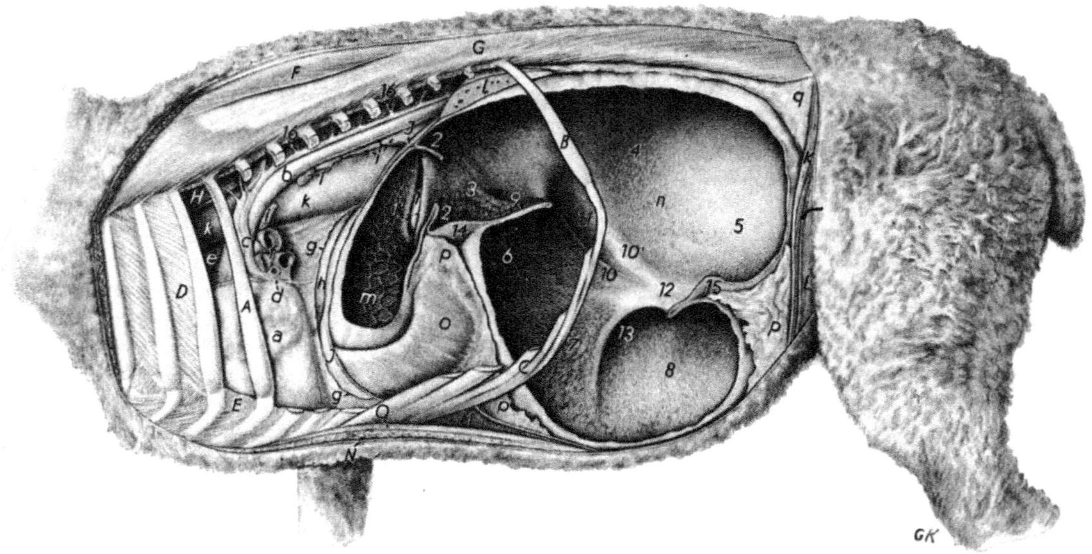

Fig. 213. Topography of the thoracic and abdominal organs of a sheep. The left lung, the left half of the diaphragm, and part of the spleen have been removed; the reticulum and rumen have been opened. Left aspect. (After Wilkens 1956b.)

A Fourth rib; *B* Thirteenth rib; *C* Costal arch; *D* Ext. intercostal muscles; *E* Int. intercostal muscles; *F* Spinalis et semispinalis thoracis et cervicis; *G* Longissimus; *H* Longus colli; *J* Diaphragm; *K* Int. abdominal oblique muscle; *L* Ext. abdominal oblique muscle; *M* Transversus abdominis; *N* Pectoral muscles; *O* Rectus abdominis

a Heart inside pericardium; *b* Aorta; *c* Left azygous vein; *d* Left phrenic nerve; *e* Trachea; *f* Root of left lung; *g* Caudal mediastinum, its cut edge; *h* Accessory lobe of right lung; *i* Caudal mediastinal lymph nodes; *k* Esophagus; *l* Cut surface of spleen; *m* Reticulum; *n* Rumen; *o* Abomasum; *p* Supf. wall of greater omentum; *q* Fat

1 Reticular groove, *1'* its lips; *2, 2* Ruminoreticular fold; *3* Cranial sac of rumen; *4* Dorsal sac of rumen; *5* Caudodorsal blind sac; *6* Recessus ruminis; *7* Ventral sac of rumen; *8* Caudoventral blind sac; *9* Cranial pillar; *10, 10'* Right longitudinal pillar; *11* Insula ruminis; *12* Caudal pillar; *13* Ventral coronary pillar; *14* Cranial groove; *15* Caudal groove, covered by the attachment of the greater omentum; *16* Dorsal intercostal artery and vein

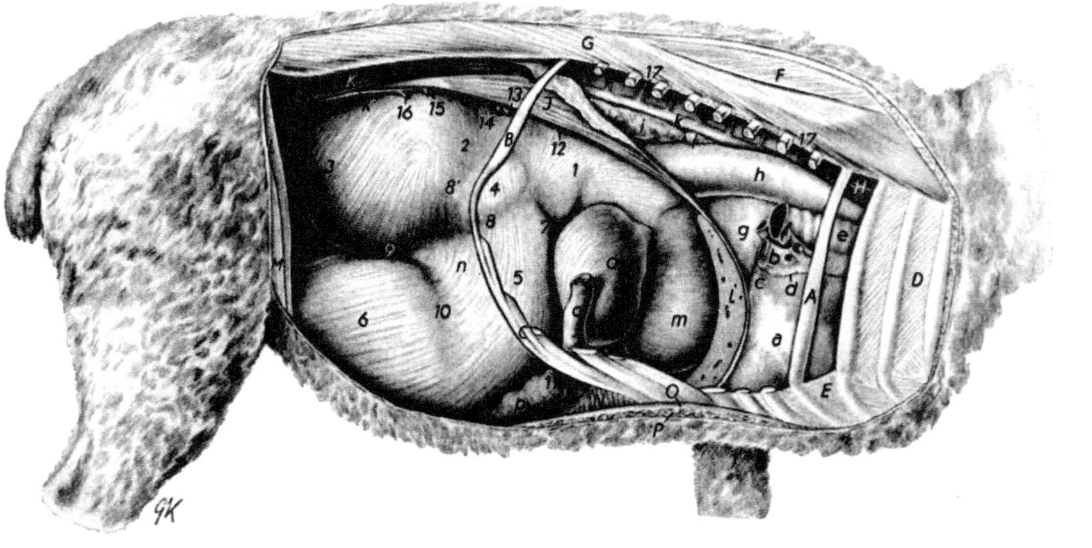

Fig. 214. Topography of the thoracic and abdominal organs of a sheep. The right lung, right half of diaphragm, omenta, intestines, kidneys, and most of the liver have been removed. Right aspect. (After Wilkens 1956b.)

A Fourth rib; *B* Thirteenth rib; *C* Costal arch; *D* Ext. intercostal muscles; *E* Int. intercostal muscles; *F* Spinalis et semispinalis thoracis et cervicis; *G* Longissimus; *H* Longus colli; *J* Right crus of diaphragm; *K* Psoas musculature; *L* Int. abdominal oblique muscle; *M* Ext. abdominal oblique muscle; *N* Transversus abdominis; *O* Rectus abdominis; *P* Pectoral muscles

a Heart inside pericardium; *b* Caudal vena cava; *c* Plica venae cavae; *d* Right phrenic nerve; *e* Trachea; *f* Root of right lung; *g* Caudal mediastinum; *h* Esophagus; *i* Caudal mediastinal lymph nodes; *k, k'* Aorta; *l* Liver; *m* Reticulum; *n* Rumen; *o* Omasum; *p* Abomasum; *q* Duodenum

1 Cranial sac of rumen; *2* Dorsal sac of rumen; *3* Caudodorsal blind sac; *4* Insula ruminis; *5* Ventral sac of rumen; *6* Caudoventral blind sac; *7* Cranial groove; *8, 8'* Right longitudinal groove; *9* Caudal groove; *10* Ventral coronary groove; *11* Pylorus; *12* Splenic artery; *13* Celiac artery; *14* Cranial mesenteric artery; *15* Right renal artery; *16* Left renal artery; *17* Dorsal intercostal artery and vein

Since the anatomy of the **INTERIOR OF THE RUMINANT STOMACH** is closely related to the physical and chemical processes in the proventricular compartments, a knowledge of these processes, by which the ingesta are prepared for chemical digestion in the abomasum, makes for a better understanding of the anatomical structure. These processes include: thorough soaking and mixing of the partly chewed food; breakdown of cellulose by bacteria and protozoa; regurgitation of the ingesta for remastication in the mouth by complex movements of the ruminoreticulum; sorting of the remasticated bolus in the ruminoreticulum and its eventual transport through the omasum into the abomasum; shunting of liquid ingesta, especially of milk in the suckling animal, directly into the abomasum; and, finally, the periodic elimination (via the esophagus) of gases that accumulate in large quantities as a result of bacterial activity in the rumen. These processes, some of which not as yet fully understood, necessitate exact coordination of the activities of the various compartments; and the ruminant stomach, despite its size and compartmentalization, has to be regarded as a

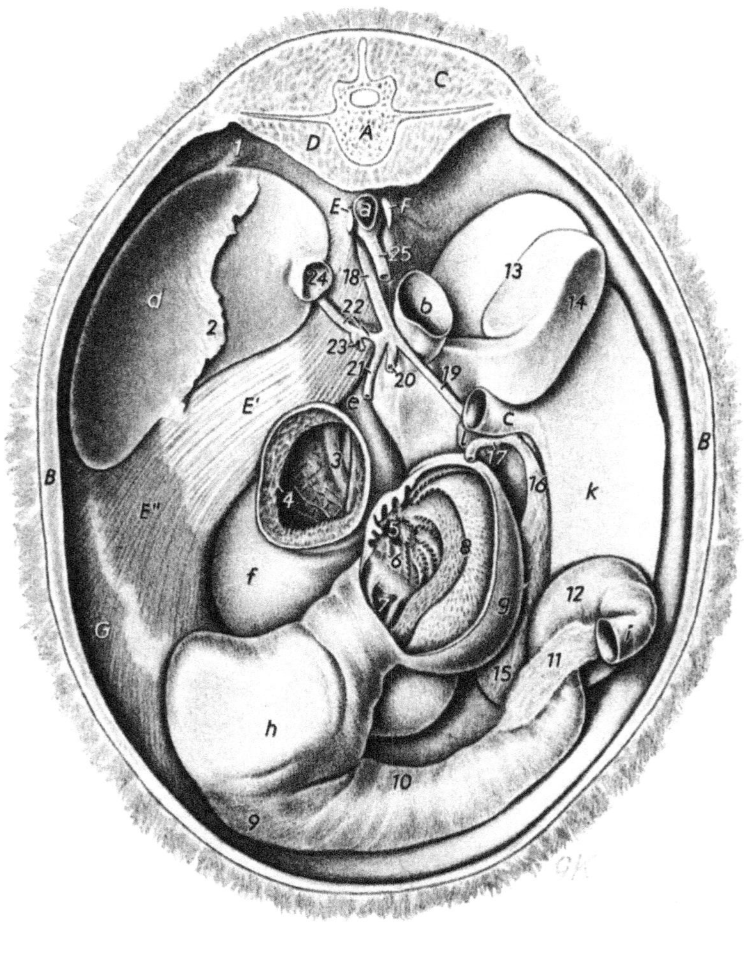

Fig. 215. Cross section through a sheep at the level of the second lumbar vertebra, showing the cranial abdominal organs. Caudal aspect. (After Wilkens 1956b.)

A Second lumbar vertebra; *B* Abdominal wall; *C* Epaxial musculature; *D* Psoas musculature; *E*–*F* Diaphragm: *E* left crus, *E'* tendinous center, *E''* costal part, *F* right crus; *G* Transversus abdominis

a Aorta; *b* Caudal vena cava; *c* Portal vein; *d* Spleen; *e* Esophagus; *f* Reticulum; *g* Omasum; *h* Abomasum; *i* Duodenum; *k* Liver; *l* Hepatic lymph nodes

1 Phrenicosplenic ligament; *2* Gastrosplenic ligament; *3* Reticular groove; *4* Ruminoreticular fold; *5* Reticulo-omasal opening; *6* Omasar groove at the base of the omasum; *7* Omasoabomasal opening, flanked by the vela abomasica; *8* Omasal laminae; *9, 10* Greater and lessel curvatures of abomasum; *11* Stump of lesser omentum; *12* Pylorus; *13* Renal impression of liver; *14* Caudate process; *15* Gall bladder; *16* Cystic duct; *17* Bile duct; *18* Celiac artery; *19* Hepatic artery; *20* Left gastric artery; *21* Left ruminal artery; *22* Splenic artery; *23* Right ruminal artery; *24* Splenic vein; *25* Cranial mesenteric artery

single functional unit. It is not surprising, therefore, that through faulty husbandry and incorrect feeding the delicate functional balances essential to this organ may be easily disturbed and cause disease.

As already mentioned, the grooves that divide the **RUMEN** externally are represented on the inside by pillars of corresponding name, e. g., the left longitudinal groove on the outside is the left longitudinal pillar on the inside. Some of these pillars are very prominent and result from a foldlike duplication of the internal muscle layer of the stomach wall; others are small and mere thickenings of the stomach wall. The **cranial pillar** (207/*9*) projects caudodorsally like a shelf into the rumen and lies between the cranial sac and the recessus ruminis. The **caudal pillar** (*12*) projects cranially between the two blind sacs. The **right longitudinal pillar** (*10, 10'*) connects the cranial and caudal pillars on the right side and, like the corresponding groove, is split into two limbs. The **left longitudinal pillar** (229/*d*) continues the left end of the cranial pillar, but does not reach the caudal pillar. The cranial, caudal, and longitudinal pillars surround the **intraruminal opening** through which the dorsal sac communicates with the ventral sac. The **dorsal** and **ventral coronary pillars** (207/*13, 14*) are branches of the caudal pillar, and like the corresponding grooves, the ventral coronary pillar extends completely around the base of the caudoventral blind sac, whereas the dorsal does not.

The mucous membrane of the rumen has no glands and is covered with a cornified stratified squamous epithelium. In the suckling animal this epithelium is light in color, but becomes greenish yellow to dark brown in older animals as a result of plant dyes and tannic acid staining the cells of the stratum corneum. The mucous membrane forms large conical or tongue-shaped **papillae** up to 1 cm. long, which give the internal surface of the rumen its characteristic pilelike appearance. The papillae are well developed in the ventral sac, the blind

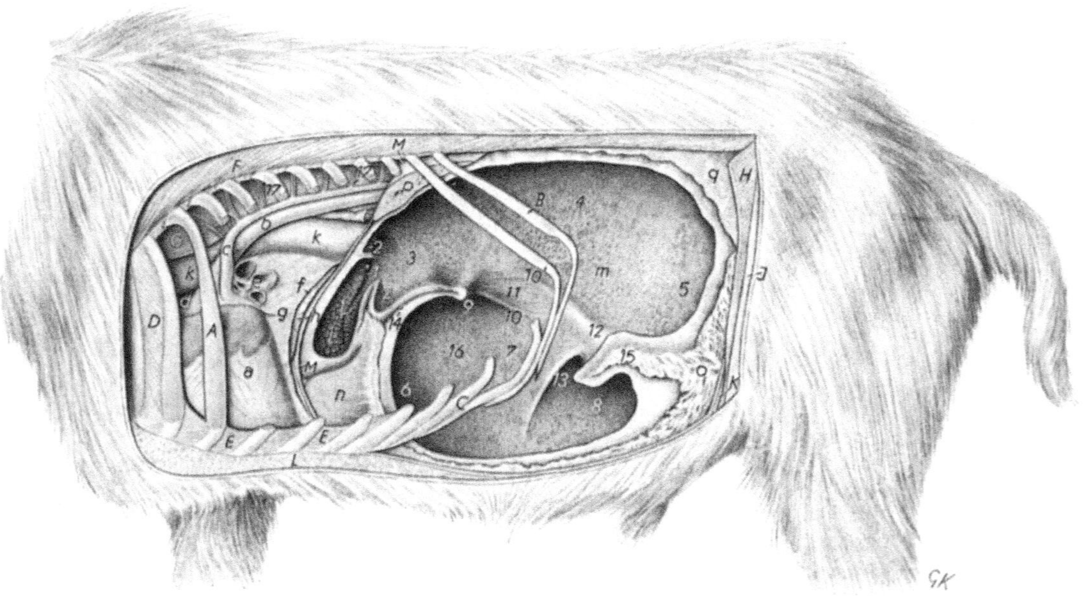

Fig. 216. Topography of the thoracic and abdominal organs of a male goat. The left lung has been removed and the reticulum and rumen have been opened. Left aspect. (After Wilkens 1956a.)

A Fourth rib; *B* Thirteenth rib; *C* Costal arch; *D* Ext. intercostal muscle; *E* Int. intercostal muscle; *F* Longissimus; *G* Longus colli; *H* Int. abdominal oblique; *J* Aponeurosis of ext. abdominal oblique; *K* Rectus abdominis; *L* Pectoral muscles; *M* Diaphragm

a Heart inside pericardium; *b* Aorta; *c* Left azygous vein; *d* Trachea; *e* Root of left lung; *f* Cut edge of caudal mediastinum; *g* Left phrenic nerve; *h* Accessory lobe of right lung; *i* Caudal mediastinal lymph node; *k* Esophagus; *l* Reticulum; *m* Rumen; *n* Abomasum; *o* Greater omentum; *p* Cut surface of spleen; *q* Fat

1 Reticular groove, *1'* its lips; *2, 2* Ruminoreticular fold; *3* Cranial sac of rumen; *4* Dorsal sac of rumen; *5* Caudodorsal blind sac; *6* Recessus ruminis; *7* Ventral sac; *8* caudoventral blind sac; *9* Cranial pillar; *10, 10'* Right longitudinal pillar; *11* Insula ruminis; *12* Caudal pillar; *13* Ventral coronary pillar; *14* Cranial groove; *15* Caudal groove, covered by attachment of greater omentum; *16* Omasal bulge; *17* Dorsal intercostal artery and vein

sacs, and in the cranial sac, but decrease in size toward the pillars, on which they are absent (223). Most of the roof of the dorsal sac also lacks papillae (224). This seems to be associated with the regular presence of a large bubble of gas on top of the ingesta, so that the roof of the rumen hardly ever comes in contact with stomach contents. The papillae greatly increase the surface area of the ruminal mucosa, through which primarily fatty acids and sodium are absorbed. Whether the papillae have a mechanical function as well is debatable. They have been thought to increase friction between the ingesta and the wall of the rumen, thereby facilitating the mixing of the ingesta during contractions of the stomach, but their primary function seems to be absorptive.

Rhythmic ruminal contractions take place 10—14 times every 5 minutes in the ox, and 7—16 times every 5 minutes in the sheep and goat; that is, about 2—3 per minute. The movement of the ingesta along the rough ruminal wall and the simultaneous rupture of the many fine gas bubbles that result from bacterial fermentation produce the characteristic sound that accompanies the contractions. The movements of the rumen can be felt by placing the hand against the left paralumbar fossa, and may also be seen, since the wall of the fossa moves with each contraction.

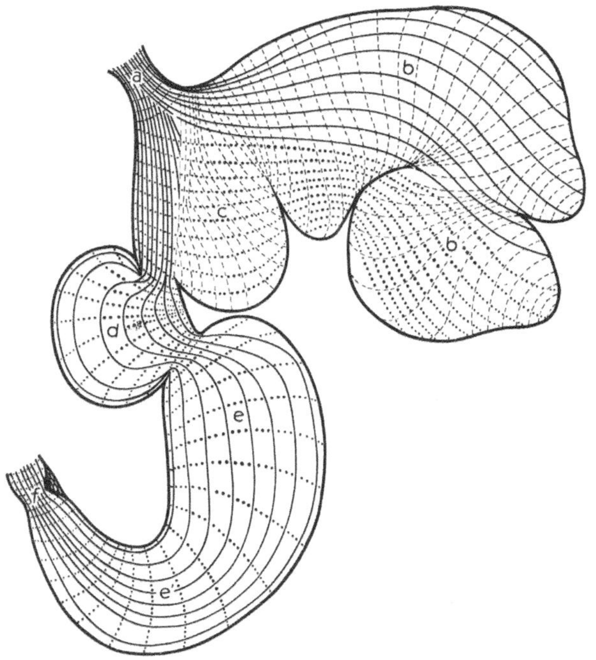

Fig. 217. The muscular coat of the ruminant stomach. Schematic. (Redrawn from Pernkopf 1930.)

a Cardia; *b*, *b'* Dorsal and ventral sacs of rumen; *c* Reticulum; *d* Omasum; *e*, *e'* Abomasum; *f* Pylorus

Solid lines: *longitudinal layer.* This layer divides at the cardia and passes as *ext. oblique fibers* onto the dorsal sac of the rumen. The other division forms the longitudinal muscle layer in the region of the reticular groove, the omasum, the abomasum, and the pylorus. Dotted lines: *circular layer.* This layer forms the *cardiac sphincter* and the circular fibers of the reticulum, omasum, and abomasum, and the *pyloric sphincter.* From the reticulum, the fibers of this layer continue onto the cranial and ventral sacs of the rumen. Broken lines: *int. oblique fibers.* They are found on the dorsal and ventral sacs of the rumen and encircle the blind sacs. They form the lips of the reticular groove and the *cardiac loop*

The rumen communicates with the reticulum through the **ruminoreticular opening,** which is about 18 cm. high (dorsoventrally) and 13 cm. wide in the ox, and is almost completely surrounded by the **ruminoreticular fold** (207, 209/*2*), which is high ventrally where it separates the cranial sac of the rumen from the reticulum. The fold gradually decreases in height as it ascends along the left side, then continues across the roof, and tapers off on the right side of the cranial sac of the rumen. Its ruminal surface is covered with papillae, while its reticular surface bears the honeycomb crests typical of the reticulum. During rumenotomy* for the removal of foreign bodies from the reticulum, the surgeon inserts his hand into the rumen and through the ruminoreticular opening into the reticulum to examine the interior and remove offending objects that frequently lodge there.

The **RETICULUM** has an especially well-developed muscular wall, which upon contraction can almost occlude the lumen and lift the ingesta into the cranial sac of the rumen. The nonglandular mucosa of the reticulum (226) forms permanent **crests** 8—12 mm. high, which intersect to form honeycomb-like **cells.** Each cell is subdivided by lower, secondary crests, and both the crests and the floor of the cells are studded with small **papillae.** Prominent cords of muscle fibers run inside the free edges of the reticular crests.

* Operation in which the rumen is opened, usually through the left flank.

The function of the reticular cells is still unknown. According to Grau (1955) finely chewed food particles settle into the reticular cells, and, upon contraction of the reticulum, are passed into the omasum, while the coarse material is lifted over the ruminoreticular fold back into the cranial sac of the rumen. In contrast, Hofmann (1969) believes that during contraction of the reticulum the cells hold on to the coarse material, while the suspension of finely chewed food particles passes through the temporarily held coarse material, like through a sieve, before entering the omasum.

The **GASTRIC GROOVE** of the ruminant stomach is well developed and of considerable physiological importance. It extends from the cardia through the reticulum, omasum, and abomasum almost to the pylorus and is customarily divided into three segments: the **reticular groove,** the **omasal groove,** and the **abomasal groove.**

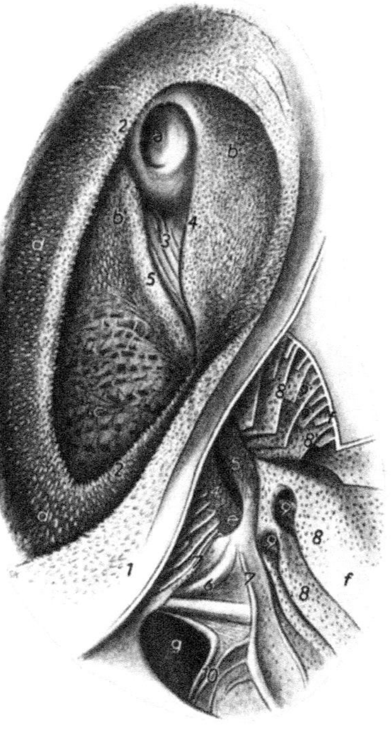

Fig. 218

Fig. 218. Reticular and omasal grooves of the ox. Caudoventral aspect.

a Cardia; *b* Cardiac part of stomach; *c* Reticulum; *d* Cranial sac of rumen; *e* Reticulo-omasal opening; *f* Omasum; *g* Omasoabomasal opening

1 Cranial pillar; *2* Ruminoreticular fold, surrounding the ruminoreticular opening; *3* Reticular groove, floor, *4* its right, *5* its left lip; *6* Omasal groove at the base of the omasum; *7,7* Mucosal folds representing the lips of the omasal groove; *8* Omasal laminae, *8'* their cut edges; *9, 9'* Interlaminar recesses; *10* Velum abomasicum parietale

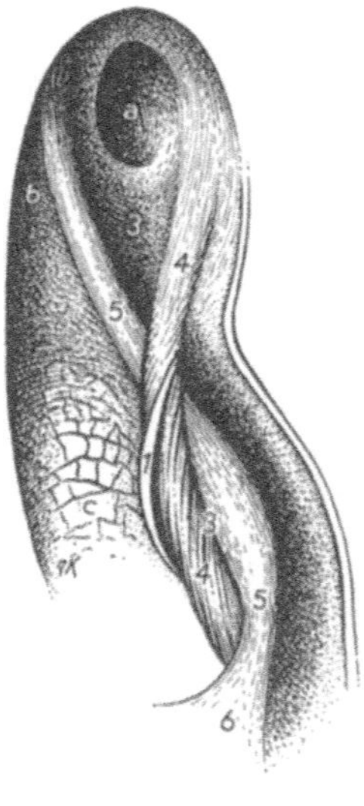

Fig. 219

Fig. 219. Schematic representation of the reticular groove of the ox. Caudoventral aspect. See Fig. 218 for legends

The **cardiac opening** is at about the level of the eighth intercostal space essentially dorsal to the fundus of the reticulum, so that solid food arriving at the cardia usually drops into that compartment (207, 213). (In the suckling animal, swallowed milk is conducted via the reticular and omasal grooves directly into the abomasum.) The **reticular groove** (sulcus reticuli, 218, 219, 225) is formed by two muscular ridges or lips extending from the cardia to the reticulo-omasal opening, and is 15—20 cm. long in the ox and 7—10 cm. long in the sheep and goat. The muscle layers of the stomach wall taking part in the formation of the lips and floor of the groove are described on page 157. The two lips meet dorsal to the cardiac opening and pass ventrally and slightly caudally along the right wall of the reticulum, the **right lip** (218/*4*) twisting around the **left lip** (*5*) in a clockwise direction when the groove is viewed from a dorsal position. At the ventral end of the groove, the right lip passes around the ventral aspect of the **reticulo-omasal opening** from left to right. The spiral twist of the lips around each other is such that the floor of the groove faces at first caudally, then to the left, and finally cranially. The floor of the reticular groove presents longitudinal folds and is marked by horny papillae at the reticulo-omasal opening, which are thin and curved (papillae unguiculiformes). A reflex, triggered by the presence of salts in liquids passing through pharynx and proximal portion of esophagus brings the lips of the reticular groove together to form a tube (Dietz et al., 1970).

The reticulum is the part involved in traumatic gastritis (hardware disease), a fairly common disease of cattle. Since coarse food and especially heavier foreign bodies enter the reticulum first, nails and pieces of wire, which are occasionally present in the ration of stable-fed animals, lodge in this compartment. The forceful and complete contraction of the reticulum causes these objects to penetrate the reticular wall and injure neighboring organs. The object commonly penetrates cranially into the diaphragm and through the right pleural cavity into the pericardium* a short distance away from the stomach (207, 208). Less frequently, the liver, omasum, abomasum, or the ventral body wall are penetrated. If the disease is diagnosed early, the offending object can often be removed before it has done serious damage, through an incision in the left flank and dorsal sac of the rumen.

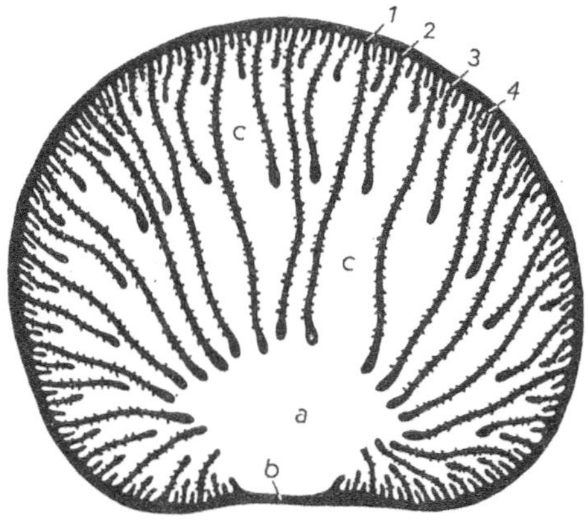

Fig. 220. Transverse section of the bovine omasum. Schematic. *a* Omasal canal; *b* Omasal groove; *c* Interlaminar recesses; *1, 2, 3, 4* Omasal laminae of various sizes

The **OMASUM** is larger than the reticulum in the ox and smaller than the reticulum in the sheep and goat. It is spherical and communicates with the reticulum through the reticulo-omasal opening and with the abomasum through the omasoabomasal opening (210/*5, 7*; 212/*12, 15*). It is nearly filled with many parallel folds of varying sizes, the **omasal laminae** (220/*1*—*4*), which arise from the wall and project with their free edges into the interior. Between the laminae are the **interlaminar recesses** (*c*). If, as in Figure 220, numbers 1 through 4 are assigned to the laminae of the same relative size, with 1 to the highest fold and 4 to the lowest, it will be found that they are arranged in the following sequence: 1, 4, 3, 4, 2, 4, 3, 4, 1, 4, etc. The omasum of the ox has 12—16 of the highest laminae and a total of 90—130. In the sheep the total number is 72—80, and in the goat 80—88.

The omasal laminae are thin muscular sheets covered with mucous membrane, and consist of two outer layers and an intermediate layer of muscle in the high laminae and only the two outer layers in the low laminae. The intermediate layer is derived from the inner circular layer of the omasal wall, with fibers directed from the attached border to the free border of the lamina. The outer layers are the muscularis mucosae of the mucous membrane on either side of the fold, the fibers of which run parallel to the free border of the lamina and thus cross those of the intermediate layer. Along the free edge of each lamina, the outer layers form a marginal thickening, which is especially prominent on the high laminae close to the reticulo-omasal opening (222). The omasal laminae are covered with papillae, which are short and stubby over most of the surface, making it rough to the touch, but which are longer and more cornified toward the reticulum.

Connecting the reticulo-omasal opening with the omasoabomasal opening at the base of the omasum is the **omasal groove** (sulcus omasi, 220/*b*), the middle segment of the gastric groove. The omasal groove is flanked by two mucosal ridges which, like the omasal laminae, are covered with papillae (222/*c'*). It lies opposite the free borders of the omasal laminae, facing caudally, dorsally, and to the right, and with the tallest of laminae surrounds and forms the **omasal canal** (220/*a*). The muscular **omasal pillar** (pila omasi, 212/*14*; 218/*below 6*) crosses the omasal groove near the **omasoabomasal opening,** through which the ingesta enter the abomasum. The omasoabomasal opening is flanked by two mucosal folds, the **vela**

* Causing traumatic pericarditis.

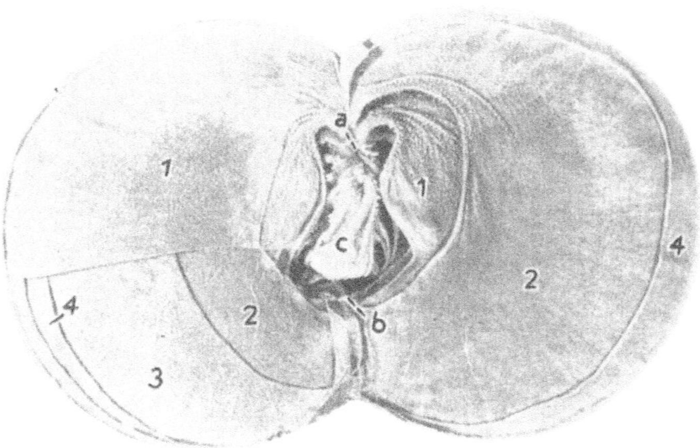

Fig. 221. Bovine omasum, opened along curvature. Photograph.

a Reticulo-omasal opening; *b* Omasoabomasal opening; *c* Omasal groove; *1, 2, 3, 4* Omasal laminae of various sizes

abomasica, (212/*15*; 222/*b'*, *b'*), which are thought to play a role in the closure of the omasoabomasal opening. In the ox, the omasal surface of the vela is covered with stratified squamous epithelium and is nonglandular, while the surface facing the abomasum has a glandular mucosa. In the sheep and goat, the omasal surface of the vela is also partly covered with glandular abomasal mucosa.

According to Stevens, Sellers, and Spurrell (1960) the omasum acts as a **two-stage pump, transferring material from the reticulum to the abomasum.** The **first stage** consists of contractions and relaxations of the omasal canal, during which ingesta are aspirated from the reticulum and the more fluid components pressed into the interlaminar recesses. Some discharge into the abomasum also takes place. The second stage consists of contractions of the omasal body by which omasal contents are discharged into the abomasum. The grid of laminae opposite the omasal groove (220/*b*) retains coarse material which appears to be returned to the reticulum from time to time. The contractions of the omasal canal are linked to the cyclic ruminoreticular contractions; the contractions of the omasal body are independent. Like the ruminoreticulum, the omasum also absorbs fatty acids, sodium, and water.

The **ABOMASUM,** in contrast to the proventricular compartments, is lined with glandular mucosa, which as in the simple stomach is divided, principally, into two regions: the region of the proper gastric glands and the pyloric gland region. The **region of the proper gastric glands** includes the lining of most of the fundus and body of the abomasum (203/*3*). The mucosa of this region is grayish red and is arranged in large permanent **spiral folds** (plicae spirales abomasi, 227), which are slightly oblique to the longitudinal axis of the body of the abomasum (210/*h*). They begin at the omasal end, at first increase in height, and then decrease again toward the pyloric part. A bandlike area along the lesser curvature is free of folds and is considered to be the **abomasal groove,** the third segment of the gastric groove. The **pyloric gland region** is roughly coextensive with

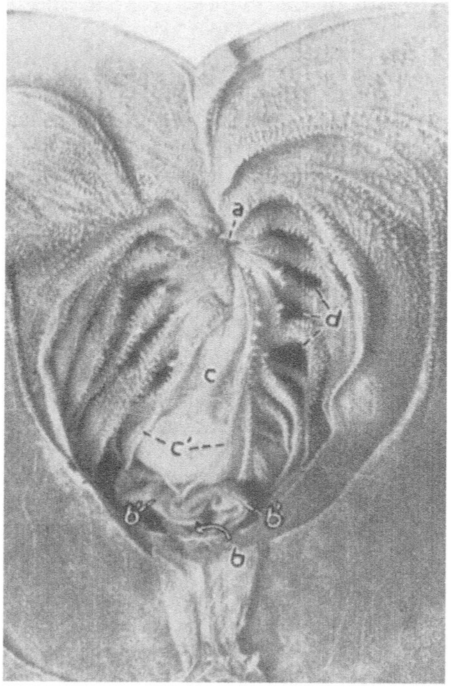

Fig. 222. Omasal groove. Detail of Fig. 221. Photograph.

a Reticulo-omasal opening; *b* Omasoabomasal opening; *b', b'* Vela abomasica; *c* Omasal groove; *c'* Mucosal folds on each side of omasal groove; *d* Interlaminar recesses; note the thick, muscular free edges of the omasal laminae

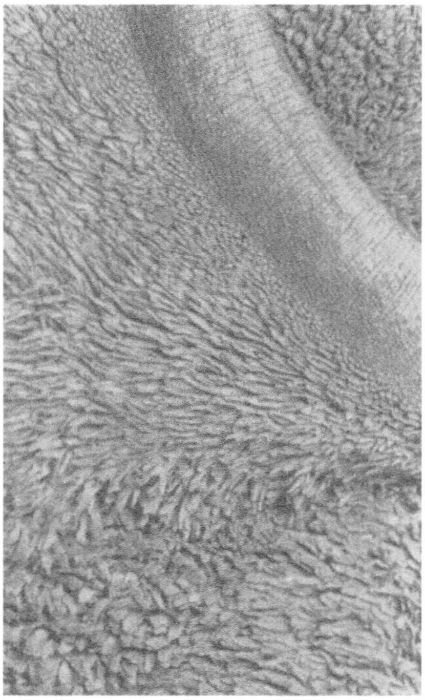

Fig. 223

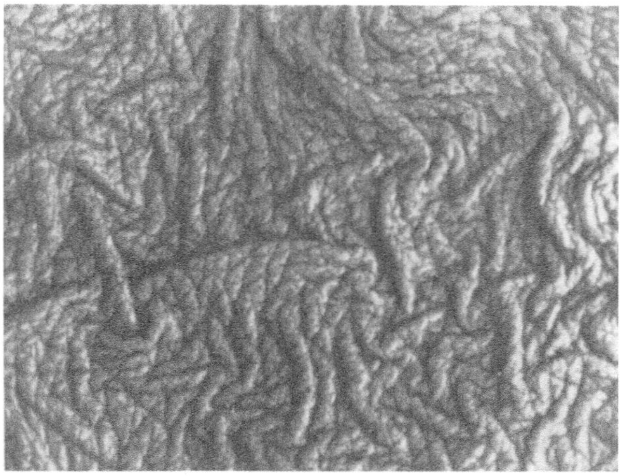

Fig. 224

Fig. 223. Mucosa of the ventral sac of the rumen and ruminal pillar. Photograph. Note the various shapes of the rumen papillae and the absence of papillae on the pillar

Fig. 224. Mucosa of the roof of the dorsal sac of the rumen. Photograph. Note the absence of papillae

Fig. 225. Reticular groove with adjacent reticular mucosa. Photograph.
1 Cardia; *2* Reticulo-omasal opening; *3* Floor of reticular groove with cornified papillae; *4* Right, *5* left lip of reticular groove

Fig. 226. Mucosa of the reticulum. Photograph. Note the papillae on the large and small crests and on the floor of the cells

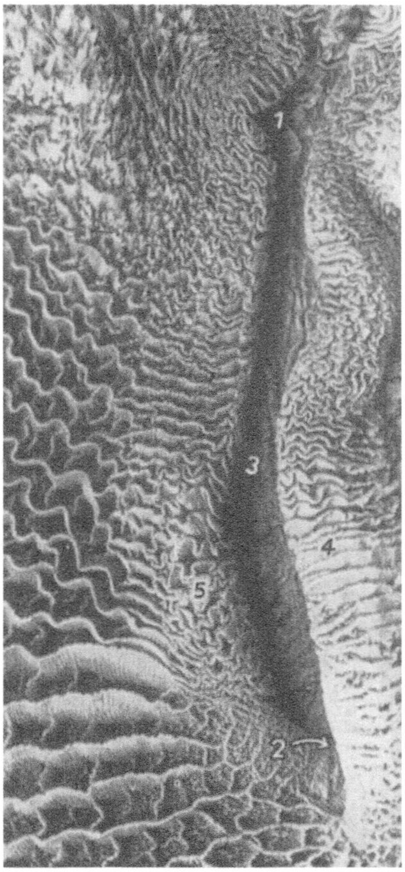

Fig. 225

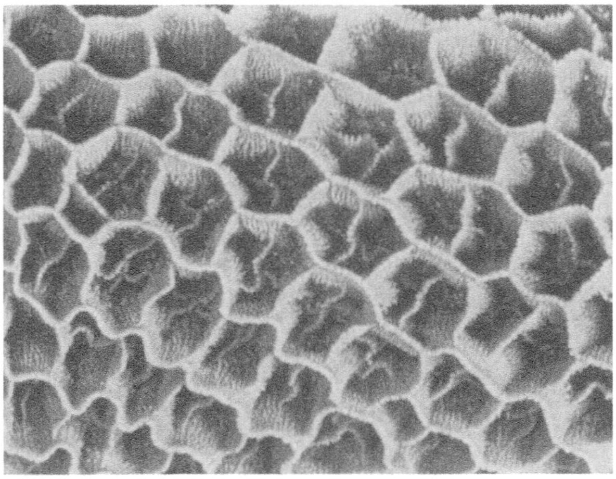

Fig. 226

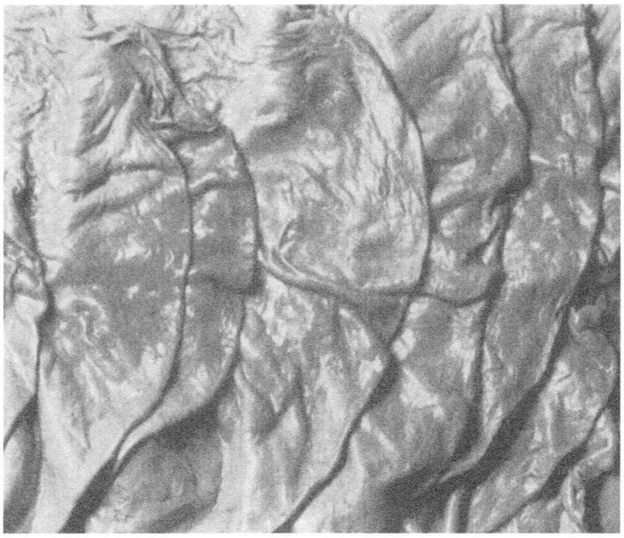

Fig. 227. Mucosa of the body of the abomasum. Photograph. Note the permanent spiral folds

the pyloric part of the abomasum (203/*4*). The mucosa here is lighter, more yellowish, than in the body, and in the vicinity of the pylorus forms irregular transient folds (228). A small **"cardiac" gland region** surrounds the omasoabomasal opening (203/*2*). The glands in this region are similar to the cardiac glands at the cardia of the simple stomach, although the cardia of the ruminant stomach is not at the omasoabomasal junction, but at the point where the esophagus enters the ruminoreticulum.

The **pyloric sphincter** is not very well developed in the ruminants, and is augmented by a large **torus pyloricus,** a round protuberance about 3 cm. in diameter on the inside of the lesser curvature (210/*8*; 228).

ATTACHMENT OF THE RUMINANT STOMACH. The dorsal sac of the rumen is firmly attached to the crura of the diaphragm and to the left psoas musculature as far caudally as the third or fourth lumbar vertebra (229). The dorsal end of the spleen is included in the area of direct attachment of the rumen, so that the parietal surface of the spleen adheres to the diaphragm and the visceral surface to the rumen without the interposition of peritoneum (216/*p*).

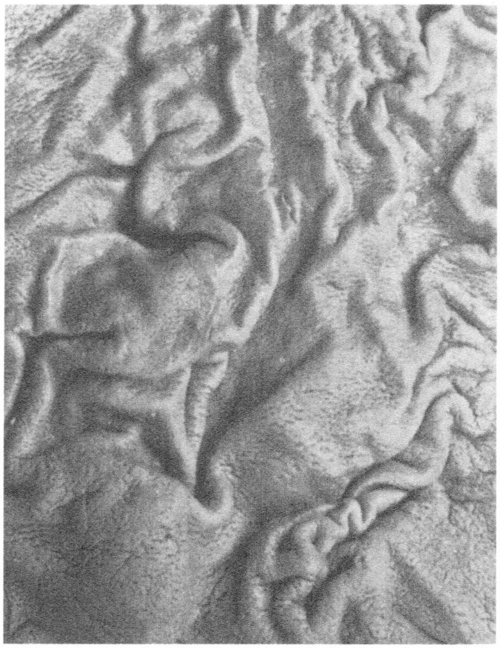

Fig. 228. Mucosa of the pyloric part of the abomasum. Photograph.
Note the large gastric folds, the torus pyloricus in the lower left corner, and the minute short grooves between the areae gastricae. The foveolae gastricae are too small to be seen

OMENTA. To understand the arrangement of the omenta associated with the ruminant stomach, it is useful to begin with a few developmental remarks. The simple, spindle-shaped primordium of the ruminant stomach, like that of the simple stomach, is suspended from the roof of the embryonic abdominal cavity by the dorsal mesogastrium, which attaches to the dorsal border (greater curvature) of the primordium and becomes, postnatally, the greater omentum. Likewise, the ventral mesogastrium attaches to the ventral border (lesser curvature) of the primordium, passes to the floor of the abdominal cavity, and parts of it form the lesser omentum (14/*A*). The transformation of the simple, spindle-shaped primordium into the definitive stomach consisting of four compartments results in complex changes in the line (greater curvature) along which the dorsal mesogastrium attaches to the stomach. The rumen, reticulum, and most of the abomasum develop from the area of the greater curvature of the primordium, while the omasum and a small portion of the abomasum come from the area of the lesser curvature. Consequently, the line of attachment of the greater omentum passes at first over the rumen, then close to the reticulum, and then over the abomasum. It begins at the esophageal hiatus, passes caudally along the right longitudinal groove, then to the left between the two blind sacs in the caudal groove, and then cranially along the left longitudinal groove. At the cranial end of the left longitudinal groove it is fairly close to the reticulum. It inclines, however, ventrally, and underneath the stomach gains the greater curvature of the abomasum along which it proceeds to the cranial part of the duodenum, where it meets the attachment of the mesoduodenum. Cranially and to the right, the developing liver comes to lie against the right face of the dorsal mesogastrium and unites with it, and the left lobe of the pancreas also becomes associated here with the dorsal mesogastrium.

The **LESSER OMENTUM** (210/*r*; 211/*q*; 234/*m*; 239/*8*, *9*), postnatally, originates on the visceral surface of the liver along a line from the porta to the esophageal notch. From this line it passes to the omasum, to the lesser curvature of the abomasum, and to the cranial border of the cranial part of the duodenum.

The **superficial wall** of the **GREATER OMENTUM** arises from the greater curvature of the abomasum, the caudal border of the cranial part of the duodenum (209/*p*), and from the ventral border of the descending duodenum (234/*x*). From the descending duodenum on the right (229/*e*, *3*), it descends along the abdominal wall, crosses the median plane ventral to the stomach, and passes dorsally between the ventral sac of the rumen and the left abdominal wall to the left longitudinal groove of the rumen (*d*).

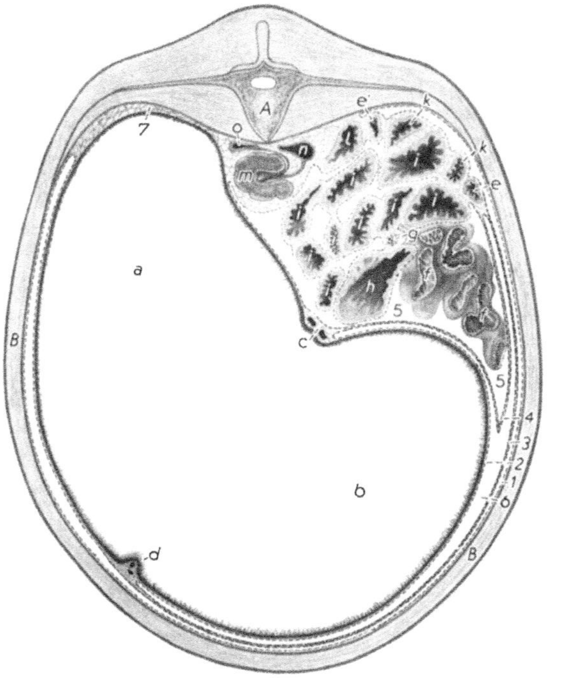

Fig. 229. Cross section of the abdomen of the ox at the level of the fourth lumbar vertebra. Semischematic. Caudal aspect. (Redrawn from Schmaltz 1895). The peritoneum is indicated by broken lines.

A Fourth lumbar vertebra; *B* Abdominal wall

a Dorsal sac of rumen; *b* Ventral sac of rumen; *c* Right longitudinal groove with right ruminal artery and vein and attachment of the deep wall of the greater omentum; *d* Left longitudinal groove and pillar with left ruminal artery and vein and attachment of the supf. wall of the greater omentum; *e* Descending duodenum; *e'* Ascending duodenum; *f* Jejunum; *g* Ileum; *h* Cecum; *i* Spiral loop of ascending colon; *k* Distal loop of ascending colon; *l* Descending colon; *m* Left kidney; *n* Caudal vena cava; *o* Aorta

1 Parietal peritoneum; *2* Visceral peritoneum; *3* Supf. wall of greater omentum; *4* Deep wall of greater omentum; *5* Supraomental recess; *6* Omental bursa, its caudal recess; *7* Attachment of the rumen to the roof of the abdomen

The **deep wall of the greater omentum** attaches to the right longitudinal groove of the rumen (229/*c*). Just as the caudal ends of the longitudinal grooves are connected around the caudal end of the rumen by the caudal groove between the blind sacs, so the sheet of omentum attaching along the left longitudinal groove and that attaching along the right longitudinal groove join in the caudal groove. The deep wall of the greater omentum leaves the right longitudinal groove and passes ventrally in contact with the visceral surface of the ventral sac of the rumen. It then passes around the ventral aspect of the intestines, which lie to the right of the rumen, and turns dorsally and ascends between the intestines and the superficial wall of the greater omentum, uniting with it on the ventral surface of the descending duodenum (229/*4*, *e*; 234/*y*). Cranially, the deep wall of the greater omentum (209, 210/*q*; 235/*u*) forms a transverse sheet that passes craniodorsally cranial to the intestines, and attaches to the pancreas, the dorsal border of the liver, and the sigmoid loop of the duodenum. Caudally, the deep and superficial walls of the greater omentum are continuous in a fold that arcs from the caudal flexure of the duodenum on the right to the caudal groove of the rumen on the left (234/*z*; 236/*v*; 237/*q*, *q'*; 244/*p*, *q*).

As in the other domestic mammals the walls of the greater omentum enclose the **caudal recess of the omental bursa** (229/*6*) which contains the ventral sac of the rumen (*b*) and its blind sac, and in the live animal, is only a capillary space. The omental bursa communicates also in the ruminants with the peritoneal cavity through the **epiploic foramen** (210/*arrow*), which is bounded dorsally by the caudal vena cava (*b*) and ventrally by the portal vein (*c*). The foramen leads into the **vestibule of the omental bursa** (*r'*), which is formed by the lesser omentum (*r*) on the right, the rumen on the left, the liver dorsocranially, and the omasum caudoventrally.

Dorsal to the sling formed by the deep wall of the greater omentum is the extensive **supraomental recess** (229/*5*), which is open caudally and contains most of the intestines. The large opening of the recess lies cranial to the pelvic inlet and slightly to the right of it, and is bounded on the left by the rumen and on the right by the caudal edge of the greater omentum. Portions of intestine usually protrude from the recess and lie at the pelvic inlet. In the pregnant cow, the gravid uterus may extend into the supraomental recess.

If the abdominal cavity of the ruminant is opened through the right flank, the descending duodenum (234/*8*) is usually the only part of the alimentary canal that is visible. Dorsal to the duodenum is the mesoduodenum (*w*) and ventral to the duodenum is the superficial wall of the greater omentum (*x*). Cutting the superficial wall of the greater omentum opens the omental bursa and exposes the deep wall of the greater ometum (*y*). Only after cutting the deep wall are the intestines in the supraomental recess exposed.

Intestines

(146, 211, 229—238)

While the stomach of the ruminants occupies more than half of the abdominal cavity, the intestines, despite their considerable length, are confined to a relatively small part of the abdominal cavity. The intestines are suspended from the roof of the abdominal cavity by a common mesentery which collects them into a large, disc-shaped mass. With its plane roughly sagittal, this mass fills the supraomental recess to the right of the rumen and caudal to the omasum (229, 235).

In the ox, the total length of the intestines is 33—63 m. of which 27—49 m. is small intestine. In the sheep, the total length is 22—43 m. and in the goat an average of 33 m., with the small intestine being 18—35 m. in both species.

The **DUODENUM** (234/*6, 7*) begins at the pylorus close to the ends of the ninth to the eleventh ribs. Its **cranial part,** related laterally to the gall bladder, passes dorsally to the porta along the visceral surface of the liver, where it forms the **sigmoid loop** (210/*i'*) and is continued at the **cranial flexure** by the descending duodenum. The **descending duodenum,** attached medially to the coils of the colon, passes caudally to about the level of the tuber coxae in the ox (not quite so far in the small ruminants), and there turns medially and then cranially, forming the sharp **caudal flexure** (234/*9*). The **ascending duodenum** is connected to the descending colon by the **duodenocolic fold** (236/*x*), and passes cranially, high in the common mesentery (229/*e'*). Ventral to the pancreas, the duodenum turns ventrally forming the **duodenojejunal flexure,** and is continued by the jejunum (235/*v, w*).

The **JEJUNUM** (230/*2*) is very long and of small diameter. It is attached to the free edge of the mesentery, which is suspended by its root from the roof of the abdominal cavity. Along this edge, the jejunum is arranged in numerous close coils surrounding the spiral loop of the ascending colon, which is applied against the left face of the mesentery. This arrangement is best seen when the intestinal tract is laid out on the dissecting table (230). In situ, the jejunal coils are more crowded and lie lateral to the spiral colon. The more cranial of them lie deep within the supraomental recess and are related to the liver, pancreas, omasum, abomasum, and rumen through the deep wall of the greater omentum. Ventrally, the jejunal coils are related through the greater omentum to the abdominal floor (235) or, if the rumen is distended, to the ventral sac of that organ (229). The caudal coils of the jejunum are more mobile because of their longer mesenteric attachment (230) and usually project from the supraomental recess. Depending on the fullness of the stomach or the intestines, these coils may be found in the pelvic inlet or on the left side caudal to the caudodorsal blind sac of the rumen. Jejunal loops from this region occasionally become incarcerated between the right deferent duct and the abdominal wall in steers.

The **ILEUM** is the straight, terminal part of the small intestine (230/*3*) passing cranially ventral to the cecum, to which it is connected by the **ileocecal fold** (*3'*), and enters the large intestine on the ventromedial surface of the cecocolic junction. The **ileal orifice** (231—233) is roughly at the level of the fourth lumbar vertebra in the ox and at the level of the caudalmost point of the costal arch in the sheep and goat.

The large intestine of the ruminants, with the exception of the cecum and the proximal parts of the colon, is only slightly wider than the small intestine.

The **CECUM** (230/*4*; 231/*b*), a slightly S-shaped blind tube with a diameter of about 12 cm., extends caudally from the ileocolic junction and protrudes with its free, blunt end from the supraomental recess (234/*r*; 236/*o*). When the cecum is distended with ingesta, the free end may extend into the pelvic cavity or curve to the left in front of the pelvic inlet (235/*x*). Its cranial portion has a more constant position, since it is firmly attached to the mesentery.

The **COLON** (146/*F, G, H*) continues cranially from the cecum. Its diameter is at first the same as that of the cecum, but soon diminishes. The **ascending colon** (*11—15*), by far the longest of the colic segments, has a peculiar arrangement and can be divided into proximal, spiral, and distal loops.

The **proximal loop** (146/*11*; 230/*5*) describes essentially an S-shaped curve, and like the letter, consists of three parts. The first part runs cranially from the cecocolic junction, continuing the direction of the cecum. Ventral to the right kidney and at about the level

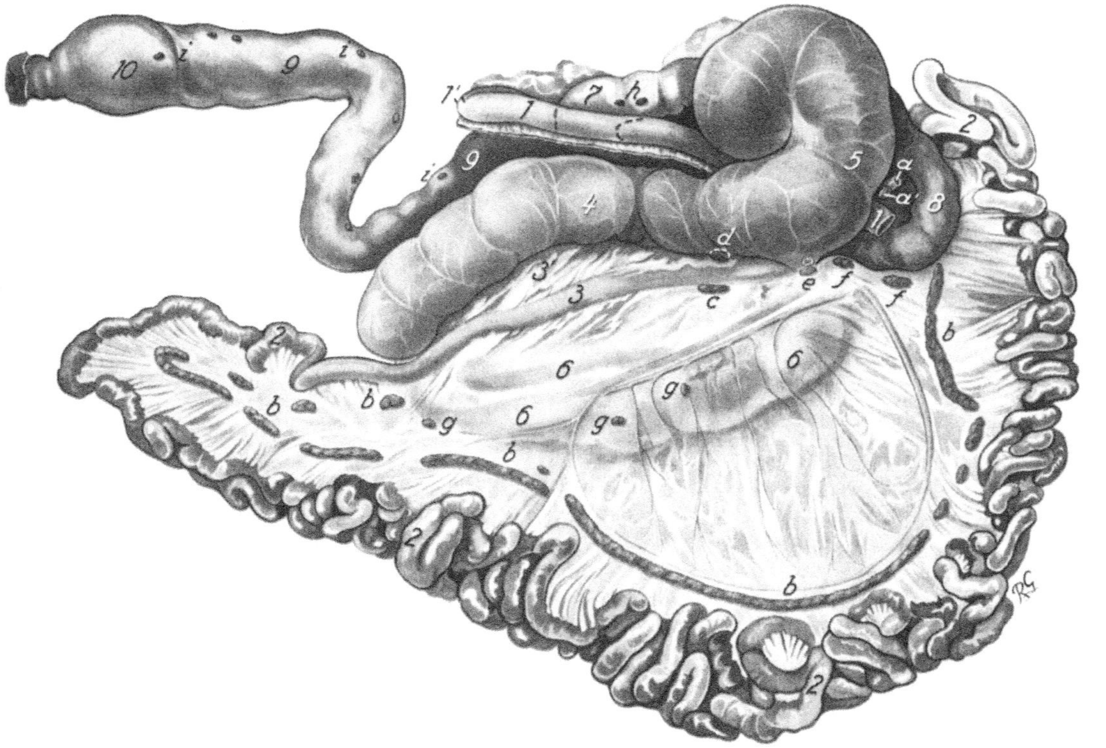

Fig. 230. Intestines of the ox. Right aspect. (After Zietzschmann, 1958.)

1 Descending duodenum with stump of the supf. and deep walls of the greater omentum; *1'* Caudal duodenal flexure; *2* Jejunum; *3* Ileum; *3'* Ileocecal fold; *4* Cecum; *5* Proximal loop of ascending colon; *6* Spiral loop seen through the mesentery; *7* Distal loop of ascending colon; *8* Transverse colon; *9* Descending colon; *10* Rectum

a Cranial mesenteric artery; *a'* One of the celiac lymph nodes; *b* Jejunal lymph nodes; *c* Ileal lymph node; *d* Cecal lymph node; *e—h* Colic lymph nodes, *e* at the ileocecocolic junction, *f* on the proximal loop, *g* on the spiral loop, *h* on the distal loop of the ascending colon; *i* Caudal mesenteric lymph nodes

of the twelfth rib, it doubles back on itself dorsolaterally and, as the second part, passes caudally and slightly laterally along the right abdominal wall, but separated from it by the greater omentum. It is parallel to the descending duodenum, which lies a short distance dorsal to it. The proximal loop then doubles on itself once more, this time mediodorsally, and passes, as the third part, cranially again. The third part lies on the left surface of the mesentery and is related to the ascending duodenum, the descending colon, and the left kidney. At this point

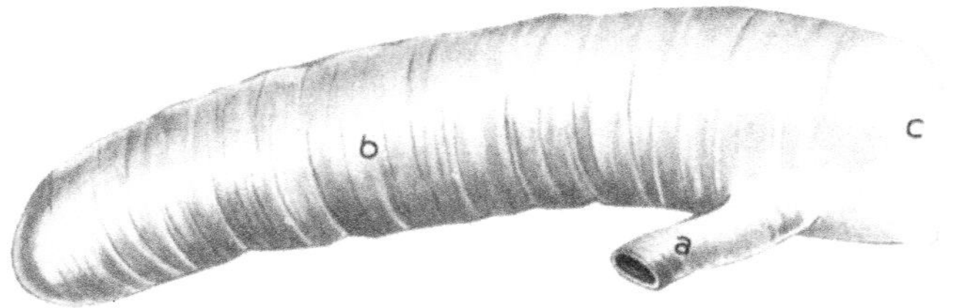

Fig. 231. Cecum of the ox.

a Ileum; *b* Cecum; *c* Ascending colon

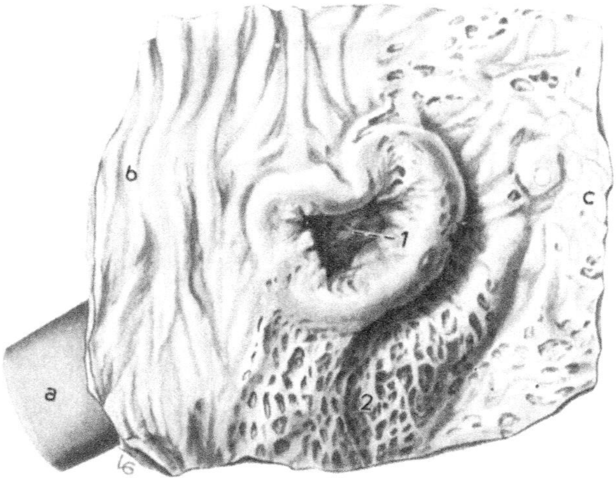

Fig. 232 (Ox)

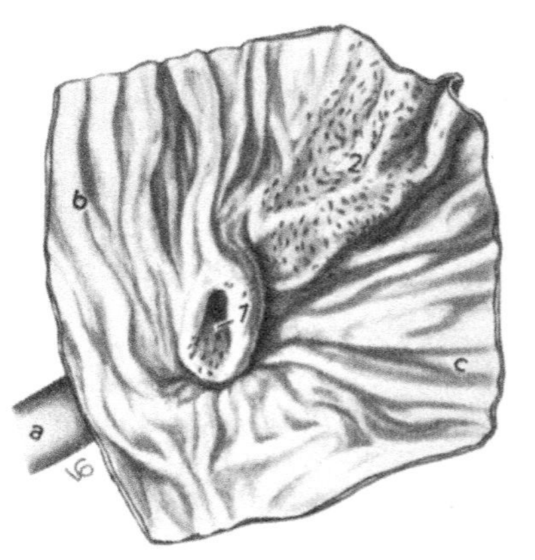

Fig. 233 (Goat)

Figs. 232 and 233. Ileal orifice of the ox and goat. *a* Ileum; *b* Mucosa of cecum; *c* Mucosa of ascending colon; *1* Ileal orifice; *2* Last patch of aggregate lymph nodules extending into the colon

the proximal loop is followed by the **spiral loop,** which rapidly decreases in diameter. During development, the coils of the spiral loop come to lie against the left surface of the mesentery and are therefore best observed in the adult from the left side (235, 237). The spiral loop is actually a long, single loop of gut, which is rolled up on itself in a sagittal plane (146/*12, 13, 14*). The coil that results is like an elliptical disc in the cow (235) but slightly more raised to form a low cone in the sheep and goat (237). Centripetal turns spiral toward the center of the coil where at a **central flexure** (*4*) the gut reverses direction and with centrifugal turns returns toward the periphery of the coil. There are 1.5—2 centripetal turns in the ox, 3 in the sheep, and 4 in the goat, and an equal number of centrifugal turns. In the ox, the turns are generally closely packed, and the last centrifugal turn, after a short caudal course, is followed at the level of the first lumbar vertebra by the distal loop of the ascending colon (146/*15*). In the sheep and goat, the turns are also closely packed, but the distal half of the last centrifugal turn spirals away from the coil and continues quite close to the attachment of the jejunal coils. The **distal loop** (230/*7*) lies medial to the proximal loop and ascending duodenum. It consists of an upper caudally directed part, a tight flexure at the level of the fifth lumbar vertebra, and a lower cranially directed part. The cranially directed part passes along the right side of the mesentery to the level of the last thoracic vertebra, where it turns sharply to the left and is continued as the short transverse colon (230/*8*).

The **transverse colon,** as in the other species, passes in front of the cranial mesenteric artery from right to left. It is suspended by the short transverse mesocolon and is related dorsally to the pancreas. The **descending colon,** which follows the transverse colon to the left of the cranial mesenteric artery, is embedded together with the ascending duodenum in the left side of the root of the mesentery and passes caudally in close proximity to the roof of the abdominal cavity (229/*l*; 235/*z*). At the level of the last lumbar vertebra the descending mesocolon becomes somewhat longer, allowing the caudal part of the descending colon (colon sigmoideum, 234/*12*) more range than the rest of the descending colon. The longer mesenteric attachment of the colon at this point increases the range of the veterinarian's arm during rectal palpation.

The **RECTUM** follows the colon into the pelvic cavity. Most of it is covered with peritoneum, the retroperitoneal portion being relatively short. The mesorectum decreases rapidly in length. Despite the well-developed and relatively thick muscular coat, the rectum of the ruminants may be considerably distended by the accumulation of feces prior to evacuation. (The feces of the small ruminants are divided in the distal part of the spiral loop and arrive as pellets at the rectum.) Inconstant **transverse folds** (plicae transversales recti, 555/*17*), which result from localized constrictions of the circular musculature, are often found on the inside of the rectal wall.

The **rectococcygeus** (490/*14*) consists of thick muscle bundles that arise dorsally from the longitudinal musculature of the rectum and are inserted on the ventral surface of the first few caudal vertebrae (in the ox to the third). In the cow, muscle bundles from the ventral surface of the rectum decussate in the perineal body and unite with the constrictor vulvae.

The **ANAL CANAL,** the terminal segment of the alimentary canal, is short and ends caudally at the anus (528/*a*). At the junction of rectum and anal canal, the mucosa for a length of about 10 cm. in the ox, and about 1 cm. in the small ruminants presents a number of longitudinal folds (columnae rectales) alternating with depressions. This plicated zone is followed directly by the **cutaneous zone** of the anal canal.

Both the voluntary **external anal sphincter** (490/*15*) and the smooth **internal anal sphincter** are present in the ruminants. The **levator ani** (*16*) arises from the ischiatic spine and adjacent sacroischiatic ligament and is inserted in the wall of the anal canal (see also pp. 339 and 365 for genital muscles associated with the anal canal). Habel (1966) in his study of the perineum of the cow describes these muscles in detail.

TOPOGRAPHY AND RELATIONS OF THE INTESTINES (SUMMARY). The ruminant stomach is so large that it occupies most of the abdominal cavity and leaves little room for the intestinal tract. The ruminoreticulum occupies the left half of the abdominal cavity and with the ventral sac of the rumen at times also a considerable portion of the right half (229). The omasum lies under cover of the ribs to the right of the median plane, and dorsocranial to it is the liver. The abomasum occupies the floor of the abdominal cavity in the xiphoid region (235). The remaining space, essentially the caudal part of the right half, contains the intestines, which form a disc-shaped mass, roughly sagittal in position, reaching from the liver to the pelvic cavity, and often into it. The intestinal mass is suspended from the roof of the abdominal cavity and, with the exception of the cecum and a few caudal jejunal coils, is contained in the supraomental recess (234). The left surface of the intestinal mass lies against the dorsal sac of the rumen and the deep wall of the greater omentum covering the ventral sac. On the right side, the intestinal mass is related to the right abdominal wall, but is separated from it by the two walls of the greater omentum (229). Cranially, the intestines extend deeply into the intrathoracic part of the abdominal cavity and are in contact here through the deep wall of the greater omentum with the omasum, the abomasum, and the visceral surface of the liver; dorsally, they are related to the kidneys and the pancreas (235). Obviously, stomach contractions—those of the ruminoreticulum in particular—and variations in the fullness of the stomach or the intestines change the position and the relations of the intestinal tract. During pregnancy, the uterus expands mainly cranioventrally and to the right, displacing the rumen and the intestinal mass craniodorsally and to the left. In the latter part of pregnancy, the soft abdominal wall caudal to the costal arch and last rib becomes distended so as to prevent undue pressure of the greatly enlarged uterus on the other abdominal organs.

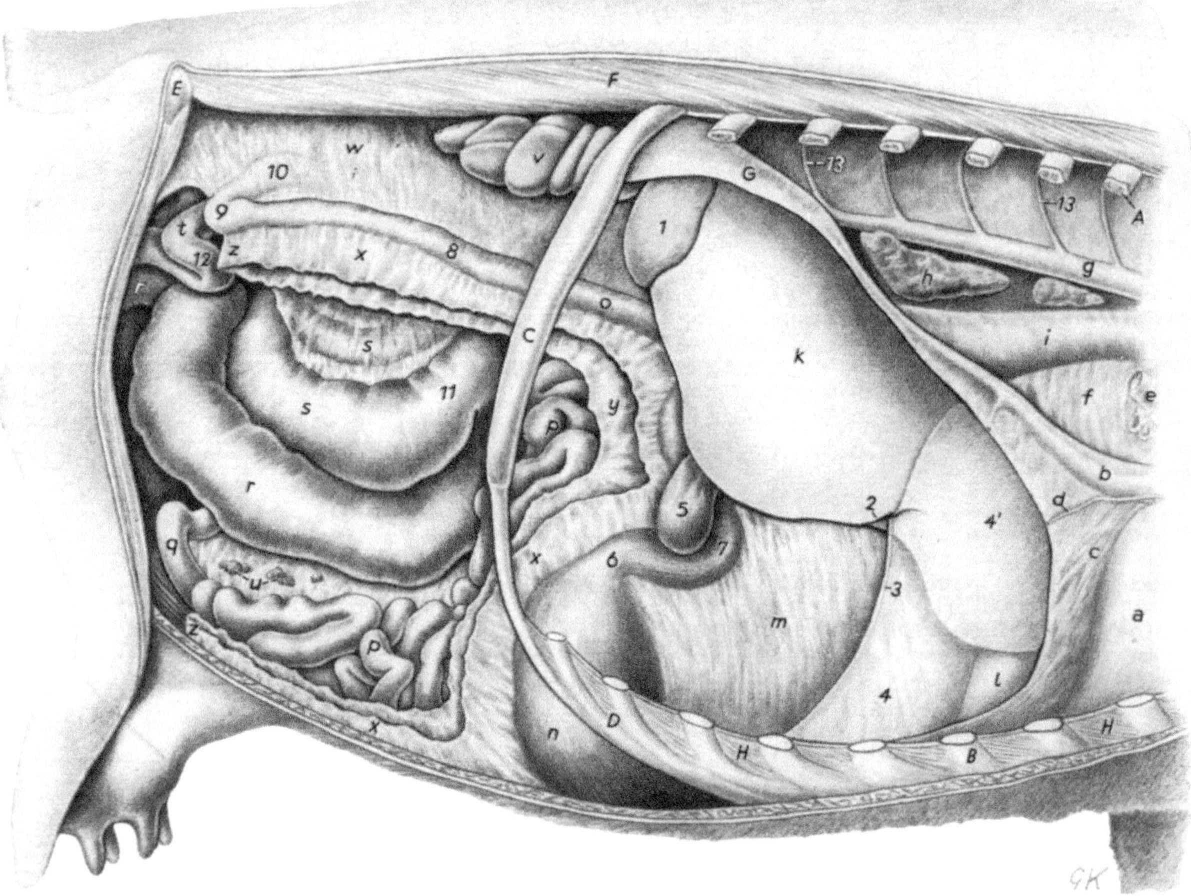

Fig. 234. Topography of the thoracic and abdominal organs of a cow. A portion of the greater omentum, diaphragm, and the right lung have been removed. Right aspect. (After Nickel and Wilkens 1955.)

A Stump of seventh rib; *B* Seventh costal cartilage; *C* Thirteenth rib; *D* Costal arch; *E* Tuber coxae; *F* Longissimus; *G* Diaphragm; *H* Intercostal muscle

a Pericardium; *b* Caudal vena cava; *c* Plica venae cavae; ***d*** Right phrenic nerve; ***e*** Root of right lung; *f* Mediastinum; *g* Aorta; ***h*** Caudal mediastinal lymph nodes; *i* Esophagus; ***k*** Liver; *l* Reticulum; ***m*** Lesser omentum, covering the omasum; ***n*** Abomasum; *o* Duodenum; *p* Jejunum; *q* Ileum; *r* Cecum; *s* Ascending colon; ***t*** Descending colon; ***u*** Jejunal lymph nodes; *v* Right kidney; ***w*** Mesoduodenum; ***x*** Supf. wall, *y* deep wall of greater omentum, enclosing the caudal recess of the omental bursa; ***z, z*** Caudal edge of greater omentum

1—4 On the liver: *1* caudate process, *2* notch for round ligament, *3* round ligament, *4* falciform ligament, covering liver at *4'*; *5* Gall bladder; *6* Pylorus; *7* Cranial part of duodenum; *8* Descending duodenum; *9* Caudal duodenal flexure; *10* Ascending duodenum; *11* Proximal loop of ascending colon; *12* Descending colon; *13* Dorsal intercostal artery

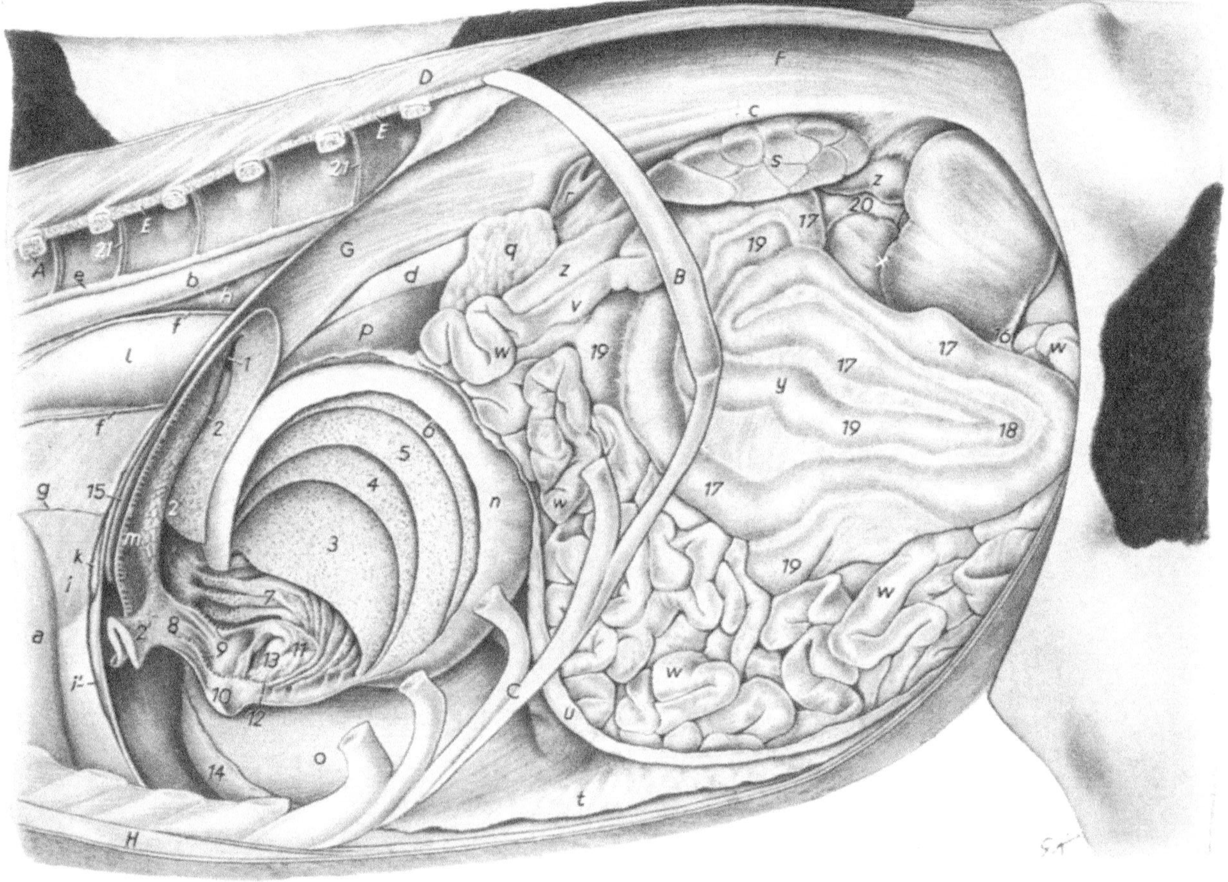

Fig. 235. Topography of the thoracic and abdominal organs of an ox. The rumen and most of the reticulum have been removed, and the omasum has been opened from the left. Left aspect. (After Nickel and Wilkens 1955.)

A Stump of seventh rib; *B* Thirteenth rib; *C* Costal arch; *D* Longissimus; *E* Intercostal muscle; *F* Psoas musculature; *G* Crura of diaphragm; *H* Rectus abdominis

a Pericardium; *b, c* Aorta; *d* Caudal vena cava; ***e*** Left azygous vein; *f, f* Dorsal and ventral vagal trunks; ***g*** Left phrenic nerve; ***h*** Caudal mediastinal lymph node; ***i*** Caudal mediastinum, ***i'*** its cut edge; ***k*** Accessory lobe of right lung; *l* Esophagus; ***m*** Reticulum, most of it removed; ***n*** Omasum, opened from the left and part of the laminae removed; *o* Abomasum; ***p*** Liver; *q* Pancreas; ***r*** Left adrenal gland; *s* Left kidney; *t* Supf. wall, ***u*** deep wall of greater omentum; *v* Ascending duodenum; *w* Jejunum; *x* Cecum; *y* Ascending colon; *z* Descending colon

1 Cardia; *2* Reticular groove; *2', 2'* Right lip of reticular groove, cut at the reticulo-omasal opening, the lower stump pulled ventrally and to the left; *3, 4, 5, 6, 7* Omasal laminae of various sizes; *8* Omasal groove; *9* Mucosal folds accompanying omasal groove; *10* Omasal pillar; *11* Omasoabomasal opening; *12* Velum abomasicum viscerale; *13* Spiral folds; *14* Greater curvature of abomasum; *15* Left lobe of liver; *16* Ileocecal fold; *17—20* Ascending colon: *17* centripetal turns, *18* central flexure, *19* centrifugal turns, *20* distal loop; *21* Dorsal intercostal artery and vein

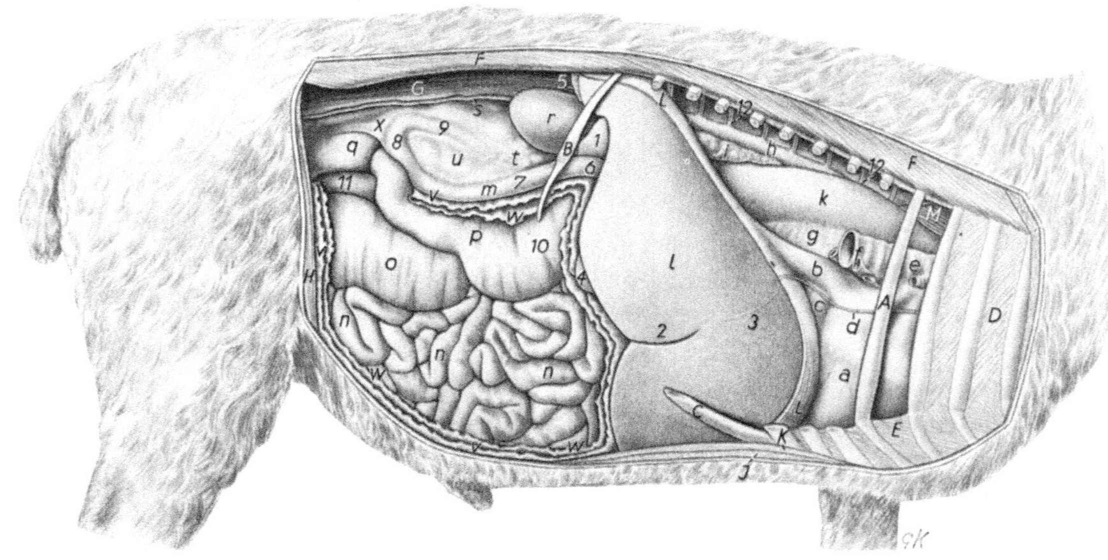

Fig. 236. Topography of the thoracic and abdominal organs of a castrated male sheep. The right lung, right half of diaphragm, costal arch, and parts of the greater omentum have been removed. Right aspect. (After Wilkens 1956b.)

A Fourth rib; *B* Thirteenth rib; *C* Eighth costal cartilage; *D* Ext. intercostal muscle; *E* Int. intercostal muscle; *F* Longissimus; *G* Psoas musculature; *H* Int. and ext. abdominal oblique muscles; *J* Pectoral muscles; *K* Rectus abdominis; *L* Diaphragm; *M* Longus colli

a Heart inside pericardium; *b* Caudal vena cava; *c* Plica venae cavae; *d* Right phrenic nerve; *e* Trachea with tracheal bronchus; *f* Root of right lung; *g* Caudal mediastinum; *h* Aorta; *i* Caudal mediastinal lymph node; *k* Esophagus; *l* Liver; *m* Duodenum; *n* Jejunum; *o* Cecum; *p* Ascending colon; *q* Descending colon; *r* Right kidney; *s* Right ureter; *t* Pancreas; *u* Mesoduodenum; *v* Supf. wall, *w* deep wall of greater omentum; *x* Duodenocolic fold

1 Caudate process of liver; *2* Notch for round ligament; *3* Falciform ligament; *4* Gall bladder; *5* Right triangular ligament; *6* Cranial duodenal flexure; *7* Descending duodenum; *8* Caudal duodenal flexure; *9* Ascending duodenum; *10* Ventrolateral part of the proximal loop of the ascending colon; *11* First centripetal turn of spiral colon; *12* Dorsal intercostal artery and vein

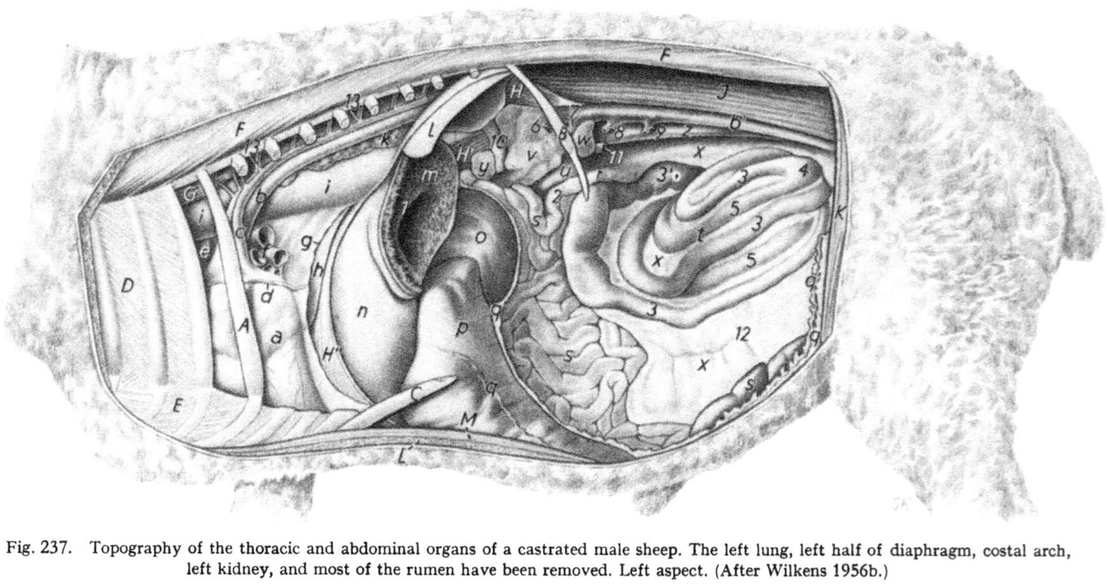

Fig. 237. Topography of the thoracic and abdominal organs of a castrated male sheep. The left lung, left half of diaphragm, costal arch, left kidney, and most of the rumen have been removed. Left aspect. (After Wilkens 1956b.)

A Fourth rib; *B* Thirteenth rib; *C* Eighth costal cartilage; *D* Ext. intercostal muscle; *E* Int. intercostal muscle; *F* Longissimus; *G* Longus colli; *H* Diaphragm, its left crus, *H'* part of its right crus, *H''* its sternal part; *J* Psoas musculature; *K* Int. and ext. abdominal oblique muscles; *L* Pectoral muscles; *M* Rectus abdominis

a Heart inside pericardium; *b*, *b'* Aorta; *c* Left azygous vein; *d* Left phrenic nerve; *e* Trachea; *f* Root of left lung; *g* Caudal mediastinum, its cut edge; *h* Accessory lobe of right lung; *i* Esophagus; *k* Caudal mediastinal lymph node; *l* Spleen; *m* Rumen, part of its cranial sac; *n* Reticulum; *o* Omasum; *p* Abomasum; *q* Supf. wall, *q'* deep wall of greater omentum; *r* Ascending duodenum; *s* Jejunum; *t* Spiral loop of ascending colon; *u* Descending colon; *v* Pancreas; *w* Left adrenal gland; *x* Mesentery; *y* Ruminal lymph nodes; *z* Caudal vena cava

1 Ruminoreticular fold; *2* Duodenojejunal flexure; *3—5* On the ascending colon: *3'* distal part of proximal loop, *3* centripetal turns, *4* central flexure, *5* centrifugal turns; *6* Celiac artery; *7* Cranial mesenteric artery; *8* Right renal artery; *9* Left renal artery; *10* Splenic vein; *11* Left renal vein; *12* Jejunal vessels; *13* Dorsal intercostal artery and vein

MESENTERIES. The intestinal mass of the ruminants is suspended as a unit from the roof of the abdominal cavity by a common mesentery (230). During development, the spiral loop of the ascending colon becomes attached to the left surface of the **mesentery,** and because of this union, the ascending colon of the ruminants is much less mobile than that of the pig or horse, for instance. The jejunum is tightly coiled along the edge of the mesentery and forms a semicircle around the spiral colon. Also, the more dorsal parts of the intestines, the descending and ascending duodenum, the proximal and distal loops of the ascending

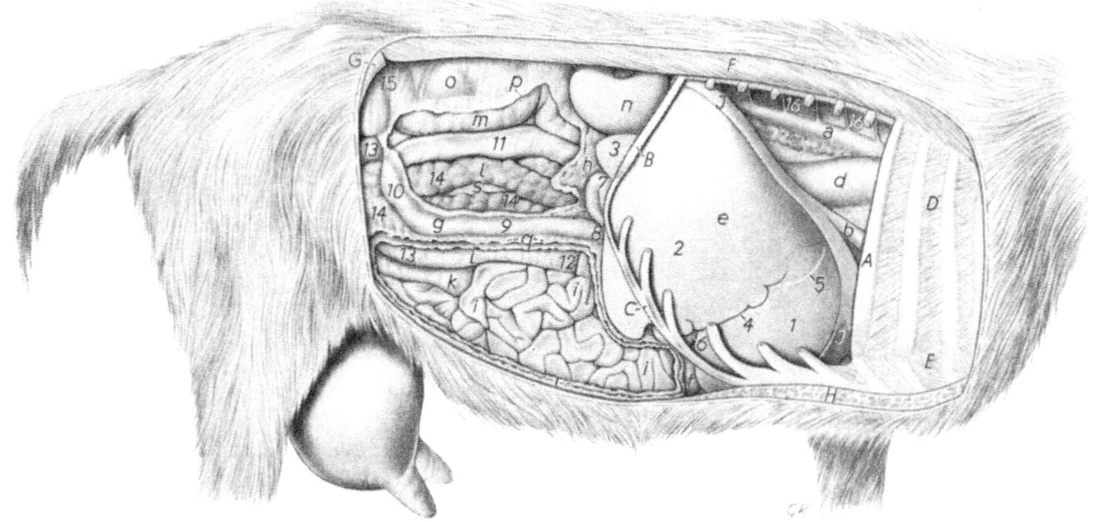

Fig. 238. Topography of the thoracic and abdominal organs of a female goat. The right lung and right half of the diaphragm have been removed, and the greater omentum and the mesoduodenum have been fenestrated. Right aspect. (After Wilkens 1956a.)

A Sixth rib; *B* Thirteenth rib; *C* Costal arch; *D* Ext. intercostal muscle; *E* Int. intercostal muscle; *F* Longissimus; *G* Int. abdominal oblique muscle; *H* Pectoral muscles; *J*, *J'* Diaphragm

a Aorta; *b* Caudal vena cava; *c* Caudal mediastinal lymph node; *d* Esophagus; *e* Liver; *f* Abomasum; *g* Duodenum; *h* Pancreas; *i* Jejunum; *k* Cecum; *l* Spiral loop of ascending colon; *m* Descending colon; *n* Right kidney; *o* Left kidney, covered by the mesentery; *p* Mesoduodenum, fenestrated; *q* Supf. wall, *r* deep wall of greater omentum, enclosing the caudal recess of the omental bursa; *s* Mesenteric fat

1—5 On the liver: *1* left lobe, *2* right lobe, *3* caudate process, *4* notch for round ligament, *5* falciform ligament; *6* Gall bladder; *7* Sigmoid loop of duodenum; *8* Cranial duodenal flexure; *9* Descending duodenum; *10* Caudal duodenal flexure; *11* Ascending duodenum; *12, 13, 13* Proximal loop and centripetal turns of ascending colon; *14* Centrifugal turns of ascending colon; *15* Descending colon; *16* Dorsal intercostal artery

colon, and the descending colon, have lost much of their initial mobility because their mesenteries have largely fused. Only the blind end of the cecum remains relatively mobile. The descending mesocolon and mesorectum appear like a caudal extension of the mesentery. In well-nourished animals large quantities of fat are associated with the more dorsal parts of the intestinal tract and often obscure the arrangement of its parts.

The intestinal mucosa of ruminants is exceptionally well supplied with **LYMPHORETICULAR TISSUE,** in the form of both solitary and of aggregate nodules, and is surpassed in this respect only by the pig. The amount of lymphatic tissue present is proportionately greater in the suckling animal than in the adult.

The **solitary lymph nodules** are visible to the naked eye and are abundant throughout the small intestine, only occasionally found in the cecum and colon of the ox, and are said to be absent in the cecum and colon of the small ruminants. They form a lymphatic ring in the columnar zone of the rectoanal junction in ox and sheep. The **patches of aggregate lymph nodules** are up to 2 cm. wide and reach a length of 25 cm. in the ox and of 15 cm. in the sheep and goat. The ox has 24—40, the sheep 18—40, and the goat 25—30. The most distal patch of aggregate nodules extends through the ileal opening a short distance into the cecum (232, 233). It is composed of large follicles, many with very distinct fossulae, which render the surface of the patch uneven and pitted. A similar patch is regularly observed at the distal end of the proximal loop of the ascending colon. It is 7—20 cm. long in the ox and 4—20 cm. long in the small ruminants, and consists of a number of deep fossulae with rounded margins. No other patches of aggregate nodules are present in the colon of ruminants.

Liver

(155, 159, 210, 215, 234, 236, 238—244)

As in the other species, the color of the ruminant liver depends on the age and the condition of the animal. In well-nourished and younger animals, it is light brown; in emaciated and older animals it is a darker reddish brown. The liver of suckling calves has a slightly yellowish tinge.

The weight of the liver varies with the age, breed, condition, and even with the sex of the animal. The relative weight of the liver in calves and lambs is always greater than in adults.

Table 10. Absolute and relative liver weights of domestic ruminants

	Weight	Per Cent of Dressed Carcass Weight
Steers	4.5—10 kg.	1.76—2.34
Bulls	4.2— 8.5 kg.	1.58—1.92
Cows	3.4— 9.2 kg.	2.26—2.53
Sheep	775 gm. (av.)	

SHAPE. The ruminant liver is not visibly divided into lobes and lacks interlobar notches, except for the **notch for the round ligament** (239/*15*), which is present on the right border.* This notch is visible also on the diaphragmatic surface (240) and is more pronounced in the liver of the small ruminants. In its vicinity the umbilical vein enters the visceral surface of the liver in the fetus. After birth, the vein degenerates into a thin cord, the **round ligament of the liver** (239/*7*), which in older animals often disappears altogether. Ventral to an imaginary line connecting the notch for the round ligament with the esophageal notch (*11*) on the opposite border is the undivided **left lobe of the liver** (*a*). The **gall bladder** (*e*) is embedded in a fossa on the visceral surface, and dorsal to the fossa is the **right lobe of the liver** (*b*), which is also undivided. Between left and right lobes are the **quadrate lobe** (*c*) and **caudate lobe** (*d, d'*). The latter has a **papillary process** (*d'*) projecting toward the porta and an exceptionally large (in the ox) **caudate process** (*d*), which protrudes from the dorsal border of the liver. The caudate process and the right lobe form the deep **renal impression** for the cranial pole of the right kidney (243/*l*). The liver of the sheep (241) is slightly narrower and has a more prominent papillary process than the liver of the goat (242).

The **POSITION** of the ruminant liver is very unusual, when compared to that of the less specialized carnivores or pig, primarily because of the presence of a forestomach, notably the rumen, which crowds the liver out of the left half of the abdominal cavity. While, in the carnivores and pig, the rounded dorsal border of the liver is oriented transversely, and in the horse obliquely from right dorsal to left ventral, in the ruminants the corresponding border follows the median plane (243; compare to 157—160); so that the ruminant liver is situated entirely in the right half of the abdominal cavity, except for a small portion ventral to the esophageal notch. The caudal vena cava (243/*10*) is partly embedded in the rounded border of the liver and receives the hepatic veins before passing through the diaphragm (240/*4*, *3*). The **diaphragmatic surface** of the liver is convex and conforms to the curvature of the diaphragm against which it lies. From its most cranial point at the level of the sixth intercostal space, the long axis of the liver extends caudodorsally to the twelfth intercostal space or the dorsal end of the thirteenth rib (234, 236, 238, 244). Below the right kidney, the right lobe and caudate process occasionally pass the caudal border of the thirteenth rib.

* The directional designations given here pertain to the liver in situ.

The rounded border of the liver, then, faces to the left and slightly cranially and is crossed by the esophagus; while the opposite border is directed chiefly caudoventrally and to the right (243). In the ox, this latter (right) border follows a line extending in a caudally slightly convex arc from the sixth costal cartilage to the junction of proximal and middle thirds of the thirteenth rib (244). In the small ruminants, the liver is more vertically placed; the right border follows the right costal arch, but may cross it opposite the ninth and tenth

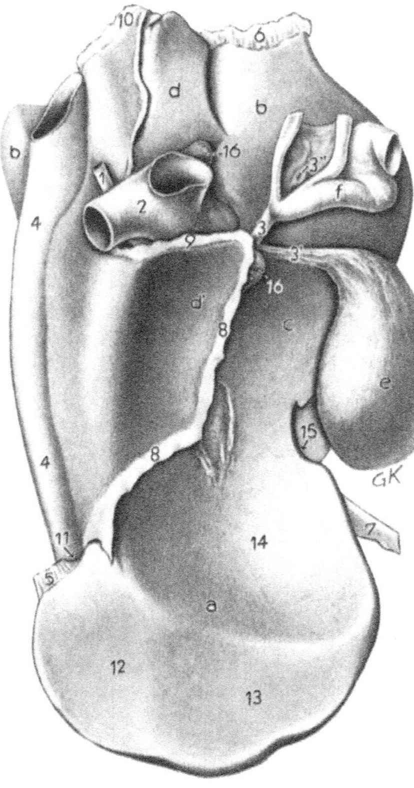

Fig. 239 (Visceral surface)

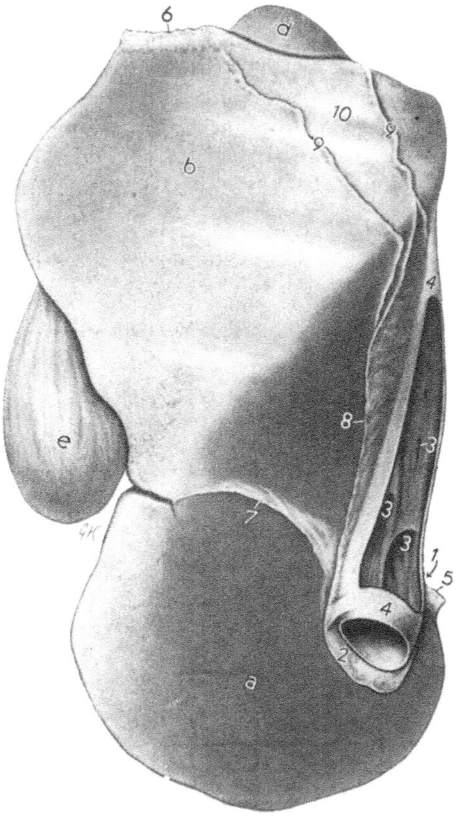

Fig. 240 (Diaphragmatic surface)

Fig. 239. Liver of the ox. Visceral surface. *a* Left lobe; *b, b* Right lobe; *c* Quadrate lobe; *d* Caudate lobe, its caudate process, *d'* its papillary process; *e* Gall bladder; *f* Sigmoid loop of duodenum; *1* Hepatic branch of hepatic artery; *2* Portal vein; *3* Bile duct; *3'* Cystic duct; *3''* Major duodenal papilla with opening of bile duct; *4* Caudal vena cava in its groove; *5* Left triangular ligament; *6* Right triangular ligament; *7* Falciform ligament with round ligament (degenerated umbilical vein) in its free border; *8, 9* Stump of lesser omentum; *8* Hepatogastric ligament; *9* Hepatoduodenal ligament; *10* Hepatorenal ligament; *11* Esophageal notch; *12* Reticular impression; *13* Abomasal impression; *14* Omasal impression; *15* Notch for round ligament *16* Hepatic lymph nodes

Fig. 240. Liver of the ox. Diaphragmatic surface. *a* Left lobe; *b* Right lobe; *d* Caudate process; *e* Gall bladder; *1* Esophageal notch; *2* Part of the tendinous center of the diaphragm; *3* Openings of the hepatic veins; *4* Caudal vena cava, fenestrated; *5* Left triangular ligament; *6* Right triangular ligament; *7* Falciform ligament; *8, 9* Coronary ligament; *10* Area nuda (area not covered with peritoneum)

intercostal spaces, the protruding portion then lying directly against the right abdominal wall (236, 238). The **visceral surface** of the liver is related to the cranial sac of the rumen, reticulum, omasum, cranial part of the duodenum, pancreas, jejunum, and right kidney (209, 210). The reticulum, omasum, and kidney leave distinct impressions on the liver when it is fixed in situ (239). In some cows the liver also touches the abomasum (*13*).

The **ATTACHMENT** of the liver to the diaphragm is accomplished chiefly by the caudal vena cava, which is firmly attached to the liver and to the right crus of the diaphragm. In addition, there is a fairly large triangular region (area nuda, 240/*10*) on the dorsomedial part of the diaphragmatic surface, which adheres directly to the diaphragm.

The **right triangular ligament** (243/*g*) leaves the dorsal border of the liver to the right of the caudate process and attaches the liver to the dorsolateral abdominal wall. Medial to the right triangular ligament is the **hepatorenal ligament,** the area of adhesion or attachment of the right kidney and the liver (239/*10*). The **left triangular ligament** (243/*f*) attaches the liver to the diaphragm in the region of the esophageal hiatus. The narrow **coronary ligament** (240/*9, 8*) extends from the right triangular ligament along the diaphragmatic surface of the liver to the right of the caudal vena cava, and passes ventral to the vein before reaching the left triangular ligament at the esophageal notch. The **round ligament** is often absent in older animals; the **falciform ligament** (211/*5, 6, 6'*; 234/*3, 4, 4'*), however, is always present as a thin serosal sheet lying on the diaphragmatic surface of the liver. It arises from this surface along a line extending from the notch for the round ligament to the esophageal notch and extends ventrally and to the right to attach on the tendinous center of the diaphragm. The **lesser omentum** (210/*r*; 234/*m*; 243/*i*) connects the visceral surface of the liver with the omasum, the lesser curvature of the abomasum, and the duodenum.

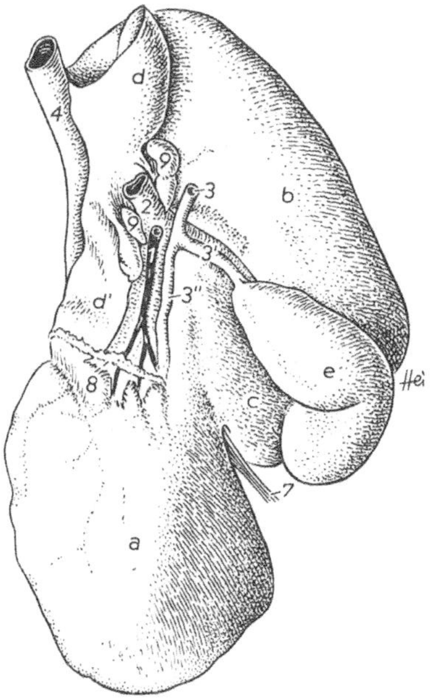

Fig. 241 (Sheep)

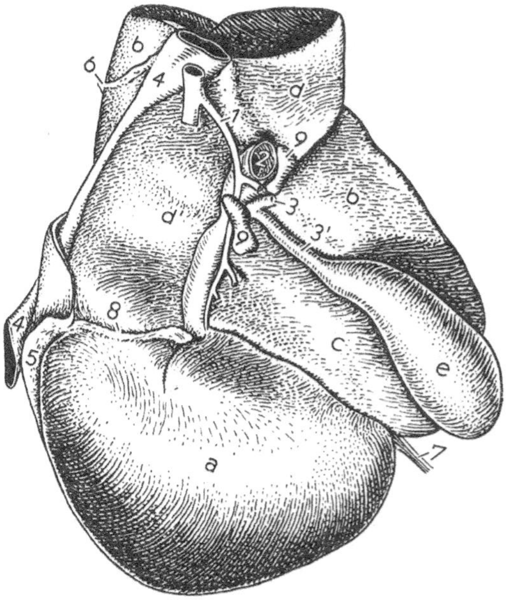

Fig. 242 (Goat)

Figs. 241 and 242. Liver of the sheep and the goat. Visceral surface. *a* Left lobe; *b* Right lobe; *c* Quadrate lobe; *d* Caudate lobe, its caudate process, *d'* its papillary process; *e* Gall bladder; *1* Hepatic branch of hepatic artery; *2* Portal vein; *3* Bile duct; *3'* Cystic duct; *3''* Common hepatic duct (shown only in the sheep); *4* Caudal vena cava in its groove; *5* Left triangular ligament; *6* Right triangular ligament (shown only in the goat); *7* Falciform ligament; *8* Stump of lesser omentum; *9* Hepatic lymph nodes

The **GALL BLADDER** of the ox is pear-shaped and may reach a diameter of about 10 cm.; in the small ruminants it is more elongated (239—242/*e*). It is attached to a fossa on the visceral surface of the liver, and its exposed surface is covered with peritoneum. It projects ventrally from the border of the liver, so that much of it lies directly against the abdominal wall (234/*5*). The hepatic ducts converge within the liver and form the common hepatic duct, which emerges at the porta. This duct unites with the **cystic duct** from the gall bladder and forms the short but wide **bile duct** (241/*3*). Short **hepatocycstic ducts,** bile channels entering the gall bladder from the liver directly through its attached wall, may be present. The bile duct enters the duodenum at the sigmoid loop about 50—70 cm. distal to the pylorus in the ox, and 30—40 cm. distal to the pylorus in the small ruminants. The **major duodenal papilla** at the end of the bile duct is not very prominent (239/*3''*). In the small ruminants the bile duct opens in common with the pancreatic duct.

Pancreas

The pancreas of the ruminants (166) is of a lighter or darker yellowish brown color and weighs on the average 550 gm. in the adult ox, and 50—70 gm. in the adult sheep. It consists of a narrow **body** (*a*) and right and left **lobes** (*c, b*) of which the right is larger. On the dorsal surface of the body is the **incisura pancreatis** (*3*), which is occupied by the portal vein. The left lobe of the pancreas (210/*k''*; 211/*r*; 237/*v*) is inserted between the dorsal sac of the rumen and the left crus of the diaphragm, and since it enters here the area of adhesion between the

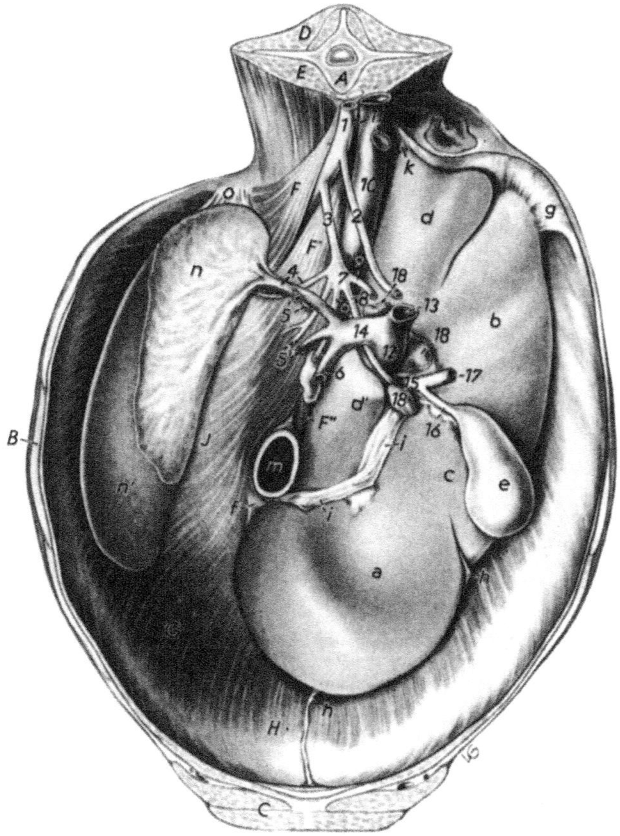

Fig. 243. Bovine liver and spleen in situ. Caudal aspect.

A Third lumbar vertebra; *B* Lateral abdominal wall with costal arch; *C* Rectus abdominis and pectoral muscles; *D* Epaxial musculature; *E* Psoas musculature; *F—H* On the diaphragm: *F* left crus, *F'*, *F''* medial part of right crus, *G* costal part, *H* sternal part, *J* left part of tendinous center

a—d' On the liver: *a* left lobe, *b* right lobe, *c* quadrate lobe, *d* caudate lobe, its caudate process, *d'* its papillary process; *e* Gall bladder; *f* Left triangular ligament; *g* Right triangular ligament; *h, h* Falciform ligament with round ligament (degenerated umbilical vein) in its free border; *i* Stump of lesser omentum; *k* Hepatorenal ligament; *l* Cut surface of right kidney; *m* Esophagus passing through diaphragm; *n* Spleen, attached portion, *n'* its free ventral end; *o* Phrenicosplenic ligament

1 Abdominal aorta, passing through aortic hiatus of diaphragm; *2* Cranial mesenteric artery; *3* Celiac artery; *4* Splenic artery and vein (see also *14*); *5* Right ruminal artery and vein; *5'* Left ruminal artery and vein; *6, 6* Left gastric artery and vein; *7* Hepatic artery; *8* Hepatic branch of hepatic artery, passing to the porta; *9* Gastroduodenal artery; *10* Caudal vena cava; *11* Left renal artery and vein; *12, 13* Portal vein; *14* Splenic vein (see also *4*); *15* Common hepatic duct; *16* Cystic duct; *17* Bile duct; *18* Hepatic lymph nodes

two, it is firmly attached to both structures. It also makes contact with the dorsal end of the spleen. The body and the right lobe of the pancreas lie against the visceral surface of the liver and the curvature of the omasum. Caudodorsally, the pancreas is in contact with the right kidney; caudoventrally, with the jejunum and transverse colon. The right lobe (244/*h*) extends caudolaterally in the mesoduodenum and accompanies the descending duodenum to the level of the second to the fourth lumbar vertebra. The bovine pancreas has only one secretory duct, the **accessory pancreatic duct,** which arises from the right lobe of the gland and enters the descending duodenum 30—40 cm. distal to the major duodenal papilla formed by the bile duct. Occasionally there is a second duct, the pancreatic duct, which enters in common with the bile duct at the major duodenal papilla. In the small ruminants, only the **pancreatic duct** is present. It enters the duodenum together with the bile duct at the major duodenal papilla.

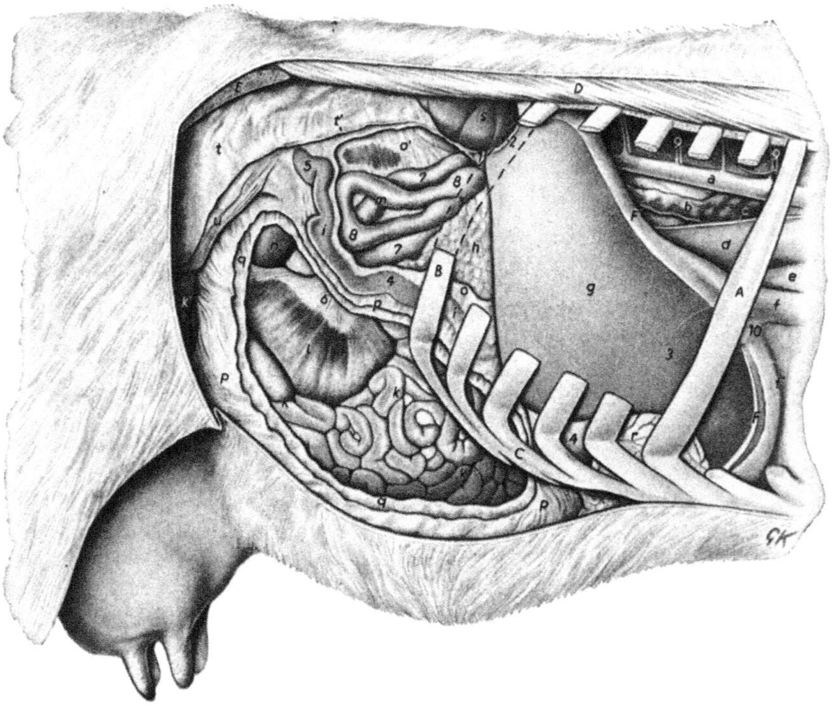

Fig. 244. Topography of the thoracic and abdominal organs of a cow. The right lung and part of the diaphragm have been removed, and the greater omentum and mesoduodenum have been fenestrated. Right aspect.

A Eighth rib; *B* Thirteenth rib, its removed portion indicated by broken lines; *C* Costal arch; *D* Longissimus; *E* Int. abdominal oblique muscle; *F*, *F'* Diaphragm;

a Aorta; *b* Caudal mediastinal lymph node; *c* Dorsal vagal trunk; *d* Esophagus; *e* Caudal mediastinum; *f* Caudal vena cava; *f'* Plica venae cavae; *g* Liver; *h* Pancreas; *i* Descending duodenum; *k* Jejunum; *k'* Caudal loops of jejunum protruding from supraomental recess; *l* Cecum; *m* Ascending colon; *n* Descending colon; *o* Mesoduodenum, *o*, *o'* lateral and medial laminae of the mesentery; *p* Supf. wall, *q* deep wall of greater omentum (between *p* and *q* is the caudal recess of the omental bursa, deep to *q* is the supraomental recess, the contents of which are visible through the fenestrated greater omentum); *r* Lesser omentum; *s* Right kidney; *t* Retroperitoneal fat; *t'* Cut edge of parietal peritoneum; *u* Cranial part of broad ligament (mesovarium)

1 Caudate process of liver; *2* Right triangular ligament; *3* Falciform ligament; *4* Gall bladder; *5* Caudal duodenal flexure; *6* Ileocecal fold; *7* Proximal loop, *8* distal loop of ascending colon; *9* Dorsal intercostal artery and vein; *10* Left phrenic vein

The Alimentary Canal of the Horse

Esophagus

The esophagus of the horse (*62*/*41*) follows the short laryngopharynx dorsal to the cricoid cartilage of the larynx and passes caudally on the dorsal surface of the trachea. In the distal third of the neck it deviates to the left and comes to lie lateral to the trachea (131) from about the fifth cervical vertebra onward, occasionally reaching the median plane ventral to the trachea (Sack and Sadler, 1979). The esophagus is accessible to surgery in the caudal third of the neck. It passes through the thoracic inlet between the trachea and the left first rib, and soon regains the dorsal surface of the trachea. It continues caudally in the mediastinum dorsal to the base of the heart and the tracheal bifurcation, crossing the aorta on the right. At the level of the fourteenth thoracic vertebra and about 12 cm. ventral to the vertebral column, it passes through the esophageal hiatus of the diaphragm slightly to the left of the median plane (257, 258/*o*). As it enters the abdominal cavity it joins the stomach at an acute angle.

At its cranial end, the esophagus receives the two **lateral longitudinal esophageal muscles** (87/*15*), which originate from the pharyngeal raphe, as well as muscle bundles from the cricoid and arytenoid cartilages. The muscular coat of the esophagus consists of striated muscle to the level of the tracheal bifurcation and of smooth muscle from there to the cardiac part of the stomach; it is 4—5 mm. thick proximally, increasing gradually toward the cardia where it reaches an extraordinary thickness of 1.2—1.5 cm. For more detail on the structure of the esophagus see page 100.

Stomach

(136, 245—248, 257)

The stomach of the horse is unusually small for an animal of its size, and has a normal capacity of only 8—15 liters. It is a simple stomach, bent sharply so that cardia and pylorus lie close to each other (245, 246), and has a very pronounced fundus (*b*) which projects as a true **blind sac** (saccus cecus) well above the level of the cardia. The stomach lies under cover of the ribs for the most part on the left side of the median plane, only the pyloric part being on the right (248/*3*). Even when greatly distended, it never reaches the ventral abdominal wall. The **greater curvature,** traced from the cardiac part over the summit of the blind sac to the pylorus, faces at first to the right, then dorsally, then to the left, and then ventrally, and over the pyloric part again to the right. The **lesser curvature** is short and faces dorsally and to the right. The **parietal surface** faces craniodorsally and slightly to the left and is applied dorsally against the diaphragm and ventrally against the left lobe of the liver, on which it produces the distinct gastric impression. Laterally, the parietal surface lies against the gastric surface of the spleen (257/*s*). The **visceral surface** of the stomach, facing caudoventrally, is in contact with jejunum, descending colon, the diaphragmatic flexure and right dorsal segment of the ascending colon, and the left lobe of the pancreas (248). The **blind sac** is the most dorsal and also the most caudal part of the stomach, and lies opposite the fourteenth or fifteenth intercostal spaces. The **body** of the stomach lies cranioventral to the blind sac and is opposite the ninth to twelfth intercostal spaces. When the stomach is distended, it expands principally to the left and caudally, but it may also advance cranially to the level of the eighth or seventh intercostal spaces. The position of the horse's stomach deep in the intrathoracic part of the abdominal cavity renders it inaccessible to clinical examination from the exterior (257/*r*); it can be reached only by means of a stomach tube introduced through the nostril and passed along the esophagus, or during acute dilatation of the stomach by means of a trocar*. Trocarization is performed high in the seventeenth intercostal space at the level of the tuber coxae (Habel 1978), but since the trocar may pass through the pleural cavity, this is a rather dangerous measure.

The **SHAPE** and **STRUCTURE** of the equine stomach fits the description of a simple stomach with composite lining as given on page 101—105. The general plan of its muscle layers is described on page 105. The **cardiac loop,** which is formed by the internal oblique fibers, and which continues along the inside of the lesser curvature, is especially well developed. It does not raise the mucous membrane sufficiently, however, to produce a gastric groove that is visible in the interior of the stomach; the mucous membrane has to be removed to see the gastric groove in the horse (247). Thick bundles of the circular muscle layer (*g*, *g'*) cross the internal oblique fibers (*f'*) externally and provide the floor of the gastric groove. In the vicinity of the cardia, these fibers (*g*) combine with the cardiac loop to form the strong **cardiac sphincter,** which keeps the cardia tightly closed even after the death of the animal. This strong sphincter and the very oblique angle at which the esophagus joins the stomach make it practically impossible for stomach contents and gases to enter esophagus. Also, being entirely within the bony thorax, the stomach is not directly influenced by ab-

* An instrument with a sharp point used inside a cannula.

dominal press. Thus the horse—with rare exceptions—is unable to relieve even a greatly distended stomach by vomiting, and if acute gastric distention is left untreated, the stomach may rupture. The circular muscle layer forms the **pyloric sphincter** (245/*8*) in the region of which the longitudinal muscle layer is also unusually thick. Proximal to the sphincter is a muscular ring (*7*), causing a constriction at which the pyloric part is divided into the pyloric antrum (*d*) and the pyloric canal (*e*).

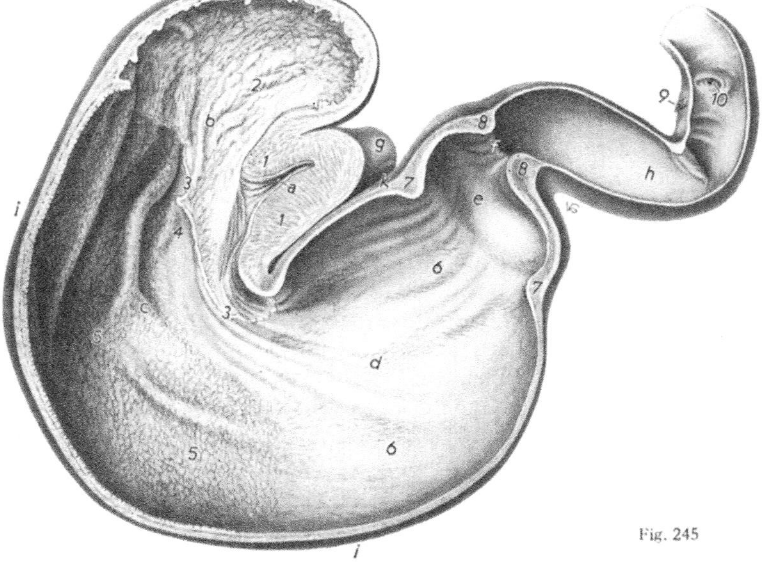

Fig. 245

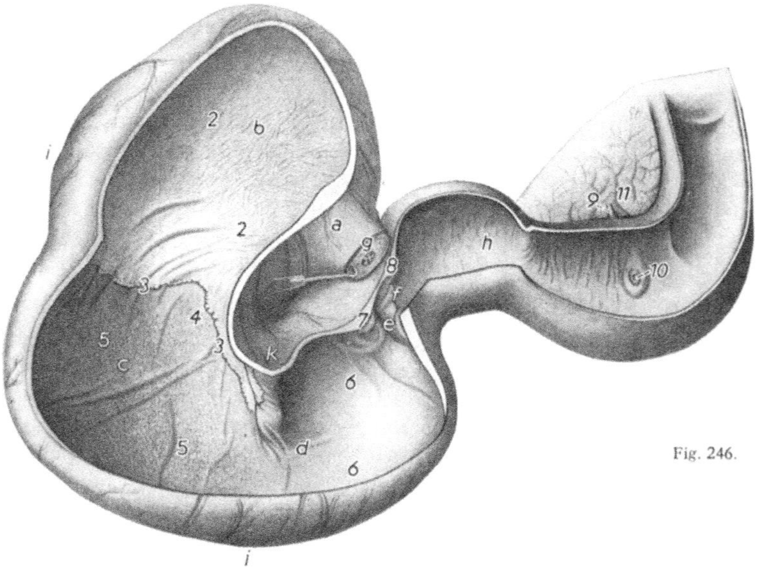

Fig. 246.

Figs. 245 and 246. Stomach of the horse, opened on the visceral surface. (Note the contracted blind sac in Fig. 245.)

a Cardia (sectioned obliquely in Fig. 245); ***b*** Blind sac (fundus); ***c*** Body; ***d, e*** Pyloric part; ***d*** Pyloric antrum; ***e*** Pyloric canal; ***f*** Pyloric orifice; ***g*** Esophagus (pulled caudally in Fig. 246); ***h*** Cranial part of duodenum (in Fig. 246 ampulla duodeni); ***i*** Greater curvature; ***k*** Lesser curvature

1, 1 Cardiac sphincter (not shown in Fig. 246); ***2*** Nonglandular proventricular part of mucosa; ***3*** Margo plicatus; ***4*** Region of mixed cardiac and pyloric glands; ***5*** Region of proper gastric glands; ***6*** Pyloric gland region; ***7*** Muscular ring separating ***d*** and ***e***; ***8*** Pyloric sphincter; ***9*** Accessory pancreatic duct; ***10*** Major duodenal papilla with openings of bile and pancreatic ducts inside the ampulla hepatopancreatica; ***11*** Body of pancreas (shown only in Fig. 246)

The extensive **proventricular part** of the gastric mucosa is whitish, nonglandular, and covered with thick, stratified squamous epithelium. It is not restricted to the blind sac but extends for a short distance also into the body of the stomach (245, 246/*2*), meeting the **glandular part** (*5*, *6*) along an irregular raised ridge, the **margo plicatus** (*3*). The moderately full stomach is often slightly constricted along the margo plicatus. The **glandular part** of the stomach is divided into three regions (136) as in other domestic mammals. The **cardiac gland region** (*4'*) is a narrow zone along the margo plicatus that does not extend to the lesser curvature; it has both pyloric and cardiac glands. The **region of the proper gastric glands** (*3*) is reddish brown and contains distinct areae and foveolae gastricae, the **pyloric gland region** (*4*) is yellowish pink; neither presents any special features.

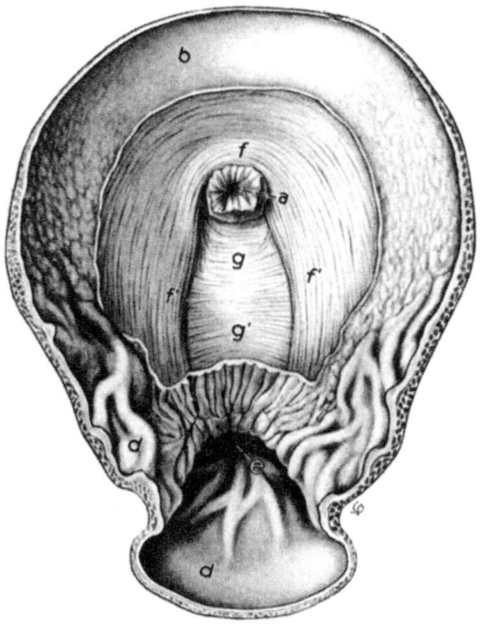

Fig. 247. Gastric groove of the horse. The mucous membrane has been removed in the area surrounding the cardiac part of the stomach.

a Cardia; *b* Blind sac lined with nonglandular mucous membrane; *d* Pyloric gland region; Between *b* and *d* is the region of the proper gastric glands; *e* Pyloric orifice; *f*, *f'* Edge of internal oblique fibers, forming the cardiac loop (*f*) and the edges of the gastric groove (*f'*, *f'*); *g*, *g'* Circular muscle layer, forming the floor of the gastric groove, *g* fibers in the vicinity of the cardia, forming the cardiac sphincter together with the cardiac loop (*f*)

ATTACHMENT OF THE STOMACH. The stomach is directly attached to the diaphragm with its cardiac part and a small area dorsal to it. The **gastrophrenic ligament** arises from the dorsal end of the area of direct attachment and passes caudolaterally to the apex of the blind sac, where it is continued by the **phrenicosplenic ligament** and ventrally by the **gastrosplenic ligament** (13/*4*).

The **GREATER OMENTUM** of the horse is thin and contains very little fat (248/*p*). It extends in irregular fashion caudoventrally between the parts of the intestines, enclosing the ample **caudal recess of the omental bursa.** It may reach the vicinity of the deep inguinal ring, and portions of it have been known to have entered the tunica vaginalis in male animals. The line of attachment of the greater omentum begins on the blind sac of the stomach as a continuation of the gastrophrenic ligament. From here it follows the greater curvature to the cranial part of the duodenum on the right (248/*f*), passes onto the adjacent right dorsal colon (*h*) and then back along the transverse colon (*h'*) to the beginning of the descending colon on left side of the body (*i*). It passes caudally on the descending colon for a short distance (*r*), reverses its course and runs cranially at first on the descending colon, and then on the left lobe of the pancreas back to the blind sac where it rejoins the gastrophrenic ligament. The structures enclosed by this circular line of attachment (visceral surface of stomach, proximal part of duodenum, left lobe of pancreas, transverse colon, part of the right dorsal and the descending colon) form the craniodorsal boundary of the omental bursa. The **epiploic foramen** (13/*11*) is a narrow slit, 4—6 cm. long, between the base of the caudate process of the liver and the right lobe of the pancreas, with the caudal vena cava (*e*) dorsally and the portal vein (*f*) ventrally. It leads into the **vestibule of the omental bursa**, which is continuous across the lesser curvature of the stomach with the large caudal recess, the principal cavity of the omental bursa.

The **LESSER OMENTUM,** consisting of **hepatogastric** and **hepatoduodenal ligaments,** connects the stomach and duodenum with the liver, and forms part of the wall of the vestibule of the omental bursa. The **vestibule** is a potential space between liver and stomach,

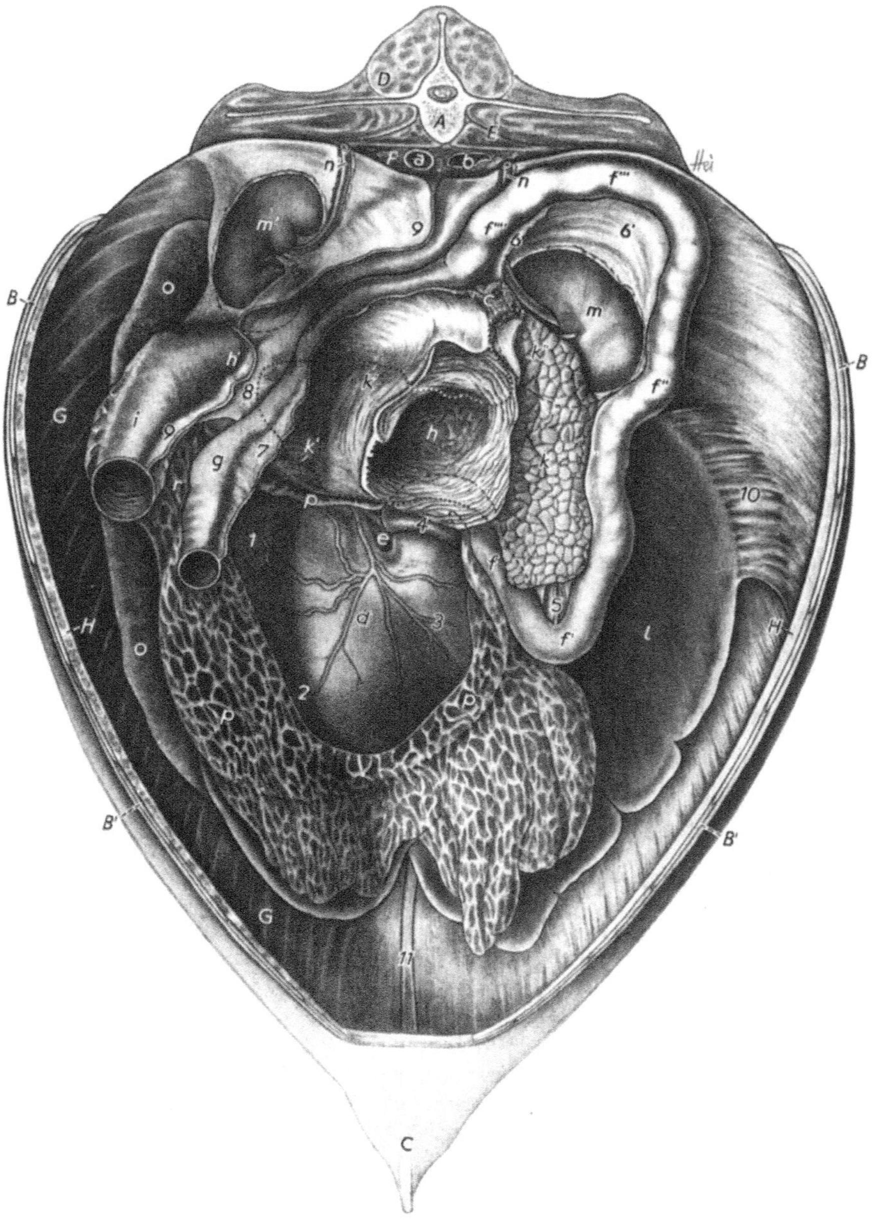

Fig. 248. Cranial abdominal organs of the horse in situ. Caudoventral aspect.

A Second lumbar vertebra; *B* Eighteenth rib; *B'* Costal arch; *C* Sternum; *D* Epaxial musculature; *E* Psoas musculature; *F* Left crus of diaphragm; *G* Costal part of diaphragm; *H* Cut surface of transversus abdominis

a Abdominal aorta; *b* Caudal vena cava; *c* Cross section of the root of the mesentery; *d* Visceral surface of stomach with visceral branch of left gastric artery; *e* Entrance to the vestibule of the omental bursa over the lesser curvature of the stomach; *f* Cranial part of duodenum; *f'* Cranial duodenal flexure; *f''* Descending duodenum; *f'''* Caudal duodenal flexure; *f''''* Ascending duodenum; *g* Jejunum; *h, h'* Transverse colon, *h* at the junction with the right dorsal colon; *i* Descending colon (small colon); *k* Body of pancreas; *k'* Outline of left lobe of pancreas (dotted line); *k''* Right lobe of pancreas; *l* Right lobe of liver; *m, m'* Right and left kidneys, their full extent is indicated by the broken lines; *n, n'* Right and left ureters; *o* Visceral surface of spleen; *p* Greater omentum, the deep wall has been fenestrated to permit a view into the caudal recess of the omental bursa; *r* Origin of greater omentum on proximal part of descending colon

1—*4* Stomach: *1* blind sac, *2* body, *3* pyloric part, *4* pylorus; *5* Bile and pancreatic ducts; *6, 6'* Mesoduodenum, *6* attaching on the base of the cecum, *6'* attaching on the right kidney; *7* Mesentery of jejunum; *8* Duodenocolic fold; *9* Descending mesocolon; *10* Right triangular ligament of liver; *11* Falciform and round ligaments of liver

leading caudoventrally into the caudal recess of the omental bursa. It is bounded on the left by the gastrophrenic ligament (13/*7*) and by a part of the hepatogastric ligament (*8*); cranially by the liver (*d*); dorsally by the left lobe of the pancreas (*g*); ventrally by the remaining portion of the hepatogastric ligament and by the hepatoduodenal ligament (*8*, *9*); and on the right by part of the hepatoduodenal ligament, the caudal vena cava (*e*), and portal vein (*f*).

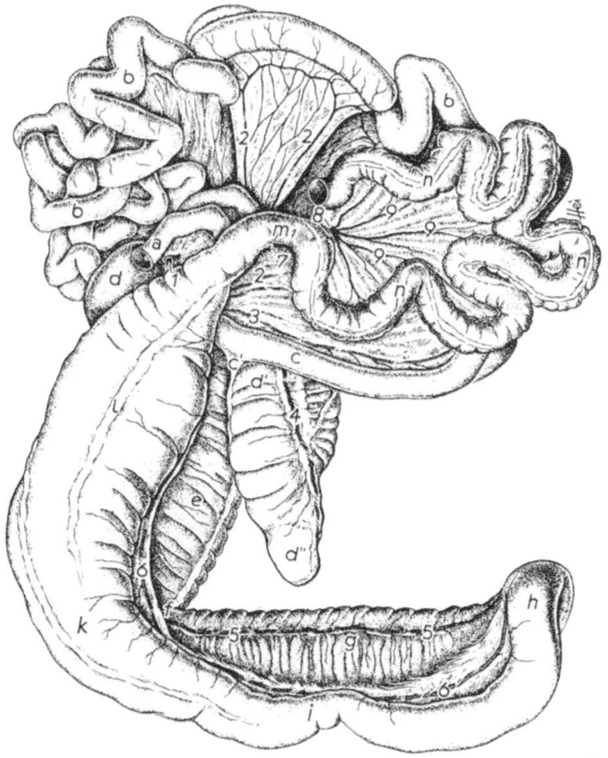

Fig. 249. Intestines of the horse.

a Distal end of descending duodenum and caudal flexure; *b* Jejunum; *c* Ileum; *c'* Ileocecal fold; *d, d', d''* Base, body, and apex of cecum; *e* Right ventral colon; *f* Sternal flexure; *g* Left ventral colon; *h* Pelvic flexure; *i* Left dorsal colon; *k* Diaphragmatic flexure, *l* Right dorsal colon; *m* Transverse colon; *n* Descending colon (small colon)

1 Cranial mesenteric artery; *2* Jejunal arteries in the mesentery; *3* Ileal artery; *4* Medial cecal artery on the medial band of the cecum; *5* Colic branch of ileocolic artery; *6* Right colic artery; *7* Middle colic artery; *8* Caudal mesenteric artery, giving rise to the left colic and cranial rectal arteries; *9* Branches of the caudal mesenteric artery to the descending colon

Intestines

In contrast to the rather small stomach, the intestines of the horse are very extensive and occupy by far the greatest part of the abdominal cavity, with the cecum and ascending colon having the greatest capacity. These are large chambers in which bacterial fermentation of cellulose takes place, much as in the rumen in the ruminants. Because of its large size, the ascending colon of the horse is also known as the great colon.

The length of the **SMALL INTESTINE** of the horse varies from animal to animal, but it also varies within the same animal in relation to digestive activity. The duodenum has an average length of 1 m., the jejunum 25 m., and the ileum about .7 m. The variation in length is associated with great variation in the lumen of the small intestine, which may have a diameter of up to 5—7 cm. or may be almost obliterated by contraction of the intestinal wall. Arrested contractions of this kind are often found in the freshly killed, unembalmed animal and give the small intestine a beaded appearance.

The **DUODENUM** (147/*B*; 248/*f—f''''*) begins at the pylorus to the right of the median plane and, as in the other domestic mammals, describes a loop around the caudal aspect of the root of the mesentery. Its **cranial part** lies against the visceral surface of the liver, forming a sharp bend (ansa sigmoidea) and a dilatation (ampulla duodeni, 246/*h*). The second curve described by the duodenum on the visceral surface of the liver is the **cranial flexure** (248/*f'*). In its concavity lies the body of the pancreas. Here, the bile and pancreatic ducts (ampulla hepatopancreatica) enter at the **major duodenal papilla** and the accessory pancreatic duct enters nearly opposite these at the **minor duodenal papilla** (167, 246). The **descending duodenum** (248/*f''*) continues from the cranial flexure caudodorsally, passing between the visceral surface of the right lobe of the liver and the right dorsal colon. Then, lying against the diaphragm, it passes the lateral border of the right kidney. With the **caudal flexure** (*f'''*) the duodenum turns medially around the base of the cecum (258/*t*) and the root of the mesentery and crosses the median plane at the level of the third or fourth lumbar vertebra. From the caudal flexure, the short **ascending duodenum** (248/*f''''*) passes cranially and to the left, is inserted between the root of the mesentery on the right (*c*) and the beginning of the descending mesocolon on the left (*9*), and is united with the transverse colon and the proximal part of the descending colon by the duodenocolic fold (*8*). At the **duodenojejunal flexure,** the duodenum is continued by the jejunum (*g*).

The duodenum has a relatively fixed position, being attached to adjacent organs by the short hepatoduodenal ligament and its continuation, the mesoduodenum. The **hepatoduodenal ligament** attaches the cranial part of the duodenum to the hepatic porta and contains the bile and pancreatic ducts. The **mesoduodenum** (248/*6'*), which follows, connects the descending duodenum to the right lobe of the liver, to the pancreas, and to the right kidney. At the caudal flexure, the mesoduodenum (*6*) connects the duodenum to the base of the cecum. The mesoduodenum attaches the ascending duodenum to the caudal aspect of the root of the mesentery (*c*) and cranial to that separates into two parts: one becomes the **duodenocolic fold,** which ends in a free edge (*8*); the other, on the opposite surface of the duodenum, continues as the mesentery of the jejunum (*7*).

The very long **JEJUNUM** (147/*C*) is suspended by an extensive mesentery (mesojejunum) from the area of the last thoracic and the first two lumbar vertebrae. The average length of the mesojejunum as measured from its **root** (248/*c*) to the attachment of the jejunum is 40—60 cm.; along the jejunum it measures 25 m. Because of the length of the mesojejunum, the jejunal coils have great mobility, and may be encountered almost anywhere in the abdominal cavity. Most of the jejunum, however, lies in the dorsal part of the left half of the abdomen, where it intermingles with the large coils of the descending colon (257/*x*). Cranially, the jejunal coils are in contact with the spleen, stomach, liver, pancreas, and the cranial parts of the colon. On the left, they extend caudally beyond the last rib into the area of the tuber coxae. They may even insinuate themselves between the left and right parts of the great colon, reaching the cecum and the ventral abdominal wall (259/*a*), and some are encountered in the inguinal region. As a result of this mobility of the jejunal coils, intestinal colic may develop because of volvulus, intussusception, or incarceration of jejunal loops in the epiploic foramen or the tunica vaginalis after castration.

The **ILEUM** (147/*D*), the terminal portion of the small intestine, when moderately contracted measures about 70 cm. in length; when fully contracted it is only 20 cm. long. When moderately contracted, the ileum can be distinguished from the jejunum by its thicker muscular wall and narrower lumen. When relaxed, these distinguishing characteristics are absent, and one has to rely on the proximal extent of the **ileocecal fold** (*18*) to determine the jejuno-ileal junction.

The ileum arises from the jejunal coils in the left flank, crosses to the right at the level of the third to the fourth lumbar vertebra, and then passes almost straight dorsally to the lesser curvature of the base of the cecum, where it ends at the ileal orifice. Although the ileum seems to join the large intestine at the junction of the base and body of the cecum, it really joins it at the cecocolic junction as in the other domestic mammals. Developmental studies of the equine intestinal tract (Martin 1921, Westerlund 1918) have shown that much of the bulbous base of the cecum is actually part of the ascending colon. In the fetus, the initial part of the ascending colon forms a tight S-shaped flexure. The proximal part of the flexure becomes greatly enlarged and forms the cranial part of the base of the cecum; the distal part becomes the proximal part of the right ventral colon of the adult. This S-shaped flexure can still be seen in the adult animal (255/*a'*, *2*, *b*). For morphological and, more importantly, physiological considerations it has become customary in the horse to consider the cecocolic junction slightly more distal than it is, and to include the proximal part of the ascending colon (*a'*) in the cecum.

The end of the ileum is partly telescoped into the cecum so that the **ileal orifice** (250, 251/*1*) is in the center of a slight elevation (papilla ilealis) formed by an annular fold of mucous membrane. Contained in this fold is a network of veins (Schummer 1953), which when engorged distend the annular fold (251, 253). This increases ileal protrusion slightly, narrows the ileal opening, and prevents reflux of cecal contents into the ileum. When the veins are not engorged, the annular fold is relaxed and the ileal opening is widened (250, 252). Although a m. sphincter ilei is often described as surrounding the ileal orifice, it is not present in the horse. However, the network of veins surrounding the orifice, and, because of its unusual contractility, the entire muscle coat of the ileum may be regarded as a functional ileal sphincter.

During contraction, the ileum shortens and thickens, and its mucous membrane becomes folded. The ileum can deliver jejunal contents into the cecum against the higher pressure resulting in that organ from contraction and fermentation. By active muscular contraction of the ileum, and closure of the ileal opening as a result of engorgement, the ileum prevents the backflow of ingesta and the equalization of pressure between jejunum and the base of the cecum. Disturbance of this sensitive balance is not uncommon and is one of the causes of colic in horses.

The ileum is attached to the most distal part of the mesentery (147/*17*), and on its antimesenteric border it is attached to the dorsal band of the cecum by the **ileocecal fold** (*18*).

Fig. 250

Fig. 251

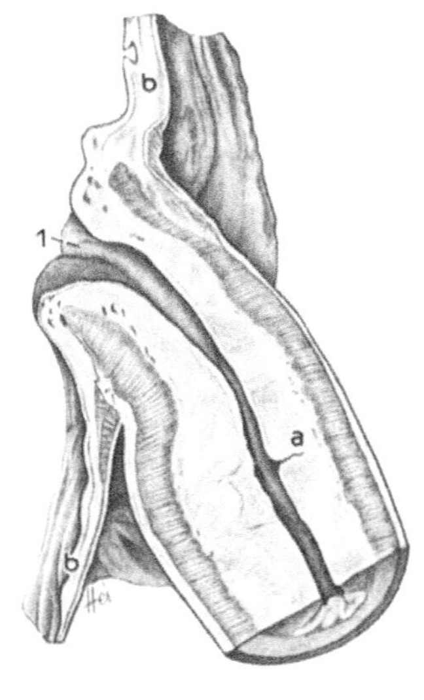

Fig. 252

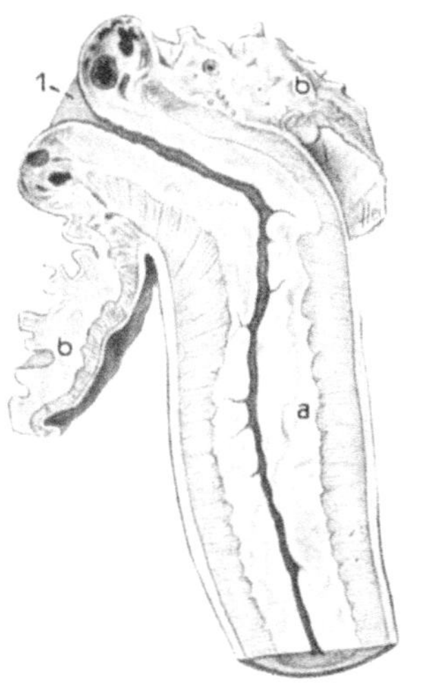

Fig. 253

Fig. 250. Ileal opening in center of papilla ilealis in which the venous plexus is not engorged.

Fig. 251. Ileal opening in center of papilla ilealis in which the venous plexus is engorged. The ileocecal fold (*a'*) is gathered into folds as a result of the contraction of the ileum.

Fig. 252. Longitudinal section of specimen shown in Fig. 250. The muscular coat of the ileum is contracted, causing high folds in the mucous membrane and a narrow lumen. The venous plexus surrounding the ileal opening is not engorged. The circular muscle layer of the ileum decreases in height toward the ileal opening.

Fig. 253. Longitudinal section of specimen of Fig. 251. The muscular coat of the ileum is contracted, causing high folds in the mucous membrane and a narrow lumen. The venous plexus surrounding the ileal opening is engorged. The circular muscle layer of the ileum decreases in height toward the ileal opening.

a Ileum; *a'* Ileocecal fold; *b* Mucosa of cecum; *1* Ileal opening in center of papilla ilealis containing a venous plexus

The **LARGE INTESTINE** of the horse is distinguished from that of the other domestic mammals chiefly by its enormous capacity, its peculiar position in the abdominal cavity, by the shape of the cecum, and the length of the descending colon. As in the pig, the large intestine is sacculated by longitudinal muscular **bands** (256).

The **CECUM** of the horse (254) is on the average about 1 m. long and has a capacity of 16—68 liters, with an average of 33 liters. The bulbous **base** of the cecum (*b*) is the most dorsal part and presents a **greater curvature** dorsally and a **lesser curvature** ventrally. Developmentally, its cranial part belongs to the ascending colon. Continuing cranioventrally from the base is the **body** (*b'*), which gradually tapers toward the **apex.** The body of the cecum has

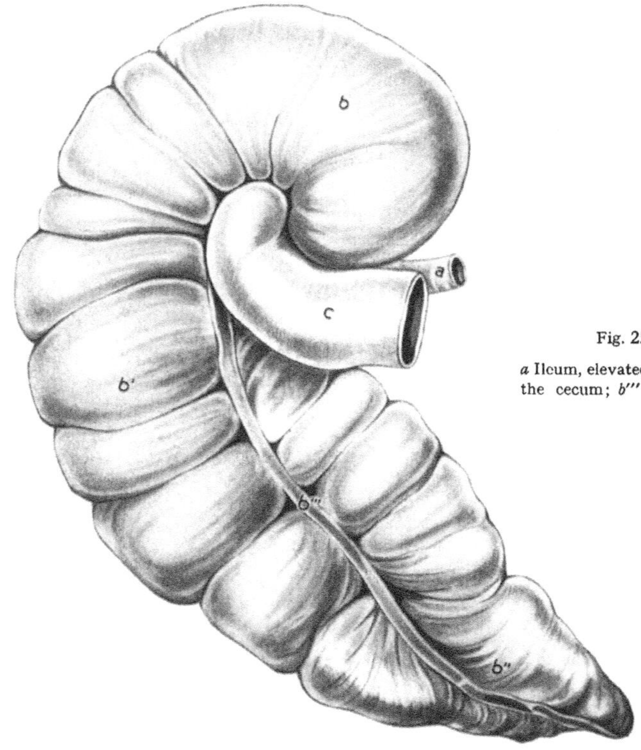

Fig. 254. Cecum of the horse. Right lateral aspect.

a Ileum, elevated; *b* Base of the cecum; *b'* Body of the cecum; *b''* Apex of the cecum; *b'''* Lateral band; *c* Proximal part of right ventral colon

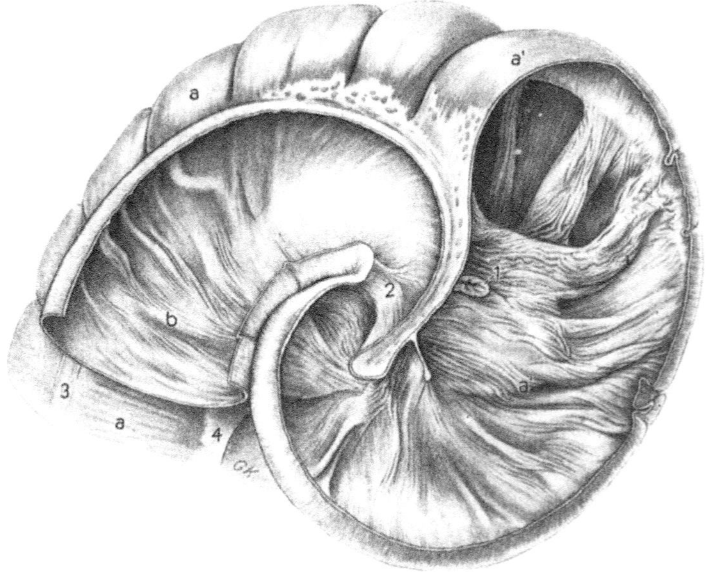

Fig. 255. Base of the cecum and proximal part of the ascending colon (right ventral colon) of the horse, opened laterally to show the ileal and cecocolic orifices. Fixed in situ. Lateral aspect.

***a* Body of the cecum; *a'* Base of the cecum; *b* Proximal part of right ventral colon**

1 Ileal orifice on papilla ilealis; *2* Cecocolic orifice; *3* Lateral band of cecum; *4* Dorsal band of cecum with ileocecal fold

longitudinal **bands** (teniae, *b'''*) on the dorsal, ventral, lateral, and medial surfaces. The dorsal and medial bands end at the apex; the ventral band usually joins the medial band near the apex or fades out before reaching it (259/*1*, *1'*, *2*, *3*). The **ileocecal fold** attaches along the dorsal band; the **cecocolic fold** (256/*1*) attaches along the lateral band; and the cecal vessels and lymph nodes accompany the medial and lateral bands. The four muscular bands make for four rows of sacculations on the cecum. Deep grooves, **semilunar folds** (255) in the interior, separate neighboring sacculations of a row.

Medial on the lesser curvature of the base of the cecum is the ileal opening (255/*1*). Distal to this is the **cecocolic opening** (*2*), which connects the base of the cecum with the ascending colon. It is caudolateral to the ileal opening, and near the center of the circle described by the greater curvature of the base of the cecum. It lies between two substantial folds, which form the **cecocolic valve** (valva cecocolica). There is no cecal sphincter controlling this opening.

The cecum occupies a large part of the right half of the abdominal cavity (258). Its base (*t*) extends from the pelvic inlet forward to about the fourteenth or fifteenth rib, lying against the paralumbar fossa, and farther cranially against the costal part of the diaphragm, the right lobe of the liver (*q*), and the right dorsal colon (*v'*). Dorsally, it is attached to the ventral surface of the right kidney (*r*) and to the right lobe of the pancreas. The left surface of the base of the cecum is attached to the root of the mesentery, and in this attachment the cecal vessels reach the organ. Caudally, the cecum is related to the loops of the jejunum and the descending colon. The cecum is related dorsally also to the duodenum (*s*), which passes with its caudal flexure around the top of the base. The body of the cecum (*t'*) follows the contour of the abdominal floor and is directed cranioventrally and slightly medially. Caudally, the body lies against the right flank, but because of its medial inclination it loses contact with the lateral abdominal wall and comes to lie between the right and left ventral colon, with the apex of the cecum about 20 cm. caudal to the xiphoid cartilage (259).

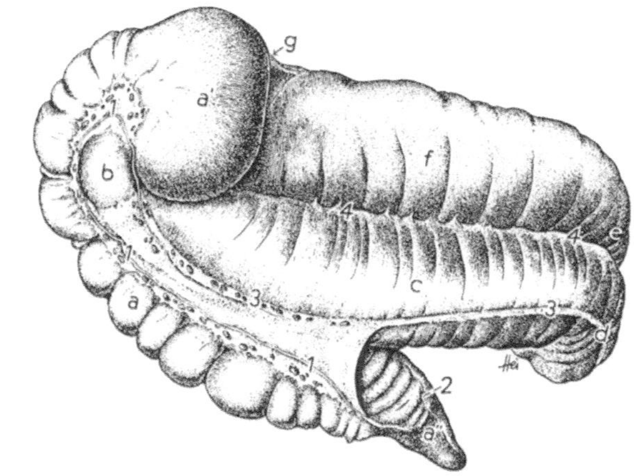

Fig. 256. Cecum and right dorsal and ventral colon of the horse. Right lateral aspect.

a, a', a'' Body, base, and apex of cecum; *b* Proximal dilated part of right ventral colon; *c* Right ventral colon; *d* Sternal flexure; *e* Diaphragmatic flexure; *f* Right dorsal colon; *g* Junction of right dorsal colon and transverse colon

1 Lateral band of cecum with lateral cecal artery and vein, cecal lymph nodes, and attachment of the cecocolic fold; *2* Dorsal band of cecum; *3* Lateral free band of right ventral colon with some colic lymph nodes and attachment of the cecocolic fold; *4* Short peritoneal bridge (ascending mesocolon) connecting dorsal and ventral colon

The **COLON** of the horse is characterized by the extraordinary capacity and peculiar position of the ascending colon and by the great length of the descending colon. The transverse colon is short, and as in the other species is located just cranial to the cranial mesenteric artery.

The **ascending colon** (147/*F*, *F'*), also known as the **great colon** (colon crassum) because of its large size, is 3—4 m. long and has a capacity of 55—130 liters with an average of about 80 liters. When it is removed from the animal it is a long, U-shaped loop consisting of two parallel limbs and a terminal flexure (*13*). The beginning and the end of the ascending colon are to the right of the root of the mesentery, to which they are attached by the ascending mesocolon (*20*), which passes distally between the two limbs of the loop and connects them. This long U-shaped loop is accomodated in the abdominal cavity by being folded once on itself, forming a double horseshoe-shaped loop. In situ, the proximal parts of the original loop pass cranioventrally from their origin on the right side of the abdominal cavity, are directed to the left by the curvature of the diaphragm, and return in caudodorsal direction on the left side of the abdominal cavity. As a result, the terminal flexure of the loop comes to lie close

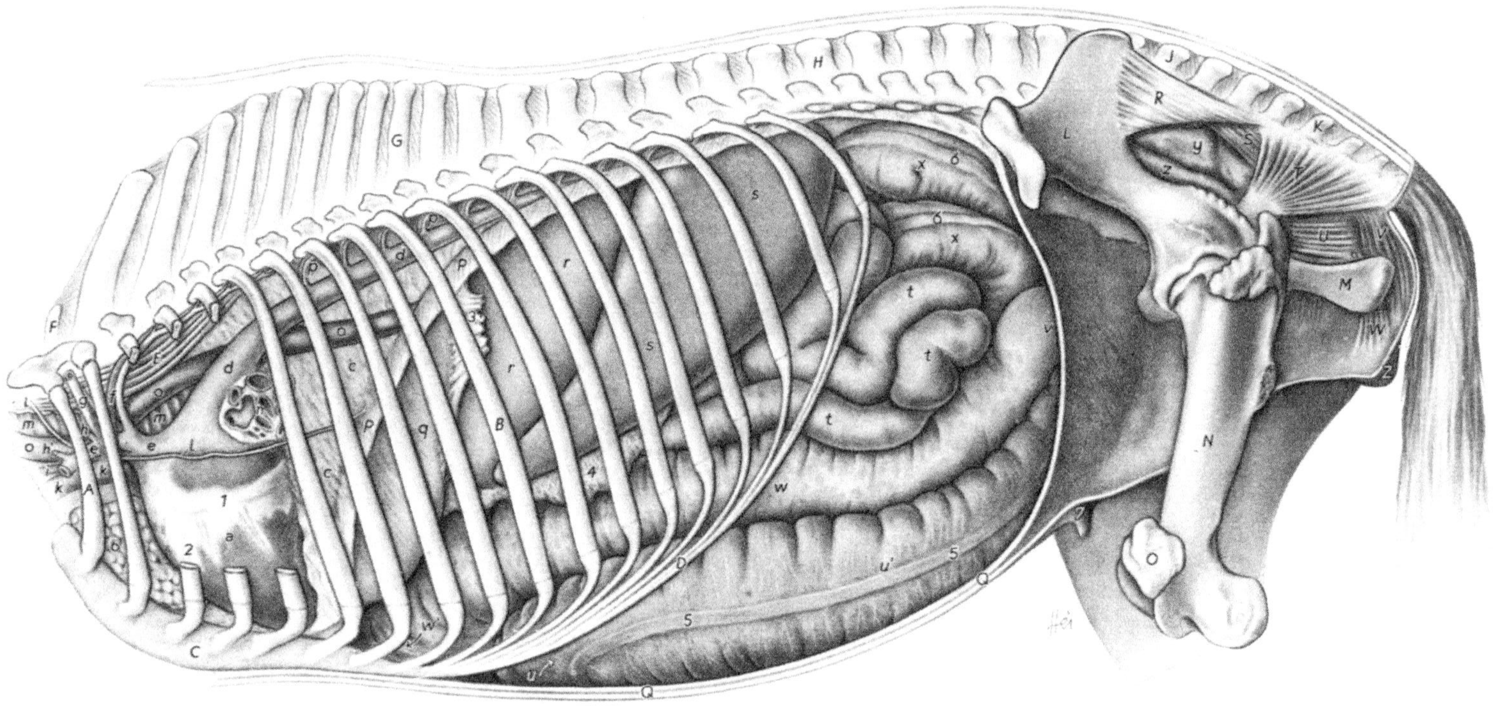

Fig. 257. Topography of the thoracic and abdominal organs of the horse. The left lung and part of the diaphragm have been removed. Left lateral aspect.

A First rib; *B* Tenth rib; *C* Sternum; *D* Costal arch; *E* Longus colli; *F* First thoracic vertebra; *G* Tenth thoracic vertebra; *H* First lumbar vertebra; *J* Sacrum; *K* First caudal vertebra; *L* Ilium; *M* Ischium; *N* Femur; *O* Patella; *P* Diaphragm; *Q* Abdominal floor; *R* Dorsal sacroiliac ligament; *S* Sacrocaudalis ventralis lateralis; *T* Coccygeus; *U* Levator ani; *V* Ext. anal sphincter; *W* Constrictor vulvae; *Z* Vulva

a Heart; ***b*** Cranial mediastinum containing fat; *c* Caudal mediastinum; ***d*** Thoracic aorta; ***e*** Brachiocephalic trunk; ***e'*** Left subclavian artery; ***f*** Costocervical artery and vein; ***g*** Deep cervical artery and vein; ***h*** Vertebral artery; ***h'*** Supf. cervical artery; ***i*** Left common carotid artery; ***k*** Left ext. jugular vein; ***l*** Left phrenic nerve; ***m*** Trachea; ***n*** Root of left lung; *o* Esophagus; ***p*** Dorsal intercostal arteries; ***q*** Left lobe of liver; ***r*** Stomach; *s* Spleen; *t* Jejunum; ***u*** Sternal flexure; ***u'*** Left ventral colon; *v* Pelvic flexure; *w* Left dorsal colon; *w'* Diaphragmatic flexure; *x* Descending colon; *y* Rectum; *z* Vagina

1 Coronary groove; *2* Paraconal interventricular groove; *3* Left triangular ligament of liver; *4* Greater omentum; *5* Lateral free band of left ventral colon; *6* Free band of descending colon; *7* Udder

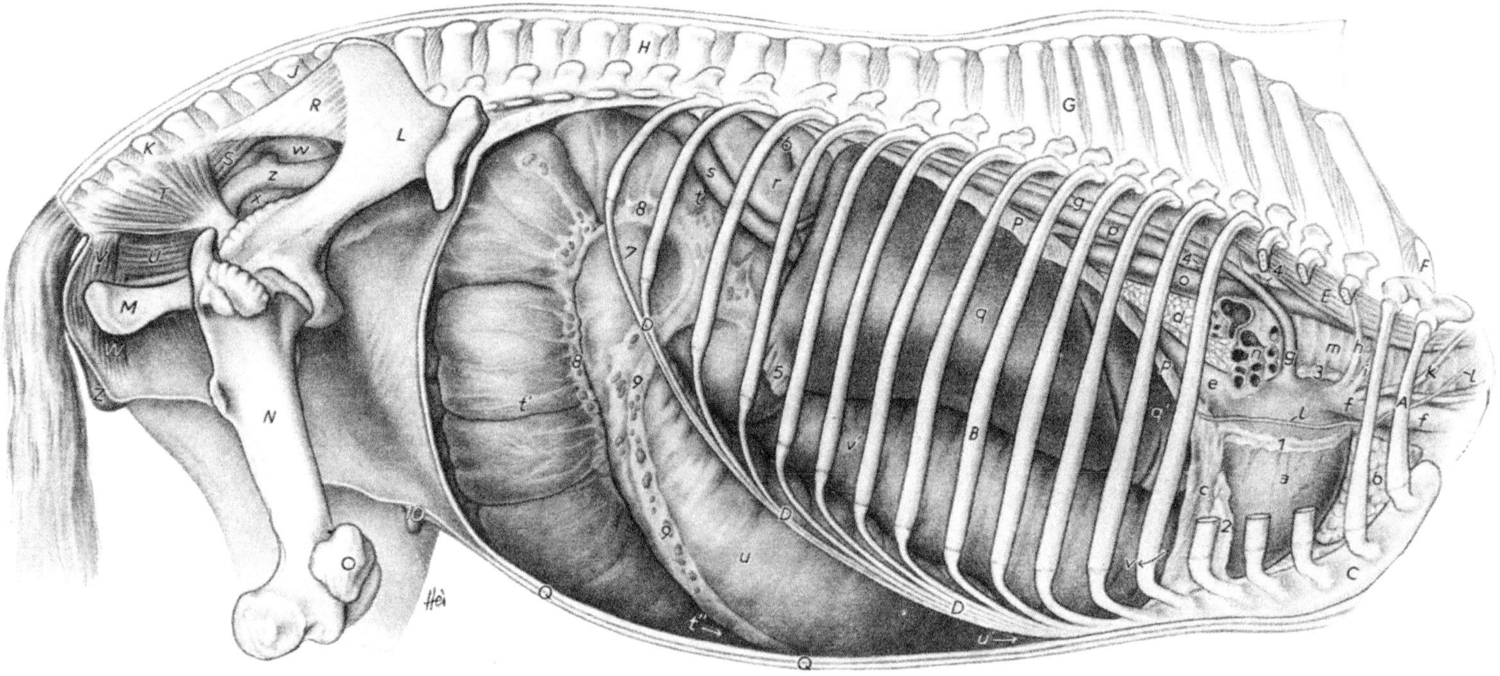

Fig. 258. Topography of the thoracic and abdominal organs of the horse. The right lung and part of the diaphragm have been removed. Right lateral aspect.

A First rib; *B* Tenth rib; *C* Sternum; *D* Costal arch; *E* Longus colli; *F* First thoracic vertebra; *G* Tenth thoracic vertebra; *H* First lumbar vertebra; *J* Sacrum; *K* First caudal vertebra; *L* Ilium; *M* Ischium; *N* Femur; *O* Patella; *P* Diaphragm; *Q* Abdominal floor; *R* Dorsal sacroiliac ligament; *S* Sacrocaudalis ventralis lateralis; *T* Coccygeus; *U* Levator ani; *V* Ext. anal sphincter; *W* Constrictor vulvae; *Z* Vulva

a Heart; *b* Cranial mediastinum containing fat; *c* Plica venae cavae; *d* Accessory lobe and part of the caudal lobe of right lung; *e* Caudal vena cava; *f* Cranial vena cava; *g* Right azygous vein; *h* Costocervical vein; *i* Deep cervical vein; *k* Vertebral vein; *l* Right phrenic nerve; *m* Trachea; *n* Root of right lung; *o* Esophagus; *p* Thoracic aorta; *q* Right lobe of liver; *q'* Quadrate lobe of liver; *r* Right kidney; *s* Descending duodenum; *t*, *t'*, *t''* Base, body, and apex of cecum; *u* Right ventral colon; *u'* Sternal flexure; *v* Diaphragmatic flexure; *v'* Right dorsal colon; *w* Rectum; *x* Urinary bladder; *z* Vagina

1 Coronary groove; *2* Subsinual interventricular groove; *3* Cranial mediastinal lymph nodes; *4* Dorsal mediastinal lymph nodes; *5* Right triangular ligament of liver; *6* Hilus of right kidney; *7* Proximal dilated portion of right ventral colon; *8* Lateral cecal band with cecal lymph nodes and attachment of the cecocolic fold; *9* Lateral free band of right ventral colon with colic lymph nodes and attachment of the cecocolic fold; *10* Udder

to the pelvic inlet, and is therefore called the pelvic flexure (*13*). The two horseshoe-shaped loops lie more or less on top of each other, and the ventral one is called the ventral colon and the other the dorsal colon. Thus the following parts of the ascending colon, listed in proximodistal sequence, can be distinguished: right ventral colon; sternal flexure; left ventral colon; pelvic flexure; left dorsal colon; diaphragmatic flexure; and right dorsal colon (249/*e*—*l*).

The **right ventral colon** (254/*c*) begins at the cecocolic orifice in the lesser curvature of the base of the cecum. It is narrow and muscular (collum coli) at its origin and is directed dorsocaudally (255/*2*). Then follows a dilated part, which, still in the lesser curvature of the cecum, bends sharply ventrally and then cranioventrally (255, 256/*b*) and is followed by a short constricted part. At this point the right ventral colon leaves the lesser curvature of the cecum, and, becoming quite wide, passes forward in the direction of the sternum (256/*c*; 258/*u*). The right ventral colon is connected with the cecum by means of the **cecocolic fold** (256/*1*), which ends at the lateral band of the cecum. It is related caudally to the body of the cecum and lies against the right abdominal wall, following more or less the curvature of the right costal arch. The transversely directed **sternal flexure** (259/*d*) lies in the xiphoid region on the floor of the abdominal cavity, curves around the apex of the cecum (*b'*), and continues on the left side as the left ventral colon. The **left ventral colon** (*e*) passes caudally and slightly dorsally following the curvature of the abdominal floor (257/*u'*) and ends in front of the pelvic inlet at the **pelvic flexure** (*v*), which is usually a little to the left of the pelvic inlet, but may be on the right side, lying transversely in front of the pelvic cavity.

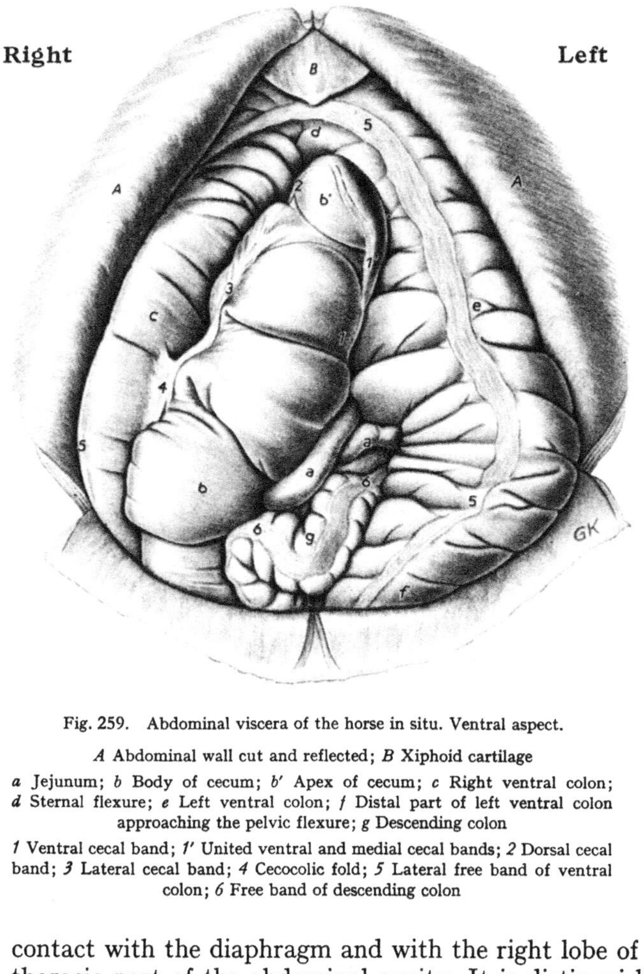

Fig. 259. Abdominal viscera of the horse in situ. Ventral aspect.

A Abdominal wall cut and reflected; *B* Xiphoid cartilage

a Jejunum; *b* Body of cecum; *b'* Apex of cecum; *c* Right ventral colon; *d* Sternal flexure; *e* Left ventral colon; *f* Distal part of left ventral colon approaching the pelvic flexure; *g* Descending colon

1 Ventral cecal band; *1'* United ventral and medial cecal bands; *2* Dorsal cecal band; *3* Lateral cecal band; *4* Cecocolic fold; *5* Lateral free band of ventral colon; *6* Free band of descending colon

The **left dorsal colon** (*w*) passes cranially along the left abdominal wall. It is dorsal and slightly lateral to the left ventral colon, and upon reaching the diaphragm is continued by the **diaphragmatic flexure** (*w'*), which lies craniodorsal to the sternal flexure and is in contact with the diaphragm, the liver, and the stomach. The **right dorsal colon** (258/*v'*), which follows the diaphragmatic flexure, passes caudodorsally on the right and makes considerable contact with the diaphragm and with the right lobe of the liver, remaining within the intrathoracic part of the abdominal cavity. It is distinguished from the other parts of the great colon by its enormous diameter and has been described as a "stomachlike dilatation" (ampulla coli) of the colon (249/*l*). On the right side of the root of the mesentery at about the level of the seventeenth or eighteenth thoracic vertebra, the right dorsal colon decreases sharply in diameter, turns medially, and is continued by the narrow transverse colon(*m*).

The **transverse colon** (248/*h, h'*) is short and passes from right to left cranial to the cranial mesenteric artery. It is connected dorsally with the pancreas (*k, k'*) and the roof of the abdominal cavity, and by a short transverse mesocolon also with the root of the mesentery (*c*).

The **descending colon,** which begins to the left of the root of the mesentery, has a smaller diameter than the ascending colon and is therefore also known as the **small colon** (colon tenue, 147/*H*). When compared to the descending colon of the other domestic mammals, it is unusually long, reaching a length of 2.5—4 m. It is suspended by the long **descending mesocolon** (249/*9*), which allows its large loops wide range. Nevertheless, most of the small colon occupies, together with the jejunum, the space caudal to the stomach and dorsal to the left parts of the great colon, extending also into the pelvic cavity (257/*x*).

DIAMETER OF THE COLON. Except for the constricted part at the beginning of the **ascending colon,** the diameter of the right and left ventral colon is very wide, measuring about 25—30 cm. (249/*e, g*). A short distance proximal to the pelvic flexure, the diameter of the colon narrows to about 6—10 cm. and retains this width to about the middle of the left dorsal colon (*i*). From here onward the diameter increases steadily and reaches its maximum of 30—50 cm. at the right dorsal colon (*l*), but decreases abruptly at the beginning of the short transverse colon (*m*). The diameter of the **descending colon** is about 7—10 cm. and remains the same throughout its length. The ampulla-like right dorsal colon lies caudoventral to the diaphragm and liver (258). When distended, it may press on the lungs on the other side of the diaphragm and cause respiratory difficulty, and chronic distension usually results in pressure atrophy of the right lobe of the liver.

At the narrow cecocolic orifice and the initial portion of the ascending colon, at the narrow pelvic flexure, and at the funnel-shaped narrowing where the ascending colon joins the transverse colon, the flow of ingesta may be impeded and impaction may result.

MESOCOLON. The **ascending mesocolon** is very long and narrow. It originates from the right surface of the root of the mesentery and, attached to the muscular bands, extends between the dorsal and ventral colon to the pelvic flexure. With dorsal and ventral colon closely applied to each other, the ascending mesocolon is represented only by a short peritoneal bridge on each side of their attachment (256/*4*). In the vicinity of the pelvic flexure, however, the dorsal and ventral colon separate and the ascending mesocolon spans the distance between them. Thus only the proximal and terminal parts of the ascending colon are fixed either directly or indirectly to the roof of the abdominal cavity; the greater part of this long loop, particularly those segments lying on the left side, including the pelvic flexure, are free and are supported only by the ventral abdominal wall. Consequently, only the left parts of the great colon are subject to displacement or volvulus, as occurs occasionally, and this displacement can as a rule be diagnosed by rectal palpation.

The **descending mesocolon,** which usually contains large amounts of fat, originates cranially from the left surface of the root of the mesentery and attaches along a straight line to the roof of the abdominal cavity. At the pelvic inlet it is continued by the mesorectum.

BANDS AND SACCULATIONS OF THE COLON. The right and left ventral colon, like the cecum, have four well-developed bands, two dorsal and two ventral. Because the dorsal bands are included in the mesocolic attachment between dorsal and ventral colon, they are known as the **medial** and **lateral mesocolic bands.** They extend along each side of the attachment to the vicinity of the pelvic flexure, where the mesocolon is attached only to the medial mesocolic band. The two ventral bands are exposed and are known as the **medial** and **lateral free bands** (259/*5*). Between these bands are four distinct rows of sacculations. The pelvic flexure and the succeeding left dorsal colon have only one mesocolic band, and therefore are not sacculated (249/*h, i*). Close to the diaphragmatic flexure, two new bands begin on the dorsal surface of the dorsal colon, which become distinct on the right dorsal colon. Thus at the diaphragmatic flexure and on the right dorsal colon there are three bands, one mesocolic facing the ventral colon and two free bands dorsally. The sacculations between these three bands are not very distinct (*k, l*).

The two unusually wide bands of the **descending colon** give it a very characteristic appearance (*n*). The mesocolic band is concealed in the attachment of the descending mesocolon; the other is exposed and runs along the antimesenteric border of the descending colon. The two bands produce two rows of regular, almost semispherical sacculations in which the feces are divided into the characteristic fecal balls.

The muscular bands and the sacculations between them can be palpated rectally and aid in the identification of the different parts of the large intestine during rectal palpation or surgery. The descending colon is the most easily recognized because of its deep sacculations.

The **RECTUM** (147/*J*) is 20—30 cm. long and continues the descending colon into the pelvic cavity. It is suspended at first by the **mesorectum** (15, 16/*4*), but as it passes at the level of the fourth or fifth sacral vertebra into the retroperitoneal part of the pelvic cavity, it is surrounded by connective tissue. It gradually increases in diameter and forms the capacious and muscular **ampulla recti** (494/*z*), which extends to the level of the second or third caudal vertebra and ends at the short anal canal. The **rectococcygeus** (559/*26''*), derived from the well-developed longitudinal muscle layer of the rectum, is a thick, paired band which attaches on the ventral surface of the fourth caudal vertebra. Some of its fibers unite with those of the other side and form a muscular loop dorsal to the rectum, rather than going to the tail. In the stallion, the rectum is related ventrally to the bladder, pelvic urethra, deferent ducts, seminal vesicles, prostate gland, and the bulbourethral glands (15; 494). In the mare, it is related to the vagina, vestibule, cervix, and to the body of the uterus; ventral to these are the urethra and the urinary bladder (16; 559).

The **ANAL CANAL** of the horse forms a rounded protuberance ventral to the root of the tail, with the **anus,** the terminal opening of the digestive tract, in the center (529/*a*). The protuberance is covered with heavily pigmented skin and fine hairs. Inside the skin the anus is surrounded by the smooth **internal** and the striated **external anal sphincter** (559/*26*). In the mare, fibers of the cutaneous part of the external anal sphincter pass ventrally into the labia and are continuous with the constrictor vulvae (*19'*). The **levator ani** (*25*) is a well-developed muscular band, originating from the deep surface of the sacroischiatic ligament and thus indirectly from the ischiatic spine, and is inserted on the lateral wall of the anal canal deep to the external anal sphincter. The **retractor penis,** or **retractor clitoridis** in the mare (*26'*), originates from the ventral surface of the second caudal vertebra as a paired grayish white muscular band, which descends under cover of the levator and external anal sphincter on each side of the rectum. The rectal part of the muscle unites with the corresponding part from the other side ventral to the rectum and forms a loop* which suspends the rectum. From this loop, the retractor penis (pars penina, 499/*a*) is continued onto the ventral surface of the penis. Similarly, the retractor clitoridis (pars clitoridica) continues into the labia, sometimes reaching the level of the body of the clitoris (Habel 1953).

The anal canal is lined with a band of whitish, nonglandular mucosa, about 3—4 cm. wide, and is covered with stratified squamous epithelium. Proximally it is continuous with the rectal mucosa at the distinct **anorectal line,** while distally at the **anocutaneous line** it is succeeded by the skin that covers the anal protuberance.

LYMPHORETICULAR TISSUE OF THE INTESTINES. Solitary lymph nodules are found in all segments of the intestines, varying in number and distribution. They are about 1—2 mm. in diameter and some have central depressions. **Patches of aggregate lymph nodules** occur in large numbers in the small intestine, where between 100 and 200 have been counted. Some are small, round plaques 5—6 mm. in diameter; most of them, however, are narrow bands about 2—6 cm. long, and some may be as long as 10—14 cm. In the large intestine, patches of aggregate lymph nodules are found at the apex of the cecum and in the pelvic flexure of the great colon. The prominent patches found at the ileocolic junction in the ruminants and pig are absent in the horse.

Liver

(156, 257, 258, 260—262)

Depending on the amount of blood it contains, the liver of the horse is light or dark brownish red. In young animals and in animals in good condition it usually has a lighter color. Its weight averages 5 kg. generally, and 2.5—3.5 kg. in older animals. Occasionally, particularly in old animals, the right lobe is atrophied as a result of the pressure of the right dorsal colon; then the left lobe is larger than the right. Sometimes the left lobe is atrophied by the pressure of the stomach; then the right lobe is the larger of the two.

* Formerly known as the suspensory ligament of the anus.

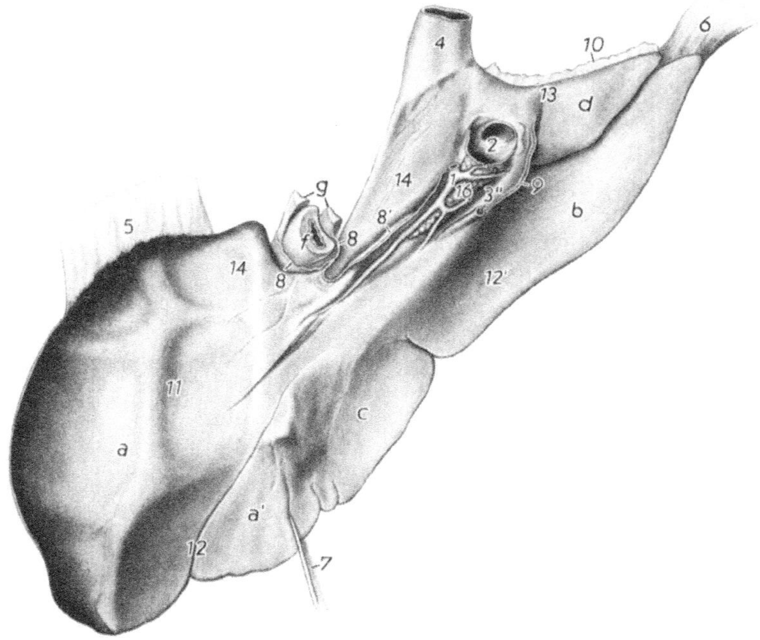

Fig. 260. Liver of the horse. Visceral surface.

a Left lateral lobe; *a'* Left medial lobe; *b* Right lobe; *c* Quadrate lobe; *d* Caudate process; *f* Esophagus in esophageal notch; *g* Part of the diaphragm, forming esophageal hiatus

1 Hepatic branch of hepatic artery; *2* Portal vein; *3''* Common hepatic duct; *4* Caudal vena cava; *5* Left triangular ligament; *6* Right triangular ligament; *7* Falciform ligament with round ligament in its free edge; *8* Gastrophrenic ligament; *8', 9* Cut edge of lesser omentum; *10* Hepatorenal ligament; *11* Gastric impression; *12* Colic impression; *12'* Duodenal and colic impressions; *13* Renal impression; *14* Area of contact with pancreas; *16* Hepatic lymph nodes

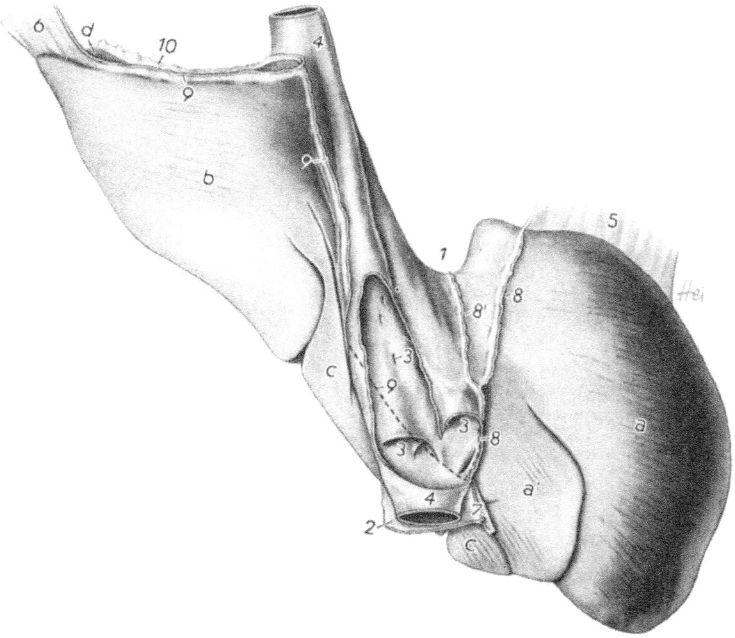

Fig. 261. Liver of the horse. Diaphragmatic surface.

a Left lateral lobe; *a'* Left medial lobe; *b* Right lobe; *c, c* Quadrate lobe; *d* Caudate process

1 Esophageal notch; *2* Diaphragm; *3* Openings of hepatic veins; *4* Caudal vena cava, fenestrated to show hepatic veins; *5* Left triangular ligament; *6* Right triangular ligament; *7* Falciform ligament with round ligament in its free edge; *8* Left, *8'* middle, and *9* right laminae of coronary ligament; *10* Hepatorenal ligament

LOBATION. The liver of the horse is only partly lobated. The absence of a gall bladder, a useful landmark between the quadrate and right lobes in the other species, poses some difficulty in determining the boundary between these lobes in the horse. Nevertheless, the same method used for the division of the liver in the other species applies. To the left of an imaginary line connecting the esophageal notch (260/*f*) with the notch for the round ligament (*7*) is the **left lobe,** which is subdivided into **left lateral** (*a*) and **left medial lobes** (*a'*). To the right of the notch for the round ligament and ventral to the porta is the **quadrate lobe** (*c*), the ventral margin of which presents a number of short notches. To the right of these notches is a deeper notch, which in the absence of a gall bladder marks the limit of the quadrate lobe. To the right of an imaginary line from this notch to where the caudal vena cava crosses the dorsal border of the liver (*4*) is the usually undivided **right lobe** (*b*). There is no papillary process, and the **caudate lobe** dorsal to the porta consists only of a slender **caudate process** (*d*), which is directed to the right. The dorsal border of the liver is thick, and includes the **renal impression,** the groove for the vena cava, and the **esophageal notch;** the remaining borders around the periphery of the liver are sharp.

The gall bladder is absent in the horse. The **hepatic ducts** unite to form the **common hepatic duct** (*3''*), which carries the bile to the duodenum. Since there is no cystic duct delimiting the beginning of the bile duct, the distal, dilated portion of the common hepatic duct

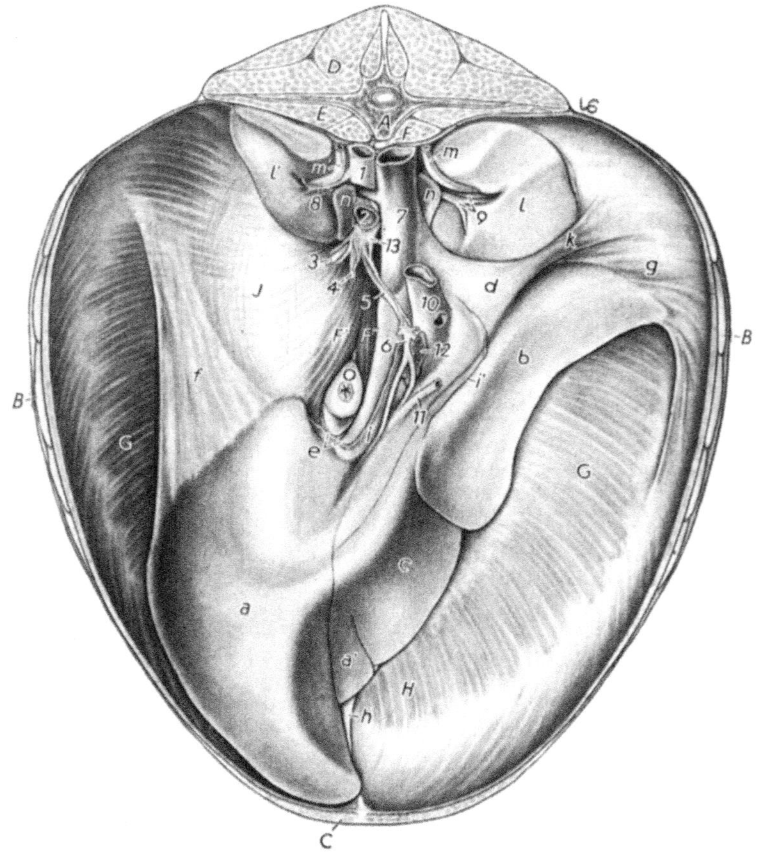

Fig. 262. Liver and kidneys of the horse in situ. Caudoventral aspect.

A First lumbar vertebra; *B* Lateral abdominal wall with costal arch; *C* Ventral abdominal wall with rectus abdominis and linea alba in cross section; *D* Epaxial musculature; *E* Psoas musculature; *F—J* Diaphragm: *F* tendons of origin of crura, *F'*, *F''* medial part of right crus, *G* costal part, *H* sternal part, *J* left portion of tendinous center

a—d On the liver: *a* left lateral lobe, *a'* left medial lobe, *b* right lobe, *c* quadrate lobe, *d* caudate process; *e* Gastrophrenic ligament at esophageal notch; *f* Left triangular ligament; *g* Right triangular ligament; *h* Falciform ligament with round ligament in its free edge; *i*, *i'* Cut edge of lesser omentum; *k* hepatorenal ligament, partly removed; *l*, *l'* Right and left kidneys, the left kidney has been transected and its perirenal fat has been removed; *m*, *m'* Right and left ureters; *n*, *n'* Right and left adrenal glands; *o* Esophagus in esophageal hiatus

1 Aorta; *2* Cranial mesenteric artery; *3* Splenic artery; *4* Left gastric artery; *5* Hepatic artery; *6* Hepatic branch of hepatic artery; *7* Caudal vena cava; *8* Left renal artery and vein; *9* Right renal artery and vein; *10* Portal vein; *11* Common hepatic duct; *12* Hepatic lymph nodes; *13* Celiac plexus

is usually called the **bile duct.** The bile duct opens with the pancreatic duct 12—15 cm. distal to the pylorus on the major duodenal papilla (167/*1'*).

POSITION AND RELATIONS. The liver lies entirely under cover of the ribs (257, 258/*q*); normally three-fifths of it is to the right and two-fifths to the left of the median plane. Its long axis is directed obliquely from right to left in a cranioventral direction (262). Its convex **diaphragmatic surface** lies against the diaphragm except where the two are separated by the caudal vena cava, which receives the hepatic veins (261/*3*) before it reaches the foramen venae cavae. Since the vein is united firmly with the liver and the diaphragm, it provides a strong attachment of the liver near the median plane.

The ventral border of the liver is dorsal to the sternum at the level of the sixth or seventh rib and at about the junction of ventral and middle thirds of the dorsoventral diameter of the thorax. The right border ascends obliquely in a dorsocaudal direction to a point about 10 cm. dorsal to the fifteenth costochondral junction. From there it runs dorsally, follows the fifteenth intercostal space, and widens to form the **renal impression** before reaching the vertebral column (258). The left border arises from the ventral end of the sixth or seventh rib, and passes dorsally and only slightly caudally to the dorsal end of the tenth intercostal space (257).

The **visceral surface** of the liver is irregularly concave and in the hardened state displays the impressions made by the various organs with which it is contact. On the left is the **gastric impression** (260/*11*) produced by the stomach, and ventral to it the **colic impression** (*12*) produced by the diaphragmatic flexure and the right dorsal colon. On the right is the **renal impression** (*13*), which receives the cranial pole of the right kidney, the **duodenal impression** (*12'*) made by the cranial part and the beginning of the descending part of the duodenum, and the **cecal impression** from the most cranial part of the base of the cecum (258/*t*). The body of the pancreas also lies against the visceral surface of the liver (248). All of these organs, with perhaps the exception of the stomach, lie ventral to the liver, so that they seem to support this large gland, but the liver's strong attachments to the diaphragm (see below) are evidence that they only partly support it.

ATTACHMENT. The attachment of the liver to the diaphragm is firm and allows the liver little movement. The relatively long **left triangular ligament** (262/*f*) originates from the left part of the tendinous center of the diaphragm and is inserted on the left lateral lobe just lateral to the esophageal notch. The **right triangular ligament** (*g*) is short and connects the caudal border of the right lobe to the costal part of the diaphragm. Medially it splits off the **hepatorenal ligament** (261/*10*). The **coronary ligament** consists of three narrow laminae visible on the diaphragmatic surface. The right lamina (*9*) is the medial continuation of the right triangular ligament and follows the caudal vena cava on the right. The left lamina (*8*) comes from the left triangular ligament and passes to the foramen venae cavae of the diaphragm on the left lobe. The middle lamina (*8'*) comes from the esophageal notch, where it is continuous with the gastrophrenic ligament (260/*8*), and passes cranioventrally to join the left lamina. The right and left laminae of the coronary ligament unite ventral to the caudal vena cava and give rise at this point to the **falciform ligament** (261/*7*), which attaches the liver to the sternal part of the diaphragm. The **round ligament of the liver** (262/*h*) passes from the ventral border of the liver to the region of the umbilicus, at first in the free edge of the falciform ligament and then retroperitoneally along the floor of the abdominal cavity. The visceral surface of the liver is connected to the lesser curvature of the stomach and cranial part of the duodenum by the hepatogastric and hepatoduodenal parts of the lesser omentum (see page 183).

Pancreas

(167, 248, 439)

Depending on the amount of blood it contains, the pancreas of the horse has a light or dark yellowish red color. The **left lobe** (167/*b*) is relatively long while the **right lobe** (*c*) is short and plump. The **body** is perforated by the portal vein (*3*). Most of the pancreas lies to the right of the median plane on the dorsal wall of the abdominal cavity opposite the seventeenth and eighteenth thoracic vertebrae. It is inserted between the visceral surface of the stomach and liver cranially, and the base of the cecum, right dorsal colon, and transverse colon caudally (248). The body of the pancreas (*k*) lies against the visceral surface of the liver in the concavity of the cranial flexure of the duodenum. The left lobe (*k'*) extends along the visceral surface of the stomach to the spleen, while the right lobe (*k''*) accompanies the descending duodenum to the right kidney. Dorsally, the pancreas is related to the crura of the diaphragm, to the aorta and caudal vena cava, to the celiac artery and its branches, to the portal vein and its branches, and to the celiac ganglion and plexus (439). Its caudoventral surface is attached on the right to the base of the cecum and, more medially, to the right dorsal colon and transverse colon. The left lobe is attached to the blind sac of the stomach.

The pancreas of the horse has two secretory ducts. The **pancreatic duct,** the larger of the two, ends on the concave surface of the cranial duodenal flexure alongside the bile duct and enters at the **major duodenal papilla** (167/*1'*). The **accessory pancreatic duct** (*2*) enters at the **minor duodenal papilla** almost opposite the major duodenal papilla.

BIBLIOGRAPHY

Esophagus and Stomach

Alexander, F., and D. Benzie: A radiological study of the digestive tract of the foal. Quart. J. exp. Physiol. **36,** 1951.

Anderson, W. D., A. F. Weber: Normal arterial supply to the ruminant (ovine) stomach. J. Animal Sci. **28,** 379—382 (1969).

Andres, J.: Über den Magen der Wiederkäuer. Schweiz. Arch. Tierheilk. **70,** 1928. Ref.: Anat. Ber. **18,** 1930.

Auernheimer, O.: Größen- und Formveränderungen der Baucheingeweide der Wiederkäuer nach der Geburt bis zum erwachsenen Zustand. Zürich, Diss. med. vet., 1909.

Baier, U.: Über Venennetze am Speiseröhreneingang bei den Haussäugetieren. Berl. Tierärztl. Wschr. **45,** 1929.

Barone, R.: Topographie des viscères du Porc et de la Truie. Rev. Méd. Vét. **103,** 1952.

Barone, R.: La topographie des viscères abdominaux, chez les Équidés domestiques. Bull. Soc. Sc. Vét. Lyon 1952/1953.

Barone, R.: La topographie des viscères abdominaux, chez la jument gravide. Rev. Méd. Vét. **113,** 1962.

Barthol, A.: Beiträge zur Anatomie und Histologie der Magenschleimhaut von Sus scrofa domestica. Leipzig, Diss. med. vet., 1914.

Bauer, K. H.: Über das Wesen der Magenstraße. Arch. Klin. Chirurg. **124,** 1930.

Becker, R. B., S. P. Marshall, and P. T. Dix Arnold: Anatomy, development, and functions of the bovine omasum. J. Dairy Sci. **46,** 1963.

Benzie, D., and A. T. Phillipson: The alimentary tract of the ruminant. Edinburgh, Oliver and Boyd, 1957.

Biedermann, F.: Metrische Untersuchungen am Pferdemagen. Berlin, Diss. med. vet., 1921.

Botha, G. S. M.: The gastro-esophageal junction. Boston: Little, Brown and Co., 1962.

Bremmer, C. G., R. G. Shorter, and F. H. Ellis: Anatomy of feline esophagus with special reference to its muscular wall and phrenoesophageal membrane. J. Surg. Res. **10,** 1970.

Brownlee, A., and J. Elliot: Studies on the normal and abnormal structure and function of the omasum of domestic cattle. Brit. Vet. J. **116,** 1960.

Carlin, J.: Studien über den Hundemagen im Röntgenbilde unter normalen Verhältnissen. Berlin, Diss., 1928.

Church, D. C., Jessup jun., L. Gordon, R. Bogart: Stomach development in the suckling lamb. Amer. J. Vet. Res. **23,** 220—225 (1962).

Comline, R. S., I. A. Silver and D. H. Steven: Physiological anatomy of the ruminant stomach. Handbook of Physiology, Section 6, Vol. V. Washington, Am. Physiol. Soc., 1968.

Cordiez, E., J. M. Bienfait, J. Mignon: Alimentation et développement volumétrique des estomacs du jeune bovin. Ann. Méd. Vét. Bruxelles **107,** 293—316 (1963).

Czepa, A., u. R. Stigler: Der Wiederkäuermagen im Röntgenbild. I. Mitt. Pflügers Arch. **212,** 1926.

Czepa, A., u. R. Stigler: Der Verdauungstrakt des Wiederkäuers im Röntgenbilde. II. Mitt. Abderhaldens Fortschr. naturwiss. Forsch. H. 6, 1929.

Delaney, J., E. Grim: A note on the weight of the dog stomach. Amer. J. Vet. Res. **25,** 1560—1561 (1964).

Derrick, R. J., B. E. Patten and D. A. Titchen: The muscle of the reticulo-omasal orifice of the sheep. J. Anat. **116,** 1973.

Dietz, et al.: Untersuchungen zur Vagusfunktion, zur Vagusbeeinflussung und zu Vagusausfällen am Verdauungsapparat des erwachsenen Rindes. Arch. Exp. Vet. Med. **24,** 1970.

Dougherty, W. R.: Physiology of eructation in ruminants. In Handbook of Physiology, Section 6, Volume 5. Washington, American Physiological Society, 1968.

Dyce, K. M.: Observations upon the gastro-intestinal tract of the living foal. Brit. Vet. J. **116,** 1960.

Dyce, K. M.: Some remarks upon the functional anatomy of the ruminant stomach. Tijdschr. Dierg. **93,** 1968.

Ellenberger, W.: Ein Beitrag zur Lehre von der Lage und Funktion der Schlundrinne der Wiederkäuer. Arch. wissensch. prakt. Tierheilk. **21.**

Ellenberger, W.: Über die Schlundrinne der Wiederkäuer und ein Modell der Wiederkäuermägen. Arch. wissensch. prakt. Tierheilk. **24.**

Florentin, P.: Anatomie topographique des viscères abdominaux du Boeuf et du Veau. Rev. Méd. Vét. **104,** 1953.

Frewein, J.: Der Anteil des Sympathicus an der autonomen Innervation des Rindermagens. Wiener Tierärztl. Mschr. 398—412 (1963).

Fröhlich, A.: Untersuchungen über die Übergangszonen und einige Eigentümlichkeiten des feineren Baues der Magenschleimhaut der Haussäugetiere. Leipzig, Diss. med. vet., 1907.

Grau, H.: Zur Funktion der Vormägen, besonders des Netzmagens, der Wiederkäuer. Berlin. Münchn. Tierärztl. Wschr. **15,** 1955.

Gross, F.: Untersuchungen über die zwischen der Fundusdrüsen- und Pylorusdrüsenzone des Pferdemagens befindliche Intermediärzone. Leipzig, Diss. med. vet., 1920.

Gruzdev, P. V., V. A. Nikodimova: Die Blutversorgung des Magens beim Schwein. Ref. Ž. Moskva, Životnovodsto i Veter. 31—36 (1971).

Haane, G.: Über die Drüsen des Oesophagus und des Übergangsgebietes zwischen Pharynx und Oesophagus. Arch. wiss. prakt. Tierheilk. **31,** 1905.

Haane, G.: Über die Cardiadrüsen und die Cardiadrüsenzone des Magens der Haussäugetiere. Gießen, Diss. med. vet., 1905.

Habel, R. E.: Applied veterinary anatomy. 2nd. ed. Ithaca, New York, Author, 1978.

Hänlein, G., B. Baumgardt: Die morphologischen und physiologischen Veränderungen des Kälbermagens bei der Frühentwöhnung. Z. Tierphysiol., Tierernähr. Futtermittelk. **21,** 327—337 (1966).

Hájovská, B.: Veränderungen in der Topographie der Vormägen und des Labmagens bei Schafen während Gravidität. Folia Vet., Košice **9,** 147—156 (1965).

Hájovská, B.: Postnatale Entwicklung des Pansenbandes bei Schafen. Folia Vet., Košice **14,** 11—17 (1970).

Hájovská, B.: Veränderungen in der Lage des Pansens beim Schaf und ihre Ursachen. Folia Vet., Košice **14,** 17—23 (1970).

Hauser, H.: Über interessante Erscheinungen am Epithel der Wiederkäuervormägen. Z. mikr. anat. Forsch. **17,** 1929.

Hellfors, J. A.: Die Verbreitung und Anordnung des elastischen Gewebes in den einzelnen Wandschichten des Ösophagus einiger Haustiere. Leipzig, Diss. med. vet., 1913.

Helm, R.: Vergleichend anatomische und histologische Untersuchungen über den Oesophagus der Haussäugetiere. Zürich, Diss. med. vet., 1907.

Hofmann, R.: Zur Topographie und Morphologie des Wiederkäuermagens im Hinblick auf seine Funktion. Nach vergleichenden Untersuchungen an Material ostafrikanischer Wildarten. Berlin und Hamburg: Paul Parey, 1969.

Jones, R. S.: The position of the bovine abomasum. An abattoir survey. Vet. Rec. **74,** 1962.

Künzel, E.: Die Speiseröhre des Schafes in funktioneller Betrachtung. Tbl. Vet. Med. A **8,** 1961.

Lagerlöf, N.: Investigations of the topography of the abdominal organs in cattle, and some clinical observations and remarks in connection with the subject. Skand. Veterinärtidskrift **19,** 1929.

Lambert, P. S.: The development of stomach in the ruminant. Vet. J. **104,** 1948.

Langer, P.: Vergleichend anatomische Untersuchungen am Magen der Artiodactyla (Owen, 1848). II. Teil: Untersuchungen am Magen der Tylopoda und Ruminantia. Gegenb. Morph. Jb. **119,** 1973.

Lauwers, H.: Morphology of the bovine forestomachs with special reference to their absorptive ability. Meded. Fac. Dierg. Gent **17,** 1973.

Lauwers, H., N. R. de Vos, and M. Sebruyns: The volumes and the anatomical development of the stomachs of calves at slaughter weight. Vlaams Dierg. Tijdschr. **38,** 1969.

Mangold, E., u. W. Klein: Bewegungen und Innervation des Wiederkäuermagens. Leipzig, Thieme, 1927.

Mann, C. V., and R. G. Shorter: Structure of the canine esophagus and its sphincters. J. Surg. Res. **4,** 1964.

Marschall, A.: Über den Einfluß des N. vagus auf die Bewegungen des Magens der Wiederkäuer. Bern, Diss., 1910.

Martin, P.: Die Entwicklung des Wiederkäuermagens. Österr. Mschr. Tierheilk. 1896.

Massig, P.: Über die Verbreitung des Muskel- und elastischen Gewebes und speziell über den Verlauf der Muskelfasern in der Wand der Wiederkäuermägen. Zürich, Diss., 1907.

Miller, M. E., G. C. Christensen, and H. E. Evans: Anatomy of the dog. Philadelphia, Saunders, 1964.

Müller, K.: Untersuchungen über die cardiale Übergangszone des Pferdemagens. Leipzig, Diss. med. vet., 1914.

Neumann-Kleinpaul, K., u. G. Schützler: Untersuchungen über Druckmessungen, Ruptur, Fassungsvermögen und Gewicht am Magen des Pferdes. Arch. Tierheilk. **75,** 1940.

Nickel, R., u. H. Wilkens: Zur Topographie des Rindermagens. Berl. Münch. Tierärztl. Wschr. **68,** 1955.

Pernkopf, E.: Beiträge zur vergleichenden Anatomie des Vertebratenmagens. Zschr. Anat. I. **91,** 1930.

Pernkopf, E.: Die Entwicklung des Vorderdarmes, insbesondere des Magens der Wiederkäuer. Zschr. Anat. **94,** 1931.

Pernkopf, E., J. Lehner: III. Vorderdarm. In: Bolk, L., E. Göppert, E. Kallius, W. Lubosch: Handbuch der vergleichenden Anatomie der Wirbeltiere III, 349—476. Berlin u. Wien, Urban & Schwarzenberg, 1937. Neudruck Asher & Co., Amsterdam, 1967.

Reetz, G.: Beiträge zur Anatomie und Histologie des dritten Magens der Wiederkäuer. Leipzig, Diss. med. vet., 1911.

Sack, W. O., and L. Sadler: The dissected horse, No. 53. In: W. O. Sack and R. E. Habel: Rooney's guide to the dissection of the horse. Ithaca, New York, Veterinary Textbooks, 1979.
Sack, W. O., P. Svendsen: Size and position of the abomasum in ten cows fed high-roughage or high-concentrate rations. Amer. J. Vet. Res. **31**, 1539—1543 (1970).
Schels, H.: Untersuchungen über Wandbau und Funktion des Netzmagens des Rindes. München, Diss. med. vet., 1956.
Schmaltz, R.: Topographische Anatomie der Körperhöhlen des Rindes. Berlin, Schoetz, 1895.
Schmidt, H.: Der funktionelle Bau der Schleimhautmuskulatur des Magens (Schwein). Morph. Jb. **86**, 1939.
Schnorr, B., B. Vollmerhaus: Das Oberflächenrelief der Pansenschleimhaut bei Rind und Ziege. (Erste Mitteilung zur funktionellen Morphologie der Vormägen der Hauswiederkäuer.) Zbl. Vet. Med. A **14**, 93—104 (1967).
Schnorr, B., B. Vollmerhaus: Das Blutgefäßsystem des Pansens von Rind und Ziege. IV. Mitt. zur funktionellen Morphologie der Vormägen der Hauswiederkäuer. Zbl. Vet. Med. A **15**, 799—828 (1968).
Schummer A.: Zur Formbildung und Lageveränderung des embryonalen Wiederkäuermagens. Gießen, Diss. med. vet., 1932.
Schwabe, F.: Anatomische und histologische Untersuchungen über den Labmagen der Wiederkäuer, insbesondere über das Muskel- und elastische Gewebe desselben. Leipzig, Diss. med. vet., 1910.
Schwarz, E.: Zur Anatomie und Histologie des Psalters der Wiederkäuer. Bern, Diss. med. vet., 1910.
Scott, A. and I. C. Gardner: Papilla form in the forestomach of the sheep. J. Anat. **116**, 1973.
Sellers, A. F., and C. E. Stevens: Motor functions of the ruminant stomach. Physiol. Rev. **46**, 1966.
Sisson, S., and J. D. Grossmann: The anatomy of the domestic animals. 4th ed. Philadelphia, Saunders, 1953.
Spörri, H.: Physiologie der Wiederkäuer-Vormägen. Schweiz. Arch. Tierheilk., Sonderheft April 1, 1951.
Stevens, C. E., A. F. Sellers, and F. A. Spurrell: Function of the bovine omasum in ingesta transfer. Am. J. Physiol. **198**, 1960.
Stigler, R.: Die Verdauungsorgane des Wiederkäuers im Röntgenbild. Tierärztl. Umschau **33**, 1929.
Stilinović, Z., J. Rac, Z. Robić: Lineare Dimensionen von Pansen und Haube, gemessen „in vivo". Vet. Arh. Zagreb **35**, 204—212 (1965).
Stojanov, I.: Untersuchung der afferenten Innervation des Wiederkäuermagens. Bulgarska akademija na naukite, Sofija **12**, 247—265 (1967).
Sussdorff, M. v.: Die Lagerung des Schlundes der Haussäugetiere im hinteren Mittelfell. Dtsch. Tierärztl. Wschr. 1896.

Tamate, H.: Development of the stomach of the goat. J. Agric. Res., Tohoku (Japan) 1957.
Tamate, H., A. D. McGilliard, N. L. Jackson, and R. Getty: Effects of various dietaries on the anatomical development of the stomach of the calf. J. Dairy Sci. **45**, 1962.
Trautmann, A.: Beitrag zur Physiologie des Wiederkäuermagens. I. Mitt. Einfluß der Nahrung auf die Ausbildung der Vormägen bei jugendlichen Wiederkäuern. Arch. Tierernährung u. Tierzucht. **7**, 1932.

Warner, E. D.: The organogenesis and early histogenesis of the bovine stomach. Am. J. Anat. **102**, 1958.
Warner, R. G., and W. P. Flatt: Anatomical development of the ruminant stomach. In: Physiology of digestion in the ruminant. Washington, D. C., Butterworth, Inc., 1965.
Weaver, A. D.: A post-mortem survey of some features of the bovine abomasum. Brit. Vet. J. **120**, 1964.
Wensing, C. J. G.: Die Innervation des Wiederkäuermagens. Tijdschr. Diergenessk. **93**, 1352—1360 (1968).
Wester, J.: Der Schlundrinnenreflex beim Rind. Berl. Tierärztl. Wschr. **46**, 1930.
Wester, J.: Das Erbrechen bei Wiederkäuern. Berl. Tierärztl. Wschr. **47**, 1931.
Wild, H.: Über den Vorgang des Rülpsens (Ructus) beim Wiederkäuer. Gießen, Diss., 1913.
Wilkens, H.: Zur Topographie der Verdauungsorgane der Ziege. Dtsch. Tierärztl. Wschr. **63**, 1956a.
Wilkens, H.: Zur Topographie der Verdauungsorgane des Schafes unter besonderer Berücksichtigung von Funktionszuständen. Zbl. Vet. Med. **3**, 1956b.
Wilkens, H., u. G. Rosenberger: Betrachtungen zur Topographie und Funktion des Oesophagus hinsichtlich der Schlundverstopfung des Rindes. Dtsch. Tierärztl. Wschr. **64**, 1957.

Ziegler, H.: Anatomie für die Praxis. I. Von den Vormägen des Rindes. Schweiz. Arch. Tierheilk. **76**, 1934.
Zietzschmann, O.: Der Darmkanal der Säugetiere, ein vergleichend anatomisches und entwicklungsgeschichtliches Problem. Anat. Anz. (Erg.-H.) **60**, 1925.
Zietzschmann, O.: Über die Form und Lage des Hundemagens. Berl. Tierärztl. Wschr. **50**, 1938.
Zietzschmann, O.: Das Mesogastrium dorsale des Hundes mit einer schematischen Darstellung seiner Blätter. Morph. Jb. **83**, 1939.

Intestines

Alexander, F., and D. Benzie: A radiological study of the digestive tract of the foal. Quart. J. exp. Physiol. **36**, 1951.

Barone, R.: Topographie des viscères du Porc et de la Truie. Rev. Méd. Vét. **103**, 1952.
Barone, R.: La topographie des viscères abdominaux, chez les Équidés domestiques. Bull. Soc. Sc. Vét. Lyon, 1952/1953.
Barone, R.: La topographie des viscères abdominaux, chez la jument gravide. Rev. Méd. Vét. **113**, 1962.
Bassett, E. G.: The anatomy of the pelvic and perineal region in the ewe. Austral. J. Zool. **13**, 1956.
Benzie, D., and A. I. Phillipson: The alimentary tract of the ruminant. Edinburgh, Oliver and Boyd, 1957.
Borelli, V., A. Fernandes Filho: Unregelmäßigkeiten in der Ansa spiralis des Kolons von Ziegen. Rev. Fac. Med. Vet., São Paulo **7**, 319—323 (1965).

Borelli, V., I. L. de Santis Prada: Unregelmäßigkeiten im Muster der Ansa spiralis des Kolons bei Schafen. Rev. Fac. Med. Vet., São Paulo **7**, 521—526 (1966/67).

Borelli, V., J. G. Lopes Pereira, O. Miguel: Veränderungen im Verhalten der Taenia libera im Colon fluctuans der Equiden. Rev. Fac. Med. Vet., São Paulo **8**, 379—384 (1970).

Carlens, O.: Studien über das lymphatische Gewebe des Darmkanals bei einigen Haustieren mit besonderer Berücksichtigung der embryonalen Entwicklung, der Mengenverhältnisse und der Altersinvolution dieses Gewebes im Dünndarm des Rindes. Zschr. Anat. **86**, 1928.

Doughri, A. M., K. P. Altera and R. A. Kainer: Some developmental aspects of the bovine fetal gut. Zbl. Vet. Med. A, **19**, 1972.

Dyce, K. M.: The ileocecocolic region of the horse. Anat. Anz. **103**, 1956.

Dyce, K. M.: Observations upon the gastro-intestinal tract of the living foal. Brit. Vet. J. **116**, 1960.

Engelmann, K.: Beitrag zur Anatomie der Baucheingeweide des Göttinger Zwergschweines unter besonderer Berücksichtigung ihrer Blutgefäßversorgung. Diss. med. vet. München, 1971.

Florentin, P.: Anatomie topographique des viscères abdominaux du Boeuf et du Veau. Rev. Méd. Vét. **104**, 1953.

Florentin, P.: Anatomie topographique des viscères abdominaux du Mouton et de la Chèvre. Rev. Méd. Vét. **106**, 1955.

Gerisch, D. und K. Neurand: Topographie und Histologie der Drüsen der Regio analis des Hundes. Anat. Histol. Embryol. **2**, 1973.

Graeger, K.: Zur Topographie der Bauchorgane des Schweines unter besonderer Berücksichtigung verschiedener Füllungszustände des Magens. Zbl. Vet. Med. **4**, 1957.

Grau, H., u. J. Schlüns: Experimentelle Untersuchungen zum zentralen Chylusraum der Darmzotten. Anat. Anz. **111**, 1962.

Grau, H.: Der After von Hund und Katze unter biologischen und praktischen Gesichtspunkten. Tierärztl. Rdsch. **41**, 1935.

Gomerčić, H., K. Babic: A contribution to the knowledge to the variations of the arterial supply of the duodenum and the pancreas in the dog (Canis familiaris). Anat. Anz., **132**, 281—288 (1972).

Greer, M. B., M. L. Calhoun: Anal sacs of the cat (Felis domesticus). Amer. J. Vet. Res. **27**, 773—781 (1966).

Habel, R. E.: The perineum of the mare. Cornell Vet. **43**, 1953.

Habel, R. E.: The topographic anatomy of the muscles, nerves, and arteries of the bovine female perineum. Am. J. Anat. **119**, 1966.

Habel, R. E.: Applied veterinary anatomy. 2nd. ed. Ithaca, New York, Author, 1978.

Habermehl, K. H.: Die Verlagerung der Bauch- und Brustorgane des Hundes bei verschiedenen Körperstellungen. Zbl. Vet. Med. **3**, 1956.

Haesler, K.: Der Einfluß verschiedener Ernährung auf die Größenverhältnisse des Magen-Darmkanals bei Säugetieren. Z. Züchtg. **17**, 1930.

Hagemeier, K.: Röntgenologische Beobachtungen am Darmkanal, insbesondere am Blinddarm der Ziege. Hannover, Diss. med. vet., 1937.

Happich, A.: Blutgefäßversorgung der Verdauungsorgane in Bauch- und Beckenhöhle einschließlich Leber, Milz und Bauchspeicheldrüse beim Schaf. Diss. med. vet. Hannover, 1961.

Hebel, R.: Untersuchungen über das Vorkommen von lymphatischen Darmkrypten in der Tunica submucosa des Darmes von Schwein, Rind, Schaf, Hund und Katze. Anat. Anz. **109**, 1960.

Heidenhain, M.: Über Zwillings-, Drillings- und Vierlingsbildungen der Dünndarmzotten, ein Beitrag zur Teilkörpertheorie. Anat. Anz. **40**, 1912.

Jakobshagen, E.: Grundzüge des Innenreliefs vom Rumpfdarm der Wirbeltiere. Anat. Anz. **83**, 1936.

Kienitz, M.: Über die Größenverhältnisse des Magens und Darmkanals bei verschiedenen Hunderassen nebst einem Beitrag zur Morphologie des Blinddarmes der Hunde. Berlin, Diss. med. vet., 1921.

Klimmeck, K.: Beiträge zur Anatomie des Darmes vom Schweine. Berlin, Diss. med. vet., 1922.

Kolda, J.: Zur Topographie des Darmes beim Schaf und bei der Ziege. Zschr. Anat. Entw. gesch. **95**, 1931.

Lagerlöf, N.: Investigations of the topography of the abdominal organs in cattle, and some clinical observations and remarks in connection with the subject. Skand. Veterinärtidskrift **19**, 1929.

Martin, P.: Zur Blind- und Grimmdarmentwicklung beim Pferd. Beitr. path. Anat. allg. Path. (Festschr. E. Bostroem) **69**, 1921.

Martin, W. D., T. F. Fletcher and W. E. Bradley: Perineal musculature in the cat. Anat. Rec. **180**, 1974.

May, H.: Vergleichend-anatomische Untersuchungen der Lymphfollikelapparate des Darmes der Haussäugetiere. Gießen, Diss., 1903.

McCance, R. A.: The effect of age on the weights and lengths of the pig's intestines. J. Anat. **117**, 1974.

Mladenowitsch, L.: Vergleichende anatomische und histologische Untersuchungen über die Regio analis und das Rectum der Haussäugetiere. Leipzig, Diss. med. vet., 1907.

Mouwen, J. M. V. M.: Structure of the mucosa of the small intestine as it relates to intestinal function in pigs. Netherl. J. vet. Sci. **3**, 34—37 (1970).

Mu, M. M. and R. Lingam: The rudimentary appendix of the cat. Acta Anat. **87**, 119—123 (1974).

Müller, L. F.: Die Bewegungserscheinungen am Darme des Pferdes nach Röntgenuntersuchungen beim Pony. Wiss. Zschr. Univ. Leipzig, H. 5, 1952/53.

Muthmann, E.: Beiträge zur vergleichenden Anatomie des Blinddarmes und der lymphoiden Organe des Darmkanals bei Säugetieren und Vögeln. Anat. Hefte, **48**, 1913.

Najbrt, R.: Sacculations and bands on the caecum of the horse. Acta vet. Brno **31**, 377—384 (1970).
Neimeier, K.: Röntgenologische Beobachtungen am Magen-Darmkanal des Schweines. Hannover, Diss. med. vet., 1939.
Nickel, R.: Über die Ermittlung der Länge und Lage des Verdauungskanals. Zschr. exper. Med. **91**, 1933.
Nitschke, T. und F. Preuss: Zur Homologie der Mm. rectococcygeus und coccygeorectalis. Anat. Anz. **130**, 170—175 (1972).

Preuss, F., H. Lange: Die Colon-Doppelwendel des Schweines. Zbl. Vet. Med., A **17**, 803—817 (1970).

Santis Prada, I. L. de, V. Borelli, J. Peduti Neto: Die Anordnung der Ansa spiralis des Corriedale-Schaf-Colons. Rev. Fac. Med. Vet., São Paulo **8**, 639—646 (1971).
Santis Prada, I. L. de, J. Peduti Neto, V. Borelli: Gesamtlänge des Darmes bei Schafen der Rasse Corriedale. Rev. Fac. Med. Vet., São Paulo **8**, 651—656 (1971).
Schaller, O.: Gibt es beim Hund einen "Musculus sphincter ani tertius"? Wien. Tierärztl. Mschr. **48**, 1961.
Schmaltz, R.: Topographische Anatomie der Körperhöhlen des Rindes. Berlin, Schoetz, 1895.
Schröder, L.: Über die Lage des Dickdarmes bei einer hochgraviden Stute. Monatshefte Vet. Med. **11**, 1956.
Schummer, A.: Morphologische Untersuchungen über die Funktionszustände des Ileums. Tierärztl. Umschau, **8**, 1953.
Smith, R. N.: The arrangement of the ansa spiralis of the sheep colon. J. Anat. (Lond.) **89**, 1955.
Smith, R. N., and G. W. Meadows: The arrangement of the ansa spiralis of the ox colon. J. Anat. (Lond.) **90**, 1956.
Smith, R. N.: The pattern of the ansa spiralis of the sheep colon. Brit. Vet. J. **113**, 1957.
Smith, R. N.: Irregular patterns of the ansa spiralis of the sheep colon. Brit. Vet. J. **114**, 1958.
Smith, R. N.: The arrangement of the ansa spiralis of the goat colon. Anat. Anz. **106**, 1959.
Srnetz, A.: Beitrag zur Topographie der Bauchorgane des Schweines. Jber. Vet. med. **56**, 1935.
Stempel, M.: Beiträge zur Anatomie des Schafdarmes. Berlin. Diss. med. vet., 1925.

Tiedemann, K.: Die Angioarchitektur der Schleimhaut des Anorektalgebietes bei Hund und Schwein. Diss. med. vet. Berlin, 1968.

Vodovar, N., J. Flanzy, and A. C. François: Intestin grêle du porc. I. Dimensions en fonction de l'age et du poids, étude de la jonction du canal cholédoque et du canal pancréatique a celui-ci. Ann. Biol. **4**, 1964.

Westerlund, A.: Om Hästens Ileo-Ceco-Kolska Tormonerâde. Lunds Universitäts Arskrift. N. F. Acad. **15**/5 Kungl. fysiolografiska Sällskapets Handlingen N. F. **30**/5, 1918.
Wetzel, G.: Weitere Veränderungen des Darmkanals bei pflanzlicher und tierischer Nahrung. Anat. Anz. **72**, Ergh., 1931.
Wiethölter, G.: Topographische Anatomie der Bauch- und Beckenorgane von Hund und Katze im Röntgenbild. Leipzig, Diss. med. vet., 1964.
Wilkens, H.: Zur Topographie der Verdauungsorgane des Schafes unter besonderer Berücksichtigung von Funktionszuständen. Zbl. Vet. Med. **3**, 1956.
Wilkens, H.: Zur Topographie der Verdauungsorgane der Ziege. Dtsch. Tierärztl. Wschr. **63**, 1956.

Zietzschmann, O.: In Ellenberger-Baums Handbuch der vergleichenden Anatomie der Haustiere, 18th ed. Berlin, Springer, 1943.
Zietzschmann, O.: In Schönberg, F., u. O. Zietzschmann: Die Ausführung der tierärztlichen Fleischuntersuchung, 5th ed. Berlin, Hamburg, Paul Parey, 1958.
Zimmermann, A.: Zur vergleichenden Anatomie des Wurmfortsatzes am Blinddarm. Berl. Tierärztl. Wschr. **38**, 1922.

Liver and Pancreas

Annunziata, M.: Beitrag zum Studium der intra- und extrahepatischen Gallengänge bei der Ziege. Systematisierung des Ramus principalis sinister. Rev. Fac. Med. Vet., São Paulo **8**, 119—138 (1969).

Becker, G.: Zur Frage des Vorkommens eines Oddi'schen Sphinkter im Mündungsgebiet des Ductus choledochus der Haussäugetiere. Hannover, Diss. med. vet., 1933.
Bevandić, M., I. Arnautović, I. Krčmar, J. Lorger: Vergleichende Übersicht über die Gallenwege der Haustiere. Vet., Sarajevo **16**, 301—315 (1967).
Borelli, V., J. Peduti Neto, I. L. de Santis Prada: Die Entfernung zwischen der Papilla duodeni hepatica und der Papilla duodeni pancreatica beim Büffel (Bubalus bubalis Linnaeus, 1758). Rev. Fac. Med. Vet., São Paulo **8**, 375—378 (1970).
Boyden, E. A.: The choledochoduodenal junction in the cat. Surgery **41**, 1957.
Braus, H.: Anatomie des Menschen, Vol. II. Berlin, Springer, 1924 and 1956.

Čalyj, A. S.: Die Gewichts- und Größenveränderungen der Leber in Abhängigkeit vom Alter des Schweines. Veterynarija, Kyïv, 50—53 (1965).
Cuq, P.: La segmentation hépatique des Carnivores. Rec. Méd. Vét. **141**, 1965.

Darany, J.: Zur vergleichenden Anatomie der Gallenblase. Budapest, Diss. med. vet., 1931.
Di Dio, L. J. A., and E. A. Boyden: The choledochoduodenal junction in the horse—a study of the musculature around the ends of the bile and pancreatic ducts in a species without a gall bladder. Anat. Rec. **143**, 1962.

Eggeling, H. v.: Leber und ventrales Magengekröse. Morph. Jahrb. **66**, 1931.
Eichel, J.: Maße, Formen und Gewichte der Lebern von Rindern und Schafen. Berlin, Diss. med. vet., 1925.
Eichhorn, E. P., and E. A. Boyden: The choledochoduodenal junction in the dog—a restudy of Oddi's sphincter. Am. J. Anat. **97**, 1955.
Elias, H.: Beobachtungen über den Bau der Säugerleber. Anat. Nachr. **1**, 1949.

Ferner, H.: Das Inselsystem des Pankreas. Stuttgart, Thieme, 1952.

Geyer, H., G. Aberger, H. Wissdorf: Beitrag zur Anatomie der Leber beim neugeborenen Kalb. Topographische Untersuchungen mit Darstellung der Gallenwege und der intrahepatischen Venen. Schweiz. Arch. Tierheilk. **113**, 577—586 (1971).

Habel, R. E.: Applied veterinary anatomy. 2nd. ed. Ithaca, New York, Author, 1978.
Hess, O.: Ausführungsgänge des Hundepankreas. Arch. Anat. Physiol. **118**, 1907.
Höcke, M.: Beiträge zur vergleichenden Histologie des Pankreas der wichtigsten Haussäugetiere (Hund, Katze, Schwein, Schaf, Ziege, Rind, Pferd) mit besonderer Berücksichtigung des „Ausführenden Apparates" und der „Pankreasinseln". Bern, Diss., 1907.
Holzapfel, R.: Die Mündung von Gallen- und Pankreasgang beim Menschen. Anat. Anz. **69**, 1930.

Jablan-Pantić, O.: Merkmale und Vergleich der intrahepatischen Gallengänge bei Haustieren. Acta Vet., Beograd **13**, 3—14 (1964).

Klöpping, E.: Die Gefäßkonfiguration des Venaportae- und des Venae-hepaticae-Systems in der Leber einiger Haustiere. Diss. Rijksuniv. Utrecht, 1968.
Krastin, L.: Lobierung und Vaskularisation der Leber der Säuger. Latvijas Biol. Biedribus Raksti **1**, 1929; Bull. Soc. Biol. Lettonie **1**, 1929.
Kretschmar, S.: Untersuchungen über die Leberzellen und Leberläppchen des Schweines während des Wachstums. Leipzig, Diss., 1914.

Lanz, A.: Wägungen und Messungen an der Pferdeleber. Arch. wiss. prakt. Tierheilk. **76**, 1941.
Löhner, L.: Über die extrahepatischen Gallenwege der Säuger in vergleichend-physiologischer Betrachtung. Biol. gen. **5**, 1929. Ref. Anat. Ber. **20**, 1930/31.

Mamedov, Ju. A.: Ähnlichkeiten und Unterschiede des Pankreas von Mensch und Tier. Doklady Akademii nauk Azerbajdžanskoj SSR, Baku **21**, 68—71 (1965).
Mann, F. C., J. P. Foster and S. D. Brimhall: The relation of the common bile duct to the pancreatic duct in common domestic and laboratory animals. J. Lab. Clin. Med. **5**, 1920.
Meyer, F.: Terminologie und Morphologie der Säugetierleber nebst Bemerkungen über die Homologie ihrer Lappen. (Eine vergleichend-anatomische, entwicklungsgeschichtliche Untersuchung.) Zürich, Diss. med. vet., 1911.
Mintzlaff, M.: Leber, Milz, Magen, Pankreas des Hundes. Leipzig, Diss. med. vet., 1909.

Nielsen, S. W., and E. J. Bishop: The duct system of the canine pancreas. Am. J. Vet. Res. **15**, 1954.

Paiva, O. M., A. Fernandes Filho, I. L. de Santis Prada: Beitrag zur Untersuchung der Gallengänge bei Felis catus domestica. Systematisierung des Ramus principalis sinister. Rev. Fac. Med. Vet., São Paulo **8**, 603—624 (1971).
Peduti Neto, J., I. L. de Santis Prada, V. Borelli: Der Abstand vom Torus pyloricus zur Papilla duodeni major bei Schafen. Rev. Fac. Med. Vet., São Paulo **8**, 635—637 (1971).
Pfuhl, W.: Die Gallenblase und die extrahepatischen Gallengänge. In: v. Möllendorffs Handbuch der mikroskopischen Anatomie des Menschen, Vol. V/2. Berlin, Springer, 1932.
Popper, H., u. F. Schaffner: Die Leber. Struktur und Funktion. Stuttgart, Thieme, 1961.

Sajonski, H., u. M. Dziadek: Form, Lage, Masse und Gewicht der Bauchspeicheldrüse des Schweines. Zbl. Vet. Med. **2**, 1955.
Santis Prada, I. L. de, H. Higashi: Über das Auftreten des Anulus pancreatis beim Rind. Rev. Fac. Med. Vet., São Paulo **7**, 535—540 (1966/67).
Santis Prada, I. L. de, V. Borelli, A. Fernandes Filho: Über die Verbindung des Ausführungsgangsystems der Bauchspeicheldrüse mit dem Zwölffingerdarm bei Pferden. Rev. Fac. Med. Vet., São Paulo **8**, 411—416 (1970).
Schreiber, H.: Zum Bau und Entleerungsmechanismus der Gallenblase (Vorläufige Mitteilung). Anat. Anz. **87**, 1939.
Schreiber, H.: Das Muskellager der menschlichen Gallenblasenwand im Vergleich zu der vierfüssiger Säuger. Zschr. Anat. Entw. gesch. **111**, 1941.
Siwe, St.: Über Onto- und Phylogenese des Pankreas. Gegenb. Morph. Jahrb. **68**, 1931.
Small, E., R. Olsen, and Th. Fritz: The canine pancreas. Vet. Med. Small Animal Clin. **59**, 1964.
Szuba, Z., St. Nogalski: Der morphometrische Zusammenhang zwischen der Leber und der Milz bei der Hauskatze. Zeszyty nauk, wyzszej Szkoly roliniczej, Szczecin, 205—211 (1971).

Vodovar, N., J. Flanzy, A. C. François: Intestin grêle du porc. I. Dimensions en fonction de l'âge et du poids, étude de la jonction du canal cholédoque et du canal pancréatique à celui-ci. Ann. Biol. Animale, Biochim., Biophys. **4**, 27—34 (1964).

Wass, W. M.: The duct systems of the bovine and porcine pancreas. Am. J. Vet. Res. **26**, 1965.

Zimmermann, A.: Zur vergleichenden Anatomie der extrahepatischen Gallenwege. Arch. wiss. prakt. Tierheilk. **68**, 1934.

Spleen

General and Comparative

(13, 263—268, also 172—174, 178—180, 186, 195, 197, 199, 207, 213, 215, 216, 243, 257)

The spleen (lien)* develops between the two serosal layers of the dorsal mesogastrium, which becomes the greater omentum in the adult. As the spleen increases in size, the outer layer of the mesogastrium proliferates until it envelops the entire organ (12, 13). In this way the spleen receives its peritoneal covering and, at the same time, forms close relations and connections with the stomach and the greater omentum.

The functions of the spleen are: during embryonic life, the production of erythrocytes; in the adult, the production of lymphocytes and, later, the destruction of erythrocytes, and storage of iron (hemosiderin). Further, the spleen is capable of storing blood and returning it to the circulation when the need arises. The spleen also forms an important link in the reticulo-endothelial system and, through its ability to repel and remove infectious agents, constitutes an important defense mechanism of the body. Although it is evident that the spleen belongs to the circulatory system, for topographic and didactic reasons, it is here considered directly after the chapter on the digestive system.

POSITION, SHAPE, AND ATTACHMENT. Although the spleen of the domestic mammals differs in shape and attachment, it always lies against the left abdominal wall and is usually covered by the ribs (197/*k*). In the carnivores, pig, and to a lesser degree in the horse, the position of the spleen depends on the fullness of the stomach. In the dog, for example, it may be displaced caudally to the flank when the stomach is greatly distended (174/*d*).

The spleen is a laterally flattened, elongated organ and presents parietal and visceral **surfaces,** cranial and caudal, mostly sharp, **borders,** and dorsal and ventral **extremities.** In the carnivores (263), the ventral extremity or end is usually a little wider than the dorsal. The spleen of the pig (264) is of uniform width and on cross section it is triangular. In the ox (265), the spleen resembles a rather long oval, while in the small ruminants it is more compact, rather triangular in the sheep, and more rectangular in the goat (266). The spleen of the horse (267) resembles a printed comma in shape; its dorsal end is wide and its ventral end is pointed and bent cranially.

The long, groovelike **hilus** (*a*) is on the visceral surface of the spleen in the carnivores, pig, and horse, and is marked by the splenic vessels and nerves, and the attachment of the **gastrosplenic ligament** (*4*). The hilus divides the visceral surface into unequal cranial and caudal surfaces; the cranial (facies gastrica) lies against the greater curvature of the stomach, the caudal (facies intestinalis) is in contact with the intestines. The hilus of the ruminant spleen is a small indentation on the dorsal end (265, 266/*a*).

Color, consistency, size, and weight of the spleen depend largely on the amount of blood it contains, but also on the age, state of nutrition, breed, and sex of the animal. In the dog, e.g., up to 16 per cent of the total blood volume may be stored in the spleen.

STRUCTURE. Deep to the peritoneal covering of the spleen is a **capsule** (268/*b*), which consists of collagenous and elastic fibers and contains many smooth muscle cells. Numerous **trabeculae** (*c*) containing, or accompanied by, vessels run from the capsule to the interior

---

* Also Gr. *splen*, hence splenic.

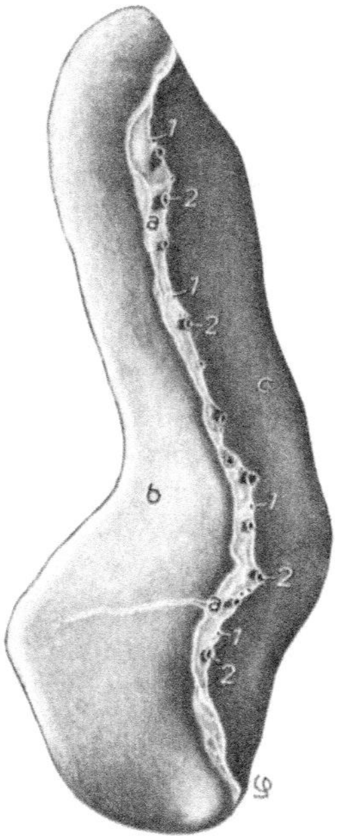

Fig. 263 (Dog)

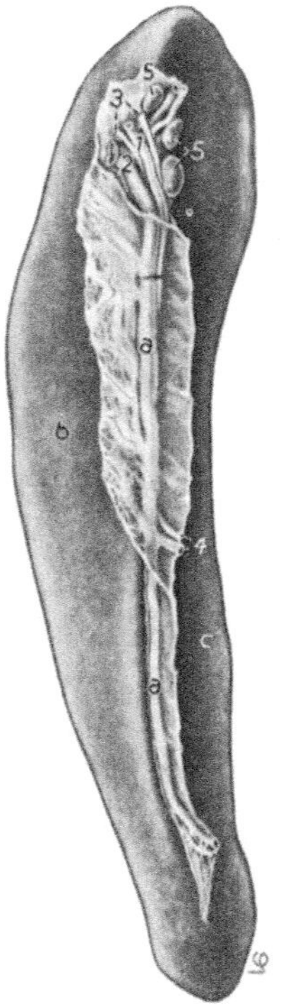

Fig. 264 (Pig)

Figs. 263—267. The spleens of the domestic mammals. Visceral surface.

a Hilus and (except ruminants) attachment of greater omentum; *b* Intestinal surface; *c* Gastric surface; *d* Area of adhesion to the dorsal sac of the rumen; *e* Area covered with peritoneum; *f* Dorsal extremity; *g* Ventral extremity

1 Splenic artery or its branches; *2* Splenic vein or its branches; *3* Nerves of splenic plexus; *4* Short gastric arteries and veins in the gastrosplenic ligament (pig, horse); *5* Splenic lymph nodes (pig, horse); *6* Phrenicosplenic ligament; *7* Renosplenic ligament

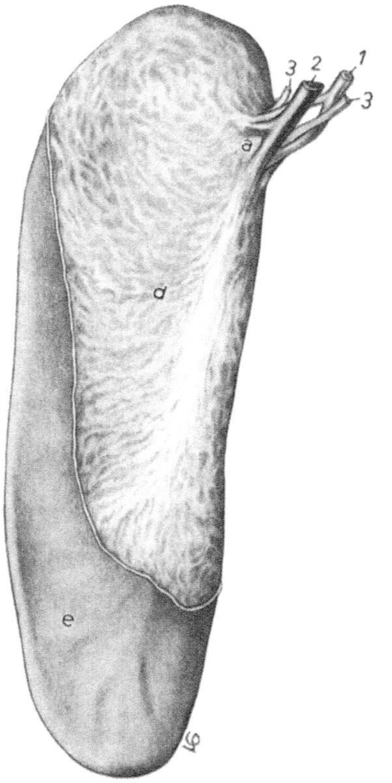

Fig. 265 (Ox)

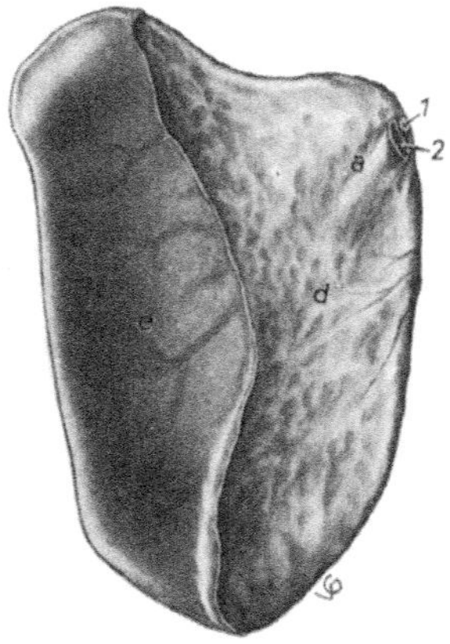

Fig. 266 (Goat)

and form a spongy framework. This supports a soft meshwork of reticular connective tissue and blood vessels, known as the splenic pulp. Most of the splenic pulp, because of the preponderance of red blood cells, is known as the **red pulp.** Scattered throughout the red splenic pulp are whitish **corpuscles** (lymphonoduli lienales, *e*), which are visible to the unaided eye, especially in the pig and ox, and are referred to, collectively, as the **white pulp.** The corpuscles are nodular accumulations of lymphoreticular tissue along the course of small pulp arteries. They produce lymphocytes and often show a lighter central area, particularly under conditions of stress as during an infection.

The red pulp consists of the fine, blood-filled meshwork of reticular connective tissue and accompanying **venous sinuses** which are of varying widths. The walls of the sinuses consist of ribbon-like endothelial cells arranged longitudinally on the inside, and circular fibers derived from the reticular connective tissue on the outside. Through adjustable openings in this peculiar latticework, blood is believed to pass from the reticular meshwork into the sinus and vice versa. The blood reaches the pulp through the pulp arteries which break up, forming little brushes (penicilli). The arteries of the penicilli then become the sheathed arteries and give rise to the capillaries. From the capillaries, the blood pours into the reticular meshwork of the red pulp (open circulation) and, after having been subjected to the influences of the reticular cells, enters the venous sinuses which channel it into the pulp veins. From here the 'cleansed' venous blood, now rich in lymphocytes, enters the trabecular veins which carry it to the splenic vein.

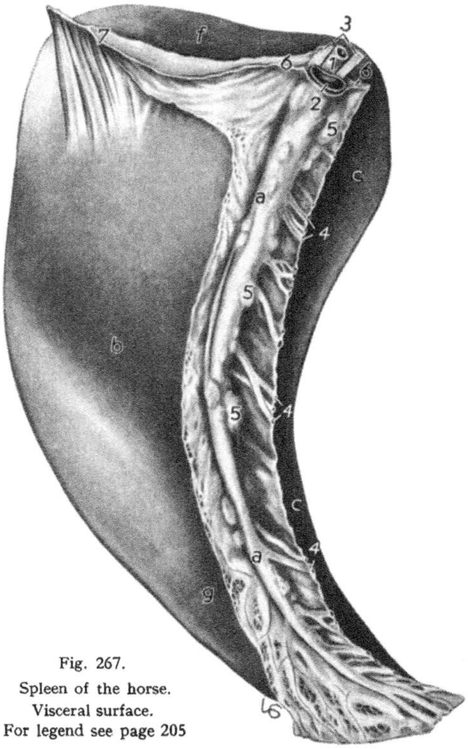

Fig. 267.
Spleen of the horse.
Visceral surface.
For legend see page 205

BLOOD VESSELS, LYMPHATICS, AND INNERVATION. The spleen receives its blood supply from the splenic artery, which is a branch of the celiac artery. The splenic vein, a branch of the portal vein, carries the blood from the spleen to the liver. The **lymphatics** of the spleen in the carnivores, pig, and horse, enter the splenic lymph nodes (264, 267/*5*) which are situated at the hilus. The spleen of the ruminants lacks its own lymph nodes; the lymphatics pass to the atrial nodes on the rumen or directly to the celiac nodes. In the ox, some of the splenic lymphatics pass through the diaphragm and enter the caudal mediastinal nodes. The unusually thick nerves of the **splenic plexus** (*3*) come from the celiac plexus, which is formed by the sympathetic and vagus (parasympathetic) nerves.

The Spleen of the Carnivores

The spleen is bright red in the fresh state but sometime after death it turns bluish red or grayish. In the dog, the spleen is flat and long, and usually slightly narrower in its middle section (263). The ventral end is the widest part and is often bent slightly cranially. The **parietal surface** of the spleen is slightly convex. The **visceral surface** presents the long, ridgelike **hilus** (*a*), through which the branches of the splenic artery and vein enter the organ. The splenic lymph nodes are somewhat removed from the hilus; they lie next to the splenic artery and vein in the greater omentum.

The spleen of the cat is rather similar to that of the dog, especially with regard to color and shape.

The spleen of the carnivores is larger, relatively, than that of the herbivores. The canine spleen is on the average 9.7—24 cm. long, 2.5—4.6 cm. wide, and weighs from 8—147 gm., thus varying widely with the size of the breed. The spleen of the cat has a length of 11.4—18.5 cm., a width of 1.4—3.1 cm., and a weight of 5.5—32 gm.

The **POSITION** of the spleen depends largely on the fullness of the stomach. In a dog with a moderately full stomach (173), the dorsal end of the spleen (*d*) is covered by the last two ribs. The ventral end projects beyond the costal arch and reaches ventrally as far as the level of the ventral ends of the seventh to the tenth ribs, and caudally to a transverse plane through the second to the fourth lumbar vertebrae. The parietal surface is related to the left crus and costal part of the diaphragm and ventrocaudally also to the flank, where the spleen can be palpated when enlarged. The visceral surface is related cranially to the stomach and caudally to the left kidney, descending colon, and jejunum. When the stomach is empty (172), the spleen is entirely intrathoracic. But when the stomach is greatly distended (174), the spleen is displaced fully into the flank and may, occasionally, reach the pelvic inlet.

The spleen of the carnivores is suspended by the part of the greater omentum that leaves the left crus of the diaphragm along a line (178/*c'*) from the esophageal hiatus to the celiac artery. Cranially, this part of the omentum passes directly to the fundus of the stomach; caudally, it is wider and passes first to the hilus of the spleen and then to the greater curvature of the stomach, forming an extensive **gastrosplenic ligament** (*d*). The relatively loose attachment of the spleen is the reason that its ventral end in particular can shift so extensively when the stomach becomes distended.

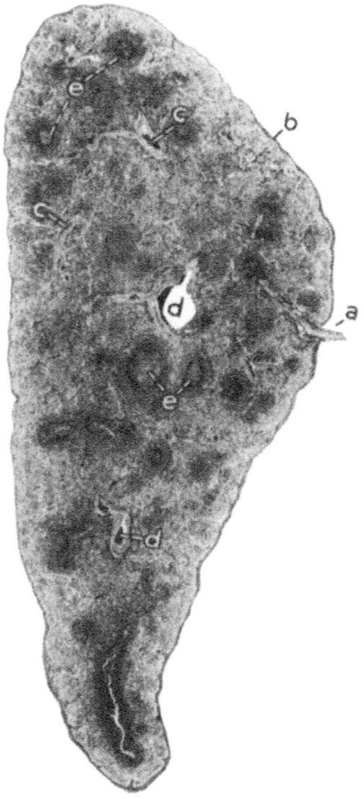

Fig. 268. Cross section of the spleen of the dog. Microphotograph.

a Hilus; *b* Capsule; *c* Trabeculae; *d* Cross sections of trabecular arteries and veins; *e* Corpuscles (white pulp), the remaining parenchyma is the red pulp

The Spleen of the Pig

The spleen is bright red, but darkens after it is exposed to the air. Splenic **corpuscles** are distinctly visible on the cut surface. The spleen is long and narrow and has a pointed ventral extremity (264). Its cross section is triangular. The **hilus** (*a*) is on a longitudinal crest on the visceral surface. Here, the numerous branches of the splenic artery and vein enter the organ. The splenic lymph nodes (*5*) are near the hilus.

The spleen is 24—45 cm. long, 3.5—12.5 cm. wide, and weighs 90—335 gm.

POSITION. When the stomach is empty or moderately full, the spleen, in nearly dorsoventral orientation, lies against the costal part of the diaphragm and its ventral end may extend beyond the costal arch (195, 197). When the stomach is full, the caudal border of the spleen reaches the last rib, and its ventral end moves across the costal arch as far as the umbilical region (199). The dorsal end of the spleen is related to the fundus of the stomach, the left kidney, and the left lobe of the pancreas. The visceral surface makes contact with the greater curvature of the stomach cranially, and with jejunum and colon caudally. The **gastrosplenic ligament** (*9*) provides the spleen with only a loose attachment to the stomach and for this reason torsion of the spleen is not a rare occurrence in the pig.

The Spleen of the Ruminants

The color of the bovine spleen differs with the age and sex of the animal. In calves it is reddish brown to bluish red, in bulls and steers dark red or reddish brown, and in cows bluish gray or gray, and is of a more solid consistency in males than in females. The splenic **corpuscles** reach a size of 1—2 mm.

The spleen of the ox (265), shaped like an elongated oval, is flat and of uniform width. The **hilus** (*a*) is restricted to a small area on the visceral surface close to the dorsal end of the cranial border. It is 41—50 cm. long, 11—14.5 cm. wide, 2—3 cm. thick, and weighs 665 to 1,155 gm. The greater splenic weights are found among male animals.

POSITION. Owing to the extraordinary expansion of the greater curvature of the gastric primordium to form the ruminoreticulum, the spleen has lost its connection with the greater omentum and is found directly applied to the left dorsocranial surface of the rumen (207). Most of its **visceral surface** (265/*d*) is firmly attached to the rumen without the interposition of peritoneum. Its **parietal surface** lies against the diaphragm to which a small dorsal area is attached. The spleen is thus firmly fixed in position. The peritoneal reflection from the parietal surface of the spleen to the diaphragm is the **phrenicosplenic ligament,** and the reflection from the visceral surface to the rumen is the **gastrosplenic ligament.** The spleen of the ox (243/*n*, *n'*) extends from the dorsal ends of the last two ribs to the costochondral junctions of the seventh and eighth ribs and is related to the dorsal and cranial sacs of the rumen and to the cranial surface of the reticulum. Its cranial border passes obliquely from the dorsal end of the last rib to the ventral end of the seventh intercostal space. The spleen is accessible for biopsy during expiration through the left eleventh intercostal space at the level of the tuber coxae.

SMALL RUMINANTS. The spleen of the sheep is reddish brown and that of the goat reddish gray; **corpuscles** are distinctly visible on the cut surface. The spleen of the small ruminants can be distinguished easily from that of the ox, because of its more compact shape. In the sheep it is roughly triangular with a somewhat pointed ventral extremity (215/*d*); its parietal surface is convex and its visceral surface concave. The small **hilus** is on the visceral surface close to the dorsal end of the cranial border. The spleen of the goat (266) is more rectangular than that of the sheep. It is difficult, however, to distinguish with certainty between the spleens of these two animals. The length of the sheep's spleen is 8.5—14 cm., its width 6—11 cm., and its weight 46—133 gm. In the goat, the spleen is 9.4—12.4 cm. long, 6.5—7 cm. wide, and weighs approximately 70 gm.

The **POSITION** of the spleen in the small ruminants is similar to that described for the ox. It is inserted between the dorsal sac of the rumen and the diaphragm and is firmly attached to both structures by connective tissue. It is found high in the abdominal cavity adjacent to the vertebral column in the region of the tenth to the thirteenth ribs (213, 216).

The Spleen of the Horse

The spleen of the horse, because of its thick capsule, is bluish in color. After the death of the animal it turns a dark reddish brown. Splenic corpuscles are not visible on the cut surface. The spleen is shaped like a printed comma with a wide dorsal end and a pointed ventral end (267). The **hilus** (*a*) is situated close to the cranial border of the slightly concave visceral surface and extends the entire length of the organ.

The spleen is 40—67 cm. long; at its broadest point it is 17—22 cm. wide; its thickness ranges from 2—6 cm.; and its weight is 950—1,680 gm. In the thoroughbred, the spleen weighs an average of 2,023 gm., while in draft horses the average is only 1,000 gm.

POSITION. The spleen of the horse, covered by the ribs, lies against the left abominal wall (257). Its wide dorsal end is lateral to the left kidney and extends caudally to the transverse process of the first lumbar vertebra. The ventral end is pointed cranioventrally and reaches the ventral halves of the ninth to the eleventh intercostal spaces. The concave cranial border of the spleen follows roughly the line between the costal part and the tendinous center of the diaphragm. The convex caudal border, from dorsal to ventral, at first runs parallel to the costal arch, then, from the sixteenth intercostal space onward, follows a line connecting the olecranon with the tuber coxae. The **parietal surface** of the spleen is in contact with the diaphragm. The deep relations of the spleen are: the stomach cranially; the left kidney and left lobe of the pancreas dorsally; and the jejunum, small colon, and occasionally, the left dorsal colon caudally. Caudal displacement of the spleen depends not only on the fullness of the stomach, but also on the degree of distention of the great colon. Only the dorsal end of the spleen can be palpated rectally.

The dorsal end of the spleen is attached to the diaphragm by the **phrenicosplenic ligament** (13, 267/*6*) which blends cranially with the **gastrophrenic ligament** (13/*7*). There is also an attachment to the left kidney by the **renosplenic ligament** (13/*5*; 267/*7*). The part of the greater omentum extending from the splenic hilus to the greater curvature of the stomach is the **gastrosplenic ligament** (*4*) and contains the splenic artery and vein. The splenic lymph nodes (*5*) lie along the hilus.

BIBLIOGRAPHY

Spleen

Barcroft, J.: Neue Milzforschungen. Naturwiss. **13,** 1925.

Curson, H. H.: Accessory spleens in a horse. 16th Report of the Director of vet. Services of South Africa, 1930.

Dorfman, R. F.: Nature of the sinus lining cells of the spleen. Nature **190,** 1961.

Dreyer, B. J. van R.: The segmental nature of the spleen. Blood **18,** 1961.

Fillenz, Marianne: The innervation of the cat spleen. Proc. Roy. Soc. London, B. **174,** 459—468 (1970).

Godinho, H. P.: Anatomical studies about blood circulation of the dog's spleen. I. Venous drainage: venous lienal zones. Arq. Esc. Vet. **15,** 1963.

Godinho, H. P.: Anatomical studies on the termination and anastomoses of the a. lienalis and arterial lienal segments in the dog. Arq. Esc. Vet. **16,** 1964.

Hartmann, A.: Die Milz. In: v. Möllendorffs Handbuch der mikroskopischen Anatomie des Menschen. VI/1. Berlin, Springer, 1930.

Hartwig, H.: Die makroskopischen und mikroskopischen Merkmale und die Funktion der Pferdemilz in verschiedenen Lebensaltern und bei verschiedenen Rassen. Z. mikrosk.-anat. Forsch. **55,** 287—410 (1949).

Herrath, E. v.: Über einige Beobachtungen bei der Durchspülung verschiedener Säugermilzen. Anat. Anz. **80,** 38—44 (1935).

Herrath, E. v.: Bau und Funktion der Milz. Zschr. Zellforsch. **23,** 375—430 (1935).

Herrath, E. v.: Vergleichend-quantitative Untersuchungen an acht verschiedenen Säugermilzen. Z. mikrosk. anat. Forsch. **37,** 389—406 (1935).

Herrath, E. v.: Einiges über die Beziehungen zwischen Bau und Funktion der Säugermilz. Anat. Anz. Erg. H. **81,** 182—186 (1936).

Herrath, E. v.: Experimentelle Untersuchungen über die Beziehungen zwischen Bau und Funktion der Säugermilz. 1. Der Einfluß des Lauftrainings auf die Differenzierung der Milz heranwachsender Tiere. a) Hunde. Z. mikrosk. anat. Forsch. **42,** 1—32 (1937).

Herrath, E. v.: Experimentelle Ergebnisse zur Frage der Beziehungen zwischen Bau und Funktion der Säugermilz. Anat. Anz. Erg. H. **85,** 196—207 (1938).

Herrath, E. v.: Zur vergleichenden Anatomie der Säugermilz und ihrer Speicher- und Abwehraufgaben. Zugleich ein Beitrag zur Typologie der Milz und zum Problem der artlichen und individuellen Milzgröße. Med. Klin. **34,** 1355—1359 (1938).

Herrath, E. v.: Die Milztypen beim Säuger. Anat. Anz. Erg. H. **87,** 247—254 (1939).

Herrath, E. v.: Zur Frage der Typisierung der Milz. Anat. Anz. **112,** 140—149 (1963).

Krzywanek, Fr. W.: Neue Ansichten über die Funktion der Milz im Blutkreislauf. Berl. Tierärztl. Wschr. **43,** 1927.

Krzywanek, Fr. W.: Weiteres über die neu erkannte Milzfunktion. Berl. Tierärztl. Wschr. **45,** 1929.

Langer, P.: Die Altersveränderungen der Milz beim Pferd mit besonderer Berücksichtigung der Gitterfasern. 5. Beitrag zur Altersanatomie des Pferdes. Diss. med. vet. Hannover, 1941.

Moore, R. D., V. R. Mumaw, and M. D. Schoenberg: The structure of the spleen and its functional implications. Exp. Molec. Path. **3,** 1964.

Obiger, L.: Untersuchungen über die Altersveränderungen der Milz bei Hunden. (2. Beitrag zur Altersanatomie des Hundes.) Diss. med. vet. Hannover, 1940.

Popescu, P.: Beitrag zum Studium der Topographie und der Punktionstechnik der Milz beim Rinde. Diss. med. vet. Bukarest, 1937.

Reissner, H.: Untersuchungen über die Form des Balkengerüstwerks der Milz bei einigen Haussäugetieren, sowie über die Verteilung von elastischem und kollagenem Bindegewebe und glatter Muskulatur in Kapsel und Trabekel. Zschr. mikr. anat. Forsch. **16**, 1929.

Riedel, H.: Das Gefäßsystem der Katzenmilz. Z. Zellforsch. **15**, 459—529 (1932).

Schulz, P.: Maße und Gewichte der Milzen unserer Schlachttiere. Dtsch. Schlacht. u. Viehhof Ztg. **56**, 1956.

Schwarze, E.: Über Bau und Leistung der Milzkapsel unserer Haussäugetiere. Berl. Tierärtzl. Wschr. 1937.

Skramlik, E. v.: Die Milz. Mit besonderer Berücksichtigung des vergleichenden Standpunktes. Erg. Biol. **2**, 1927.

Snook, T.: A comparative study of the vascular arrangements in mammalian spleens. Amer. J. Anat. **87**, 31—78 (1950).

Steger, G.: II. Beitrag zur ,,Anatomie für den Tierarzt''. Zur Biologie der Milz der Haussäugetiere. Dtsch. Tierärztl. Wschr. **46**, 1938.

Steger, G.: Die Artmerkmale der Milz der Haussäugetiere. Morph. Jb. **83**, 125—157 (1939).

Steger, G.: Die Artmerkmale der Milz der Haussäugetiere (Pferd, Rind, Schaf, Ziege, Schwein, Hund, Katze, Kaninchen und Meerschweinchen). Leipzig, Diss. med. vet. (1939).

Steger, G.: IV. Beitrag zur ,,Anatomie für den Tierarzt''. Die tierartlichen Merkmale der Haussäugermilzen bezüglich Form, Hilus und Gefäßen. Dtsch. Tierärztl. Wschr. **47**, 1939.

Tischendorf, F.: Beobachtungen über die feinere Innervation der Säugermilz. Klin. Wschr. **26**, 125 (1948).

Tischendorf, F.: Milz. In: Kükenthals Handbuch der Zoologie (hrsg. v. J.-G. Helmcke und H. v. Lengerken). VIII/5 (2). 1—32. De Gruyter & Co., Berlin 1956.

Tischendorf, F.: Die Milz. In: v. Möllendorf-Bargmanns Handbuch der mikroskopischen Anatomie des Menschen. VI/6. Erg. VI/I. Berlin, Heidelberg, New York, Springer, 1969.

Vereby, K.: Vergleichende Untersuchungen über die Kapsel, Trabekel und Gefäße der Milz. I. Die Milz des Schafes und Rindes. Z. Anat. Entw.-gesch. **112**, 634—652 (1943).

Wagemeyer, M.: Über den Einbau des Gefäßsystems der Milz in die Trabekelarchitektur und dessen funktionelle Bedeutung. Mainz, Diss. med. 1956.

Respiratory System

General and Comparative

The respiratory organs provide for the exchange of gases between the blood and the atmosphere and, within limits, improve the quality of the inspired air and regulate its flow. The respiratory system begins at the **nostrils,** through which the air enters the **nasal cavities,** and is continued by the **nasopharynx, larynx,** and **trachea** to the **lungs.**

The exchange of gases takes place in the pulmonary alveoli where the alveolar blood capillaries make contact with the air through the extremely thin alveolar wall. In its passage from the nostrils to the alveoli the air is usually purified, moistened, and warmed, and its volume is regulated by the nostrils and the larynx. The diaphragm and the other respiratory muscles, by increasing or diminishing the size of the thoracic cavity, govern the respiratory volume.

The respiratory passages are lined with a mucus-producing, ciliated, pseudostratified epithelium containing large numbers of goblet cells. In regions where the mucosa is exposed to wear, such as at the nostrils and the larynx, the lining consists of stratified squamous epithelium. The wavelike movements of the cilia transport the fine dust particles, that have been trapped by the moist epithelium, either toward the nostrils, and thus out of the system, or toward the pharynx, where the respiratory passages cross those of the digestive system and the particles are eliminated by swallowing. Seromucous glands in the walls of the respiratory passages serve to add moisture to the air. The inspired air is warmed by extensive vascular plexuses in the respiratory mucosa of the nasal cavities. The amount of blood flowing through these plexuses can be regulated, and the warmth thus added to the air stream facilitates the evaporation of glandular secretions and saturates the air with water (which is important for olfaction). The respiratory air stream, while usually passing through the nose, may at times be directed through the mouth, though this is the case only in labored breathing, for the oral cavity lacks the special features for preparing the air to enter the lungs. Dogs frequently breathe through their mouths (panting) because it provides for evaporating fluids, thus cooling the body.

The olfactory region is located in the caudal part of the nasal cavity, while the organ of phonation is lodged in the larynx. The olfactory region registers the presence of harmful substances in the air and triggers a reflex that closes the air passage in the larynx. The voice is produced in the larynx chiefly by expired air. The air passages in the head, including the oral cavity, determine the shape of the air column and increase the resonance of the voice. The shape of the human oral cavity can be altered in order to produce articulated sound (speech); the oral cavity of domestic mammals is, within limits, also capable of being changed, so that different sounds can be produced.

Nose

The human nose projects distinctly from the face; in domestic mammals it is incorporated in the face and forms the large dorsal and lateral areas (dorsum nasi) rostral to the eyes. The **apex** of the nose in the carnivores and pig, however, does protrude to some extent from the face (18, 20). The **nostrils** in the apex lead into the **nasal cavity,** to which are connected, directly or indirectly, several **paranasal sinuses.** The **nasal septum** (277—281/*a*) forms a

Fig. 269 (Cat)

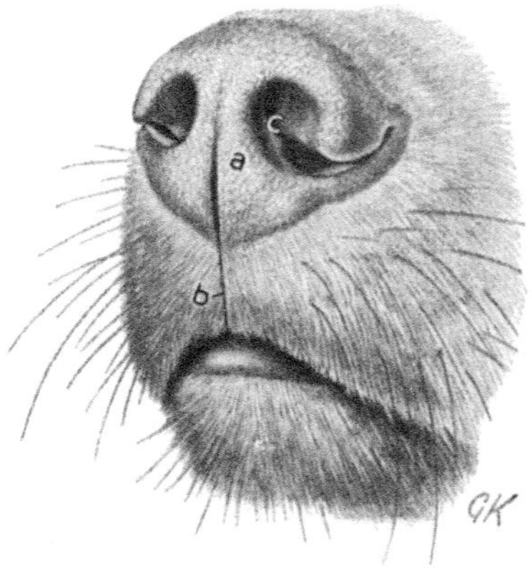

Fig. 270 (Dog)

Fig. 271 (Pig)

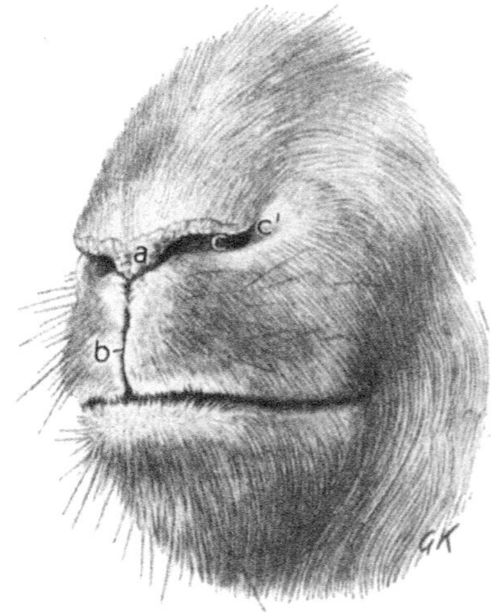

Fig. 272 (Goat)

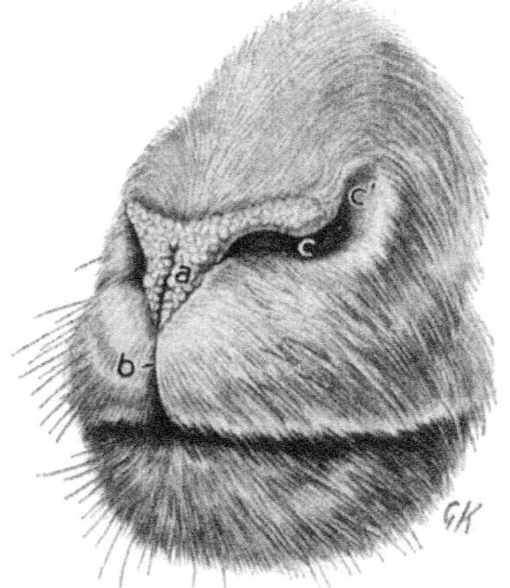

Fig. 273 (Sheep)

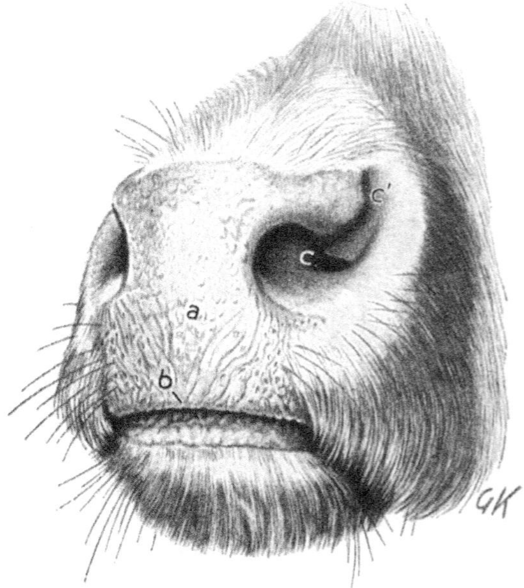

Fig. 274 (Ox)

Figs. 269—274. The muzzles of the domestic mammals. Craniolateral aspect.
a Planum nasale (cat, dog, goat, sheep), planum rostrale (pig), planum nasolabiale (ox); *b* Philtrum; *c* Nostril; *c'* Alar groove

partition between the nostrils and divides the nasal cavity into right and left halves. The caudal portion of the septum, which is formed by the perpendicular plate of the ethmoid, is osseous, while rostrally the septum consists of cartilage and becomes progressively more flexible toward the apex.

The **WALL OF THE NOSE** consists of skin externally, a middle supporting layer of bone or (rostrally) cartilage, and a mucous membrane which lines the nasal cavity. The osseous support of the wall is formed by the nasal, maxillary, incisive, frontal, lacrimal, and zygomatic bones, and by the perpendicular plate of the palatine bone. The free borders of the nasal and incisive bones (apertura nasi ossea) provide attachment for the nasal cartilages which support the nostrils. Associated with the bones and cartilages of the supporting layer are the nasal muscles which regulate the width of the nostrils.

The **FLOOR OF THE NASAL CAVITY** is the roof of the oral cavity. It is an osseous shelf consisting of portions of the incisive bones, the palatine processes of the maxillary bones, and the horizontal plates of the palatine bones, and it is covered with nasal mucosa dorsally and oral mucosa ventrally. The vomer (25/*C*) is attached to the dorsal surface of these bones and supports the nasal septum.

The partition between nasal and cranial cavities is formed by the ethmoid, the nasal part of the frontal bone, and the rostrum of the presphenoid (283).

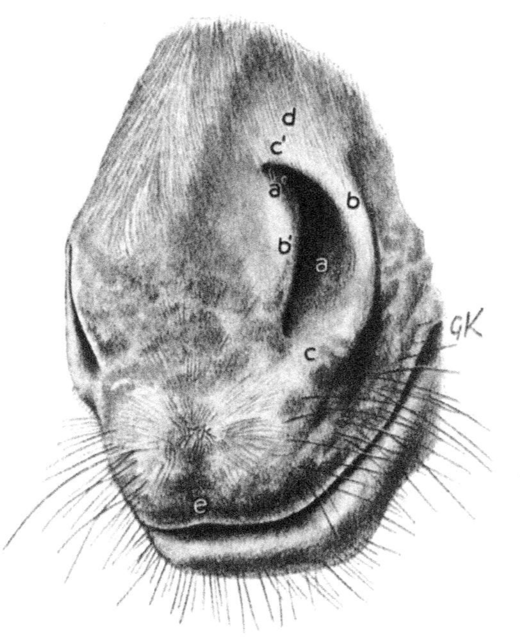

Fig. 275 with narrow nostrils

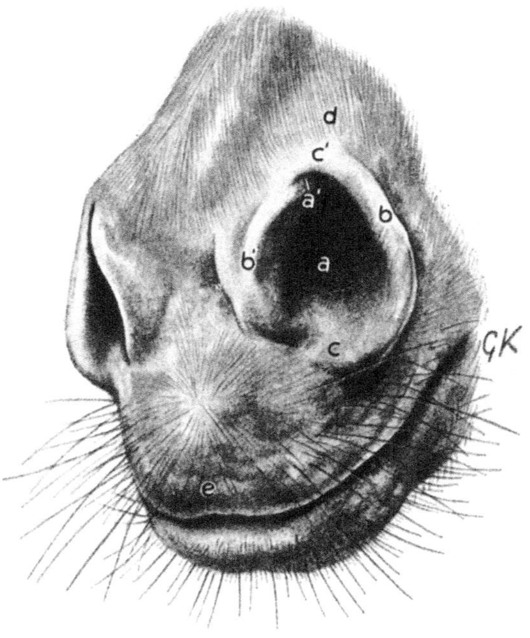

Fig. 276 with dilated nostrils

Figs. 275, 276. The muzzle of the horse. Craniolateral aspect. *a* Nostril; *a'* Dorsal part of nostril leading into nasal diverticulum; *b* Lateral border of nostril (ala nasi); *b'* Medial border of nostril; *c, c'* Junctions of lateral and medial and borders of nostril; *d* Nasal diverticulum; *e* Depression in upper lip (philtrum)

Apex of the Nose

The apical segment of the nose and the **NASAL CARTILAGES** present here differ considerably in the domestic mammals (277—281). The rostral part of the nasal septum widens along its dorsal and ventral margins to form the **dorsal and ventral lateral nasal cartilages** (*b, c, d*). These support the lateral wall of the apex, and in the carnivores, pig, and ruminants come together lateral to the nostril. The dorsal cartilage in the horse does not project very far laterally, and the narrow ventral cartilage covers only the palatine suture or may be absent. A further difference in the formation of the nasal cartilages in the horse is the presence of **alar cartilages** (*g—g''*), which are attached, sometimes by the interposition of a joint, to the rostral border of the nasal septum. They support the nostrils dorsally, medially, and ventrally, and consist of a **lamina*** dorsally and a **cornu**** ventrally. The medial support of the nostrils in the other animals is provided by the nasal septum and in the pig, in addition, by the **rostral bone** (*a'*) which is joined to the septum and the incisive bones. Dorsally, the nostrils are supported by the rostral part of the dorsal lateral cartilages (*c*). The ventral and lateral support of the nostrils is provided by the **lateral accessory cartilages** (*e*), which in the ruminants are attached to the dorsal lateral cartilages, in the carnivores to the ventral lateral cartilages, and in the pig to the rostral bone. In the horse there is no lateral support

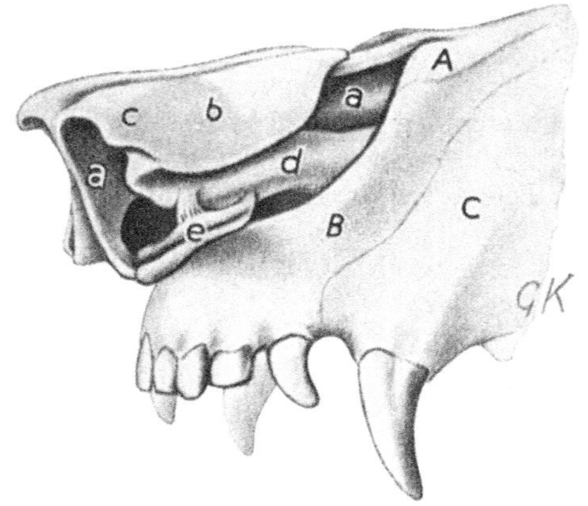

Fig. 277 (Dog)

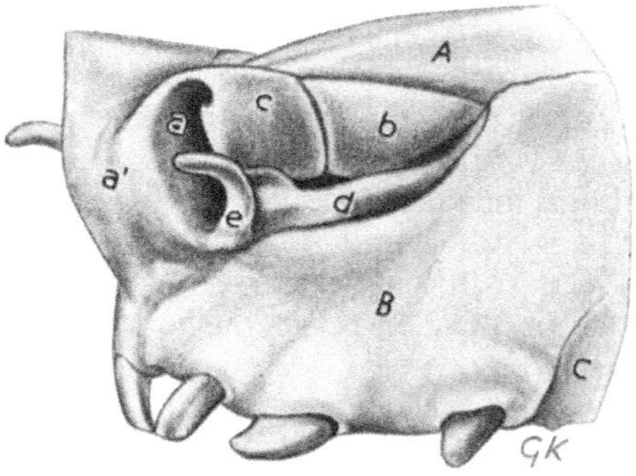

Fig. 278 (Pig)

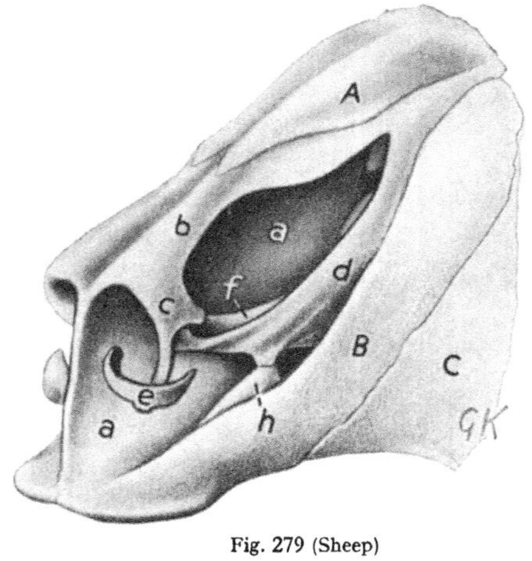

Fig. 279 (Sheep)

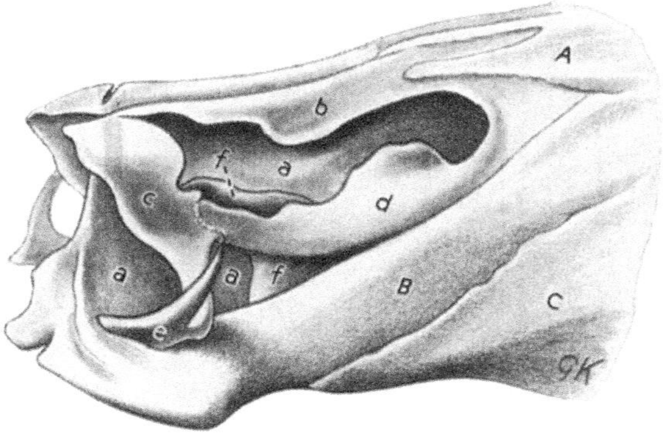

Fig. 280 (Ox)

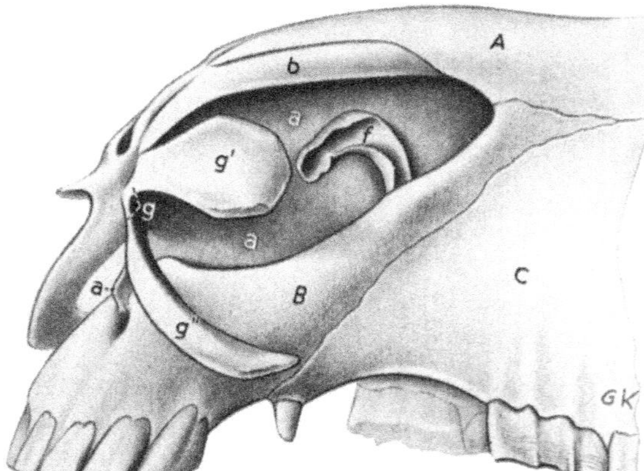

Fig. 281 (Horse)

Figs. 277—281. Nasal cartilages of the domestic mammals Craniolateral aspect. *A* Nasal bone; *B* Incisive bone; *C* Maxilla; *a* Nasal septum; *a'* Rostral bone (pig); *b*, *c* Dorsal lateral nasal cartilage, *c* rostral portion; *d* (sheep; *d*, *h*) Ventral lateral nasal cartilage (except horse); *e* Lateral accessory nasal cartilage (except horse); *f* Medial accessory nasal cartilage (not visible in dog and pig); *g* Alar cartilage, *g'* its lamina, *g''* its cornu (horse)

* L., plate.
** L., horn,

for the nostril. The **medial accessory cartilage** (*f*) lies inside the alar fold (see p. 216) and originates with its wide base from the ventral nasal concha and the ventral lateral nasal cartilage. In the horse, the medial accessory cartilage is large and S-shaped, while in the other species it is small.

The appearance of the **NOSTRILS** (nares) and the area of modified skin that surrounds them varies considerably from species to species. In the carnivores and small ruminants, this specialized skin area (planum nasale) lacks hair, is rather narrow in the goat and sheep (272, 273) and more extensive in the cat and dog (269, 270). In the pig, the area (planum rostrale, 271) forms part of the **rostrum,** or **snout,** fuses centrally with the upper lip, and bears short tactile hairs. In the ox, it involves the upper lip more fully (planum nasolabiale, 274) and has tactile hairs along its lateral borders. The corresponding area in the horse is normal skin covered with a coat of short hair interspersed with tactile hairs (275).

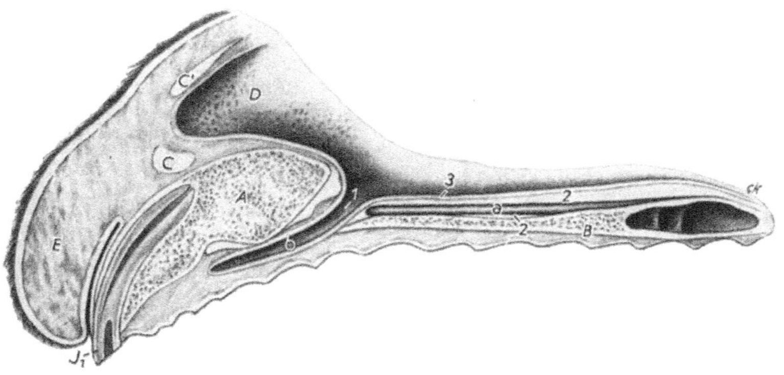

Fig. 282. Left vomeronasal organ of the horse. Sagittal section. Left lateral aspect. *A* Incisive bone; *B* Hard palate; *C, C'* Alar cartilage, *C* its cornu, *C'* its lamina; *D* Nasal sectum; *E* Upper lip; J_1 First incisor; *a* Vomeronasal duct; *b* Incisive duct; *1* Opening of vomeronasal duct; *2* Vomeronasal cartilage, opened; *3* Edge of nasal mucosa

The planum nasale of the cat consists of small cobblestone-like elevations, while in the other species the surface is divided by fine grooves into small polygonal fields (areae). This cutaneous relief remains unchanged throughout life and, since it is distinctive for the individual animal, "noseprints" may be taken for purposes of identification (Horning et al. 1926, Salomon 1930, Schulz 1930). The area is kept moist in the pig and ruminants by **glands** which secrete through pores (foveolae) in the small polygonal fields. These glands are absent in the carnivores, but the glands in the mucosa of the nasal septum, and the lateral nasal and lacrimal glands perform this function.

The **philtrum** (269—274/*b*) in the carnivores and small ruminants is deep and extends to the nostrils. In the pig, ox, and horse, it is shallow or absent.

While the medial wall of the **nostrils** is uniform, the lateral wall (ala nasi) presents a more or less deep groove in most species. In the carnivores, this groove is dorsolateral to the nostril and is between the dorsal lateral and the lateral accessory cartilages. The nostrils of the pig tend to be oval, conforming to the shape of the supporting cartilages. In the small ruminants the nostrils are narrow slits, in the ox they are oval, and they are drawn out laterally by an alar groove (sulcus alaris, 272—274/*c'*). This groove is bounded medially by the dorsal lateral nasal cartilage, which is covered with mucosa, and laterally by the lateral accessory cartilage, which is covered with skin. In the horse, the lamina of the alar cartilage and the medial accessory cartilage support the **alar fold** (62/*2*), which ends in the dorsal part of the nostril. The alar fold forms a nearly horizontal shelf which divides the nostril into dorsal and ventral passages (275, 276/*a, a'*); the dorsal leads into a blind cutaneous pouch, the **nasal diverticulum**; the ventral leads into the nasal cavity.

The **muscles** of the nose and upper lip act together to dilate the nostrils. This is particularly noticeable during labored breathing or when the animal is scenting. These muscles are well developed in the horse and can transform the normally semilunar nostrils to become circular (275, 276). They are poorly developed in the pig and carnivores.

Nasal Cavity

(25, 57—62, 283—291, 400, 401)

The narrow rostral portion of the nasal cavity (vestibulum nasi) is usually lined with mucous membrane covered with stratified squamous epithelium, but in the horse, skin with fine hairs extends into it for a short distance. The middle portion of the nasal cavity is the largest part and contains the **nasal conchae*** (287/*O*, *P*). In the small caudal part of the nasal cavity are the much more numerous **ethmoidal conchae** (*R*). Caudoventrally, the nasal cavity communicates through the **choanae** (62/*13*), two openings separated by the vomer, with the nasopharynx (*b*).

Most of the middle portion and septum of the nasal cavity (regio respiratoria) is lined with mucosa that is covered with ciliated pseudostratified epithelium and a varying number of goblet cells and contains mostly serous glands. In the caudal part of the nasal cavity, the mucosa is specialized for olfaction (regio olfactoria) and contains the olfactory nerve endings and glands. In the submucosa of the respiratory region are numerous vascular plexuses (25/*1*, *2*, *3*) consisting primarily of veins, which because of their muscular walls are capable of dilation and constriction. This extensive vascularization warms the air and, by causing evaporation of the glandular secretions, adds moisture to the inhaled air. When engorged, these plexuses swell like erectile tissue and impede the flow of air.

The dorsal and ventral nasal conchae project from the lateral wall and divide the nasal cavity into three meatuses. The **dorsal nasal meatus** (25/*b*) is a narrow passage between the roof of the nasal cavity and the dorsal concha and leads into the caudal part of the nose. The **middle nasal meatus** (*c*) is between the dorsal and ventral conchae and it also leads into the caudal part. In the carnivores and ruminants, this meatus is split caudally by the middle nasal concha into dorsal and ventral channels (288/*2*, *2'*, *2''*). The principal aperture to the paranasal sinuses is found in the middle meatus. The **ventral nasal meatus** (25/*d*) is the largest. It lies between the ventral concha and the floor of the nasal cavity and leads into the nasopharynx. Most of the respiratory air passes through this meatus. The **common nasal meatus** (*a*), the narrow space between the nasal septum and the conchae, extends from the roof of the nasal cavity to the floor and is continuous laterally with the other meatuses. Similar air spaces between the ethmoidal conchae (287/*R*) are the **ethmoidal meatuses.**

In the rostral portion of the nasal cavity, the mucosa of the lateral wall forms a number of folds which extend from the nasal conchae to the nostril. The **straight fold** (plica recta, 287/*5*) is the most dorsal of these and is continuous with the dorsal concha. It is double in the horse, but is united rostrally. Ventral to the straight fold is the **alar fold** (plica alaris, *6*), which is continuous with the ventral nasal concha and contains the supporting medial accessory cartilage. In the horse it also contains the lamina of the alar cartilage. The **basal fold** (plica basalis, 61, 62/*3*) is the most ventral. In the horse it originates from the ventral concha, but in the other species it is independent of the concha and unites rostrally with the alar fold.

The **NASAL AND ETHMOIDAL CONCHAE** are thin osseous scrolls that are covered on each side with mucous membrane, and originate with a **basal lamella** (401/*a*, *b*) from the lateral wall of the nasal cavity. This lamella projects medially like a shelf and is continued by one, two, or, rarely, more **spiral lamellae** (*a'*, *b'*) which roll up on themselves and form the scrolls. The spiral lamellae enclose airfilled **recesses** (*4*, *5*) which communicate extensively with the nasal meatuses (*1*, *2*, *3*). Along their free border, the spiral lamellae may form **bullae**** (*a''*, *b''*) which may, in turn, be subdivided by small transverse septa into **cells** (cellulae, 288/*c*). If the free border of a spiral lamella unites with its basal lamella or with an adjacent facial bone, a conchal sinus results. Bullae, cells, and conchal sinuses are never entirely sealed off and communicate through small apertures (*8*, *9*, *10*) with the nasal cavity. In addition to the conchal sinuses there are other large paranasal sinuses (see p. 223).

The large dorsal, middle, and ventral nasal conchae (*N*, *O*, *P*) are located in the large middle portion of the nasal cavity, while the smaller and more numerous ethmoidal conchae (*Q*, *R*) are

* L. *concha*, a shell or conch.

** L. *bulla*, a bubble.

in the caudal portion of the nasal cavity. The caudal ends of the dorsal and middle nasal conchae are part of the ethmoid labyrinth of scrolls (ethmoidal conchae) and are known as

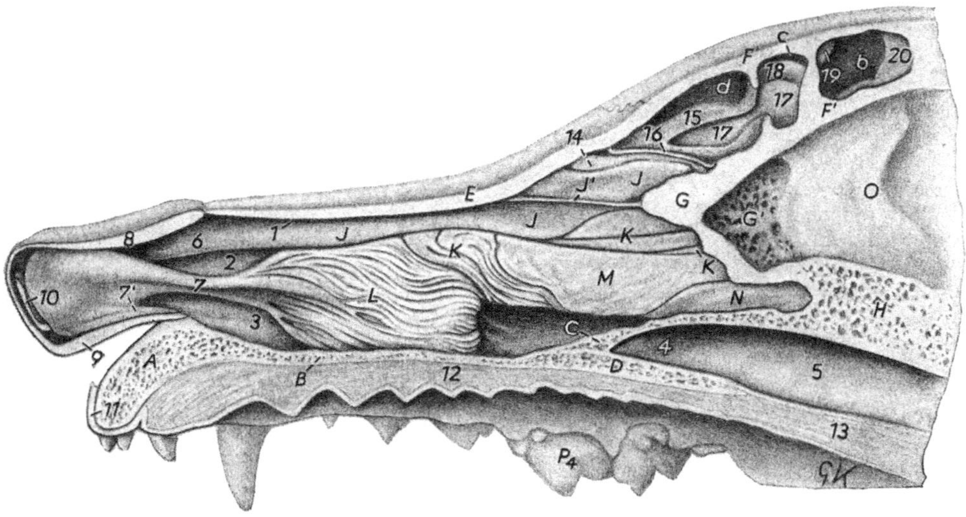

Fig. 283. Nasal cavity of the dog after removal of the nasal septum. Sagittal section. Medial aspect of right half. (After Graeger 1958.)

A Incisive bone; *B* Maxilla; *C* Vomer; *D* Palatine bone; *E* Nasal bone; *F, F'* Frontal bone; *G* Perpendicular plate, *G'* cribriform plate of ethmoid; *H* Body of presphenoid; *J* Dorsal nasal concha, *J'* cut edge of the caudal portion of its basal lamella; *K* Middle nasal concha; *L* Ventral nasal concha; *M, N* Ethmoidal conchae; *M* Endoturbinate III; *N* Endoturbinate IV; *O* Cranial cavity; P_4 Fourth premolar (sectorial); *b* Lateral frontal sinus; *c* Medial frontal sinus; *d* Rostral frontal sinus

1 Dorsal nasal meatus; *2* Middle nasal meatus; *3* Ventral nasal meatus; *4* Choana; *5* Nasopharynx; *6* Straight fold; *7* Alar fold; *7'* Basal fold; *8* Dorsal lateral nasal cartilage; *9* Cut edge of nasal septum; *10* Nostril; *11* Oral mucosa; *12* Hard palate; *13* Soft palate; *14—19* Further ethmoidal conchae: *14* Ectoturbinate 1; *15—18* Ectoburbinate 2: *15* basal lamella (fenestrated), *16* cut edge of dorsomedial spiral lamella; *17* medial portion, *18* lateral portion of ventrolateral spiral lamella; *19* Ectoturbinate 3, its dorsal spiral lamella; *20* Osseous lamella in lateral frontal sinus

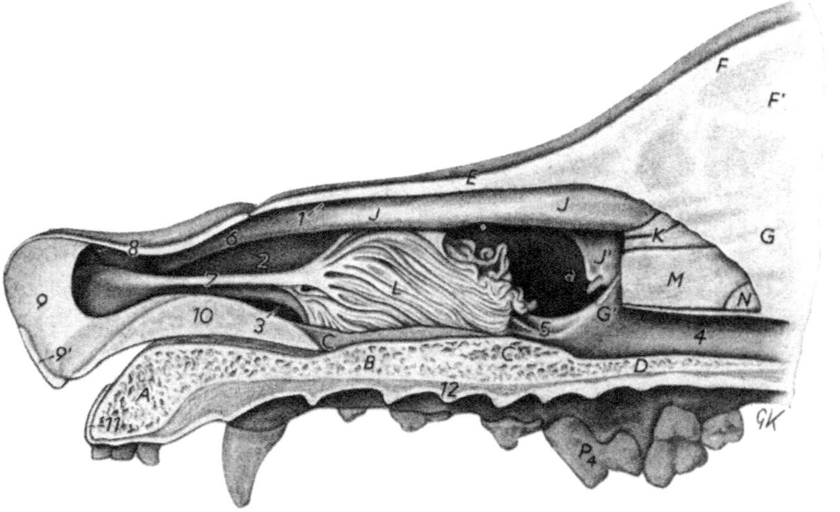

Fig. 284. Nasal cavity of the dog after partial removal of the nasal septum, middle nasal concha, endoturbinate III, and orbital plate of the ethmoid to expose the nasomaxillary opening. Sagittal section. Medial aspect of right half. (After Graeger 1958.)

A Incisive bone; *B* Maxilla; *C* Vomer; *D* Palatine bone; *E* Nasal bone; *F* Frontal bone; *F'* Median septum between frontal sinuses; *G, G'* Perpendicular plate and orbital plate of ethmoid; *J* Dorsal nasal concha; *J'* Uncinate process; *K* Middle nasal concha; *L* Ventral nasal concha; *M, N* Ethmoidal conchae; *N* Endoturbinate III; *M* Endoturbinate IV; P_4 Fourth premolar (sectorial); *a* Maxillary recess;

1 Dorsal nasal meatus; *2* Middle nasal meatus; *3* Ventral nasal meatus; *4* Choana; *5* Mucosal fold, with nasomaxillary opening and uncinate process (*J'*) above; *6* Straight fold; *7* Alar fold; *8* Dorsal lateral nasal cartilage; *9, 10* Nasal septum, *9'* its cut edge; *11* Oral mucosa; *12* Hard palate

endoturbinates I and II respectively*. The nasal conchae, like the paranasal sinuses, vary greatly in shape and size with the species, and, within a species, with the age of the animal. Here follows only a general description of the conchae. For more detail see pages 248, 255, 261, and 271 in the special chapters. The apertures through which the sinuses, recesses, bullae, and cells communicate with the nasal and ethmoidal meatuses are shown in Figures 287—291, and 400.

The **DORSAL NASAL CONCHA** is the longest concha in the domestic mammals. It extends from the cribriform plate to the nasal vestibule and is attached to the ethmoidal crest of the nasal bone. In the carnivores (284/*J*), the middle portion of the concha is coiled, enclosing a recess, while the rostral and short caudal portions consist only of the basal lamella and have the appearance of a shelf or a longitudinal swelling respectively (285, 286/*1*). In the pig and ruminants (287, 290/*0*), the caudal two-thirds of the concha is coiled and encloses the **dorsal conchal sinus**; the rostral third is platelike and consists only of the basal lamella.

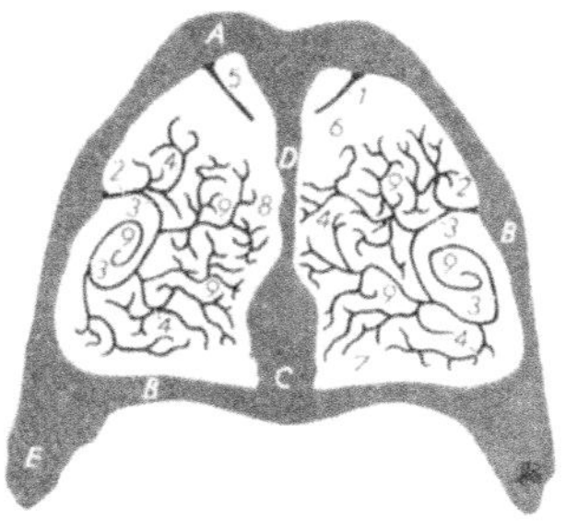

Fig. 285. Transverse section of the nasal cavity of the dog, about 2 cm. rostral to the infraorbital foramen. Rostral aspect. Semischematic. (After Graeger 1958.)

A Nasal bone; *B* Maxilla; *C* Vomer; *D* Nasals eptum; *E* Second premolar; *1* Dorsal nasal concha; *2—4* Ventral nasal concha; *2* basal lamella, *3* spiral lamella, *4* secondary lamellae; *5* Dorsal nasal meatus; *6* Middle nasal meatus; *7* Ventral nasal meatus; *8* Common nasal meatus; *9* Recesses in the ventral nasal concha

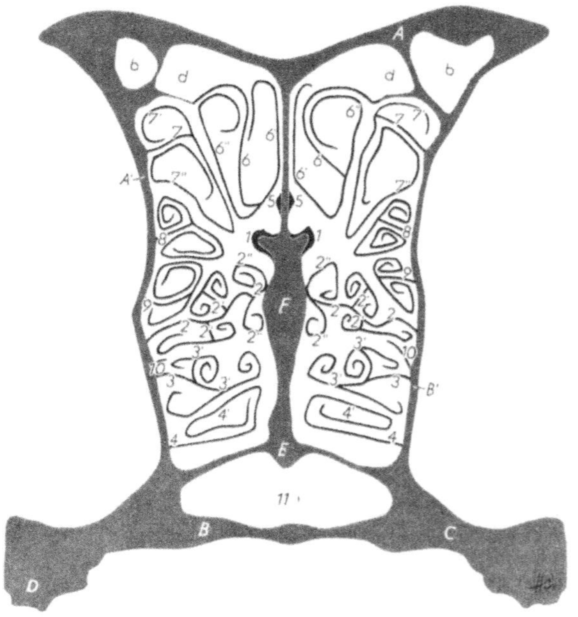

Fig. 286. Transverse section of the nasal cavity of the dog just rostral to the cribriform plate, showing the labyrinth formed by the ethmoidal conchae. Rostral aspect. Semischematic. (After Graeger 1958.)

A, A' Frontal bone; *A'* Orbital part of frontal bone, fused with orbital plate of ethmoid; *B, B'* Palatine bone: *B* its horizontal plate, *B'* its perpendicular plate, fused with orbital plate of ethmoid; *C* Maxilla; *D* Molar tooth; *E* Vomer; *F* Perpendicular plate of ethmoid *b* Lateral frontal sinus; *d* Rostral frontal sinus

1 Dorsal nasal concha; *2* Middle nasal concha, its basal lamella, *2'* secondary lamellae with secondary spiral lamellae, which originate from the periphery of the cribriform plate, *2''* spiral lamellae, which originate from the center of the cribriform plate; *3* Endoturbinate III, its basal lamella, *3'* its double spiral lamellae; *4* Endoturbinate IV, its basal lamella, *4'* its dorsal spiral lamella (its ventral spiral lamella is not present at this level); *5* Ectoturbinate 1; *6* Ectoturbinate 2, its basal lamella, *6'* its dorsomedial spiral lamella, *6''* its ventrolateral spiral lamella; *7* Ectoturbinate 3, its basal lamella, *7'* its dorsal spiral lamella, *7''* its ventral spiral lamella; *8* Ectoturbinate 4, its basal lamella with a dorsolateral and a ventromedial spiral lamella; *9* Ectoturbinate 5, its basal lamella with a dorsal and a ventral spiral lamella; *10* Ectoburbinate 6; *11* Nasopharynx

In the horse, the entire dorsal concha (401/*A*) coils toward the middle nasal meatus, and is divided by a septum (400/*small star*) into rostral and caudal parts. The rostral part encloses a recess and bulla, which is divided into several cells (*i*). The caudal part encloses the **dorsal conchal sinus** (*e*), which combines with the frontal sinus (*d—d'''*) to form the conchofrontal sinus.

---

* The ethmoidal conchae that reach the nasal septum are also known as endoturbinates, and are numbered with Roman numerals; those that do not reach the septum are ectoturbinates, and are numbered with Arabic numerals (Fig. 286).

The **VENTRAL NASAL CONCHA** arises from the conchal crest of the maxilla and is independent of the ethmoidal conchae. In the dog (284/*L*; 285/*2—4*) it presents several branching secondary lamellae that arise from the spiral lamella, creating multiple narrow recesses. In the cat, the ventral concha is very small. In the pig (287/*Q*), two spiral lamellae arise from the basal lamella, one coiling dorsally and the other coiling ventrally. The dorsal encloses a recess (*a*), while the ventral is divided into a recess (*b*) rostrally and the **ventral conchal sinus** (*i*) caudally. Ruminants also have dorsal and ventral spiral lamellae which, in the sheep and goat, enclose recesses (288/*a, b*). Both spiral lamellae in the goat, but only the ventral one in the sheep, have subdivided bullae (*c*) in their free borders. In the ox, dorsal and ventral spiral lamellae enclose recesses and subdivided bullae (290/*m, n, q*) only rostrally, while caudally the two spiral lamellae unite and enclose the **ventral conchal sinus** (*o*). The ventral concha in the horse coils toward the middle meatus and is, like the dorsal concha, divided into rostral and caudal parts by a septum (400/ ○). The rostral part encloses a recess and subdivided bulla (*k, l*), while the caudal forms the **ventral conchal sinus** (*b*) which communicates widely with the rostral maxillary sinus (*a*).

The **MIDDLE NASAL CONCHA** of the carnivores (284/*K*) is long and presents many secondary lamellae internally. In the pig (287/*P*), the concha is small, but there are dorsal and ventral spiral lamellae which enclose recesses. In the ox (290/*l*), the dorsal spiral lamella forms the **middle conchal sinus,** while the ventral encloses a recess. In the sheep and goat, the dorsal spiral lamella of the middle concha has the same arrangement as in the ox; the ventral lamella, however, is divided into a short ventral sinus in front (288/*e'*) and a recess behind (*e''*). In the horse, the concha is small and forms the **middle conchal sinus** (400/*c''*).

Incisive Duct, Vomeronasal Organ, and Lateral Nasal Gland

The **INCISIVE DUCT*** (282/*b*) is a paired tube in the floor of the nasal cavity, which is directed rostroventrally and connects the nasal cavity with the oral cavity. Its nasal opening

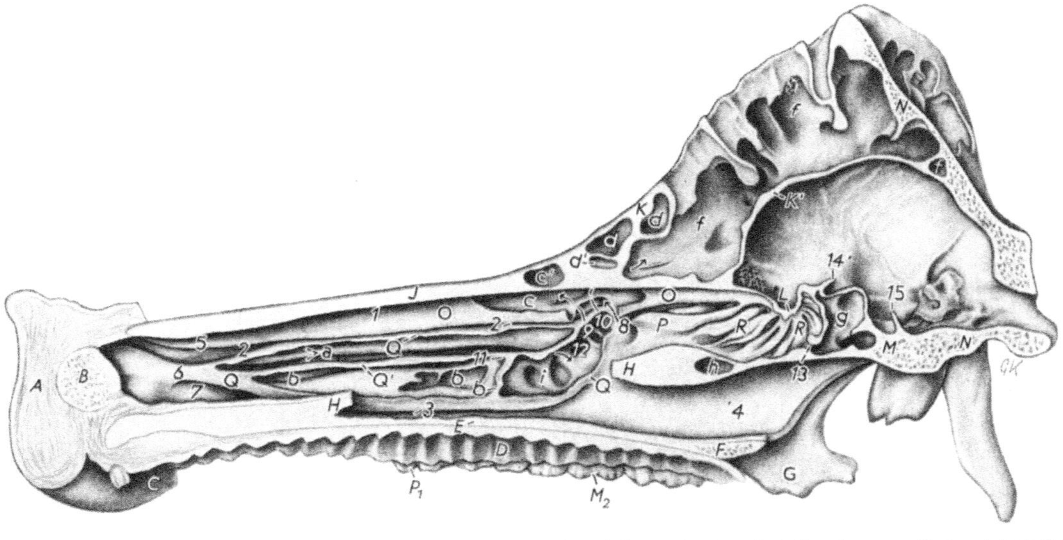

Fig. 287. Nasal cavity of the pig. The nasal conchae have been opened, and the nasal septum including the perpendicular plate of the ethmoid have been removed. Median section. Medial aspect of right half. (After Loeffler 1959a.)

A Rostral portion of nasal septum; *B* Rostral bone; *C* Upper lip; *D* Hard palate; *E* Palatine process of maxilla; *F* Horizontal plate of palatine bone; *G* Pterygoid process of palatine bone; *H, H* Vomer (partly removed); *J* Nasal bone; *K, K'* Frontal bone; *L* Ethmoid; *M* Sphenoid; *N* Occipital bone; *O* Dorsal nasal concha; *P* Middle nasal concha; *Q* Ventral nasal concha, *Q'* its basal lamella; *R* Ethmoidal conchae; P_1 First premolar; M_2 Second molar

a, b Dorsal and ventral recesses of ventral concha; *c, c'* Dorsal conchal sinus; *d* Right medial rostral frontal sinus; *d'* Left medial rostral frontal sinus; *f* Caudal frontal sinus; *g* Sphenoid sinus; *h* Ethmoidal cell; *i* Ventral conchal sinus

1 Dorsal nasal meatus; *2* Middle nasal meatus; *3* Ventral nasal meatus; *4* Nasopharynx; *5* Straight fold; *6* Alar fold; *7* Basal fold; *8* Nasomaxillary aperture; *9* Aperture of dorsal conchal sinus; *10* Aperture of caudal frontal sinus; *11* Aperture of ventral conchal sinus; *12* Aperture of a lateral recess of the ventral conchal sinus (*i*); *13* Aperture of sphenoid sinus; *14* Orbitosphenoid crest; *15* Dorsum sellae turcicae

---

* Formerly nasopalatine duct.

is hidden in the ventral meatus at the level of the canine tooth. The oral opening of the duct is on the **incisive papilla** (26/*1*) just caudal to the upper incisors. In the horse, the duct does not open into the oral cavity, but ends blindly under the oral epithelium.

The **VOMERONASAL ORGAN** (282/*a*) consists of a pair of blind ducts which lie in the floor of the nasal cavity on each side of the nasal septum. The epithelial lining of the ducts resembles that of the nasal cavity and contains elements of both the respiratory and the olfactory regions. The ducts are supported by thin-walled cartilaginous sleeves, but also by the vomer. The ducts extend caudally from their openings in the incisive ducts at about the level of the canine tooth to a level of the second to fourth cheek tooth where they, as mentioned, end blindly. They are 15—20 cm. long in the horse and ox, and 2—7 cm. long in the smaller species. Contrast-radiography (Frewein 1972) shows the vomeronasal organ of the pig to be relatively the smallest. The vomeronasal organ performs special olfactory functions notably the investigation of urinary pheromones. This seems to be related to the "Flehmen" reaction, a peculiar sustained retraction of the upper lip.

The **LATERAL NASAL GLAND,** which is absent in the ox, is not very large in the other domestic mammals. In the pig and carnivores, it is in the maxillary sinus (recess), while in the horse and small ruminants it lies close to the nasomaxillary aperture. Its duct

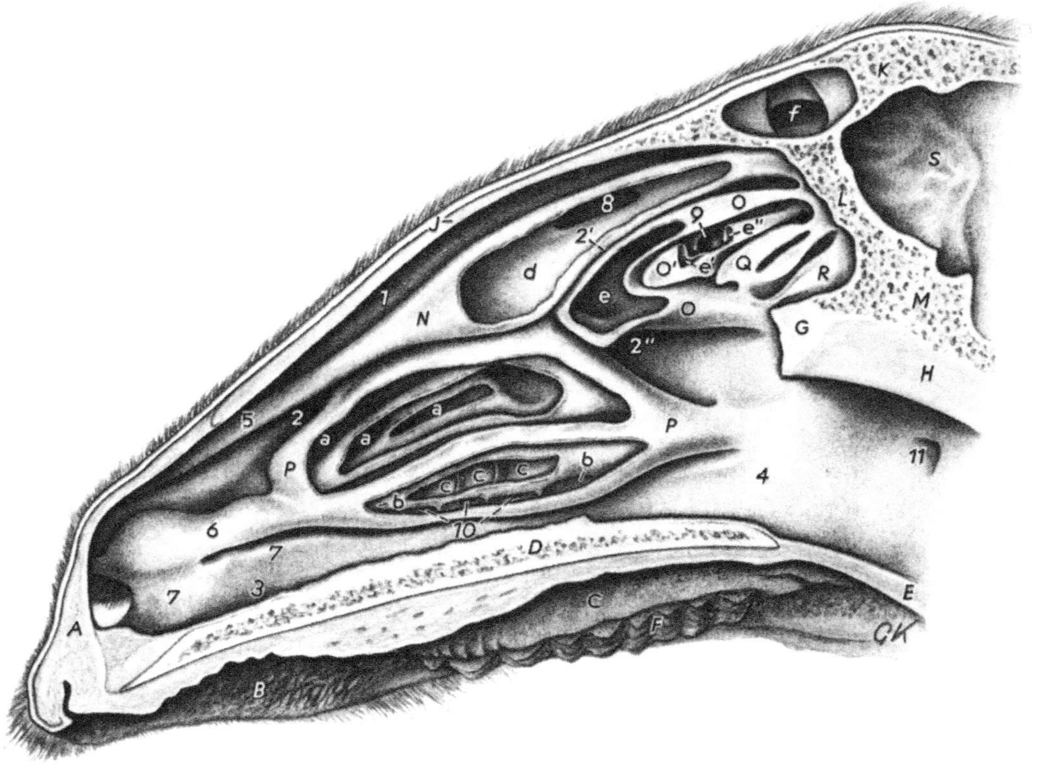

Fig. 288. Nasal cavity of the sheep. The nasal conchae have been opened and the nasal septum and perpendicular plate of the ethmoid have been removed. Median section. Medial aspect of right half. (After Loeffler 1958.)

A Rostral portion of nasal septum; *B* Upper lip; *C* Hard palate; *D* Palatine process of maxilla and horizontal plate of palatine bone, dorsal to them cut edge of vomer (the palatine sinus remains unopened since it does not quite reach the median plane); *E* Soft palate; *F* Second molar; *G* Caudal portion of vomer; *H* Pharyngeal septum; *J* Nasal bone; *K* Frontal bone; *L* Ethmoid; *M* Sphenoid; *N* Dorsal nasal concha; *O, O'* Middle nasal concha, *O'* its ventral spiral lamella; *P* Ventral nasal concha; *Q, R* Ethmoidal conchae; *Q* Endoturbinate III; *R* Endoturbinate IV; *S* Cranial cavity

a, b Dorsal and ventral recesses of ventral concha; *c* Subdivided bulla of ventral concha; *d* Dorsal conchal sinus; *e* Dorsal sinus of middle concha; *e'* Ventral sinus and *e''* recess of middle concha; *f* Lateral frontal sinus

1 Dorsal nasal meatus; *2* Middle nasal meatus, split caudally (*2', 2''*); *3* Ventral nasal meatus; *4* Choana; *5* Straight fold; *6* Alar fold; *7* Basal fold; *8* Aperture of dorsal conchal sinus; *9* Aperture of middle conchal sinus; *10* Apertures of cells in the ventral concha; *11* Pharyngeal opening of auditory tube runs along the middle meatus and opens inside the nostril close to the straight fold or at the end of it; in the horse, the opening is at the level between the first and second cheek tooth. Except in the horse, the secretion of the lateral nasal gland passes through the incisive duct into the oral cavity. The secretion helps moisten the inhaled air, and in the dog also the nose; and it is also believed to play a role in the functioning of the vomeronasal organ.

The **opening of the nasolacrimal duct** is located in the floor of the nostril at the junction of skin and mucosa. There is a second opening of the nasolacrimal duct in the pig—and often in the dog—on the lateral surface of the ventral nasal concha near its caudal end.

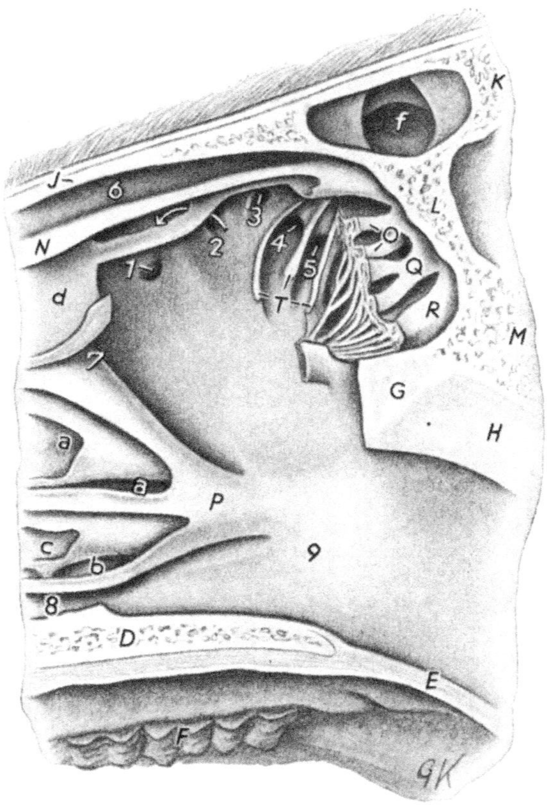

Fig. 289. Apertures of the paranasal sinuses in the sheep. Enlarged section of Figure 288. The dorsal nasal concha has been fenestrated, and the ethmoidal conchae have been removed, leaving their caudal segments and basal lamellae. (After Loeffler 1958.)

D Palatine process of maxilla and horizontal plate of palatine bone; *E* Soft palate; *F* Second molar; *G* Caudal portion of vomer; *H* Pharyngeal septum; *J* Nasal bone; *K* Frontal bone; *L* Ethmoid; *M* Sphenoid; *N* Dorsal nasal concha; *O* Middle nasal concha; *P* Ventral nasal concha; *Q*, *R* Ethmoidal conchae; *Q* Endoturbinate III; *R* Endoturbinate IV; *T* Basal lamellae of further ethmoidal conchae

a, *b* Dorsal and ventral recesses of ventral concha; *c* Subdivided bulla of ventral concha; *d* Dorsal conchal sinus; *f* Lateral frontal sinus

1 Nasomaxillary aperture; *2* Aperture of dorsal conchal sinus; *3* Aperture of medial frontal sinus; *4* Aperture of lateral frontal sinus; *5* Aperture of lacrimal sinus; *6* Dorsal nasal meatus; *7* Middle nasal meatus; *8* Ventral nasal meatus; *9* Choana

Nasopharynx

(57, 58/*c*; 59—62/*b*)

After leaving the nasal cavity, the inspiratory air passes through the **choanae** (60/*13*), two openings separated by the vomer, into the nasopharynx which is dorsal to the soft palate (*36*). Dorsally, the nasopharynx is separated from the caudal part of the nasal cavity by a horizontal partition formed by the ethmoid and palatine bones and by the vomer, and more caudally it is related to the base of the skull. In the pig, the rostral part of the nasopharynx is divided into left and right channels by the median **pharyngeal septum** (59/*12*) which is absent in the carnivores and horse, and in the ruminants (60, 61/*12*) is incomplete and projects ventrally from the roof of the nasopharynx. Caudoventrally, the nasopharynx communicates through the **intrapharyngeal opening** with the laryngopharynx, and is it is here where the respiratory pathway intersects the digestive pathway (see also p. 46). The intrapharyngeal opening (79/*11*) is formed by the free border of the soft palate rostrally and by the palatopharyngeal arches laterally and caudally.

The nasopharynx communicates dorsolaterally with the middle ear through the **auditory tube.** Close to the pharyngeal opening of this tube (59—62/*14*) are the **pharyngeal** and **tubal tonsils** which differ in shape and size with the species. In the horse, and in fact in all Equidae, the auditory tube is greatly dilated to form the **guttural pouch** (62/*14'*).

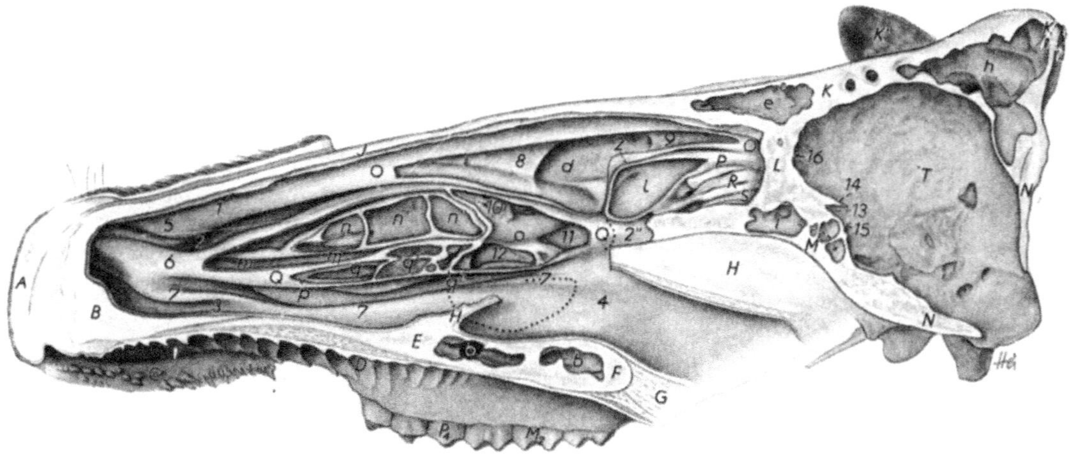

Fig. 290. Nasal cavity of the ox. The nasal conchae have been fenestrated and the nasal septum and perpendicular plate of the ethmoid removed. Median section. Medial aspect of right half. (After Wilkens 1958.)

A Nose; *B* Rostral portion of nasal septum; *C* Upper lip; *D* Hard palate; *E* Palatine process of maxilla; *F* Horizontal plate of palatine bone; *G* Soft palate; *H, H* Vomer; *J* Nasal bone; *K* Frontal bone; *K'* Intercornual protuberance; *K''* Right horn; *L* Ethmoid; *M* Sphenoid; *N* Occipital bone; *O* Dorsal nasal concha; *P* Middle nasal concha; *Q* Ventral nasal concha; *R, S* Ethmoidal conchae; *R* Endoturbinate III; *S* Endoturbinate IV; *T* Cranial cavity; P_4 Fourth premolar; M_2 Second molar

b Palatine sinus; *d* Dorsal conchal sinus; *e* Medial rostral frontal sinus; *h* Caudal frontal sinus; *i* Sphenoid sinus; *l* Middle conchal sinus; *m* Dorsal recess of ventral concha; *n—n''* Dorsal subdivided bulla of ventral concha; *o* Ventral conchal sinus; *p* Ventral recess of ventral concha; *q—q''* Ventral subdivided bulla of ventral concha

1 Dorsal nasal meatus; *2* Middle nasal meatus, split caudally (2', 2''); *3* Ventral nasal meatus; *4* Choana; *5* Straight fold; *6* Alar fold; *7* Basal fold; *8, 9* Osseous lamellae in dorsal nasal concha; *10, 11, 12* Dorsal, middle, and ventral recesses of ventral conchal sinus; *13* Sulcus chiasmatis; *14* Orbitosphenoid crest; *15* Groove to the foramen orbitorotundum; *16* Ethmoid fossa. (The dotted lines indicate the bony defect in the roof of the palatine sinus which is covered with mucosa.)

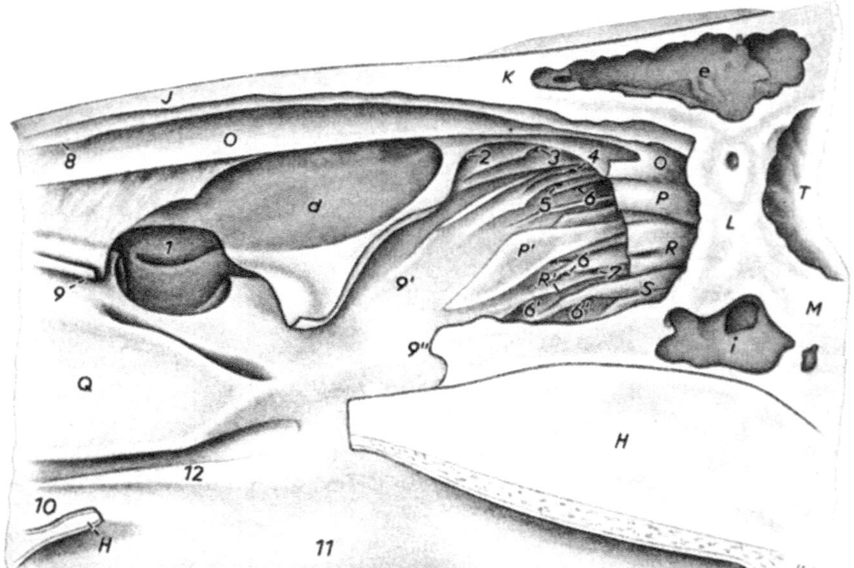

Fig. 291. Apertures of the paranasal sinuses in the ox. Enlarged section of Figure 290. The nasal conchae have been fenestrated, and the ethmoidal conchae have been removed, leaving their caudal segments and basal lamellae. (After Wilkens 1958.)

H, H Vomer; *J* Nasal bone; *K* Frontal bone; *L* Ethmoid; *M* Sphenoid; *O* Dorsal nasal concha; *P, P'* Middle nasal concha, *P'* its basal lamella; *Q* Ventral nasal concha; *R, S* Ethmoidal conchae; *R, R'* Endoturbinate III, *R'* its basal lamella; *S* Endoturbinate IV; *T* Cranial cavity; *d* Dorsal conchal sinus; *e* Medial rostral frontal sinus; *i* Sphenoid sinus

1 Nasomaxillary aperture; *2* Aperture of dorsal conchal sinus; *3* Aperture of medial rostral frontal sinus; *4* Aperture of caudal frontal sinus; *5* Aperture of lateral rostral frontal sinus; *6* Apertures of ethmoidal cells; *6'* Recesses in the lateral wall of the ethmoid; *7* Aperture of sphenoid sinus; *8* Dorsal nasal meatus; *9* Middle nasal meatus, split caudally (9', 9''); *10* Ventral nasal meatus; *11* Choana; *12* Basal fold

Paranasal Sinuses

(292—303, 364—366, 377, 378, 385—387, 389, 400, 402)

The paranasal sinuses are air-filled spaces lined with mucosa that communicate with the nasal cavity. They originate in the embryo from sprouts of the nasal epithelium that grow into the adjacent cranial bones and into some nasal conchae, and by their enlargement hollow out these bones. The sinuses are not fully developed at birth, but continue to enlarge for a long time as the skull matures.

The function of the paranasal sinuses is still not fully understood; a number of theories have been advanced, but none of them is entirely convincing (Blanton 1969). The sinuses do tend to lessen the weight of the head of the large herbivorous animals, especially in those in which extensive surfaces for attachment of the large masticatory muscles and sufficient room for the large teeth are needed. The regional differences in the mucosal lining of the sinuses suggest functions that are as yet unknown. Figures 292—303 illustrate the topography and extent of the paranasal sinuses and show, by direct comparison, how widely they differ in the domestic mammals.

In addition to the smaller **conchal sinuses** described on page 218 the following paranasal sinuses are present: the **maxillary sinus,** the **frontal sinus,** the **palatine sinus,** the **sphenoid sinus,** and the **lacrimal sinus.**

The way in which the paranasal sinuses are connected to the nasal cavity and to each other varies with the species. In general, two types of sinus systems can be distinguished: the first comprises the sinuses that communicate collectively with the middle nasal meatus; the second comprises those that communicate, each with its own opening, with the ethmoidal meatuses in the caudal part of the nasal cavity. In the first type the sinuses are arranged in series, while in the second they are parallel. In the horse, all paranasal sinuses communicate either directly or indirectly with the middle nasal meatus. The rostral and caudal maxillary sinuses open directly into the nasal cavity. The remaining conchofrontal and sphenoid sinuses open into the caudal maxillary sinus. In the ox, the palatine, maxillary, and lacrimal sinuses communicate with the middle nasal meatus; the palatine and maxillary directly and the lacrimal via the maxillary sinus. The remaining frontal, sphenoid, and conchal sinuses in the ruminants open independently into ethmoidal meatuses in the caudal part of the nasal cavity. In the pig, only the maxillary sinus communicates with the middle nasal meatus; the remaining lacrimal and frontal sinuses open separately into ethmoidal meatuses. In the carnivores, the maxillary sinus is a recess which communicates through a large opening with the middle nasal meatus. The frontal and sphenoid sinuses are joined to the ethmoidal meatuses.

The **MAXILLARY SINUS** is represented in the carnivores by the **maxillary recess** (364/*a*; 365/*a, a'*). In contrast to a true sinus, which lies between the external and internal laminae of the facial bones, the recess is bounded laterally by the compact maxillary, lacrimal, and palatine bones, and medially by the orbital plate of the ethmoid. Its wide **nasomaxillary opening** (364/*3*) is narrowed by the **uncinate process** (*J'*), which divides the opening as well as the recess into rostral and caudal parts. The maxillary sinus in the pig invades both the maxillary and the zygomatic bones, and consists of rostral and caudal portions (377, 378/*a, a'*). The **nasomaxillary aperture** (377/*1*) leads into the middle nasal meatus. In the ruminants, the maxillary sinus (387/*a*) is in the maxillary, zygomatic, and in the bulla of the lacrimal bones, and is in wide communication with the palatine sinus (*b*), with which it shares the **nasomaxillary aperture** (*1*) into the middle meatus. The horse has rostral and caudal maxillary sinuses, separated by an osseous septum (402/*9*); both sinuses communicate with the nasal cavity through a common **nasomaxillary aperture** (*1*). The caudal maxillary sinus is subdivided by a ventral partition into a larger ventrolateral (*c*) and a smaller dorsomedial part (*c'*), and communicates widely with the conchofrontal sinus (*d, e*) dorsomedially and with the sphenopalatine sinus (*f, g*) ventromedially.

The **FRONTAL SINUSES** in the carnivores, small ruminants, and horse occupy the dorsal part of the skull between the nasal cavity, the cranial cavity, and the orbits. In the pig and ox, the frontal sinuses extend caudally also into the parietal, interparietal, occipital, and temporal bones, and thus surround the cranial cavity also dorsally, laterally, and caudally. The cornual process of the horned ruminants is also excavated by the frontal sinus. Right and left frontal sinuses are separated by a median **septum,** which deviates slightly from the median plane (389/*6*). The frontal sinuses open separately into the ethmoidal meatuses except in the horse, where the sinus communicates via the caudal maxillary sinus with the middle nasal meatus.

The frontal sinus is undivided in the **cat** (58/*a'''*), while in the **dog** there are three: lateral, medial, and rostral frontal sinuses (365/*b, c, d*), each with its own opening into ethmoidal meatuses. This **pig** has also three frontal sinuses on each side: the large caudal (378/*f*), and the smaller medial and lateral rostral sinuses (*d, e*). They vary considerably in shape and are incompletely subdivided by numerous osseous lamellae. The differences of the frontal sinuses in the **ruminants** are illustrated in Figures 296—300. A transverse septum (389/*7*) divides the frontal sinuses of the **ox** into rostral and caudal sinuses. The rostral are the lateral rostral (*g*), the medial rostral (*e*), and the inconstant intermediate rostral (*f*) frontal sinuses. The much larger caudal frontal sinus (*h, h'*) is subdivided by an incomplete oblique lamina (*8*) into rostromedial and caudolateral parts and extends also into the cornual process of the frontal bone (*10*). The **small ruminants** have only two frontal sinuses, a small medial and a large lateral (386/*e, f*), which extend caudally only to the level of the zygomatic process of the frontal bone. The shape of the medial frontal sinus of the **sheep** varies considerably from one side to the other; in the **goat** this sinus is more uniform. The lateral sinus extends into the cornual process in horned animals. In the **horse,** the frontal sinus occupies mainly the frontal bone (402/*d—d'''*). Transverse lamellae partly divide the sinus into rostral, medial, and caudal recesses which communicate with one another. Sometimes there is a nasal recess extending into the nasal bone. Rostromedially, the frontal sinus communicates widely with the dorsal conchal sinus forming the conchofrontal sinus. The frontal sinus of the horse communicates via the caudal maxillary sinus with the nasal cavity.

The **LACRIMAL SINUS** is present only in the pig and ruminants. In the pig (378/*b*), it occupies the lacrimal bone and, as a rule, has its own access to an ethmoidal meatus, but may be connected to the lateral rostral frontal sinus (*e', e' on right side of skull*). In the ox, the lacrimal sinus communicates with the maxillary sinus (387/*3*), and in the small ruminants it is either independent with its own aperture leading into an ethmoidal meatus (386/*c, 4*), or it may be a lateral recess of the lateral frontal sinus (296/*c, f*).

The **PALATINE SINUS** is absent in the carnivores and pig. In the ruminants it occupies the horizontal lamina of the palatine bone and the palatine process of the maxilla. Right and left sinuses of the ox extend to the median plane, but are separated by a septum; they do not extend so far medially in the small ruminants. Because the osseous roof of the palatine sinus is incomplete, the sinus is separated from the nasal cavity only by a double layer of mucosa (290/*dotted line*). The palatine and maxillary sinuses communicate with one another across the infraorbital canal through the wide **maxillopalatine aperture** (387/*2*) and share a common **nasomaxillary aperture** (*1*) for access to the nasal cavity. The palatine sinus of the horse occupies the perpendicular plate of the palatine bone, but is bounded dorsally also by the ethmoid and the vomer. It opens rostrally into the caudal maxillary sinus, and, since it is continuous caudally with the sphenoid sinus, it is commonly referred to as the **sphenopalatine sinus** (301/*f, g*).

Right and left **SPHENOID SINUSES** are relatively small, of unequal size, and are separated by a median septum; they are not present in the dog and small ruminants. In the cat, the sphenoid sinus occupies the presphenoid and contains part of endoturbinate IV, which projects into it from the nasal cavity. The sphenoid sinus is relatively large in the pig and opens into an ethmoidal meatus (287/*13, g*). In the fully developed pig, the central cavity of the sinus is in the body of the presphenoid and basisphenoid bones. Opening into the central cavity are three recesses (295/*g—g''*) of which the lateral (*g*) is the largest. In the ox (298/*7, 8*), the sphenoid sinus is present in over fifty per cent of cases and occupies the body and wing of the presphenoid. The ductlike rostral portion (*6*) lies ventrolateral to the ethmoidal labyrinth and makes the connection with the ethmoidal meatuses (387/*i, 5*). In the horse, the sphenoid sinus is located in the presphenoid, but it may be absent. As a rule it communicates

PLATE V

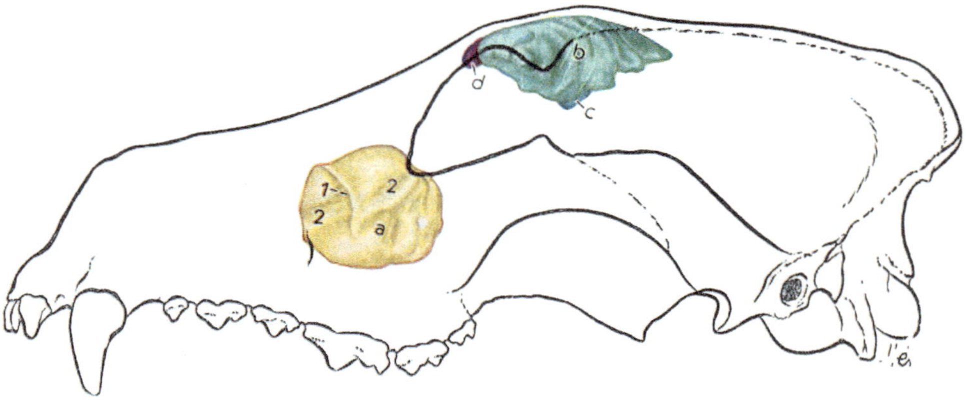

Fig. 292. "Plastoid" cast of the paranasal sinuses of the dog. Lateral aspect. (After Graeger 1958.) *a* Maxillary recess; *b* Lateral frontal sinus; *c* Medial frontal sinus; *d* Rostral frontal sinus. *1* Groove for the uncinate process; *2* Groove for the bony lacrimal canal

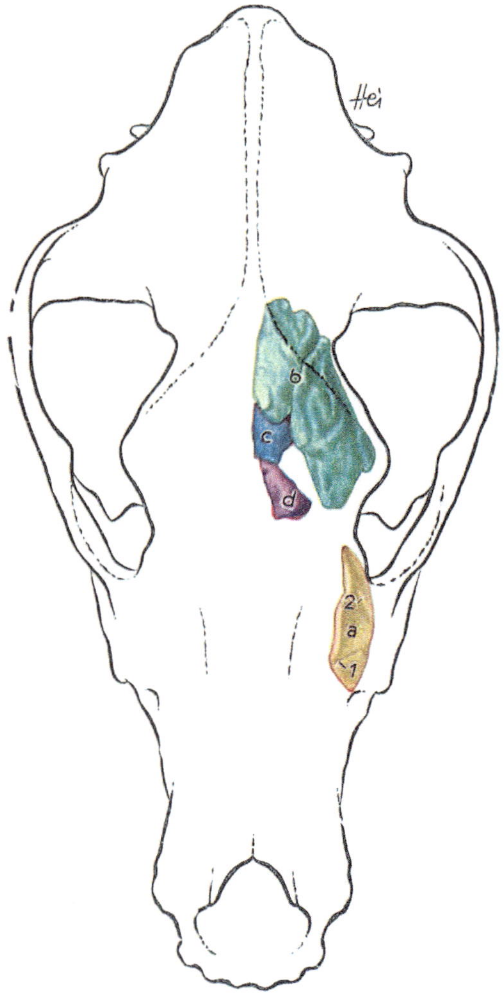

Fig. 293. "Plastoid" cast of the paranasal sinuses of the dog. Dorsal aspect. (After Graeger 1958.) *a* Maxillary recess; *b* Lateral frontal sinus; *c* Medial frontal sinus; *d* Rostral frontal sinus; *1* Groove for the uncinate process; *2* Groove for the bony lacrimal canal

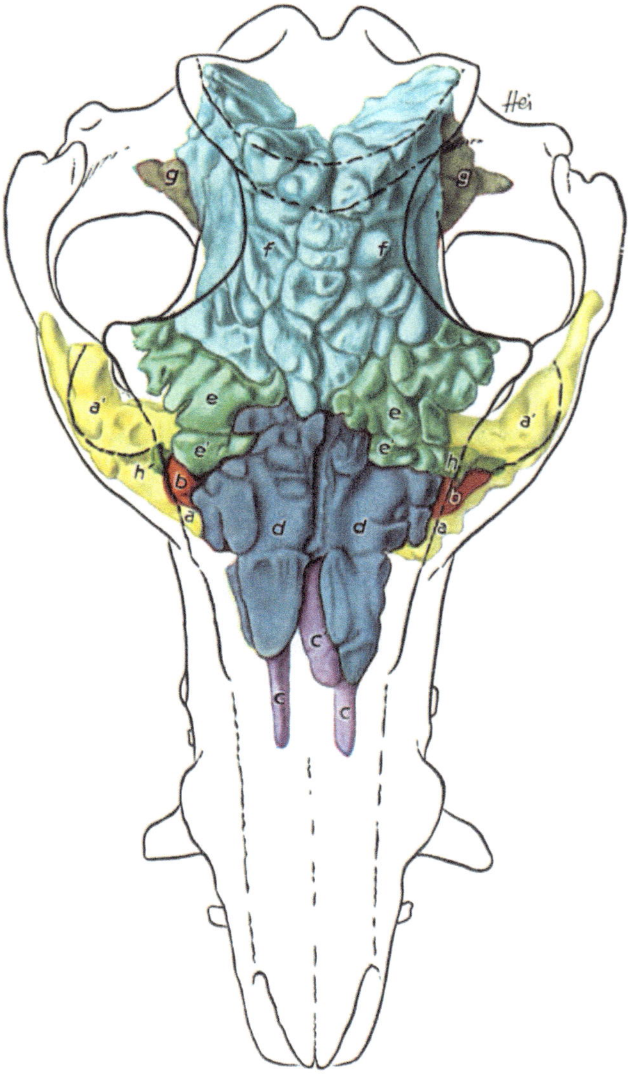

Fig. 294 (Pig, dorsal aspect)

Figs. 294, 295. "Plastoid" casts of the paranasal sinuses of a two year old pig. (After Loeffler 1959a.)

a Maxillary sinus, *a'* its zygomatic portion; *b* Lacrimal sinus; *c* Dorsal conchal sinus, its conchal portion, *c'* its nasal portion; *d* Medial rostral frontal sinus; *e* Lateral rostral frontal sinus, *e'* portion of this sinus with its own aperture into the nasal cavity, *e''* portion of this sinus in the lateral wall of the nose; *f* Caudal frontal sinus; *g* Sphenoid sinus, *g'* its diverticulum in the pterygoid process of the basisphenoid, *g''* its caudal recess; *h* Ethmoidal cells; *1* Nasomaxillary aperture

Fig. 295 (Pig, lateral aspect)

PLATE VII

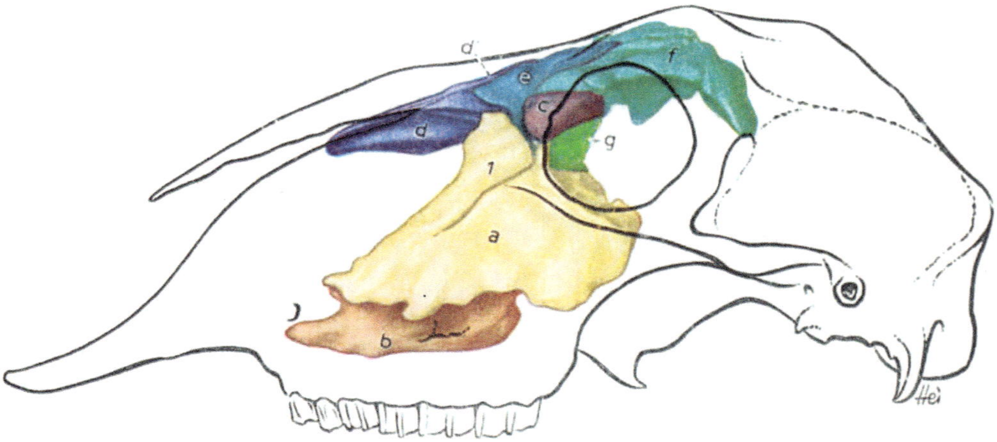

Fig. 296 (Sheep, lateral aspect)

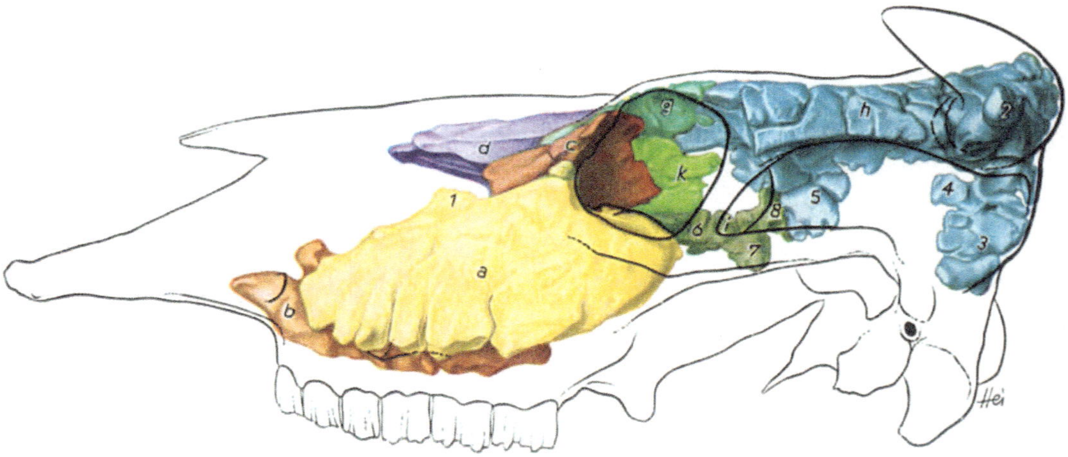

Fig. 297 (Ox, lateral aspect)

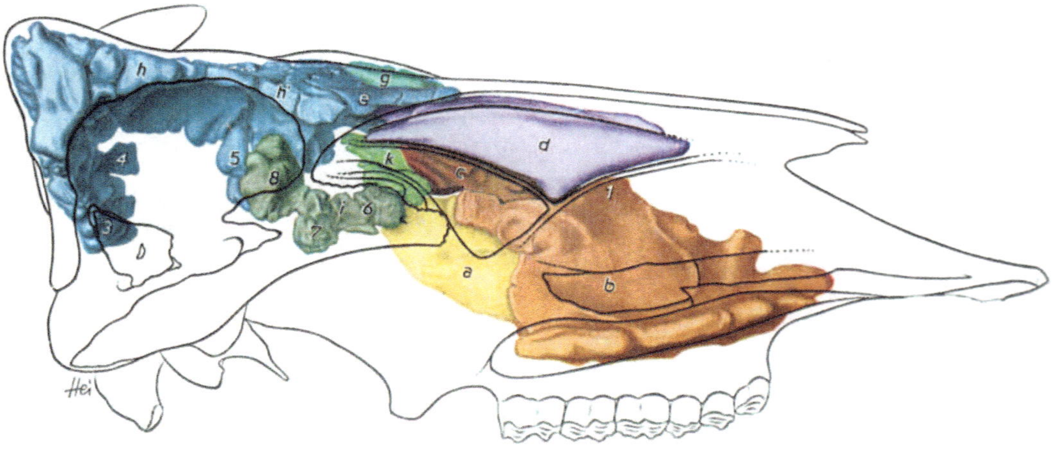

Fig. 298 (Ox, medial aspect)

PLATE VIII

Figs. 296, 299. "Plastoid" casts of the left paranasal sinuses of a sheep. (After Loeffler 1958.)

a Maxillary sinus; *b* Palatine sinus; *c* Lacrimal sinus, a diverticulum of the lateral frontal sinus in this specimen; *d* Dorsal conchal sinus, its conchal portion, *d'* its frontal portion; *e* Medial frontal sinus; *f* Lateral frontal sinus; *g* Ethmoidal cells;

1 Nasomaxillary aperture

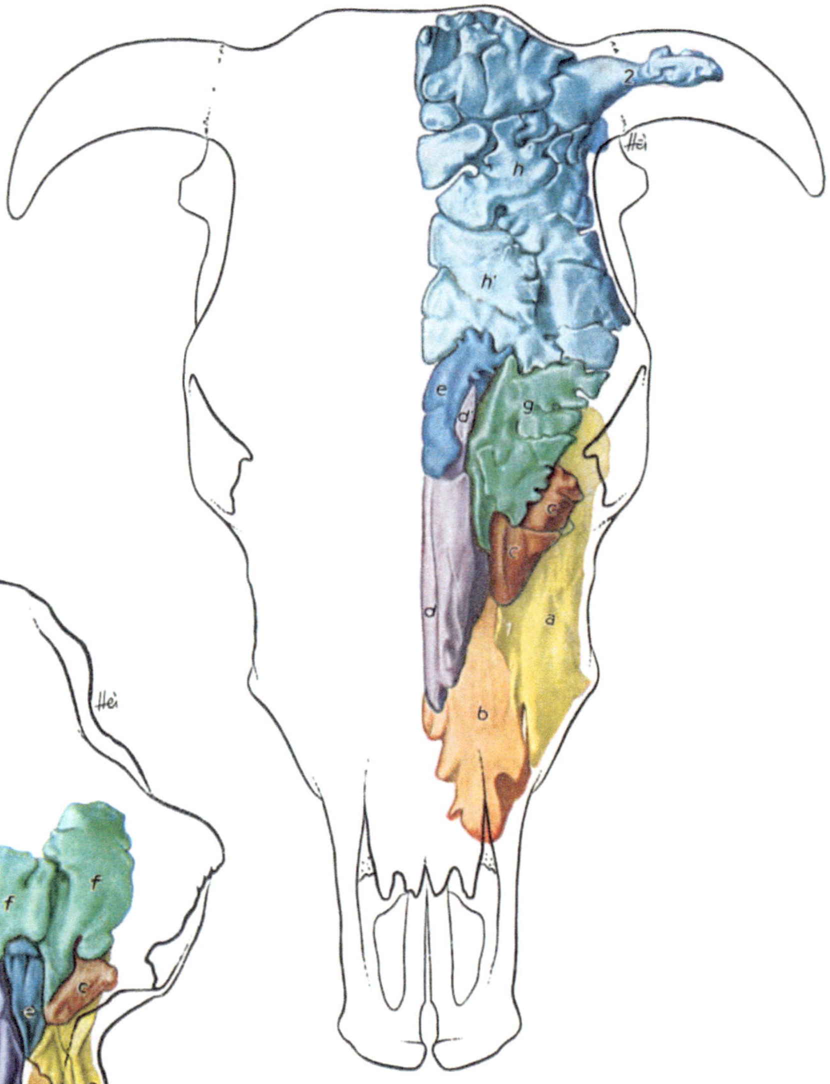

Fig. 300 (Ox, dorsal aspect)

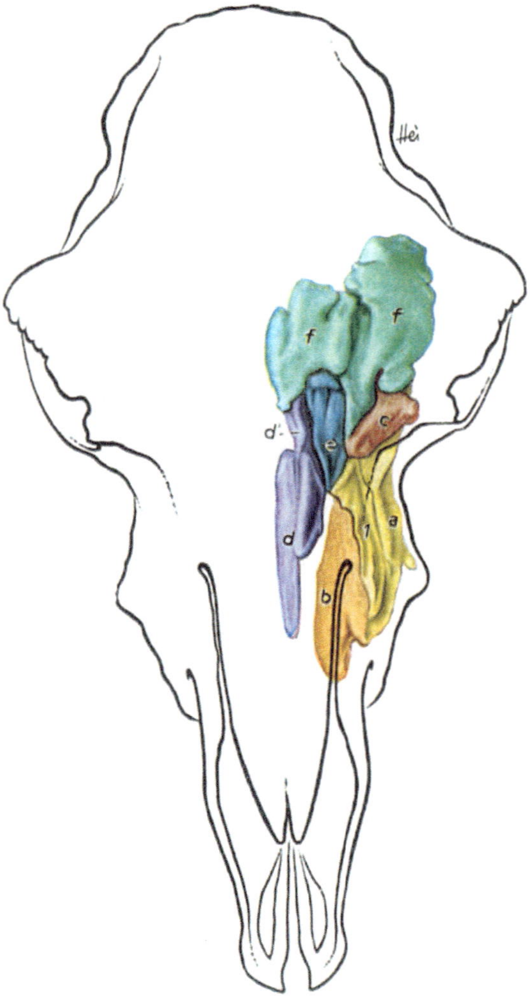

Fig. 299 (Sheep, dorsal aspect)

Figs. 297, 298, 300. "Plastoid" casts of the left paranasal sinuses of an eight-year-old cow. (After Wilkens 1958.)

a Maxillary sinus; *b* Palatine sinus; *c* Lacrimal sinus; *d* Dorsal conchal sinus, *d'* its frontal portion; *e* Medial rostral frontal sinus; *g* Lateral rostral frontal sinus; *h* Caudal frontal sinus, its caudolateral portion, *h'* its rostromedial portion; *i* Sphenoid sinus; *k* Ethmoidal cells

1 Nasomaxillary aperture; *2* Cornual, *3*, *4* nuchal, and *5* postorbital diverticula of caudal frontal sinus; *6*–*8* Sphenoid sinus, *6* rostral portion, *7*, *8* caudal portion, *7* ventral recess, *8* dorsal recess

PLATE IX

Fig. 301. "Plastoid" cast of the left paranasal sinuses of an 18-year-old horse. Lateral aspect. (After Nickel and Wilkens 1958.)

a Rostral maxillary sinus; *b* Ventral conchal sinus; *c* Caudal maxillary sinus, its ventrolateral portion, *c'* its dorsomedial portion; *d, d', d''* Frontal sinus; *e* Dorsal conchal sinus; *f, g* Sphenopalatine sinus

1 Infraorbital foramen; *2* Facial crest; *3* Maxillary tuberosity; *4* Pterygoid process of basisphenoid; *5* Septum of the maxillary sinuses; *6* Nasolacrimal canal

Fig. 302. "Plastoid" cast of the left paranasal sinuses of an 18-year-old horse. Medial aspect. (After Nickel and Wilkens 1958.)

a Rostral maxillary sinus; *b* Ventral conchal sinus; *c* Caudal maxillary sinus, its ventrolateral portion, *c'* its dorsomedial portion; *c''* Middle conchal sinus; *d, d', d''* Frontal sinus; *e* Dorsal conchal sinus; *f, g* Sphenopalatine sinus; *i* Bulla and cells of dorsal nasal concha; *l* Bulla of ventral concha

1 Position of dorsal nasal concha; *2* Position of ventral nasal concha; *3* Rostroventral boundary of ethmoid labyrinth; *4* Basisphenoid; *5* Basilar part of occipital bone; *6* Sphenopalatine foramen; *7* Palatine canal

Fig. 303. "Plastoid" cast of the paranasal sinuses of an 18-year-old horse. Dorsal aspect. (After Nickel and Wilkens 1958.)

a Rostral maxillary sinus; *b* Ventral conchal sinus; *c* Caudal maxillary sinus; *d, d', d''* Frontal sinus; *e* Dorsal conchal sinus; *f* Palatine sinus

1 Infraorbital foramen; *1'* Infraorbital canal; *2* Facial crest; *3* Septum of the maxillary sinuses; *4* Nasolacrimal canal; *5* Supraorbital foramen; *6* Orbit

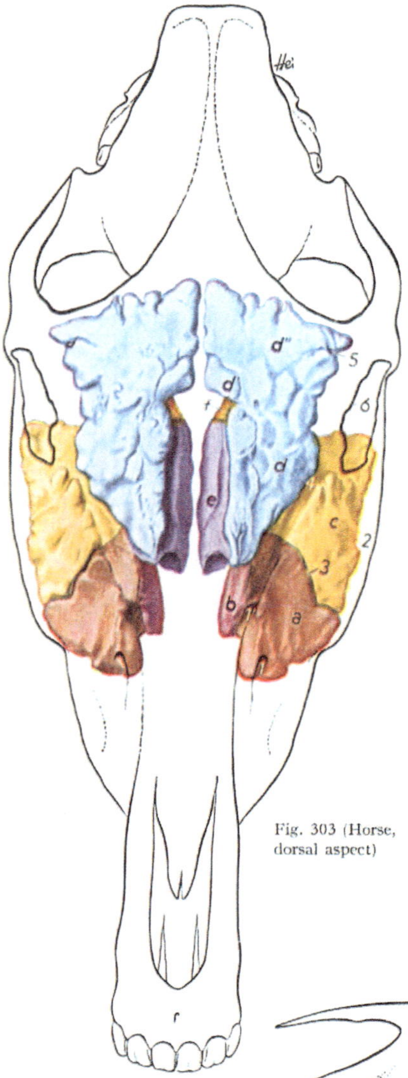

Fig. 303 (Horse, dorsal aspect)

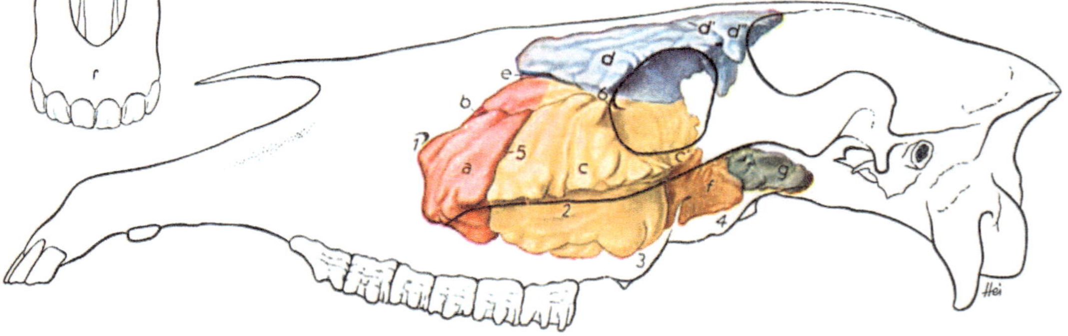

Fig. 301 (Horse, lateral aspect)

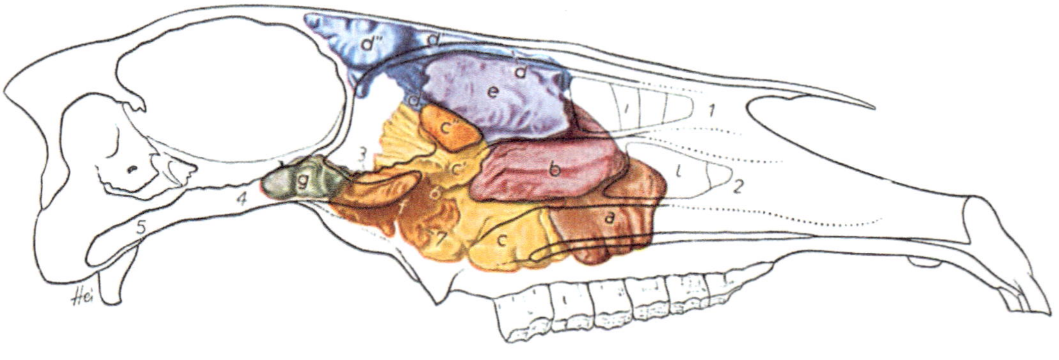

Fig. 302 (Horse, medial aspect)

rostrally with the palatine sinus forming the **sphenopalatine sinus** (402/*f*, *g*), but it may open independently into an ethmoidal meatus.

The part of the sphenoid sinus occupying the presphenoid is closely related to the optic nerve and chiasma, so that inflammation of the sinus may affect the vision of the animal.

A varying number of small sinuses (cellulae ethmoidales) are present in the pig and ruminants. They are related to the ethmoid bone and open separately into ethmoidal meatuses. In the ruminant they lie in the medial wall of the orbit (387/*k*), in the pig they lie in the skull rostral to the orbit (377/*h*).

Larynx

The larynx is a short cartilaginous tube that connects the lower part of the pharynx with the trachea, and contains the organ of phonation. Its rostral opening can be closed for the protection of trachea and lungs, especially during deglutition. The larynx of the domestic mammals is at the level of the base of the cranium ventral to the laryngopharynx and the beginning of the esophagus, still within the intermandibular space in the ruminants and horse, but more caudal in the other species.

The larynx consists of several cartilages lined on the inside with mucous membrane. The cartilages are connected to each other, to the hyoid bone, and to the trachea by ligaments and muscles, and become partly ossified with age.

The cartilages, ligaments, muscles, nerves, and vessels of the larynx develop from branchial arch material, while the mucous lining is derived from the lung bud, which leaves the ventral surface of the foregut. The origin of this bud from the ventral surface of the foregut brings about the crossing of the digestive and the respiratory pathways in the pharynx. Respiratory air entering the nasal cavities and nasopharynx *dorsal* to the digestive tube must pass to the *ventral* side of the pharynx in order to reach the larynx and trachea.

The branchial arches that take part in the formation of the larynx each receive a branch from the vagus, the **cranial** and **caudal laryngeal nerves.** The caudal laryngeal nerves pass to the primordium of the larynx caudal to the last (sixth) pair of aortic arches. During the development and relocation of the heart, these aortic arch arteries—and with them the caudal laryngeal nerves—are carried into the thorax. The right artery, while being transformed into the right pulmonary artery, loses its connection with the dorsal aorta and releases the right caudal laryngeal nerve to be caught up by the fourth arch artery, which becomes the right subclavian artery. On the left side, the connection between the last aortic arch and the dorsal aorta remains in the form of the ductus and later the ligamentum arteriosum, and the nerve on that side is not released. Thus, the caudal laryngeal nerves, which in the embryo pass directly from the vagus to the larynx, postnatally descend in the neck, loop around the subclavian artery on the right and the ligamentum arteriosum on the left, and return to the larynx. For this reason they are also known as the **recurrent laryngeal nerves.**

The cranial laryngeal nerves, which pass directly to the larynx, innervate the mucous membrane of the larynx, while the caudal (recurrent) laryngeal nerves innervate the laryngeal muscles except the cricothyroideus.

Cartilages of the Larynx

(306—329)

The skeleton of the larynx is composed of the following cartilages: the unpaired cricoid cartilage caudally, the unpaired thyroid cartilage ventrally and laterally, the paired arytenoid cartilages dorsally, and the unpaired epiglottis rostrally. The epiglottis fits like a lid over the entrance of the larynx and closes it during swallowing. In addition to these there are the interarytenoid cartilages and the inconstant sesamoid cartilages. The corniculate and cuneiform cartilages of man are processes of the arytenoid or epiglottic cartilages in the domestic mammals. The cricoid, thyroid, and the body of the arytenoid cartilages consist of hyaline cartilage, while the epiglottis, cuneiform, and the vocal and corniculate processes of the arytenoid cartilages consists of elastic cartilage.

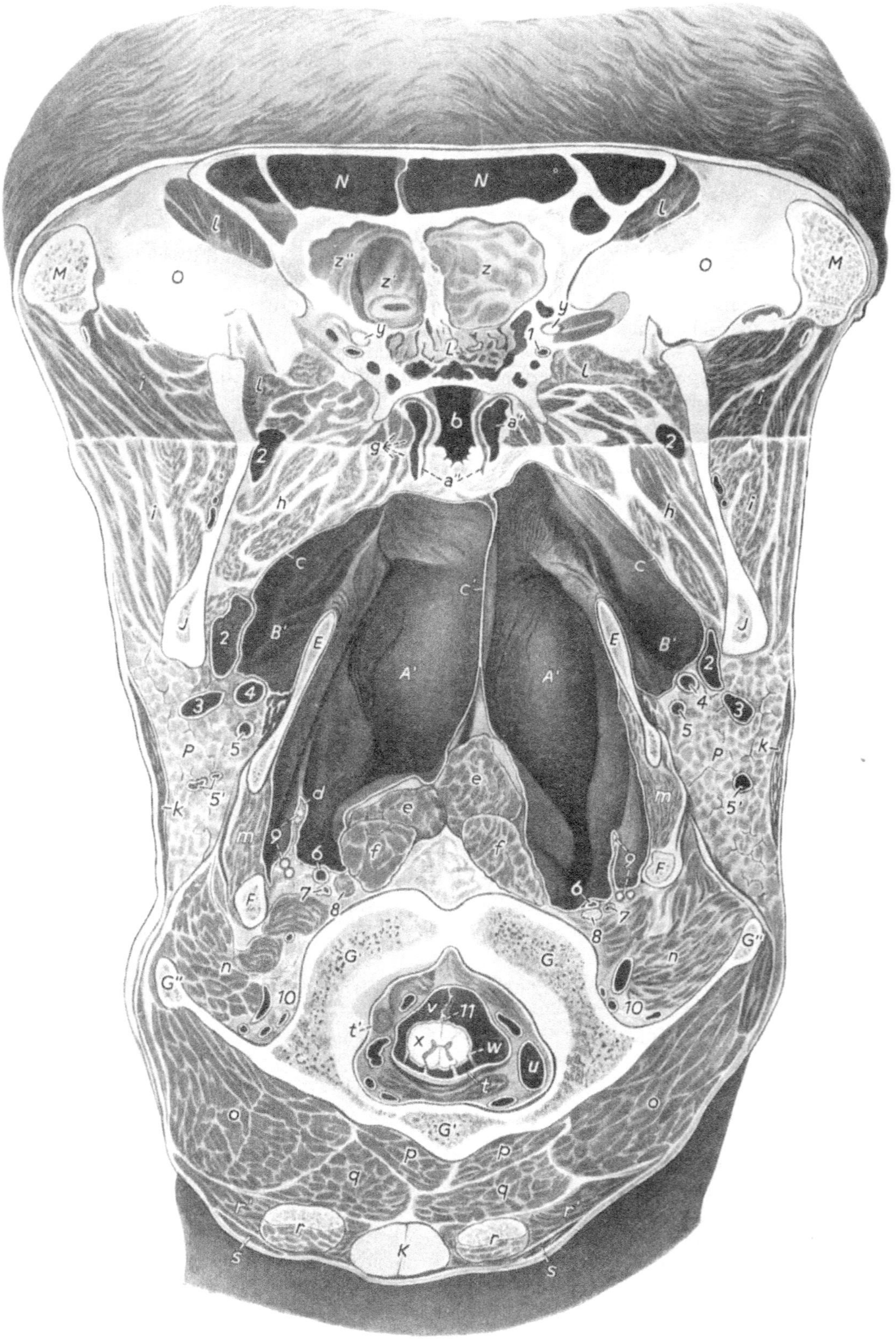

Fig. 304 (Caudodorsal aspect)

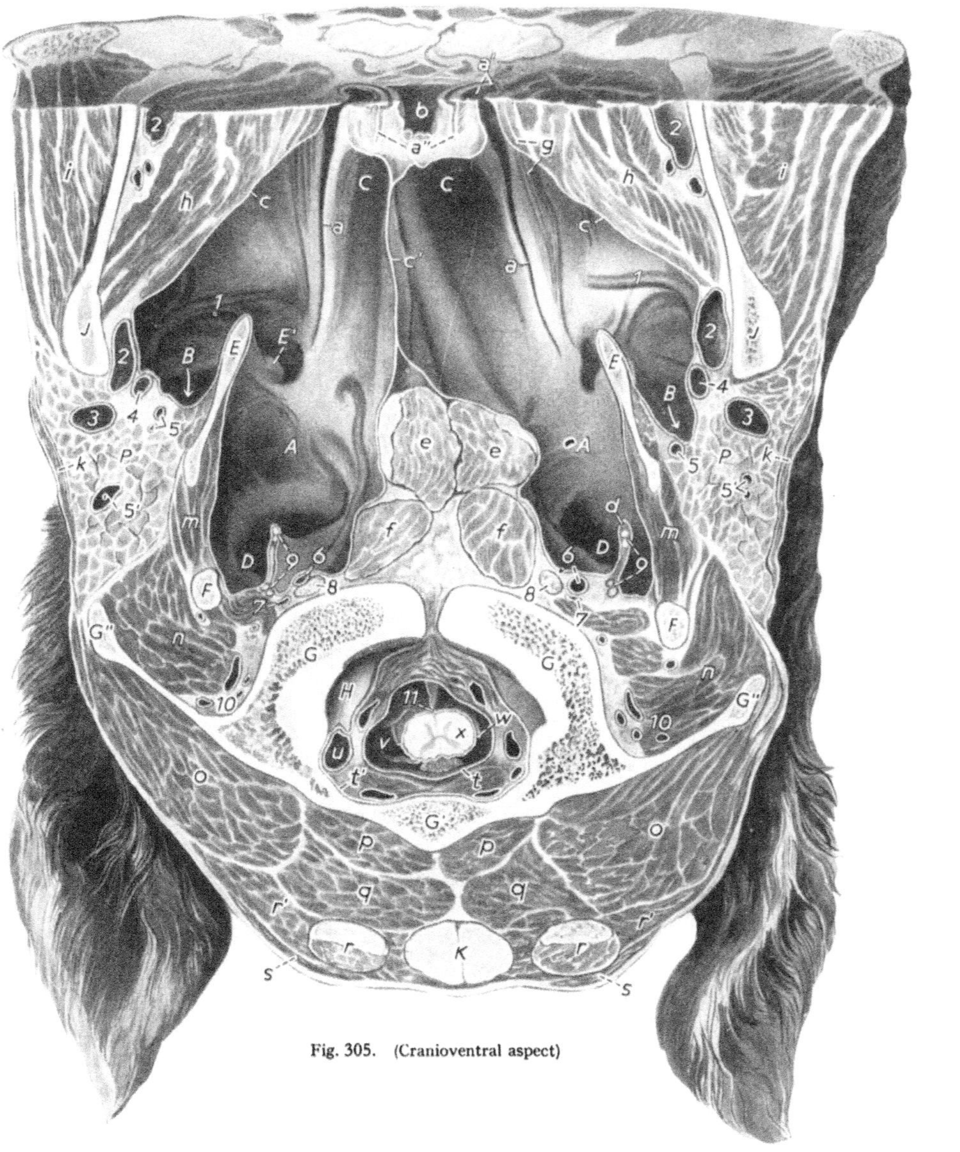

Fig. 305. (Cranioventral aspect)

Figs. 304 and 305. Guttural pouch of the horse. Opened by dorsal section. (After Zietzschmann 1943.)

A Medial, *B* lateral recess of guttural pouch; *A'* Floor of medial recess, related to the pharynx; *B'* Floor of lateral recess; *C* Rostral, *D* caudal recess on the roof of the guttural pouch; *E* Stylohyoid; *E'* Tympanohyoid; *F* Paracondylar process of occipital bone; *G* Atlas, ventral part of its cranial articular processes, *G'* dorsal tubercle, *G''* transverse process (wing); *H* Occipital condyle; *J* Ramus of mandible; *K* Ligamentum nuchae; *L* Ethmoid; *M* Zygomatic arch; *N* Frontal bones and sinuses; *O* Retro-orbital fat; *P* Parotid salivary gland

a Auditory tube; *a'* Transverse section of auditory tube near its pharyngeal opening, surrounded by cartilage of auditory tube; *a''* Medial lip of cartilage of auditory tube; *b* Pharyngeal recess; *c, c'* Mucosa of guttural pouches; *d* Fold of mucosa containing the ninth to twelfth cranial nerves; *e* Longus capitis; *f* Rectus capitis ventralis; *g* Levator and tensor veli palatini; *h* Pterygoideus; *i* Masseter; *k* Parotidoauricularis; *l* Temporalis; *m* Occipitohyoideus and digastricus; *n, o* Obliquus capitis cranialis and caudalis; *p, q* Rectus capitis dorsalis major, its deep and superficial part; *r, r'* Semispinalis capitis, *r* biventer cervicis, *r'* complexus; *s* Splenius; *t* Dura mater; *t'* Endostium; *u* Blood vessel in epidural space; *v* Subarachnoid space; *w* Spinal roots of accessory nerve; *y* Optic nerve; *z* Telencephalon; *z'* Olfactory bulb; *z''* Falx rhinencephali

1 Maxillary artery medial to pterygoideus on the roof of the guttural pouch; *2* Maxillary vein; *3* Common trunk of the supf. termporal, dorsal cerebral, and transverse facial veins; *4* Ext. carotid artery; *5, 5'* Caudal auricular artery and vein; *6* Int. carotid artery; *7* Ventral cerebral vein; *8* Cranial cervical ganglion; *9* Ninth to twelfth cranial nerves; *10* Branches of occipital artery and vein under the wing of the atlas; *11* Ventral spinal artery

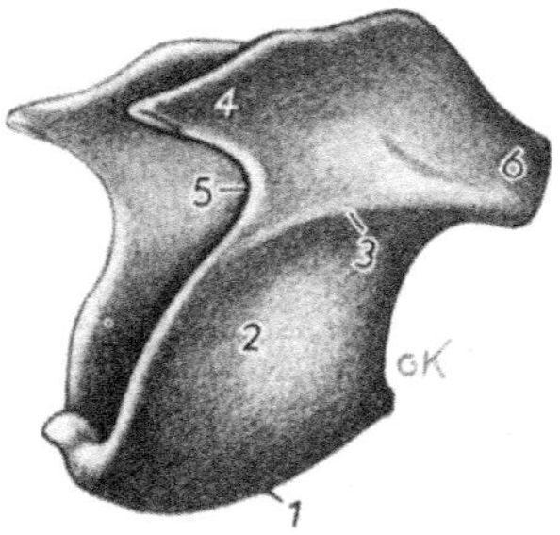

Fig. 306 (Dog)

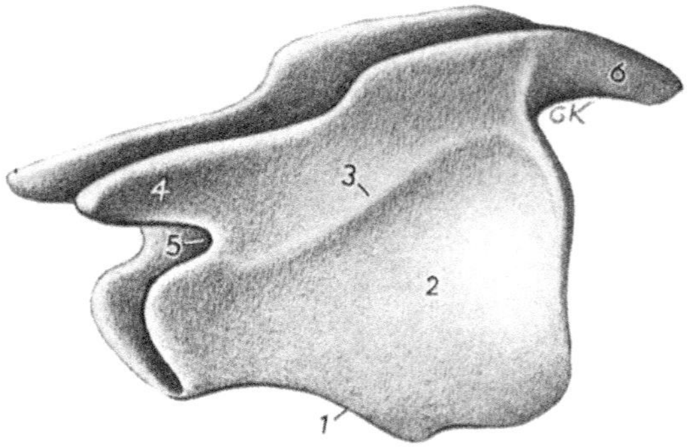

Fig. 308 (Ox)

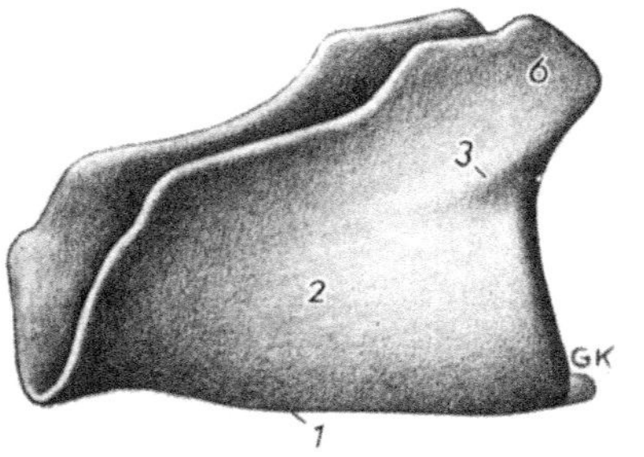

Fig. 307 (Pig)

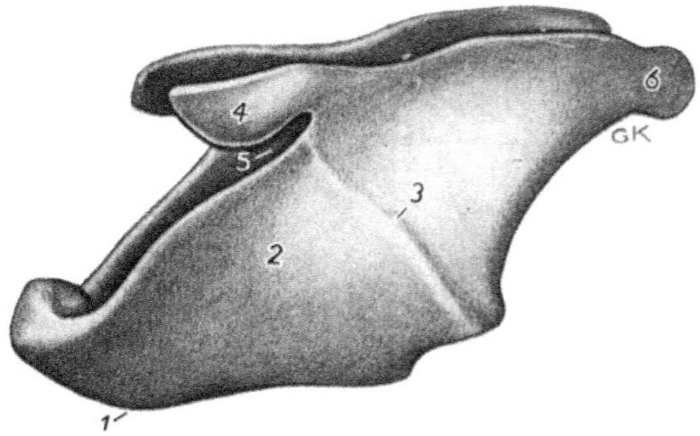

Fig. 309 (Horse)

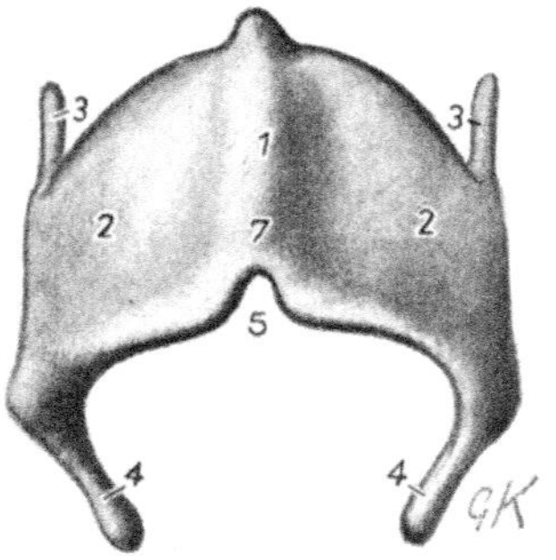

Fig. 310 (Dog)

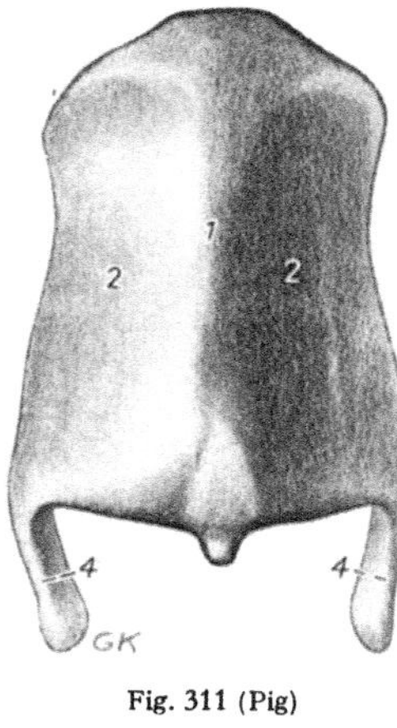

Fig. 311 (Pig)

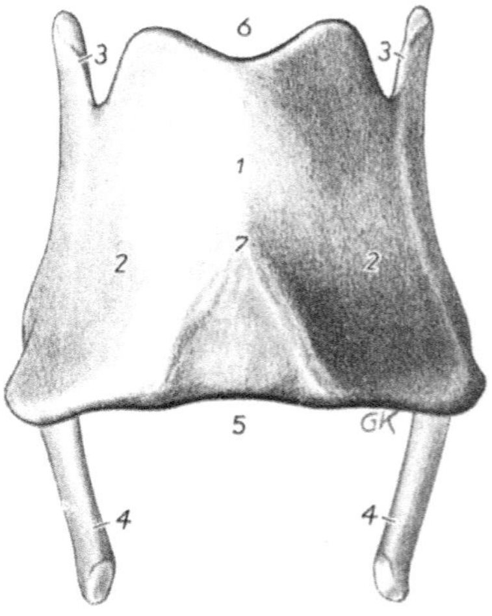

Fig. 312 (Ox)

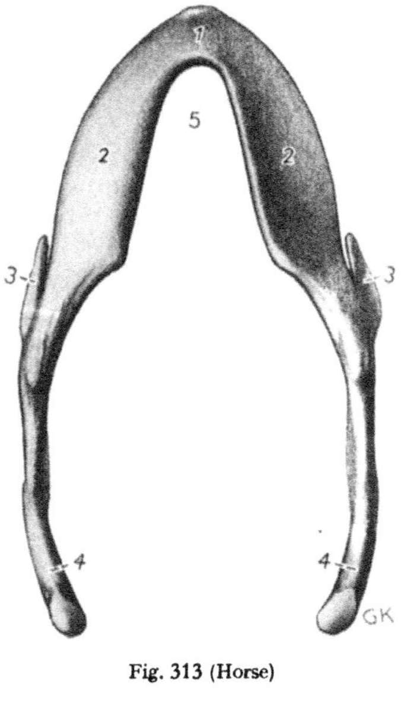

Fig. 313 (Horse)

Figs. 306—309. Thyroid cartilage. Lateral aspect.

1 Junction of laminae; *2* Left lamina; *3* Oblique line; *4* Rostral cornu (except in the pig); *5* Thyroid fissure (except in the pig); *6* Caudal cornu

Figs. 310—313. Thyroid cartilage. Ventral aspect.

1 Junction of laminae; *2* Lamina; *3* Rostral cornu (except in the pig); *4* Caudal cornu; *5* Caudal thyroid notch (except in the pig); *6* Rostral thyroid notch; *7* Laryngeal prominence

The **THYROID CARTILAGE** (306—313) consists of right and left **laminae,** which are united ventrally, and partly encloses the cricoid and arytenoid cartilages from below and from the sides. The slightly convex external surface of each lamina is divided by an **oblique line** (306—309/*3*) into two areas, which serve for the attachment of muscles. Dorsally, the thyroid

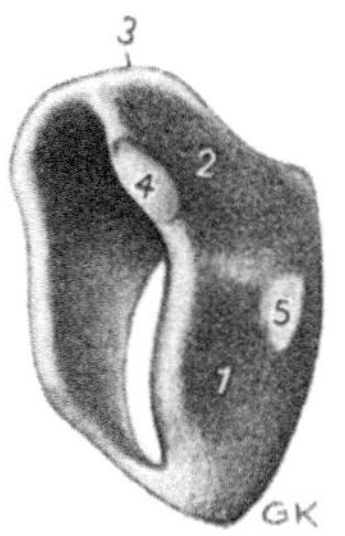

Fig. 314 (Dog)

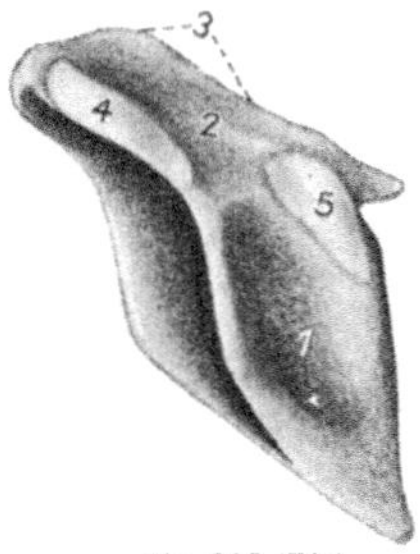

Fig. 315 (Pig)

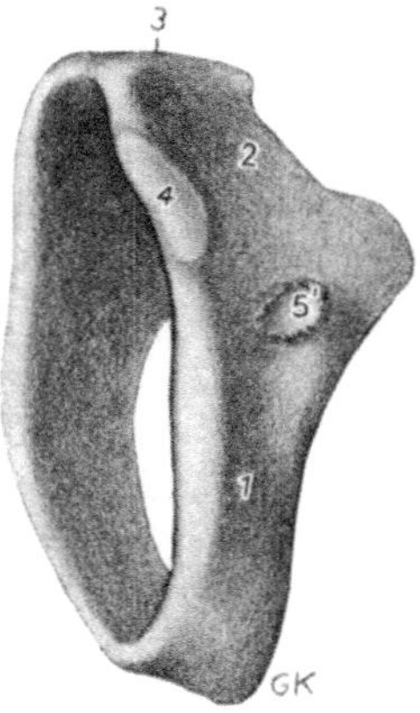

Fig. 316 (Ox)

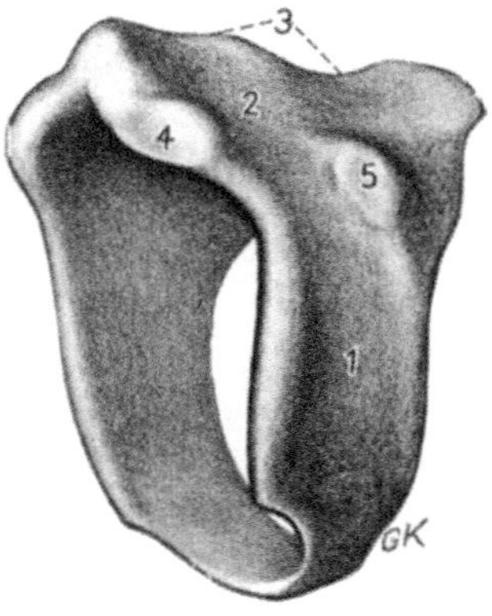

Fig. 317 (Horse)

Figs. 314—317. Cricoid cartilage. Rostrolateral aspect.

1 Cricoid arch; *2* Cricoid lamina; *3* Median crest; *4* Articular surface for arytenoid cartilage; *5*, *5'* Articular surface for thyroid cartilage (roughened area in the ox)

lamina is expanded to form rostral and caudal cornua (*4*, *6*). The **caudal cornu** articulates with the cricoid cartilage, and the **rostral cornu,** which is absent in the pig, articulates with the thyrohyoid. The **thyroid fissure** (*5*) separates the rostral cornu from the rostral border of the cartilage and, except in the carnivores, is bridged over by connective tissue, leaving a small foramen at the depth of the fissure. This is the **thyroid foramen** which transmits the sensory fibers of the cranial laryngeal nerve to the interior of the larynx for the innervation of the mucous membrane. In the ruminants, a shallow **rostral thyroid notch** is formed at the union of the two laminae (312/*6*). An exceptionally deep **caudal thyroid notch** is found in the horse (313/*5*). Through it the surgeon gains access to the interior of the larynx. The

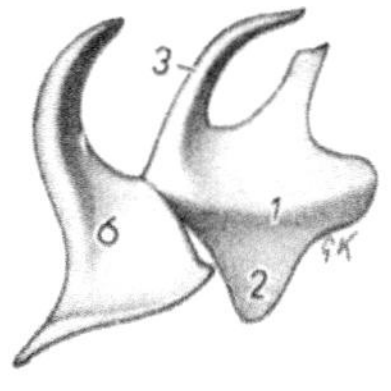

Fig. 318 (Dog)

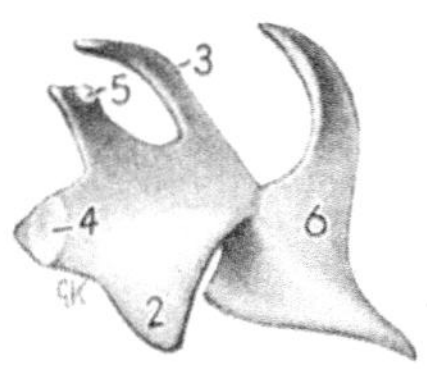

Fig. 319 (Dog)

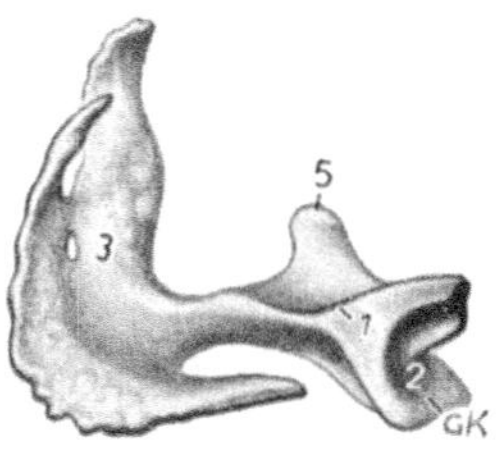

Fig. 320 (Pig)

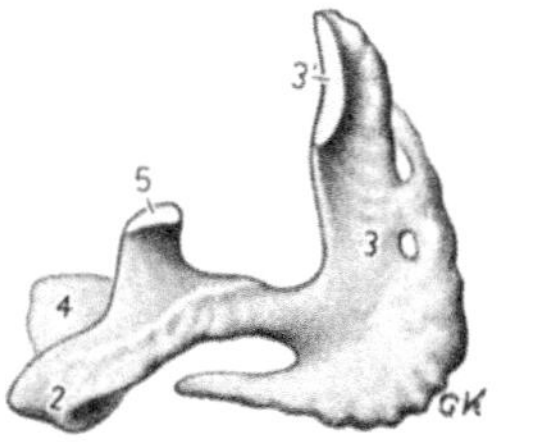

Fig. 321 (Pig)

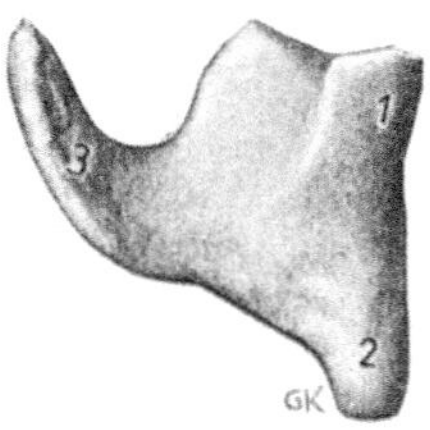

Fig. 322 (Ox)

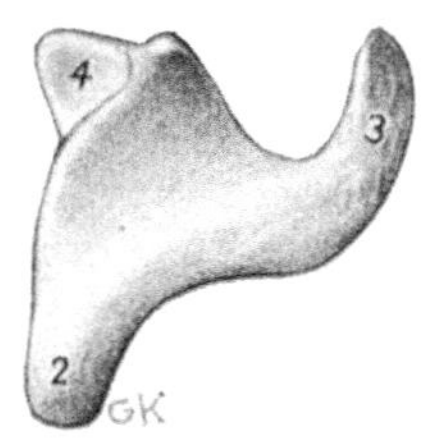

Fig. 323 (Ox)

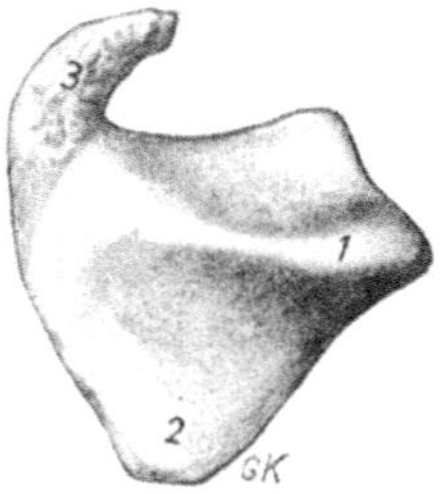

Fig. 324 (Horse)

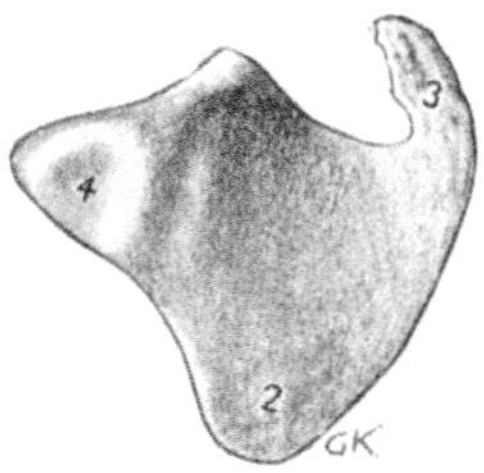

Fig. 325 (Horse)

Figs. 318—325. Left arytenoid cartilage.

Figs. 318, 320, 322, 324. Lateral aspect. Figs. 319, 321, 323, 325. Medial aspect. (Right and left cartilages of the pig have been separated.)

1 Muscular process; *2* Vocal process; *3* Corniculate process, *3'* cut surface; *4* Articular surface; *5* Interarytenoid cartilage (dog and pig); *6* Cuneiform process (dog)

caudal thyroid notch is deep also in the cat, but shallow in the dog and ruminants, and absent in the pig. The slight **laryngeal prominence** (310, 312/*7*) found on the ventral surface of the thyroid cartilage in the dog, older pig, and ruminants, is more caudally placed than the laryngeal prominence (Adam's apple) of man.

The **CRICOID CARTILAGE** (314—317) is joined to the caudal parts of the thyroid cartilage and is partly covered by the thyroid laminae. It has the appearance of a signet ring, and consists of a wide **lamina** (*2*) dorsally and a narrow **arch** (*1*) laterally and ventrally. The dorsal surface of the lamina is marked by a **median crest** (*3*) for the attachment of muscles. On the cranial border of the lamina are surfaces (*4*) for articulation with the arytenoid cartilages, and at the junction of lamina and arch is the surface (*5*) for articulation with the caudal cornu of the thyroid cartilage. The cricoid arch is narrowest ventrally, and its lateral surfaces, as far as they are covered by the thyroid laminae, are slightly concave.

The paired **ARYTENOID CARTILAGE** (318—325) has the shape of a three-sided pyramid; its apex points rostrodorsally and its base faces the cricoid cartilage. The base is narrow from side to side and is formed by the crest-shaped **muscular process** laterally (*1*), the **vocal process** ventrally (*2*), and the **articular surface** for the cricoid dorsomedially (*4*). The vocal process, in contrast to the otherwise hyaline nature of the arytenoid, consists of elastic cartilage. Medial, lateral, and dorsal surfaces connect the base of the cartilage with the apex, which consists of the elastic **corniculate process** (*3*). The latter, as its name indicates, is shaped like a horn; it is absent in the cat. In the pig, the corniculate process has a

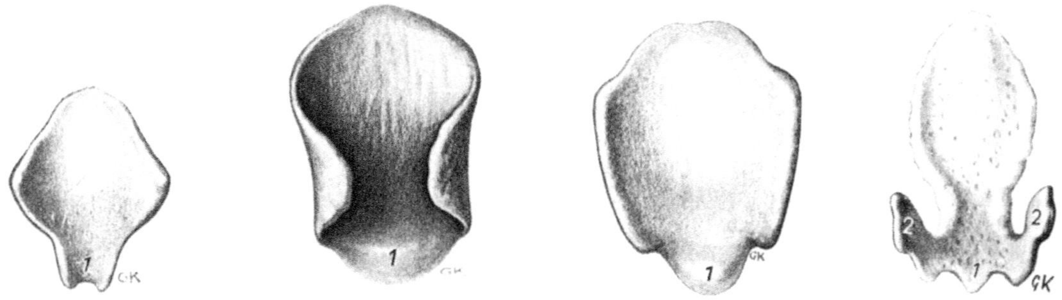

(Dog) (Pig) (Ox) (Horse)

Figs. 326—329. Epiglottic cartilage. Dorsal aspect. *1* Petiolus; *2* Cuneiform process (horse)

lateral semilunar extension. In the dog, the cuneiform process (*6*) arises from the base of the corniculate process. A small **interarytenoid cartilage** is present between the arytenoid cartilages in both the dog and pig (*5*).

The elastic **EPIGLOTTIS** (326—329) is the most rostral of the laryngeal cartilages and closes the laryngeal entrance during deglutition. Its **base** is next to the thyroid cartilage and is continued by a short process (petiolus, *1*) with which the epiglottis is connected to the thyroid cartilage or the thyrohyoid ligament. Except in the dog, a strong cushion of fat (334—338) adheres firmly to the lingual surface of the base, and changes its shape with the movements of the epiglottis. The **apex** of the epiglottis is directed rostrally and is pointed in the dog, small ruminants, and horse; but is rounded in the pig and ox. In the horse, the elastic **cuneiform processes** (329/*2*) project dorsally from each side of the base of the epiglottis.

Ligaments and Articulations of the Larynx

(330—333, 404)

The laryngeal cartilages are connected to each other, to the trachea, and to the hyoid bone by a number of ligaments and by the articulations between the cricoid and thyroid cartilages, between the cricoid and arytenoid cartilages, and between the thyroid cartilage and hyoid bone (except in the carnivores and pig).

The **CRICOTRACHEAL CONNECTION** is ligamentous and formed by the **cricotracheal ligament** (330—333/*10*), an annular band of fibroelastic tissue connecting the caudal border of the cricoid cartilage with the rostral border of the first tracheal cartilage.

The **CRICOTHYROID CONNECTION** is both articular and ligamentous. To form the articulation (404/*n*), the cricoid cartilage is furnished on each side with a shallow facet for the caudal cornu of the thyroid cartilage. These joints permit both dorsal and ventral flexion of the cartilages in respect to each other; in ventral flexion, for example, the vocal folds are tensed.

The elastic **cricothyroid ligament** connects the two cartilages ventrally and laterally (330—333/*9*). The ligament is well developed ventrally and extends here also onto the internal surface of the thyroid cartilage. In the horse, the ventral part of the cricothyroid ligament (404/*3'*) closes the deep caudal thyroid notch and is incised to gain access to the interior of the larynx as, for instance, in the correction of laryngeal hemiplegia (roaring). From the origin on the cricoid cartilage, the cricothyroid ligament splits off elastic fibers (membrana fibroelastica, 330/*9'*), which pass forward in the submucosa and attach on the vocal ligament. These fibers correspond to the conus elasticus of man.

The **CRICOARYTENOID CONNECTION** is both articular and ligamentous. For the articulation, the cricoid cartilage is furnished with a cylindrical articular surface, which is parallel to the rostral margin of the cricoid lamina; on the arytenoid cartilage there is a correspondingly concave facet. The **cricoarytenoid ligament** (330—333/*8*) is ventromedial to the joint and functions prominently during phonation. Since the capsule of the cricoarytenoid joint is loose and since there are not ligaments lateral to the joint, the arytenoid cartilages are capable of a variety of movements, namely, short dorsal and ventral movements around the longitudinal axis of the articular cylinder, gliding movements parallel to the same axis, and rotary movements around an axis that connects the cricoarytenoid joint with the apex of the arytenoid cartilage. The latter involve partial loss of contact between the articular surfaces. In the pig, the corniculate processes of the arytenoid cartilages are fused dorsomedially, limiting especially the rotary movements.

The **INTERARYTENOID CONNECTION** is ligamentous and formed by the **transverse arytenoid ligament.** This ligament connects the dorsocaudal angles of the two cartilages, but also sends thin fibers to the rostral border of the croicoid lamina. A small **interarytenoid cartilage** is found at this point in the dog and pig.

The **CONNECTION BETWEEN THE THYROID CARTILAGE AND THE THYROHYOID** is both articular and ligamentous. The thyrohyoid articulation (404/*m*) connects the rostral cornu of the thyroid cartilage with the thyrohyoid. In the dog, this takes the form of a synchondrosis*, and in the pig, in which the cornu is absent, there is a fibrous union between the thyrohyoid and the lateral surface of the thyroid lamina. The **thyrohyoid membrane** (*2*) extends from the rostral border of the thyroid cartilage to the basihyoid and to the two thyrohyoids.

The **THYROEPIGLOTTIC CONNECTION** is ligamentous and is formed by the elastic **thyroepiglottic ligament** (331—333/*3*) which passes from the base of the epiglottis to the ventral part of the thyroid cartilage. As Figures 330—333 show, this ligament differs in shape from species to species, because the relationships of the two cartilages it connects differ. They further illustrate that the movements of the laryngeal cartilages, and particularly those of the epiglottis, are not exactly the same in each species.

The **HYOEPIGLOTTIC CONNECTION** is formed by the **hyoepiglottic ligament** (330—332/*1*). This ligament is an elastic band which, together with the hyoepiglotticus, extends from the base of the epiglottis to the hyoid bone. It, too, varies with the species.

The arytenoid cartilage is connected to the floor of the larynx by two ligaments: the vestibular ligament rostrally and the vocal ligament caudally (*6*, *7*). These ligaments vary greatly in shape and attachment, once again pointing to the species-bound differences in laryngeal dynamics. The entrance to the lateral laryngeal ventricle in the dog and horse (335, 338/*1*) lies between these ligaments (see also p. 235).

---

* Gr. *syn-*, together; *chondros*, cartilage: a cartilaginous union.

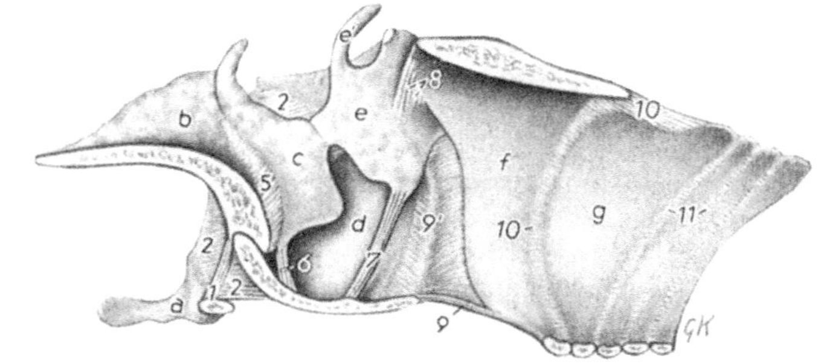

Fig. 330 (Dog)

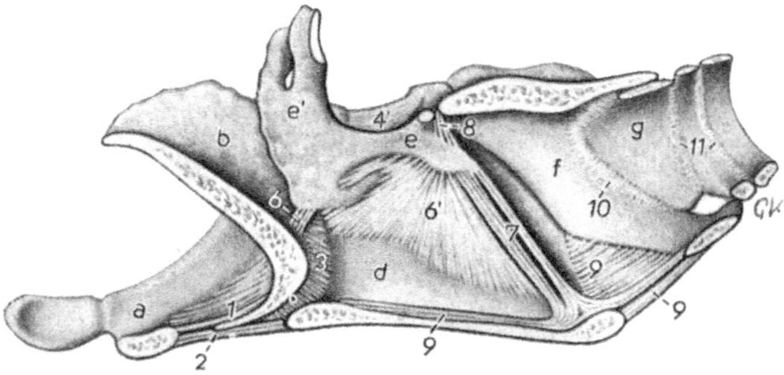

Fig. 331 (Pig)

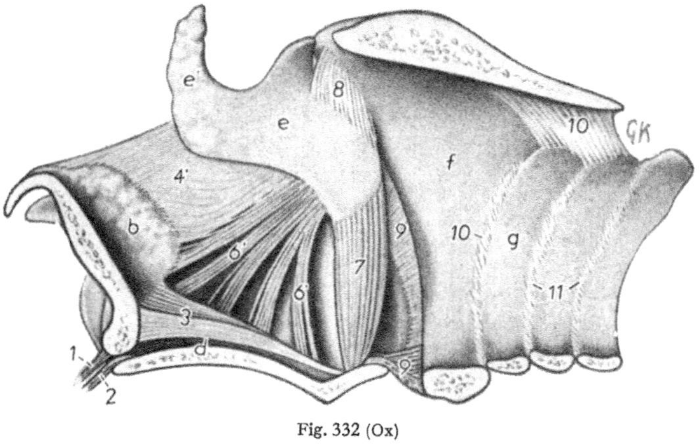

Fig. 332 (Ox)

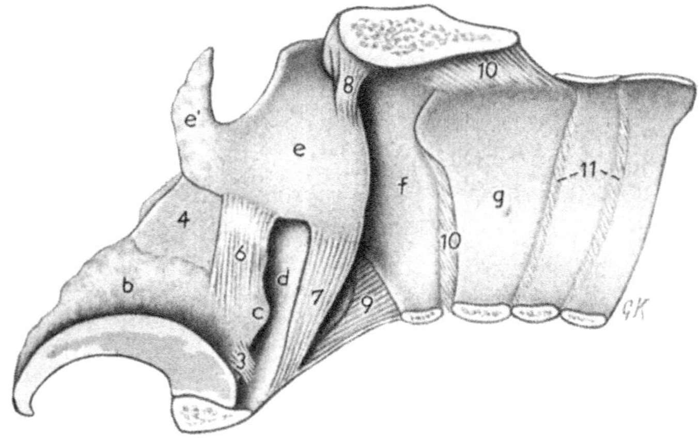

Fig. 333 (Horse)

Figs. 330—333. Ligaments of the larynx of the domestic mammals. Median section. Medial aspect.

a Hyoid bone (shown only in the dog and pig); *b* Epiglottic cartilage; *c* Cuneiform process (dog, horse); *d* Thyroid cartilage; *e* Arytenoid cartilage; *e'* Corniculate process; *f* Cricoid cartilage; *g* First tracheal cartilage

1 Hyoepiglottic ligament (not shown in horse); *2* Thyrohyoid membrane (not shown in horse); *3* Thyroepiglottic ligament (not visible in the dog); *4, 4'* Aryepiglottic fold; *5* Fibers belonging to aryepiglottic fold (dog); *6* Vestibular ligament; *6'* Fibers for the support of the lateral laryngeal ventricle (pig), fibers that correspond to the vestibular ligament (ox); *7* **Vocal ligament (split in the pig)**; *8* **Cricoarytenoid ligament;** *9* Cricothyroid ligament; *9'* Membrana fibroelastica (shown only in the dog); *10* Cricotracheal ligament; *11* Annular ligaments of trachea

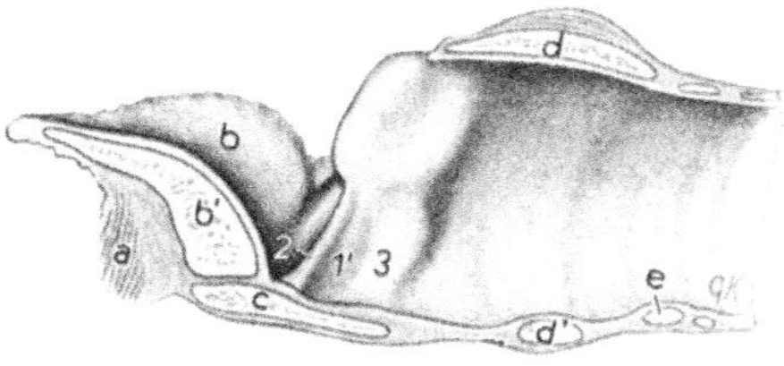

Fig. 334 (Cat)

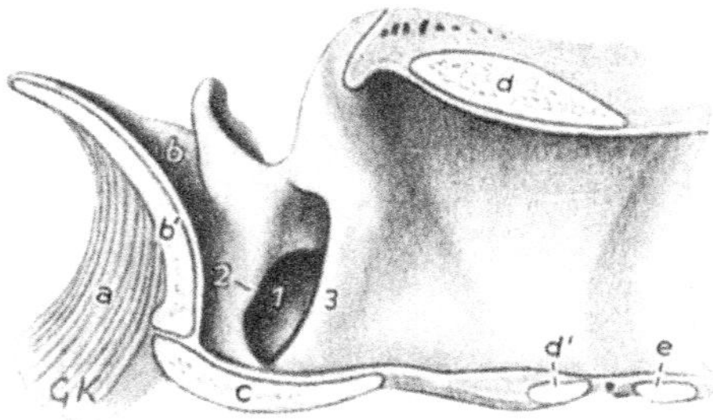

Fig. 335 (Dog)

Figs. 334—338. Larynx of the domestic mammals. Median section. Medial aspect.

a Hyoepiglotticus; *b* Epiglottis; *b'* Epiglottic cartilage; *c* Thyroid cartilage; *d* Cricoid lamina; *d'* Cricoid arch; *e* First tracheal cartilage

1 Entrance to laryngeal ventricle; 1' Depression rostral to vocal fold; 2 Vestibular fold; 3 Vocal fold; 4 Median laryngeal recess (pig, horse)

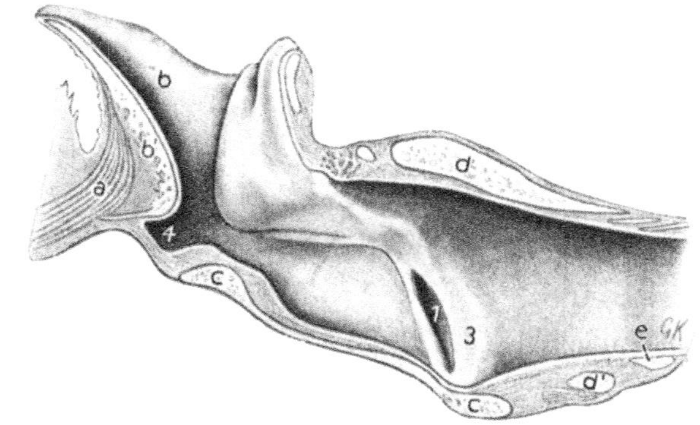

Fig. 336 (Pig)

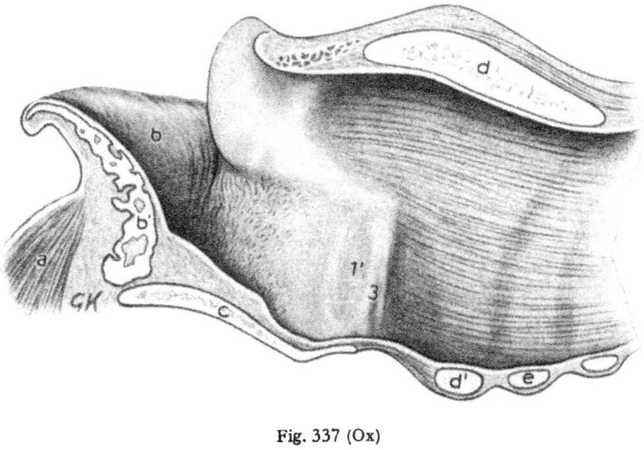

Fig. 337 (Ox)

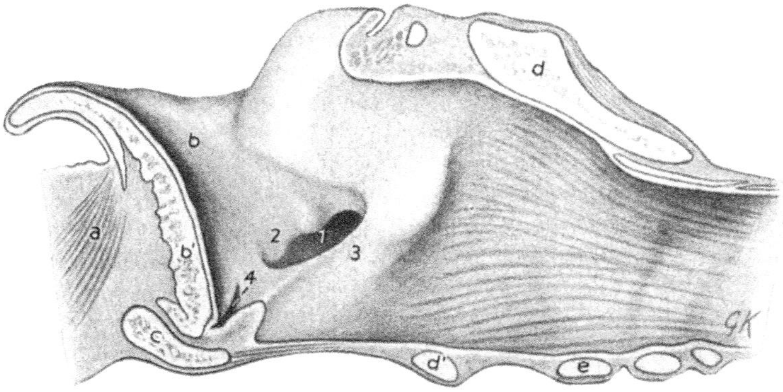

Fig. 338 (Horse)

The **vestibular ligament,** which is absent in the cat, extends in the dog from the floor of the thyroid cartilage to the cuneiform process of the arytenoid cartilage. In the pig, this ligament extends from the base of the epiglottis to the lateral surface of the arytenoid and its corniculate process and does not enter in the formation of the vestibular fold. In the ruminants, the vestibular ligament is represented by a number of radiating fibers (332/*6'*). These pass in the submucosa from the base of the epiglottis and the floor of the thyroid cartilage to the lateral surface of the arytenoid cartilage, and are relatively better developed in the small ruminants than in the ox. In the horse, the vestibular ligament extends from the cuneiform process at the base of the epiglottis to the lateral surface of the arytenoid cartilage.

The elastic **vocal ligament** (330—333/*7*) arises from the ventral part of the thyroid cartilage or from the cricothyroid ligament and passes to the vocal process of the arytenoid cartilage, forming different angles with the laryngeal floor in the different species. In the pig, the vocal ligament is split longitudinally, and between the two parts is the entrance to the laryngeal ventricle. There is, in addition, a fibrous sheet (331/*6'*) between the vocal and the vestibular ligaments in the pig. It extends from the thyroid lamina to the ventral border of the arytenoid cartilage and is lateral to the ventricle.

Muscles of the Larynx

The muscles associated with the larynx are striated and are either extrinsic or intrinsic. The *EXTRINSIC MUSCLES* are the **thyrohyoideus** (369, 370/*3*; 380/*3*; 391/*3*; 405/*3*), **hyoepiglotticus** (59—62/*33*; 380/*2*; 391/*2*; 405/*2*) and the **sternothyroideus** (described on p. 34). They, together with the hyoid muscles (p. 32) move the entire larynx, especially during swallowing.

The *INTRINSIC MUSCLES* pass from one laryngeal cartilage to another and move them in relation to one another. At this point a brief comparative description of their morphology and attachments is given. Their functions, principally in phonation and reflex closing of the glottis, will be considered on page 238 in the section on the dynamics of the larynx.

The **cricothyroideus** (369, 370, 380, 391, 405/*5*) arises from the lateral surface of the cricoid arch and passes rostrodorsally to the caudal border or the lateral surface of the thyroid cartilage and its caudal cornu. Its dorsal extent varies with the species. The **cricoarytenoideus dorsalis** (*6*) originates from the cricoid lamina lateral to the median crest and passes rostrolaterally to the muscular process of the arytenoid cartilage. The **arytenoideus transversus** (*7*) arises from the muscular process of the arytenoid cartilage and passes medially over the dorsal border of the cartilage, to which it is loosely attached; it ends together with its fellow from the other side on a median raphe. The **cricoarytenoideus lateralis** (*8*) lies on the internal surface of the thyroid lamina. It passes rostrodorsally from the rostroventral border of the cricoid arch to the caudal portion of the muscular process of the arytenoid cartilage. The **thyroarytenoideus** (*9*) of the dog and horse is divided into a rostral ventricularis and a caudal vocalis, while in the other species it is a uniform triangular plate. In the cat and ruminants, its wide portion originates on the base of the epiglottis, on the thyroid cartilage, and on the cricothyroid ligament (in the pig only on the thyroid cartilage), and in these three species is inserted mainly on the muscular process, but with a few rostral fibers also on the vocal process, of the arytenoid cartilage. The **ventricularis** of the dog and horse, together with the vestibular ligament, is contained in the vestibular fold (340/*8*). The muscle arises ventrally from the thyroid lamina (in the horse also from the cricothyroid ligament) and ends on the rostral part of the muscular process of the arytenoid cartilage. Its fibers are directed dorsally and, in the horse, are related medially to the cuneiform process. In the horse a rostral slip of the ventricularis continues over the muscular process and unites dorsal to the arytenoideus transversus with the muscle from the other side. The **vocalis** in the dog and horse lies inside the vocal fold (*9*) and is lateral as well as caudal to the vocal ligament (*5*). The muscle arises ventrally from the thyroid cartilage and ends largely on the muscular process of the arytenoid cartilage, but with a few rostral fibers also on the vocal process.

Laryngeal Cavity and its Lining

(334—340)

MUCOSA. Because of the exposure to wear, the mucosa lining the entrance, vestibule, and the rostral border of the vocal folds of the larynx, and in the dog and pig also the ventricles, is covered with mostly noncornified stratified squamous epithelium. The remainder of the larynx, including the ventricles in the horse, is lined, like the other air passages, with respiratory mucosa which is covered with ciliated, pseudostratified, cylindrical epithelium containing goblet cells. The beat of the cilia moves mucus and foreign particles toward the pharynx. Taste buds have been found in the epithelium on the laryngeal surface of the epiglottis, except in the horse. Serous, mucous, and seromucous **laryngeal glands** are present in most of the interior of the larynx. Their distribution, however, is not the same in each species. Glands are always absent in the free margin and immediate vicinity of the vocal folds, but are regularly present in the mucosa of the epiglottis, the aryepiglottic folds, the corniculate processes and the vestibular folds. The glands in the vestibular folds provide the moisture for the adjacent glandless vocal folds. The **lymphatic tissue** present takes the form of solitary nodules and tonsillar follicles and is usually better developed in older animals.

WALL OF THE LARYNGEAL CAVITY. In the lateral wall of the larynx of the dog, pig, and horse the muscosa forms lateral evaginations, the **laryngeal ventricles** (335, 336, 338/*1*; 340/*13*), which lie against the medial surfaces of the thyroid cartilage. The cat and the ruminants lack the ventricles and instead have shallow depressions (334, 337/*1'*) which result from the fact that the underlying thyroarytenoideus is quite thin here. The entrance to the laryngeal ventricle in the dog and horse lies between the vestibular fold and the vocal fold (340). The **vestibular fold** (*8*) encloses the vestibular ligament, the ventricularis (*1*), and the cuneiform process (*c*), which is located in the dorsal end of the fold in the dog and in the ventral end of the fold in the horse. The **vocal fold** (*9*), which projects farther medially than the vestibular fold, contains the vocal ligament (*5*) and the vocalis (*2*). Because the thyroarytenoideus of the pig is undivided and forms an unbroken surface together with a wide sheet of fibers (331/*6'*), a raised vestibular fold as found in the dog and horse does not exist. The entrance to the laryngeal ventricle in the pig is bounded rostrally by a narrow fold that contains the rostral half of the split vocal ligament (*7*), while the fold caudal to the ventricle is the vocal fold (336/*3*). Because of the absence of a ventricle, the vocal folds of the cat and ruminants are not covered with mucosa laterally and are, therefore, not true folds, but rather wide ridges (334, 337/*3*). Ruminants do not have a vestibular fold. The cat, however, despite the absence of a ventricle, has a vestibular fold (334/*2*); but the fold contains neither ligament nor muscle. At the base of the epliglottis in the pig, sheep and horse is a ventral evagination of the mucosa, the **median laryngeal recess** (336, 338/*4*); it is absent in the other species.

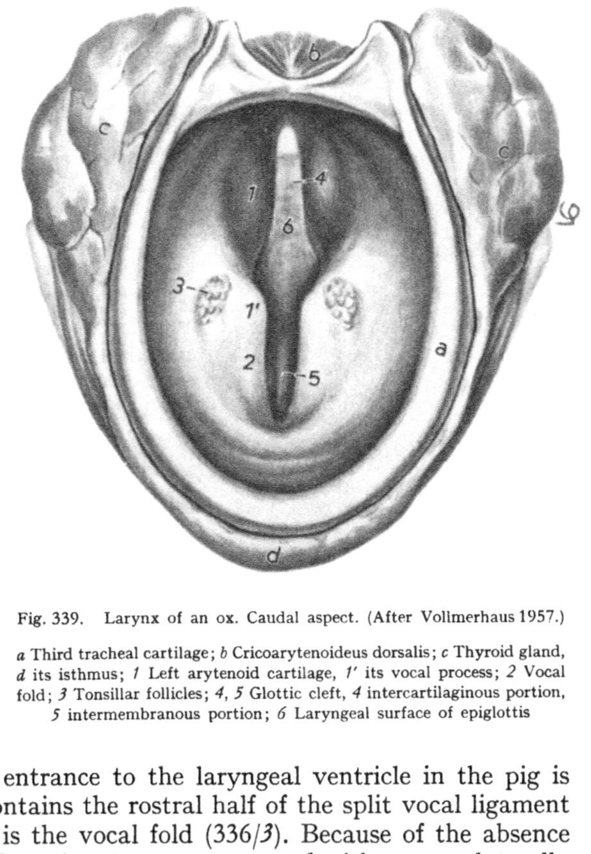

Fig. 339. Larynx of an ox. Caudal aspect. (After Vollmerhaus 1957.)

a Third tracheal cartilage; *b* Cricoarytenoideus dorsalis; *c* Thyroid gland, *d* its isthmus; *1* Left arytenoid cartilage, *1'* its vocal process; *2* Vocal fold; *3* Tonsillar follicles; *4*, *5* Glottic cleft, *4* intercartilaginous portion, *5* intermembranous portion; *6* Laryngeal surface of epiglottis

On each side of the entrance to the larynx is the **aryepiglottic fold,** which arises from the lateral margin of the epiglottis, and ends at the arytenoid cartilage in the dog and horse

(88/*10*), at the cricoid cartilage in the cat (58/*0'*), and dorsal to the arytenoid and cricoid cartilages in the pig and ruminants. The aryepiglottic folds, the epiglottis, and the corniculate processes enclose the entrance to the laryngeal cavity and project from the floor of the laryngopharynx, often through the intrapharyngeal opening into the nasopharynx (86). The piriform recesses (88/*7*) are two channels of the floor of the laryngopharynx, bounded medially by the epiglottis and aryepiglottic folds and laterally by the thyroid laminae and the hyoid bones. Soft food or fluid passing through the laryngopharynx is diverted into these channels by the rostral surface of the epiglottis. Solid food, however, will often pass directly over the epiglottis.

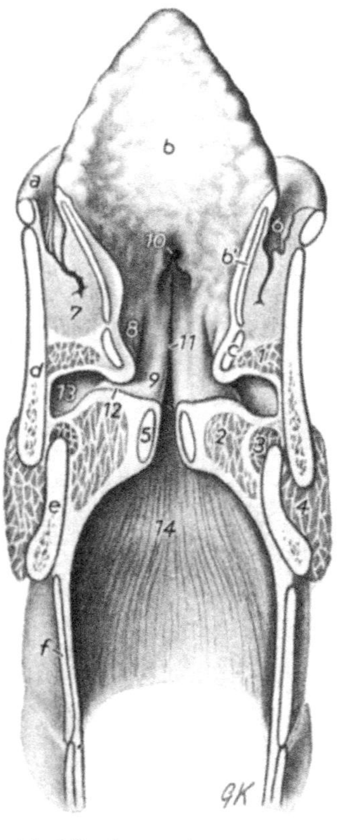

Fig. 340. Larynx of the horse. Dorsal section.

a Thyrohyoid; *b* Epiglottis; *b'* Epiglottic cartilage; *c* Cuneiform process; *d* Thyroid cartilage; *e* Arch of cricoid cartilage; *f* First tracheal cartilage

1 Ventricularis; *2* Vocalis; *3* Cricoarytenoideus lateralis; *4* Cricothyroideus; *5* Vocal ligament; *6* Piriform recess; *7* Palatine tonsil; *8* Vestibular fold; *9* Vocal fold; *10* Median laryngeal ventricle; *11* Glottic cleft; *12* Entrance to lateral ventricle; *13* Lateral ventricle; *14* Infraglottic cavity

The **LARYNGEAL CAVITY** may be divided into three sections: the vestibule rostrally, the infraglottic cavity caudally (340/*14*), and, between the two, the narrow glottic cleft (340/to the right of *5*).

The **vestibule of the larynx** is compressed laterally and extends from the wide laryngeal entrance (aditus laryngis) to the glottis. A short distance rostral to the glottis, the vestibular folds (*8*), enclosing a cleft (rima vestibuli), project into the lumen of the vestibule. Immediately rostral to the vocal folds in the dog, pig, and horse, the vestibule communicates with the **laryngeal ventricles** (*12*, *13*), and in the pig, sheep and horse it communicates ventrally with the **median laryngeal recess** (*10*). In the cat and ruminants there are small recesses in the lateral wall of the vestibule between the aryepiglottic folds and the arytenoid cartilages.

The glottis encloses the narrowest part, the **glottic cleft** (rima glottidis) of the laryngeal cavity. The glottis consists of the two vocal folds (*9*) and the arytenoid cartilages with their vocal processes and mucosal covering and functions mainly in phonation. The ventral or intermembranous part of the glottic cleft (339/*5*) is bounded by the vocal folds, while the dorsal or intercartilaginous part of the cleft (*4*) lies between the arytenoid cartilages.

The **infraglottic cavity** (340/*14*) extends from the glottic cleft to the beginning of the trachea. It is contained entirely within the ringlike cricoid cartilage and is continuous with the lumen of the trachea.

Taken as a whole, the cavity of the larynx can be likened to an hourglass. The upper funnel of the hourglass is the vestibule, the lower funnel is the infraglottic cavity, and the narrow aperture between the funnels is the glottic cleft. When laryngeal ventricles are present, as in the dog, pig, and horse, the vocal folds project more freely into the laryngeal lumen. This should be kept in mind when attempting intratracheal intubation in these species. In the horse, a paralysis of mostly the left caudal laryngeal nerve and the muscles it supplies causes the left vocal fold to be slack. Inspiratory air is caught by the loose fold and produces the characteristic sound of a roarer. This condition is known as laryngeal hemiplegia and can lead to extreme respiratory distress.

Movements of the Larynx and its Cartilages

It is necessary to distinguish between the movements that involve the entire larynx and cause a change of its position, and movements of the individual laryngeal cartilages in relation to each other. The latter are concerned with the constriction or dilatation of the laryngeal cavity, principally at the glottis.

MOVEMENTS OF THE ENTIRE LARYNX are caused mainly by muscles that originate on the sternum, the hyoid bone, and, via the pharyngeal wall, from the base of the cranium. Some of the hyoid and lingual muscles also influence the position of the larynx, although only indirectly. In general, these muscles move the larynx only during swallowing (see p. 56).

POSITION OF THE LARYNX DURING BREATHING. The rostral end of the larynx of the domestic mammals is related to the free edge of the soft palate, and the larynx in the normal breathing position forms a direct continuation of the nasopharynx (65A). The angle between the soft palate and the larynx is obtuse so that only a slight deflection of the air stream occurs in the pharynx. In man, the larynx is more caudally placed, has no relation to the soft palate, and the two structures are at right angles to each other. As a result, expiratory air is directed against the palate, which is essential for the production of articulated sound.

During the more common **nasal breathing** in the domestic mammals, the nasopharynx is dilated, the soft palate lies against the root of the tongue and closes the oropharynx (*1*), and the epiglottis (*c*) projects through the intrapharyngeal opening into the nasopharynx. On inspiration, therefore, air passes directly from the nasal cavity through the nasopharynx into the larynx and from there into the trachea and lungs, and vice versa on expiration. In **oral breathing,** the animal has to raise the soft palate. This closes off the nasopharynx and opens the oropharynx, and air passes from the oral cavity through the dilated oropharynx into the larynx. The horse cannot breathe through the mouth with ease and will do so only in distress. The reason for this is that it cannot elevate its soft palate high enough to obliterate the intrapharyngeal opening (see also page 73).

The **POSITION OF THE LARYNX DURING SWALLOWING** (65B) has been discussed on page 56 in the chapter dealing with the digestive tract. It was shown how the hyoid and lingual muscles pull the larynx rostroventrally to bring its entrance under the root of the tongue. At the same time, the bolus forces the oropharynx and the laryngopharynx open and, in so doing, obliterates the nasopharynx. With the larynx pulled rostroventrally, the epiglottis is laid over the laryngeal entrance, preventing food from entering the air passages. The movement of the epiglottis to close the larynx is thus brought about by the action of extrinsic muscles, because special intrinsic muscles for this purpose are not present in the domestic mammals. It has not been established whether the epiglottis, while the larynx is being returned to the breathing position, reverts to its original position by virtue of its elasticity or because of the action of the hyoepiglotticus which is always present in domestic mammals, but not in man. The role of the omohyoideus, sternohyoideus, and sternothyroideus muscles in returning the larynx after swallowing, and the dangers of overextending the animal's head and neck when administering liquid medication, were discussed on page 35. Overextension of the head and neck stretches these muscles and impedes the normal sequence of movements the larynx undergoes during swallowing. It is equally important not to overextend head and neck of the anesthetized animal, as accumulated saliva may enter the larynx and trachea and cause serious complications.

The **MOVEMENTS OF THE LARYNGEAL CARTILAGES** are effected by the intrinsic muscles of the larynx. These muscles, with the exception of the hyoepiglotticus which acts on the epiglottis, cause the dilatation or constriction of the glottis and the attendant changes in the length, tension, and thickness of the vocal folds. Complete closure of the glottis may be the result of a reflex, which prevents harmful substances from entering the lungs, or it may be voluntary as in coughing, abdominal press, etc. Partial closure of the glottis occurs during phonation and as an additional means of regulating the respiratory volume.

In **PHONATION,** the vocal folds are caused to vibrate, like the lips of a trumpet player, by the more or less forceful explusion of air from the lungs. This may also occur, although less often, during inspiration. Little puffs of air are pressed through the glottic cleft as the vibrating vocal folds open and close. The pitch of the sound produced depends on the frequency of the vibrations, and this, as with a violin string, depends on the thickness, the length, and the tension of the vocal folds. The thinner, shorter, and tenser the vocal folds, the higher the pitch, and vice versa for the low tones. The loudness of the sound depends on the force with which the air is pressed through the glottis.

While the larynx produces only the raw sound, the succeeding chambers that contain the upper air column, namely, the laryngeal vestibule, the pharynx, and the nasal and oral cavities, act as resonators and change the character of the sound. Man's ability to produce articulated sound (speech) is the result of the manifold changes which are possible in the shape and length of the upper resonators. These changes are produced by the action of the labial, lingual, and palatine muscles and are governed by a highly developed speech center in the central nervous system.

The intrinsic muscles of the larynx act on an elastic system of ligaments, which is altered in shape as long as the muscles contract, but which returns to its resting state when the muscles relax. This elastic system connects the laryngeal cartilages and is composed principally of the cricothyroid, cricoarytenoid, vocal, and vestibular ligaments. When the larynx is at rest and its elastic components are in a state of equilibrium, the glottic cleft is about half open. This is the intermediary or resting position. Action of the intrinsic muscles moves the laryngeal cartilages away from, or out of, this intermediate position, either toward complete dilatation or complete closure of the glottis. In either direction, the muscles cause the elastic elements to be tensed, so that upon relaxation of the muscles the elastic elements contract and return the cartilages to their original position.

Dilatation of the glottis is caused by the contraction of the **cricoarytenoideus dorsalis,** which pulls the muscular process of the arytenoid cartilage dorsolaterally and thus lifts the vocal process away from the median plane. The remainder of the intrinsic muscles close the glottis. The **cricoarytenoideus lateralis** pulls the muscular process of the arytenoid cartilage ventrolaterally. This apposes the vocal processes and results in a narrowing of the inter-ligamentous part of the glottic cleft when the vocal folds are relaxed. The **arytenoideus transversus** draws the arytenoid cartilages together and assists the cricoarytenoideus lateralis in closing the glottic cleft, particularly its intercartilaginous part. The arytenoideus transversus is thought, in addition, to augment the action of the cricoarytenoideus dorsalis in dilating the glottis. The **cricothyroideus,** when the thyroid cartilage is fixed, moves the arch of the cricoid cartilage rostrally and the lamina caudally. This increases the distance between the floor of the laryngeal cavity and the rostral border of the cricoid lamina. The arytenoid cartilages, which are attached to the rostral border of the cricoid lamina, are therefore also raised and the vocal folds become tense. With the vocal folds thus fixed, the **vocalis** can act and increase their thickness and tension. The **ventricularis** in the dog and horse constricts the vestibular cleft and, in cat, pig, and ruminants, assists the vocalis from which it is not separated. The positioning and fixation of the thyroid cartilage that is essential for the movements of the other laryngeal cartilages is brought about by the muscles that act on the hyoid bone (see p. 32) and by the sternothyroideus and hyothyroideus.

When the animal is breathing quietly, the glottic cleft changes very little in size and remains more or less in its intermediary position. During rapid or labored breathing, however, it opens wider. The movements of the glottis during respiration are produced mainly by the cricoarytenoideus dorsalis in interaction with the elastic system of ligaments. The muscle causes the arytenoid cartilage to move up and down around the longitudinal axis of the cylindrical articular surface on the cricoid lamina (314—317/*4*), and also laterally and medially parallel to the same axis. In phonation, the glottic cleft is constricted much more than in respiration. The arytenoid cartilages are caused to rotate medially with a resultant loss of contact between the articular surfaces of the cricoarytenoid joint. After the arytenoids have been fixed in position, the vocalis muscles contract and alter the thickness and the tension—but not the length—of the vocal folds, and thus determine the character of their vibration.

Trachea

(131, 132, 341—352, 357—363)

The trachea is a noncollapsible tube which continues the respiratory pathway from the cricoid cartilage of the larynx to the root of the lung where it bifurcates to form the right and left principal bronchi.

POSITION. In the neck, the trachea is related dorsally to the esophagus and the longus colli and longus capitis muscles, which cover the ventral surface of the vertebral column (131).

Fig. 341 (Cat)

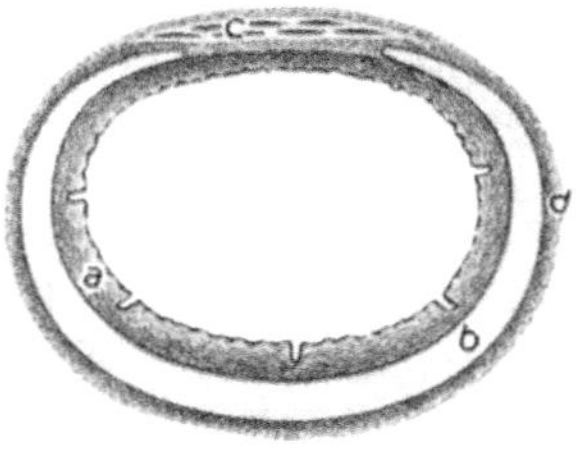

Fig. 342 (Dog)

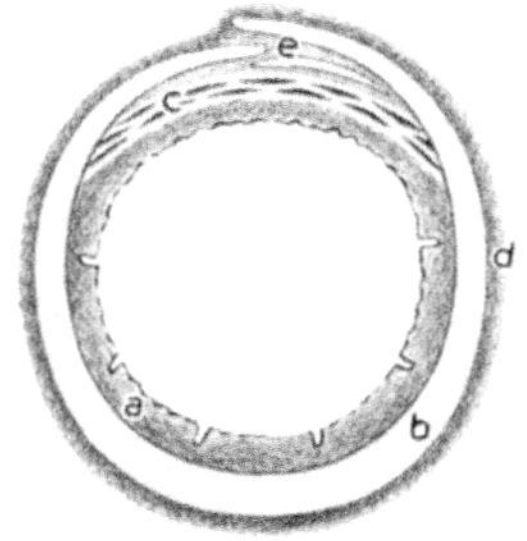

Fig. 343 (Pig)

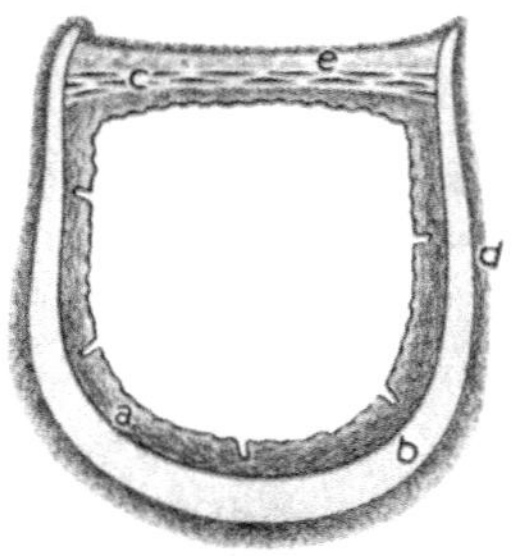

Fig. 344 (Goat)

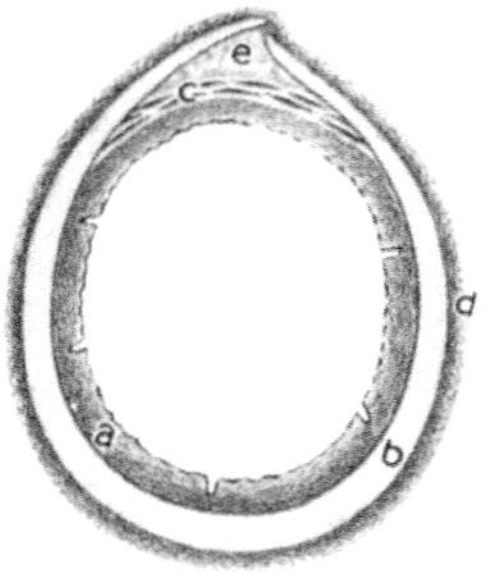

Fig. 345 (Sheep)

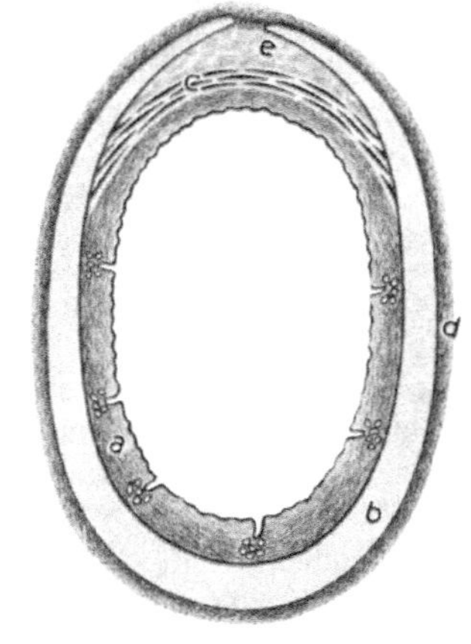

Fig. 346 (Ox)

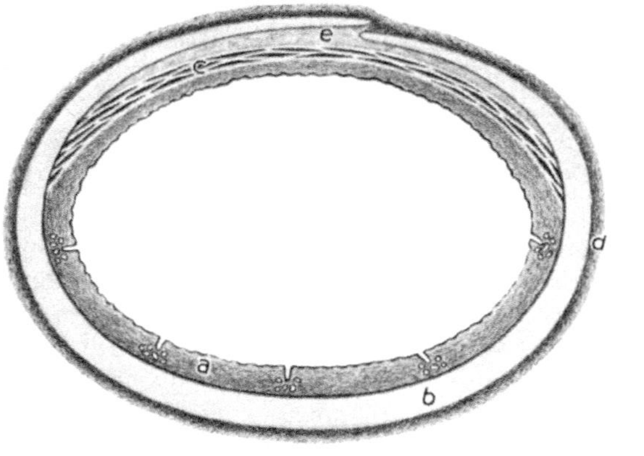

Fig. 348 (Horse)

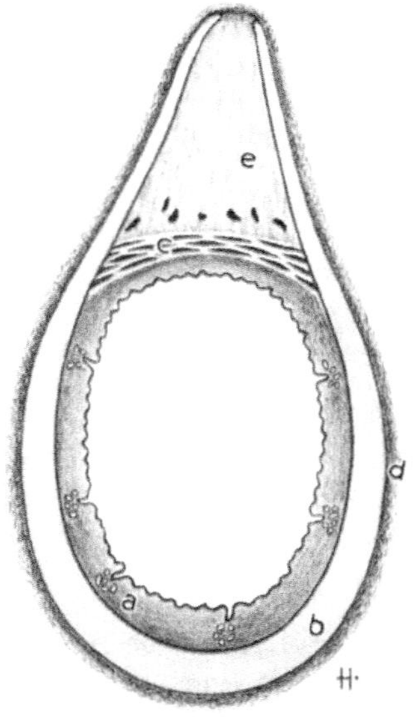

Fig. 347 (Ox)

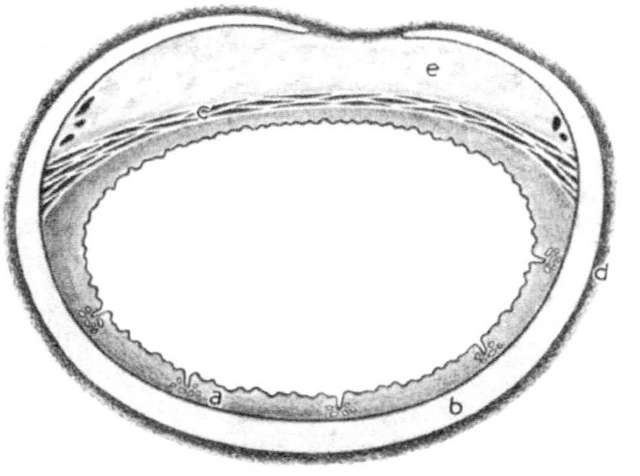

Fig. 349 (Horse)

Figs. 341—346, 348. Cross sections of the cervical portion of the trachea of the domestic mammals. Fixed in situ. Schematic.

Figs. 347, 349. Cross sections of the cervical portion of the unpreserved trachea of the ox and of the horse. Schematic.

a Mucous membrane with tracheal glands and their excretory ducts; *b* Tracheal cartilage; *c* Trachealis muscle; *d* Adventitia; *e* Loose connective tissue

Ventrally and laterally, it is surrounded by the straplike cervical muscles (*17*) that pass from the sternum to the head. The sternohyoidei, the ventralmost of these, are fused in the midline and have to be separated to expose the trachea, e.g., in the tracheotomy. Dorsolaterally, the trachea is accompanied by the common carotid arteries (*f*), vagosympathetic trunks (*m*), internal jugular veins (often absent in the horse and always absent in the goat and sheep), caudal laryngeal nerves (*n*), and the tracheal lymphatic trunks (*k*). In the caudal third of the neck, the esophagus and the vessels and nerves that accompany the trachea on the left gradually descend to the left lateral surface of the trachea (132) so that the trachea passes through the thoracic inlet in contact with the ventral vertebral muscles (352).

In the thoracic cavity, the trachea continues caudally in the mediastinum, dorsal to the cranial vena cava (*20*). The esophagus (*u*) returns to the median plane and assumes its original position dorsal to the trachea. The trachea then crosses the aortic arch on the right side and, dorsal to the base of the heart and at the level of the fourth to the sixth intercostal spaces, divides into the two principal bronchi. A short distance cranial to its bifurcation, the trachea in the pig and ruminants gives off the **tracheal bronchus** (*v'*) for the cranial lobe of the right lung.

STRUCTURE. The trachea consists of a series of incomplete cartilaginous rings (341—349/*b*) of the hyaline type, which prevent the collapse of the tube and are covered with an **adventitia** (*d*) and lined with a **mucous membrane** (*a*). The **tracheal cartilages** are united by **annular ligaments** (330—333/*11*) which fuse with the perichondrium and consist mainly of fibrous but also of elastic tissue. On the dorsal surface, the connective tissue is loose (341—349/*e*), contains lymphoid tissue, and fills the space between the free ends of the cartilages. Between this connective tissue and the mucous membrane is the **trachealis** (*c*), a smooth muscle with mainly transversely oriented fibers. In the carnivores, the trachealis lies external to the cartilages.

The **MUCOUS MEMBRANE** of the trachea is covered with respiratory epithelium, consisting of ciliated pseudostratified columnar cells and goblet cells. The sweep of the cilia, as elsewhere in the respiratory passages, is directed outward. In the deep layers of the propria and in the submucosa are the mostly seromucous **tracheal glands** (*a*). They produce the tracheal mucus, and are present mainly in the ventral and lateral walls of the trachea. Increased production of mucus during inflammation of the trachea gives rise to abnormal respiratory sounds, known as moist rales. The **submucosa** is generally thin, but is well developed dorsally where the cartilages are incomplete. Numerous longitudinal elastic fibers present in the mucosa help the trachea to return to its normal length after it has been stretched by the extension of the neck.

The number of the **TRACHEAL CARTILAGES** is not constant in all species and varies even in animals of the same species. Occasionally, adjacent cartilages become partly or completely fused. This is seen most often in the pig and least often in the ruminants. Equally variable is the cross sectional appearance of the trachea (341—349). This depends on the shape of the cartilages or, if it occurs in the same animal, on the state of contraction of the trachealis. In the carnivores and in the goat (as in man) there is a considerable gap between the ends of each tracheal cartilage with the result that part or all of the dorsal wall of the trachea is membranous (paries membranaceus) and is without cartilaginous support. The ends of the tracheal cartilages in the sheep overlap in the cranial portion of the trachea (345), have a position as those of the goat in the middle third (344), and in the caudal third the left ends of the cartilages extend farther dorsally than the right ends. In the horse, several thin plates of cartilage are found between the ends of the tracheal cartilages, just cranial to the bifurcation (363/*2'*).

The Lungs

The lungs originate in the embryo from the floor of the foregut, usually as a single median bud, which soon splits into right and left lung buds. In the dog, the lungs have a paired origin; each bud develops independently from the lateral wall of the flattened foregut (Sack 1964). The two lung buds develop into the right and left lungs (pulmo dexter, sinister).

PLATE X

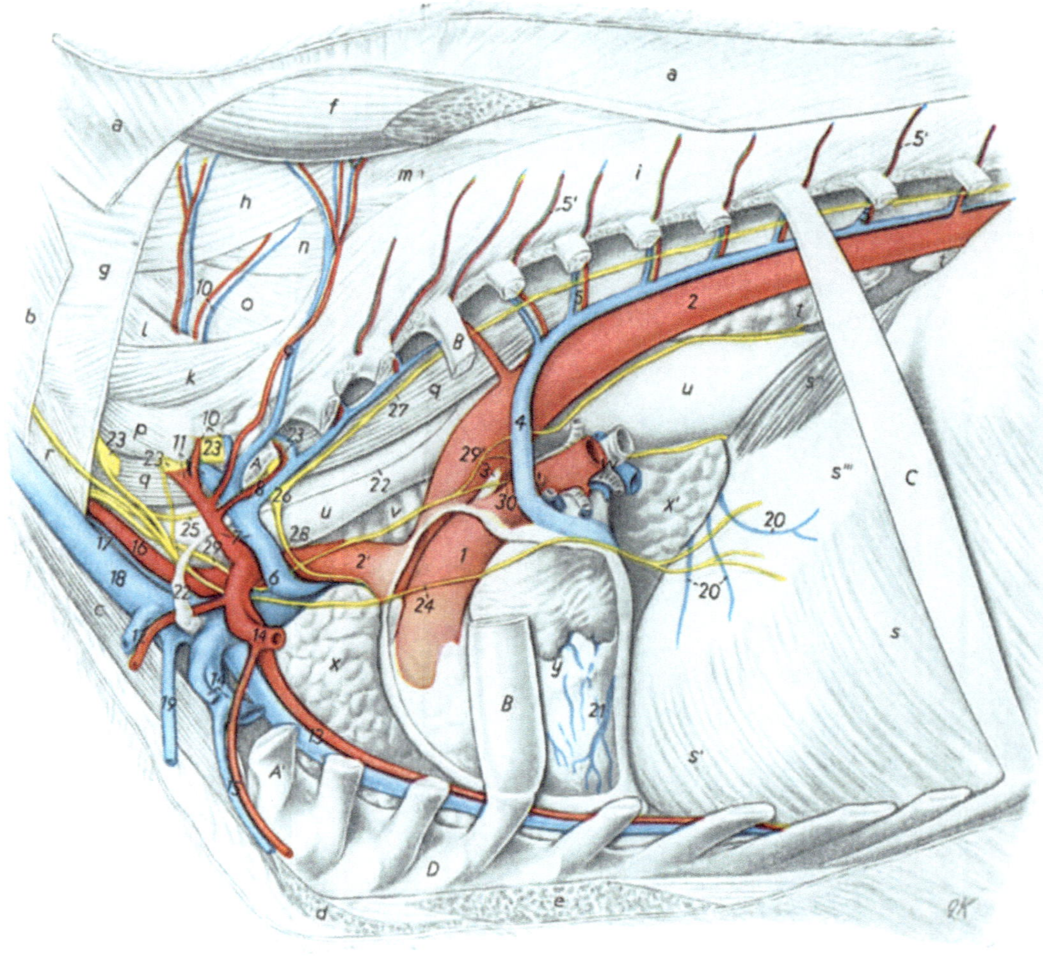

Fig. 351. Thoracic cavity of the ox, showing vessels and nerves. The mediastinal pleura has been removed. Left aspect. (Arteries red, veins blue, nerves yellow).

A First rib; *A'* Cartilage of first rib; *B* Fourth rib; *C* Ninth rib; *D* Sternum

a **Trapezius;** ***b*** **Brachiocephalicus;** ***c*** **Sternocephalicus;** ***d*** **Supf. pectoral muscles;** ***e*** **Deep pectoral muscle;** ***f*** **Rhomboideus cervicis;** ***g*** **Serratus ventralis cervicis;** ***h*** **Splenius;** ***i*** **Longissimus thoracis;** ***k*** **Longissimus cervicis;** ***l*** **Longissimus capitis et atlantis;** ***m*** **Semispinalis cervicis;** *n*, *o* Semispinalis capitis; *n* Biventer cervicis; *o* Complexus; *p* Intertransversarius; *q*, *q* Longus colli; *r* Scalenus dorsalis; *s* Diaphragm, costal part, *s'* its sternal part, *s''* its left crus, *s'''* its tendinous center; *t*, *t* Caudal mediastinal lymph nodes; *u* Esophagus; *v* Trachea; *w* Root of lung; *x* Right lung, cranial lobe, *x'* accessory lobe; *y* Heart, the pericardium has been opened

1 Pulmonary artery; *2* Aorta; *2'* Brachiocephalic trunk; *3* Lig. arteriosum; *4* Left azygous vein; *5* Dorsal intercostal arteries and veins; *5'* Dorsal branches of intercostal vessels; *6* Left subclavian artery and vein; *7* Costocervical trunks; *8* Supreme intercostal artery and vein; *9* Descending scapular artery and vein; *10* Deep cervical artery and vein; *11* Vertebral artery and vein; *12* Supf. cervical artery and vein; *13* Int. thoracic artery and vein; *14* Axillary artery, displaced dorsally; *14'* Axillary veins; *15* Ext. thoracic artery and vein; *16* Common carotid artery; *17* Int. jugular vein; *18* Ext. jugular vein; *19* Cephalic vein; *20* Branches of left phrenic vein; *21* Intermediate branch of great coronary vein; *22* Thoracic duct; *23* Sixth, seventh, and eighth cervical and first thoracic nerves, stumps of their ventral branches; *24* Left phrenic nerve (note its origin); *25* Cervical part of sympathetic trunk; *26* Cervicothoracic ganglion; *27* Symphathetic trunk; *28* Cardiac branches of sympathetic trunk; *29* Vagus; *29'* Division of vagus into dorsal and ventral vagal trunks; *30* Caudal laryngeal nerve

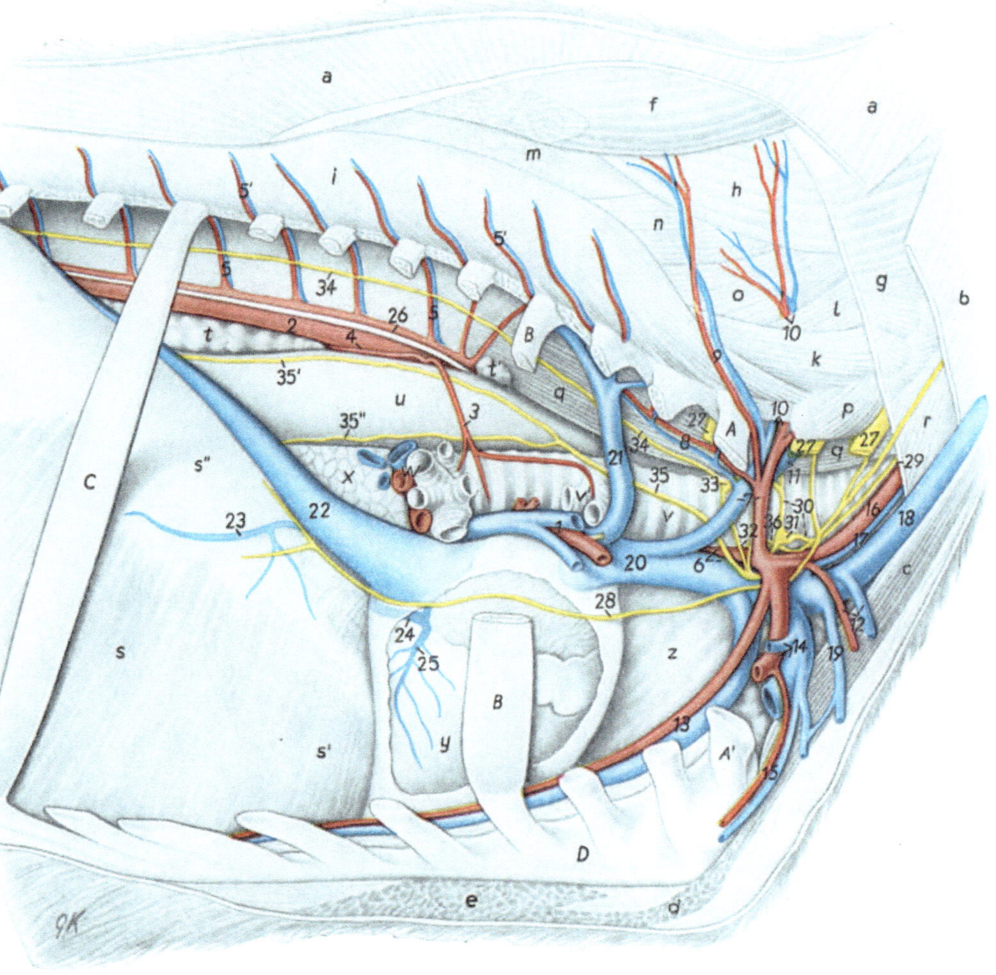

Fig. 352. Thoracic cavity of the ox, showing vessels and nerves. The mediastinal pleura has been removed. Right aspect. (Arteries red, veins blue, nerves yellow) (After Wilkens and Rosenberger 1957.)

A First rib; *A'* Cartilage of first rib; *B* Fourth rib; *C* Ninth rib; *D* Sternum

a Trapezius; *b* Brachiocephalicus; *c* Sternocephalicus; *d* Supf. pectoral muscles; *e* Deep pectoral muscle; *f* Rhomboideus cervicis; *g* Serratus ventralis cervicis; *h* Splenius; *i* Longissimus thoracis; *k* Longissimus cervicis; *l* Longissimus capitis et atlantis; *m* Semispinalis cervicis; *n, o* Semispinalis capitis; *n* Biventer cervicis; *o* Complexus; *p* Intertransversarius; *q, q* Longus colli; *r* Scalenus dorsalis; *s* Diaphragm, costal part, *s'* its sternal part, *s''* its tendinous center; *t* Caudal mediastinal lymph node; *t'* Middle mediastinal lymph node; *u* Esophagus; *v* Trachea; *v'* Tracheal bronchus; *w* Root of lung; *x* Accessory lobe of right lung; *y* Heart, the pericardium has been opened; *z* Cranial mediastinum

1 Pulmonary artery and vein, branches for the right cranial lobe; *2* Aorta; *3, 4* Bronchial and esophageal branches of the bronchoesophageal artery; *5* Dorsal intercostal arteries and veins; *5'* Dorsal branches of intercostal vessels; *6* Right subclavian artery and vein; *7* Costocervical trunks; *8* Supreme intercostal artery and vein; *9* Descending scapular artery and vein; *10* Deep cervical artery and vein; *11* Vertebral artery and vein; *12* Supf. cervical artery and vein; *13* Int. thoracic artery and vein; *14* Axillary artery and veins; *15* Ext. thoracic artery and vein; *16* Common carotid artery; *17* Int. jugular vein; *18* Ext. jugular vein; *19* Cephalic vein; *20* Cranial vena cava; *21* Right azygous vein; *22* Caudal vena cava; *23* Right phrenic vein; *24* Coronary sinus; *25* Middle vein of heart; *26* Thoracic duct; *27* Sixth and seventh cervical and first thoracic nerves, stumps of their ventral branches; *28* Right phrenic nerve (note its origin); *29* Vagosympathetic trunk; *30* Cervical portion of sympathetic trunk; *31* Middle cervical ganglion; *32* Ansa subclavia; *33* Cervicothoracic ganglion; *34* Sympathetic trunk; *35* Vagus; *35'* Dorsal vagal trunk; *35''* Ventral vagal trunk; *36* Caudal laryngeal nerve

PLATE XII

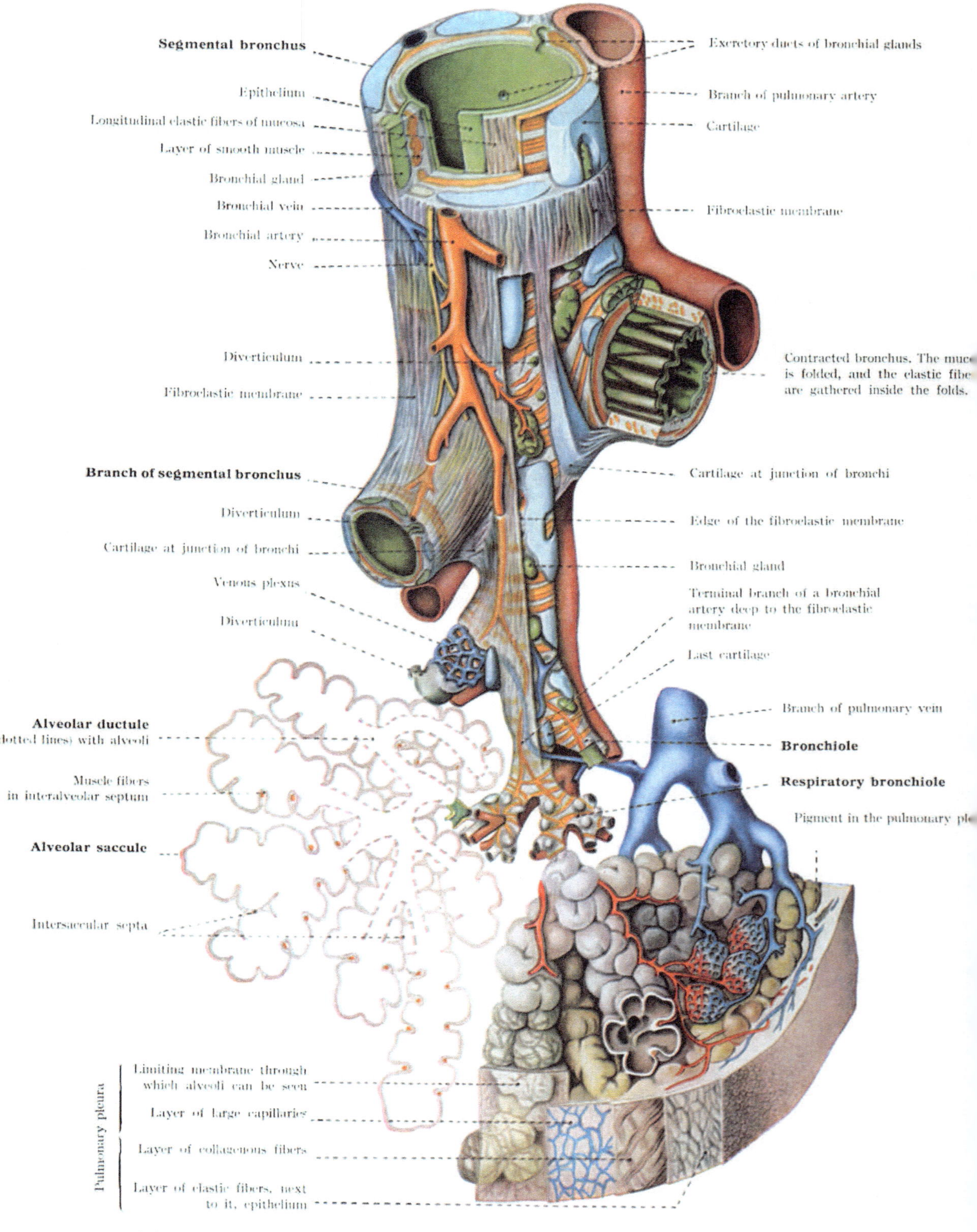

Fig. 353. Segmental bronchus and its branches with accompanying vessels and nerves. Schematic. (After Braus 1956.) On some of the alveoli the capillary network is shown, on others the network of elastic fibers is shown. On the left, the alveolar ductules and saccules are shown schematically in longitudinal section.

The lungs are located in the **pleural sacs** which come together medially to form the **mediastinum.** The walls of the pleural sacs adhere laterally to the ribs as the **costal pleura,** caudally to the diaphragm as the **diaphragmatic pleura,** and medially to the organs in the mediastinum or to the other pleural sac as the **mediastinal pleura** (see also pp. 4 and 5). The organs between right and left mediastinal pleurae are exposed in Figure 351 and 352. The caudal vena cava and its fold (plica venae cavae), passing between the heart and the diaphragm on the right side, form a recess, which communicates freely with the right pleural cavity and contains the accessory lobe of the right lung (352/*x*; 7/*10*).

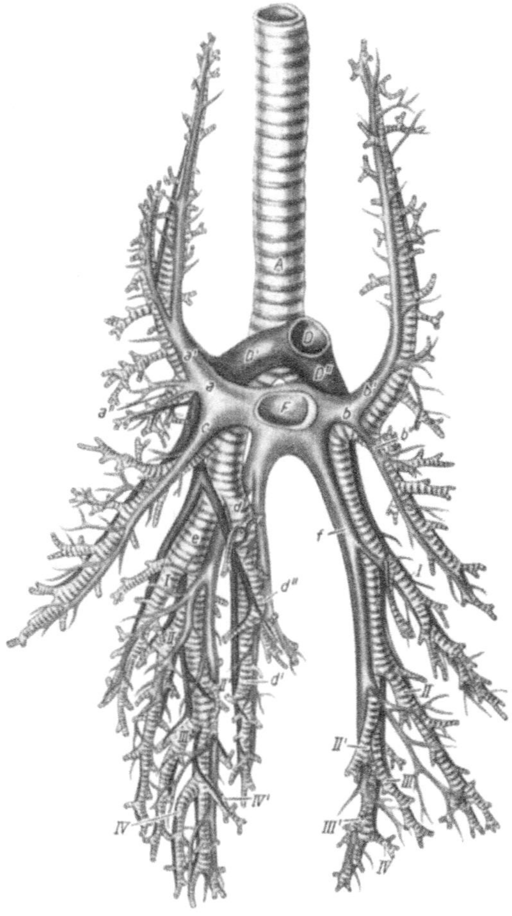

Fig. 350. The bronchial tree and accompanying blood vessels of the cat. Ventral aspect. (After Adrian 1964.) Drawn from several "Plexiglas" casts. Bronchi: gray with dark stripes; Arteries: black; Veins: light gray.

A Trachea; *D* Pulmonary trunk; *D'* Right pulmonary artery; *D''* Left pulmonary artery; *F* Portion of left atrium

a Right cranial bronchus and vessels, *a'* its cranial, *a''* its caudal segmental bronchus and vessels; *b* Left cranial bronchus and vessels, *b'* its cranial, *b''* its caudal segmental bronchus and vessels; *c* Middle bronchus and vessels; *d* Accessory bronchus, *d'* its dorsal, *d''* its ventral segmental bronchus; *e* Right caudal bronchus and vessels; *f* Left caudal bronchus and vessels

I, II, III, IV Ventral segmental bronchi of caudal lobe, beginning at the second bronchus on the left and the third bronchus on the right, the vein has left the bronchus and runs intersegmentally; *II' III', IV'* Medial segmental bronchi of caudal lobe

The lungs are covered with **pulmonary pleura,** which is continuous at the **hilus** and **pulmonary ligament** (356/*9, 14*) with the mediastinal pleura. They fill their respective pleural sacs completely, leaving only a capillary space which in the healthy animal contains a small amount of serous fluid. The fluid facilitates frictionless movement of the lungs as they expand and contract in respiration.

Each lung is shaped like a semicone and has an **apex** (356/*12*), which is directed cranially and lies at the thoracic inlet, and an oblique **base** (*8*), which faces caudoventromedially and lies on the diaphragm. The surfaces of the lungs correspond to the walls of the pleural sacs. The **costal surface** lies against the ribs and bears their impressions when the lung was fixed in situ (354). The **medial surface** lies against the bodies of the thoracic vertebrae and the mediastinum and displays the impressions of the organs located in the mediastinum (356). The **diaphragmatic surface** (*8*), which is at the base of the lung, is concave and lies on the diaphragm. The medial surface meets the costal surface dorsally to form the rounded **dorsal border** (*4*), and ventrally to form the sharp **ventral border** (*5* near *3*). The ventral border presents the **cardiac notch** (*3*), which permits the heart and pericardium to make contact with the lateral thoracic wall (376/*5*). The costal and the diaphragmatic surfaces of the lung meet at the **basal border** (356/*5* near *8*) which is of considerable clinical importance. During respiration the basal border moves in and out of the costodiaphragmatic recess, but never opens it fully under normal conditions (see also p. 6).

The **medial surface** of the lung contains the **hilus** (*9*), through which the principal bronchus and the pulmonary and bronchial vessels and nerves pass from the mediastinum to the lung. The aggregate of these structures is known as the **root of the lung** (radix pulmonis). The deepest of the above-mentioned impressions on the medial surface is the **cardiac impression** (356/*ventral to the hilus*) in which are lodged the heart and pericardium. The cardiac impression is deeper on the left lung than on the right because the heart lies slightly more to the left. Most of the other impressions are dorsal to the cardiac impression and are shown in Figure 356. The most prominent are the **aortic impression** (*10*), the **esophageal impression** (*13*), and on the right lung the **groove for the caudal vena cava** (355/*4*).

The **COLOR** of the lung depends on the amount of blood it contains. If the animal is bled completely, the lung is pink, but if blood remains in the lung after death, it is a darker red. The human lung is often gray, grayish blue, or even black because of the accumulation of dust and carbon particles. This is true also of the lungs of cats and dogs that have been kept for the most part in large cities or indoors.

WEIGHT. Because of the considerable amount of air contained in the lung, it will float in water. Lungs of stillborn animals will sink. If they float, the animal has breathed and is considered to have been alive and therefore should not be called stillborn. This flotation test is of importance in some forensic cases. Diseased lungs may also sink due to the presence of exudate replacing the residual air in the alveoli. The absolute weight of the lung differs with the species. On the average, it represents 1—1.5 per cent of the body weight.

The **SIZE** of the lung depends on the amount of air it contains and is considerably larger after inspiration than after expiration. The **collapsed lung** is smaller than the functional lung after expiration, and results from the introduction of air into the pleural sacs (pneumothorax) in the live animal, or, after death, when the pleural cavities are opened. The right lung is always larger than the left, the proportion being 4:3.

BRONCHIAL TREE. The bifurcation of the trachea gives rise to the thick, but short, **principal bronchi** (357—363/*3*) which divide into lobar bronchi upon entering the lung. The **lobar bronchi** (*4—8*) pass into the dorsal part of the lung and each of them enters and ventilates one lobe (*a—d*). The tracheal bronchus (*4*) present in the pig and ruminants is also a lobar bronchus. The number and distribution of the lobar bronchi is not the same in all domestic species and differs especially between right and left lungs. The lobar bronchi give rise to a large number of **segmental bronchi** (*9—11*) each of which enters and ventilates a bronchopulmonary segment. A **bronchpulmonary segment** is a cone-shaped, self-contained section of lung tissue within a lobe. The base of the segment lies under the pleura and its apex points toward the hilus. The existence of bronchopulmonary segments may be demonstrated by inflation of an individual segmental bronchus or by making plastic casts of the bronchial tree. Adjacent bronchopulmonary segments are thought to communicate in the dog (Miller, Christensen, and Evans 1964). The branches of the segmental bronchi give rise to bronchioles. The **bronchioles** (bronchuli) are very small tubes of which many are less than 1 mm. in diameter. In contrast to the bronchi, the bronchioles have no glands and their walls are without cartilaginous support (353). The bronchioles are the last branches of the bronchial tree concerned solely with the conduction of air. The terminal subdivisions of the bronchioles give rise usually to two **respiratory bronchioles** (353) whose walls contain some alveoli*. The respiratory bronchioles branch once or twice and are followed by the **alveolar ductules,** which are completely surrounded by alveoli. The alveolar ductules terminate in **alveolar saccules.** The exchange of gases takes place in the walls of the alveoli. The sum of the alveoli constitutes the respiratory surface of the lung.

* L. *alveolus*, a small hollow or cavity. Respiratory bronchioles are not present in ruminants.

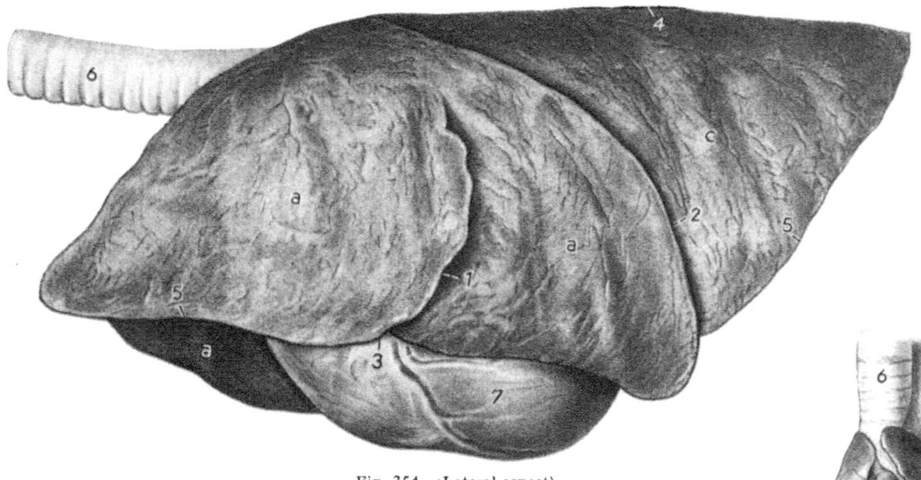

Fig. 354 (Lateral aspect)

Fig. 354. Costal surface of the left lung of a dog. (After Baum, unpublished.)

a, a (black letters) Cranial and caudal parts of left cranial lobe; *a (white letter)* Undivided right cranial lobe; *c* Caudal lobe; *1* Intralobar fissure; *2* Caudal interlobar fissure; *3* Cardiac notch; *4* Dorsal border; *5 (near a)* Ventral border; *5 (near c)* Basal border; *6* Trachea; *7* Heart

Fig. 355. Lungs and heart of a dog, fixed in situ. Caudoventral aspect. (After Baum, unpublished.)

a (black letter) Right cranial lobe; *a, a (white letters)* Cranial and caudal parts of left cranial lobe; *b* Middle lobe; *c* Caudal lobe; *d* Accessory lobe; *1, 1* Cranial interlobar fissure on the right side of animal, and intralobar fissure on the left side; *2* Caudal interlobar fissure; *3* Cardiac notch; *4* Groove in the accessory lobe for the caudal vena cava; *5* Basal border; *6* Trachea; *7* Heart; *8* Esophagus; *9* Esophageal impression

Fig. 356. Left lung of a dog, fixed in situ. Medial aspect. (After Baum 1943.)

a, a Cranial and caudal parts of cranial lobe; *c* Caudal lobe; *1* Intralobar fissure; *2* Caudal interlobar fissure; *3* Cardiac notch; *4* Dorsal border; *5 (near 3)* Ventral border; *5 (near 8)* Basal border; *6* Border between medial and diaphragmatic surfaces; *7* Medial surface; *8* Base of lung, and diaphragmatic surface; *9* Hilus; *10* Aortic impression; *11* Impression made by left subclavian artery; *12* Apex of lung, and impression made by int. thoracic artery; *13* Esophageal impression; *14* Pulmonary ligament; *15* Principal bronchus; *16* Pulmonary artery; *17* Pulmonary veins

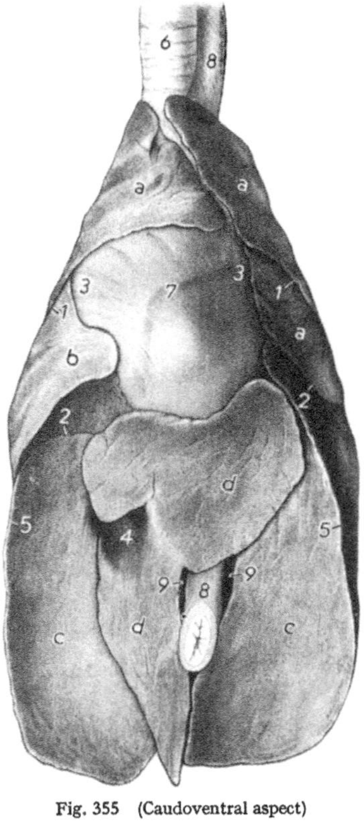

Fig. 355 (Caudoventral aspect)

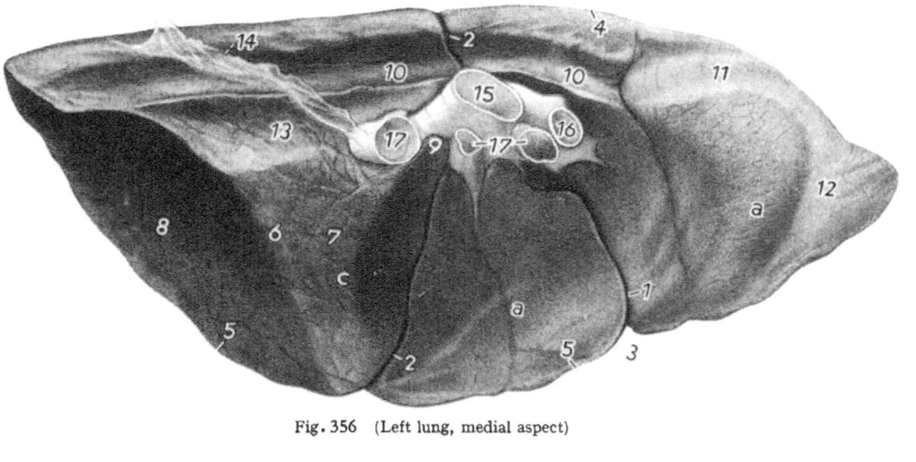

Fig. 356 (Left lung, medial aspect)

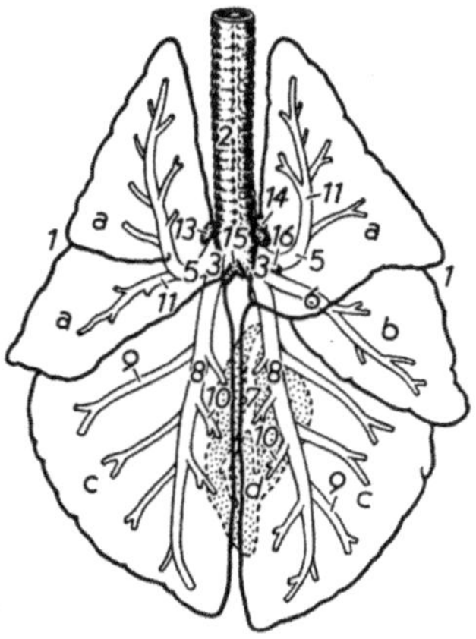

Fig. 357 (Cat)

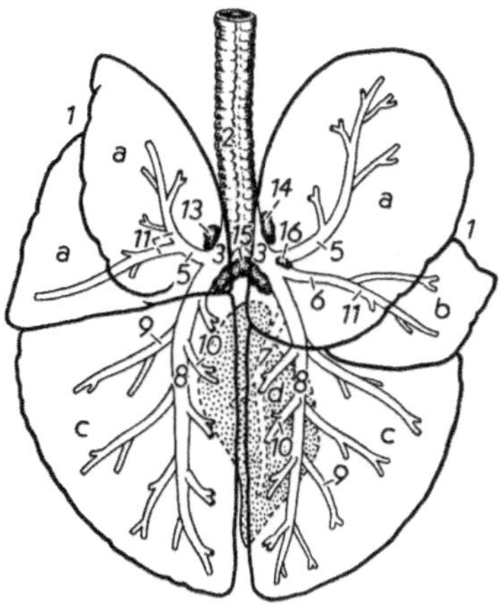

Fig. 358 (Dog)

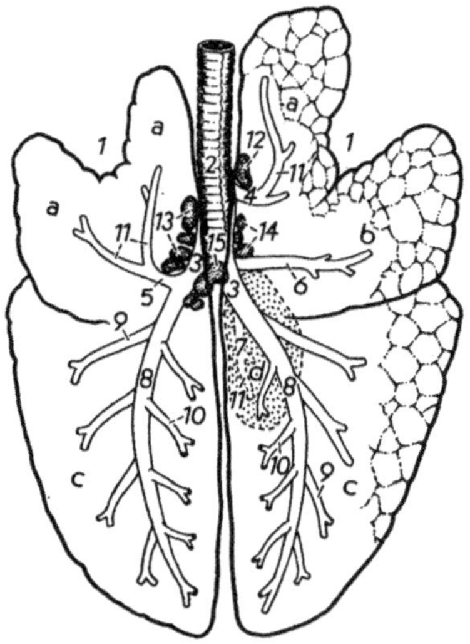

Fig. 359 (Pig)

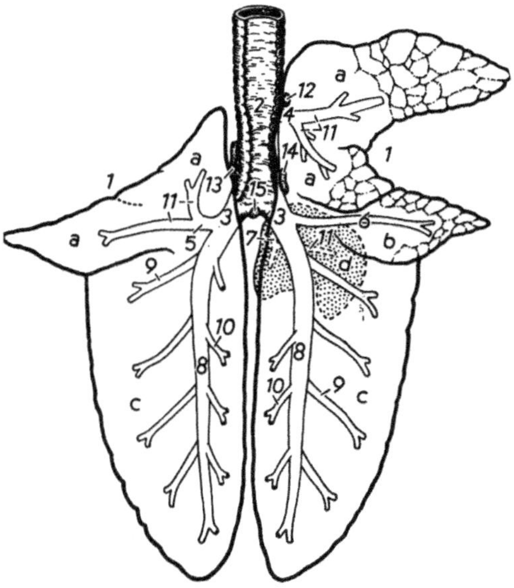

Fig. 360 (Goat)

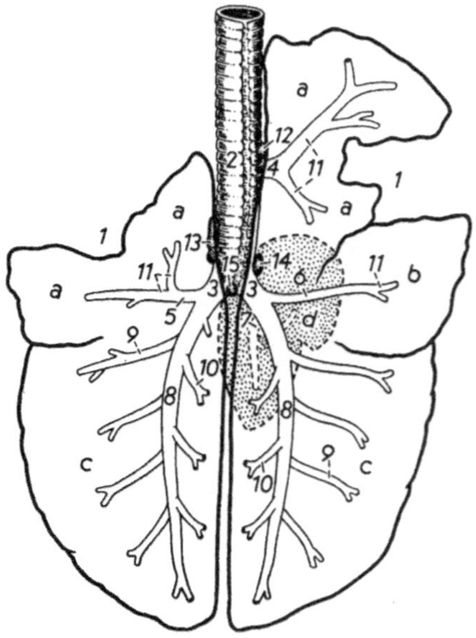

Fig. 361 (Sheep)

THE LOBES OF THE LUNG. The naming of the lobes of the lung was based formerly on the external fissures and on the relations of the lobes to neighboring structures. The great differences in the depth of these fissures and the absence of fissures in the equine lung have led to much confusion in homology and nomenclature. Basing the names on the division of the bronchi (Seiferle 1956) gets around these difficulties and provides a workable scheme for the homology of the lobes among domestic mammals. According to this scheme each lung has a **cranial lobe** (357—363/*a*) ventilated by the cranial bronchus (the right cranial bronchus in the pig and ruminants is the tracheal bronchus) and a **caudal lobe** (*c*) ventilated by the caudal bronchus. The right lung, in addition, has a **middle lobe** (*b*) ventilated by the middle bronchus, and an **accessory lobe** (*d*) ventilated by the accessory bronchus. In some species the cranial lobe is divided into cranial and caudal parts. The actual lobation of the lungs of the domestic mammals is given in the table on page 246.

Figs. 357—363. Lobation, bronchial tree, and lymph nodes of the lungs of the domestic mammals. Schematic. Dorsal aspect. (After Müller 1938.)

a Cranial lobe; *b* Middle lobe (except in the horse); *c* Caudal lobe; *d* Accessory lobe (stippled)

1 Cardiac notch; *2* Trachea; *2'* Cartilaginous plates in dorsal wall of trachea (horse); *3* Principal bronchus; *4—8* Lobar bronchi; *4* Tracheal bronchus (pig and ruminants); *5* Cranial bronchus; *6* Middle bronchus (except in the horse); *7* Accessory bronchus; *8* Caudal bronchus; *9* First (left) and third (right) ventral segmental bronchi of caudal lobe; *10* Second dorsal segmental bronchus of caudal lobe; *11* Further segmental bronchi; *12* Cranial tracheobronchial lymph node(s); *13* Left tracheobronchial lymph node(s); *14* Right tracheobronchial lymph node(s); *15* Middle tracheobronchial lymph node(s); *16* Pulmonary lymph node

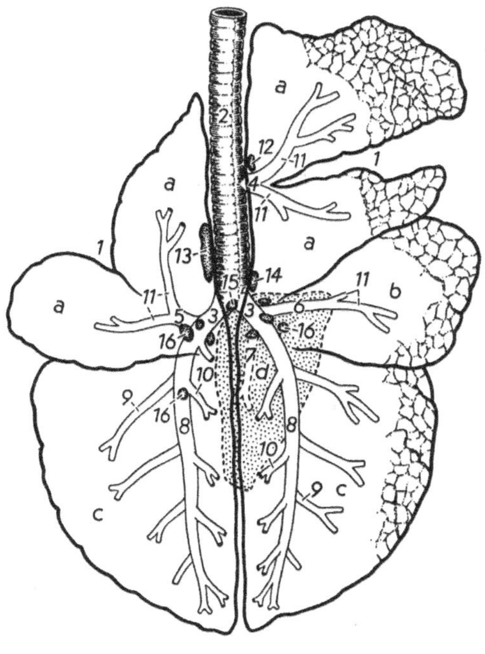

Fig. 362 (Ox)

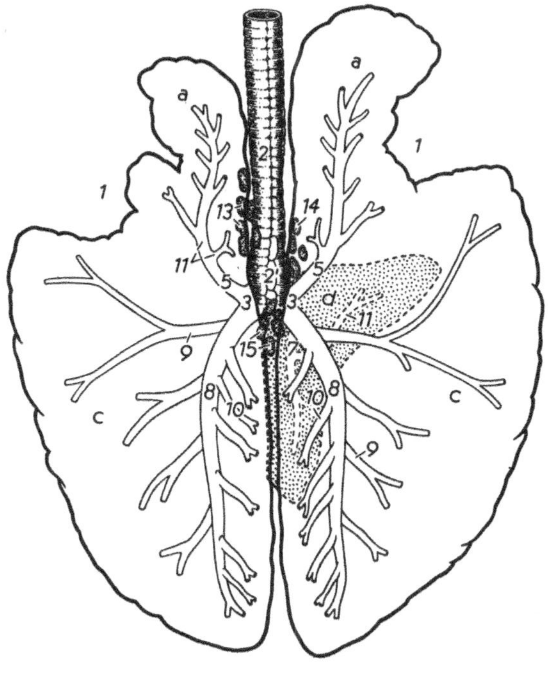

Fig. 363 (Horse)

The **MICROSCOPIC STRUCTURE OF THE LUNGS** will be considered here only as far as is useful for the understanding of the gross anatomy. The lungs, like the large glands, consist of an interstitial connective tissue framework, or stroma, and a parenchyma.

THE INTERSTITIAL FRAMEWORK. The lungs are covered with **pleura,** which, like all serosal membranes, consists of a simple squamous epithelium on a thin lamina propria. The collagenous and elastic fibers of the lamina propria are continuous with the elastic **interlobular and intralobular connective tissue,** which constitutes, with the deeper peribronchial and perivascular tissues, the interstitial framework of the lungs. It is this elastic connective tissue framework that enables the lungs to expand and contract in respiration. The fibers are stretched when the lungs increase in size during inspiration. Inspiration occurs when the respiratory muscles, by enlarging the thoracic cavity, create a vacuum in the lungs so that air can enter through the unobstructed trachea and bronchi. When, during expiration, the thoracic cavity decreases in size, the elastic elements of the pulmonary stroma contract and expell the air.

The pulmonary connective tissue framework is fairly evenly distributed in the carnivores, sheep, and horse, while in the pig and ox it causes a more or less distinct lobulation of the parenchyma (359, 362). In the goat, only the cranial and middle lobes are lobulated, but the lobulation is not very distinct if the lungs are still warm. **Pulmonary lobules** are small units of lung tissue of very irregular shape and size, surrounded by interlobular connective tissue. A lobule may be so large as to be ventilated by a small bronchus, or it may be so small as to accommodate only a bronchiole and its branches. The division of the lungs into these small lobular units facilitates the constant changes in shape which the lungs undergo during respiration.

The organization of the **PARENCHYMA** is similar to that of a tubuloalveolar gland. The excretory duct system of the gland corresponds to the air-conducting bronchial tree, and the secreting endpieces of the gland correspond to the alveoli. To assure the patency of the air-conducting system, the walls of the bronchi, like the trachea, contain incomplete rings of hyaline cartilage. As the bronchi diminish in diameter toward the periphery of the lung, the cartilages decrease in number and are gradually replaced by small plates or strips of elastic

Table 11. Lobation of the lungs of the domestic mammals.

	Left Lung	*Right Lung*
Carnivores	Divided cranial lobe Caudal lobe	Cranial lobe Middle lobe Caudal lobe Accessory lobe
Pig	Divided cranial lobe Caudal lobe	Cranial lobe Middle lobe Caudal lobe Accessory lobe
Ruminants	Divided cranial lobe Caudal lobe	Divided cranial lobe Middle lobe Caudal lobe Accessory lobe
Horse	Cranial lobe Caudal lobe	Cranial lobe Caudal lobe Accessory lobe

cartilage. The last cartilages are found at the junctions of the bronchioles, which, as a rule, are without cartilaginous support (353). The **mucosa** that lines the bronchi has a ciliated pseudostratified epithelium containing goblet cells. In its propria are the seromucous **bronchial glands** and accumulations of lymphocytes. The mucosa contains an elastic network of mostly longitudinal fibers, which are stretched during inspiration and contract during expiration, returning the tissues to their original position. Distally (toward the alveoli), the epithelium gradually becomes thinner and the goblet cells and cilia disappear. In the respiratory bronchioles, the epithelium consists of only a single layer of cells. The bronchial glands also diminish in number and are absent in the bronchioles. External to the mucosa is a very elastic **layer of smooth muscle** which consists proximally of circular fibers and distally of more oblique fibers. The muscular layer also diminishes distally and is absent in the smaller bronchioles. Isolated muscle fibers are found, however, in the free margins of the interalveolar septa (353). The peribronchial connective tissue is loose and joins the bronchi to the interstitial framework of the lung. For details about the structure of the alveoli, histology texts should be consulted. The available evidence suggests that the alveolar wall is composed of alveolar cells, which contain osmiophilic inclusions in the vicinity of their nuclei. Peripherally, the cytoplasm of the alveolar cells is very thin and forms the limiting membrane of the alveolus. This is followed externally by the basement membrane and this, in turn, by the cytoplasm of the endothelial cells of the alveolar capillaries, which always serve two adjacent alveoli. Reticular and elastic fibrils in the meshes of the capillaries complete the alveolar wall. The layers of the alveolar wall must be penetrated in the respiratory exchange of gases. Before the introduction of the electron microscope, it was thought that the alveolar capillaries were exposed directly to the air in the alveoli.

The **BLOOD VESSELS** of the lung may be divided into those of the functional blood supply and those of the nutritional blood supply. The **functional blood supply** comes from the right ventricle of the heart and enters the lungs as venous blood through the **pulmonary arteries** and its branches. It passes through the alveolar capillaries (where it is oxygenated) and leaves the lungs through the **pulmonary veins** to reenter the heart at the left atrium. The functional blood supply nourishes the lung tissue distal to the bronchioles. The **nutritional blood supply** arises from the bronchoesophageal artery and follows as **bronchial arteries** the ramifications of the bronchial tree, supplying the bronchi proximal to the bronchioles. The **bronchial veins,** after leaving the lungs, enter the azygous vein.

Considering that the same amount of blood that enters the systemic circulation of the body must also pass through the very much smaller lungs, it becomes clear that the pulmonary circulation is a very efficient one, and, with the constantly varying demands of the body, one that is readily regulated. Diseases of the lungs impede pulmonary blood flow and, therefore, make great demands on the heart and on the systemic circulation.

The pattern of distribution of the pulmonary vessels is not the same in all domestic mammals. In the sheep and horse (as in man) the pulmonary arteries follow the bronchi, while the veins run intersegmentally. In the ox, both pulmonary arteries and veins accompany the bronchi. In the carnivores and pig, the vessels of the cranial and middle lobes follow the bronchi, as in the ox; but in the caudal lobes only the arteries follow the bronchi and the veins run between the bronchopulmonary segments, as in the sheep and horse (350).

The **LYMPHATICS** of the lung form superficial and deep networks which are, however, difficult to differentiate. They collect the lymph for the tracheobronchial, pulmonary, and mediastinal nodes (351, 352/*t*, *t'*; 357—363/*12—16*). The **NERVES** of the lungs are derived from the vagus and the sympathetic trunks and accompany the bronchi and blood vessels.

The Respiratory Organs of the Carnivores

The apex of the **NOSE** is joined to the upper lip, is movable, and protrudes somewhat beyond the skull (18, 19). It is bisected by the **philtrum** (269, 270/*b*), which extends to about the level of the nostrils. In some breeds the philtrum is a deep cleft giving rise to the so-called double-nosed appearance.

NASAL CARTILAGES. The rostral part of the **nasal septum** (277/*a*) is split ventrally and projects beyond the incisive bones. Both **dorsal** and **ventral lateral nasal cartilages** (*b, c, d*) are expansive and unite with each other laterally for some distance to form two cartilaginous tubes, one for each nostril. Projecting forward from the ventral lateral cartilage, and shaped like an anchor, is the **lateral accessory cartilage** (*e*). It helps to support the nostril ventrally, and with its distal part forms a groove which underlies the slit in the lateral border of the nostril (270). The **medial accessory cartilage** arises from the ventral nasal concha and the ventral lateral nasal cartilage, and extends rostrally inside the alar fold (283/*7*).

The circular **NOSTRILS** are supported medially and ventrally by the nasal septum, and dorsally by the dorsal lateral nasal cartilages. The lateral accessory cartilages, which also contribute to the ventral support, and the dorsal lateral cartilages form the lateral wall of the nostril. The modified skin surrounding the nostrils, the **planum nasale**, is without hair and has a tuberculate surface in the cat. In the dog (270/*a*), it is divided by fine grooves into small plaquelike fields. The nose, having no glands of its own, is kept moist by the glands in the nasal septum, by the lateral nasal gland, and also because of the constant flow of lacrimal fluid from the orifice of the nasolacrimal duct in the floor of the nostril by the lacrimal gland. In 88 per cent of dogs, the **nasolacrimal duct** has a second opening at the level of the canine tooth, lateral to the ventral concha.

The **NASAL CAVITY** (57, 58, 283—286) is almost completely filled with the large nasal and smaller ethmoidal conchae. In the caudal part of the nasal cavity of the dog is a recess, occupied by endoturbinate IV (283/*N*), that extends caudally to the presphenoid, but it does not correspond to the sphenoid sinus of the other species.

The **dorsal nasal concha** may be divided into rostral, middle, and caudal portions. The rostral portion (285/*1*) is a long, curved shelf supported by the basal lamella which projects toward the nasal septum from the roof of the nasal cavity. The caudal portion (286/*1*) is only about 1 cm. long and forms an elongated swelling overlying a low basal lamella. The scroll-like middle portion has a basal lamella (366/*J'*), which originates from the perpendicular plate of the ethmoid (*G*) as well as from the nasal septum, and a spiral lamella. The spiral lamella (*J*) coils first ventrally, then laterally, and then dorsally, and encloses a recess. It releases several secondary lamellae and forms the **uncinate process** (284/*J'*).

The **ventral nasal concha** of the dog occupies the rostral half of the nasal cavity and is extensively folded. Its caudal portion fills almost the entire lumen of the nasal cavity (285) and is arranged in many longitudinal folds which unite rostrally to form the alar fold (284/*7*). The ventral concha of the cat is similar to that of the dog, but is much smaller.

The **middle nasal concha** (283/*K*; 286/*2*) has numerous secondary lamellae which give rise to secondary scrolls. Its rostral end lies between the ventral concha and endoturbinate III in the dog, and in the cat it extends quite far rostrally. The **ethmoidal conchae** (283/*M*, *N*) of the carnivores extend considerably forward in the nasal cavity and also invade the frontal sinus (*15*—*19*).

The **incisive ducts** connect the nasal cavity with the oral cavity and receive the ducts of the **vomeronasal organ** (see also p. 219). The **lateral nasal gland** is located in the maxillary recess (364/*a*). Its duct runs along the middle nasal meatus und opens in the nasal vestibule near the straight fold from where its secretion reaches the nose to keep it moist. The mucosal folds in the nasal vestibule are described on page 216.

The **NASOPHARYNX** (57, 58/*c*) is relatively long and is not divided by a septum. The **auditory tubes** (*v*) open into it on each side, and dorsally is the platelike **pharyngeal tonsil** (*w'*).

PARANASAL SINUSES. Both dog and cat have frontal sinuses and a maxillary recess, and the cat, in addition, has a sphenoid sinus. The **MAXILLARY RECESS** (364, 365/*a*, *a'*) is not a true sinus, because it does not lie between internal and external plates of a cranial bone. Instead, it is bounded laterally by the maxilla, lacrimal, and palatine bones; and medially it is bounded by the orbital plate of the ethmoid. The entrance to the maxillary recess, the **nasomaxillary opening** (364/*3*) is reduced in size by the **uncinate process** (*J'*) of the dorsal nasal concha. The uncinate process enters the recess and divides it into a large rostral and a small caudal part (365/*a'*, *a*). The size of the maxillary recess is relatively constant and does not vary with increasing age of the animal. In the cat it is relatively small. There is a lateral, a medial, and a rostral **FRONTAL SINUS** in the dog, and they are separated from those of the other side by a **median septum** (365). The three frontal sinuses do not communicate with each other; each opens separately into the nasal cavity. By far the largest is the **lateral frontal sinus** (*b*), which is subdivided into several recesses by osseous lamellae. Its aperture (366/*1*) to the nasal cavity is narrowed by ectoturbinate 3 (*6*) which also invades the sinus. The **medial frontal sinus** (*c*) is usually small, but may extend to the caudal limit of the lateral frontal sinus by crowding the lateral sinus away from the median septum. The medial frontal sinus is invaded through its aperture (*2*) by ectoturbinate 2 (365/*c*). The **rostral frontal sinus** (*d*) is at the junction of the maxilla, the frontal, and the nasal bones; and part of its floor is formed by ectoturbinate 2 (*9*, *10*). Its aperture (366/*3*) to the nasal cavity is a narrow cleft on the medial wall of the sinus. (For more illustrations of the frontal sinuses see also Figures 283, 292, and 293.) The size of the frontal sinuses of the dog does not usually vary with age. They do, however, vary from animal to animal and from side to side in the same animal. In the cat there is only one **frontal sinus** on each side (58/*a'''*). It is located in the frontal bone and communicates rostrally with the nasal cavity. The **SPHENOID SINUS** excavates the presphenoid and is present only in the cat. Right and left sinuses are usually of

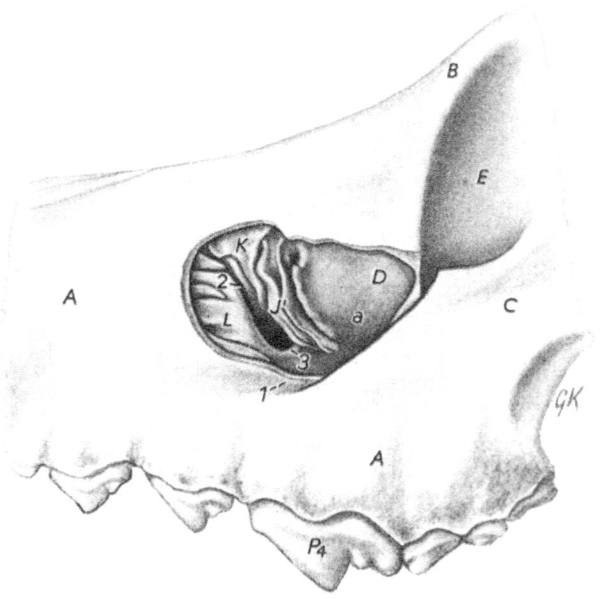

Fig. 364

Fig. 364. Nasomaxillary opening and uncinate process of the dog. Lateral aspect. (After Graeger 1958.)

A Maxilla; *B* Frontal bone; *C* Zygomatic bone; *D* Ethmoid, its orbital plate; *E* Orbit; *J'* Uncinate process; *K* Middle nasal concha; *L* Ventral nasal concha; P_4 Fourth premolar (sectorial); *a* Maxillary recess; *1* Infraorbital canal (opened); *2* Middle nasal meatus; *3* Mucosal fold, above it the nasomaxillary opening through which the uncinate process projects into the maxillary recess

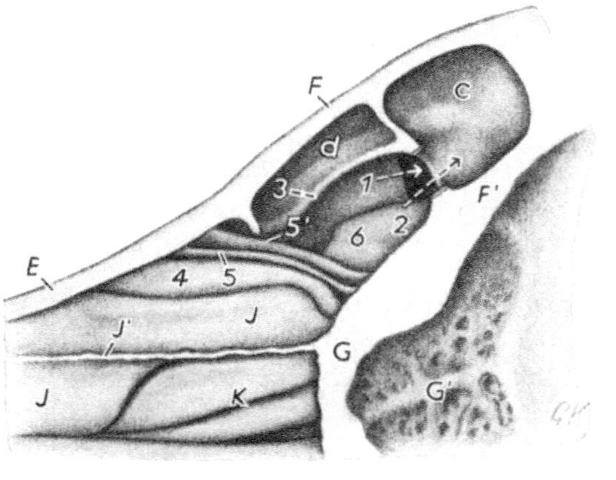

Fig. 366

Fig. 366. Nasal apertures of the frontal sinuses of the dog. The nasal septum and portions of ectoturbinate 2 and 3 have been removed. Sagittal section. Medial aspect of right half. (After Graeger 1958.)

E Nasal bone; *F*, *F'* Frontal bone; *G* Ethmoid, its perpendicular plate, *G'* its cribriform plate; *J*, *J'* Dorsal nasal concha, *J'* cut edge of the caudal portion of its basal lamella; *K* Middle nasal concha; *c* Medial frontal sinus; *d* Rostral frontal sinus

1 Aperture of lateral frontal sinus; *2* Aperture of medial frontal sinus; *3* Ventral boundary of aperture of rostral frontal sinus; *4* Ectoturbinate 1; *5* Cut edge of dorsomedial spiral lamella of ectoturbinate 2; *5'* Cut edge of ventrolateral spiral lamella of ectoturbinate 2; *6* Ventral spiral lamella of ectoturbinate 3

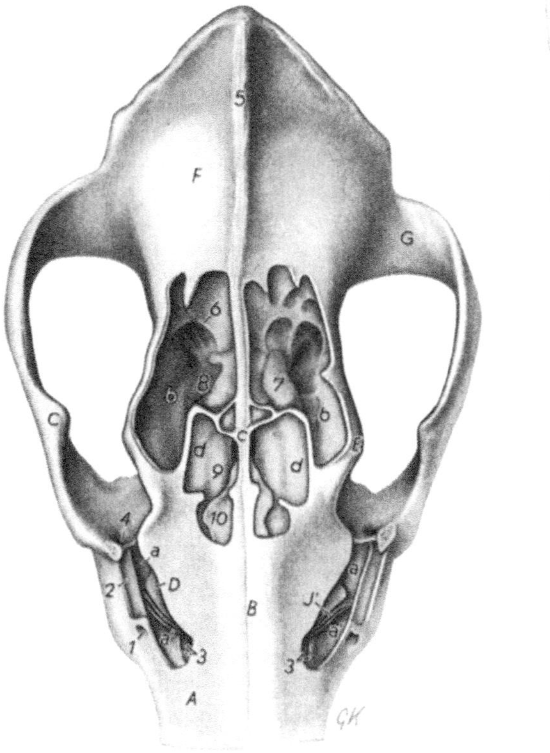

Fig. 365

Fig. 365. Paranasal sinuses of the dog. Dorsal aspect. (After Graeger 1958.)

A Maxilla; *B* Nasal bone; *C* Zygomatic bone, its frontal process; *D* Ethmoid, its orbital plate; *E* Frontal bone, its zygomatic process; *F* Parietal bone; *G* Temporal bone, its zygomatic process; *J'* Uncinate process (cut edge)

a, *a'* Caudal and rostral parts of maxillary recess, separated by the uncinate process (*J'*); *b* Lateral frontal sinus; *c* Medial frontal sinus with ventrolateral spiral lamella of ectoturbinate 2; *d* Rostral frontal sinus

1 Infraorbital foramen; *2* Infraorbital canal (opened); *3* Mucosal fold; *4* Maxillary foramen; *5* Ext. sagittal crest; *6* Osseous lamella; *7* Dorsal, *8* ventral spiral lamella of ectoturbinate 3; *9* Basal lamella, *10* dorsomedial spiral lamella of ectoturbinate 2

unequal size because the septum that separates them is generally not exactly median in position. Through the aperture of the sphenoid sinus enters endoturbinate IV.

LARYNX. The laryngeal cartilages of the dog differ in shape from those of the cat. The **thyroid cartilage** in the cat is longer than it is high, whereas in the dog it is higher than it is long (367, 368/*b'*). Its oblique line (*3*) is a low elevation in the cat, but is prominent in the dog. The rostral cornu (*1*) in the cat is long and straight; in the dog it is short and is bent medially. The thyroid fissure ventral to the rostral cornu is usually not bridged by connective tissue, with the result that a thyroid foramen is not formed in the carnivores. The caudal cornu (*2*) is wide at its base; it projects caudally only slightly in the cat, but quite far in the dog. Ventrally, the thyroid cartilage presents a caudal thyroid notch which is deep in the cat and shallow in the dog. The laryngeal prominence is present only in the dog. The median crest of the **cricoid cartilage** (367/*4*; 368/*5*) is low and rounded in both species. The arytenoid and thyroid articular surfaces on the cricoid present no special features (see p. 230). The cricoid arch is reduced in rostrocaudal width along the caudal border in the cat and along the cranial border in the dog. The **arytenoid cartilages** of the carnivores differ in that those of the dog have corniculate processes (368/*4*) which are shaped like hooks projecting dorsocaudally, and those of the cat have not. The muscular and vocal processes and the articular surface of the arytenoid cartilage present no special features. A small **interarytenoid cartilage** (319/*5*) is present in the dog between the two arytenoid cartilages; it is absent in the cat. Miller, Cristensen, and Evans (1964) also describe a **sesamoid cartilage** which is sometimes paired and appears to be intercalated in the arytenoideus transversus. The **cuneiform process** (368/*c*) of the arytenoid lies in the aryepiglottic fold in the dog, and is absent in the cat. The **epiglottis** (*a*) of the carnivores has a pointed apex. Its petiolus is absent in the cat, and in the dog it is wide and slightly forked. (For further illustrations of the laryngeal cartilages see Figures 306, 310, 314, 318, 319, and 326.)

THE CONNECTIONS OF THE LARYNGEAL CARTILAGES (330). The crico-thyroid connection is articular. From the medial surface of the sheetlike **cricothyroid ligament** (*9*), which in the cat closes the deep caudal thyroid notch, the membrana fibroelastica (*9'*) to the vocal ligament is given off. The joint capsule of the **cricoarytenoid joint** is strengthened ventromedially by the short **cricoarytenoid ligament** (*8*). The two arytenoid cartilages are held together by the **transverse arytenoid ligament** associated with which is the small interarytenoid cartilage. The connection between the thyrohyoid and the rostral cornu is a synchondrosis and not a diarthrodial joint as in the ruminants and horse. The **thyrohyoid membrane** (*2*) extends as a wide sheet from the hyoid bone to the thyroid cartilage. The epiglottis rides with its petiolus on the rostral border of the thyroid cartilage, the two cartilages being held together by the **thyroepiglottic ligament.** The **hyoepiglottic ligament** (*1*) extends from the basihyoid to the rostral border of the base of the epiglottis. The **vestibular ligament** (*6*) of the dog extends from the floor of the thyroid cartilage to the cuneiform process; it is absent in the cat. The **vocal ligament** (*7*) is directed cranioventrally and attaches on the floor of the thyroid cartilage. The **cricotracheal ligament** (*10*) connects the cricoid cartilage with the first tracheal cartilage.

MUSCLES OF THE LARYNX (369, 370). In the carnivores, the laryngeal muscles, except the thyroarytenoideus, conform in general to the description given on page 234. The **thyroarytenoideus** of the cat (369/*9*) is a uniform triangular sheet, which has an extensive origin on the base of the epiglottis, the floor of the thyroid cartilage, and the adjacent cricothyroid ligament, and, becoming progressively narrower, is inserted on the muscular and vocal processes of the arytenoid cartilage. In the dog it consists of ventricularis and vocalis muscles. The **ventricularis** (370/*9'*) enclosed with the vestibular ligament in the vestibular fold, arises ventrally from the thyroid lamina and is inserted on the muscular process of the arytenoid cartilage. The **vocalis** (*9''*) lies lateral and caudal to the vocal ligament inside the vocal fold, and extends from the floor of the thyroid cartilage to the muscular process of the arytenoid cartilage, but a few rostral fibers of the vocalis are attached to the vocal process.

LARYNGEAL CAVITY. The **aryepiglottic fold** at the entrance to the larynx connects in the dog (80/*8*) the epiglottis with the arytenoid cartilage and also encloses the cuneiform process; in the cat (58/*o'*), in which the cuneiform and corniculate processes are absent, it connects the epiglottis with the cricoid cartilage. **Laryngeal ventricles** (335/*1*) are present between the vestibular and the vocal folds in the dog. The ventricles are absent in the cat, but their position is marked by shallow depressions (334/*1'*). In addition to the ventral part of the cuneiform process, the **vestibular fold** of the dog (335/*2*) contains the ventricularis and

Fig. 367. Laryngeal cartilages of the cat. Lateral aspect.

a Epiglottic cartilage; *b* Thyroid cartilage, *b'* its lamina; *c* Left arytenoid cartilage; *d* Cricoid cartilage, its arch, *d'* its lamina; *e* Trachea; *1* Rostral cornu; *2* Caudal cornu; *3* Oblique line on thyroid cartilage; *4* Median crest of cricoid

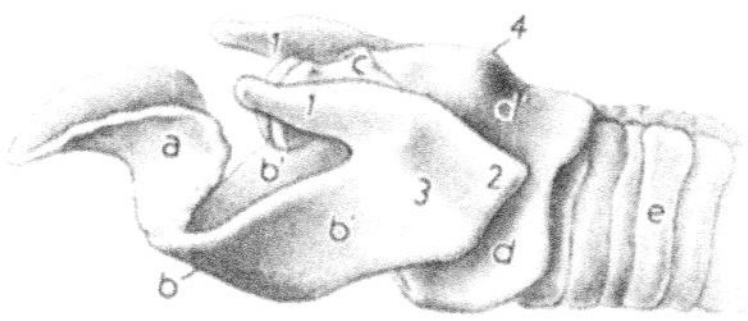

Fig. 367 (Cat)

Fig. 368. Laryngeal cartilages of the dog. Lateral aspect.

a Epiglottic cartilage; *b* Thyroid cartilage, *b'* its lamina; *c* Cuneiform process of arytenoid cartilage; *d* Arytenoid cartilage; *e* Cricoid cartilage, its arch, *e'* its lamina; *f* Trachea; *1* Rostral cornu; *2* Caudal cornu; *3* Oblique line on thyroid cartilage; *4* Corniculate process of arytenoid cartilage; *5* Median crest of cricoid

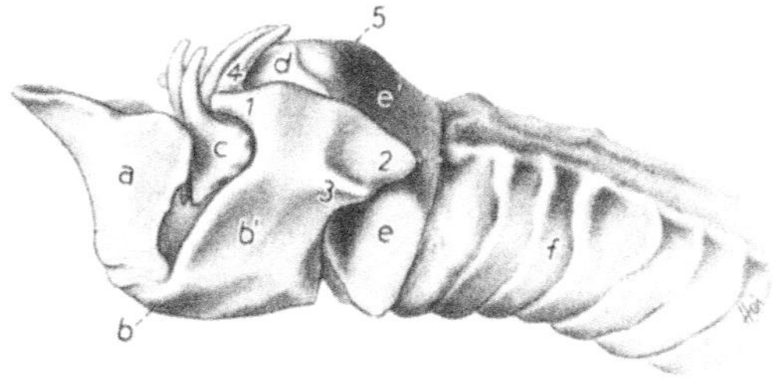

Fig. 368 (Dog)

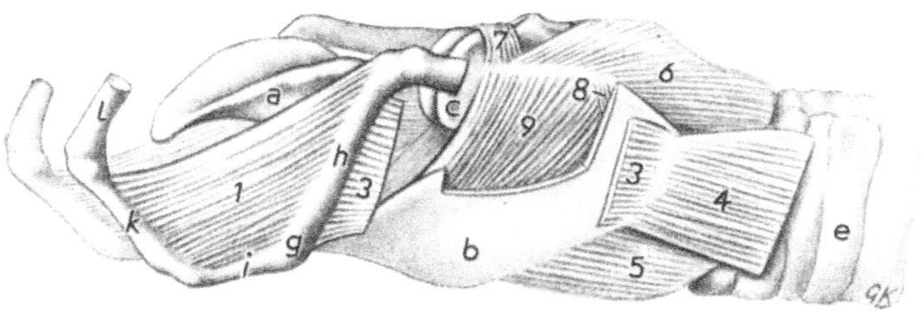

Fig. 369 (Cat)

Figs. 369, 370. Laryngeal muscles of the carnivores. Left lateral aspect.

a Epiglottis; *b* Thyroid cartilage; *c* Arytenoid cartilage, in the dog its corniculate process; *c'* Cuneiform process (dog); *d* Cricoid cartilage (not visible in the cat); *e* Trachea; *f* Lateral laryngeal ventricle; *g* Basihyoid; *h* Thyrohyoid; *i* Ceratohyoid; *k* Epihyoid; *l* Stylohyoid (not shown in the dog)
1 Ceratohyoideus; *3* Thyrohyoideus; *4* Sternothyroideus; *5* Cricothyroideus; *6* Cricoarytenoideus dorsalis; *7* Arytenoideus transversus; *8* Cricoarytenoideus lateralis; *9* Thyroarytenoideus (cat); *9'* Ventricularis; *9''* Vocalis

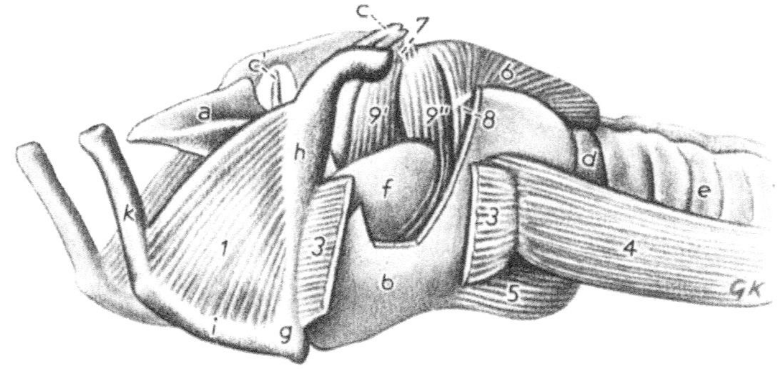

Fig. 370 (Dog)

the vestibular ligament. In the cat, the vestibular fold, which is attached dorsally to the arytenoid cartilage, contains neither the ligament nor the muscle and consists only of mucosa. The **vocal fold** (*3*) is directed cranioventrally and in addition to the vocal ligament contains the vocalis in the dog, and the entire thyroarytenoideus in the cat.

LYMPHATIC TISSUE in the form of solitary nodules is present in the dog in the lateral laryngeal ventricles and occasionally on the laryngeal surface of the epiglottis. In the cat, solitary nodules are always present on the laryngeal surface of the epiglottis, and in this species there is an aggregate of nodules (tonsilla paraepiglottica, 37/*8*) located along the edge of the aryepiglottic fold.

The **TRACHEA** consists of 38—43 cartilages in the cat and of 42—46 cartilages in the dog. The ends of the tracheal cartilages do not meet dorsally, and the **trachealis** is attached to the outside of the cartilages (341, 342).

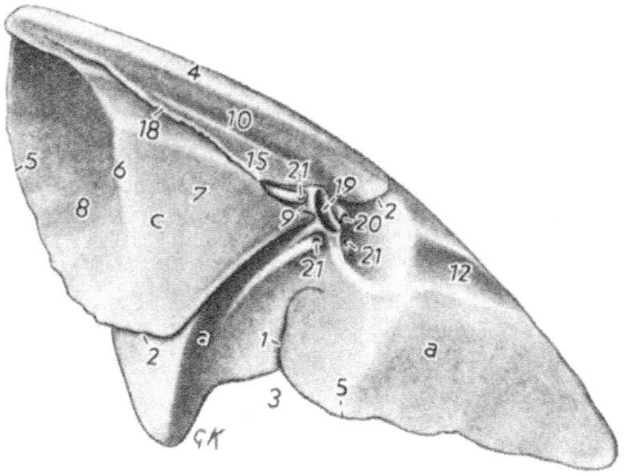

Fig. 371 (Cat, left lung)

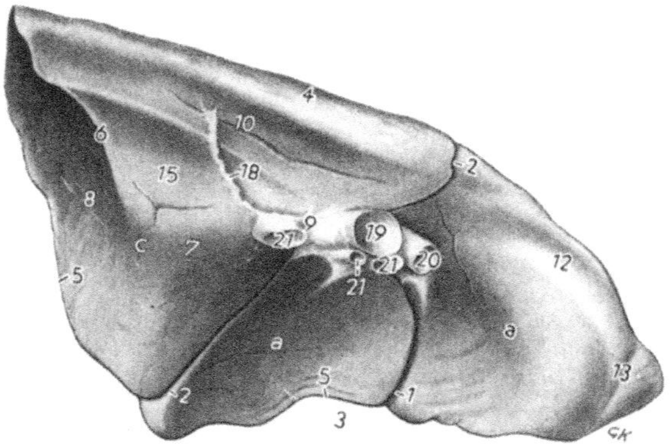

Fig. 373 (Dog, left lung)

Figs. 371, 373. Left lung of the carnivores, fixed in situ. Medial aspect.

a, *a* Divided cranial lobe; *c* Caudal lobe

1 Intralobar fissure; *2* Caudal interlobar fissure; *3* Cardiac notch; *4* Dorsal border; *5* (near *8*) Basal border; *5* (near *a*) Ventral border; *6* Border between diaphragmatic and mediastinal surfaces; *7* Mediastinal surface; *8* Diaphragmatic surface; *9* Hilus; *10* Aortic impression; *12* Impression made by left subclavian artery; *13* Impression made by int. thoracic artery (dog); *15* Esophageal impression; *18* Pulmonary ligament; *19* Principal bronchus; *20* Pulmonary artery; *21* Pulmonary veins

Figs. 372, 374. Right lung of the carnivores, fixed in situ. Medial aspect.

a Cranial lobe; *b* Middle lobe; *c* Caudal lobe; *d* Accessory lobe

1 Cranial interlobar fissure; *2* Caudal interlobar fissure; *3* Cardiac notch; *4* Dorsal border; *5* (near *8*) Basal border; *5* (near *a*) Ventral border; *6* (cat) Border between diaphragmatic and mediastinal surfaces; *6* (dog) Cranial border of accessory lobe; *7* Mediastinal surface; *8* Diaphragmatic surface; *9* Hilus; *10* Aortic impression; *12* Impression made by int. thoracic artery and vein (dog); *13* Groove for caudal vena cava; *14* Impression made by cranial vena cava; *15* Impression made by right azygous vein (cat); *16* Impression made by costocervical vein (dog); *21* Esophageal impression; *22* Impression made by trachea (cat); *24* Impressions made by ribs; *25* Pulmonary ligament; *26* Principal bronchus; *28* Pulmonary artery; *29* Pulmonary veins

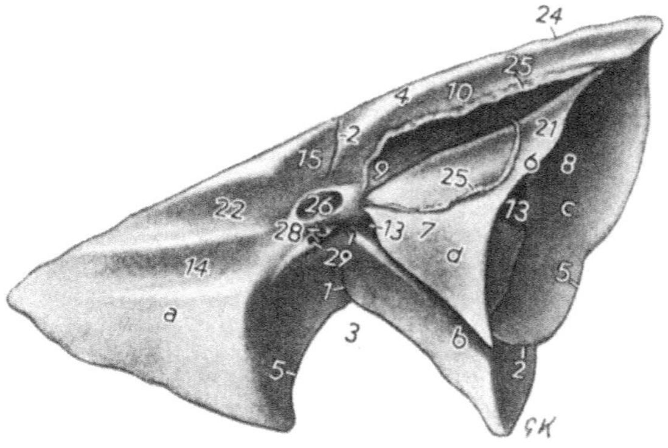

Fig. 372 (Cat, right lung)

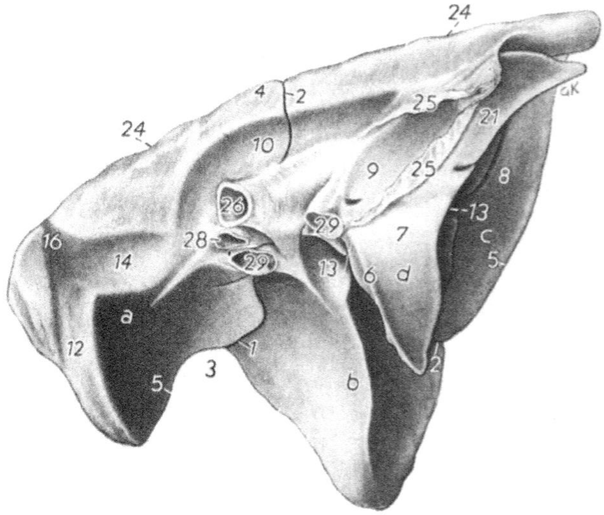

Fig. 374 (Dog, right lung)

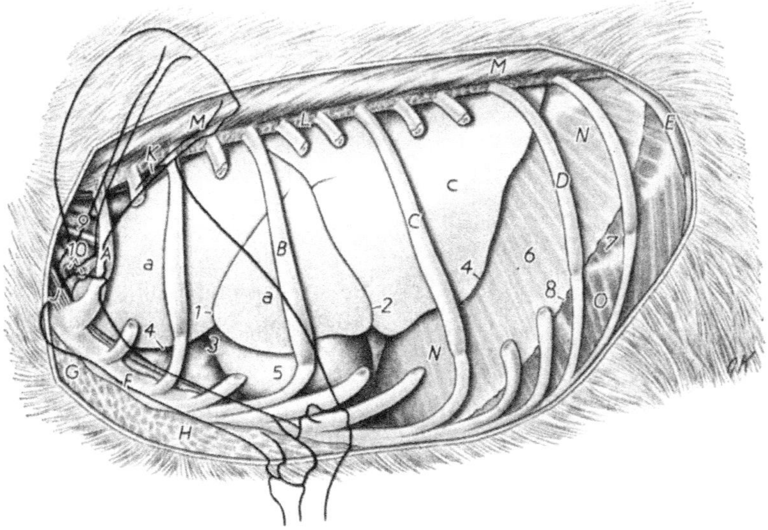

Fig. 375. Thoracic organs of the dog. Left aspect.

A First rib; *B* Fifth rib; *C* Eighth rib; *D* Eleventh rib; *E* Thirteenth rib; *F* Sternum; *G* Supf. pectoral muscles; *H* Deep pectoral muscle; *J* Sternocephalicus; *K* Longus colli; *L* Intercostal muscles; *M* Longissimus thoracis; *N* Diaphragm; *O* Transversus abdominis

a, a Divided cranial lobe; *c* Caudal lobe

1 Intralobar fissure; *2* Caudal interlobar fissure; *3* Cardiac notch; *4* (near *c*) Basal border; *4* (near *a*) Ventral border; *5* Heart inside pericardium; *6* Diaphragm covered with pleura; *7* Diaphragm not covered with pleura; *8* Line of pleural reflection; *9* Trachea (dorsal to the trachea is the esophagus, ventral to the trachea is the common carotid artery); *10* Axillary artery and vein

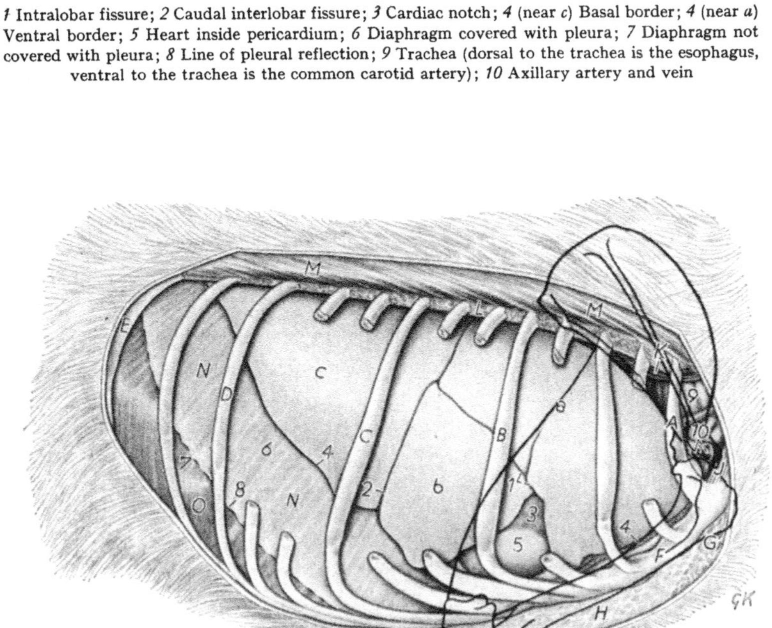

Fig. 376. Thoracic organs of the dog. Right aspect.

A First rib; *B* Fifth rib; *C* Eighth rib; *D* Eleventh rib; *E* Thirteenth rib; *F* Sternum; *G* Supf. pectoral muscles; *H* Deep pectoral muscle; *J* Sternocephalicus; *K* Longus colli; *L* Intercostal muscles; *M* Longissimus thoracis; *N* Diaphragm; *O* Transversus abdominis

a Cranial lobe and caudal border of triceps; *b* Middle lobe; *c* Caudal lobe

1 Cranial interlobar fissure; *2* Caudal interlobar fissure; *3* Cardiac notch; *4* (near *c*) Basal border; *4* (near *F*) Ventral border; *5* Hear inside pericardium; *6* Diaphragm covered with pleura; *7* Diaphragm not covered with pleura; *8* Line of pleural reflection; *9* Trachea (ventral to the trachea is the common carotid artery); *10* Axillary artery and vein

The **LUNGS** of the carnivores present very deep **fissures** which reach almost to the bronchi; only the left intralobar fissure (375/*1*) is not quite so deep as the others. The **left lung** consists of the divided cranial lobe (*a*) and the caudal lobe (*c*). The **right lung** consists of the undivided cranial lobe, the middle lobe, and the caudal lobe, to which the accessory lobe is attached (374). The apex of the right cranial lobe is rounded in the dog (*a*) and is pointed in the cat (372/*a*). Animals kept indoors for most of their lives have gray, bluish gray, or even black lungs owing to the inhalation of carbon particles (anthracosis). As a rule, **pulmonary lobules** cannot be discerned with the naked eye on the surface of the lung. The position of the lungs when fixed in situ in its expiratory positions is shown in Figures 375 and 376.

BRONCHIAL TREE. On the left side, the **cranial bronchus** (357, 358/*5*) divides into cranial and caudal segmental bronchi to serve the cranial and caudal parts of the cranial lobe. The **caudal bronchus** (*8*), after entering the caudal lobe, gives off dorsal and ventral segmental bronchi. On the right side, the **cranial bronchus** (*5*) ventilates the undivided cranial lobe by means of dorsal and ventral segmental bronchi. The **middle bronchus** (*6*) arises a short distance caudal to the origin of the cranial bronchus and passes into the middle lobe. The **caudal bronchus** (*8*) enters the caudal lobe and gives off dorsal and ventral segmental bronchi. The **accessory bronchus** (*7*) arises from the right caudal bronchus at the cranial end of the accessory lobe. (See also Fig. 350.) The division of the bronchi may vary between dog and cat or in the same species. For more detail see Tucker and Krementz (1957).

BLOOD VESSELS OF THE LUNGS. In the carnivores, right and left **bronchial arteries** for the respective halves of the bronchial tree are present. The right artery of the dog arises from the right sixth dorsal intercostal artery, and the left artery arises directly from the aorta close to the origin of the sixth dorsal intercostal artery. Right and left **bronchial veins** follow the arteries and empty into the azygous vein. The **pulmonary arteries** and **veins** follow the bronchi in the cranial and middle lobes (350). In the caudal lobes, only the arteries follow the bronchi, while the veins run intersegmentally. The following **anastomoses** are present: (1) between branches of the bronchial artery and those of the pulmonary artery, with the formation of intima cushions in the pulmonary arteries, (2) in the area of the hilus between branches of the bronchial vein and branches of the pulmonary vein, (3) between branches of the bronchial artery and those of the pulmonary vein, and (4) between the capillaries of the bronchial arteries and the alveolar capillary networks formed by the pulmonary arteries.

The Respiratory Organs of the Pig

The disc-shaped apex of the **NOSE** is fused centrally with the upper lip and projects from the face as the **rostrum** or snout (20). The **nostrils** are circular and are supported by the rostral bone and the nasal cartilages.

NASAL CARTILAGES (278). Associated with the nasal cartilages of the pig is the **rostral bone** (os rostrale, *a'*) which is situated between the nostrils and is attached to the rostral border of the nasal septum. It is a thick transverse plate strengthening the snout in adaptation to the burrowing and rooting habits of the pig. The **dorsal** and **ventral lateral nasal cartilages** (*c*, *b*, *d*) extend well into the lateral walls of the nostrils and are united along most of their length; the dorsal cartilage is divided by a deep fissure into rostral and caudal parts. The rostral part supports the nostril in much the same fashion as the lamina of the alar cartilage does in the horse. The **lateral accessory cartilage** (*e*) is a curved rod, which arises from the ventral part of the rostral bone and extends into the lateral wall of the nostril. The **medial accessory cartilage** connects the ventral concha with the ventral lateral nasal cartilage and supports the alar fold (287/*6*).

The modified skin surrounding the nostrils, the **planum rostrale** (271/*a*), is covered with short tactile hairs and is divided by grooves into small convex fields. It is kept moist by glands opening in the center of these fields. The **nasolacrimal duct** opens on the floor of the nostril at the junction of skin and nasal mucosa. The pig has a second nasolacrimal opening which is on the lateral surface of the ventral nasal concha near its caudal end. This opening becomes patent after birth and remains functional throughout life, while the rostral portion of the nasolacrimal duct, with its opening in the nostril, degenerates.

The **NASAL CAVITY** is relatively long and narrow (287). Right and left ventral nasal meatuses lead through the **choanae** (59/*13*) into the **nasopharynx.**

The **dorsal nasal concha** (287/*O*) may be divided into rostral, middle, and caudal portions. The rostral portion is long and extends caudally to the level of the third or fourth cheek tooth; the caudal 2 cm. of the concha is the caudal portion. Both rostral and caudal portions are plate-like and consist merely of the basal lamella of the concha that projects ventrally from the roof of the nasal cavity. In the middle portion there is a spiral lamella which turns first laterally and then dorsally and unites with the nasal bone to form the **dorsal conchal sinus** (*c, c'*). The small dorsal part (*c'*) of this sinus lies between the external and internal plates of the nasal bone, while the larger ventral part occupies the concha.

The basal lamella of the **ventral nasal concha** (*Q*) arises from the lateral wall of the nasal cavity and divides into dorsal and ventral spiral lamellae. The dorsal turns dorsally and then laterally and encloses a recess (*a*). The ventral curves first ventrally and then laterally, and encloses a recess (*b*) rostrally and the **ventral conchal sinus** (*i*) caudally.

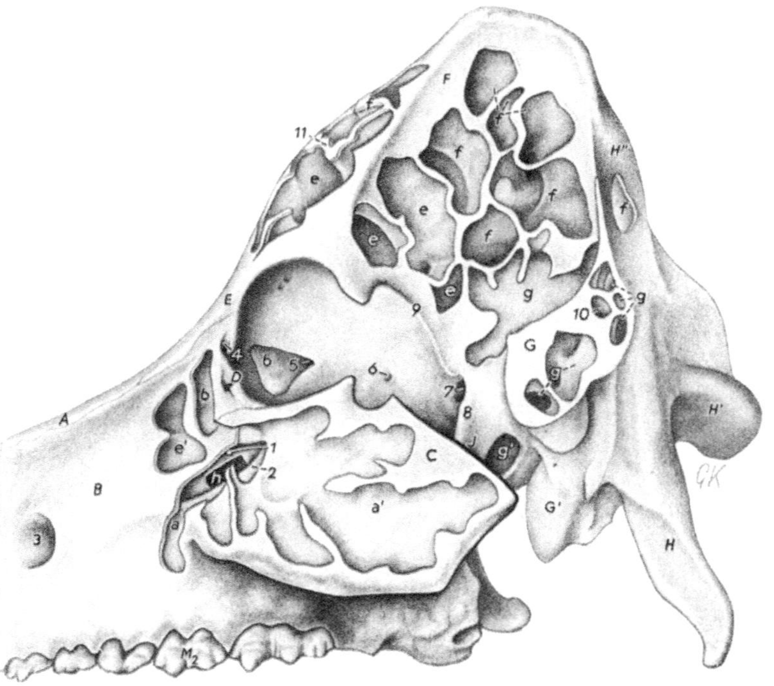

Fig. 377. Skull of a pig with opened paranasal sinuses. Lateral aspect. (After Loeffler 1959a.)

A Nasal bone; *B* Maxilla; *C* Zygomatic bone; *D* Lacrimal bone; *E* Frontal bone; *F* Parietal bone; *G* Temporal bone, cut surface of zygomatic process; *G'* Tympanic bulla; *H* Paracondylar process; *H'* Occipital condyle; *H''* Squamous part of occipital bone; *J* Presphenoid, pterygoid process; M_2 Second molar

a, a' Maxillary sinus; *b, b* Lacrimal sinus; *e* Lateral rostral frontal sinus, *e'* its rostral recess in the lateral nasal wall; *f* Caudal frontal sinus; *g* Sphenoid sinus, *g'* its recess in the pterygoid process of the presphenoid; *h* Ethmoidal cell

1 The arrow points to the nasomaxillary aperture; *2* Partition between lateral and medial parts of maxillary sinus; *3* Infraorbital foramen; *4* Lacrimal foramen; *5* Aperture of lacrimal sinus; *6* Ethmoid foramen; *7* Optic canal; *8* Pterygoid crest; *9* Orbitotemporal crest; *10* Ext. acoustic meatus; *11* Transverse septum of frontal sinuses

The **middle nasal concha** (*P*) is small and projects very little from the ethmoid labyrinth. Its basal lamella gives rise to dorsal and ventral spiral lamellae which contain resecces.

The **incisive duct** and the **vomeronasal cartilage** are sometimes split. The **lateral nasal gland** is located in the maxillary sinus. A description of the folds in the nasal vestibule (287/*5, 6, 7*) may be found on page 216.

The rostral part of the **NASOPHARYNX** is divided by the median **pharyngeal septum** (59/*12*) which extends ventrally from the vomer. The **pharyngeal openings of the auditory tubes** (*14*) appear as vertical slits on the lateral walls of the nasopharynx, and in the roof of the nasopharynx is the uneven **pharyngeal tonsil** (*15*).

PARANASAL SINUSES. In addition to the dorsal and ventral conchal sinuses, the pig has maxillary, frontal, lacrimal, and sphenoid sinuses.

The **MAXILLARY SINUS** (377/*a*, *a'*) occupies the caudal part of the maxilla and the zygomatic bone, and in fully grown pigs extends considerably into the zygomatic arch. Rostrally, the sinus is uniform, while caudally it is divided into medial and lateral recesses by a substantial bony lamella (*2*). This lamella arises from the floor of the sinus and contains an ethmoidal cell (see below) and the infraorbital canal. The **nasomaxillary aperture** (*1*) is found in the middle nasal meatus at the level of the sixth cheek tooth and is covered medially by the dorsal nasal concha. At birth, the maxillary sinus is less than 1 cm. in diameter. With the development of the skull it reaches the dimensions shown in Figures 294, 295, 377, and 378 in mature pigs.

The **FRONTAL SINUSES** are usually not present at birth. In fully grown pigs, however, on which this description is based, they constitute large air spaces which occupy several of the dorsal and lateral cranial bones. Right and left frontal sinuses are separated from each other by a sagittal septum (378/*6*) which usually deviates from the median plane. A variably placed transverse septum (*7*) divides the frontal sinuses of a side into rostral and caudal sinuses. Rostral to the transverse septum are the medial and lateral rostral frontal sinuses, and caudal to the septum

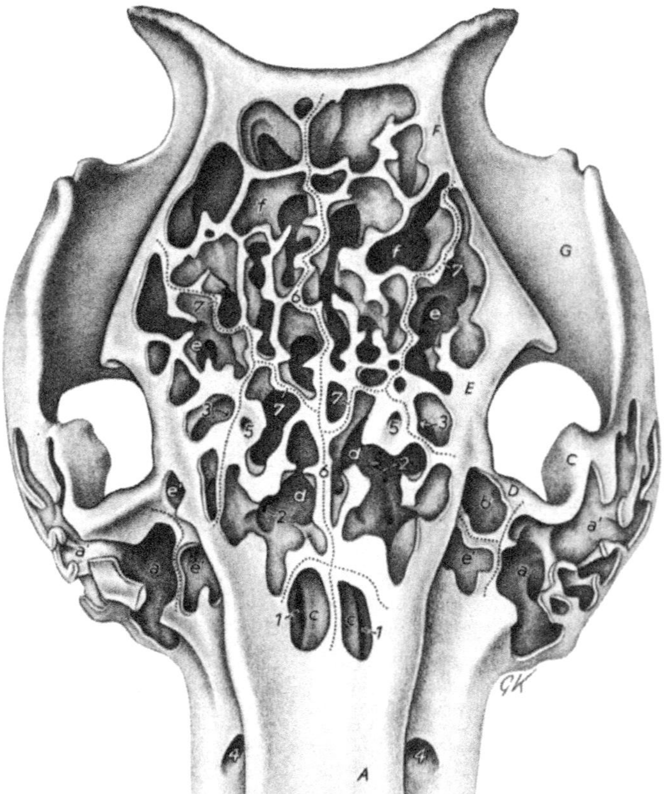

Fig. 378. Skull of a pig with opened paranasal sinuses. Rostrodorsal aspect. (The dotted lines indicate the boundaries between the sinuses.) (After Loeffler 1959a.)

A Nasal bone; *B* Maxilla; *C* Zygomatic bone; *D* Lacrimal bone; *E* Frontal bone; *F* Parietal bone; *G* Zygomatic process of temporal bone

a, a' Maxillary sinus; *b* Lacrimal sinus; *c* Dorsal part of dorsal conchal sinus; *d* Medial rostral frontal sinus; *e* Lateral rostral frontal sinus, *e'* its rostral recess in the lateral nasal wall; *f* Caudal frontal sinus

1 Communication between dorsal and ventral parts of dorsal conchal sinus (see Fig. 287/*c, c'*); *2* Aperture of medial rostral frontal sinus; *3* The arrow points in the direction of the aperture of the lateral rostral frontal sinus; *4* Infraorbital foramen; *5* Supraorbital foramen; *6* Sagittal septum of frontal sinuses; *7* Transverse septum of frontal sinuses

is the caudal frontal sinus. The **caudal frontal sinus** (*f*) is the largest paranasal sinus in the pig. It occupies the frontal and the occipital bones, and in older animals it extends also into parts of the temporal bone, thus surrounding the cranial cavity dorsally, laterally and caudally. It is separated from the sphenoid sinus only by a thin osseous plate. The aperture of the caudal frontal sinus (287/*10*) is in the middle nasal meatus close to the aperture of the dorsal conchal sinus. The two **rostral frontal sinuses** (378/*d*, *e*) excavate the frontal bone medial, but also rostral and caudal (377), to the orbit, and vary in their extent considerably from one animal to another. Their apertures (378/*2*, *3*) connect them to dorsal ethmoidal meatuses in the caudal part of the nasal cavity. The frontal sinuses are all variously, but incompletely, subdivided by bony lamellae to form a number of recesses or diverticula. For more detail see the study by Loeffler (1959a).

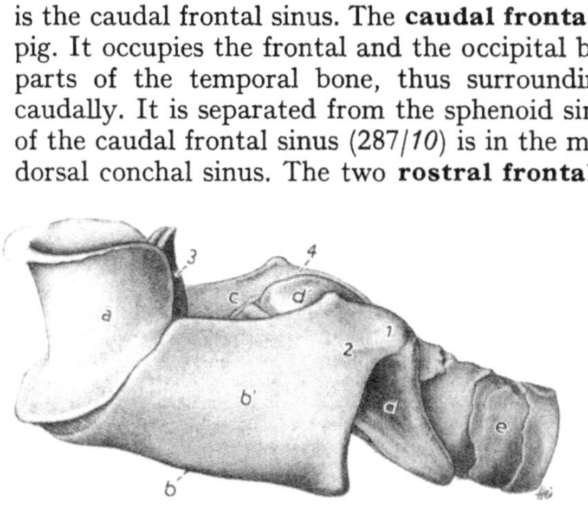

Fig. 379. Laryngeal cartilages of the pig. Left lateral aspect.

a Epiglottic cartilage; *b* Thyroid cartilage, *b'* its lamina; *c* Left arytenoid cartilage; *d* Cricoid cartilage, arch, *d'* lamina; *e* Trachea *1* Caudal cornu; *2* Oblique line; *3* Corniculate process of arytenoid; *4* Median crest of cricoid

The **LACRIMAL SINUS** starts to develop when the pig is about six months old. It is usually an independent sinus, but occasionally it is a diverticulum of the lateral rostral frontal sinus. The lacrimal sinus (377/*b*) is related ventrally to the maxillary sinus, medially and dorsally to the rostral frontal sinuses, and caudally to the orbit, and makes contact also with ethmoidal cells. Unless the sinus is a diverticulum of the lateral rostral frontal sinus, the aperture of the lacrimal sinus opens into a lateral ethmoidal meatus.

The **SPHENOID SINUS** (377/*g*, *g'*) of the adult pig is very large compared to that of the other domestic mammals. In the eight-month-old pig, however, it is only about 1 cm. in diameter. When fully developed, the sphenoid sinus consists of a central cavity and three recesses. The central cavity lies in the bodies of the presphenoid and basisphenoid and underlies the optic chiasma and the floor of the cranial cavity 287/*g*). The small caudal recess (295/*g''*) extends into the basilar part of the occipital bone; the rostral recess (*g'*) extends into the pterygoid process of the basisphenoid and into the sphenoid process of the palatine bone; and the lateral recess (*g*), the largest of the three, occupies the squamous part temporal bone, extending as far as the caudal frontal sinus in old animals. The aperture of the sphenoid sinus (287/*13*) opens into a ventral ethmoidal meatus.

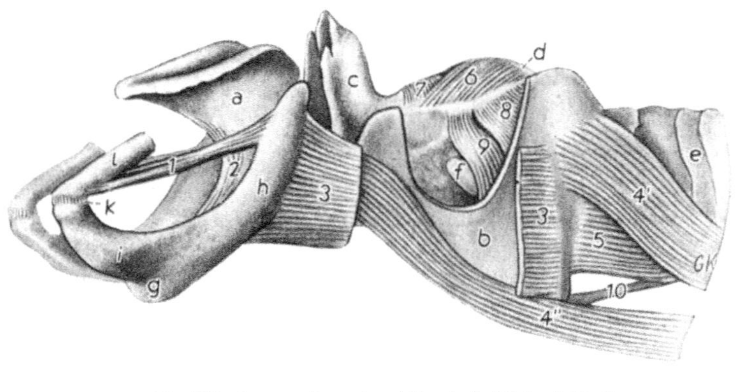

Fig. 380. Laryngeal muscles of the pig. Left lateral aspect.

a Epiglottis; *b* Thyroid cartilage; *c* Corniculate process of arytenoid; *d* Cricoid cartilage; *e* Trachea; *f* Lateral laryngeal ventricle; *g* Basihyoid; *h* Thyrohyoid; *i* Ceratohyoid; *k* Epihyoid ligament; *l* Stylohyoid

1 Ceratohyoideus; *2* Hyoepiglotticus; *3* Thyrohyoideus; *4'*, *4''* Dorsal and ventral parts of sternothyroideus; *5* Cricothyroideus; *6* Cricoarytenoideus dorsalis; *7* Arytenoideus transversus; *8* Cricoarytenoideus lateralis; *9* Thyroarytenoideus; *10* Cricothyroid ligament

Several **ethmoidal cells** surrounding the ethmoid labyrinth are present in the pig. They open into ethmoidal meatuses and are labeled with the letter (*h*) in Figures 287, 294, 295, and 377.

LARYNX. The **thyroid cartilage** (379/*b*) is very long; its wide and relatively high laminae are smooth and present short oblique lines (*2*) only caudally. The rostral cornu, thyroid fissure, and thyroid foramen are absent, and the caudal cornu (*1*) is plump and short. Ventrally, the rostral margin of the thyroid cartilage is convex, while caudally there is a short median projection (311). A laryngeal promince on the ventral surface of the thyroid cartilage may be present in older pigs. The lamina of the **cricoid cartilage** (379/*d'*) is long, and its median crest is prominent. Attached to the caudal end of the lamina are one or two small cartilaginous plates which sometimes fuse with the lamina. The articular surface for the arytenoid is convex, while that for the thyroid is concave. The arch of the cricoid cartilage (*d*) is narrow, directed

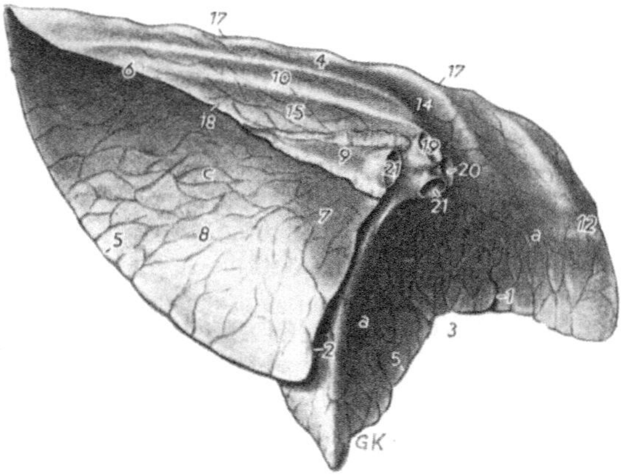

Fig. 381. Left lung of the pig, fixed in situ. Medial aspect.

a, a Divided cranial lobe; *c* Caudal lobe
1 Intralobar fissure; *2* Caudal interlobar fissure; *3* Cardiac notch; *4* Dorsal border; *5* (near *8*) Basal border; *5* (near *a*) Ventral border; *6* Border between diaphragmatic and mediastinal surfaces; *7* Mediastinal surface; *8* Diaphragmatic surface; *9* Hilus; *10* Aortic impression; *12* Impression made by left subclavian artery; *14* Impression made by left azygous vein; *15* Esophageal impression; *17* Impressions made by ribs; *18* Pulmonary ligament; *19* Principal bronchus; *20* Pulmonary artery; *21* Pulmonary veins

caudoventrally, and drawn out to a point. The **arytenoid cartilages** of the pig are unusual because of their large corniculate processes (320, 321/*3*). These processes are fused dorsomedially (*3'*) so as to form an unpaired structure, and on each side they have a large semilunar appendage. The muscular process of the arytenoid is high, the articular surface is deeply concave, and the vocal process is large. Between the arytenoid cartilages is a small interarytenoid cartilage (*5*). The **epiglottis** (379/*a*) has high sides and forms a deep trough. Its dorsal border is rounded, and its ventral border is turned rostrally and related to the hyoid bone.

Fig. 382. Right lung of the pig, fixed in situ. Medial aspect.

a Cranial lobe; *b* Middle lobe; *c* Caudal lobe; *d* Accessory lobe
1 Cranial interlobar fissure; *2* Caudal interlobar fissure; *3* Cardiac notch; *4* Dorsal border; *5* (near *3*) Ventral border; *5* (near *c*) Basal border; *6* Border between diaphragmatic and mediastinal surfaces; *7* Mediastinal surface; *8* Diaphragmatic surface; *9* Hilus; *10* Aortic impression; *12* Impression made by int. thoracic vessels; *13* Groove for caudal vena cava; *14* Impression made by cranial vena cava; *16* Impression made by costocervical vein; *21* Esophageal impression; *24* Impressions made by ribs; *25* Pulmonary ligament; *26* Principal bronchus; *27* Tracheal bronchus; *28, 28'* Pulmonary arteries; *29* Pulmonary veins

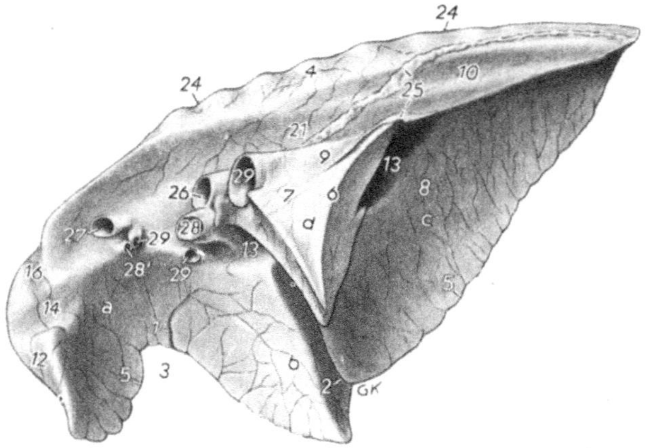

THE CONNECTIONS OF THE LARYNGEAL CARTILAGES (331). The cricoid and thyroid cartilages come together in the **cricothyroid joint** and are connected by the **cricothyroid ligament** (*9*). The ligament bridges the large gap between the caudally-directed cricoid arch and the caudal border of the thyroid cartilage, and is continued for a considerable distance on the inside of the thyroid cartilage. The fibroelastic membrane (330/*9'* shown here is the dog) is short in the pig. The weak **cricoartenoid ligament** (331/*8*) is a ventromedial thickening in the capsule of the **cricoarytenoid joint.** Because of the fusion of the corniculate processes, this joint, compared with that of the other domestic mammals, is very restricted in its movements. The **transverse arytenoid ligament** connects the arytenoid cartilages to one another and contains the small interarytenoid cartilage. The connection between the thyrohyoid and the thyroid lamina is not articular, as in the ruminants and horse, but is syndesmotic*. The epiglottis rests with its wide base on the **thyrohyoid membrane** (*2*) and is firmly attached to the hyoid bone by the **hyoepiglottic ligament** (*1*). The epiglottis is loosely connected to the thyroid cartilage by the **thyroepiglottic ligament** (*3*). The **vestibular ligament** (*6*) extends from the base of the epiglottis to the lateral surface of the corniculate process of the arytenoid cartilage. The **vocal ligament** (*7*) is obliquely placed and consists of rostral and caudal parts, which lie next to each other. It arises from the caudal part of the floor of the thyroid cartilage and from the adjacent cricothyroid ligament, and passes rostrodorsally to the vocal process. The wide space on each side between the ventricular and vocal ligaments is filled by a sheet of fibers (*6'*) which is lateral to the lateral laryngeal ventricle, and passes from the thyroid lamina to the rostroventral border of the arytenoid. The **cricotracheal ligament** (*10*) connects the cricoid cartilage to the first tracheal cartilage.

MUSCLES OF THE LARYNX (380). In the pig, the laryngeal muscles, except the thyroarytenoideus, conform in general to the description given on page 234. The **thyroarytenoideus** (*9*) is not divided into ventricularis and vocalis, but is uniform and somewhat fan-shaped. It has a wide origin on the thyroid cartilage and is inserted on the muscular process, and with a few rostral fibers also on the vocal process, of the arytenoid cartilage.

LARYNGEAL CAVITY. The **aryepiglottic folds** (38/*m*) at the entrance of the larynx connect the lateral borders of the epiglottis with the dorsal wall of the larynx, by-passing the arytenoid cartilages laterally. A **median laryngeal recess** (336/*4*) is present in the floor of the vestibule. The large corniculate processes of the arytenoids project medially from the lateral walls of the vestibule. The vestibular folds are absent. At the caudal end of the relatively long laryngeal vestibule are the **laryngeal ventricles** (*1*), the entrances to which are between the rostral and caudal parts of the vocal ligaments. The caudal part of the vocal ligament and the thyroarytenoideus form the bulk of the **vocal fold** (*3*).

Only small amounts of **LYMPHATIC TISSUE** are found in the laryngeal mucosa of young pigs. In older animals, solitary nodules are present in the aryepiglottic folds, in the lateral ventricles, and in the mucosa of the epiglottis. Larger aggregations of lymphatic tissue lateral to the base of the epiglottis are the **paraepiglottic tonsils.**

The **TRACHEA** of the pig is circular in cross section (343) and consists of 32—36 **cartilages.** The ends of the tracheal cartilages always overlap each other and the **trachealis muscle** is attached to their deep surface.

In the fresh state, the **LUNGS** of the pig are bluish pink, and their fissures are moderately deep. The **left lung** is composed of a divided cranial lobe (381/*a*, *a*) and a caudal lobe (*c*). The **right lung** consists of an undivided cranial lobe (382/*a*), which is served by the tracheal bronchus (*27*), a middle lobe, a caudal lobe, and an accessory lobe (*b*, *c*, *d*). The position of the lungs when fixed in situ in the state of expiration is shown in Figures 383 and 384. **Lobulation** is not quite as distinct as in the ox, but is visible over the entire surface of the lungs.

BRONCHIAL TREE (359). On the left side, the **cranial bronchus** (*5*) divides into cranial and caudal segmental bronchi (*11*), each serving one of the parts of the cranial lobe. The **caudal bronchus** (*8*) gives off four dorsal and four ventral segmental bronchi which ventilate the caudal lobe. On the right side, the **tracheal bronchus** (*4*) serves the undivided right cranial lobe. Close to the tracheal bifurcation arises the **middle bronchus** (*6*) which serves the middle lobe. Exceptionally, a branch of the tracheal bronchus may supply the cranial portion of the middle lobe. The **accessory bronchus** (*7*) arises just caudal to the middle bronchus and passes ventrally into the small accessory lobe. The right **caudal bronchus** (*8*), like the one on the left, gives off four dorsal and four ventral segmental bronchi.

BLOOD VESSELS OF THE LUNGS. The branches of the **pulmonary arteries** and **veins** follow the bronchi in the cranial and middle lobes of the lungs. In the caudal lobes only the arteries follow the bronchi, while the veins run intersegmentally, receiving blood from adjacent bronchopulmonary segments. The **bronchial artery** arises from the bronchoesophageal artery, and the **bronchial vein** comes from the left azygous vein. Each divides at the tracheal bifurcation to supply the two lungs. There are many anastomoses between the pulmonary and the bronchial vessels.

---

* Gr. *syn-*, together; *desmos*, ligament: connected by ligaments.

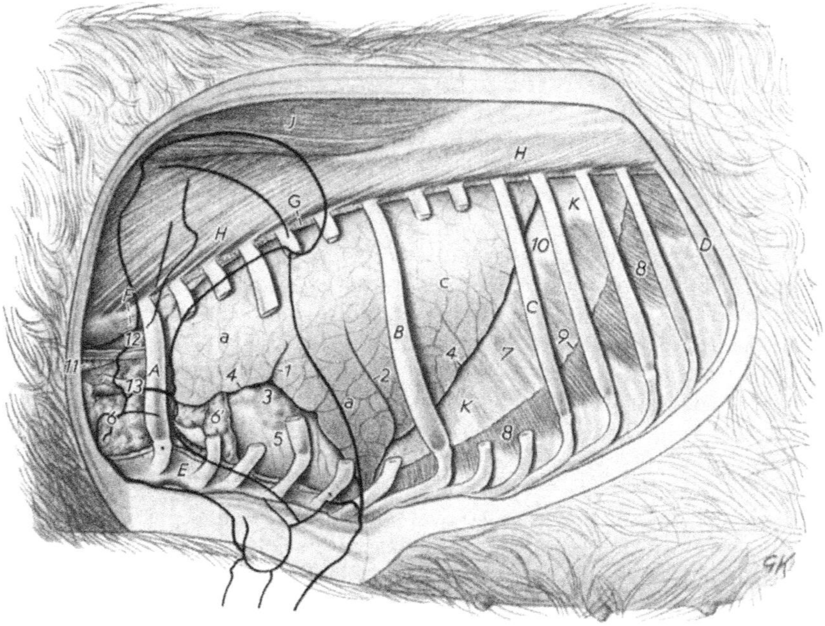

Fig. 383. Thoracic organs of the pig. Left aspect.

A First rib; *B* Seventh rib; *C* Tenth rib; *D* Fourteenth rib; *E* Sternum; *F* Longus colli; *G* Intercostal muscles; *H* Longissimus thoracis; *J* Spinalis thoracis et cervicis; *K* Diaphragm; *a, a* Divided cranial lobe; *c* Caudal lobe

1 Intralobar fissure; *2* Caudal interlobar fissure; *3* Cardiac notch; *4* (near *a*) Ventral border; *4* (near *c*) Basal border; *5* Heart inside pericardium; *6* Cervical part, *6'* thoracic part, of thymus (ventral to the thoracic part of the thymus is the cranial lobe of the right lung); *7* Diaphragm covered with pleura; *8* Diaphragm not covered with pleura; *9* Line of pleural reflection; *10* Tendinous center of diaphragm; *11* Trachea; *12* Esophagus (between trachea and esophagus is the common carotid artery accompanied by the vagosympathetic trunk); *13* Ext. jugular vein

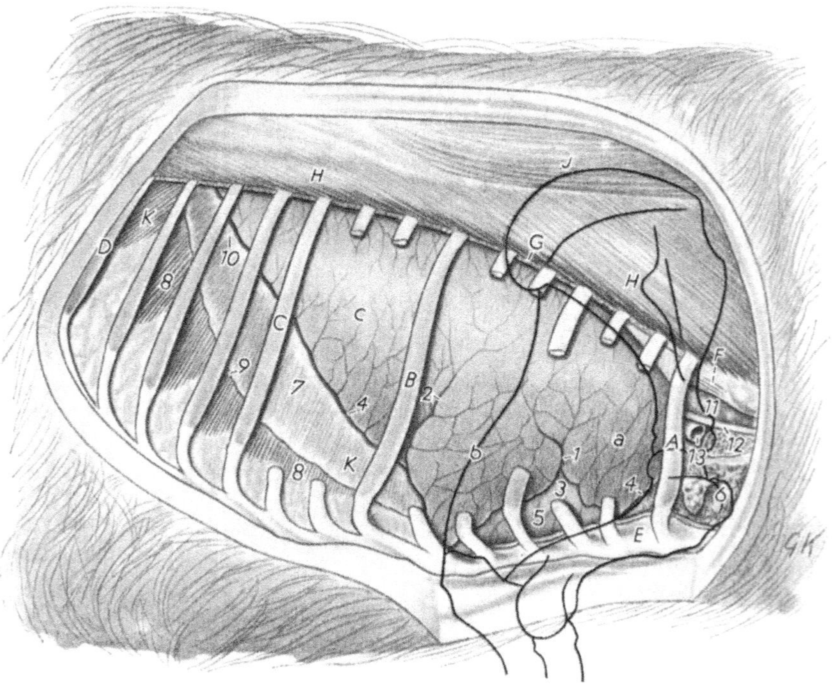

Fig. 384. Thoracic organs of the pig. Right aspect.

A First rib; *B* Seventh rib; *C* Tenth rib; *D* Fourteenth rib; *E* Sternum; *F* Longus colli; *G* Intercostal muscles; *H* Longissimus thoracis; *J* Spinalis thoracis et cervicis; *K* Diaphragm; *a* Cranial lobe; *b* Middle lobe and caudal border of triceps; *c* Caudal lobe

1 Cranial interlobar fissure; *2* Caudal interlobar fissure; *3* Cardiac notch; *4* (near *a*) Ventral border; *4* (near *c*) Basal border; *5* Heart inside pericardium; *6* Thymus; *7* Diaphragm covered with pleura; *8* Diaphragm not covered with pleura; *9* Line of pleural reflection; *10* Tendinous center of diaphragm; *11* Trachea; *12* Common carotid artery; *13* Ext. jugular vein

The Respiratory Organs of the Ruminants

NOSE. There is little difference between the **NASAL CARTILAGES** of the small ruminants and those of the ox. The **nasal septum** (279, 280/*a*) reaches to the rostral border of the incisive bone, and in old cows is occasionally strengthened in front by a transverse plate of bone (os rostrale). The rostral part of the **dorsal lateral nasal cartilage** (*c*) extends all the way into the lateral wall of the nostril. The **ventral lateral nasal cartilage** (*d*) is fused with the dorsal lateral cartilage only rostrally and caudally, and the gap that remains between the two cartilages is bridged by connective tissue. The **lateral accessory nasal cartilage** (*e*) is shaped like an anchor and arises from the dorsal lateral cartilage. Its branches are directed rostrally and caudally and support the ventral part of the lateral wall of the nostril. The **medial accessory nasal cartilage** (*f*) is large in the ox. It originates from the ventral nasal concha and the ventral lateral cartilage, is situated inside the alar fold (290/*6*, and p. 216), and has a ligamentous connection also with the rostral part of the dorsal lateral cartilage.

The **NOSTRILS** of the small ruminants are nearly horizontal slits; the nostrils of the ox are wider and more oval (272—274). They are supported medially by the nasal septum, dorsally by the dorsal lateral cartilage, and laterally as well as ventrally by the lateral accessory cartilage. The **alar groove** (*c'*) is between the dorsal lateral nasal cartilage and the lateral accessory cartilage. The areas of modified skin (*a*) between the nostrils, the **planum nasale** in the sheep and goat and the much larger **planum nasolabiale** of the ox, are without hair and are divided by fine grooves into numerous plaquelike fields. These surface markings are characteristic for each individual animal and do not change with age. Glands in the subcutis keep the areas moist. The **nasolacrimal duct** opens close to the nostril on the medial surface of the alar fold.

The **NASAL CAVITY** of the ruminants is relatively long. A small area (290/*dotted line*) of its lateroventral wall that overlies the palatine sinus has no osseous support and is formed by mucosa only.

The rostral third of the **dorsal nasal concha** (60, 61/*8*) is a thick shelf supported by the basal lamella, which is porous in the ox and of compact bone in the small ruminants. In the caudal two-thirds of the concha a spiral lamella is present. It coils first ventrally, then laterally, and then dorsally, and encloses the **dorsal conchal sinus** (288/*d*), which also extends a short distance into the rostral end of the frontal bone. In the ox, the dorsal conchal sinus is divided into several recesses by sagittal lamellae (290/*8*, *9*).

The **ventral nasal concha** of the ox differs considerably from that of the small ruminants. The structure of the concha in the sheep and goat is simpler and will be described first. The basal lamella, which projects into the nasal cavity from the lateral nasal wall, gives rise to two spiral lamellae, of which the dorsal coils toward the middle meatus and the ventral coils toward the ventral meatus, each enclosing a recess (288/*a*, *b*). The free borders of the spiral lamellae in the goat form subdivided bullae, which communicate through small openings with their respective recesses. In the sheep a subdivided bulla is present only on the ventral spiral lamella (*c*). For purposes of description, the ventral nasal concha of the ox can be divided into rostral and caudal parts. The rostral part (388/*left diagram*) is similar in structure to the entire ventral concha of the goat; that is to say, there is a basal lamella (*A*), and dorsal and ventral spiral lamellae (*B*, *C*) each of which form subdivided bullae (*b'*, *c'*). The basal lamella continues into the caudal part of the concha and splits to form dorsal and ventral basal lamellae (*A'*, *A''*). These remain connected to each other laterally by a porous bony plate (*D*). The two basal lamellae give rise to spiral lamellae (*B'*, *C'*) which unite medially and enclose the **ventral conchal sinus** (290/*o*). This sinus has three recesses (*10*, *11*, *12*) in its lateral wall.

The **middle nasal concha** consists of a basal lamella and dorsal and ventral spiral lamellae. In the ox, the dorsal spiral lamella extends rostrally for a considerable distance and encloses the **middle conchal sinus** (290/*l*). The ventral spiral lamella is shorter and encloses a recess. In the sheep and goat, the dorsal spiral lamella encloses a sinus (288/*e*), while the ventral lamella encloses a short second sinus (*e'*) rostrally and a recess (*e''*) caudally.

The **incisive duct** is 6 cm. long in the ox and about 1 cm. long in the small ruminants. It is directed obliquely oroventrally, and connects the ventral nasal meatus with the oral cavity. The **vomeronasal duct** joins the incisive duct close to its oral opening (see also page 220). The folds in the lateral wall of the nasal vestibule (288/*5, 6, 7*) are described on page 216. The **lateral nasal gland** is absent in the ox; in the sheep and goat it is located close to the nasomaxillary aperture.

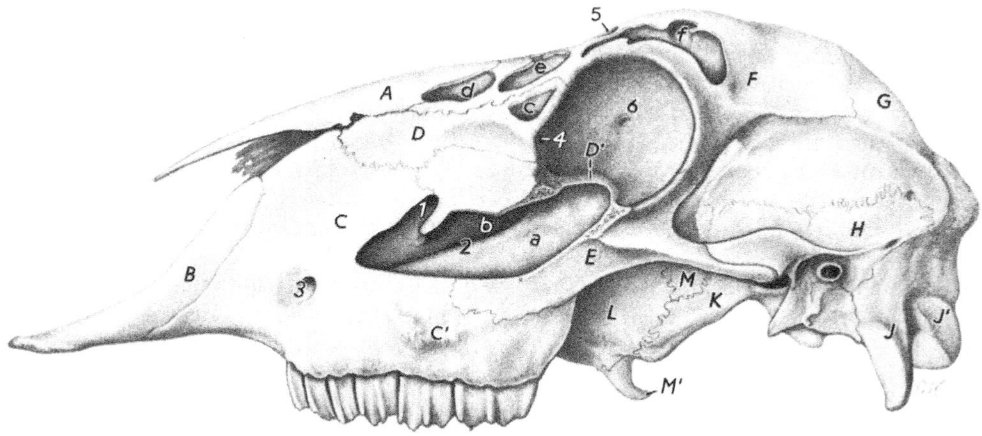

Fig. 385. Skull of a sheep with opened paranasal sinuses. Lateral aspect. (After Loeffler 1958.)

A Nasal bone; *B* Incisive bone; *C* Maxilla; *C'* Facial tuberosity; *D* Lacrimal bone; *D'* Lacrimal bulla; *E* Zygomatic bone; *F* Frontal bone; *G* Parietal bone; *H* Temporal bone; *J* Occipital bone, paracondylar process; *J'* Occipital condyle; *K* Sphenoid; *L* Palatine bone; *M* Pterygoid bone; *M'* Hamulus

a Maxillary sinus; *b* Palatine sinus; *c* Lacrimal sinus; *d* Dorsal conchal sinus; *e* Medial frontal sinus; *f* Lateral frontal sinus *1* Nasomaxillary aperture; *2* Infraorbital canal; *3* Infraorbital foramen; *4* Lacrimal foramen; *5* Supraorbital foramen; *6* Ethmoid foramen

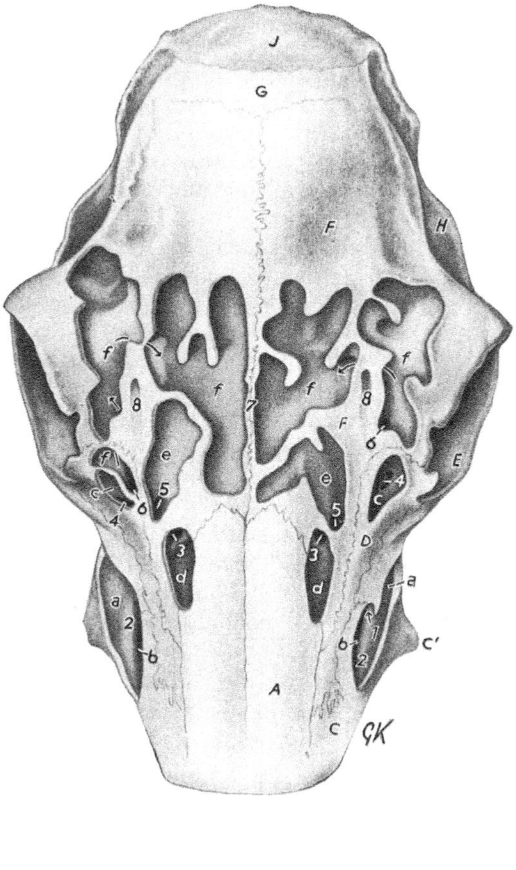

The dorsal part of the **NASOPHARYNX** is divided by the median **pharyngeal septum** (60, 61/*12*), which hangs down from above and contains the pharyngeal tonsil in its caudal end (*15*). The **auditory tubes** (*14*) open on the lateral walls of the nasopharynx.

PARANASAL SINUSES. In addition to the dorsal, ventral, and middle conchal sinuses (see above), the ruminants have maxillary, palatine, lacrimal, frontal, and sphenoid sinuses. These sinuses surround the nasal cavity almost completely, are in the rostral and dorsal walls of the orbit, and, in the ox, extend

Fig. 386. Skull of a sheep with opened paranasal sinuses. Dorsal aspect. (After Loeffler 1958.)

A Nasal bone; *C* Maxilla; *C'* Facial tuberosity; *D* Lacrimal bone; *E* Zygomatic bone; *F* Frontal bone; *G* Parietal bone; *H* Temporal bone; *J* Occipital bone

a Maxillary sinus; *b* Palatine sinus; *c* Lacrimal sinus; *d* Dorsal conchal sinus; *e* Medial frontal sinus; *f* Lateral frontal sinus (the arrows show where the diverticula of the sinus communicate)

1 Arrow pointing to nasomaxillary aperture; *2* Infraorbital canal; *3* Aperture of dorsal conchal sinus; *4* Aperture of lacrimal sinus; *5* Aperture of medial frontal sinus; *6* Aperture of lateral frontal sinus; *7* Sagittal septum of frontal sinuses; *8* Opened supraorbital canal

into the caudal part of the skull and surround much of the cranial cavity (296—300). In general, the maxillary, palatine, and lacrimal sinuses communicate with the middle nasal meatus through the nasomaxillary aperture; while the frontal, sphenoid, and most of the conchal sinuses open separately into ethmoidal meatuses in the caudal part of the nasal cavity.

The size and the boundaries of the **MAXILLARY SINUS** are shown in Figures 296-300/*a*. The maxillary sinus excavates the maxilla (387/*C*), the zygomatic bone (*E*), and the bulla of the lacrimal bone (*D'*). It communicates with the nasal cavity through the **nasomaxillary aperture** (*1*), which is in the caudal part of the middle meatus and has a different shape in

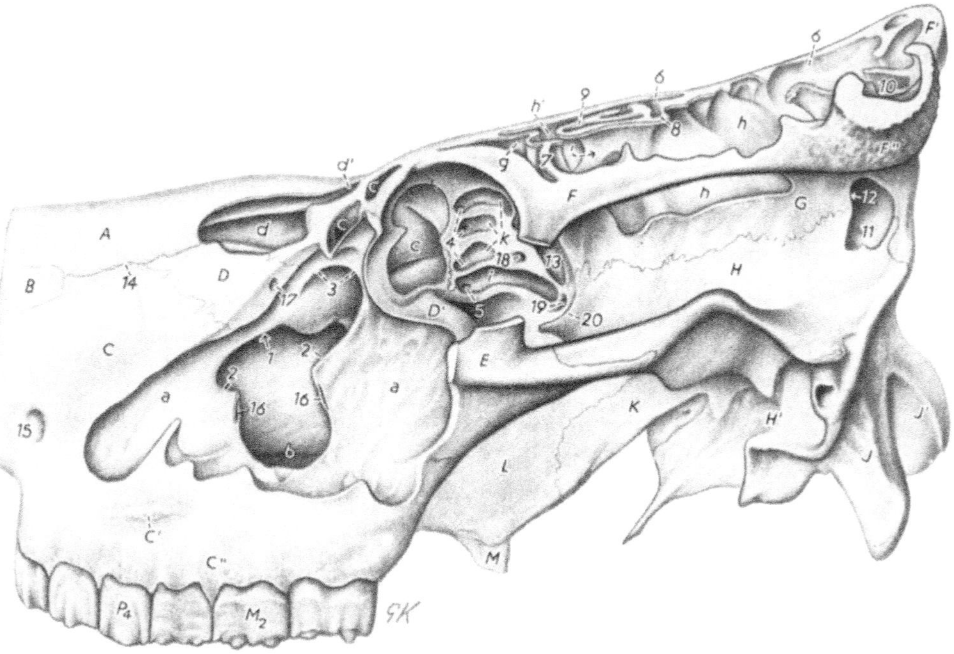

Fig. 387. Paranasal sinuses of the ox. Lateral aspect. (After Wilkens 1958.)

A Nasal bone; *B* Incisive bone; *C* Maxilla; *C'* Facial tuberosity; *C''* Alveolar process of maxilla; *D* Lacrimal bone; *D'* Lacrimal bulla; *E* Zygomatic bone; *F* Frontal bone; *F'* Intercornual protuberance; *F''* Stump of cornual process; *G* Parietal bone; *H, H'* Squamous and petrous parts of temporal bone; *J* Occipital bone, paracondylar process; *J'* Occipital condyle; *K* Sphenoid; *L* Palatine bone; *M* Pterygoid bone, hamulus; P_4 Fourth premolar; M_2 Second molar

a Maxillary sinus, its medial wall has been fenestrated; *b* Palatine sinus; *c* Lacrimal sinus; *d, d'* Dorsal conchal sinus; *d'* Caudal part of dorsal conchal sinus extending into frontal bone; *g* Lateral rostral frontal sinus; *h, h'* Caudal frontal sinus; *i* Rostral part of sphenoid sinus; *k* Ethmoidal cells

1 The arrow points in the direction of the nasomaxillary aperture; *2* Maxillopalatine aperture; *3* Entrance to lacrimal sinus; *4* Apertures of ethmoidal cells; *5* Aperture of sphenoid sinus; *6* Sagittal septum of frontal sinuses; *7* Transverse septum of frontal sinuses; *8* Incomplete oblique septum dividing caudal frontal sinus; *9* Supraorbital groove; *10* Cornual, *11*, *12* nuchal, and *13* postorbital diverticula of caudal frontal sinus; *14* Nasomaxillary fissure; *15* Infraorbital foramen; *16* Infraorbital canal; *17* Lacrimal canal; *18* Ethmoid foramen; *19* Optic canal; *20* Pterygoid crest

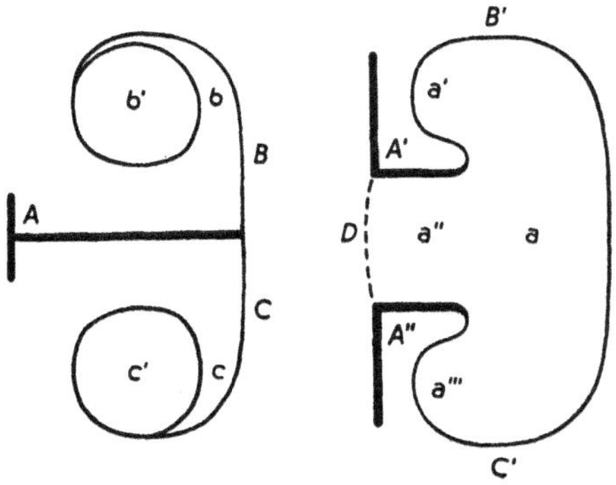

Fig. 388. Ventral nasal concha of the ox. Transverse sections through its rostral (left) and caudal (right) parts. Schematic. (After Wilkens 1958.)

A Rostral part of basal lamella; *A', A''* Split caudal parts of basal lamella; *B, B'* Dorsal spiral lamella; *C, C'* Ventral spiral lamella; *D* Porous bony plate between the two basal lamellae; *a—a'''* Ventral conchal sinus with three recesses in its lateral wall; *b, b'* Dorsal recess and bulla; *c, c'* Ventral recess and bulla

the ox than in the small ruminants. In the young animal, the maxillary sinus is close to the rostral and ventral walls of the orbit. With increasing age it expands in the direction of the cheek teeth and the infraorbital foramen. The maxillary sinus communicates widely with the palatine sinus (*b*) over the infraorbital canal (*16*).

The **PALATINE SINUS** is contained mainly in the palatine process of the maxilla and in the horizontal plate of the palatine bones, i.e., inside the hard palate (290/*b*). In the ox, right and left palatine sinuses are separated from each other by a bony median septum. In the sheep and goat, the two sinuses do not reach as far medially (299/*b*). Defects in the osseous roof of the palatine sinuses (290/*dotted line*) permit the mucous membrane of the nasal cavity to lie back to back against that of the palatine sinus. Several bony lamellae on the floor of the palatine sinus divide it into shallow recesses.

The **LACRIMAL SINUS** is small. In the ox, it is a caudodorsal diverticulum of the maxillary sinus and lies rostromedial to the orbit in the lacrimal and frontal bones (387/*3, c*). Bony lamellae protruding into the sinus form several recesses. The lacrimal sinus of the small ruminants is not associated with the maxillary sinus, but is an independent sinus that opens separately into one of the ethmoidal meatuses (289/*5*). Occasionally it is a lateral diverticulum of the lateral frontal sinus.

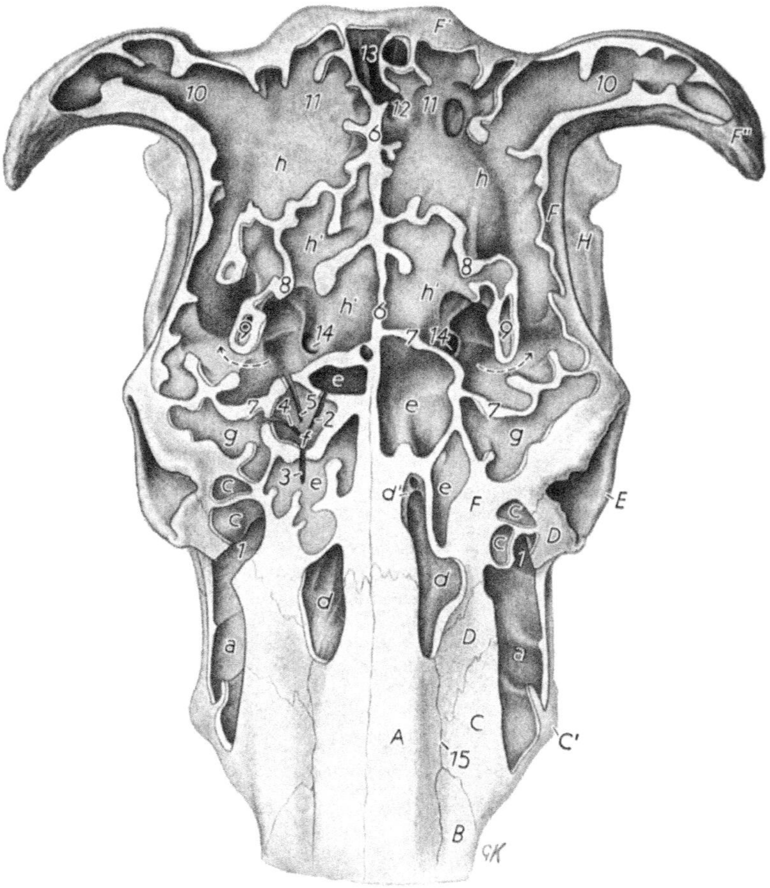

Fig. 389. Paranasal sinuses of the ox. Dorsal aspect. (After Wilkens 1958.)

A Nasal bone; *B* Incisive bone; *C* Maxilla; *C'* Facial tuberosity; *D* Lacrimal bone; *E* Zygomatic bone; *F* Frontal bone; *F'* Intercornual protuberance; *F''* Cornual process; *H* Temporal bone

a Maxillary sinus; *c* Lacrimal sinus; *d, d'* Dorsal conchal sinus; *d'* Caudal part of dorsal conchal sinus extending into frontal bone; *e* Medial rostral frontal sinus; *f* Intermediate rostral frontal sinus (only present on the left side); *g* Lateral rostral frontal sinus; *h, h'* Caudal frontal sinus

1 Entrance to lacrimal sinus; *2* Aperture of medial rostral frontal sinus (*e*); *3* Aperture of intermediate rostral frontal sinus (*f*); *4* Aperture of lateral rostral frontal sinus (*g*); *5* Aperture of caudal frontal sinus (*h, h'*); *6* Sagittal septum of frontal sinuses; *7* Transverse septum of frontal sinuses; *8* Incomplete oblique lamina subdividing caudal frontal sinus; *9* Opened supraorbital canal; *10* Cornual diverticulum, *11* floor, *12, 13* nuchal diverticulum, *14* postorbital diverticulum of caudal frontal sinus; *15* Nasomaxillary fissure

In the adult ox, the **FRONTAL SINUSES** excavate the frontal, parietal, and interparietal bones, and, in part, also the temporal and occipital bones. Thus they almost completely surround the cranial cavity (298). Right and left frontal sinuses are separated from each other by a **median septum** (389/*6*) which deviates slightly at the caudal end. A **transverse septum** (*7*) separates the large caudal frontal sinus from the smaller rostral frontal sinuses: the **medial,** the inconstant **intermediate,** and the **lateral rostral frontal sinuses** (*e, f, g*), each of which opens separately into ethmoidal meatuses (*2, 3, 4*).

The **caudal frontal sinus** is divided by an incomplete oblique septum (*8*) into caudolateral and rostromedial parts which communicate with each other (*h, h'*). It has three diverticula. The **cornual diverticulum** (*10*) extends into the cornual process of the frontal bone and is opened in the dehorning of adult animals. The **nuchal diverticulum** (*13*) extends into the caudal wall of the skull (387/*11*) and sometimes communicates with the nuchal diverticulum of the other side. When this occurs, the two diverticula form a common cavity, which often communicates with only one of the caudal frontal sinuses, the left in Figure 389. The **postorbital diverticulum** (*14*; 387/*13*) is between the orbit and the cranial cavity. Numerous bony lamellae form smaller diverticula or recesses. The aperture of the caudal frontal sinus (389/*5*) is at the rostral end and opens into an ethmoidal meatus. The different depths of the frontal sinuses are shown in Figure 298, and these should be remembered when trephining*. The frontal sinuses of the calf are similar in extent to those of the adult sheep or goat (296). As the calf grows older and its head grows larger, the sinuses gradually excavate the roof of the skull and expand into the lateral and nuchal walls.

The frontal sinuses of the small ruminants occupy only the frontal bone and consist of a small medial and a larger lateral sinus. The **medial frontal sinuses** (386/*e*) are rather asymmetrical in the sheep and do not always reach the median plane. In the goat, right and left medial sinuses lie against the median septum. The **lateral frontal sinus** (*f*) extends caudally to the level of the caudal border of the zygomatic process, and in horned animals extends also into the cornual process. The apertures of these two sinuses open into dorsolateral ethmoidal meatuses (289/*3, 4*). The **lacrimal sinus** of the sheep and goat may by a diverticulum of the lateral frontal sinus; otherwise, it is an independent sinus and has its own aperture (386/*4*).

The **SPHENOID SINUS** is absent in the sheep and goat, and develops in only about 50 per cent of bovine animals. If present, the right and left sphenoid sinuses are separated by a **septum** which seldom coincides with the median plane. Each sinus consists of a tubelike rostral part and a somewhat wider caudal part. The former (297/*6*) is situated ventrolateral to the ethmoid labyrinth and contains the aperture (291/*7*) with which the sinus opens into one of the lateral ethmoidal meatuses. The latter (297/*7, 8*) lies in the body and wing of the presphenoid ventral to the rostral cranial fossa, and partly surrounds the optic canal (290/*i and arrow*).

In the medial wall of the orbit are several **ethmoidal cells** (387/*k*), which vary considerably in size, shape, and number (2—3 in the small ruminants, up to 10 in the ox). Their medial walls are formed by the ethmoid, while their lateral walls are formed by the frontal and palatine bones, and by the wings of the presphenoid. The ethmoidal cells open into ethmoidal meatuses (291/*6*). Their relation to the paranasal sinuses is shown in Figures 296/*g* and 297, 298/*k*.

LARYNX. The laminae of the **thyroid cartilage** (390/*b'*) are long and tall and present rostroventrally-inclined oblique lines (*3*). The rostral cornu is short and straight, while the caudal cornu is long and curves ventrally. The thyroid fissure ventral to the rostral cornu (*1*) is bridged by connective tissue to form the thyroid foramen. Rostral and caudal thyroid notches (312/*5, 6*) are shallow and a laryngeal prominence (*7*) is present. The lamina of the **cricoid cartilage** forms a high median crest (390/*6*). Caudoventral to the articular surface for the arytenoid is a small, rough area (316/*5'*) for the syndesmotic attachment of the caudal cornu of the thyroid cartilage. The slender **arytenoid cartilages** (322, 323) present a long and narrow vocal process ventrally, a thick muscular process laterally, a caudally curved corniculate process craniodorsally, and a concave articular surface for the cricoid caudodors-

* Removing a disc of bone with a trephine (a crown saw).

ally. The lateral surfaces of the **epiglottis** (390/*a*) are wide and their free edges are rounded. The petiolus on the base of the epiglottis is short and slightly curved rostrally for attachment on the thyroid cartilage.

THE CONNECTIONS OF THE LARYNGEAL CARTILAGES (332). The connection between cricoid and thyroid cartilages is entirely ligamentous in the ruminants because a syndesmosis replaces the cricothyroid joint found in the other domestic mammals. The **cricothyroid ligament** (*9*) fills the narrow space between the cartilages. Ventrally, the cricothyroid ligament extends only a short distance onto the internal surface of the thyroid cartilage, and laterally, a short fibroelastic membrane (330/*9'*, shown here in the dog) is split off to the vocal ligament. The joint capsule of the **cricoarytenoid joint** is strengthened ventromedially by the heavy **cricoarytenoid ligament** (*8*). The two arytenoid cartilages are united by the **transverse arytenoid ligament** without the interposition of an interarytenoid cartilage. The thyrohyoid and the rostral cornu of the thyroid cartilage form the **thyrohyoid joint.** The base of the epiglottis is thick and rests on the **thyrohyoid membrane** (*2*), which bridges the space between the hyoid bone and the thyroid cartilage and is quite thick laterally. The **hyoepiglottic ligament** (*1*) extends rostrally from the base of the epiglottis to the basihyoid. The **thyroepiglottic ligament** (*3*) which also originates from the base of the epiglottis passes caudally and covers the entire floor of the thyroid cartilage. The **vocal ligament** (*7*) is oriented dorsoventrally, and passes from the vocal process of the arytenoid to the floor of the thyroid cartilage. In the ruminants, a fan-shaped sheet of fibers (*6'*) located in the submusoca takes the place of the vestibular ligament found in the other domestic mammals. These fibers arise from the epiglottis and the floor of the thyroid cartilage and pass to the lateral surface of the arytenoid cartilage. The larynx is connected to the trachea by the **cricotracheal ligament** (*10*).

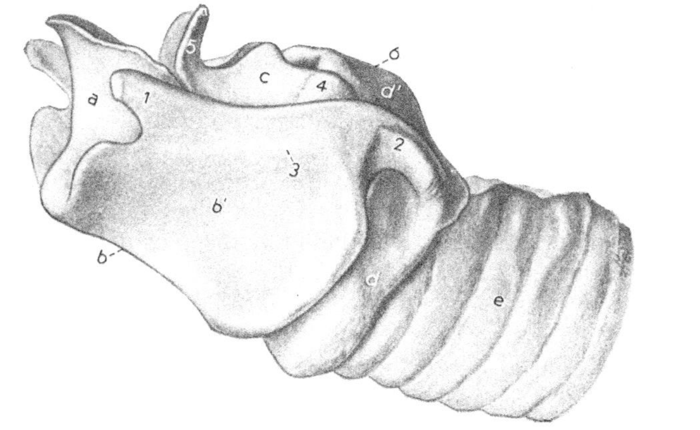

Fig. 390. Laryngeal cartilages of the ox. Left lateral aspect.

a Epiglottic cartilage; *b* Thyroid cartilage; *b'* its lamina; *c* Left arytenoid cartilage; *d* Cricoid cartilage, arch, *d'* its lamina; *e* Trachea; *1* Rostral cornu; *2* Caudal cornu; *3* Oblique line on thyroid cartilage; *4* Muscular process of arytenoid cartilage; *5* Corniculate process of arytenoid cartilage; *6* Median crest of cricoid cartilage

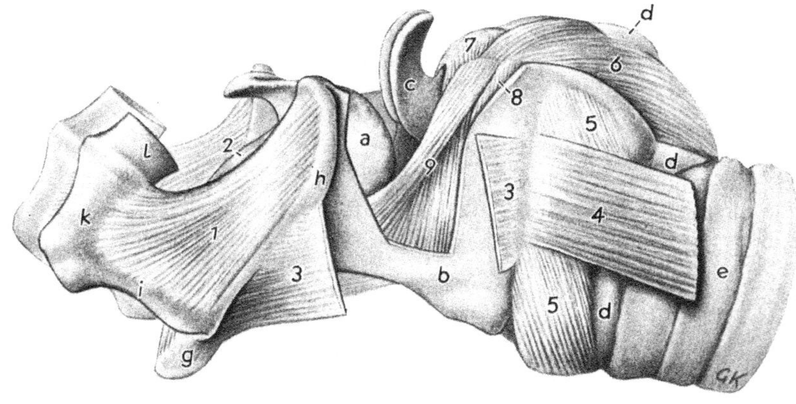

Fig. 391. Laryngeal muscles of the ox. Left lateral aspect.

a Epiglottis; *b* Thyroid cartilage; *c* Corniculate process of arytenoid; *d* Cricoid cartilage; *e* Trachea; *g* Basihoid, its lingual process; *h* Thyrohyoid; *i* Ceratohyoid; *k* Epihyoid; *l* Stylohyoid

1 Ceratohyoideus; *2* Hyoepiglotticus; *3* Thyrohyoideus; *4* Sternothyroideus; *5* Cricothyroideus; *6* Cricoarytenoideus dorsalis; *7* Arytenoideus transversus; *8* Cricoarytenoideus lateralis; *9* Thyroarytenoideus

MUSCLES OF THE LARYNX (391). The laryngeal muscles of the ruminants, except the thyroarytenoideus, conform in general to the description given on page 234. The **thyroarytenoideus** (*9*) is not divided into ventricularis and vocalis (dog and horse), but is a uniform fan-shaped muscle. It originates from the base of the epiglottis, the floor of the thyroid cartilage, and the thyroepiglottic ligament and is inserted on the muscular and vocal processes of the arytenoid.

LARYNGEAL CAVITY. At the entrance to the larynx the **aryepiglottic folds** (332/*4'*) pass from the lateral margins of the epiglottis to the dorsal wall of the larynx, by-passing the arytenoid cartilages laterally (39, 40/*m, l*). The vestibular folds and the laryngeal ventricles are absent. The shallow fossa (337/*1'*) in the medial surface of the thyroarytenoideus corresponds in location to the ventricle of the other species. Because of the absence of the ventricles, the nearly vertical **vocal folds** (*3*) are covered with mucosa only medially and rostrally and, consequently, appear more like heavy ridges than true folds.

LYMPHATIC TISSUE is regularly present in the laryngeal mucosa of the ox. It appears in the form of **solitary lymph nodules** on the epiglottis, on the medial surfaces of the aryepiglottic folds, and in the form of two small patches of nodules, about 5 mm. in diameter, bilaterally on the floor of the vestibule. Two further patches of 8—40 nodules each are located in the mucosa covering the vocal processes (339/*3*). According to Vollmerhaus (1957) the number of nodules in these patches seems to grow larger with age and as a result of increased lymphatic activity. The sheep and goat, like the pig, have **paraepiglottic tonsils** which extend from the free edge of the aryepiglottic folds to the floor of the vestibule. Solitary nodules are present on the epiglottis and the vocal folds.

The **TRACHEA** consists of 48—60 cartilages, to the insides of which the **trachealis muscle** is attached. The trachea of the ox displays marked differences in shape in live and in dead animals, due principally to changes in tension of the trachealis and the annular ligaments. In the live animal, the trachea is more or less oval (346), but after death it becomes compressed laterally (347). The trachea of the goat is U-shaped on cross section with a variable distance between right and left ends of the cartilages (344). In the sheep, the ends of the tracheal cartilages overlap in the cranial third of the trachea (345); have a position as those of the goat in the middle third; and in the caudal third, the left ends of the cartilages extend farther dorsally than the right ends.

LUNGS. The **left lung** of the ruminants consists of a divided cranial lobe and a caudal lobe (398). The **right lung** consists of a divided cranial lobe*, which is supplied by the tracheal bronchus, a middle lobe, a caudal lobe, and an accessory lobe (399). The accessory lobe (397/*d*) is attached mainly to the middle lobe rather than to the caudal lobe as in the other domestic mammals. The **cranial lobes** are more distinctly divided by intralobar fissures in the ox than in the sheep and goat. In the goat especially, the left cranial lobe shows no clear evidence of division (360, 392/*a*). **Lobulation** of the lungs is conspicuous in the ox, almost completely absent in the sheep, and visible only on the cranial and middle lobes in the goat (360) after the lung has cooled following death. In the sheep's lungs, lobules may be discerned along the ventral and basal borders after formalin fixation. Fresh lungs of the sheep may be differentiated from those of the goat by their color and by the shape of their lobes. Sheep's lungs are a bright orange, while those of the goat tend to be pink. The caudal part of the left cranial lobe and the middle lobe are wide in the sheep, but narrow in the goat. The topography of the bovine lungs fixed in expiratory position is shown in Figures 398 and 399.

BRONCHIAL TREE (360—362). The **left cranial bronchus** (*5*) immediately divides into cranial and caudal segmental bronchi (*11*) which serve the two parts of the cranial lobe. The bronchus for the right cranial lobe, the **tracheal bronchus** (*4*), arises directly from the trachea a short distance cranial to the tracheal bifurcation. Like the left cranial bronchus, the tracheal bronchus divides immediately into cranial and caudal segmental bronchi (*11*) which serve the two parts of the right cranial lobe. Both **right** and **left caudal bronchi** (*8*) give off four ventral (*9*) and usually five larger dorsal (*10*) segmental bronchi. The **middle bronchus** (*6*) originates from the right principal bronchus a short distance caudal to the tracheal bifurcation. It ventilates the middle lobe by a small dorsal and a larger ventral segmental bronchus. The **accessory bronchus** (*7*) originates from the principal bronchus

* Much of which is to the left of the median plane.

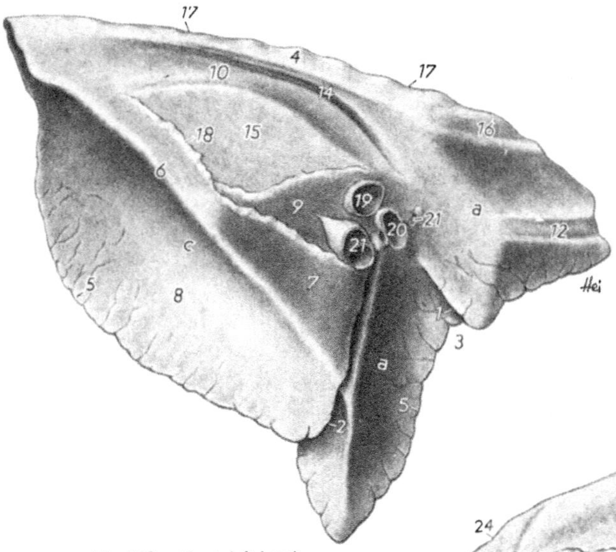

Fig. 392. (Goat, left lung)

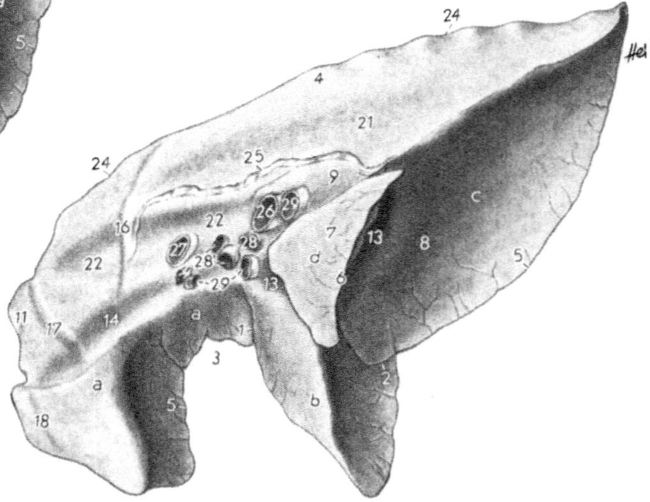

Fig. 393. (Goat, right lung)

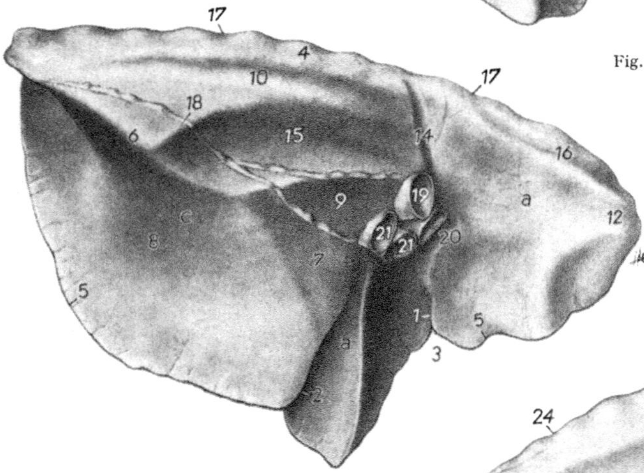

Fig. 394. (Sheep, left lung)

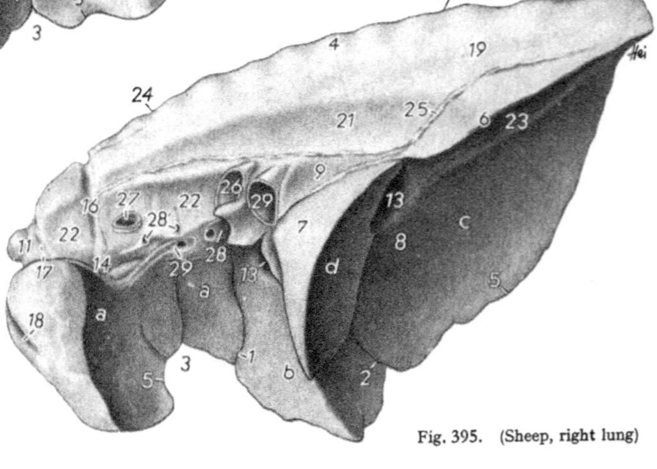

Fig. 395. (Sheep, right lung)

ventromedial to the middle bronchus. It enters the accessory lobe and divides into dorsal and ventral branches. For more detail and variations see the studies of Guzsal (1955), Hare (1955), and Barone (1956).

BLOOD VESSELS OF THE LUNGS. In the ox, the **pulmonary vessels** accompany the bronchi. In the sheep, only the arteries accompany the bronchi, while the veins run intersegmentally. The **bronchial artery** is unpaired and arises from the bronchoesophageal

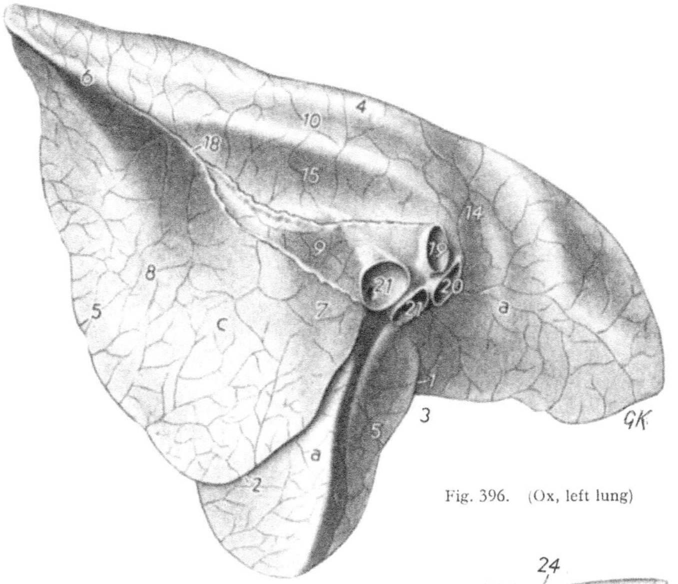

Fig. 396. (Ox, left lung)

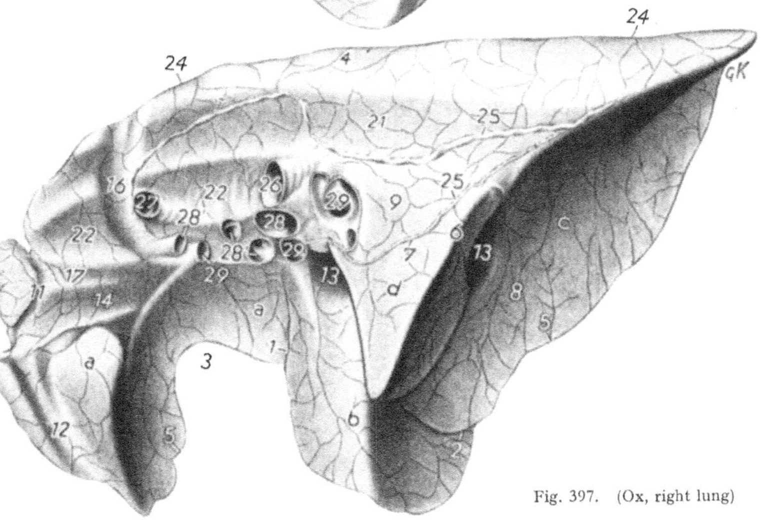

Fig. 397. (Ox, right lung)

Figs. 392, 394, 396. Left lung of the ruminants, fixed in situ. Medial aspect.

a, a Divided cranial lobe; *c* Caudal lobe; *1* Intralobar fissure; *2* Caudal interlobar fissure; *3* Cardiac notch; *4* Dorsal border; *5* (near *8*) Basal border; *5* (near *a*) Ventral border; *6* Border between diaphragmatic and mediastinal surfaces; *7* Mediastinal surface; *8* Diaphragmatic surface; *9* Hilus; *10* Aortic impression; *12* Impression made by left subclavian artery; *14* Impression made by left azygous vein; *15* Esophageal impression; *16* Impression made by the longus colli; *17* Impressions made by ribs; *18* Pulmonary ligament; *19* Principal bronchus; *20* Pulmonary artery; *21* Pulmonary veins

Figs. 393, 395, 397. Right lung of the ruminants, fixed in situ. Medial aspect.

a, a Divided cranial lobe; *b* Middle lobe; *c* Caudal lobe; *d* Accessory lobe; *1* Cranial interlobar fissure; *2* Caudal interlobar fissure; *3* Cardiac notch; *4* Dorsal border; *5* (near *3*) Ventral border; *5* (near *c*) Basal border; *6* Border between mediastinal and diaphragmatic surfaces; *7* Mediastinal surface; *8* Diaphragmatic surface; *9* Hilus; *11* Impression made by costocervical artery; *12* Impression made by int. thoracic vessels; *13* Groove for caudal vena cava; *14* Impression made by cranial vena cava; *16* Impression made by costocervical vein; *17* Impression made by deep cervical vein; *18* Impression made by int. thoracic vein; *19* Impression made by thoracic duct; *21* Esophageal impression; *22* Impression made by trachea; *23* Impression made by the right medial part of the right crus of the diaphragm; *24* Impressions made by ribs; *25* Pulmonary ligament; *26* Principal bronchus; *27* Tracheal bronchus; *28, 28'* Branches of pulmonary artery; *29* Pulmonary veins

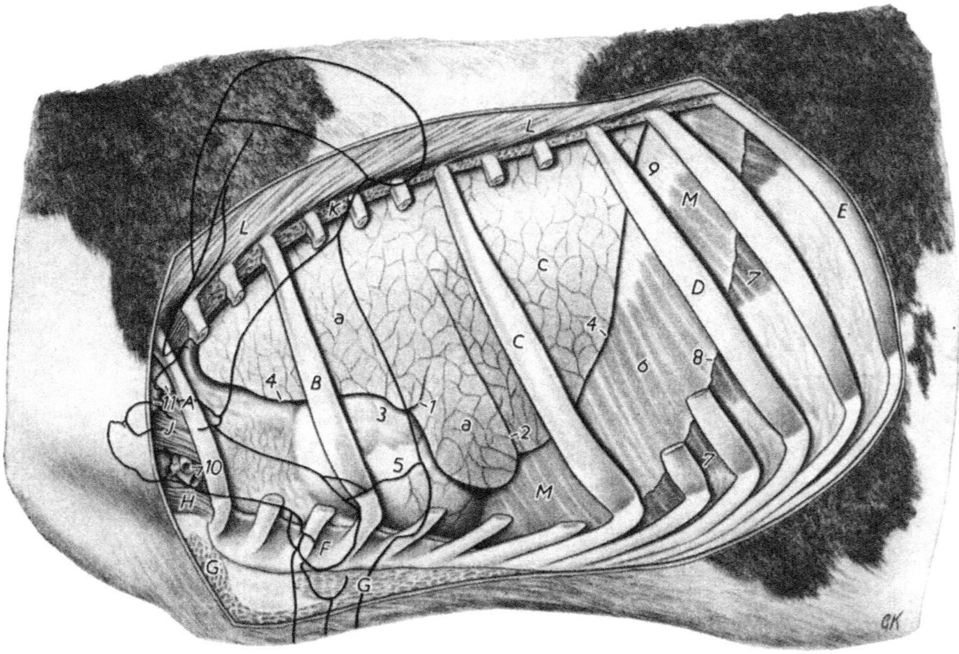

Fig. 398. Thoracic organs of the ox. Left aspect.

A First rib; *B* Fourth rib; *C* Eighth rib; *D* Eleventh rib; *E* Thirteenth rib; *F* Sternum; *G* Pectoral muscles; *H* Sternocephalicuse *J* Scalenus ventralis; *K* Intercostal muscles; *L* Longissimus thoracis; *M* Diaphragm; *a, a* Divided cranial lobe; *c* Caudal lobe

1 Intralobar fissure; *2* Caudal interlobar fissure; *3* Cardiac notch; *4* (near *B*) Ventral border; *4* (near *c*) Basal border; *5* Heart insidd pericardium (cranial to the heart is the cranial part of the right cranial lobe); *6* Diaphragm covered with pleura; *7* Diaphragm not covered with pleura; *8* Line of pleural reflection; *9* Tendinous center of diaphragm; *10* Axillary artery and vein; *11* Stumps of brachial plexus

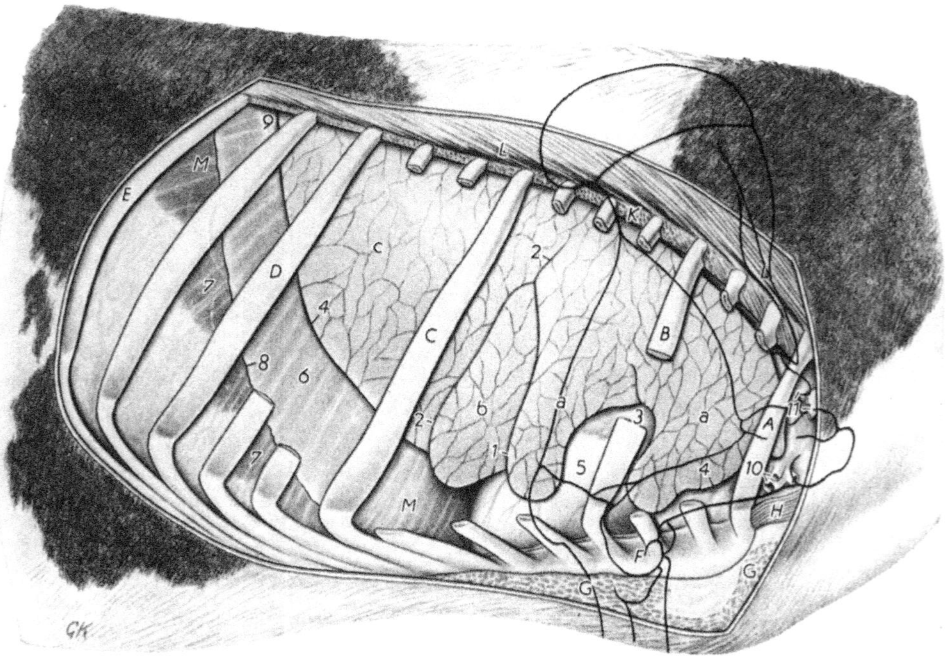

Fig. 399. Thoracic organs of the ox. Right aspect.

A First rib; *B* Fourth rib; *C* Eighth rib; *D* Eleventh rib; *E* Thirteenth rib; *F* Sternum; *G* Pectoral muscles; *H* Sternocephalicus; *K* Intercostal muscles; *L* Longissimus thoracis; *M* Diaphragm; *a, a* Divided cranial lobe; *b* Middle lobe; *c* Caudal lobe

1 Cranial interlobar fissure; *2* Caudal interlobar fissure; *3* Cardiac notch; *4* (near *c*) Basal border; *4* (near *a*) Ventral border; *5* Heart inside pericardium; *6* Diaphragm covered with pleura; *7* Diaphragm not covered with pleura; *8* Line of pleural reflection; *9* Tendinous center of diaphragm; *10* Axillary artery and vein; *11* Stumps of brachial plexus

artery in the small ruminants or, in the ox, often directly from the aorta. It passes to the tracheal bifurcation, where it divides to supply the right and left halves of the bronchial tree. A bronchial vein is not present. Blood entering the lungs by the bronchial artery passes through the alveolar capillary network and is returned to the heart by the pulmonary veins. Precapillary anastomoses between bronchial and pulmonary arteries are present.

The Respiratory Organs of the Horse

NOSE. The **NASAL CARTILAGES** of the horse differ in several respects from those of the other animals. The absence of lateral accessory cartilages leaves the lateral walls of the nostrils without support. The **alar cartilage** (281/*g—g''*) is attached to the rostral border of the nasal septum and consists of a **lamina** dorsally (*g'*) and of a **cornu** ventrally (*g''*). The **dorsal lateral nasal cartilage** (*b*) is narrow, while the ventral lateral cartilage is either absent or, if it is present, it is very narrow and covers only the palatine suture. The **medial accessory cartilage** (*f*) is S-shaped and is connected to the ventral nasal concha; it gives cartilaginous support to the alar fold (62/*2*).

The **NOSTRILS** are semilunar in outline during normal breathing (275). But when the respiratory volume increases, they are dilated and become circular (276). The **alar fold,** supported by the lamina of the alar cartilage and by the medial accessory cartilage, projects laterally into the dorsal part of the nostril and divides the nostril into dorsal and ventral passages. The dorsal passage (*a'*) leads into a blind, cutaneous pouch, the **nasal diverticulum** (*d*); and the ventral passage leads into the nasal cavity. The skin surrounding the nostrils is covered with fine hair and is furnished with long tactile hairs, which are also found on the upper lip. The orifice of the **nasolacrimal duct** is easily visible through the nostril. It is located on the floor of the nasal vestibule at the junction of the skin and the mucous membrane. The nasolacrimal duct of the horse may have several orifices.

NASAL CAVITY. The **choanae** (62/*13*) that connect the ventral nasal meatus with the nasopharynx are short and undivided in the horse. The straight, alar, and basilar folds (*2, 3*) found in the vestibule are described on page 216.

The **dorsal** and **ventral nasal conchae** (400/*K, M*) consist of rostral and caudal parts which are separated by oblique **conchal septa** (∗,⊙). The basal lamellae of the two chonchae project into the nasal cavity from the lateral wall and are continued by spiral lamellae, which coil toward the middle nasal meatus and form subdivided bullae at their free margins (401). Rostral to the conchal septa, the conchae enclose recesses (400/*h, k,*), while caudal to the septa the spiral lamellae fuse with the adjacent facial bones and dorsal and ventral conchal sinuses, and a bulla (*b'*) which communicates with the ventral sinus, are formed. Occasionally, small air spaces are also present in the conchal septa. The **dorsal conchal sinus** (*e*) communicates freely with the frontal sinus (*d—d'''*), and the single cavity so formed is called the **conchofrontal sinus.** The conchofrontal sinus communicates via the caudal maxillary sinus (*c*) with the middle nasal meatus (*1, caudal arrow*). The **ventral conchal sinus** (*b*) communicates over the infraorbital canal (*13*) with the rostral maxillary sinus (*a*) and through this sinus with the middle nasal meatus (*1, rostral arrow*).

The **middle nasal concha** of the horse (*L*) is very small compared to that of the other species. It encloses the **middle conchal sinus** (*c''*) which communicates with the caudal maxillary sinus. The **ethmoidal conchae** (*N*) are small and numerous and do not protrude from the caudal part of the nasal cavity.

The **incisive duct** (282/*b*), in contrast to that of the other species, does not open into the oral cavity, but ends blindly under the oral mucosa. See page 220 for a description of the vomeronasal organ and the lateral nasal gland.

The short choanae lead into the **NASOPHARYNX** (62/*b*) which, because of the absence of a pharyngeal septum (pig and ruminants), is not divided into two longitudinal channels. The **pharyngeal opening of the auditory tube** (*14*) is located on the lateral wall of the nasopharynx at the level of the lateral angle of the eye. On the medial lip of this opening is the triangular **tubal tonsil.**

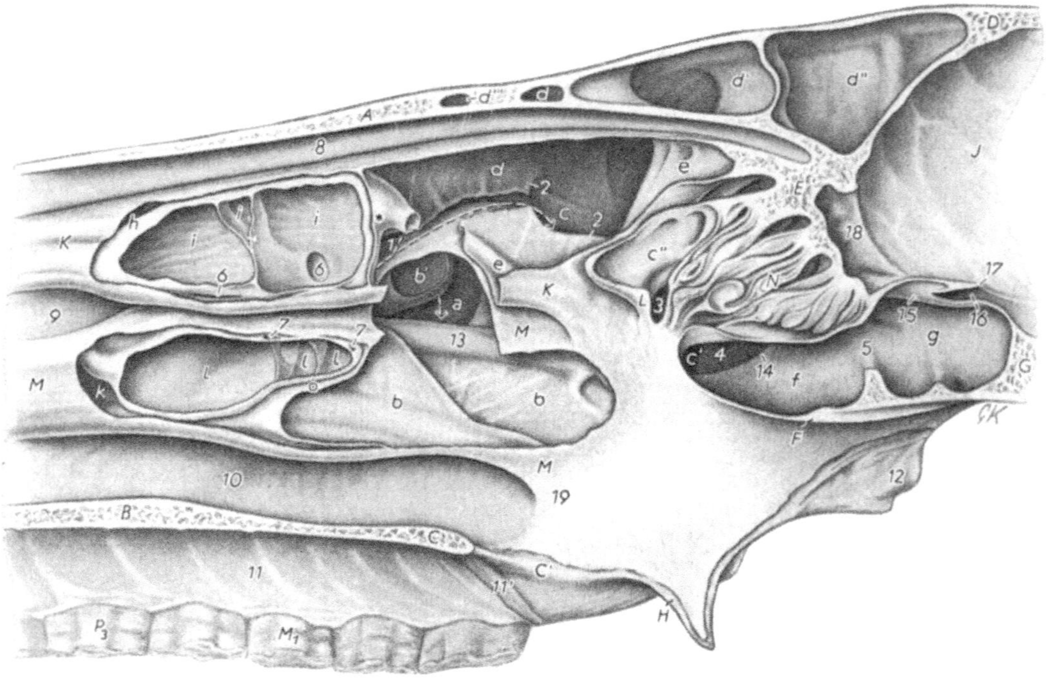

Fig. 400. The paranasal sinuses and nasal conchae of the horse, opened on the medial side. Medial aspect after median section of the head (After Nickel and Wilkens 1958.)

A Nasal bone; *B* Palatine process of maxilla; *C* Palatine bone, its horizontal plate, *C'* its perpendicular plate; *D* Frontal bone; *E* Ethmoid; *F* Vomer; *G* Sphenoid; *H* Hamulus of pterygoid bone; *J* Cranial cavity; *K* Dorsal nasal concha; *L* Middle nasal concha; *M* Ventral nasal concha; *N* Ethmoidal conchae; P_3 Third premolar; M_1 First molar

a Rostral maxillary sinus; *b* Ventral conchal sinus; *b'* Bulla in caudolateral free edge of ventral nasal concha; *c*, *c'* Caudal maxillary sinus; *c''* Middle conchal sinus; *d—d'''* Frontal sinus; *e* Dorsal conchal sinus; *f*, *g* Sphenopalatine sinus; *h* Recess in dorsal nasal concha; *i* Subdivided bulla of dorsal concha; *k* Recess in ventral nasal concha; *l* Subdivided bulla of ventral concha; * Septum of dorsal concha; ⊙ Septum of ventral concha

1 Nasomaxillary aperture, the rostral branch of the arrow points into the rostral maxillary sinus (*a*), the caudal branch of the arrow points into the caudal maxillary sinus (*c*); *2* Edge of frontomaxillary aperture; *3* Aperture of middle conchal sinus into caudal maxillary sinus (*c*); *4* Aperture of sphenopalatine sinus into caudal maxillary sinus (*c*), bounded laterally by wall of caudal nasal foramen (*14*); *5* Sphenopalatine sinus; *6*, *7* Apertures of subdivided bullae into conchal recesses; *8* Dorsal nasal meatus; *9* Middle nasal meatus; *10* Ventral nasal meatus; *11* Mucosa on hard palate, *11'* its cut edge; *12* Pterygoid process of palatine bone; *13* Infraorbital canal; *14* Wall of caudal nasal foramen (lateral boundary of *4*); *15* Optic canal; *16* Chiasmatic groove; *17* Orbitosphenoid crest; *18* Ethmoid fossa; *19* Lateral wall of choana

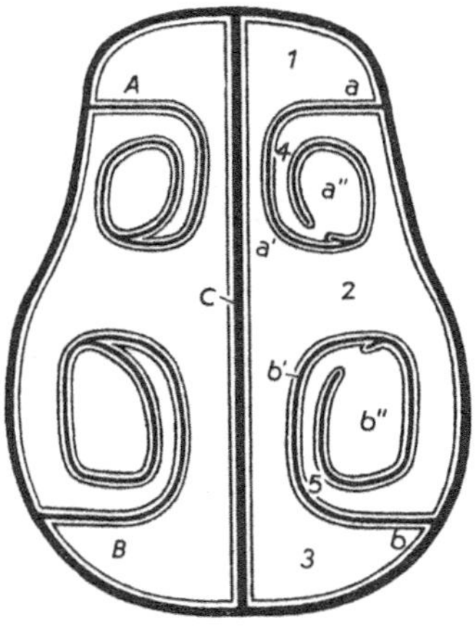

Fig. 401. Cross section of the nasal cavity of the horse taken through the rostral parts of the nasal conchae. Schematic. The thin lines represent the nasal mucosa. (After Nickel and Wilkens 1958.)

A Dorsal nasal concha; *B* Ventral nasal concha; *C* Nasal septum; *a* Basal lamella; *a'* Spiral lamella; *a''* Bulla with its aperture; *b* Basal lamella; *b'* Spiral lamella; *b''* Bulla with its aperture; *1* Dorsal nasal meatus; *2* Middle nasal meatus; *3* Ventral nasal meatus; *4* Recess of dorsal concha; *5* Recess of ventral concha

The paired **AUDITORY TUBES,** which connect the nasopharynx with the middle ear, are greatly modified in the Equidae. The auditory tubes (305/*a*) run inside trough-shaped cartilages (*a'*, *a''*) which are positioned in such a way that their slitlike openings (*a*) face caudoventrally. From the edges of these slits the mucous membrane of the auditory tubes has expanded caudoventrally and formed the two large guttural pouches (*A*, *B*, *C*, *D*). The **GUTTURAL POUCHES** occupy the space between the base of the cranium and the atlas dorsally and the pharynx ventrally, and are in contact with one another medially where their thin mucous membranes lie back to back to form a median septum (*c'*). Dorsocaudally, the two layers of the median septum separate to pass on each side of the longus capitis (*e*) and the rectus capitis ventralis (*f*), which lie ventral to the atlas (*G*) and the base of the cranium. The slender stylohyoid bones (304/*E*) protrude rostrodorsally into the guttural pouches and partly divide each pouch into a small lateral part (*B'*) and a large medial part (*A'*) which extends much more caudally than the lateral part.

PARANASAL SINUSES (302). In addition to the dorsal, ventral, and middle conchal sinuses (see above), the horse has rostral and caudal maxillary, frontal, and sphenopalatine sinuses. All paranasal sinuses of the horse communicate either directly or indirectly with the mid-

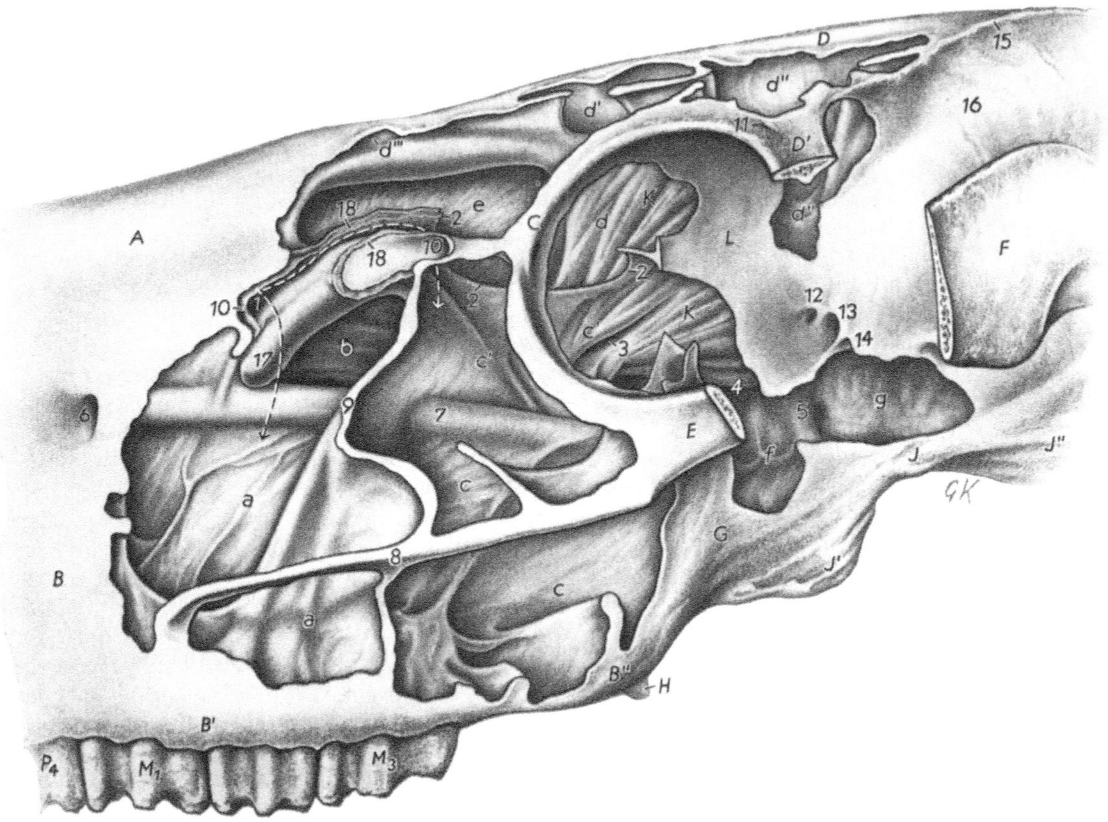

Fig. 402. Paranasal sinuses of the horse opened laterally. Left lateral aspect. (After Nickel and Wilkens 1958.)

A Nasal bone; *B* Maxilla, *B'* its alveolar process; *B''* Maxillary tuberosity; *C* Lacrimal bone; *D* Frontal bone, *D'* its zygomatic process; *E* Temporal process of zygomatic bone; *F* Zygomatic process of temporal bone; *G* Perpendicular plate of palatine bone; *H* Hamulus of pterygoid bone; *J* Presphenoid, *J'* its pterygoid process; *J''* Basisphenoid; *K* Int. plate of frontal bone, fused with the orbital plate of ethmoid and covered with mucosa; *L* Orbit, part of its medial wall; P_4 Fourth premolar; M_1 First molar; M_3 Third molar

a Rostral Maxillary sinus; *b* Ventral conchal sinus; *c*, *c'* Caudal maxillary sinus; *d—d'''* Frontal sinus; *e* Dorsal conchal sinus; *f*, *g* Sphenopalatine sinus

1 Nasomaxillary aperture, the rostral branch of the arrow points into the rostral maxillary sinus (*a*), the caudal branch of the arrow points into the caudal maxillary sinus (*c*); *2* Edge of frontomaxillary aperture; *3* Aperture of middle conchal sinus into caudal maxillary sinus; *4* Aperture of sphenopalatine sinus into caudal maxillary sinus; *5* Sphenopalatine sinus; *6* Infraorbital foramen; *7* Infraorbital canal; *8* Facial crest; *9* Septum of maxillary sinuses; *10, 10* Lacrimal canal, opened laterally and interrupted; *11* Supraorbital foramen; *12* Ethmoid foramen; *13* Optic canal; *14* Lateral boundary of orbital fissure; *15* Temporal line; *16* Temporal fossa; *17* Bulla in caudolateral free edge of ventral nasal concha, it communicates medially with the ventral conchal sinus; *18* Cut edge of mucosa

dle nasal meatus through the nasomaxillary aperture, which is common to the rostral and caudal maxillary sinuses. The ventral conchal sinus (*b*) communicates with the rostral maxillary sinus (*a*) and through it with the middle meatus. The dorsal conchal (*e*), frontal (*d*), sphenopalatine (*f*, *g*), and middle conchal (*c''*) sinuses communicate with the caudal maxillary sinus (*c*, *c'*) and through it with the middle meatus.

The rostral and caudal **MAXILLARY SINUSES** (402/*a*, *c*) are separated by an oblique bony septum (*9*) which may vary slightly in position. The two maxillary sinuses are the largest of the paranasal sinuses in the horse, and they excavate the maxilla, lacrimal, zygomatic, and in part also the ethmoid bones. In young horses, the maxillary sinuses are dorsal to the facial crest (*8*) and a considerable distance caudal to the infraorbital foramen (*6*). As the skull matures and the cheek teeth are gradually pushed from their sockets, the maxillary sinuses can expand ventrally below the facial crest, and in old horses may reach the infraorbital foramen, and extend caudally to the level of the lateral angle of the eye. The two maxillary sinuses communicate with the middle nasal meatus through the **nasomaxillary aperture** (*1*) which is a dorsoventrally flattened cleft between the dorsal and ventral nasal conchae. The nasomaxillary aperture, if probed from the nasal cavity, will be found to bifurcate about midway, one branch (*1*, *rostral arrow*) leading laterally into the rostral maxillary sinus and the other (*1*, *caudal arrow*) continuing caudally into the caudal maxillary sinus. The **rostral maxillary sinus** (*a*), the smaller of the two, is inside the maxilla. Its roof is continuous medially with the dorsal wall of the ventral conchal sinus (*b*) with which the rostral maxillary sinus communicates over the infraorbital canal. The **caudal maxillary sinus** (*c*, *c'*) excavates the bones mentioned above, and borders on the cranial and medial walls of the orbit. An osseous partition, containing the infraorbital canal (*7*) in its free margin, rises from the floor of the sinus and divides it into a large ventrolateral and a small dorsomedial part. The caudal maxillary sinus communicates dorsally (*2*) with the conchofrontal sinus, medially (*3*) with the middle conchal sinus, and caudally (*4*) with the sphenopalatine sinus.

The **FRONTAL SINUS** (303/*d*, *d'*, *d''*) is located largely between the internal and external plates of the frontal bone. In the suckling foal it extends caudally only to the orbit, while in the adult horse it extends beyond the orbit. Right and left frontal sinuses are separated by a median septum, which however does not exactly coincide with the median plane. Transverse lamellae divide the frontal sinuses into rostral, medial, and caudal recesses, which communicate with one another, and of which the rostral (402/*d'''*) occasionally extends into the nasal bone. Rostromedially, the frontal sinus communicates freely with the **dorsal conchal sinus** (*e*), with which it combines to form the **conchofrontal sinus.** The conchofrontal sinus communicates through the large **frontomaxillary aperture** (*2*) with the caudal maxillary sinus.

The **SPHENOPALATINE SINUS** (400/*f*, *g*) is an elongated cavity which communicates rostrally with the caudal maxillary sinus (*c'*). Its palatine part (*f*) primarily occupies the perpendicular plate of the palatine bone, but borders also on the ethmoid dorsally and the vomer dorsomedially. Laterally, it is related to the wall of the pterygoid fossa through which important nerves and vessels pass to the nasal cavity, palate, upper teeth, and face. Inflammation of this part my therefore affect these regions. The sphenoid part (*g*) excavates the presphenoid, and is related dorsally to the optic canal (*15*) and nerve; hence the vision of the animal may be affected when this part of the sinus is diseased. The septum between right and left sphenoid parts varies in position so that the two sides are never of equal size. The spenoid parts may be absent or, rarely, have their own apertures into ventral ethmoidal meatuses.

LARYNX. The **thyroid cartilage** of the horse is characterized by the very deep caudal thyroid notch (313/*5*) through which the laryngeal cavity may be entered surgically in the roarer. The laminae of the thyroid cartilage are long and high, and the oblique lines (403/*3*), which are not very prominent, are directed caudoventrally. The cornua are long. The thyroid fissure ventral to the rostral cornu is bridged by connective tissue, leaving an opening, the thyroid foramen, for the passage of the cranial laryngeal nerve. The lamina of the **cricoid cartilage** (*e'*) is wide and slightly concave on each side of the median crest. The articular surfaces for the arytenoids on the cranial border of the lamina are slightly convex. The rostral border of the cricoid arch is notched ventrally. The articular surface for the thyroid (317/*5*) is at the junction of arch and lamina. The corniculate process of the **arytenoid cartilage** (403/*5*) curves caudodorsally, the vocal process is short and rounded, and the muscular process (*4*) is prominent. The articular surface of the arytenoid is slightly convex. The apex of the **epiglottis** (*a*) is pointed, while at the base the epiglottis is thick and has three processes. The central of these (petiolus, 329/*1*) is short, while the lateral ones, the cuneiform processes (*2*), project dorsocaudally into the vestibular folds.

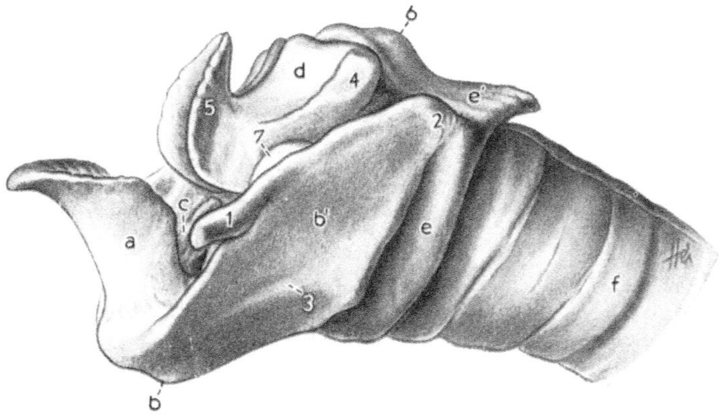

Fig. 403. Laryngeal cartilages of the horse. Left lateral aspect.

a Epiglottic cartilage; *b* Thyroid cartilage, *b'* its lamina; *c* Right cuneiform process of epiglottis; *d* Left arytenoid cartilage; *e* Cricoid cartilage, its arch, *e'* its lamina; *f* Trachea

1 Rostral cornu; *2* Caudal cornu; *3* Oblique line on thyroid cartilage; *4* Muscular process of arytenoid cartilage; *5* Corniculate process of arytenoid cartilage; *6* Median crest of cricoid cartilage; *7* Lateral laryngeal ventricle

THE CONNECTIONS OF THE LARYNGEAL CARTILAGES (333, 404). The cricoid and thyroid cartilages articulate in the **cricothyroid joint** (404/*n*). The medial portion of the **cricothyroid ligament** (*3'*) fills the deep caudal thyroid notch, while, laterally, it splits off the fibroelastic membrane (330/*9*, shown here in the dog) to the vocal ligament. The **cricoarytenoid ligament** (333/*8*) is long and strengthens the capsule of the cricoarytenoid joint ventromedially. The arytenoid cartilages are held together by the **transverse arytenoid ligament.** The **thyrohyoid membrane** (404/*2*) passes from the rostral border of the thyroid cartilage to the caudal border of the basihyoid and thyrohyoid bones. The rostral cornu of the thyroid and the thyrohyoid bone unite in the **thyrohyoid joint** (*m*). The base of the epiglottis rests directly on the ventral part of the thyroid cartilage, and is held in place by the narrow **thyroepiglottic ligament** (333/*3*). Rostrally, it is anchored to the lingual process of the basihyoid by the long **hyoepiglottic ligament** (404/*1*). The vestibular and the vocal

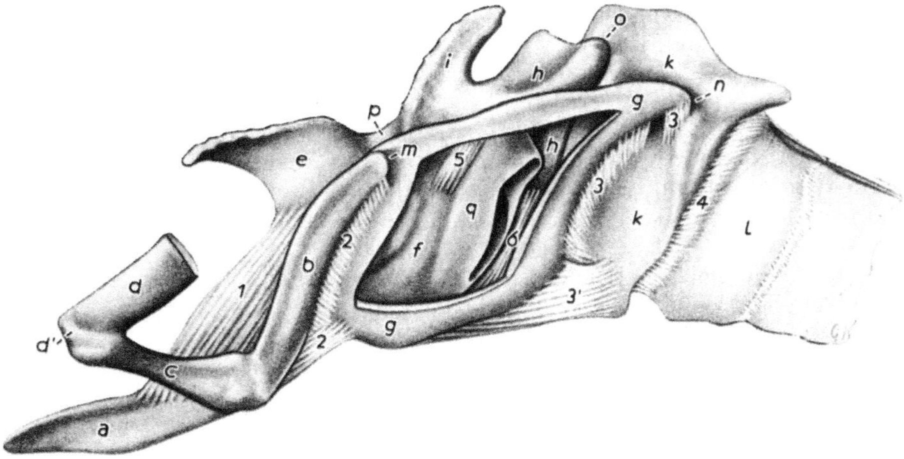

Fig. 404. The ligaments of the larynx of the horse. Left lateral aspect.

a Lingual process of basihyoid; *b* Thyrohyoid; *c* Ceratohyoid; *d* Stylohyoid; *d'* Epihyoid fused with stylohyoid; *e* Epiglottic cartilage; *f* Cuneiform process of epiglottis; *g* Thyroid cartilage, its lamina has been fenestrated; *h* Arytenoid cartilage; *i* Corniculate process of arytenoid; *k* Cricoid cartilage; *l* First tracheal cartilage; *m* Thyrohyoid joint; *n* Cricothyroid joint; *o* Cricoarytenoid joint; *p* Aryepiglottic fold; *q* Lateral laryngeal ventricle, opened

1 Hyoepiglottic ligament; *2* Thyrohyoid membrane; *3*, *3'* Cricothyroid ligament; *4* Cricotracheal ligament; *5* Vestibular ligament; *6* Vocal ligament

ligaments are well developed. The **vestibular ligament** (*5*) extends from the cuneiform process of the epiglottis (*f*) to the lateral surface of the arytenoid cartilage. The **vocal ligament** (*6*) passes rostroventrally from the base of the vocal process to the **cricothyroid ligament.** The **cricotracheal ligament** (*4*) joins the cricoid cartilage to the trachea.

MUSCLES OF THE LARYNX (405). The laryngeal muscles of the horse conform in general to the description given on page 234. The thyroarytenoideus is divided into ventricularis and vocalis. The **ventricularis** (*9'*), together with the vestibular ligament, is contained

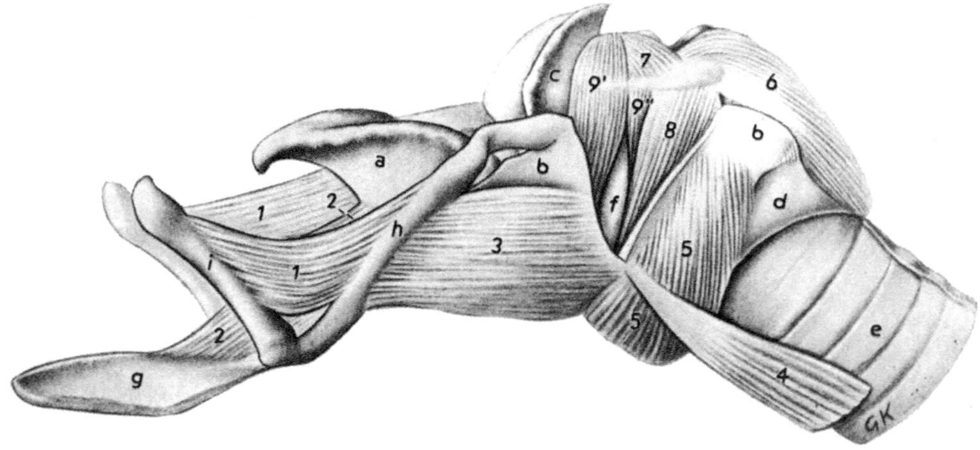

Fig. 405. Laryngeal muscles of the horse. Left lateral aspect.

a Epiglottis; *b* Thyroid cartilage; *c* Corniculate process of arytenoid; *d* Cricoid cartilage; *e* Trachea; *f* Laryngeal ventricle; *g* Lingual process of basihyoid; *h* Thyrohyoid; *i* Ceratohyoid
1 Ceratohyoideus; *2* Hyoepiglotticus; *3* Thyrohyoideus; *4* Sternothyroideus; *5* Cricothyroideus; *6* Cricoarytenoideus dorsalis; *7* Arytenoideus transversus; *8* Cricoarytenoideus lateralis; *9'* Ventricularis; *9''* Vocalis

in the vestibular fold. It arises from the cricothyroid ligament and the ventral part of the thyroid cartilage and is inserted on the muscular process of the arytenoid cartilage. Some of its fibers, however, continue over the top of the arytenoideus transversus (*7*) and join similar fibers from the other side. The ventricularis is oriented dorsoventrally and is related medially to the cuneiform process of the epiglottis. The **vocalis** (*9''*) lies caudolateral to the vocal ligament and is enclosed with it in the vocal fold. It is related laterally to the cricoarytenoideus lateralis (*8*). The vocalis originates on the ventral part of the thyroid cartilage and is inserted on the muscular process and, with a few oral fibers, also on the vocal process of the arytenoid. The position of the ventricularis and vocalis in the wall of the equine larynx is shown also in Figure 340.

LARYNGEAL CAVITY. The **aryepiglottic folds** (88/*10*) at the entrance to the larynx are short and extend caudally from the lateral borders of the epiglottis to the arytenoid cartilages. A **median laryngeal recess** (340/*10*) is present in the floor of the vestibule. The dorsal part of the **vestibular fold** (*8*) contains the vestibular ligament and the ventricularis; in the ventral part is the cuneiform process. The entrance to the **laryngeal ventricles** (*12, 13*) lies between the vestibular and the vocal folds. The **vocal fold** (*9*) is directed rostroventrally and projects well into the laryngeal cavity. It contains the vocal ligament and the vocalis. (See also Figures 86 and 338.)

LYMPHATIC TISSUE is present, also in young horses, on the arytenoid cartilages, in the floor of the vestibule, and in, and dorsal to the entrance of, the lateral laryngeal ventricle.

The **TRACHEA** of the horse is flattened dorsoventrally and consists of 48—60 cartilages, the ends of which overlap slightly (348). The **trachealis** (*c*) is attached to the inside of the cartilages. In the caudal part of the trachea, small overlapping cartilaginous plates (363/*2'*) form part of the dorsal wall where the ends of the tracheal cartilages do not meet.

The **LUNGS** of the horse are characterized by the absence of fissures (408, 409). Each lung has a **cranial** and a **caudal lobe** (*a, c*) and the right lung, in addition, has an **accessory lobe** (407/*d*) which is attached to the medial surface of the caudal lobe. The topography of the lungs fixed in expiratory position is shown in Figures 408 and 409. Lobulation is not very distinct. On closer inspection, however, small irregular lobules can be distinguished through the pulmonary pleura and on the cut surface.

BRONCHIAL TREE (363). Both **right** and **left cranial bronchi** (*5*) divide into dorsal and cranial segmental bronchi (*11*). The cranial segmental bronchus is considerably larger than the dorsal, and gives off four to five dorsal and ventral branches, which ventilate bronchopulmonary subsegments. The **caudal bronchi** (*8*) give off four large ventral segmental bronchi (*9*) and five to seven shorter dorsal ones (*10*). The **accessory bronchus** (*7*) arises a short distance caudal to the tracheal bifurcation and divides into the dorsal and ventral segmental bronchi.

BLOOD VESSELS OF THE LUNGS. The **pulmonary arteries** accompany the bronchi, while the **pulmonary veins** run intersegmentally and collect blood from adjacent bronchopulmonary segments. Three **bronchial arteries** arise from the bronchoesophageal artery. The right and left cranial bronchial arteries supply the cranial lobes as well as the tracheobronchial lymph nodes. The unpaired middle bronchial artery divides at the tracheal bifurcation to supply the two caudal lobes. Since a bronchial vein is absent, the blood carried to the lungs by the bronchial arteries passes through the bronchial and the alveolar cappillary networks and is returned to the heart by the pulmonary veins. A number of precapillary anastomoses are present between branches of the bronchial arteries and those of the pulmonary arteries.

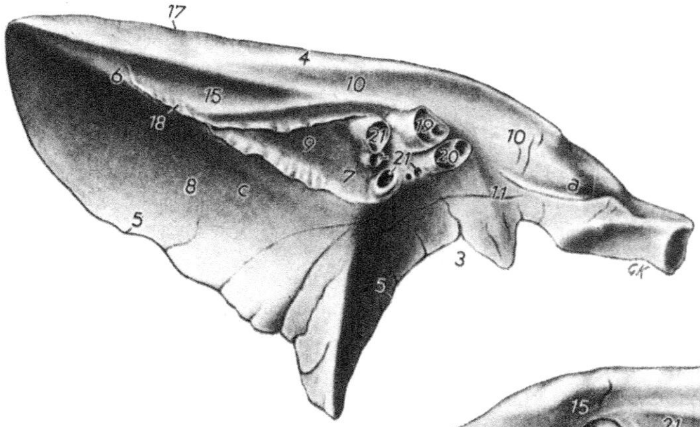

Fig. 406. Left lung of the horse, fixed in situ. Medial aspect.

a Cranial lobe; *c* Caudal lobe

3 Cardiac notch; *4* Dorsal border; *5* (near *8*) Basal border; *5* (near *3*) Ventral border; *6* Border between diaphragmatic and mediastinal surfaces; *7* Mediastinal surface; *8* Diaphragmatic surface; *9* Hilus; *10* Aortic impression; *11* Impression made by pulmonary artery; *15* Esophageal impression; *17* Impression made by ribs; *18* Pulmonary ligament; *19* Principal bronchus; *20* Pulmonary artery; *21* Pulmonary veins

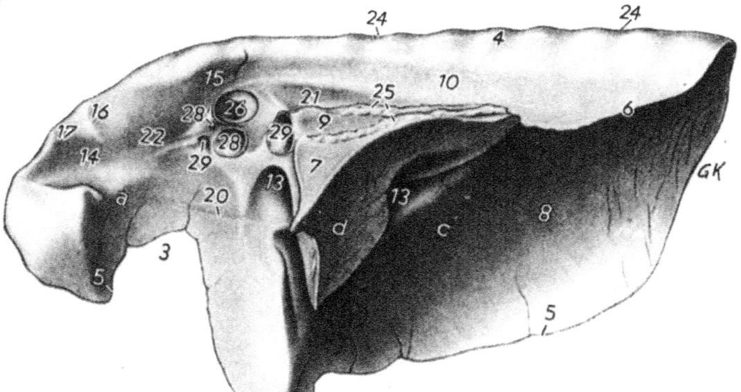

Fig. 407. Right lung of horse, fixed in situ. Medial aspect.

a Cranial lobe; *c* Caudal lobe; *d* Accessory lobe

3 Cardiac notch; *4* Dorsal border; *5* (near *3*) Ventral border; *5* (near *8*) Basal border; *6* Border between diaphragmatic and mediastinal surfaces; *7* Mediastinal surface; *8* Diaphragmatic surface; *9* Hilus; *10* Aortic impression; *13* Groove for caudal vena cava; *14* Impression made by cranial vena cava; *15* Impression made by right azygous vein; *16* Impression made by costocervical vein; *17* Impression made by deep cervical vein; *20* Impression made by right phrenic nerve; *21* Impression made by the esophagus; *22* Impression made by trachea; *24* Impression made by ribs; *25* Pulmonary ligament; *26* Principal bronchus; *28, 28'* Pulmonary arteries; *29* Pulmonary veins

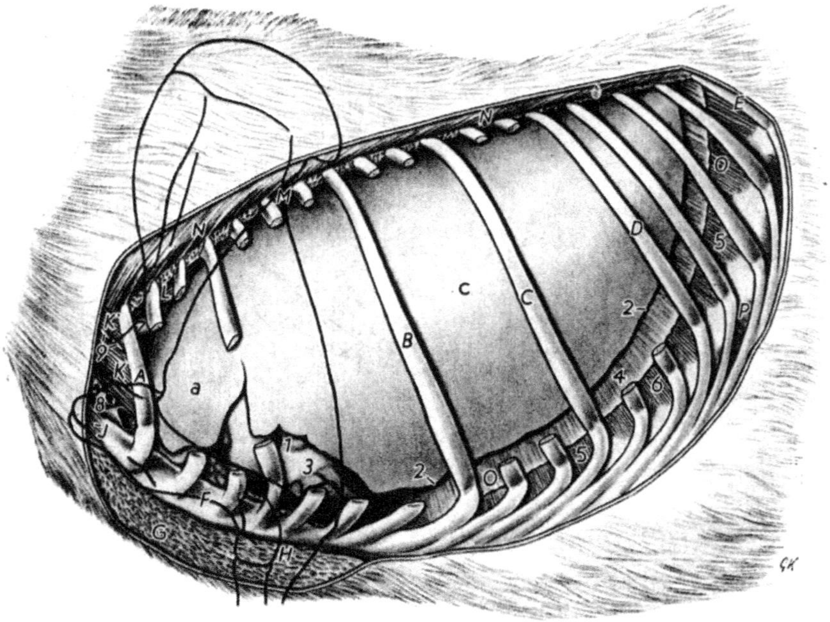

Fig. 408. Thoracic organs of the horse. Left aspect.

A First rib; *B* Eight rib; *C* Eleventh rib; *D* Fourteenth rib; *E* Eighteenth rib; *F* Sternum; *G* Supf. pectoral muscles; *H* Deep pectoral muscle; *J* Sternomandibularis; *K, K'* Scalenus ventralis; *L* Longus colli; *M* Intercostal muscles; *N* Longissimus thoracis; *O* Diaphragm; *P* Transversus abdominis; *a* Cranial lobe; *c* Caudal lobe

1 Cardiac notch; *2* Basal border; *3* Heart inside pericardium; *4* Diaphragm covered with pleura; *5* Diaphragm not covered with pleura; *6* Line of pleural reflection; *7* Esophagus; *8* Axillary artery and vein; *9* Stumps of brachial plexus

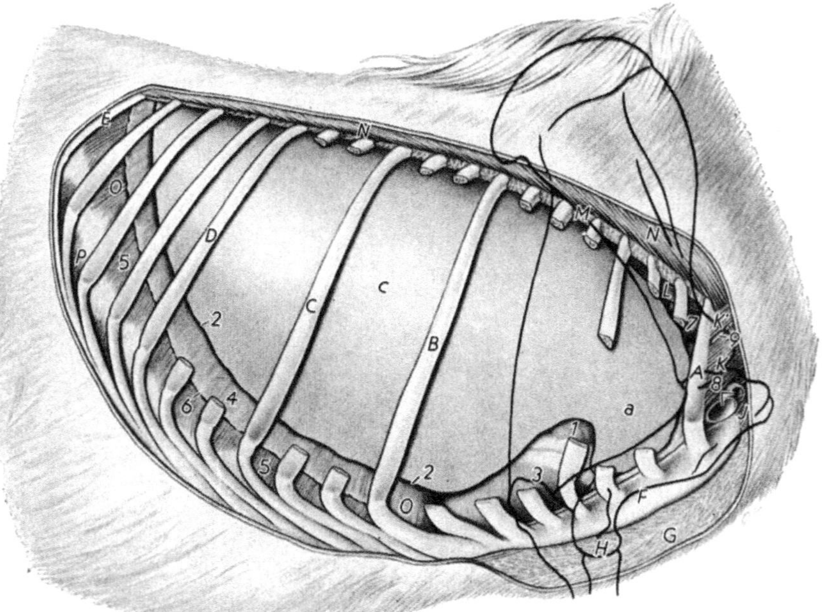

Fig. 409. Thoracic organs of the horse. Right aspect.

A First rib; *B* Eighth rib; *C* Eleventh rib; *D* Fourteenth rib; *E* Eighteenth rib; *F* Sternum; *G* Supf. pectoral muscles; *H* Deep pectoral muscle; *J* Sternomandibularis; *K, K'* Scalenus ventralis; *L* Longus colli; *M* Intercostal muscles; *N* Longissimus thoracis; *O* Diaphragm; *P* Transversus abdominis; *a* Cranial lobe; *c* Caudal lobe

1 Cardiac notch; *2* Basal border; *3* Heart inside pericardium; *4* Diaphragm covered with pleura; *5* Diaphragm not covered with pleura; *6* Line of pleural reflection; *7* Trachea; *8* Axillary artery and vein; *9* Stumps of brachial plexus

BIBLIOGRAPHY

Respiratory System

Adrian, R. W.: Segmental anatomy of the cat's lung. Am. J. Vet. Res. **25,** 1964.

Alexander, A. F., and R. Jensen: Normal structure of bovine pulmonary vasculature. Am. J. Vet. Res. **24,** 1963.

Bakker, Tj.: Het orgaan van Jacobson bij onze huisdieren. Utrecht, Diss. 1939.

Barone, R.: Arbre bronchique et vaisseaux sanguins des poumons chez les Équides domestiques. Rec. Méd. Vét. **129,** 1953.

Barone, R.: La projection pariétale des plèvres et des poumons chez les Équidés domestiques. Rev. Méd. Vét. **105,** 1954.

Barone, R.: Bronches et vaisseaux pulmonaires chez le Boeuf (Bos taurus). C. R. Assoc. Anat., Lisbonne, 1956.

Barone, R.: Arbre bronchique et vaisseaux pulmonaires chez le Chien. C. R. Assoc. Anat., Leiden, 1957

Barone, R.: La projection pariétale des plèvres et des poumons, chez les Bovins. Rev. de Méd. Vét. **112,** 1961.

Barone, R.: Les images radiologiques normales des poumons et de leur arbre broncho-vasculaire chez le chien. Rev. Méd. Vét. **121,** 1970.

Barone, R., et al.: Organe de Jacobson, nerf vomero-nasal et nerf terminal du Chien. Bull. Soc. Sci. Vét. Méd. comp. **68,** 1966.

Baum, H.: In Ellenberger—Baum, Handbuch der vergleichenden Anatomie der Haustiere. 18th ed. Berlin, Springer, 1943.

Bitter, H.: Aufzweigung und Knorpelverteilung am Bronchialbaum der Katze. Hannover, Diss. med. vet., 1974.

Blanton, P. L., and N. L. Biggs: Eighteen hundred years of controversy: The paranasal sinuses. Am. J. Anat. **124,** 1969.

Bölck, G.: Ein Beitrag zur Topographie des Rinderhalses. Diss. med. vet. Berlin, 1961.

Boyden, E. A., and D. H. Tompsett: The postnatal growth of the lung in the dog. Acta Anat. **47,** 1961.

Bonfert, W.: Untersuchungen über den Lobus intermedius der Hundelunge. Anat. Anz. **103,** 1956.

Braus, H.: Anatomie des Menschen, Vol. II. Berlin: Springer, 1924, 1956.

Broman, I.: Das Organon vomeronasale Jacobsoni — ein Wassergeruchsorgan. Anat. Hefte **58,** 1920.

Broman, I.: Über die Ursache der Asymmetrie der Lungen und der Herzlage bei den Säugetieren. Anat. Anz. **57,** 1923.

Bromann, I.: Zur Kenntnis der Lungenentwicklung. Anat. Anz. Erg. Ber. 1923.

Bucher, O.: Beitrag zum funktionellen Bau des menschlichen Kehlkopfgerüstes. Gegenbaurs Morph. Jb. **87,** 1942.

Bürgi, J.: Das grobe Bindegewebsgerüst in der Lunge einiger Haussäuger (Rind, Schwein, Pferd, Ziege, Schaf, Hund und Katze) mit besonderer Berücksichtigung der Begrenzung des Lungenläppchens. Zürich, Diss. med. vet., 1953.

Burow, W.: Beiträge zur Anatomie und Histologie des Kehlkopfes einiger Haussäugetiere. Berlin, Diss. phil. nat., 1902.

Calhoun, M. L., and K. Kartawiria: The gross and microscopic anatomy of the postnatal epiglottis of seven domestic animals. Anat. Rec. **151,** 1965, Abstr.

Calka, W.: Präkapilläre Anastomosen zwischen der A. bronchalis und der A. pulmonalis in den Lungen von Hausrindern. Folia morphol., Warszawa **28,** 65—74 (1969).

Cook, W. R.: Clinical observations on the anatomy and physiology of the equine upper respiratory tract. Vet. Rec. **79,** 1966.

Corondan, Gh., C. Radu, L. Bejan, L. Radu: Beziehungen zwischen dem bronchovaskulären Apparat der höheren Wirbeltiere und Menschen und das Problem der Gleichartigkeit ihrer Nomenklatur. Inst. agron. Timişoara, Lucrări stiint. Ser. Med. Vet. **10,** 101—110 (1967).

Diaconescu, N., and C. Veleanu: Die Rolle der Brustwirbelsäulendynamik bei der Lobierung des Lungenparenchyms. Anat. Anz. **117,** 1965.

Dscherov, D.: Über die Struktur des Terminalgefäßnetzes der Lunge beim Hund. Anat. Anz. **127,** 450—456 (1970).

Ehrsam, H.: Die Lappen und Segmente der Pferdelunge und ihre Vaskularisation. Zürich, Diss. med. vet., 1957.

Ernstmeyer, D.: Beitrag zur topographischen Anatomie des Halses des Schweines. Diss. med. vet. Zürich, 1962.

Espersen, G.: Cellulae conchales hos hest og aesel. Nord. Vet. Med. **5,** 1953.

Felder, G.: Beitrag zur Segmentanatomie der Hundelunge. Zürich, Diss. med. vet., 1962.

Franke, H.-R.: Zur Anatomie des Organum vomeronasale des Hundes. Diss. med. vet. Berlin, 1970.

Frewein, J.: Röntgenanatomie des Organum vomeronasale bei den Haussäugetieren. Zbl. Vet. Med. C, **1,** 1972.

Gehlen, H. v.: Neuere Auffassungen über die Retraktionskraft der Lunge und ihre anatomischen Grundlagen. Anat. Anz. Erg. H., 1938.

Gehlen, H. v.: Der Acinus der menschlichen Lunge als elastisch-muskulöses System. Gegenbaurs Morph. Jahrb. **85,** 1940.

Ghetie, V.: Die Lufthöhlen des Schweinekopfes. Anat. Anz. **92,** 1941.
Gigov, Z., W. Wassiliv: Die Topographie des Zwerchfells, der Pleurasäcke und einiger Brustkorborgane bei neugeborenen Kälbern. Berl. Münch. Tierärztl. Wschr. **84,** 286—290 (1971).
Graeger, K.: Die Nasenhöhle und die Nasennebenhöhlen beim Hund unter besonderer Berücksichtigung der Siebbeinmuscheln. Dtsch. Tierärztl. Wschr. **65,** 1958.
Gräper, L.: Zwerchfell, Lunge und Pleurahöhlen in der Tierreihe. Anat. Anz. Erg.-H. **66,** 1928.
Gräper, L.: Lungen, Pleurahöhlen und Zwerchfell bei den Amphibien und Warmblütern. Morph. Jb. **60,** 1928.
Graubmann, H.-D.: Zur topographischen Anatomie des Brustkorbes beim Rinde. Berlin (Humboldt-Univ.), Diss. med. vet., 1961.
Guzsal, E.: Contribution to the comparative anatomy of the bronchial tree of domestic animals. The pulmonary segments. Acta Vet. Budapest **2,** 1952.
Guzsal, E.: The topography of blood vessels and of the bronchial tree of domestic animals. Acta. Vet. Budapest **5,** 333—365 (1955).

Hare, W. C. D.: The broncho-pulmonary segments in the sheep. J. Anat. **89,** 1955.
Hegner, D.: Das Blutgefäßsystem der Nasenhöhle und ihrer Organe von Canis famillaris, gleichzeitig ein Versuch der funktionellen Deutung der Venenplexus. Diss. med. vet. Gießen, 1962.
Hillmann, D. J.: Macroscopic anatomy of the nasal cavities and paranasal sinuses of the domestic pig *(Sus scrofa domestica)*. Thesis, Iowa State Univ. Ames, Iowa, 1971.
Horning, J. G., A. J. McKee, H. E. Keller, and K. K. Smith: Nose printing your cat and dog patients. Vet. Med. **21,** 1926.

Jacobi, W.: Zur Topographie der Brusthöhlenorgane des Hausschweines (Sus scrofa domestica). Berlin (Humboldt-Univ.) Diss. med. vet., 1962.

Keilbach, R.: Das knorplige Nasenskelett einiger Säugergruppen. Zschr. Säugetierk. **21,** 1956.
Koch, H., U. Tröger und B. Vollmerhaus: Zur makroskopischen Anatomie der Lunge des Göttinger Miniaturschweines. Zbl. Vet. Med. C, **2,** 1973.
Koch, T., R. Berg: Die mediastinalen Pleuraumschlaglinien am Sternum und das Lig. sterno- bzw. phrenicopericardiacum bei einigen Säugetieren. Anat. Anz. **110,** 116—126 (1961).
Kormann, B.: Vergleichende makroskopische Untersuchungen über das Nasenloch und den Nasenvorhof der Haussäugetiere. Arch. wissensch. prakt. Tierheilk. **34.**
Krause, R.: Vergleichende anatomisch-histologische Untersuchungen über den Bau der Schaf- und Ziegenlunge. Hannover, Diss. med. vet., 1921.
Künzel, E., G. Luckhaus u. P. Scholz: Vergleichend-anatomische Untersuchungen der Gaumensegelmuskulatur. Zschr. Anat. Entwickl.-gesch. **125,** 1966.

Lassoie, L.: Les sinus osseux de la tête, chez la bête bovine. An. Méd. Vét. **96,** 1952.
Lechner, W.: Über die Nasenhöhle und deren Nebenhöhlen bei der Katze. Morph. Jb. **71,** 1932.
Lechner, W.: Über die ventralen Kehlkopfventrikel bei Pferd und Tapir. Morph. Jb. **73,** 1933.
Leppert, F.: Beitrag zur funktionellen Struktur der Trachea und des Kehlkopfes des Pferdes. Morph. Jb. **74,** 1934.
Loeffler, K.: Zur Topographie der Nasenhöhle und der Nasennebenhöhlen bei den kleinen Wiederkäuern. Berl. Münch. Tierärztl. Wschr. **71,** 1958.
Loeffler, K.: Zur Topographie der Nasenhöhle und der Nasennebenhöhlen beim Schwein. Dtsch. Tierärztl. Wschr. **66,** 1959a.
Loeffler, K.: Zur Topographie der Nasenhöhle und der Nasennebenhöhlen bei der Katze. Berl. Münch. Tierärztl. Wschr. **72,** 1959b.
Lodge, D.: A survey of tracheal dimensions in horses and cattle in relation to endotracheal tube size. Vet. Rec. **85,** 1969.

Mathea, Kl.: Beitrag zur makroskopischen Anatomie der Lunge von Schaf und Ziege. Berlin, Diss. med. vet., 1963.
Mattay, B.: Das Organum vomeronasale des Schweines. Diss. med. vet. Berlin, 1968.
McLaughlin, R. F., W. S. Tyler, and R. O. Canada: A study of the subgross pulmonary anatomy in various mammals. Am. J. Anat. **108,** 1961.
Miller, M. E., G. C. Christensen, and H. E. Evans: Anatomy of the dog. Philadelphia: W. B. Saunders Co., 1964.
Moorhead, P. D., and R. F. Cross: The subgross vascular anatomy of the feline lung. Am. J. Vet. Res. **26,** 1965.
Moskoff, M.: Beitrag zur Mechanik des Trachealskeletts des Pferdes. Beobachtungen an Frakturen der Trachelknorpel. Zschr. Anat. Entwickl. gesch. **99,** 1932.
Müller, F.: Die Eigenform der Lunge als Artdiagnostikum bei Katze und Hund. Anat. Anz. **83,** 1937.
Müller, F.: Von der Lunge. I. Beitrag „Anatomie für den Tierarzt". Dtsch. Tierärztl. Wschr. **46,** 1938.
Muratori, G.: Peribronchiale Mikroparaganglien und arteriovenöse sowie pulmo-bronchiale Anastomosen bei der Katze. Arch. ital. Anat. Embriol. 73, 133—154 (1968).

Nanda, B. S., M. R. Patel, J. S. Makhani: Bronchial tree and bronchopulmonary segments in goats. Indian Vet. J. **44,** 926—933 (1967).
Nanda, B. S., M. R. Patel: Normal pattern of the bronchopulmonary segments in goat. Indian vet. J. **45,** 124—127 (1968).
Nanda, B. S., M. R. Malik: Bronchial tree in buffalo. Indian. Vet. J. **45,** 127—130 (1968).
Negus, V. E.: The organ of Jacobson. J. Anat. **90,** 1956.

Nickel, R., u. H. Wilkens: Zur Topographie der Nasenhöhle und der Nasennebenhöhlen beim Pferd. Dtsch. Tierärztl. Wschr. **56**, 1958.
Nietz, K.: Zur Anatomie der Tuba pharyngotympanica beim Rind, Schaf, bei der Ziege, beim Schwein, Hund und Sumpfbiber. Diss. med. vet. Berlin, 1961.

Paulli, S.: Ein Os rostri bei Bos taurus. Anat. Anz. **56**, 1923.
Paulli, S.: Pneumatizität des Säugetierschädels. Festschr. Bernhard Bang, 1928.
Picco, G.: Beitrag zur Segmentanatomie der Lungen. Zürich, Diss., 1956.
Pichler, F.: Über die Gaumenkeilbeinhöhle des Rindes. Wien. Tierärztl. Mschr. **28**, 1941.
Piérard, J.: Anatomie comparée du larynx du chien et d'autres carnivores. Can. Vet. J. **6**, 1965.
Popović, S.: Eine Darstellung der morphologischen Eigentümlichkeiten des knorpeligen Nasengerüstes bei Haussäugetieren. Anat. Anz. **114**, 1964.
Popović, S. A.: Anatomische und radiologische Untersuchungen der Vaskularisation der Nasenschleimhaut des Schweines. Acta Vet. Beogard **17**, 445—458 (1967).
Prodinger, F.: Die Artmerkmale des Kehlkopfes der Haussäugetiere (Pferd, Rind, kleine Wiederkäuer, Schwein, Hund, Katze, Kaninchen). Zschr. Anat. Entwickl. gesch. **110**, 1940.

Quinlan, T., B. E. Goulden and A. S. Davies: Bilateral asymmetry of equine laryngeal muscles. N. Z. Vet. J. **23**, 1975.

Ramser, R.: Zur Anatomie des Jakobson'schen Organs beim Hund. Berlin, Diss. med. vet., 1935.
Ruoss, E.: Zur Kenntnis der Segmentanatomie der Lunge. Zürich, Diss., 1955.

Sack, W. O.: The early development of the embryonic pharynx of the dog. Anat. Anz. **115**, 1964.
Salomon, S.: Untersuchungen über das Nasolabiogramm des Rindes. Hannover, Diss. med. vet., 1930.
Schaake, R.: Aufzweigung und Knorpelverteilung am Bronchialbaum des Hundes. Hannover, Diss. med. vet., 1974.
Scholz, O.: Identifizierung vom Hund durch Nasenspiegelabdruck. Tierärztl. Rdsch. **36**, 1930.
Schorno, E.: Die Lappen und Segmente der Rinderlunge und deren Vaskularisation. Zürich, Diss. med. vet., 1955.
Schwieler, G. H., and S. Skoglund: Individual variations in the bronchial tree in cats of different ages. Acta Anat. **56**, 1964.
Seiferle, E.: Grundsätzliches zu Bau und Benennung der Haussäugerlunge. Okajimas Folia Anat. Japonica (Festschr. Nishi), **28**, 1956.
Seki, M.: Über den Bau und die Durchlässigkeit der Siebbeinplatte. Gegenbaurs Morph. Jb. **86**, 1941.
Sis, R. F., J. T. Yoder, and C. J. Starch: Devocalization of cats by median laryngotomy and dissection of the vocal folds. Vet. Med.-SAC. **62**, 1967.
Speranskij, V. S.: Bronchopulmonale Segmente und Lungenblutgefäße der Haustiere. Československ. Morfol. **12**, 373—388 (1964).
Stahl, U.: Das knorpelige Nasenskelett des Hundes unter Berücksichtigung der rassenbedingten morphologischen Unterschiede. Diss. med. vet. Berlin, 1961.
Stamp. J. T.: The distribution of the bronchial tree in the bovine lung. J. comp. Path. **58**, 1948.

Talanti, S.: Studies on the lungs in the pig. Anat. Anz. **106**, 1959.
Tompsett, D. H.: Marco resin cast of the upper respiratory tract, paranasal sinuses and guttural pouches of a horse. Proc. Anat. Soc. Gr. Brit. and Ireland, Nov., 1963.
Töndury, G.: Zur Segment-Anatomie der Lungenlappen. Schweiz. Zschr. Tuberkulose **11**, 227—236 (1954).
Töndury, G., G. Picco: Zur Anatomie der Schweinelunge. Acta anat. **16**, 436 (1952).
Tucker, J. L., and E. T. Krementz: Anatomical corrosion specimens; II. Bronchopulmonary anatomy in the dog. Anat. Rec. **127**, 1957.

Vollmerhaus, B.: Über tonsilläre Bildungen in der Kehlkopfschleimhaut des Rindes. Berl. Münch. Tierärztl. Wschr. **70**, 1957.
Vuillard, P. et al.: Anatomie des pedicules pulmonaires chez le chien. Bull. Ass. Anat. **155**, 1972.

Weber, H. W.: Zur Frage der Segmenteinteilung der Lungen. Verhdlg. Deutsch. Ges. Path. **33**, 1949.
Weber, H. W.: Die anatomischen Grundlagen und die Bedeutung der Lungensegmente. Der Tuberkulosearzt **4**, 1950.
Wilkens, H.: Zur Topographie der Nasenhöhle und der Nasennebenhöhlen beim Rind. Dtsch. Tierärztl. Wschr. **65**, 1958.
Wilkens, H. u. G. Rosenberger: Betrachtungen zur Topographie und Funktion des Oesophagus hinsichtlich der Schlundverstopfung des Rindes. Dtsch. Tierärztl. Wschr. **64**, 1957.

Ziegler, H.: Die Entwicklung des Jakobson'schen Organs beim Schäferhunde nach der Geburt. Berlin, Diss. med. vet., 1936.
Zietzschmann, O.: In Ellenberger—Baum, Handbuch der vergleichenden Anatomie der Haustiere, 18th ed. Berlin, Springer, 1943.
Zimmermann, A.: Der Stimmbandfortsatz des Gießkannenknorpels des Pferdes. Zschr. ges. Anat. **100**, 1932.
Zsebök, Z., A. Székely und E. Nagy: Beiträge zur Anatomie des Bronchialsystems und der Lungenangioarchitektur des Rindes. Acta. ved. acad. sci. Hung. **5**, 1955.

Urogenital System

The urinary and the genital organs are intimately related in development, and remain closely related morphologically throughout the life of the animal. They are derived principally from mesoderm, but short end-sections of their duct systems originate also from ectoderm. The function of the **URINARY ORGANS** is excretion of fluid wastes, whereas the function of the male and female **GENITAL ORGANS** is reproduction.

Urinary Organs

General and Comparative

The urinary organs consist of the **KIDNEYS,** which secrete urine, and the **RENAL PELVIS, URETER, BLADDER,** and **URETHRA** which function in transporting the urine to the outside. The urethra differs in the two sexes. In the male it is closely related to the genital organs.

Kidneys

The kidneys (renes*) are paired excretory glands in which waste products are continuously eliminated from the blood. They regulate the fluid and salt balance of the body and thus maintain normal osmotic pressures in the blood and tissues. They are also capable of removing foreign substances from the blood. The kidneys, therefore, have considerable controlling and regulatory influence over the blood, and the amount of blood that, with certain fluctuations, flows continuously through them is large. In man it has been calculated that about 1,500 liters of blood must flow through the kidneys for the production of 1,500 ml. of urine; this occurs about once every 24 hours.

The color of the kidneys varies from brownish red to dark bluish red depending upon the amount of blood they contain. They are basically bean-shaped. The carnivores (417—420) and small ruminants (434, 435) have thick, well-rounded kidneys; the pig has rather flat ones (426, 427). Exceptions to this basic shape are the right kidney of the horse, which is heart-shaped (437), and those of the ox, which are lobated irregular ovals (430, 431).

The kidney has **dorsal** and **ventral surfaces,** a convex **lateral** and a concave **medial border,** and **cranial** and **caudal extremities,** or **poles.** The **hilus** is an indentation in the medial border, where the renal blood vessels, lymphatics, nerves, and the ureter enter the organ (417). The hilus leads into a recess (sinus renalis) which is situated in the center of the kidney and contains the **renal pelvis** and, embedded in fat, the proximal branches of the renal vessels and nerves (410).

The kidneys lie in the lumbar area to the right and left of the median plane. The medial border of the right kidney is related to the caudal vena cava, and that of the left kidney is related to the aorta (439). The renal arteries and veins arise from these large vessels opposite the kidneys and pass to the hilus by the shortest route (*11, 12*). The right kidney is usually somewhat more cranial than the left, more so in the dog and horse (1; 248), and less so in the cat and pig (541; 545). Because of the presence of the rumen, the left kidney of the ruminants has become pendulous and almost entirely invested with peritoneum. The rumen has pushed

---

* Also Gr. *nephros;* hence nephritis, nephron, etc.

it caudally and over the median plane, so that it lies in contact with the left surface of the spiral colon, lower than, and caudal to, the right kidney (554). In the carnivores, the left kidney is loosely attached and may therefore vary slightly in position.

Generally, the kidneys are applied with their dorsal surface to the crura of the diaphragm and the iliac fascia, which covers the psoas musculature, and are held in place by connective tissue and fat. They are retroperitoneal in position and covered with peritoneum on the ventral surface which faces the abdominal cavity. Depending on the species and on the condition of the animal, the kidneys are embedded in a mass of **perirenal fat** (capsula adiposa) of varying thickness. Usually, this does not cover the ventral surface, but in a fat animal it may conceal the kidney entirely. The perirenal fat helps to protect the kidney and to keep it in position. It is well developed in the pig and ruminants, less so in the carnivores, and least in the horse. If the amount perirenal fat decreases, the left kidney especially may become pendulous and draw out its peritoneal covering like a "mesentery". In man, a migration of the kidney toward the pelvic inlet (floating kidney) is sometimes observed, and is related to his erect position.

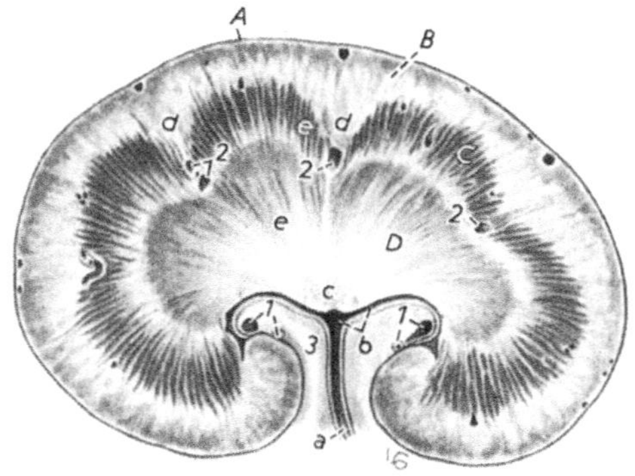

Fig. 410. Kidney of the cat, sectioned through poles and hilus.

A Fibrous capsule; *B* Cortex; *C, D* Medulla, *C* zona intermedia, *D* zona basalis; *a* Ureter entering hilus; *b* Renal pelvis; *c* Renal crest; *d* Renal columns; *e, e* Renal lobe *1* Branch of renal artery and vein; *2, 2, 2* Interlobar arteries (between each two of these is a renal lobe); *3* Fat

The kidney itself is covered with a **fibrous capsule** (410, 411, 412/*A*), which consists superficially of interweaving bundles of collagenous fibers and small amounts of elastic fibers. The capsule is loosely connected to the kidney by loose connective tissue, which may contain smooth muscle fibers. Because of its loose attachment, the fibrous capsule is easily stripped from the healthy kidney, and only where blood vessels reach or enter the deep surface of the capsule is it more closely united to the parenchyma. At the hilus, the fibrous capsule blends with the adventitia of the pelvis, ureter, and blood vessels. Because the capsule is fibrous and rather inelastic, the kidneys cannot swell as other organs if the internal pressure rises, due to an increase in arterial tension for instance.

The **MACROSCOPIC ORGANIZATION OF THE KIDNEY PARENCHYMA** is best examined on a kidney sectioned through the poles and hilus (410). The parenchyma is divided into an outer, paler cortex (*B*) and an inner, darker medulla (*C, D*). The **cortex** is brownish red and has a granular appearance. In the fresh state it contains large numbers of barely visible red dots, the renal corpuscles. The external part of the **medulla** (zona intermedia, *C*) is dark red, almost purple, which sets it off sharply from the cortex. The internal part (zona basalis, *D*) is lighter, grayish red, and shows distinct radial striations. Narrow spikes of medullary tissue radiating into the cortex are known as **medullary rays.**

The kidney of the domestic mammals consists of a varying number of radially arranged **lobes** (lobi renales), which are not readily apparent, except in the bovine kidney (432). The underlying pattern of renal lobation becomes clear, however, when the more primitive kidneys of some aquatic mammals (whale, seal, also Polar bear and river otter) are considered. In these animals, the kidney is composed of many separate kidney units (renculi) which in one of the Whalebone whales, for example, number up to 3,000. Each separate unit, or lobe, consists of a caplike cortex, enclosing the base and sides of a pyramid-shaped medulla. The apex of the pyramid, the **renal papilla,** is inserted into the cup-shaped endpiece (calix renalis) of a branch of the ureter. Such a primitive kidney resembles a bunch of grapes, with the ureter representing the stem, and is called a composite or lobated kidney.

In domestic mammals, the cortical and medullary substances of the lobes are fused to varying degrees, with the result that, outwardly, the kidney has become a uniform, compact organ. It is still possible, however, to recognize its lobation when the parenchyma of the sectioned kidney is examined, particularly by following the typical course of the interlobar vessels (410, 414/*2*).

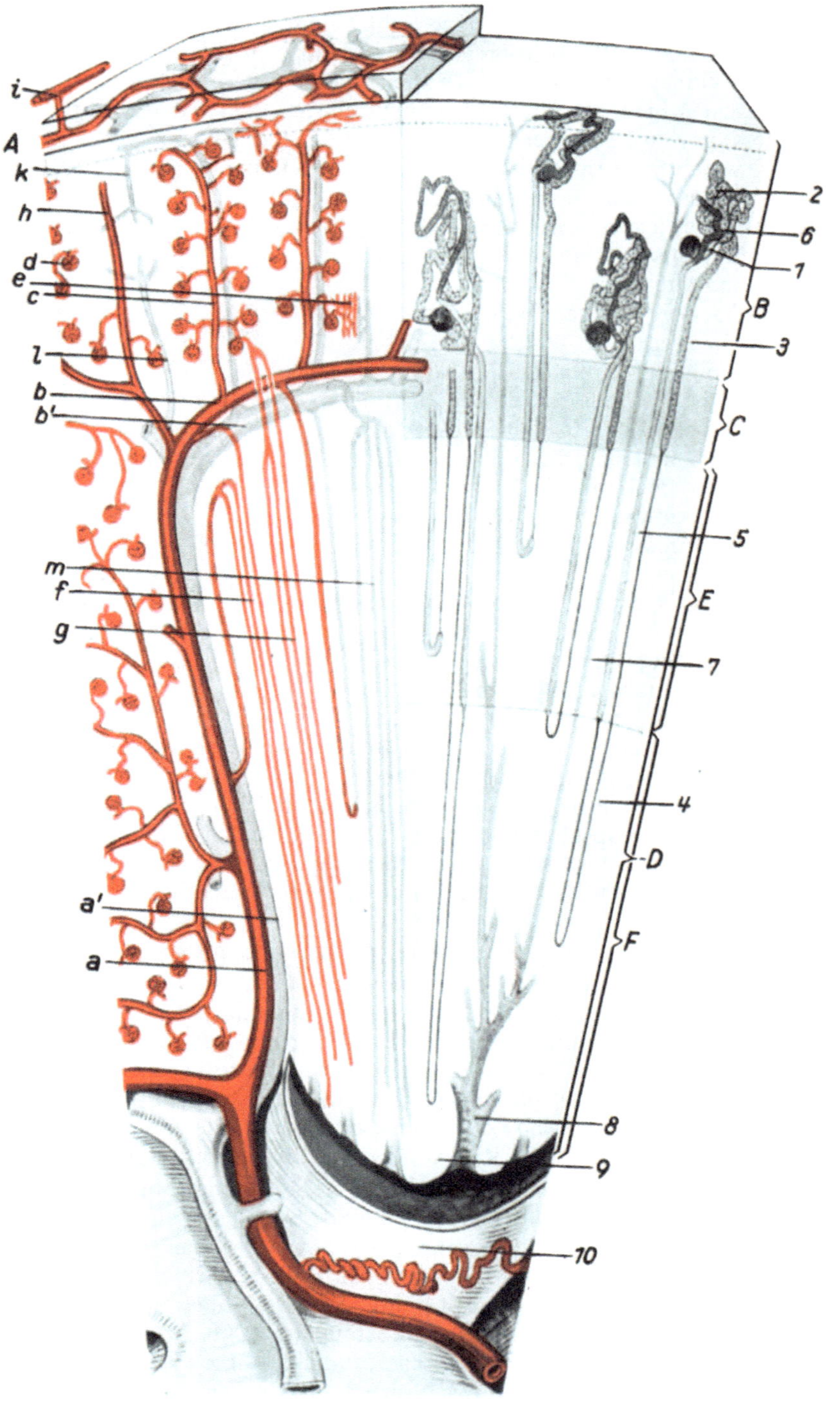

Fig. 411. Structure of the kidney. Schematic. (After Clara 1938.)

A Fibrous capsule; *B* Cortex (to the left of the interlobar vessels *a*, *a'* a renal column); *C*, *D* Medulla; *C* Stratum subcorticale; *E* Zona intermedia; *F* Zona basalis

a Interlobar artery; *a'* Interlobar vein; *b* Arcuate artery; *b'* Arcuate vein; *c* Interlobular artery; *d* Glomerulus with its afferent and efferent arterioles; *e* Cortical capillary network; *f* Straight arterioles (directly from interlobar or arcuate artery); *g* Straight arterioles (from an efferent glomerular arteriole); *h* Capsular branch; *i* Artery in adipose capsule; *k*, *l* Interlobular veins; *m* Straight venules

1 Renal corpuscle; *2* Proximal convoluted tubule; *3*, *4*, *5* Loop of nephron; *6* Distal convoluted tubule; *7* Collecting tubule; *8* Papillary duct; *9* Renal papilla with area cribrosa; *10* Renal calix

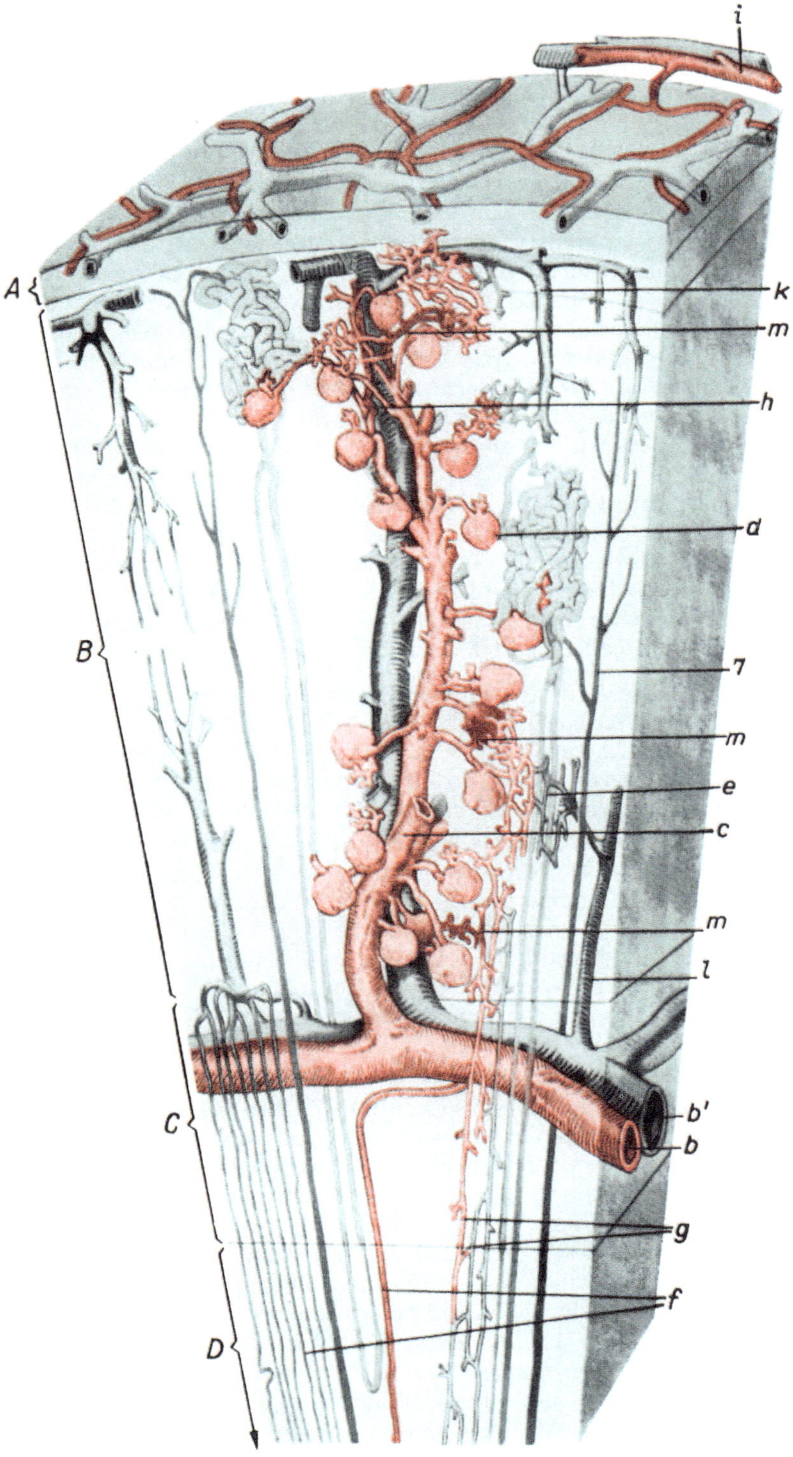

Fig. 412. Structure of the renal cortex. Semischematic. (After Clara 1938.)

A Fibrous capsule; *B* Cortex; *C*, *D* Medulla; *C* Stratum subcorticale; *b* Arcuate artery; *b'* Arcuate vein; *c* Interlobular artery; *d* Glomerulus; *e* Cortical capillary network; *f* Straight arterioles and venules (from arcuate vessels); *g* Straight arterioles and venules (arterioles from efferent glomerular arterioles); *h* Capsular branch; *i* Artery in adipose capsule; *k*, *l* Interlobular veins; *m* Branches of afferent arteriole to cortical capillary network; *7* Collecting tubule

Complete fusion of the cortical tissue of neighboring lobes results in a kidney with a smooth surface. Incomplete cortical fusion results in a kidney that is superficially divided by fissures of varying depths, as in the ox (430—432). The fact that a kidney is smooth externally does not mean that the cortex is a layer of uniform thickness on the outside of the medullary substance. On the contrary, when the deep surface of the cortex is exposed (424/*d*) it will be found to be rather irregular, because of the presence of recesses of varying sizes, one for each lobe. Each recess is bounded by irregular crests of cortical tissue (*c*), and overlies the convex base of a medullary pyramid. In section, these crests appear as wedges of cortical tissue between adjacent pyramids, and are known as the **renal columns** (410/*d*). Each lobe, therefore, consist of a medullary pyramid—capped at its base by a part of the renal cortex—and an apex, the renal papilla, which is directed toward the center of the kidney.

The fusion of the lobes is not limited to the cortex, however; it may also involve the medulla. Incomplete fusion of the medullary pyramids preserves individual renal papillae, which project separately into the renal sinus (papillated kidney). With complete fusion, the apices of the medullary pyramids also fuse and form a ridgelike common papilla, known as the **renal crest.** The renal pelvis of the papillated kidney presents a number of cup-shaped calices, one for each papilla. In kidneys with a renal crest, the pelvis is more uniform and no calices are present.

The kidneys of the carnivores, small ruminants, and horse show the highest degree of fusion, and have a smooth surface and a renal crest. The kidneys of the ox show the least fusion. Their lobation is visible on the surface, and they have separate papillae, each projecting into a calix located at the end of a branch of the ureter. They therefore lack a renal pelvis (432). The kidney of the pig is partly fused; it is smooth externally, but it retains individual papillae which shed the urine into calices that form parts of a renal pelvis (428).

The gross features described are of course closely related to the **MICROSCOPIC STRUCTURE OF THE RENAL PARENCHYMA.** Functionally, the uriniferous tubules of the kidney consist of two types: those in which the urine is produced, and those which conduct the urine to the renal pelvis. The latter are known as collecting tubules. Each urine-producing tubule, together with a renal corpuscle, constitutes a **NEPHRON.**

The **renal corpuscles*** (411/*1*) have a diameter of .1—.3 mm. and consist of a microscopic cluster of blood capillaries, the **glomerulus,** surrounded by a double-walled envelope, the **glomerular capsule****. The glomerulus is supplied and drained by small arterioles, and constitutes a capillary network inserted in the arterial pathway. The inner layer of the glomerular capsule (413/*b*), consisting of branching epithelial cells, closely invests the glomerulus. At the vascular pole of the corpuscle (*d*) the inner layer of the capsule blends with the outer layer (*c*), which consists of a continuous sheet of flat epithelial cells. At the urinary pole of the corpuscle (*e*), the capillary space between the internal and external layers of the glomerular capsule is continuous with the lumen of the tubular part of the nephron.

The tubular part of the nephron consists essentially of three segments: the proximal convoluted tubule, the loop of the nephron***, and the distal convoluted tubule. The **proximal convoluted tubule** (411/*2*), very tortuous initially, has cuboidal cells with striated cytoplasm basally and a brush border at the narrow lumen. It straightens out distally and continues into the U-shaped **loop of the nephron** (*3, 4, 5*) which consists of a thin descending limb and a thicker ascending limb. The descending limb extends from the cortex into the medullary tissue, doubles back on itself, and, as the ascending limb, passes close to its corpuscle in the cortex. Here it is continued by the **distal convoluted tubule** (*6*) which ends at the junction with the collecting tubules (*7*). These are largely directed toward the center of the kidney, becoming progressively larger in diameter as adjacent ones join at acute angles. As they approach the renal papilla (*9*) they are succeeded by larger **papillary ducts** (*8*), which open on the surface of the papilla at tiny slitlike openings, the **papillary foramina,** through which the urine enters the renal pelvis. The papillary foramina perforate the surface of the renal

---

* Formerly Malpighian corpuscles.
** Formerly Bowman's capsule.
*** Formerly loop of Henle.

papilla (area cribrosa, 423/*1*) and can be seen with the unaided eye. Proximally, the collecting tubules are lined with cuboidal cells. As they increase in diameter, their lining changes to columnar cells, and at the renal papilla to strafified columnar cells.

The topography of the nephrons (of which there are from 186,000 to 378,000 in the kidneys of the dog) and of the collecting tubules is as follows. The renal corpuscles and the proximal and distal convoluted tubules are in the cortex. The rays of medullary tissue that project into the cortex consist of the loops of the more superficially placed nephrons and the proximal segments of the collecting tubules. In the medulla proper lie the loops of the more deeply placed nephrons and the remaining collecting tubules and, apically, the papillary ducts (411).

The **BLOOD VASCULAR SYSTEM OF THE KIDNEY** is as intricate as the system of uriniferous tubules it supplies. The kidney receives its blood supply through the **renal arteries** (414/*1*), which are remarkably large branches of the abdominal aorta. They pass directly to the hilus of the kidney where they divide into a number of branches (*2*). From these branches originate the **interlobar arteries** (411/*a*), which enter the parenchyma between adjacent lobes and radiate toward the cortex. At the corticomedullary junction, each interlobar artery gives rise to a number of **arcuate arteries** (411, 412/*b*), which follow the curvature of the base of the medullary pyramids. From the convex surface of the arcuate arteries arise the small **interlobular arteries** (*c*), which enter the cortex. In the cortex, the nearly straight interlobular arteries give off afferent arterioles to the nearby renal corpuscles. Each **afferent arteriole** (vas afferens) enters a corpuscle at the vascular pole and breaks up into a spherical cluster of capillary loops, the **glomerulus.** The smaller **efferent arteriole** (vas efferens) leaves the corpuscle also at the vascular pole and immediately enters the capillary network surrounding the adjacent uriniferous tubules.

The uriniferous tubules of the medulla are supplied by the **straight arterioles** (arteriolae rectae, *f*, *g*), which come from efferent arterioles of glomeruli that lie close to the medulla (*g*), or directly from the arcuate or interlobar arteries (*f*). Many of the interlobular arteries reach the surface of the kidney and give off branches (rami capsulares, *h*) to the fibrous capsule and, occasionally, to the adipose capsule. Afferent arterioles may give off branches (*m*) directly to the cortical capillary network, which means that not all blood reaching the convoluted tubules has passed through a glomerulus.

After the blood has traversed the capillaries of the cortex, it is collected by **interlobular veins,** or by the **arcuate veins** directly. The interlobular veins enter the arcuate veins, which are branches of the **interlobar veins.** The capillaries of the medulla are drained by **straight venules** (venulae rectae) which enter the arcuate veins. From the arcuate veins, the blood passes through the interlobar and renal veins into the caudal vena cava.

The subcapsular branches (venulae stellatae,* 412) of the interlobular veins are often visible grossly, and appear like small vascular "stars" on the surface of the kidneys in some species. The capillaries of the fibrous capsule are drained largely by these branches.

Renal Pelvis

The renal pelvis is the flared proximal end of the ureter located inside the renal sinus. It differs in shape from species to species, and is absent in the ox because of the lack of fusion of the renal lobes. The pelvis collects the urine, which reaches it through the papillary foramina, and like a funnel conveys it directly into the ureter.

The wall of the pelvis consists of three layers: an external connective tissue adventitia, a thin muscle layer, and a mucous membrane that lines the pelvis. The **mucous membrane** has the transitional epithelium characteristic of all urinary passages. In the horse, the mucous membrane of the renal pelvis and of the proximal part of the ureter contains large mucous glands.

The wall of the renal pelvis extends to the base of each papilla or, in the nonpapillated kidney, to the base of the renal crest. Adventitia and muscle layer end here; only the mucous membrane continues over the papilla or crest. In the horse, two tubular **terminal**

* L., star-shaped.

recesses (440—442/*b'*) enter the relatively small renal pelvis from the poles. The renal pelvis of the dog (425) is very similar to that of the small ruminants (436), both consisting of a common cavity (*b*) which receives the renal crest. Attached to the sides of the common cavity are a number of **recesses** (*b''*), into each of which project columns of kidney tissue (so-called pseudopapillae, 423/*3*) which nearly bisect each recess. The interlobar arteries and veins ascend toward the cortex in the narrow clefts between neighboring recesses (414). The renal pelvis of the pig (429) has a number of short-stemmed, cup-shaped calices which surround single or partly fused aggregates of papillae. In the ox, the ureter first divides into two main branches directed toward the poles. These then subdivide into up to 22 smaller ones, which end by forming calices at the base of individual or partly fused aggregates of papillae (432, 433).

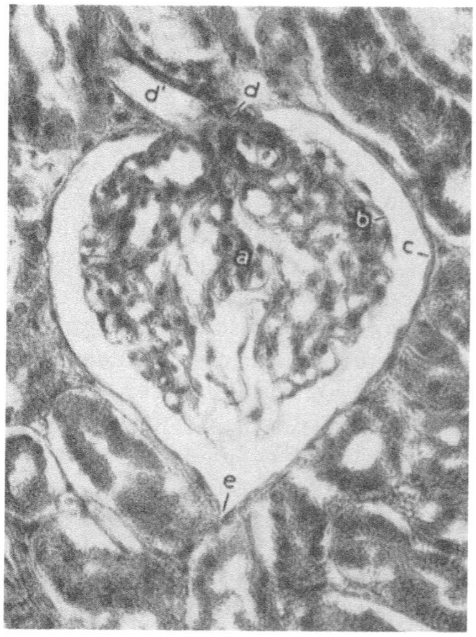

Fig. 413. Renal corpuscle of canine kidney. Microphotograph.

a Glomerulus; *b, c* Internal and external layers of glomerular capsule; *d* Vascular pole; *d'* Afferent arteriole; *e* Beginning of proximal convoluted tubule, urinary pole

Ureter

The ureter (559/*4*) is a narrow tube which conducts the urine in a continuous trickle from the renal pelvis to the bladder. It appears at the hilus of the kidney and curves caudally toward the pelvic inlet, assuming a slightly convex (medially), retroperitoneal course. Its **abdominal part** lies on the psoas muscles and crosses the large terminal branches of the aorta and caudal vena cava ventrally. Its **pelvic part** enters the genital fold or broad ligament (15, 16), and in the male crosses the dorsal surface of the ductus deferens (490/*3*). It pierces the dorsal wall of the bladder at an acute angle, near the neck (416/*e, f, g*).

As in the renal pelvis, the wall of the ureter consists of a connective tissue **adventitia,** a mostly triple-layered **muscular coat** (outer and inner longitudinal and middle circular layers), and a **mucous membrane.** The latter is arranged in longitudinal folds and is covered with transitional epithelium. In the horse, the mucous membrane contains large mucous glands in the proximal part of the ureter.

Urinary Bladder

The urinary bladder (415) is an organ capable of great distension, and when necessary it can store considerable amounts of urine. When the bladder is empty and contracted, it recedes, especially in the horse, into the pelvic cavity to varying degrees. When filled, it extends over the brim of the pelvis and lies on the ventral abdominal wall. In the female, the bladder is related dorsally to the uterus and broad ligaments. In the male, because of the short genital fold, the bladder is in contact with the rectum, and is therefore more easily palpated rectally.

The relations of the bladder are illustrated in the following Figures: **Dog:** 17/*c*; 179/*h*; 539/5. **Cat:** 186/*k*; 471/*1*; 541/*5*. **Pig:** 195/*p*; 197/*q*; 545/5. **Ox:** 548/11; 554/5. **Horse:** 15, 16/*d*; 525/*g*; 559/5.

The bladder has a cranial blind end (apex vesicae); a **neck,** which is the narrow caudal part leading into the urethra; and a **body** (corpus vesicae) in the middle. At the apex, especially in young animals, is a mass of scar tissue (415/*a'*), which is the remnant of the caudal part of the urachus. The **urachus** is the tube that connects the primitive bladder with the allantoic sac in the fetus, and is included in the umbilical cord. When the umbilical cord ruptures at birth, the urachus degenerates along its intra-abdominal course.

The wall of the bladder consists of a covering of peritoneum, a muscular coat, and a mucous membrane. The **peritoneum** covers only the exposed surfaces. The parts of the bladder not so invested are covered with a connective tissue **adventitia.** The **muscular coat** is composed of three layers of coarse, mostly S-shaped, smooth muscle bundles. In the outer layer, the bundles have a longitudinal or oblique course. They do not form a continuous sheet and may even be absent in the horse and pig (416). In the middle layer, the fibers run transversely, and in the inner layer they run longitudinally. Muscle bundles of both inner and outer layers enter, and interweave in, the middle layer. The apex and neck of the bladder are surrounded by loops of muscle bundles. At the neck, the loops are derived from both inner and outer longitudinal layers and may act in the closure of the internal urethral orifice. It is possible to isolate dorsal and ventral loops (*3*, *4*).

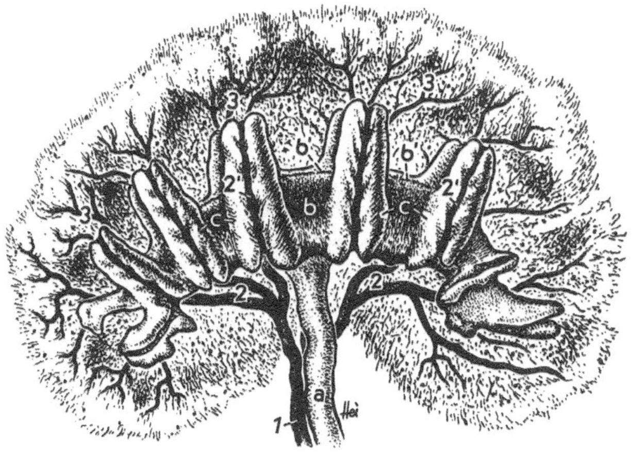

Fig. 414. "Plastoid" cast of renal pelvis and arteries of a dog.
a Ureter; *b* Renal pelvis; *b'* Partly injected papillary ducts of the area cribrosa; *c* Recesses; *1* Renal artery; *2* Branches of renal artery; *2'* Interlobar arteries passing between adjacent recesses; *3* Arcuate arteries giving off interlobular arteries. The dots are injected glomeruli

When the muscle bundles constrict the neck of the bladder, the urine is retained; when the muscle layers on body and apex contract, the urine is discharged. The muscular coat also fulfills what might be called static functions: by its tonus it renders the bladder wall sufficiently firm to prevent undue fluctuations of shape when the bladder is full.

The **mucous membrane** of the bladder rests on a thick submucosa, which allows adequate stretching when the bladder expands. The epithelial lining is of the transitional type.

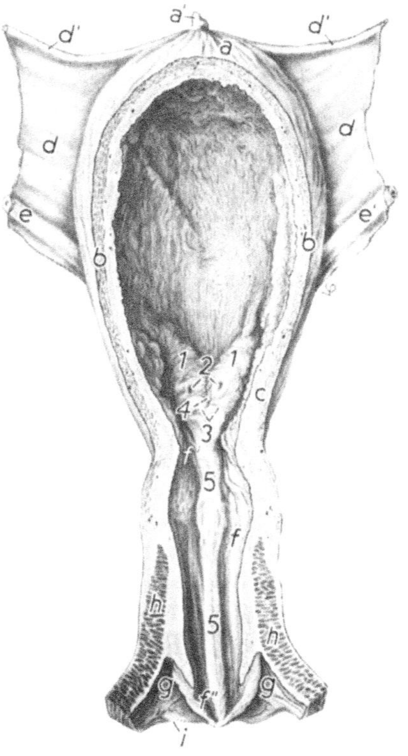

Fig. 415

Fig. 415. Bladder and urethra of the cow, opened ventrally. Ventral aspect.
a Apex, *b* body, *c* neck of bladder; *a'* Vestige of urachus; *d* Lateral ligament of bladder; *d'* Round ligament of bladder (vestige of umbilical artery); *e*, *e'* Left and right ureters; *f* Urethra, opened ventrally; *f'* Internal urethral orifice; *f''* External urethral orifice; *g*, *g* Suburethral diverticulum, opened ventrally; *h* Urethralis; *i* Mucosal fold (rudimentary hymen) at the junction of vagina and vestibule. *1* Columnae uretericae; *2* Ureteral orifices; *3* Plicae uretericae; *4* Trigonum; *5* Urethral crest

The **ureters** enter the bladder on the dorsal surface and pass through the wall at an acute angle. After penetrating the muscular coat, they continue for a short distance in the

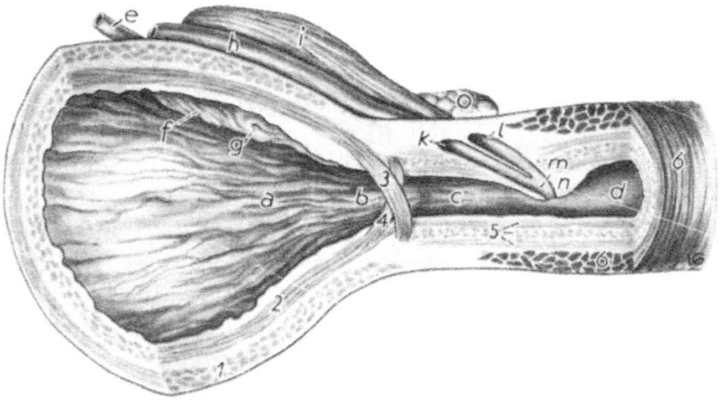

Fig. 416. Bladder and urethra of the male horse. Craniolateral aspect. Semischematic. (Modified from Gräning 1937.)
a Neck of bladder; *b* Internal urethral orifice; *c*, *d* Pelvic part of urethra; *c* Preprostatic part of urethra; *e* Ureter; *f* Columna ureterica; *g* Ureteral orifice; *h* Ductus deferens (ampulla); *i* Seminal vesicle; *k* Ductus deferens; *l* Excretory duct of seminal vesicle; *m* Ejaculatory duct and orifice; *n* Colliculus seminalis; *o* Prostate gland
1, *2* Muscular coat of bladder; *3*, *4* Muscular loops at internal urethral orifice; *5* Smooth muscle coat of urethra; *6* Urethralis

submucosa, producing two converging ridges (columnae uretericae, 415/*1*; 416/*f*) in the interior before terminating at their espective slitlike **ureteric orifice** (415/*2*). Two converging epithelial folds (plicae uretericae, *3*) continue caudally from the ureteric orifices. After they meet in the median pane, they continue as the **urethral crest** (*5*), which projects into the urethra from above and terminates in the male at the **colliculus seminalis** (416/*n*).

LIGAMENTS OF THE BLADDER. There are two lateral ligaments and one median ligament. During intrauterine life, they are related functionally to embryonic structures; after birth, they serve to support the bladder.

The **lateral ligaments of the bladder** are drawn out prenatally as vascular folds by the large umbilical arteries, which pass from the pelvic inlet to the umbilicus on each side of the median plane. In the newborn, only the caudal portions of the arteries remain, and their supporting folds become the lateral ligaments of the bladder as it becomes functional. The ligaments (15, 16/*6*) arise from the lateral pelvic wall and extend medially to the sides of the bladder. Their cranial, free edges, the **round ligaments of the bladder** (415/*d'*), are formed by the thick-walled umbilical arteries, which originate from the internal iliac arteries and end blindly at the apex of the bladder.

The **median ligament of the bladder** in prenatal life is the supporting fold of the urachus and extends along the ventral abdominal wall from the pelvis to the umbilicus. Most of it degenerates with the urachus after birth, and only a small median fold between the pelvic floor and the ventral surface of the bladder remains (15, 16/*7*). In carnivores it does not degenerate much, and reaches in the adult to the umbilicus as a narrow falciform fold (179/*i*).

Urethra

The urethra is a muscular tube, through which the urine is discharged from the bladder. It differs markedly in the two sexes. Both male and female urethrae are associated anatomically with the genital organs, but in the male, where this association is much more pronounced, there is also close functional relationship. Therefore, the male urethra is described in detail with the male genital organs (p. 319), and only a few general remarks are made here.

The **FEMALE URETHRA** (525/*f*) begins at the neck of the bladder and extends caudally along the floor of the pelvis. The urine enters it at the **internal urethral orifice** and leaves at the **external urethral orifice** (*f'*). The latter is situated in the floor of the genital tract at the junction of vagina and vestibule, and in the cow and pig shares a common opening with the **suburethral diverticulum** (555/*10*). The urethra is median in position and related dorsally to the vagina and ventrally to the pelvic symphysis. It is 6—8 cm. long in the dog, pig, and horse, and 10—12 cm. in the cow. The **muscular coat** of the urethra is continuous with the muscular coat of the bladder, and consists of an outer longitudinal and an inner circular layer of smooth muscle. Lateral and ventral to the smooth muscle is the striated **urethralis** (559/*7'*) which with other muscles forms the **sphincter urethrae.** The **mucous membrane** is similar to that of the bladder. In its submucosa it contains numerous cavernous veins (stratum cavernosum) which increase in number toward the external urethral orifice and may play a supportive role in the temporary closure of the urethra.

The preprostatic part (416/*c*) of the **MALE URETHRA** that is homologous to the female urethra is short and extends from the internal urethral orifice at the neck of the bladder to the **colliculus seminalis.** At the colliculus, the urinary and genital ducts of the male join. From the colliculus to the external urethral orifice, which is at the tip of the penis, the male urethra conducts both urine and the products of the genital glands and is therefore an integral part of both urinary and genital systems (see p. 319).

The Urinary Organs of the Carnivores

The **KIDNEYS OF THE DOG** (417, 418) are bean-shaped and have a dark, brownish red or bluish red color. Their size and weight vary widely with the size of the breed. On the average, the canine kidney weighs 40—60 gm., with the left kidney slightly heavier than the right in about two-thirds of cases.

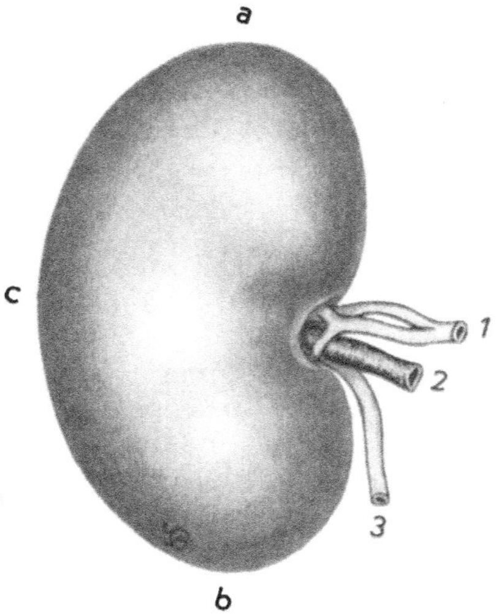

Fig. 417

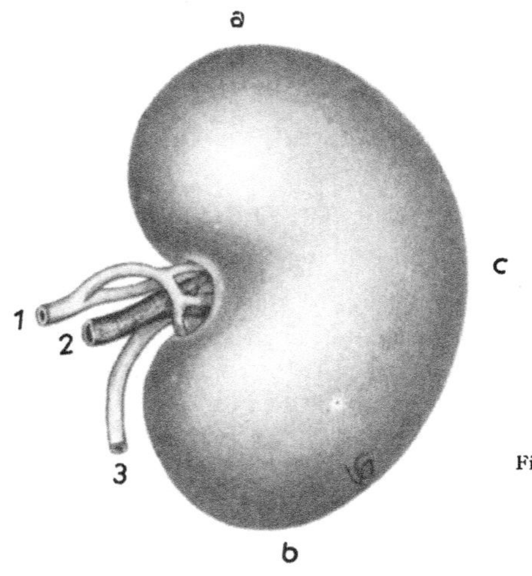

Fig. 418

Fig. 417 Right and 418 left kidneys of the dog. Ventral aspect. *a* Cranial pole; *b* Caudal pole; *c* Lateral border; *1* Renal artery; *2* Renal vein; *3* Ureter

The cranial pole of the right kidney (175—177/*e*) lies at the level of the twelfth or thirteenth rib, and the caudal pole is opposite the second or third lumbar vertebra. The left kidney is slightly more caudal. The right kidney is related dorsally to the crus of the diaphragm and psoas musculature. Cranially, it lies in the renal impression of the liver, and, medially, it is in contact with the caudal vena cava. Ventrally, it is related to the descending duodenum and pancreas and, more cranially, to the ascending colon and the stomach when the latter is distended. Laterally, it is related to the abdominal wall. The left kidney (172—174/*e*) is less firmly fixed than the right. It is related laterally to the spleen and abdominal wall, and ventromedially to the descending colon. Cranially, it makes contact with the left lobe of the pancreas and the stomach when the latter is distended. Both kidneys may be palpated through the abdominal wall.

The **KIDNEYS OF THE CAT** (419, 420) are relatively large and, depending on the amount of blood they contain, light or dark yellowish red. They are thick dorsoventrally, and their dorsal surface is slightly flattened. They are on the average 38—44 mm. long, 27—31 mm. wide, and 20—25 mm. thick. There is no appreciable weight difference between right and left kidney; each weighs from 7—15 gm.

The kidneys of the cat are more caudal in position than those of the dog. The right lies ventral to the first to fourth lumbar transverse processes and the left lies ventral to the second to fifth transverse processes (185, 186). Their relations to other organs are as described for the dog. They also can be palpated through the abdominal wall.

The amount of **perirenal fat** in the carnivores depends on the nutritional state of the animal. The fat covering the ventral surface of the kidneys is thin or absent. At the hilus and at the caudal pole the fat capsule is usually well developed.

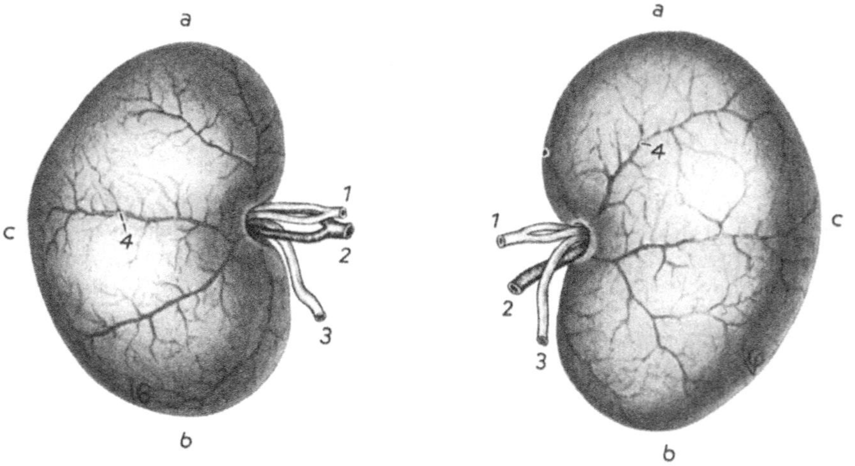

Fig. 419 Right and 420 left kidneys of the cat. Ventral aspect. *a* Cranial pole; *b* Caudal pole; *c* Lateral border; *1* Renal artery; *2* Renal vein; *3* Ureter; *4* Capsular veins

On a section made through the hilus and the poles, the **cortex** and **medulla** are easily distinguished (410, 421). In the dog, the cortex is 3—8 mm., in the cat 2—5 mm., thick. The **medullary rays** are also distinct.

The **renal crest** of the dog's kidney (423/*1*) is long and narrow. From it project 12—16 curved columns of kidney tissue (so-called pseudopapillae, *3*) which are at right angles to the longitudinal axis of the kidney. They are the central portions of the **medullary pyramids**

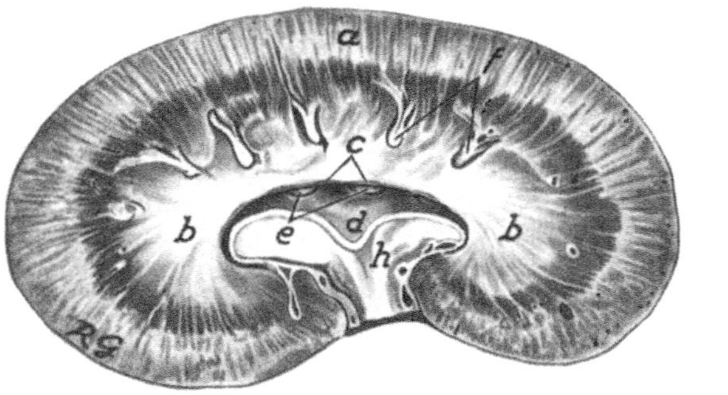

Fig. 421. Kidney of the dog, sectioned through poles and hilus. (After Zietzschmann, unpublished.)

a Cortex with medullary rays; *b* Medulla; *c* Renal crest; *d* Renal pelvis; *e* Pseudopapillae; *f* Interlobar arteries and veins breaking up to form arcuate vessels; *h* Fat

in which the **collecting tubules** from the dorsal and ventral surfaces of the kidney run to the **papillary ducts** in the renal crest. The internal architecture of the feline kidney is similar to that of the dog. The middle of the renal crest is drawn out into a cone on which the area cribrosa forms a small depression (410/*c*).

With the exception of the veins in the renal capsule, the arrangement of the **blood vessels** in the kidney of the carnivores (414) conforms to the description given on page 287. In the cat, the arrangement of the **capsular veins** (419, 420/*4*) is very characteristic. They are long, arborizing and anastomosing vessels occupying shallow grooves on the dorsal and ventral surfaces of the kidney and converging toward the hilus where they join the renal veins. In the dog, the corresponding **venulae stellatae** are distributed over the entire surface and are almost always visible grossly. Christensen (1952) has given a detailed account of the blood vascular system of the canine kidney.

The **RENAL PELVIS** in the dog (422, 425) consists of an elongated central cavity into which projects the renal crest. On the sides of the common cavity are several **recesses** (425/*b''*), which are indented (and nearly bisected) by the columns of kidney tissue (423/*3*).

Between neighboring recesses are narrow grooves in which the interlobar vessels pass to the cortex. In the cat, the central cavity of the renal pelvis is shorter from pole to pole and is more funnel-shaped (410/*b*). Similar recesses as described for the dog are present.

The **URETER** presents no special features, and the description on page 288 applies.

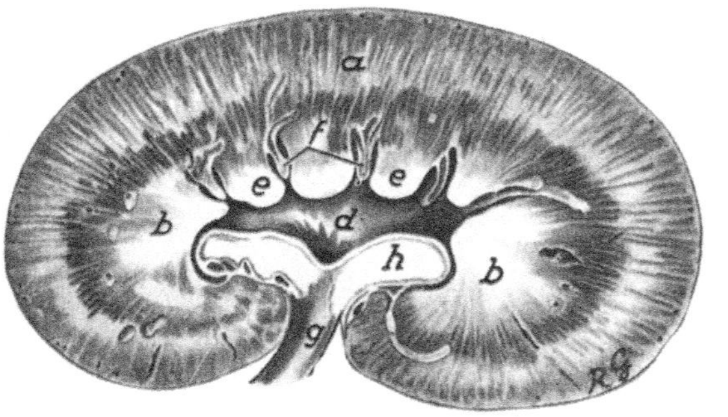

Fig. 422. Kidney of the dog, sectioned parallel to renal crest. (After Zietzschmann, unpublished.)

a Cortex; *b* Medulla; *d* Renal pelvis; *e* Pseudopapillae projecting into recesses of pelvis; *f* Interlobar arteries between adjacent recesses; *g* Ureter; *h* Fat

The **BLADDER** of the carnivores, whether full or empty, lies on the ventral abdominal wall cranial to the pubis and, with the exception of the neck which is in the pelvic cavity, is invested with peritoneum (471; 539). The bladder is one of the few organs that is not covered by the extensive greater omentum, and thus is in direct contact with the abdominal wall (179). It is related cranially to the loops of the jejunum, and when fully distended may reach to the umbilicus. When sufficiently distended, the bladder is palpable and easily accessible for surgery through the abdominal wall. The **lateral ligaments of the bladder** (539/*6*) present no special features. However, the **median ligament** (179/*i*) is rather narrow and extends from the pelvic inlet to the umbilicus.

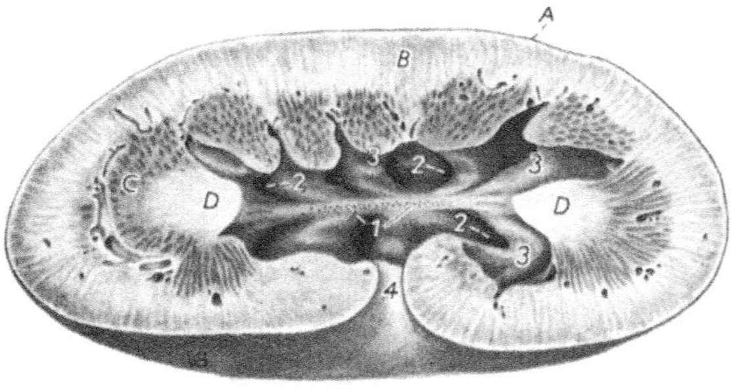

Fig. 423

Fig. 423. Kidney of the dog, sectioned parallel to medial border. The renal pelvis has been removed.

A Fibrous capsule; *B* Cortex; *C*, *D* Medulla; *1* Renal crest with area cribrosa and a slitlike papillary foramen at each end of the crest; *2* Recesses between adjacent pseudopapillae, the interlobar vessels have been removed; *3* Pseudopapillae attached to the sides of the renal crest; *4* Part of hilus

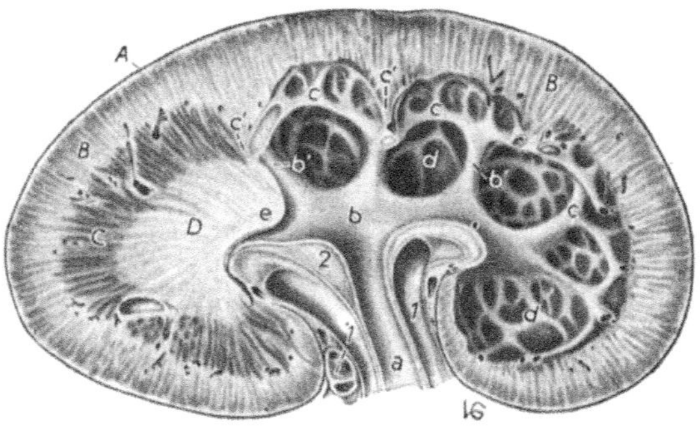

Fig. 424

Fig. 424. Kidney of the dog, sectioned through poles and hilus. The medullary substance has been removed in the right half of the picture.

A Fibrous capsule; *B* Cortex with medullary rays (light gray lines); *C*, *D* Medulla; *a* Ureter passing through hilus; *b* Renal pelvis; *b'* Recesses of renal pelvis; *c* Renal columns containing the arcuate vessels; *c'* Renal columns in section; *d* Recesses in the deep surface of the cortex from which the bases of the medullary pyramids have been removed, the arcuate vessels which lie at the corticomedullary junction are exposed; *e* Part of renal crest; *1* Branches of the renal artery and vein; *2* Fat

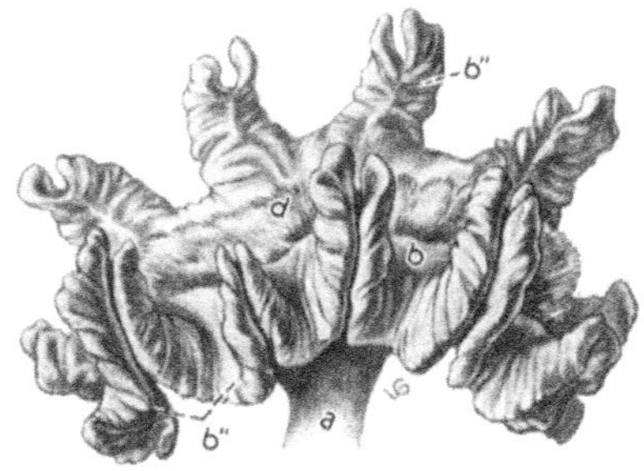

Fig. 425. "Plastoid" cast of the renal pelvis of the dog.

a Ureter; *b* Central cavity of renal pelvis; *b''* Recesses; *d* Injected papillary ducts

The **URETHRA** of the bitch (539/*7*) has a prominent circular smooth muscle layer. The inner and outer longitudinal smooth muscle layers are incomplete. The striated **urethralis** is present. The **male urethra** is described with the genital organs on p. 319.

The Urinary Organs of the Pig

The **KIDNEYS** of the pig (426, 427) are smooth externally and have several papillae, indicating partial fusion of their lobes. They are bean-shaped, flattened dorsoventrally with slightly pointed poles, and may occasionally have shallow grooves (lobation) on the surface. The **hilus** is about in the middle of the medial border and differs in width. The kidneys are brown—in bled animals often grayish brown. Their average weight is 200—280 gm. each, both kidneys being of about equal weight.

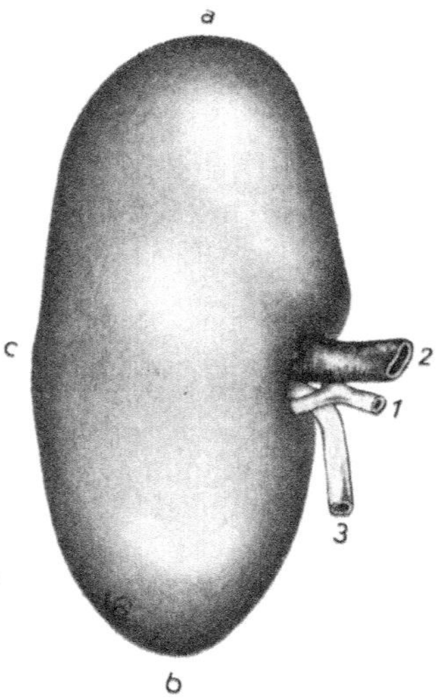

Fig. 426

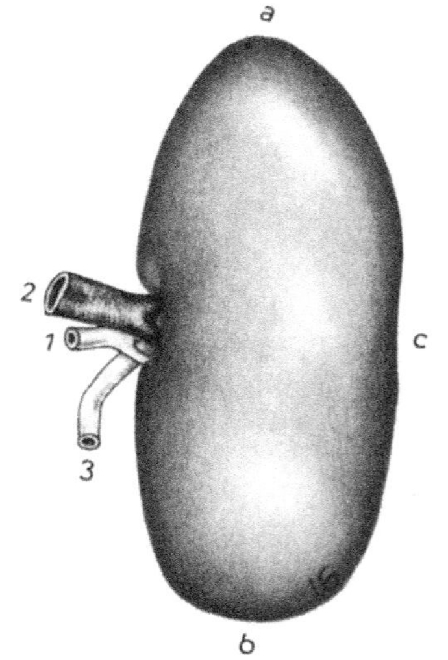

Fig. 427

Fig. 426 Right and 427 left kidneys of the pig. Ventral surface. *a* Cranial pole; *b* Caudal pole; *c* Lateral border; *1* Renal artery; *2* Renal vein; *3* Ureter

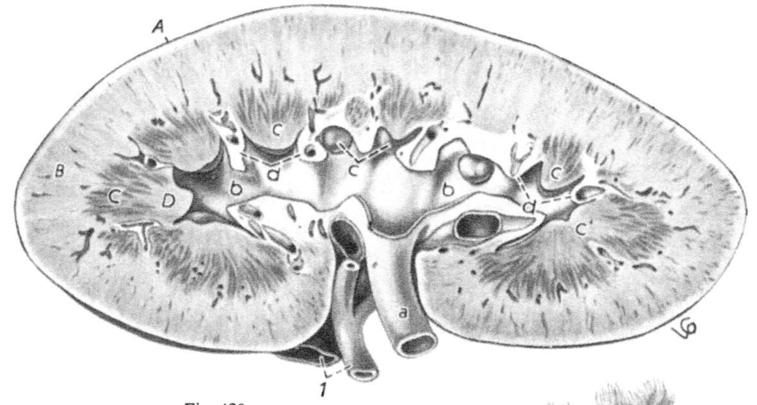

Fig. 428

Fig. 428. Kidney of the pig, sectioned through poles and hilus.

A Fibrous capsule; *B* Cortex; *C, D* Medulla

a Ureter passing through hilus; *b* Pelvis; *c* Single papilla; *c'* Large aggregate papilla; *d* Calix
1 Renal artery and vein

Fig. 429. "Plastoid" cast of renal pelvis of the pig.

a Ureter; *b* Pelvis, central cavity; *c* Calices with injected papillary ducts

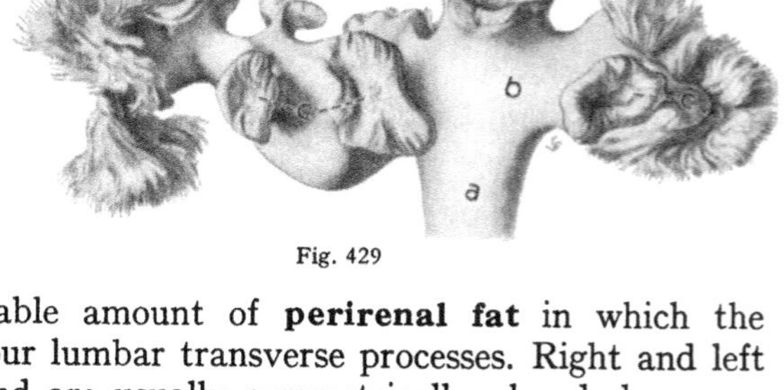

Fig. 429

Pigs in good condition have a considerable amount of **perirenal fat** in which the kidneys are embedded ventral to the first four lumbar transverse processes. Right and left kidneys lie flat against the psoas muscles and are usually symmetrically placed; however, either one may lie slightly cranial to the other (545/*1*, *2*). The right kidney is related to the intestines and to the pancreas, but does not contact the liver as in the other domestic mammals (200). The left kidney is related to the intestines, the spleen, and also to the pancreas (197). The renal vessels arise from aorta and caudal vena cava opposite the hilus of the kidneys.

The **renal cortex** in the pig (428/*B*) is 5—25 mm. thick, and the **medulla** (*C*, *D*) is often only one-half to two-thirds the thickness of the cortex. The **convoluted tubules** and **corpuscles** render the cut surface of the cortex dull and granular. The **medullary rays** are easily seen, but there is only a suggestion of **renal columns.**

The apices of the **medullary pyramids** form either single (*C*) or aggregate (*C'*) **papillae.** The single papillae are small and conical; the larger aggregate papillae are irregular elevations resulting from the fusion of 2—5 adjacent pyramids. There are 8—12 papillae per kidney. The outlines of the **areae cribrosae** differ from papilla to papilla.

Venulae stellatae, which are present in the dog and small ruminants, are absent in the pig.

The **RENAL PELVIS** consists of a central cavity (429/*b*), which leads into the ureter, and of two spacious recesses (calices majores) directed toward the poles. Both the central cavity and the recesses bear the **calices minores** (*c*), of which some are on short tubular stems. The minor calices surround the base of each papilla and differ according to the shape and size of their papilla.

The **URETERS** (545/*4*) leave the kidney at the hilus and turn sharply caudally to pass to the bladder. They enter the bladder at the neck, penetrating its muscular coat at almost right angles (476/*2*), and pass obliquely through the submucosa, raising the mucosa slightly before ending at the **ureteric orifices.** The two **ureteric folds** continue caudally from the orifices and outline the **trigonum.**

The relatively large **BLADDER** is capable of considerable distension and when inspected in unembalmed specimens may differ greatly in size. It is oval when empty or moderately full, but as it increases in size it becomes more and more spherical. With the exception of the neck, the bladder of the pig is in the abdominal cavity and rests on the ventral abdominal wall (545). The **ligaments** of the bladder conform to the general description given on page 290.

The long **FEMALE URETHRA** (545/*7*) has poorly developed outer and inner longitudinal muscle layers. The middle circular layer is thick, and together with the striated **urethralis** constricts the urethra. In the floor of the slitlike **external urethral orifice** is a small **suburethral diverticulum.** The preprostatic urethra of the male has a similar muscular arrangement. The remainder of the **male urethra** is described with the genitalia on page 319 and 331.

The Urinary Organs of the Ruminants

The **KIDNEYS OF THE OX** are brownish red and divided externally into lobes. The right kidney (430) is flattened dorsoventrally and has the shape of an irregular oval. The left kidney (431) has a rounded caudal and a pointed cranial end and appears more like a pyramid. Its left surface is flat from contact with the dorsal sac of the rumen. The **hilus** of the bovine kidney is large. The left kidney is 19—25 cm. long and the right kidney is 18—24 cm. long. The total weight of both kidneys is 1.2—1.5 kg., with the left kidney usually slightly heavier than the right. In general, the kidneys of bulls and steers weigh more than those of cows.

The right kidney (234/*v*; 554/*1*) extends from the last rib to the third lumbar vertebra and is applied with its dorsal surface against the crus of the diaphragm and the sublumbar musculature. The hilus is on the ventral surface near the medial border, and the cranial pole lies in the renal impression of the liver. Ventrally, the right kidney is related to the pancreas, colon, and cecum. The left kidney (235/*s*; 554/*2*) is suspended from the roof of the abdominal cavity by the renal vessels and is enclosed in a large fatfilled fold of peritoneum. As the dorsal sac of the rumen develops in the fetus and grows to adult size in the calf, it gradually pushes the left kidney into a median position (229/*m*), so that the kidney is related on the left to the rumen and on the right to the colon. The displacement of the left kidney is accompanied by a rotation of 45 degrees or more around its longitudinal axis, which causes the hilus to lie dorsally. The left kidney extends from the second or third lumbar vertebra to the fifth, and thus lies for the most part caudal to the right kidney (554). The sagittal fold of peritoneum (*2''*) that encloses the left kidney and its perirenal fat is of considerable height. It extends beyond the caudal pole of the kidney, curves to the left, and ends high in the left flank, blending with the broad ligament of that side. The recess between the fold and the lateral abdominal wall is occupied by the caudodorsal blind sac of the rumen. There is much **perirenal fat** in cattle in good condition. The fat, known as suet in the meat trade, is of hard consistency.

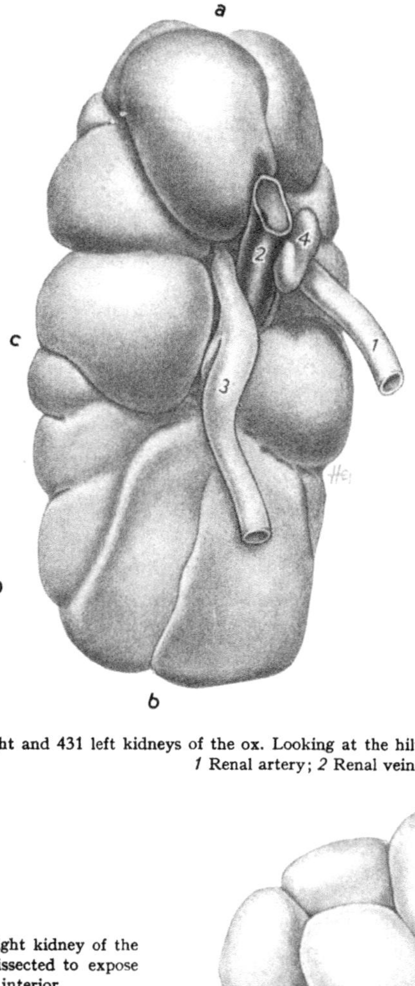

Fig. 430

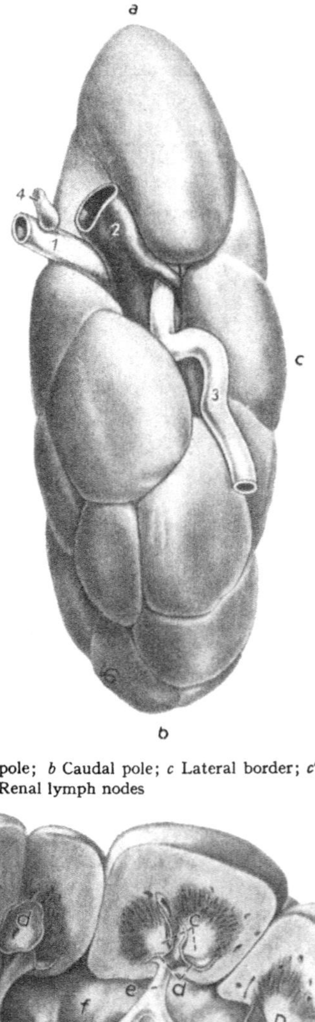

Fig. 431

Fig. 430 Right and 431 left kidneys of the ox. Looking at the hilus. *a* Cranial pole; *b* Caudal pole; *c* Lateral border; *c'* Ventral border *1* Renal artery; *2* Renal vein; *3* Ureter; *4* Renal lymph nodes

Fig. 432. Right kidney of the ox, partly dissected to expose the interior.

A Fibrous capsule; *B* Cortex; *C*, *D* Medulla

a Ureter; *b*, *b* Cranial and caudal branches of ureter; *c* Renal papilla; *d* Calix; *e* Secondary branches of ureter; *f* Interior of kidney (sinus renalis), dissected free of fat to show branching of ureter

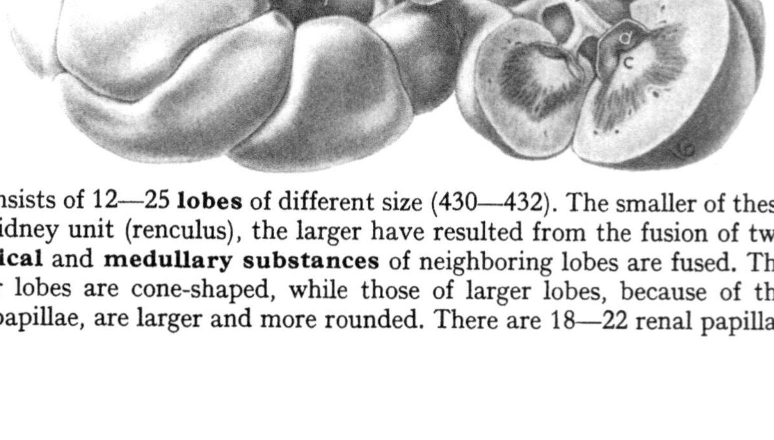

The bovine kidney consists of 12—25 **lobes** of different size (430—432). The smaller of these represent one original kidney unit (renculus), the larger have resulted from the fusion of two to five such units. **Cortical** and **medullary substances** of neighboring lobes are fused. The **papillae** of the smaller lobes are cone-shaped, while those of larger lobes, because of the fusion of several single papillae, are larger and more rounded. There are 18—22 renal papillae per kidney.

There is no renal pelvis in the ox. Instead of flaring out, the ureter of the ox after passing through the hilus divides into two branches directed toward the poles of the kidney (432/*b*, *b*). These primary branches subdivide to form 18—22 secondary branches each of which carrying a **calix** (*d*). The latter are funnel-shaped structures which embrace the base of each renal papilla (*c*). The system of tubes between ureter and calices is embedded in an ample quantity of fat in the center of the kidney.

The right **ureter** (548/*15*) of the ox follows a course similar to that seen in the other species. The left ureter (554/*4*) begins on the right of the median plane, passes along the right surface of its kidney, and then lies ventral to the right ureter for a short distance before crossing to the left and entering the bladder on the left side of the body (483/*2*).

The **KIDNEYS OF THE SHEEP AND GOAT** are similar to each other, but are very different from the bovine kidneys. The similarity between the kidneys of the small ruminants is not surprising considering that these two species are closely related and of about equal size. Remarkable, however, is that the kidneys of the sheep and goat are very much like the kidneys of the dog; and it is very difficult to distinguish them.

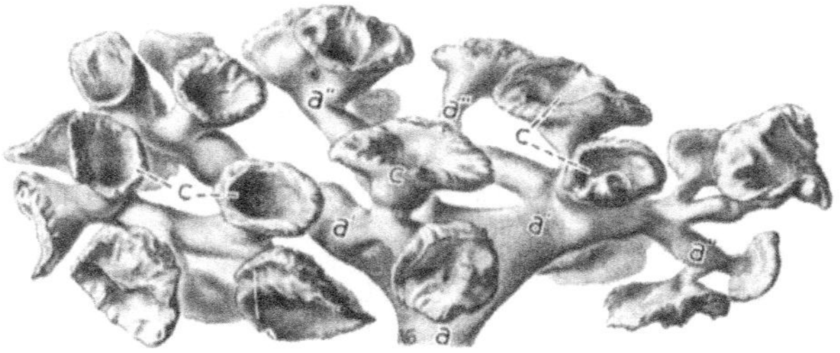

Fig. 433. "Plastoid" cast of bovine ureter and its branches and calices.

a Ureter; *a'*, *a'* Cranial and caudal branches of ureter; *a''* Secondary branches of ureter; *c* Calices

The kidneys of the small ruminants (434—436) are smooth externally and have a common papilla and a renal pelvis. They are bean-shaped, short, thick, and almost circular on cross section, and possess a distinct but rather shallow **hilus.** Their color in animals in good condition is light brown, otherwise they appear reddish brown. Their average length is from 5.5—7 cm., and their weight lies between 100—160 gm. each. In both species there is much **perirenal fat,** particularly around the left kidney.

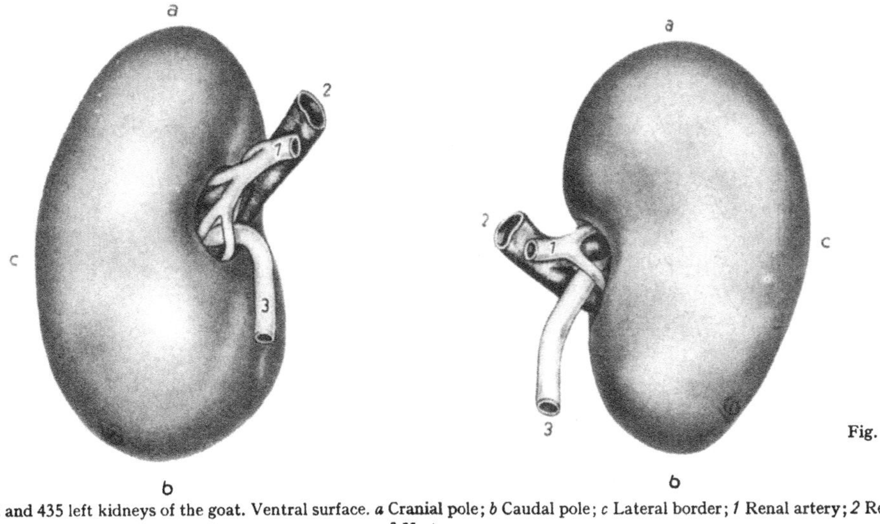

Fig. 434 Fig. 435

Fig. 434 Right and 435 left kidneys of the goat. Ventral surface. *a* Cranial pole; *b* Caudal pole; *c* Lateral border; *1* Renal artery; *2* Renal vein; *3* Ureter

The right kidney (236/*r*; 238/*n*) extends from the last rib to the level of the second lumbar vertebra. The left (*o*), as in the ox, lies to the right of the dorsal sac of the rumen and the median plane, but it is more caudal than in the ox, namely under the fourth and fifth lumbar vertebrae. Cranially, the right kidney is in contact with the renal impression of the liver; medially and ventrally it is related to the pancreas and to parts of the intestines. The left kidney is related to the rumen, ascending duodenum, and descending colon.

The **renal crest** in the small ruminants is also remarkably similar to that of the dog (423/*1*, canine kidney). The crest is joined by columns of renal tissue (so-called pseudopapillae), numbering 10—16 in the sheep and 10 in the goat. The **renal pelvis** invests the crest and pseudopapillae, and when isolated, as in a cast (436), appears like a rough negative of them. As in the carnivores, the pelvis consists of a central elongated cavity (*b*) for the renal crest, and recesses (*b''*) for the pseudopapillae.

The right **uretcr** of the small ruminants, like that of the ox, follows the vena cava, and for part of its course is dorsal to the left kidney. The first part of the left ureter lies to the right of the median plane. After leaving the kidney, the left ureter crosses the median plane ventral to the right ureter, and passes to the bladder on the left side of the body.

The **BLOOD SUPPLY** of the ruminant kidney follows the pattern described on page 287.

The **BLADDER OF THE OX** (490/*1*) is relatively large, has the shape of a long oval. and protrudes into the abdominal cavity even when only moderately full (554/*5*; 555/*8*). The full bladder occupies the ventral abdominal wall and thus, with the exception of the neck, is completely covered with peritoneum. The bladder is supported by **lateral** and **median ligaments,** which present no special features (see p. 290). In the cow, the **neck** of the bladder and the urethra are united with the ventral wall of the vagina by tough connective tissue (555). The **ureteric orifices** (415/*2*) lie close to the median plane at about the middle of the neck; the **trigonum** (*4*), therefore, is narrow. The **urethral crest** (*5*) continues caudally from the trigonum and extends to the external urethral orifice. It is up to 5 mm. high and protrudes from the dorsal wall into the lumen of the **URETHRA,** which in the cow is 10—13 cm. long. The **external urethral orifice** (547/*g*) is a narrow slit bounded laterally by mucosal folds. Ventral to the orifice is the **suburethral diverticulum** (555/*10*), a small pouch directed cranioventrally and up to 2 cm. in diameter. The entrance to this diverticulum lies exactly ventral to the urethral orifice and must be avoided when attempting to catheterize the bladder of a cow. The **musculature** of the female urethra consists of inner and outer longitudinal layers and a middle circular layer. In its caudal third, the urethra as well as the suburethral diverticulum are covered ventrally by the striated **urethralis.** The vessels in the submucosa (stratum cavernosum) are well developed in the ox, especially in the vicinity of the external urethral orifice.

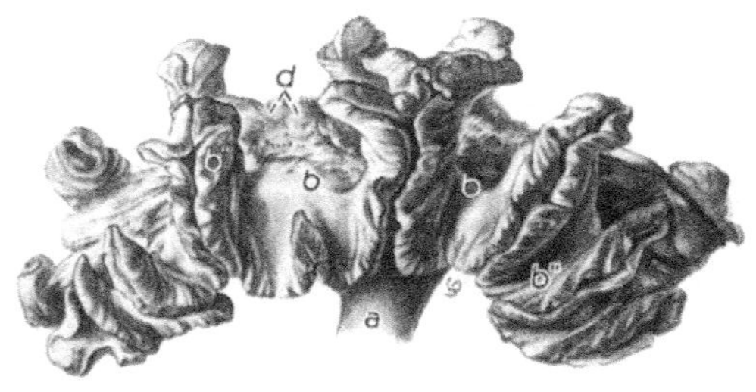

Fig. 436. "Plastoid" cast of the renal pelvis of a sheep.

a Ureter; *b* Central cavity of renal pelvis; *b''* Recesses; *d* Injected papillary ducts of renal crest

The **BLADDER OF THE SMALL RUMINANTS** is similar to the bovine bladder as far as shape, position, and attachment are concerned. The **URETHRA** of the ewe is 4—5 cm. long and that of the goat is 5—6 cm. long. Both have a stratum cavernosum between mucosa and muscular coat. The **urethral crest** extends caudally only to the middle of the urethra. A **suburethral diverticulum,** about 1—1.5 cm. long, is present in both species.

The **male urethra** is described with the genitalia on pages 319 and 336.

The Urinary Organs of the Horse

The **KIDNEYS** of the horse are smooth externally, have a common papilla and a pelvis, and are dark yellowish red to brown in the fresh state. Right and left kidneys always differ in shape (437, 438). The left kidney is bean-shaped, 15—20 cm. long, and 11—15 cm. wide. Its caudal pole is often wider than its cranial pole, and its ventral surface is more convex than the dorsal. The **hilus** is deep and is on the ventral surface close to the medial border. Radiating from the hilus are several grooves occupied by large branches of the renal artery. The right kidney is roughly triangular, resembling the heart on a playing card. Its transverse diameter, the distance from medial to lateral border, is 15—18 cm. and is longer than the longitudinal diameter, the distance from pole to pole, which is 13—15 cm. The dorsal surface of the right kidney is convex, the ventral surface is slightly concave. The **hilus** is deep and, as in the left kidney, on the ventral surface close to the medial border. The cranial border (pole, 437/*a*) of the right kidney is convex while the caudal border (pole) is nearly straight (*b*). The lateral border forms an angle (*c*). This is the apex of the "heart", the base of the "heart" is the medial border. The right kidney is always heavier than the left, and its average weight is 625 gm. with a range of 480—840 gm. The average weight of the left kidney is 602 gm. with a range of 425—780 gm.

RELATIONS. The **right kidney** is almost entirely covered by the ribs (258). It extends cranially to the fifteenth to seventeenth ribs (most commonly to the sixteenth, but in rare cases to the fourteenth). Caudally, it reaches the first lumbar vertebra, never beyond. The right kidney makes contact cranially with the renal impression of the liver. Dorsally, it is related to the crus of the diaphragm and the iliac fascia which covers the psoas musculature (262). Much of its ventral surface is attached to the base of the cecum so that the right kidney cannot be palpated rectally (258). Its medial relations are with the pancreas and the right adrenal gland, and occasionally also with the vena cava (439). Its caudal pole makes contact with the descending duodenum (248).

The longer **left kidney** (559/*2*) extends from the sixteenth to eighteenth ribs (most commonly the seventeenth) to the second or third lumbar vertebra. Occasionally, it is entirely caudal to the last rib, lying ventral to the first, second, and third lumbar transverse processes. The left kidney has the same dorsal relations as the right. Ventrally, it is related to the

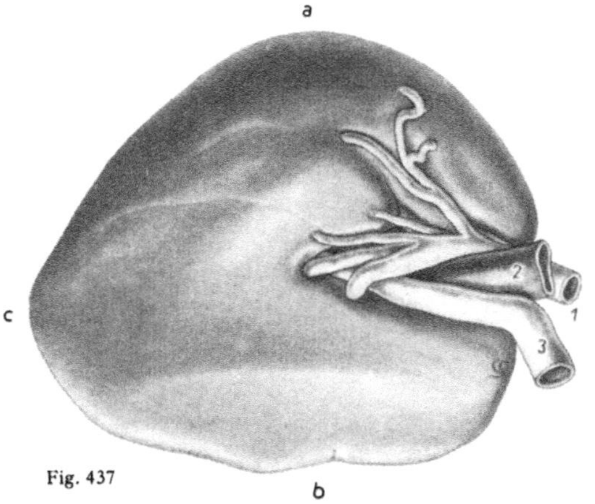

Fig. 437

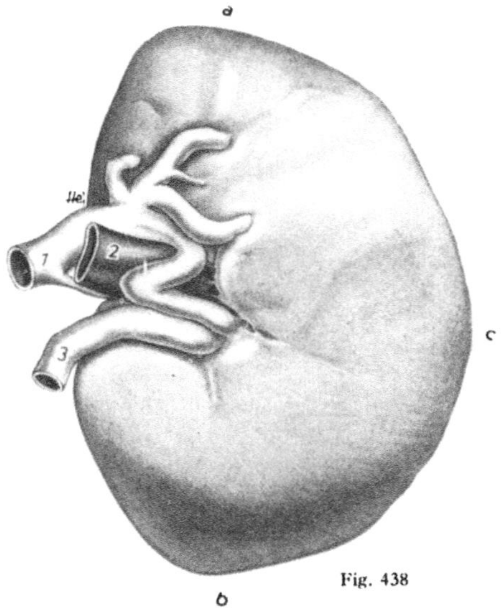

Fig. 438

Fig. 437 Right and 438 left kidneys of the horse. Ventral surface.
a Cranial pole; *b* Caudal pole; *c* Lateral border; *1* Renal artery; *2* Renal vein; *3* Ureter

jejunum, descending colon, and ascending duodenum (248). Medially, it has contact with the left adrenal gland and the left lobe of the pancreas (439). The dorsal extremity of the spleen is inserted between the lateral border of the left kidney and the abdominal wall, and is connected to the kidney by the **renosplenic ligament** (*k*). The left kidney is somewhat more loosely embedded than the right in the relatively small amount of **perirenal fat** present in the horse. Its ventral surface is covered with peritoneum, while the ventral surface of the right kidney, because of its extensive attachment to the cecum, has peritoneum only along the lateral border.

On the cut surface (440), the **cortex, medulla,** and **medullary rays** can be differentiated without difficulty. The **renal columns,** however, are not very distinct.

The kidneys of the horse have resulted from the complete fusion of 40—64 lobes. Fusion of the centrally located medullary pyramids produced the relatively small **renal crest** (*c*) on which the papillary ducts from the central part of the kidney open, forming an **area cribrosa** of varying size. The papillary ducts from the poles of the kidney open into two **terminal recesses** (*b'*) of the renal pelvis.

The **RENAL PELVIS** (440/*b*) has a small central part with which it is attached to the base of the renal crest and which is not much larger than the crest itself. The two **terminal recesses** (*b*′) proceeding from it toward the poles of the kidney are on the average 5 mm. in diameter and 6—10 cm. long. They are in effect large urine-collecting ducts lined with the same epithelium that covers the renal crest. The mucous membrane lining the central part of the pelvis forms thick high folds and contains tubular glands which secrete a thick mucus. The mucus, colored yellow by the urine, is regularly found in variable amounts in the renal pelvis of the horse.

The **URETER** (559/*4*) is about 70 cm. long and takes the usual retroperitoneal course to the bladder. With the exception of mucous glands close to the renal pelvis, it presents no special features (see p. 288).

The position of the **BLADDER** varies considerably, depending on the amount of urine it contains. When empty, it is a contracted ball of tissue, 8—10 cm. in diameter, lying deep within the pelvic cavity so that only its apex and a small part of its body is covered with peritoneum (525). As the bladder fills, it expands cranially (559/*5*) and hangs over the brim of the pelvis, and when filled to capacity it may extend to the umbilicus. The **lateral**

Right

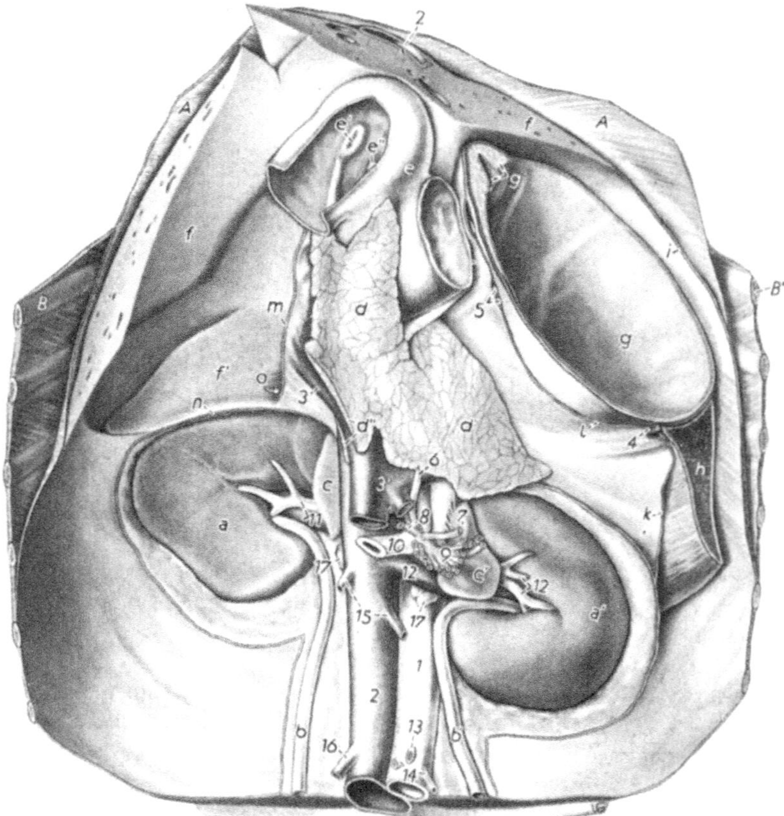

Left

Fig. 439. Kidneys, adrenal glands, and pancreas of the horse in situ. Ventral aspect.

A Diaphragm; *B*, *B*′ Right and left thirteenth ribs; *C* Third lumbar vertebra

a, *a*′ Right and left kidneys, exposed by removal of peritoneum and perirenal fat; *b*, *b*′ Right and left ureters; *c*, *c*′ Right and left adrenal glands; *d* Body of pancreas; *d*′ Left lobe of pancreas; *d*‴ Right lobe of pancreas, partly removed to expose portal vein; *e* Cranial part of duodenum, fenestrated; *e*′ Major duodenal papilla with openings of pancreatic and bile ducts; *e*″ Minor duodenal papilla with opening of accessory pancreatic duct; *f* Liver, partly removed; *f*′ Caudate process of liver; *g* Blind sac (fundus) of stomach; *g*′ Cardiac part of stomach; *h* Spleen, transected; *i* Left triangular ligament; *k* Renosplenic ligament; *l* Gastrosplenic ligament; *m* Mesoduodenum, cut close to descending duodenum; *n* Hepatorenal ligament; *o* Epiploic foramen, entrance to omental bursa

1 Aorta; *2* Vena cava; *3* Portal vein; *3*′ Branches of portal vein; *4* Splenic vessels; *5* Branches of left gastric vessels; *6* Caudal pancreaticoduodenal artery; *7* Middle colic artery; *8* Right colic artery; *9* Jejunal arteries; *10* Ileocolic artery; *11* Right renal artery and vein; *12* Left renal artery and vein; *13* Stump of caudal mesenteric artery; *14* Testicular arteries; *15* Testicular veins; *16* Right deep circumflex iliac vein; *17* Renal lymph nodes

ligaments of the bladder (*6*), which carry the obliterated umbilical arteries, show no special features. The caudal portion of the **median ligament of the bladder,** connecting the bladder with the floor of the pelvis, contains strands of smooth muscle (m. pubovesicalis).

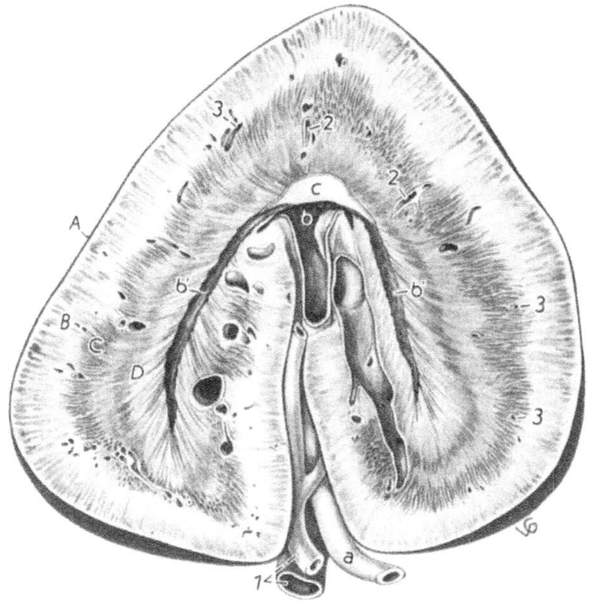

Fig. 440

Fig. 440. Right kidney of the horse, sectioned through poles and hilus.
A Fibrous capsule; *B* Cortex; *C, D* Medulla; *a* Ureter; *b* Central part of pelvis, *b'* Terminal recess of pelvis; *c* Renal crest; *1* Renal artery and vein; *2* Interlobar vessels; *3* Arcuate vessels

Fig. 441 and 442. "Plastoid" casts of right (441) and left (442) renal pelves and terminal recesses of the equine kidney.
a Ureter; *b* Renal pelvis; *b'* Terminal recess with injected papillary ducts; *d* Injected papillary ducts of renal crest

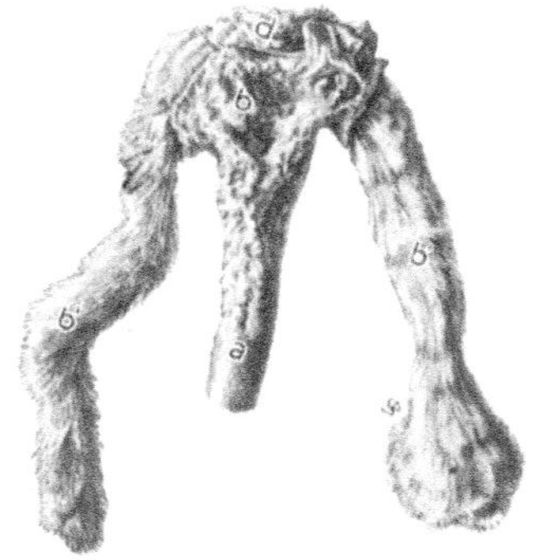

Fig. 441

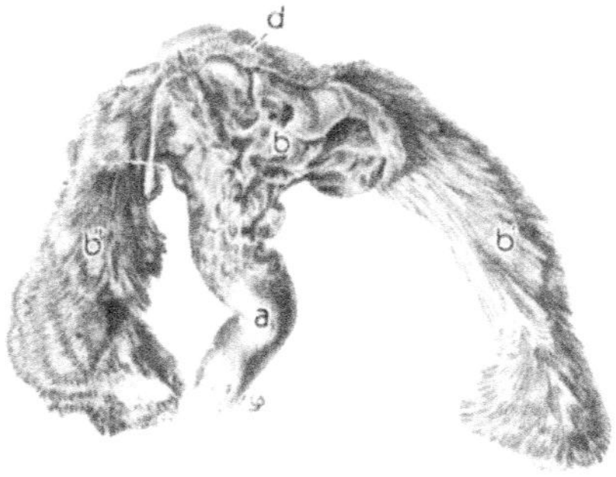

Fig. 442

The **URETHRA OF THE MARE** is about 6—8 cm. long and presents the usual triple-layered muscular coat, which is supplemented by the circular, striated **urethralis** (525/*r*). The erectile **stratum cavernosum** begins at the **internal urethral orifice** and gradually increases in thickness caudally. The **urethral crest** reaches from the internal urethral orifice, where it arises from the ureteric folds, to about the middle of the urethra.

The **male urethra** is described with the genitalia on pages 319 and 345.

Comparative Anatomy of the Kidney

In this section, the characteristic features, by which the kidneys of the domestic mammals may be identified, are summarized. These features were studied by Marschner (1937) and include color, shape, size, appearance of the surface, number and shape of the papillae, the shape of the pelvis, and the appearance of the parenchyma on the cut surface.

The kidneys of the horse, ox, pig, and cat are easily identified because of features that are characteristic for only one species. It is possible, moreover, to distinguish the right from the left kidney in the horse and ox. It is not possible, however, to distinguish the kidney of the sheep from that of the goat. This is not surprising, because these two species are close relatives. It is remarkable, however, that the dog, a species certainly not closely related to the ruminants, has kidneys so similar to those of the sheep and goat that it is virtually impossible to distinguish them. It is advisable, therefore, to be very careful when attempting to identify kidneys that may have come from these species.

HORSE (437, 438). The kidneys are smooth externally and have a renal crest. The left is bean-shaped, the right is heart-shaped, and the hilus is deep. On the ventral surface, large branches of the renal artery run in grooves before entering the parenchyma. The renal crest is relatively small—about 5 cm. long. Opening into the small pelvis are two terminal recesses, about 5 mm. in diameter and 6—10 cm. in length (440/*b'*). The pelvis always contains a yellowish mucus which is produced by the glands in its wall. Venulae stellatae are present over the entire surface of the kidney.

OX (430—432). The kidneys are very large, divided externally into lobes, and have separate papillae. The right is an irregular oval, flattened dorsoventrally; the left is shaped like a pyramid, pointed cranially and thick caudally. Instead of a pelvis the ureter forms branches, which end in 18—22 calices for an equal number of papillae.

PIG (428). The kidney is smooth externally and has 8—12 papillae. It is bean-shaped, markedly flattened dorsoventrally, and when well bled often grayish brown. The cortex is unusually thick, and the medulla is only one-half to two-thirds the thickness of the cortex. Renal columns are indistinct or missing. The renal pelvis is long and bears 8—12 calices, some on short stems (429).

CAT (419, 420). The kidney has the shape of a thick bean, is smooth externally, has a renal crest, and is often yellowish brown. The characteristic capsular veins lie in shallow grooves which radiate from the hilus. The renal crest is joined on each side by columns of renal tissue (pseudopapillae). The pelvis consists of a central cavity with several recesses on each side.

SHEEP AND GOAT (434, 435). The kidneys have the shape of the thick bean, are smooth externally, and have a renal crest. According to Grundmann (1922), the kidney of the goat is slightly thinner and longer than that of the sheep, and the hilus of the sheep's kidney is about twice as deep as that in the goat. The left surface of the left kidney is often flat from contact with the rumen. The renal crest is joined on each side by columns of renal tissue (pseudopapillae). The renal pelvis consists of a central cavity with several recesses on each side. Neither the crest nor the pelvis present features by which sheep and goat kidneys can be distinguished.

DOG (417, 418, 421—425). The kidneys are smooth externally, have a renal crest, are often bluish red in color, and are very similar in shape to the kidneys of the sheep and goat. The canine kidney, however, is less rounded on cross section and on the whole a little more slender than the kidneys of the small ruminants. The hilus does not furnish reliable distinguishing criteria. The poles are bent slightly more toward the medial border than in the small ruminants. This is seen best on a section through the poles and hilus. The renal crest is sharper than that in the small ruminants and, in addition to the papillary openings, has a small slitlike opening on each end. The pseudopapillae and the renal pelvis furnish no reliable distinguishing criteria. Venulae stellatae are present on the canine kidney.

BIBLIOGRAPHY

Urinary Organs

Ackerknecht, E.: Die Harnorgane. In: Ellenberger-Baums Handbuch der vergleichenden Anatomie der Haustiere. 18. Auflage. Springer, Berlin, 1943.

Badawi, A., E. Schenk: Innervation of the abdomino-pelvic ureter in the cat. Amer. J. Anat. **126,** 103—120 (1969).

Barone, R.: Les vaisseaux des reins chez les Équidés. Bull. Soc. Sci. Vét. Lyon **58,** 1956.

Barone, R., and B. Blavignac: Les vaisseaux sanguins des reins chez le Boeuf. Bull. Soc. Sci. Vét. Lyon **66,** 1964.

Baum, M.: Vergleichende anatomische und histologische Untersuchungen über die Harnblase der Haustiere. Leipzig, Diss. med. vet., 1911.

Becher, H.: Über besondere Zellengruppen und das Polkissen am Vas afferens in der Niere des Menschen. Zschr. wiss. Mikrosk. mikrosk. Tech. **53,** 1936.

Brasch, E.: Über die papilla renalis der Haussäugetiere. Österr. Mschr. Tierheilk., 1909.

Broger, J. B. A.: Über das Epithel des Harnleiters, der Harnblase und der Harnröhre von Pferd, Rind und Hund mit besonderer Berücksichtigung seiner Zellformen bei der künstlichen Trennung von der Propria mucosae. Bern, Diss. med. vet., 1925.

Chomiak, M., S. Szteyn, Z. Milart: Concerning the innervation of the kidney of the sheep. Zbl. Vet. Med. A **16,** 754—756 (1969).

Christensen, G. C.: Circulation of blood through the canine kidney. Am. J. Vet. Res. **13,** 1952.

Clara, M.: Anatomie und Biologie des Blutkreislaufes in der Niere. Arch. Kreisl.forsch. **3,** 1938.

Collin, B.: Les vaisseaux sanguins du rein chez le chien. Ann. Méd. Vét., Bruxelles **116,** 631—646 (1972).

Dianova, E. V.: Structure, vascularisation et innervation des organes internes des animaux domestiques (Stalingrad 1957). Rec. Méd. Vét. **136,** 407 (1960). Zit. nach Barone und Blavignac (1964).

Dorn, F. K.: Untersuchungen über den Bau der Urethra feminina von Canis familiaris, Felis domestica, von Equus caballus. Leipzig, Diss. med. vet., 1923.

Dumont, A.: Vergleichende Untersuchungen über das Nierenbecken der Haustiere. Bern, Diss. med. vet., 1909.
Eliška, O.: The perforating arteries and their role in the collateral circulation of the kidneys. Acta anat., **70**, 184—201 (1968).
Ferke, F.: Nieren, Harnleiter und Harnblase der Hauskatze (Felis domestica Briss.) Germ. Summary. Budapest, Diss. med. vet., 1933.
Fletcher, T. F., W. F. Bradley: Comparative morphologic features of urinary bladder innervation. Amer. J. Vet. Res. **30**, 1655—1662 (1969).
Fuller, P. M., D. F. Huelke: Kidney vascular supply in the rat, cat and dog. Acta anat. **84**, 516—522 (1973).
Gordon, N.: Surgical anatomy of the bladder, prostate gland, and urethra of the dog. J. Amer. Vet. Med. Assoc. **136**, 215—221 (1960).
Grahame, T.: The pelvis and calyces of the kidneys of some mammals. Brit. Vet. J. **109**, 51—55 (1953).
Grandage, J.: Some effects of posture on the radiographic appearance of the kidneys of the dog. J. Am. Vet. Med. Assoc. **166**, 1975.
Gräning, W.: Zum Bau der Harnblase des Hundes. Zschr. Anat. Entw. gesch. **103**, 1934.
Gräning, W.: Beitrag zur vergleichenden Anatomie der Muskulatur von Harnblase und Harnröhre. Zschr. Anat. Entw. gesch. **106**, 1936.
Grundmann, R.: Vergleichende anatomische Untersuchungen über die Nieren von Schaf und Ziege. Hannover, Diss., 1922.
Guzsal, E.: Über die Nierenbecken von Pferd, Hund, Schaf und Katze. Acta vet. acad. sci. Hung. **8**, 1958.
Habermehl, K. H. und E. Tuor-Zimmermann: Die Blutgefäßversorgung der Katzenniere unter besonderer Berücksichtigung der Rindenvenen. Berl. Münchn. Tierärztl. Wschr. **85**, 466—469 (1972).
Hartmann, W. and C. C. van de Watering: The function of the bladder neck in female goats. Zbl. Vet. Med. A, **21**, 1974.
Havlicek, H.: Vasa privata und Vasa publica. Neue Kreislaufprobleme. Hippokrates **2**, 105—127 (1929/30).
Hicks, R. M.: The mammalian urinary bladder: an accomodating organ. Biol. Rev. **50**, 1975.
Holle, Ute: Das Blutgefäßsystem der Niere von Schaf (Ovis aries) und Ziege (Capra hircus). Diss. med. vet. Gießen, 1964.
Hyrtl, J.: Das Nierenbecken der Säugethiere und des Menschen. Denkschr. Wiener Acad. **31**, 1872.
Kainer, R. A., L. C. Faulkner, R. M. Abdel-Raouf: Glands associated with the urethra of the bull. Amer. J. Vet. Res. **30**, 963—974 (1969).
Kriz, W.: Der architektonische und funktionelle Aufbau der Rattenniere. Zschr. Zellforsch. **82**, 1967.
Kügelgen, A. v., B. Kuhlo, W. Kuhlo, Kl.-J. Otto: Die Gefäßarchitektur der Niere. — Untersuchungen an der Hundeniere. Zwangl. Abh. Geb. norm. path. Anat. Herausgeb. Bargmann, W., u. W. H. Doerr, Stuttgart, Thieme, 1959.
Lange, W.: Über die Abhängigkeit der Glomerulusgröße der Niere bei verschiedenen großen Hunderassen. Anat. Anz. **117**, 1965.
Lantzsch, F.: Vergleichende Untersuchungen über den Bau der Urethra feminina bei Bos taurus, Ovis aries, Capra hircus und Sus scrofa. Leipzig, Diss. med. vet., 1922.
Lenk, H. J.: Zur Anatomie und Histologie der Harnblase und der Pars pelvina der Harnröhre der Säugetiere. Leipzig, Diss. med. vet., 1913.
Marschner, H.: Art- und Altersmerkmale der Nieren der Haussäugetiere (Pferd, Rind, Schwein, Schaf, Ziege, Hund, Katze, Kaninchen, Meerschweinchen). Zschr. Anat. Entw. gesch. **107**, 1937.
Meinertz, T.: Eine vergleichende Untersuchung über die Säugetierniere, besonders im Hinblick auf die Nierentypen, das Nierenbecken und die Verzweigungen der größeren Gefäße. Gegenb. Morph. Jb. **113**, 1969.
Meyer, O.: Das Verhalten der einzelnen Schichten des Pelvis renalis, speziell bei ihrem Ansatz an die Nierensubstanz bei Schwein, Schaf, Hund und Katze. Bern, Diss. med. vet., 1922.
Moffat, D. B.: The mammalian kidney. London, Cambridge Univ. Press, 1975.
Möllendorff, W. v.: Der Exkretionsapparat. In: v. Möllendorff's Handbuch der mikroskopischen Anatomie des Menschen. VII/1. Berlin, Springer, 1930.
Negrea, A., H. E. König: Beiträge zur Untersuchung der Vaskularisation der Niere bei einigen Haussäugetieren. Iaşi Lucrări ştiintifice, II, 159—164, 1970.
Petry, G. u. H. Amon: Licht- und elektronenmikroskopische Studien über Struktur und Dynamik des Übergangsepithels. Zschr. Zellforsch. **69**, 1966.
Pfeiffer, E. W.: Comparative anatomical observations of the mammalian renal pelvis and medulla. J. Anat. **102**, 1968.
Ramchandra, P. Y., and M. L. Calhoun: Comparative histology of the kidney of domestic animals. Am. J. Vet. Res. **19**, 1958.
Roost, W.: Über Nierengefäße unserer Haussäugetiere mit spezieller Berücksichtigung der Nierenglomeruli. Bern, Diss. med. vet., 1912.
Schilling, E.: Vergleichend morphologische Betrachtungen an den Nieren von Wild- und Haustieren. (Ein Beitrag zur Phylogenie der Säugernieren.) Zool. Anz. **147**, 1951.
Schilling, E.: Metrische Untersuchungen an den Nieren von Wild- und Haustieren. Zschr. Anat. Entw. gesch. **116**, 1951.
Schmeer, K.: Die Berechnung der Nierenkörperchenzahl beim Hunde. Anat. Anz. **89**, 1940.
Schwarze, E.: V. Beitrag zur „Anatomie für den Tierarzt". Von den Nieren. Dtsch. Tierärztl. Wschr. **47**, 1939.
Simić, V., u. S. Popović: Morphologische Grundmerkmale und Verschiedenheiten der Nieren bei den kleinen Wiederkäuern und Fleischfressern (Ovis aries, Canis familiaris et Felis domestica). Anat. Anz. **113**, 1963.

Sobocinski, M.: Veränderungen in der Topographie der Seitenbänder der Harnblase in den ersten Lebenswochen der Kälber. Zeszyty naukowe wyższej Szkoly rolniczej we Wroclawiu, Weteryn. 71—75 (1964).
Steigleder, G. Kl.: Konstruktionsanalytische Untersuchungen an den ableitenden Harnwegen. Bruns' Beitr. Klin. Chir. **178,** 1949.
Tennille, N. B., and G. W. Thornton: Intravenous urography studies in the unanesthetized dog. Vet. Med. **53,** 1958.
Tour-Zimmermann, Esther: Das Blutgefäßsystem der Niere der Katze (Felis catus L.). Diss. med. vet. Zürich, 1972.
Weller, U.: Das Blutgefäßsystem der Niere des Pferdes (Equus caballus). Diss. med. vet. Gießen, 1964.
Wille, K.-H.: Das Blutgefäßsystem der Niere des Hausrindes (Bos primigenius f. taurus, L., 1758). Diss. med. vet. Gießen, 1966.
Wille, K.-H.: Gefäßarchitektonische Untersuchungen an der Nierenkapsel des Rindes. (Bos primigenius f. taurus, L.). Zbl. Vet. Med. A. **15,** 372—381 (1968).
Wittmann, E.: Das absolute und relative Gewicht der Nieren von Pferd, Rind, Kalb, Schwein, Schaf und Ziege. Berlin (Humboldt-Univ.), Diss. med. vet., 1959.
Woodburne, R. T.: The sphincter mechanism of the urinary bladder and the urethra. Anat. Rec. **141,** 1961.
Wrobel, K.-H.: Das Blutgefäßsystem der Niere von Sus scrofa dom. unter besonderer Berücksichtigung des für die menschliche Niere beschriebenen Abkürzungskreislaufes. Diss. med. vet. Gießen, 1961.
Ziegler, H.: Über den Ansatz des Nierenbeckens bzw. der Nierenkelche an die Niere bei Pferd und Rind, sowie die Auskleidung der Recessus renales beim Pferd. Bern, Diss. med. vet., 1921.
Zimmermann, A.: Über die Niere der Hauskatze (Felis domestica Briss.). Dtsch. Tierärztl. Wschr. **43,** 1935.
Zimmermann, G.: Neuere Angaben über die regionale Anatomie des Harnleiters. Magyar Allatorvosok Lapja (Ung. Vet. Med. Bl.) **13,** 236—237 (1958).
Zimmermann, K. W.: Über den Bau des Glomerulus der Säugerniere. Weitere Mitteilungen. Zschr. mikroskop. anat. Forsch. **32,** 1933.
Zingel, S.: Untersuchungen über die renkulare Zusammensetzung der Rinderniere. Zool. Anz. **162,** 1959.
Zingel, S.: Metrische Untersuchungen an Rindernieren. Zool. Anz. **163,** 1959.

Reproductive Organs

The reproductive or genital organs in both sexes consist of the **GONADS,** which produce the male or female germ cells (spermatozoa and ova), the **DUCTS** that transport the germ cells, the **ACCESSORY GENITAL GLANDS** found only in the male, and the **COPULATORY ORGANS.** The genital organs differ greatly in morphology and internal organization and perform different functions in the two sexes. They are therefore described separately.

Male Genital Organs

General and Comparative

General Organization (448—452).

The two **TESTES,** which are enveloped in several layers of tissue and are carried in the **scrotum,** produce the spermatozoa. Attached to the testis is the **EPIDIDYMIS** which stores and transports the spermatozoa after they leave the testis. From the epididymis arises the **DUCTUS DEFERENS,** which enters the urethra close to the neck of the bladder. The **URETHRA** passes caudally on the floor of the pelvis and at the pelvic outlet becomes incorporated in the penis. The **ACCESSORY GENITAL GLANDS,** consisting of the **vesicular, prostate,** and **bulbourethral glands,** are grouped around the pelvic part of the urethra and open into it. At the time of ejaculation, the secretions of these glands mix with the spermatozoa, and the fluid thus emitted is the **semen.** During coitus the **PENIS,** an organ of great complexity, deposits the semen in the female genital tract.

Testis

(443—447, 454, 455, 457, 464, 471, 474, 479—482, 491—494)

The **testis***, or **testicle,** is one of a pair of organs that produces the male germ cells, and is contained in the scrotum together with the epididymis. Depending on the species, the testis is oval to nearly spherical and varies considerably in size. Thus the testes of the ram, billy goat, and boar are relatively large, while those of the carnivores are relatively small. The

* Also Gr. *orchis;* hence orchitis, cryptorchid, mesorchium, etc.

seasonal change in testicular size observed in related wild species is absent in domesticated mammals. Males of wild species exhibit libido only at certain seasons, during rut; their domesticated relatives are able to copulate at any time of the year.

The end of the testis associated with the head (caput) of the epididymis is the **extremitas capitata** (443/*a*), while the other end, the one near the tail (cauda) of the epididymis, is the **extremitas caudata** (*a'*). The border of the testis along which the epididymis is attached is the **attached** or **epididymal border** (margo epididymalis, *a''*). Opposite the attached border is the **free border** (margo liber, *a'''*). There are also **medial** and **lateral surfaces.** The orientation of the testis within the scrotum varies with the species. The extremitas caudata may point ventrally (451), caudally (452) or even slightly dorsally (450).

STRUCTURE OF THE TESTIS. As an organ originating in the body cavity, the testis has a serosal covering. This is firmly attached to the tough, fibrous capsule of the testis, the **tunica albuginea*** (444/*1*). The tunica albuginea is fairly thick and contains the superficial branches of the testicular arteries and veins, which form species-specific patterns (443). Not being elastic, it exerts pressure on the yellowish-brown parenchyma, so that when the gland is cut the parenchyma is everted from the cut surface. From the deep surface of the tunica albuginea, small connective tissue **septa** (444/*2*) of varying thickness extend into the interior and unite to form the **mediastinum testis** (*3*), a mass of fibrous tissue containing numerous fine tubules (rete testis, 445/*c*) in the central part of the gland. In the stallion, such an accumulation of fibrous tissue is present only in the extremitas capitata. The connective tissue septa contain blood vessels and nerves and divide the gland into many lobules. Each **lobule** consists of a small number of **seminiferous**** **tubules** which, peripherally, are very tortuous (tubuli seminiferi contorti, *a*). As they approach the mediastinum, they unite and form larger **straight tubules** (tubuli seminiferi recti, *b*). (In man, a seminiferous tubule is about 30—50 mm. long.) The straight tubules become continuous with the tubules of the **rete testis** in the mediastinum. The rete carries the spermatozoa into the extremitas capitata and channels them through the tunica albuginea to enter the **efferent ductules** (*d, d'*). The spermatozoa are formed from cells in the walls of the seminiferous tubules. This formation of spermatozoa (spermatogenesis) is an uninterrupted process beginning at sexual maturity and ending when the animal becomes impotent.

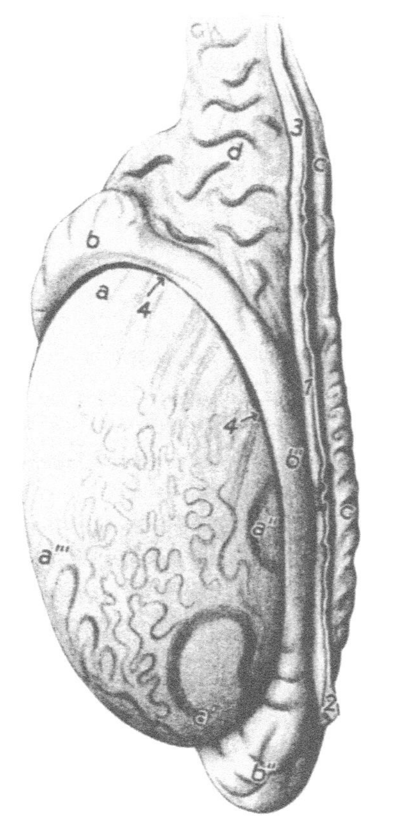

Fig. 443. Left testis and epididymis of a bull. Caudomedial aspect.
a—a''' Testis, note testicular vessels in tunica albuginea; *a* Extremitas capitata; *a'* Extremitas caudata; *a''* Epididymal border; *a'''* Free border; *b* Head of epididymis; *b'* Body of epididymis; *b''* Tail of epididymis; *c, d* (in upper part of picture) Spermatic cord; *c* Ductus deferens; *d* Testicular vessels, the veins forming the pampiniform plexus
1 Mesorchium, cut; *2* Ligament of tail of epididymis, cut; *3* Mesofuniculum, cut; *4* Testicular bursa

STRUCTURE OF THE SEMINIFEROUS TUBULES (447). Externally, the wall of the seminiferous tubule is composed of a distinct layer of connective tissue (*1*), on the inside of which is a thin basement membrane supported by reticular fibers. On the basement membrane lies a thick layer of spermatogenic and sustentacular*** cells, which surround the narrow lumen of the tubule. The spermatogenic cells are more numerous and are in various stages of division and differentiation, resulting in the production of spermatozoa. The spermatogenic cells are stratified in the wall of the tubule, so that the **spermatogonia** (*2*) are generally close to the basement membrane, while their successive descendants are progressively closer to the lumen. The spermatogonia

* L., whitish tunic. ** Literally: semen-carrying. *** Formerly Sertoli cells.

are round cells with a large nucleus. Some of them remain as stem cells (for later use), while others undergo mitosis and enter a period of growth, at the end of which they are known as **primary spermatocytes** (*3*). These cells still contain the full, diploid number of chromosomes typical of the species (dog 78, cat 38, pig 38—40, ox and goat 60, sheep 54, and horse 64). Each primary spermatocyte divides into two **secondary spermatocytes** (*4*), and each of these in turn divides into two **spermatids** (*5*). These two divisions (meiosis) follow in rapid succession, and in their course the chromosomal material of one primary spermatocyte is distributed in such a way that the four spermatids derived from it have only one-half the number of chromosomes typical of the species. The spermatids develop into **spermatozoa** (*6*) through a process of transformation, involving a repositioning and reshaping of the nucleus, cytoplasm, and the centrioles. The spermatozoa are highly differentiated, motile cells with an average length of 50—70 μ among domestic mammals. They consist of a head, neck, connecting piece, and tail, though they vary slightly in structure from species to species.

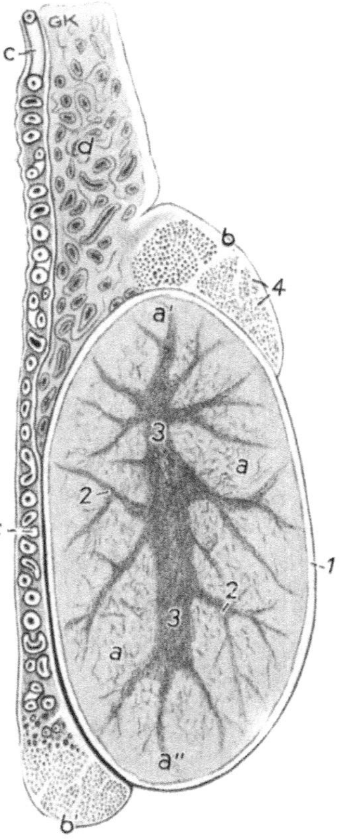

Fig. 444

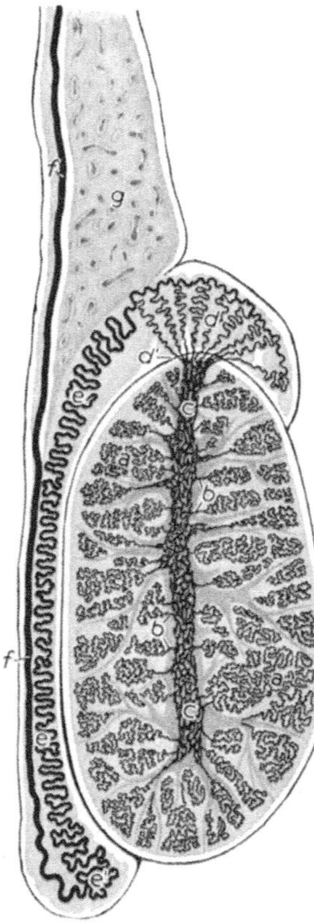

Fig. 445

Fig. 444. Longitudinal section of testis and epididymis of a bull.

a Testis; *a'* Extremitas capitata; *a''* Extremitas caudata; *b* Head of epididymis; *b'* Tail of epididymis; *c, d* (in upper part of picture) Spermatic cord; *c* Ductus deferens; *d* Testicular vessels, the veins forming the pampiniform plexus

1 Tunica albuginea covered with tunica vaginalis visceralis; *2* Small connective tissue septa dividing parenchyma into lobules; *3* Mediastinum testis, containing the rete testis; *4* Lobules of epididymis

Fig. 445. Longitudinal section of testis and epididymis of a bull. Schematic.

a Seminiferous tubules; *b* Straight tubules; *c* Rete testis; *d* Efferent ductules; *d'* Rete testis penetrating tunica albuginea and giving rise to the efferent ductules (*d*); *e, e'* Duct of epididymis at *e'* forming the tail of the epididymis; *f* Ductus deferens; *g* Testicular vessels, the veins forming the pampiniform plexus

The other type of cell present in the seminiferous epithelium is the **sustentacular cell** (*7*), which is supportive (sustentacular) in function. The sustentacular cells are attached to the basement membrane by a wide basal part, which contains the nucleus, are interconnected to form a spongelike syncytium, and extend fingerlike cytoplasmic processes toward the lumen of the tubule. The spermatogenic cells lie in the meshes of the syncytium, while the spermatids during their transformation to form spermatozoa become attached to the cytoplasmic processes and appear to receive nourishment from them.

Spermatogenesis takes place under the influence of the gonadotropic hormones, particularly those of the hypophysis, and can be disturbed, and under certain conditions halted, by lack of vitamins, starvation, fattening, psychological and climatic disturbances, and disease, to name only a few.

It has been shown that the testes, in addition to producing spermatozoa, produce male sex hormones (androgens) which with other hormones govern both the development of the typical male appearance (primary and secondary sex characteristics) and the behavior of the animal. Evidence of this is seen when the testicles are removed. This operation (castration), when performed in young animals, not only prevents the development of secondary sex characteristics, but also the onset of sexual competence, and thus influences the appearance as well as the behavior and temperament of the castrated animal. The androgens seem to be produced by the **interstitial endocrine cells*** (*8*), which are found in groups in the interlobular connective tissue of the testis. They are polygonal epithelial cells containing fat and lipoid droplets, and protein crystalloids. There is controversial evidence that the spermatogenic cells also produce hormones.

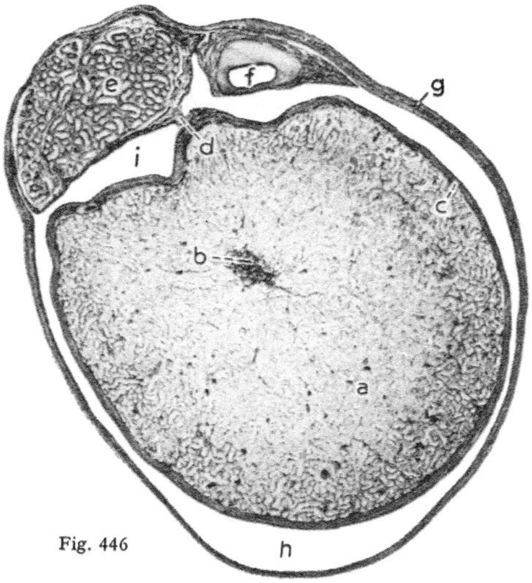

Fig. 446

Fig. 446. Transverse section of left testis, epididymis, and coverings in the dog. Caudal aspect. Microphotograph.

a Parenchyma of testis; *b* Mediastinum testis, containing the rete testis; *c* Tunica albuginea covered with tunica vaginalis visceralis; *d* Mesorchium; *e* Body of epididymis; *f* Ductus deferens; *g* Tunica vaginalis parietalis, int. spermatic and cremasteric fasciae; *h* Vaginal cavity, distended; *i* Testicular bursa

Fig. 447. Seminiferous tubule of stallion in cross section.

1 Connective tissue layer and basement membrane; *2* Spermatogonia; *3* Primary spermatocytes; *4* Secondary spermatocytes; *5* Spermatids; *6* Spermatozoa; *7* Sustentacular cells; *8* Glandular cells in intralobular connective tissue

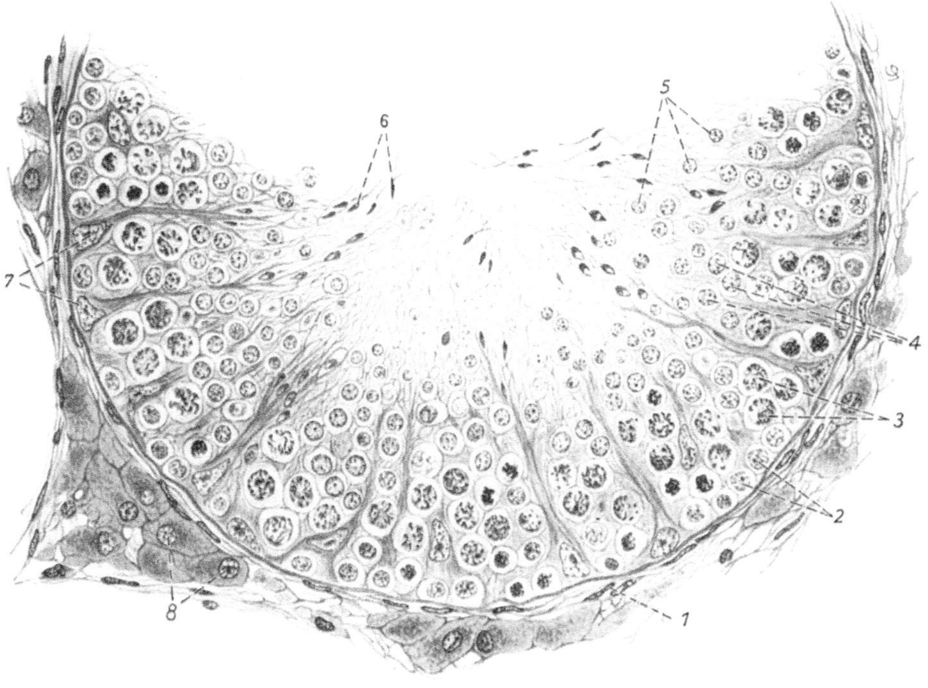

Fig. 447

* Formerly interstitial cells of Leydig.

As mentioned, the seminiferous tubules are connected by straight tubules to the centrally located network of tubules, the **rete testis.** Both the straight tubules and the tubules of the rete are lined with a single layer of cuboidal cells. The numerous tubules give the rete, when cut, a spongelike appearance. In these tubules (445/*c*) spermatozoa pass into the extremitas capitata and leave the testis through a small cribriform area in the tunica albuginea. The rete usually continues for a few mm. after having penetrated the tunic (extratesticular rete, *d'*) and gives rise to a varying number of **efferent ductules** (*d*) which occupy the head of the epididymis (Amann et al. 1977; Hemeida et al. 1978).

Epididymis

(443—446, 448—454, 474, 491)

The epididymis consists of a **head,** a **body,** and a **tail** (caput, corpus, and cauda epididymidis, 443/*b*, *b'*, *b''*). It is an elongated structure and is attached along the epididymal border of the testis. The tail of the epididymis is attached by the tough **proper ligament of the testis** (454/*11*) to the extremitas caudata of the testis. Between the body of the epididymis and the testis is the **testicular bursa** (443/*4*), an elongated recess lined with serosa and open laterally or caudally depending on the species. The spatial relationship between epididymis and testis is not same in all domestic mammals and is described for the different species in the special chapters.

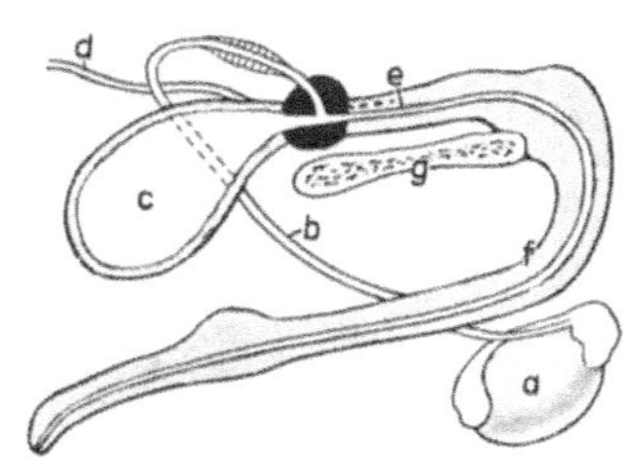

Fig. 448 (Dog)

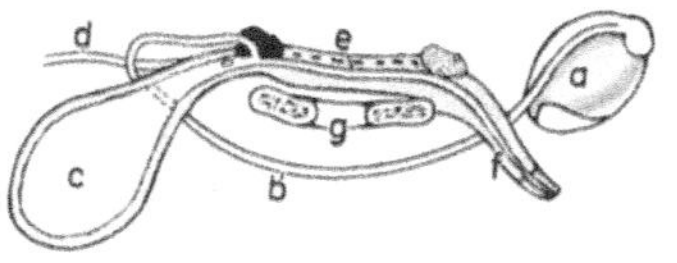

Fig. 449 (Cat)

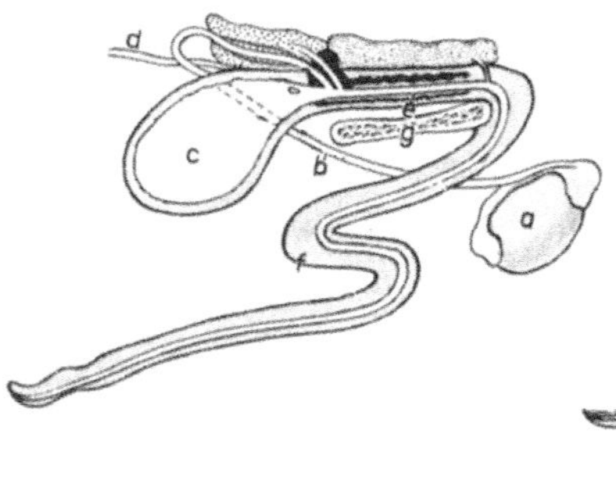

Fig. 450 (Pig)

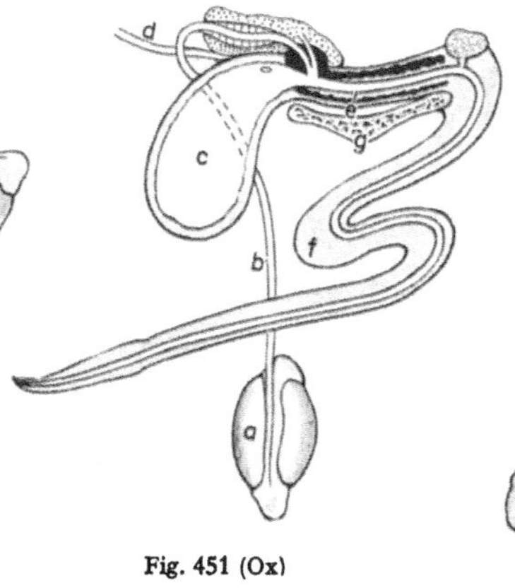

Fig. 451 (Ox)

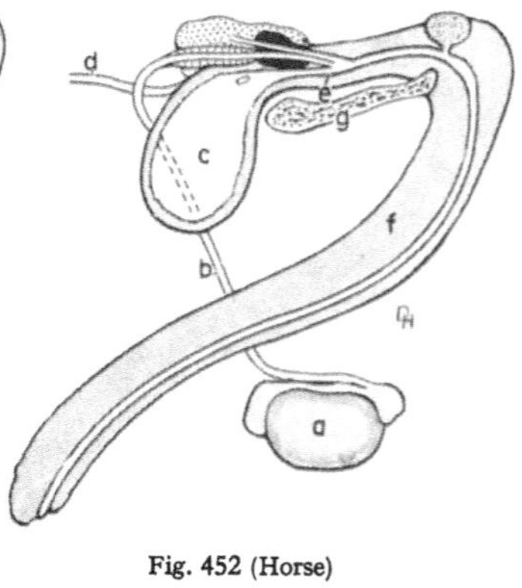

Fig. 452 (Horse)

Figs. 448—452. Male genital organs of the domestic mammals showing accessory genital glands and other structural differences characteristic for a species. Schematic.

a Right testis and epididymis; *b* Ductus deferens; *c* Urinary bladder; *d* Ureter; *e* Pelvic part of urethra surrounded by accessory genital glands (note the different ways in which the ductus deferens and the excretory duct of the vesicular gland open into the pelvic urethra); *f* Penis and extrapelvic urethra; *g* Pelvic symphysis. Crosshatched: Ampulla of ductus deferens (in the pig: glandular part of ductus deferens); Black: Prostate gland; Coarse stippling: Vesicular gland; Fine stippling: Bulbourethral gland

The **head of the epididymis** consists of a varying number of **efferent ductules** (445/*d*) and the beginning of the duct of the epididymis (*e*). Based on a large number of animals, the most often encountered ranges for the numbers of efferent ductules present in the domestic species were given by Hemeida et al. (1978): dog 13—15, cat 14—17, pig 14—16, ox 13—16, sheep 17—20, goat 18—19, and horse 14—17. The efferent ductules develop from the mesonephric tubules and in the adult consist of three segments of which the middle ones are very compact and tortuous and give the tubular contents of the head of the epididymis its lobulated appearance (444/*4*). The less tortuous distal segments of the efferent ductules join in various manners to form the **duct of the epididymis** (445/*e*). The latter passes to the tail of the epididymis where it is continuous with the **ductus deferens** *f*). The duct of the epididymis is also very tortuous and increases in diameter toward the ductus deferens. It is arranged in tightly packed coils and thus makes up the bulk of the body and tail of the epididymis. The duct of the epididymis is very long and varies widely with the species (dog 5—8 m., cat 1.5—3 m., pig 17—18 m., ox 40—50 m., small ruminants 47—52 m. and horse 72—81 m.).

Both the duct of the epididymis and the ductus deferens develop from the caudal part of the mesonephric duct. A small pear-shaped remnant (appendix epididymidis) of the cranial end of the mesonephric duct is occasionally seen on the head of the epididymis. A similar remnant (appendix testis) seen occasionally on the extremitas capitata is from the cranial end of the paramesonephric duct. Blind-ending ductules (aberrant ductules and blind efferent ductules) are frequently found between the proximal and distal segments of normal efferent ductules. They are thought to play a role in spermiostasis and granuloma formation, one of the causes of male infertility in domestic mammals (Hemeida et al. 1978).

The **efferent ductules** in the lobules of the epididymis are embedded in highly vascularized connective tissue. The ductules are thin epithelial tubes surrounded by a basement membrane and smooth muscle cells. The epithelium consists of tall, ciliated, columnar cells, which are stratified in places. Nonciliated cells that appear to be secretory are also present.

The tightly packed coils of the **duct of the epididymis** are held together by connective tissue. The wall of the duct consists of a layer of smooth muscle cells, and is lined by a double layer of very tall prismatic cells with tufts of secretory villi on the free surface. The spermatozoa, which are continuously entering the duct, are not yet motile. While slowly traversing the enormous length of the duct, they complete their maturation, using the epididymal secretion. Billions of spermatozoa are thus stored in the epididymis until ejaculation, at which time they are forced into the ductus deferens by peristaltic contractions of the duct of the epididymis.

Ductus Deferens

(416, 457, 464, 465, 474—476, 483—485)

The ductus deferens (448—452/*b*) is the continuation of the duct of the epididymis and connects the epididymis with the pelvic part of the urethra. It begins at the tail of the epididymis and runs, rather flexuous at first, along the testis medial to the epididymis. After passing the head of the epididymis it continues with the testicular vessels and nerves, and with them forms the **spermatic cord.** The ductus deferens is enclosed in a narrow serosal fold, the **mesoductus deferens,** and lies for the most part on the medial surface of the spermatic cord. The duct passes through the inguinal canal and upon entering the abdominal cavity at the vaginal ring (485/*c*) turns toward the pelvic inlet. In the cranial part of the pelvis, right and left deferent ducts enter the **genital fold** (15/*5*; 495/*3*), a horizontal sheet of peritoneum inserted between the rectum and bladder. Inside this fold, the two ducts along with the ureters (*2*) converge toward the neck of the bladder. A remnant of the paramesonephric ducts, known as **uterus masculinus** (*12*), is found occasionally in the genital fold between the deferent ducts. The wall of the terminal part of the ductus deferens thickens and forms the **ampulla** (*4*), which in the stallion may reach a diameter of 2 cm. (15/*c*). In the ruminants and dog, the ampulla is relatively thinner, while in the cat and pig it is absent. The terminal part of the ductus deferens always contains glands. Beyond the ampulla, the duct decreases in size, and in horse and ruminants unites with the excretory duct of the vesicular gland, forming the short **ejaculatory duct** (416/*m*), which ends in the dorsal wall of the urethra at the **ejaculatory orifice.** This orifice is on a small mucosal elevation, the **colliculus seminalis** (*n*), which is continuous caudally with the urethral crest (415/*5*). In the pig (450), the ductus deferens and the excretory duct of the vesicular gland usually open separately, but may end together in a small mucosal recess of the urethra. There is no vesicular gland in the carnivores, so that the ductus deferens enters the pelvic urethra alone (448).

The ductus deferens has a tunica muscularis, which consists of three layers of smooth muscle, and is lined with simple or pseudostratified columnar cells. Branched tubular **glands** with saclike dilatations are present in the thick mucous membrane of the ampulla. Their viscous mucoid secretion is released into the lumen of the duct and becomes part of the seminal fluid.

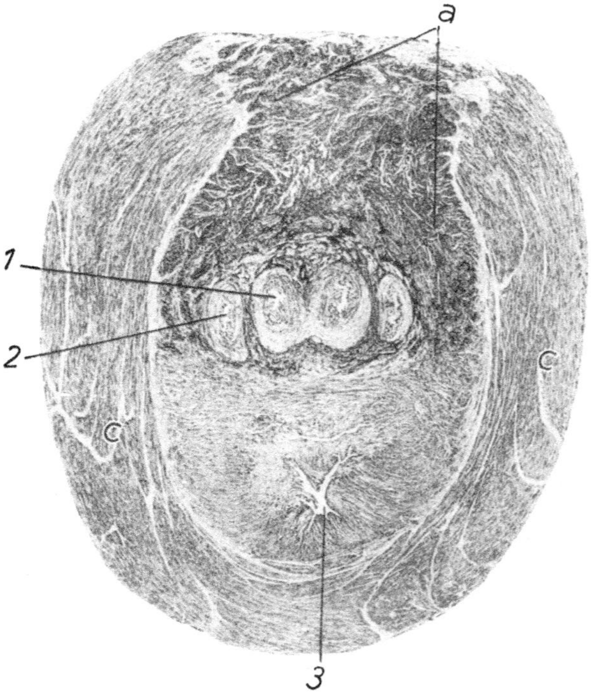

Fig. 453 A

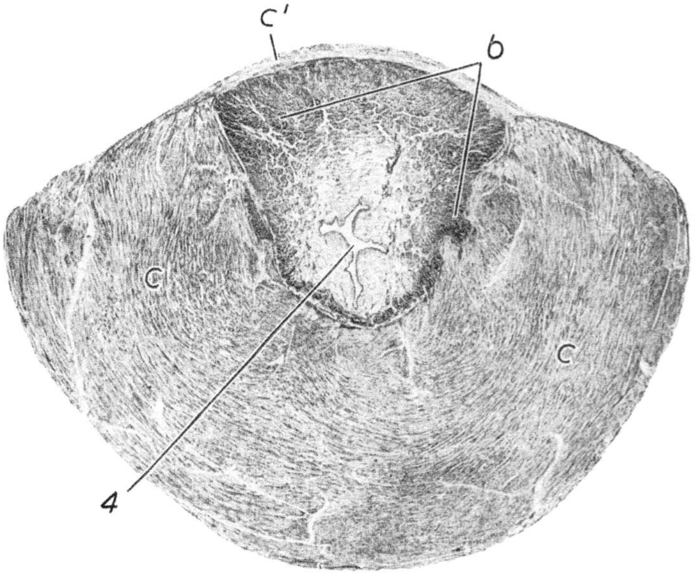

Fig. 453 B

Fig. 453. Cross sections of the pelvic part of the urethra of a bull. Microphotograph. A: at the level of numeral 6 in Fig. 483; B: at the level of numeral 7 in Fig. 483.

a Body of prostate gland; *b* Disseminate part of prostate gland; *c* Urethralis, *c'* its wide dorsal raphe

1 Ductus deferens; *2* Excretory duct of vesicular gland; *3* Preprostatic part of urethra; *4* Urethra surrounded by stratum cavernosum

Coverings of the Testis and of the Spermatic Cord

(454, 457, 464, 471, 474, 479—481, 492—494)

These coverings are derived from the various layers of the abdominal wall and more or less completely surround the testis, epididymis, and the spermatic cord (454). The layers of the abdominal wall, beginning on the outside, are: (1) Skin; (2) Superficial and deep fasciae;

(3) Muscle layer; (4) Transverse fascia; and (5) Parietal peritoneum. The corresponding coverings of the testis consist of the following layers: (1) Skin and tunica dartos, which is derived from the tela subcutanea; (2) External spermatic fascia, which may, like the fascia of the abdominal wall, consist of two layers; (3) Cremaster muscle and the cremasteric fascia that covers it, both split off from the internal abdominal oblique; (4) Internal spermatic fascia; and (5) Tunica vaginalis parietalis, a fingerlike evagination of the parietal peritoneum.

SCROTUM. The testicular coverings that have been enumerated above can be separated relatively easily between layers (2) and (3), or within layer (2), into the wall of the scrotum externally and those layers that stay with the testis. (This separation is regularly made in the "closed" method of castration and is shown in a dissection on the right side of figures 479—481.) The wall of the scrotum, thus separated from the scrotal contents, consists of skin, tunica dartos, and parts, if not all, of the external spermatic fascia (Hartig 1955).

The **skin of the scrotum** (454/*1*) is realtively thin and contains many sweat and sebaceous glands. Depending on the species and breed, it is pigmented (black) and more or less covered with hair, although the pigmentation may vary from animal to animal.

The **tunica dartos** is closely adherent to the deep surface of the skin and cannot be separated from it. It is a modified tela subcutanea and consists of smooth muscle bundles connected by

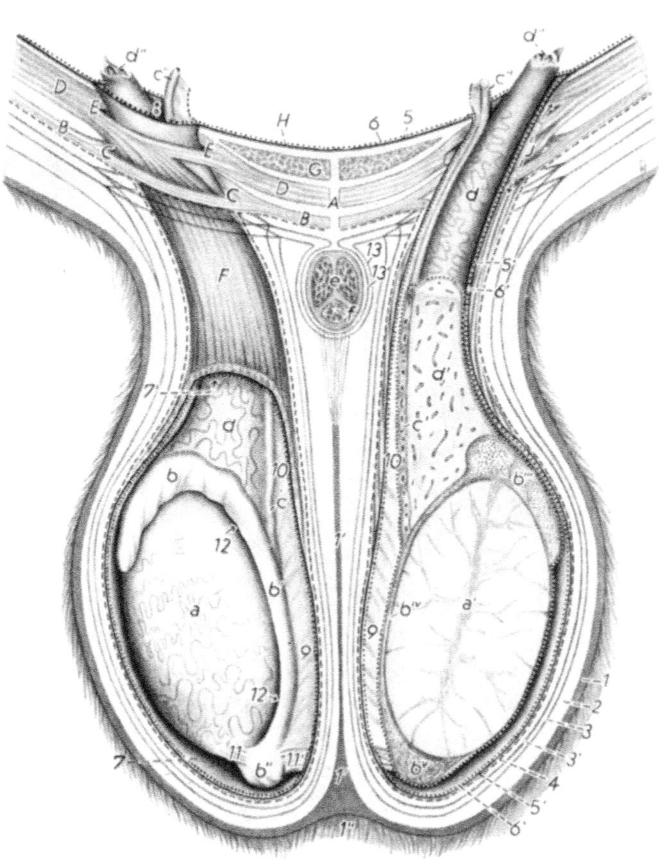

Fig. 454. Testes of a bull in situ. Schematic. Caudal aspect. The scrotum, right testis, and penis have been sectioned transversely; and the left cremaster muscle, internal spermatic fascia, and tunica vaginalis parietalis have been fenestrated to expose the left testis.

A Linea alba; *B* Ext. abdominal oblique; *C, C* Supf. inguinal ring; *D* Int. abdominal oblique; *E, E* Deep inguinal ring; *F* Cremaster muscle, fenestrated at insertion; *G* Rectus abdominis; *H* Combined transverse fascia and aponeurosis of transversus abdominis; *J* Inguinal canal

a Left testis with testicular vessels in the tunica albuginea; *a'* Right testis sectioned; b—b^{V} Epididymis; *b, b'''* head, *b'*, b^{iv} body, *b''*, b^{V} tail; *c, d, c', d'* Spermatic cord; *c, c', c''* Ductus deferens; *d, d', d''* Testicular artery and vein in spermatic cord; *d, d''* and *9* Mesorchium; *e* Penis; *f* Urethra and corpus spongiosum penis

1 Skin and tunica dartos; *1'* Scrotal septum; *1''* Scrotal raphe; *2* Loose subdartoic tissue; *3, 3'* Ext. spermatic fascia; *4* Cremasteric fascia; *5'* Int. spermatic fascia; *6'* Tunica vaginalis parietalis; *5* Transverse fascia; *6* Parietal peritoneum; *7, 7'* Vag. cavity; *8* Vaginal ring; *9* Mesorchium (also *d* and *d''* belong to the mesorchium); *10* Mesofuniculum; *11* Proper lig. of testis; *11'* Lig. of tail of epididymis; *12* Testicular bursa; *13, 13'* Supf. and deep fasciae of penis

collagenous and elastic fibers. Some of the muscle bundles enter the corium of the overlying skin. The tunica dartos forms the **scrotal septum** (*1'*), a median partition dividing the scrotum into two compartments, one for each testis. The position of the septum is marked externally by the **scrotal raphe** (492/*n*), an epithelial "seam" in the midline and continuous cranially with the **preputial raphe** (*n'*). The tunica dartos contracts under the influence of cold or as a result of mechanical stimulation, wrinkling the scrotal skin and thus accommodating the scrotum to the position of the testes. Its contraction also helps to elevate the testes, although the cremaster muscle (*e*) is principally responsible for that.

The **external spermatic fascia** (454/*3*, *3'*) is deep to the tunica dartos and is connected to it by a loose layer of connective tissue (*2*). Its two layers can be demonstrated by careful dissection. The external spermatic fascia is equally loosely attached to the underlying cremaster muscle and cremasteric fascia that covers it, and seems to be responsible for the ease with which the testes slide along inside the scrotal wall. Adhesions between the external spermatic fascia and the tunica dartos in the vicinity of the tail of the epididymis are referred to as the scrotal ligament.

As mentioned, the remaining testicular coverings are those that are more closely associated with the testis.

The **cremaster muscle** (*F*) is derived from the internal oblique, and is thus a striated muscle. It is covered externally by the thin **cremasteric fascia** (*4*) and is attached to the lateral or dorsal surface of the fingerlike process formed by the tunica vaginalis. Under certain stimuli, it raises the tunica vaginalis and thus brings testis and epididymis closer to the external inguinal ring.

The **internal spermatic fascia** (*5'*) is deep to the cremaster and is fused with the underlying tunica vaginalis. It is difficult to demonstrate.

The arrangement of the **tunica vaginalis** is best understood by considering its development. Therefore, an account is given of its origin and of the closely associated descent of the testis into the scrotum.

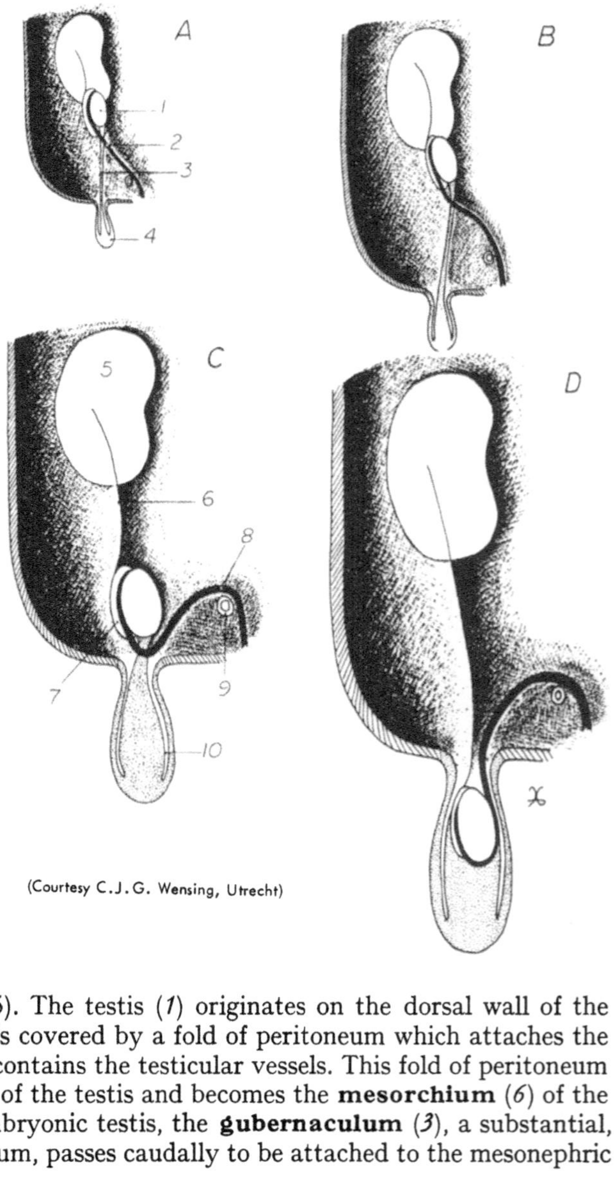

(Courtesy C.J.G. Wensing, Utrecht)

Fig. 455. Descent of the testis. Schematic. Ventral views. (After Wensing 1968.)

1 Testis; *2* Mesonephric duct; *3*, *4* Gubernaculum; *3* Intraabdominal part of gubernaculum; *4* Extra-abdominal part of gubernaculum; *5* Kidney; *6* Mesorchium; *7* Epididymis; *8* Ductus deferens; *9* Right umbilical artery; *10* Processus vaginalis

DESCENT OF THE TESTIS (455). The testis (*1*) originates on the dorsal wall of the embryo medial to the mesonephros. It is covered by a fold of peritoneum which attaches the testis to the abdominal wall and which contains the testicular vessels. This fold of peritoneum is drawn out by the subsequent descent of the testis and becomes the **mesorchium** (*6*) of the adult. From the caudal pole of the embryonic testis, the **gubernaculum** (*3*), a substantial, jellylike cord also covered with peritoneum, passes caudally to be attached to the mesonephric duct (*2*) of the same side, at a point where the duct turns medially at the pelvic inlet. This point on the duct in later development becomes the tail of the epididymis, so that this short proximal part of the gubernaculum is the forerunner of the **proper ligament of the testis** which, in the adult, connects the extremitas caudata of the testis with the tail of the epididymis. The gubernaculum, however, is continued, beyond its attachment on the mesonephric duct, to the site of the future inguinal canal. This longer distal part of the gubernaculum (*3*) becomes the **ligament of the tail of the epididymis** in the adult.

According to Wensing (1968), who studied the descent of the testis in 150 pig, 50 ox, and 50 dog fetuses, the distal end of the gubernaculum forms a free, knoblike expansion which comes to lie between the differentiating internal and external oblique muscles, i.e., in the primitive inguinal canal. This expansion is the extra-abdominal part of the gubernaculum (*4*), and at no time did he find a connection between it and the overlying scrotal wall. Parietal peritoneum evaginates in a sleevelike manner into the substance of the extra-abdominal part of the gubernaculum and keeps pace with its gradual enlargement and elongation. This peritoneal evagination (processus vaginalis, *10*) also lies in the inguinal canal and grows distally with the gubernaculum into the mesenchyme of the overlying scrotal ridge. (The scrotal ridges are the paired elevations which later fuse to form the scrotum.)

The testis, epididymis, and ductus deferens gradually move caudally, and like the gubernaculum and vaginal process before them, pass through the inguinal canal into the scrotum. Just before this occurs, the proximal part of the gubernaculum, which is enclosed by the vaginal process, enlarges (Backhouse and Butler 1960), and in so doing dilates the canal sufficiently to allow the testis to pass through. The testis passes through the inguinal canal rapidly, sliding with its own peritoneal covering along the nearly cylindrical lining (tunica vaginalis parietalis) of the vaginal process.

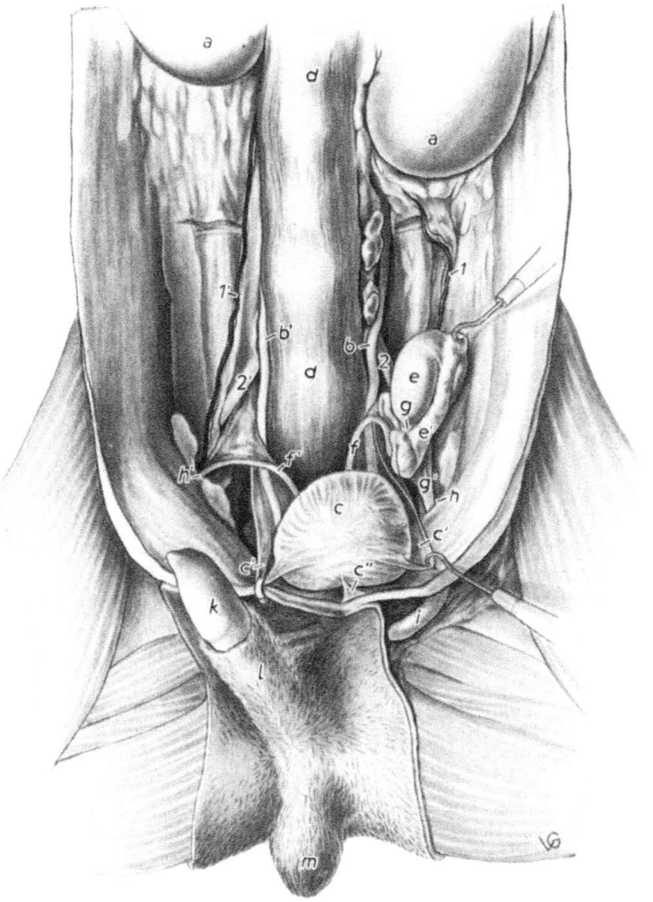

Fig. 456. Undescended left testis in a dog (abdominal cryptorchid). Ventral aspect. (Drawn from a dissection of the Dept. of Veterinary Surgery, Giessen.)

a Kidneys; *b*, *b'* Left and right ureters; *c* Urinary bladder; *c'* Lateral lig. of bladder; *c''* Median lig. of bladder; *d* Descending colon; *e*, *e'* Undescended left testis and epididymis; *f*, *f'* Left and right deferent ducts; *g* Proper ligament of testis; *g'* Ligament of tail of epididymis; *h*, *h'* Left and right vaginal rings; *i* Rudimentary left tunica vaginalis; *k* Glans penis; *l* Prepuce; *m* Scrotum containing right testis

1, *1'* Left and right testicular arteries; *2*, *2'* Left and right ext. iliac arteries

What forces are involved in the descent of the testis, and why it takes place at all, is not sufficiently understood. The following factors have been mentioned as playing a role: elongation of the body; degeneration of the mesonephros caudal to the testis, allowing space for the caudal movement of the testis; growth of the metanephros (*5*) and adrenal gland cranial to the testis, forcing the testis caudally; respiratory attemps on the part of the fetus with attendant increase of abdominal pressure; and failure of the gubernaculum to elongate with the general body growth. Wensing (1968) states that the expansion of the gubernaculum distal to the apparently unyielding inguinal canal draws the proximal end of the gubernaculum and the testis out of the abdominal cavity.

Once the testis is outside, the gubernaculum degenerates and makes room for the testis to sink into the scrotum and to acquire its full serosal investment (tunica vaginalis visceralis) particularly at its extremitas caudata and the tail of the epididymis.

In the carnivores, the testes are still in the lumbar area at the time of birth; they descend into the scrotum shortly after birth. In the pig, descent occurs in utero, and is completed shortly before birth. In ruminants, descent takes place early and is completed at three months fetal age. In the horse, the testes leave the abdominal cavity close to birth; Bergin et al. (1970), who have studied testicular descent and cryptorchidism in this species, give a range from about 30 days before birth to 10 days after birth.

*Cryptorchidism**, particularly common in pigs and horses, is the failure of one or both testes to descend into the scrotum. The affected testis (and epididymis) may remain in the abdominal cavity (456/*e*, *e'*), may partly enter the deep inguinal ring, or may be lodged in the inguinal canal. Such a testis does not complete its development and therefore produces no spermatozoa. Since this makes the animal infertile, cryptorchidism is of considerable economic importance.

The **tunica vaginalis,** as has been shown, is an evagination of the parietal peritoneum, passing through the inguinal canal into the scrotum, and when fully developed has the shape of a long flask with a narrow proximal part (471/*7'*) and a distended distal part (*7*). The distal part contains the testis and epididymis (*5, 6*) while the proximal part contains the spermatic cord (*9*). These organs, covered with tunica vaginalis visceralis, completely fill the tunica vaginalis parietalis so that only a narrow capillary space remains as in the other serosal cavities. This capillary space (cavum vaginale, 454/*7, 7'*) is continuous with the peritoneal cavity** at the **vaginal ring** (anulus vaginalis, *8*) which overlies the deep inguinal ring. At the vaginal ring, the tunica vaginalis parietalis (*6'*) is continuous with the parietal peritoneum of the abdominal cavity (*6*); and the testicular vessels, lymphatics, and nerves, and the ductus deferens pass through the ring into the abdominal cavity. In older stallions, the vaginal ring (493/*d*) may be a slit several centimeters long. Through it a loop of small intestine or parts of the greater omentum may enter the tunica vaginalis, and if the tunic has not been tied off, these structures may appear at the operative site after castration.

The **SPERMATIC CORD** (funiculus spermaticus, 454/*c, d*) fills the narrow proximal part of the tunica vaginalis parietalis. It resembles a slender cone and is attached with its wider distal end to the extremitas capitata of the testis, while its narrow proximal end is at the vaginal ring. The spermatic cord, like the testis and epididymis, is invested by serosa (tunica vaginalis visceralis). This tunic is continuous with the tunica vaginalis parietalis along a narrow band of serosa (mesofuniculum, 443/*3*, 454/*10*) which runs along the caudal aspect of the cord***.

The spermatic cord consists of the testicular artery and vein, lymphatics, testicular plexus of nerves accompanying the vessels, bundles of smooth muscle, and the ductus deferens, which is enclosed in a serosal fold of its own, the **mesoductus.**

The **testicular artery** and **vein** (494/*h*) originate from the abdominal aorta and caudal vena cava, respectively, and pass, enclosed in a serosal fold (mesorchium), to the vaginal ring (*g*), through which they enter the spermatic cord. The testicular artery, in its extra-abdominal course, is very tortuous and forms numerous coils. The testicular vein breaks up and forms a rich network, the **pampiniform plexus,** which surrounds the coils of the artery (444/*d*). This convolute of vessels constitutes the bulk of the spermatic cord (457/*c*). At the testis the many vessels of the pampiniform plexus unite again and form several larger veins which

---

* From Gr., literally: hidden testis.

** This communication is lost in man, as the narrow proximal part of the tunica vaginalis degenerates and only the part surrounding the testis remains.

*** In species with a craniocaudally directed cord, the mesofuniculum runs along the dorsal aspect.

enter the extremitas capitata. The lymphatics found in the cord drain the testis, epididymis, and tunica vaginalis, and pass to the medial iliac and lumbar aortic nodes. The **ductus deferens** (*b*; 494/*i*) lies on the caudomedial (dorsomedial) surface of the spermatic cord, and after entering the abdominal cavity leaves the vessels and passes dorsocaudally to the pelvic urethra (*q*). The mesoductus also passes through the vaginal ring, attaches the ductus to the abdominal wall, and at the pelvic inlet takes part in the formation of the genital fold.

The length of the spermatic cord depends on the position of the testis, or, more presicely, on the distance between the extremitas capitata and the vaginal ring. This distance differs individually and with the species. The horse (494) has a relatively short spermatic cord, while in the bull (485), because of the pendulous scrotum, the cord is longer. It is relatively the longest in the carnivores and pig, and because of the caudal position of the scrotum it must pass forward between the thighs to reach the external inguinal ring.

The **MESORCHIUM** is a long serosal fold which, like the mesentery suspending the gut, suspends the testis. Its organization is best understood with reference to its development. The embryonic mesorchium is short and attaches the undescended testis to the dorsal abdominal wall. It contains the testicular artery and vein and the nerve plexus that accompanies these vessels. During the descent of the testis, the mesorchium is drawn out and becomes elongated, so that in the adult the mesorchium extends from the origin of the testicular vessels in the lumbar area to the testicle in the scrotum, still enclosing the testicular vessels (494/*h*, *c*), lymphatics, and nerves with its two layers. Also belonging to the adult mesorchium is the narrow fold (454/*9*) connecting the epididymal border of the testis to the tunica vaginalis parietalis. From this narrow fold of mesorchium another narrow fold, known as the **mesepididymis,** separates and passes laterally to the body of the epididymis. Between the testis and the body of the epididymis is a recess, the **testicular bursa** (*12*) which is open laterally in most species and which in its depth is bounded by the narrow fold of mesorchium just mentioned and by the mesepididymis. The part of the mesorchium between the vaginal ring and the testis is attached to the tunica vaginalis parietalis by the **mesofuniculum** (*10*).

The **LIGAMENTS AND SEROSAL FOLDS ASSOCIATED WITH THE TESTIS, EPIDIDYMIS, AND SPERMATIC CORD** are of considerable clinical importance in castration, and are therefore summarized at this point. In the embryo, the gubernaculum extends from the caudal pole of the testis to the inguinal canal, and is divided by attachment on the mesonephric duct into proximal and distal parts. After the descent of the testis, the proximal part gives rise to the **proper ligament of the testis** (454/*11*), while the distal part becomes the **ligament of the tail of the epididymis** (*11'*), which at the same time is the distal free edge of the **mesorchium.** This part of the mesorchium (*9*) attaches the testis to the tunica vaginalis parietalis. Its proximal continuation, the **mesofuniculum** (*10*), is an equally narrow fold attaching the spermatic cord to the tunica vaginalis parietalis. Transecting the ligament of the tail of the epididymis (*11'*), and continuing the incision proximally along the narrow fold of mesorchium (*9*) and mesofuniculum (*10*), detaches the testis, epididymis, and distal part of the spermatic cord from the tunica vaginalis parietalis. This cut has been made in Figure 443, starting at (*2*) and passing (*1*) and (*3*). Ligation and transection of the spermatic cord (*c*, *d*) permits these organs to be removed from the body. The tunica vaginalis visceralis, closely investing these organs, remains undisturbed except proximally at the transection of the spermatic cord.

The **SHAPE AND POSITION OF THE TESTES AND SCROTUM** differ among the domestic mammals. In the embryo, the scrotum arises from bilateral scrotal ridges just ventral to the anus and, depending on the species, migrates cranially to a greater or lesser extent.

In the **CAT** (471), the scrotum does not migrate far from its site of origin. It is immediately ventral to the anus, and between it and the prepuce (*18*), which is directed caudally. It is relatively small and is largely hidden by dense hair. The testes are small and almost spherical, and the extremitas caudata and the tail of the epididymis (*6*) point caudally. The epididymis is dorsolateral to the testis. The spermatic cord (*9*) is very long, crosses the penis (*16*), and passes forward between the thighs to enter the superficial inguinal ring.

In the **DOG** (473), the scrotum is not as close to the anus as in the cat, and protrudes farther from the body. It has a distinct median groove and raphe. The testes are relatively small and oriented so that the extremitas caudata points caudally (464). The spermatic cord is similar to that of the cat.

The scrotum of the **BOAR** (474) is situated a short distance from the anus and does not protrude very much from the surface of the body. The testes are relatively large, with the extremitas caudata and tail of the epididymis directed dorsocaudally toward the anus (450). The free border of the testes is directed caudally, while the epididymal border and the epididymis lie against the muscles of the thighs. The long spermatic cords pass forward between the thighs to the external inguinal rings.

The **BULL** (485), **RAM** (480), and **BILLY GOAT** (481) have a pendulous scrotum, located more cranial than in the other species. When not raised, there is a well-marked neck proximal to the testes. The scrotum is relatively large and compressed so that it has cranial and caudal surfaces. The longitudinal axis of the testes is dorsoventral, with the extremitas caudata pointing ventrally. The head of the epididymis lies against the caudolateral surface of the extremitas capitata; the large tail protrudes considerably from the extremitas caudata and is visible and palpable on the scrotum; and the body of the epididymis descends along the medial surface

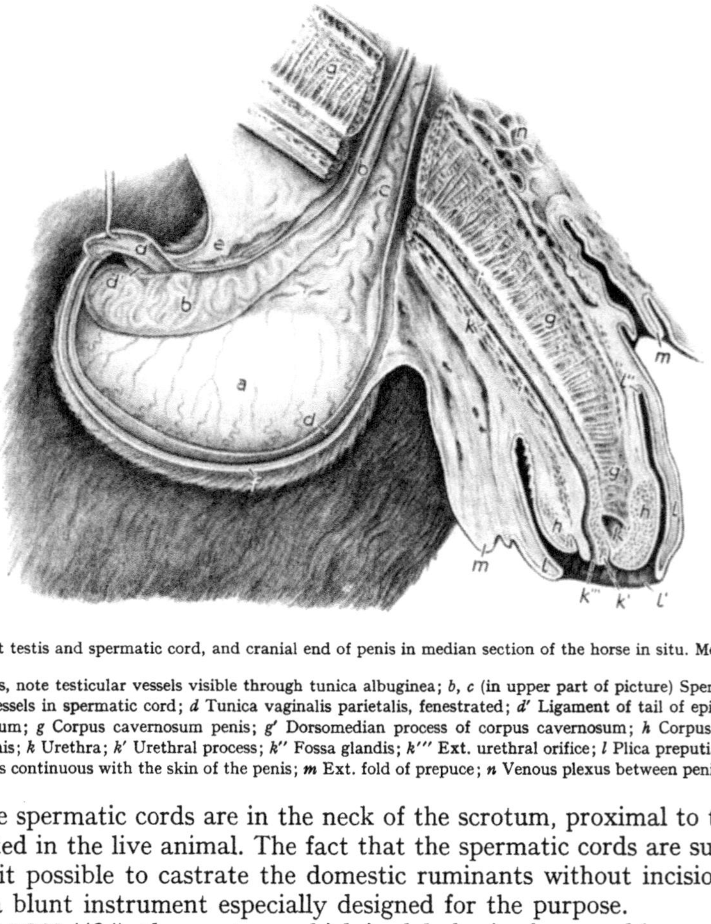

Fig. 457. Left testis and spermatic cord, and cranial end of penis in median section of the horse in situ. Medial aspect.

a Medial surface of testis, note testicular vessels visible through tunica albuginea; *b*, *c* (in upper part of picture) Spermatic cord; *b* Ductus deferens; *c* Testicular vessels in spermatic cord; *d* Tunica vaginalis parietalis, fenestrated; *d'* Ligament of tail of epididymis; *e* Cremaster muscle; *f* Wall of scrotum; *g* Corpus cavernosum penis; *g'* Dorsomedian process of corpus cavernosum; *h* Corpus spongiosum glandis; *i* Corpus spongiosum penis; *k* Urethra; *k'* Urethral process; *k''* Fossa glandis; *k'''* Ext. urethral orifice; *l* Plica preputialis; *l'* Preputial ring; *l''* Point where prepuce is continuous with the skin of the penis; *m* Ext. fold of prepuce; *n* Venous plexus between penis and abdominal wall

of the testis. The spermatic cords are in the neck of the scrotum, proximal to the testes, and are easily palpated in the live animal. The fact that the spermatic cords are surrounded here by skin, makes it possible to castrate the domestic ruminants without incision by crushing the cords with a blunt instrument especially designed for the purpose.

In the **STALLION** (494), the scrotum, which is globular in shape and has a median groove and raphe, is located between the thighs. The long axis of the testes is oblique, with the free border directed slightly cranioventrally. The epididymis follows the opposite border of the testis and slightly overlaps the lateral surface, with the tail next to the extremitas caudata and directed caudally. The spermatic cords ascend steeply to the superficial inguinal rings, cross the lateral surface of the penis, and are easily palpated.

Accessory Genital Glands

(416, 448—452, 453, 465, 471, 472, 475, 476, 483, 484, 490, 494—496)

The accessory genital glands are grouped around the pelvic urethra and differ greatly from species to species. Their growth and function are influenced by the sex hormones. If animals are castrated early in life, the accessory genital glands do not develop fully; if castration is delayed until after sexual maturity, they atrophy and cease to secrete. The accessory genital glands can be palpated rectally, more readily in the horse and ox, because the entire hand can be introduced into the rectum, than in the smaller species where only digital palpation is possible.

The **VESICULAR GLAND** (450—452) is paired and lies dorsolateral to the neck of the bladder. Its cranial end lies in the genital fold lateral to the ductus deferens. In the horse (495/*5*), it takes the form of a thick-walled pouch 10—15 cm. long and 3—6 cm. wide and is known as the **seminal vesicle.** In the ruminants (483/*5*), the vesicular gland is a compact gland of moderate size with a lobulated surface, and is 7—12 cm. long in the ox. In the pig (475/*5*), it is also compact and with an uneven surface, but it is very large, 7—12 cm. long, shaped like a three-sided pyramid with the apex directed caudally. The carnivores have no vesicular gland.

STRUCTURE. The vesicular gland of the ruminants, but more so that of the pig, is distinctly lobulated. It is a branched tubulo-alveolar gland, the secretion of which is stored in large intralobular collecting spaces. At the time of ejaculation, the smooth muscle present in the interstitial connective tissue and capsule of the gland contracts and rapidly delivers the abundant secretion into the pelvic urethra. The seminal vesicle of the horse has a tough muscular wall lined with a mucous membrane that forms many folds and crypts. Also present in the wall are numerous branched secretory tubules with alveolar pockets. In the horse and ruminants, the excretory duct of the gland joins the terminal part of the ductus deferens to form the short **ejaculatory duct** (416/*m*), which opens on the colliculus seminalis in the dorsal wall of the pelvic urethra. In the pig (450), the excretory duct opens into the urethra usually by itself, or with the ductus deferens into a small mucosal recess.

The **PROSTATE GLAND** (448—452) is present in all domestic mammals and is intimately fused with the pelvic urethra. It is largest in the carnivores, then smaller in decreasing relative size in the horse, ox, pig, and small ruminants. The **body** of the prostate (465/*5*) is a compact gland visible on the outside of the pelvic urethra. The **disseminate part** forms a glandular layer in the wall of the urethra and is visible only when the urethra is sectioned (453/*b*). In some species, the body of the prostate is divided into right and left **lobes** (495/*6'*).

In the dog (465/*5*), the prostate is globular and divided by a median groove into right and left lobes which completely surround the urethra. The prostate of the cat (471/*14*) is similar, but does not cover the ventral surface of the urethra. In both species, the disseminate part consists of a few scattered lobules in the wall of the urethra. In the pig and ox (475, 483/*6*), the body of the prostate is a small, flat mass on the dorsal surface of the urethra, and the disseminate part is covered by the urethralis muscle. The small ruminants have only the disseminate part of the gland, which in sheep is present only in the dorsal and lateral walls of the urethra. In the horse, the prostate gland consists of right and left lobes, 5—9 cm. long and 3—6 cm. wide, which lie on the dorsolateral surface of the urethra and are connected by an **isthmus** (495/*6*). The gland has numerous small excretory ducts, opening in groups on either side of the colliculus seminalis (496/*7*). There is no disseminate part.

STRUCTURE. The prostate is enclosed by a tough, fibrous **capsule** which, like the interstitial connective tissue of the gland, contains smooth muscle fibers. The disseminate part is surrounded by the urethralis muscle. The prostatic fluid is produced in branched secretory tubules and in the cranivores and horse is stored in larger collecting spaces. The disseminate part of the gland in the pig and ruminants consists of lobules radially arranged around the lumen of the urethra.

The paired **BULBOURETHRAL GLAND** (449—452) lies on the dorsal surface of the caudal end of the pelvic urethra and is closely related to the bulb of the penis. It is absent in the dog, and very small in the cat (471/*15*). In the pig, the gland consists of two large cylindrical lobes (17—18 cm. long and about 5 cm. in diameter) that lie on either side of the pelvic urethra and are completely invested by the bulboglandularis muscle (475/*8*, *8'*). In the ox and horse, the bulbourethral glands are spherical and cylindrical respectively, about 3—4 cm. in size, and lie dorsally on the urethra (483, 495/*8*). In the small ruminants, they are round and about 1.5 cm. in diameter (484/*3*). The right and left glands each have one **excretory duct** except in the horse, which has three to four ducts for each gland (496/*k*, *8*). The lobules of the branched, tubular bulbourethral gland contain collecting spaces for the storage of the secretion. Except in the ox, capsule and interlobular connective tissue contain large amounts of smooth muscle.

SEMEN. This is the mixture of spermatozoa and secretions of the accessory genital glands, which is discharged from the penis at the time of ejaculation. The secretions of the accessory glands are considered to be the substrate or vehicle for the spermatozoa, stimulating them to increased motility and enabling them to move about freely and to carry on metabolism. In short, they assure the sperm a long enough life span to reach the ovum. Color and viscosity of the semen differ from species to species; so does the amount of the ejaculate collected under physiological conditions: dog 7—15 ml., boar 200—250 ml. (amounts up to 500 ml. have been collected), bull 2—8 ml., ram 1 ml., billy goat .5—1 ml., and stallion 50—150 ml.

The number of spermatozoa in the ejaculate also varies widely among the species. Within the same species, and under normal frequency of services, the number of spermatozoa per ejaculate remains fairly constant. On the average, one cubic millimeter of semen contains 100,000 spermatozoa in the dog and boar, 1,000,000 in the bull, 3,000,000 in the ram, 2,500,000 in the billy goat, and 120,000 in the stallion. With the advent of artificial insemination, microscopic examination of semen has become an important tool for judging the breeding quality of male animals. Besides the number of spermatozoa present in the ejaculate, their vitality and appearence are important criteria for semen evaluation.

Penis and Urethra

(184, 186, 448—454, 456—472, 474—478, 483—490, 492—500)

The male copulatory organ, the penis, is composed mainly of erectile tissue and is firmly attached to the ischiatic arch of the pelvis by two **crura** (see below). Except in the cat, whose short penis is directed caudally (449), the penis extends cranially from the ischiatic arch and passes between the thighs ventral to the pelvis (485/*h*). It is flanked by the spermatic cords, surrounded by superficial and deep fasciae which are derived from the fasciae of the trunk, and covered by skin ventrally and laterally. The **free part of the penis** (pars libera penis, *k*) lies in a cutaneous sheath known as the **prepuce** (*l*). This is a reserve fold of skin that covers the shaft of the penis when it protrudes from the prepuce during erection.

In the dog and cat, the penis is cylindrical and has an elongated bone (os penis) in the center. In the pig and ruminants, the penis is long, thin, and firm even when not erect. It is folded upon itself between the thighs forming the **sigmoid flexure** (474/*n'*), which is more cranial in the ruminants than in the pig. The free part of the pig's penis is pointed and spirally twisted (*n''*). In the small ruminants, the urethra projects beyond the tip of the penis about 4 cm. (487). The penis of the horse is laterally compressed and of relatively wider diameter; its glans is well developed (494, 497).

The **shaft,** or **body, of the penis** consists of the paired corpus cavernosum penis (494/*s*), the urethra (*r*) with its corpus spongiosum (*r'*), and the glans (*t*), which is a cushion of erectile tissue at the tip; it consists further of the skin covering the free part, musculature, blood and lymph vessels, and nerves.

The **CORPUS CAVERNOSUM** constitutes the bulk of the penis. It is a paired, elongated body of erectile tissue enclosed by a thick fibroelastic capsule known as the **tunica albuginea** (459, 460, 461/*b*). Numerous connective tissue septa and trabeculae pass inward from the tunica albuginea, forming an internal framework that supports the cavernous blood spaces of the erectile tissue. Proximally, right and left corpora cavernosa separate and attach as the **crura penis** (499/*1*, *2*) to the ischiatic arch. Distal to the separation, the two corpora cavernosa are fused in the median plane and only a thin septum separates them. This septum is intact only in the carnivores; in the other species, numerous slitlike openings interrupt it, so that the

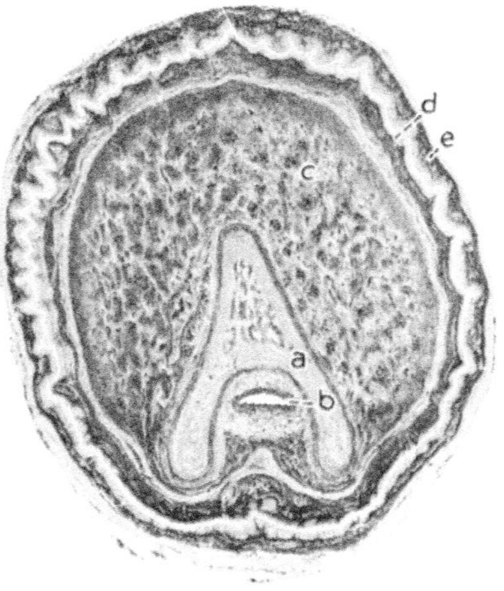

Fig. 458. Cross section of the pars longa glandis and prepuce of the dog's penis. Microphotograph.

a Os penis; *b* Urethra surrounded by corpus spongiosum; *c* Corpus spongiosum glandis; *d* Skin of glans; *e* Int. lamina of prepuce

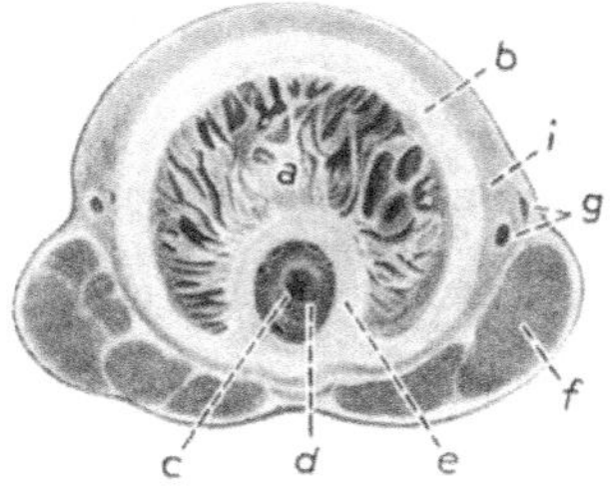

Fig. 459. Cross section of the boar's penis distal to the sigmoid flexure. (After Grabowski 1937.)

a Corpus cavernosum; *b* Tunica albuginea; *c* Urethra; *d* Corpus spongiosum; *e* Tunica albuginea; *f* Retractor penis; *g* Dorsal artery and vein of penis

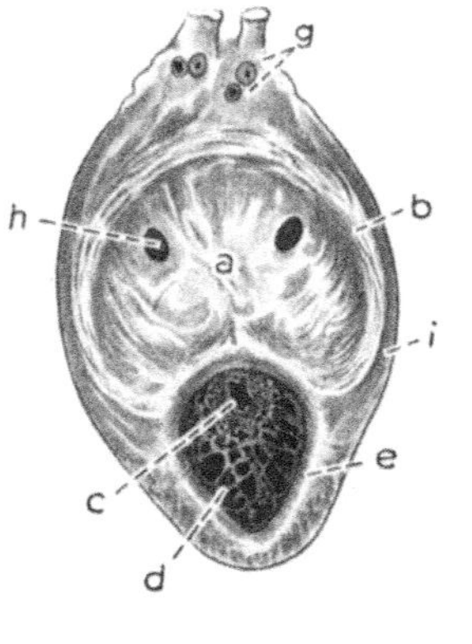

Fig. 460. Cross section of the bull's penis near the caudal end. (After Grabowski 1937.)

a Corpus cavernosum; *b* Tunica albuginea; *c* Urethra; *d* Corpus spongiosum; *e* Tunica albuginea of corpus spongiosum; *g* Dorsal arteries and veins of penis; *h* Dorsal blood channels in corpus cavernosum; *i* Connective tissue covering

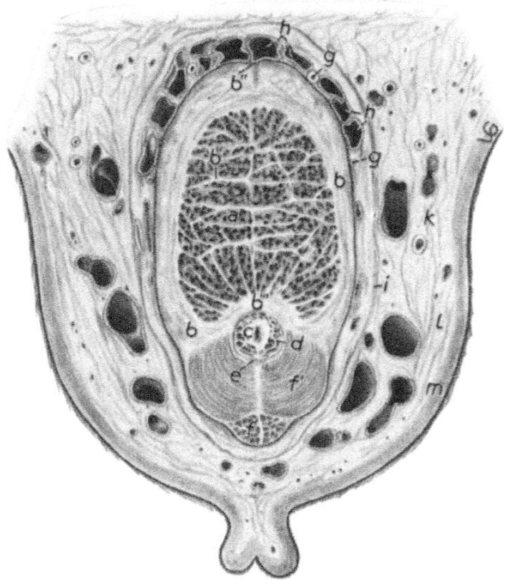

Fig. 461. Cross section of the horse's penis and surrounding structures cranial to the scrotum.

a Corpus cavernosum; *b* Tunica albuginea; *b′* Trabeculae; *b″* Dorsal groove of penis; *b‴* Urethral groove of penis; *c* Urethra; *d* Corpus spongiosum; *e* Tunica albuginea of corpus spongiosum; *f* Retractor penis; *f′* Bulbospongiosus; *g* Middle artery (and branches) of penis; *h* Dorsal veins of penis; *i* Deep fascia of penis; *k* Loose connective tissue with vessels; *l* Supf. fascia; *m* Skin

two corpora are confluent and form a uniform corpus cavernosum surrounded by a tubular tunica albuginea (462/*b*). A shallow longitudinal groove (sulcus dorsalis penis) present in some species on the dorsal surface is a reminder that the corpus cavernosum is derived from the union of an originally paired structure. Along the ventral surface of the corpus cavernosum is a deeper groove (sulcus urethralis) that contains the urethra and the corpus spongiosum surrounding it. Caudally, the shaft of the penis is attached also by a short, paired **suspensory ligament** to the ventral surface of the pelvis.

The **MALE URETHRA** (494/*q*, *r*) begins at the internal urethral orifice at the neck of the bladder, and ends at the **external urethral orifice** at the tip of the penis. In the small ruminants, it projects beyond the tip of the penis, forming the urethral process (487/*2*). The short proximal part of the urethra (pars preprostatica), which extends from the internal urethral orifice to the entrance of the two ductus deferentes, transports only urine; the remaining, much longer part transports both urine and semen.

For descriptive purposes, the male urethra is commonly divided into a pelvic part and a spongy part; most of the latter is incorporated in the shaft of the penis. The accessory genital glands are grouped around the **pelvic part** and their excretory ducts enter this part. The **mucosa** of the pelvic part is folded and covered with transitional epithelium. External to the mucosa is a more or less distinct (depending on the species) **vascular layer** (stratum cavernosum) which is followed by a thin layer of smooth muscle. Surrounding this is the striated, circular **urethralis muscle** (453B).

After leaving the pelvis through the pelvic outlet, the **spongy part** (pars spongiosa) of the urethra turns ventrally around the ischiatic arch, is flanked for a short distance by the crura of the penis, and enters the urethral groove on the ventral surface of the corpus cavernosum. The spongy part is associated with an unpaired body of erectile tissue, the **corpus spongiosum** (hence pars spongiosa) and forms an integral part of the penis (460). The mucosa of the spongy part of the urethra forms longitudinal folds which give it a rosette-like appearance on cross section. The mucosa is covered with transitional epithelium which changes at the external urethral orifice to stratified squamous epithelium. It contains lymph nodules in the ox and small glands in the horse and pig.

The **CORPUS SPONGIOSUM** more or less surrounds the urethra and is continuous proximally with the stratum cavernosum of the pelvic part of the urethra. The corpus spongiosum begins at the pelvic outlet with an enlarged part, the **bulb of the penis,** which is divided by a median septum into right and left halves (472/*c*). The bulb lies against the caudal.and lateral surfaces of the urethra as the latter turns around the ischiatic arch and is covered by the striated, circular **bulbospongiosus muscle** (473/*a*). (The bulb and the two crura penis on either side of it form the **root** (radix) **of the penis.**) The bulb of the penis lies under the skin of the perineal region and projects dorsally into the pelvic cavity, where it is related, except in the dog, to the bulbourethral gland (475).

Distal to the bulb, the corpus spongiosum becomes narrower and from here to the tip of the penis envelops the urethra and with it occupies the urethral groove. At the tip of the penis, the corpus forms vascular connections with the **corpus spongiosum glandis** (457/*h*), the erectile tissue cushion of the glans.

The **ERECTILE TISSUE OF THE CORPUS CAVERNOSUM** is enveloped by a thick **tunica albuginea** and penetrated by numerous connective tissue **trabeculae** emanating from the tunica albuginea. The **cavernous spaces** (cavernae) formed between the trabeculae are lined with endothelium, and in the unerected state of the penis contain only small amounts of blood. Depending on the relative amount of connective tissue present, two types of penises can be distinguished in the domestic mammals. In the **fibroelastic type** found in the pig and ruminants (459, 460/*a*), the trabeculae predominate over the cavernous spaces. Therefore, the consistency of the penis is firm even when not erect. In the **musculocavernous type** found in the dog and horse (462/*a*), the cavernous spaces, which have considerable amounts of smooth muscle in their walls, predominate over the trabeculae. Therefore, the penis is soft and compressible in the quiescent state.

ERECTION. The erectile tissue, trabecular framework, and tunica albuginea of the **corpus cavernosum** form a functional unit. Upon suitable stimulation these function together and bring about the extension, and in the musculocavernous type also the rigidity of the penis. For this to be achieved, the cavernous spaces are connected by special blood vessels to the blood-vascular system. When the penis is not erect, the cavernous spaces are collapsed and contain very little blood. Upon sexual excitement, the spaces fill rapidly with arterial blood released from the **helicine* arteries.** These are coiled branches of the deep artery of the penis, characterized by the presence of myoepithelial cushions capable of reducing their lumina. The helicine arteries respond to sexual stimulation with dilatation and straightening of their coils, allowing the blood to pass through them into the cavernous spaces. At the same time that the flow of blood produces a distension of the erectile tissue, outflow is reduced by pressure on the veins that drain the erectile tissue through the tunica albuginea.

---

* From Gr. *helix*, coil.

As sexual stimulation subsides after coitus and ejaculation, the helicine arteries return to their coiled state, thus restricting the inflow of arterial blood. The smooth muscle in the walls of the cavernous spaces contracts and forces the blood out of them, so that the penis returns to its flaccid state. In the fibroelastic type in which erection consists of the straightening of the sigmoid flexure the retractor penis and the retraction of the tensed elastic elements in the flexure return the penis to its S-shaped, quiescent state. (See p. 339)

The tunica albuginea of the **CORPUS SPONGIOSUM** is thin and elastic and merges distally with the skin of the glans. The cavernous spaces between the trabeculae, which in dog and horse contain smooth muscle, are not as irregular as those of the corpus cavernosum and have the appearance of an elongated venous network that surrounds the urethra. Venous blood from capillaries in the trabeculae traverses the cavernous spaces at any time, so that they are filled with blood also when the penis is not erect. During copulation—that is after erection of the corpus cavernosum has taken place—the corpus spongiosum receives an increased amount of arterial blood via the artery of the bulb of the penis. This leads to a considerable distention of the glans, and to a dilatation of the urethral lumen which facilitates the passage of the ejaculate. (Christensen [1954] has made a detailed study of the angioarchitecture and erection of the penis of the dog.)

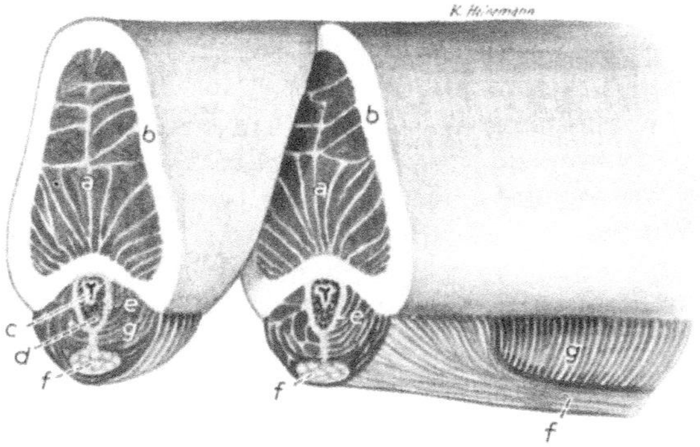

Fig. 462

Fig. 462. Segment of the horse's penis. (After Heinemann 1937.)
a Corpus cavernosum; *b* Tunica albuginea; *c* Urethra; *d* Corpus spongiosum; *e* Tunica albuginea of corpus spongiosum; *f* Retractor penis; *g* Bulbospongiosus

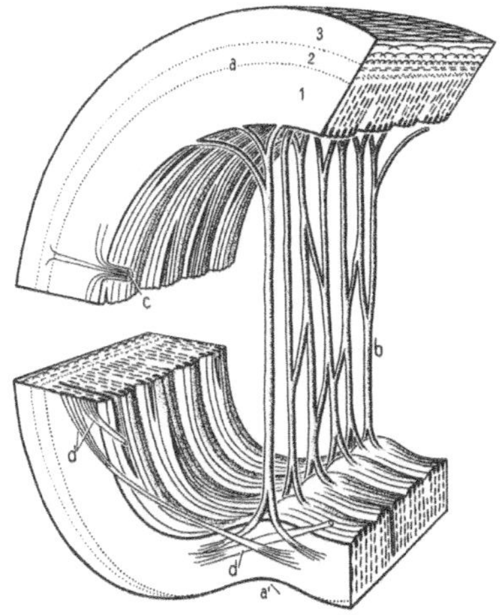

Fig. 463

Fig. 463. Tunica albuginea and trabeculae of the horse's penis. Schematic. (After Buchholz 1951, and Preuss 1954.)
a Tunica albuginea divided into internal, middle, and external layers (*1, 2, 3*); *a'* Urethral groove; *b* Median trabeculae; *c* Dorsal (horizontal) trabeculae; *d* Oblique trabeculae

The **GLANS PENIS** differs greatly from species to species. A glans similar in structure to that of man, is present only in the horse (497/*a*). It caps the distal end of the corpus cavernosum and has a circular edge, **corona glandis** (*3*), proximal to which is a constriction, **collum glandis** (*4*). At the front is the deep **fossa glandis** (*1*) which contains the **urethral process** (*2*). The glans of the dog (468, 469) is supported by the **os penis** and consists of two parts. The **pars longa** (*2*) is conical and surrounds the distal part of the bone, while the more proximal part of the glans, the **bulbus glandis** (*1*), globular when erect, is associated with the proximal part of the bone. The **erectile tissue** of the glans (corpus spongiosum glandis) is similar to a dense network of veins and is continuous with the cavernous spaces of the corpus spongiosum. Compared to horse and the dog, the cat, pig, and ruminants have a small glans, consisting of a thin layer of cavernous veins (470, 477, 486—488/*a*).

PREPUCE. In the quiescent state, the glans and the free part of the penis lie inside a cutaneous **sheath,** or **prepuce** (preputium). In contrast to the penis in man, which is entirely surrounded by skin, the penis of domestic mammals is applied to the ventral abdominal wall and is covered with skin only ventrally and laterally. The prepuce, therefore, is not a complete cylindrical fold as in man. It consists of an **external lamina,** which is continuous with the skin of the abdominal wall, and an **internal lamina,** which is in contact with the penis, and is continuous at the proximal end of the preputial cavity with the skin of the penis. External and internal laminae meet at the **preputial orifice** (464/*m''*). The prepuce of the pig and ruminants has a considerable length. In the horse, when the penis is retracted, the internal lamina gives rise to an annular fold (plica preputialis, 492/*l*) so that two folds of preputial skin surround the glans. When the penis protrudes from the prepuce during erection, both folds are drawn out and cover the shaft of the penis.

The external lamina of the prepuce has the same structure as the surrounding skin and is covered with hair. On its ventral surface is the more or less distinct preputial raphe (*n'*), a continuation of the scrotal raphe. The pig and ruminants have long, bristlelike hairs at the preputial orifice. Hairs, generally, extend only a short distance inside the preputial orifice. In the horse, however, fine hairs are present on the internal lamina including the external surface of the plica preputialis. The internal lamina of the dog, pig, and ruminants contains **lymph nodules** which in dog and ox are present also on the free part of the penis. In all domestic mammals, the skin covering the glans and the free part of the penis contains large numbers of sensory end-organs. An opening (474/*p''*) in the dorsal wall of the boar's prepuce leads into a recess, known as the **preputial diverticulum** (*p*, *p'*), which is incompletely divided by a median septum.

MUSCLES OF THE PENIS. Associated with the prepuce of the dog, pig, and ruminants are the paired cranial and caudal preputial muscles, derivatives of the cutaneus trunci. The **cranial preputial muscle** arises lateral to the xiphoid cartilage and ends in the internal lamina of the prepuce. Here right and left muscles unite and form a loop around the caudoventral aspect fo the preputial orifice; they pull the prepuce cranially. The **caudal preputial muscle** is absent in the carnivores, occasionally absent in the pig, but always present in the ruminants (489/*e*). It originates from the abdominal fascia, lateral to the tunica vaginalis, and ends in the external lamina of the prepuce near the preputial orifice, acting as a retractor of the prepuce.

The muscles directly associated with the penis and urethra vary in number and form according to the species. Therefore, only a general description is given here. The **URETHRALIS** (465/*6*) is a striated, circular muscle, which in the carnivores, goat, and horse surrounds the pelvic part of the urethra. In the pig, ox, and sheep, it covers only the ventral and lateral surfaces of the urethra (453B/*c*). Small muscles associated with the pelvic part of the urethra and accessory genital glands are present in some species and are described in the special chapters (see pp. 328, 332 and 339). In most species, the **BULBOSPONGIOSUS** (490/*8*) is a thick circular muscle, which is continuous with the urethralis at the level of the bulbourethral gland. It surrounds the urethra and bulb of the penis and extends to the proximal end of the shaft of the penis. In the horse, it continues to the glans, covering the urethra ventrally and laterally (461/*f'*; 462/*g*). The paired **ISCHIOCAVERNOSUS** (473/*b*; 485/*f*) originates from the ischiatic arch and tuber ischiadicum and, enclosing the crura of the penis, extends also to the proximal end of the shaft of the penis. It is a well-developed muscle which raises the erect penis and brings it into a position favorable for intromission before coitus. The paired **RETRACTOR PENIS** (490/*10'*, *10*) is a smooth muscle. It originates on the ventral surface of the first few caudal vertebrae and passes ventrally on each side of the rectum. Its **rectal part** passes under the rectum to become continuous with the rectal part of the opposite side. The **penile part** of the retractor continues caudoventrally along the midline, and is attached to the urethral surface of the penis. In the pig, the penile part arises independently from the sacrum and is not connected to the rectal part. The penile part in the dog (464/*g*) extends forward to the bulbus glandis, in the pig and ruminants (485/*g*) it is inserted on the distal curve of the sigmoid flexure, while in the horse (462/*f*) it passes through the bulbospongiosus and ends in the vicinity of the glans.

Blood Vessels, Lymphatics, and Innervation of the Male Genital Organs

The pelvic part of the urethra and the accessory genital glands are supplied by branches of the internal pudendal artery. Scrotum, tunica vaginalis, and prepuce are supplied by branches of the external pudendal artery and by the cremasteric artery. Testis, epididymis, and ductus deferens receive arterial blood from the testicular and deferential arteries. The pattern of the testicular vessels embedded in the tunica albuginea of the testis is species-specific and may serve to identify a testis of unknown origin. The penis of the pig and ruminants is supplied by the internal pudendal artery; that of the carnivores is supplied by the internal and external pudendal arteries; and that of the horse by the two pudendal arteries, and also the obturator artery. The arteries are accompanied by satellite veins which in certain areas form extensive plexuses.

The **lymphatics** of the accessory genital glands go to the medial iliac and the internal and external sacral nodes. Penis and prepuce are drained by the superficial and deep inguinal lymph nodes, while the lymphatics of testis and epididymis pass along the spermatic cord to the medial iliac and lumbar aortic lymph nodes.

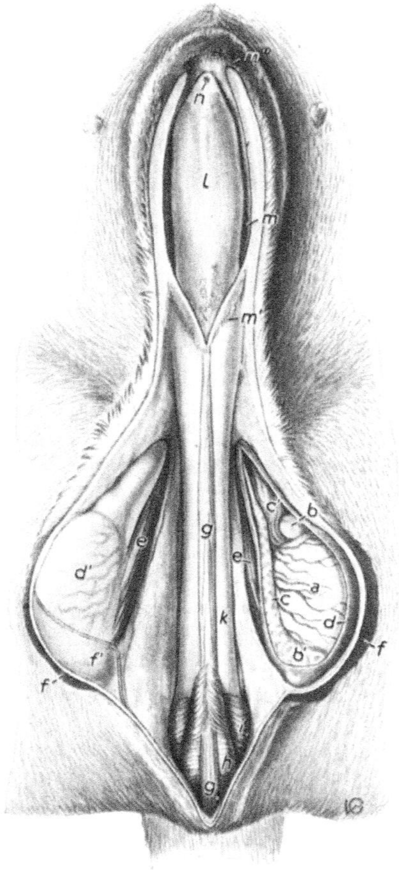

Fig. 464

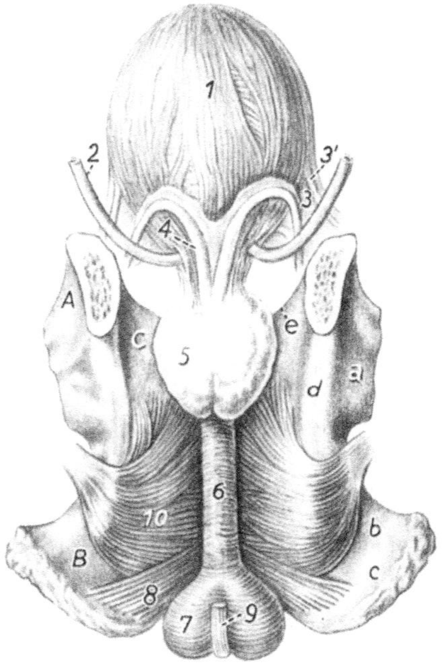

Fig. 465

Fig. 464. Male genital organs of the dog in situ. Scrotum and prepuce have been split to expose the penis. Ventral aspect.

a Medial surface of left testis; *b, b'* Head and tail of epididymis; *c* Ductus deferens; *c'* Spermatic cord; *d* Left tunica vaginalis parietalis, fenestrated; *d'* Right testis, covered by tunica vaginalis parietalis; *e* Cremaster; *f* Scrotum; *f'* Scrotal septum; *g* Retractor penis; *h* Bulbospongiosus; *i* Ischiocavernosus; *k* Body of penis; *l* Glans penis (pars longa), note lymph nodules on proximal part; *m* Preputial cavity; *m'* Caudal extent of prepuce; *m''* Preputial orifice; *n* Ext. urethral orifice

Fig. 465. Male urogenital organs associated with the pelvis of the dog. Dorsal aspect.

A Body of ilium, the wing has been removed; *B* Ischium; *C* Pubis; *a* Acetabulum; *b* Lesser ischiatic notch; *c* Tuber ischiadicum; *d* Ischiatic spine; *e* Brim of pelvis
1 Bladder; *2* Ureter; *3* Lat. ligament of bladder; *3'* Round ligament of bladder; *4* Ductus deferens (ampulla); *5* Prostate gland; *6* Pelvic part of urethra, enclosed by urethralis; *7* Bulb of penis, enclosed by bulbospongiosus; *8* Ischiourethralis; *9* Penile part of retractor penis; *10* Obturator internus

The **innervation** of the male genital organs is from the spinal nerves and from portions of the autonomic nervous system. Scrotum and prepuce are innervated by branches of the lumbar plexus, the iliohypogastric, ilioinguinal, and genitofemoral nerves. The genitofemoral nerve also innervates the tunica vaginalis and the cremaster muscle. The pudendal nerve (from the sacral plexus) supplies the penis and prepuce; and the deep perineal and caudal rectal nerves supply bulbospongiosus, ischiocavernosus, and retractor penis.

Branches of the autonomic nervous system reach the testis via the testicular plexus which accompanies the testicular vessels. The genital organs in the pelvic cavity and the penis receive autonomic innervation via the pelvic plexus and the parasympathetic pelvic nerves. (Larson and Kitchell [1958] have given a detailed description of the innervation of the genitalia in the bull and ram.)

Early castration of male animals disturbs neurohormonal processes and thus inhibits full development of the male reproductive organs, both morphologically and functionally. Such animals show no libido and are unable to produce erection of the penis. In the small ruminants and the pig, early castration inhibits the full excavation of the prepuce so that the penis cannot be protruded from the preputial orifice.

Male Genital Organs of the Carnivores

(1, 179, 184—187, 446, 448, 449, 456, 458, 464—473)

Testis, Spermatic Cord, and Coverings

The **SCROTUM** of the dog is situated between the thighs and is visible from behind (473). In the cat, it is just ventral to the anus and lies on the short, caudally directed penis (471). The skin of the scrotum is thin. In the dog it is usually pigmented (black) and covered with a few hairs, while in the cat it is densely covered with hair. The median groove and **raphe** visible on the scrotum correspond to the position of the **scrotal septum** (473/*2'*) in the interior.

The nearly spherical **TESTES** are relatively small in the dog and are enclosed in the **tunica vaginalis** and **spermatic** and **cremasteric fasciae** (446/*g*), which are connected to the tunica dartos of the scrotum by the external spermatic fascia. The **cremaster muscle** (464/*e*) is poorly developed in the cat. The testes are positioned obliquely so that the extremitas capitata points cranioventrally and the free border caudoventrally.

The **EPIDIDYMIS** (446/*e*) is attached to the dorsolateral surface of the testis and is continuous at its tail with the **DUCTUS DEFERENS** (*f*), which passes cranially medial to the epididymis to be incorporated in the slender spermatic cord. The **proper ligament of the testis** attaches the tail of the epididymis to the extremitas caudata, while the **ligament of the tail of the epididymis** connects the tail of the epididymis to the tunica vaginalis parietalis. The distance, particularly in the cat, between the scrotum and the superficial inguinal ring necessitates a relatively long **SPERMATIC CORD** and tubular tunica vaginalis that encloses it (471). The cord passes cranially between the thighs, and through the inguinal canal to the **vaginal ring,** through which its components enter the abdominal cavity. The testicular vessels (*13*) ascend in the mesorchium to the aorta and caudal vena cava, while the ductus deferens (*12*) makes a sharp caudal turn into the pelvis. An **ampulla** is present on the ductus deferens only in the dog, but is not very prominent (465/*4*). The ductus deferens passes through the substance of the prostate gland and opens into the pelvic part of the urethra on the indistinct **colliculus seminalis.** Since the carnivores have no vesicular gland, the ductus enters the urethra alone. The **genital fold** enclosing the caudal parts of the deferent ducts is small and in the dog may contain a **uterus masculinus.** This is a small vesicle partly imbedded in the prostate, and is a remnant of the paramesonephric ducts.

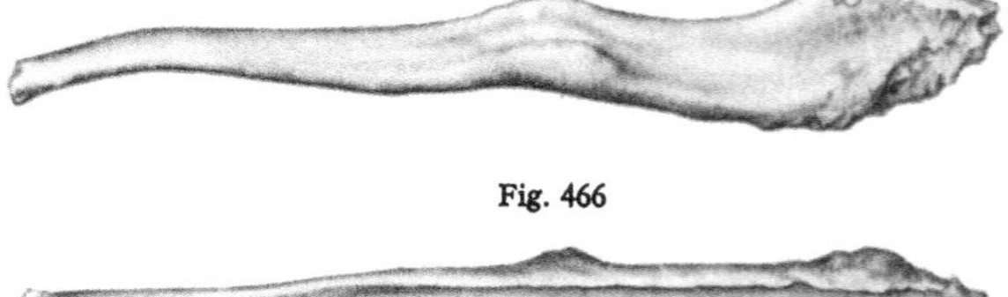

Fig. 466

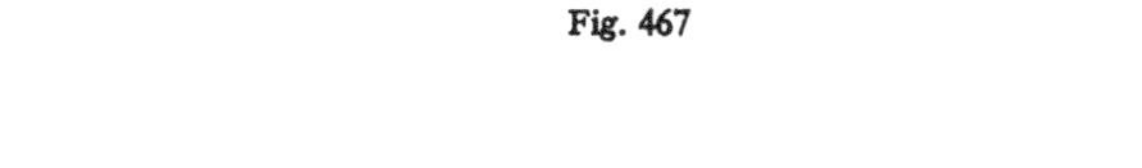

Fig. 467

Figs. 466 and 467. Os penis of the dog. Left lateral aspect (466), ventral aspect (467).

Right: thick proximal part; left: tapering distal part; urethral groove on ventral surface

Accessory Genital Glands

In the dog only the prostate gland, and in the cat only the prostate and the bulbourethral glands are present (448, 449). The pelvic part of the urethra has only a thin stratum cavernosum, and is enclosed by the striated **urethralis muscle** (465/*6*).

The **PROSTATE GLAND** of the dog (465/*5*) is spherical and divided into **right** and **left lobes** which completely surround the urethra. It is hard in consistency, varies in diameter with the breed and age from 1.5—3 cm., and can be examined by digital palpation per rectum in the live animal. The **disseminate part** is represented by a small number of prostatic lobules in the wall of the urethra. The numerous excretory ducts of the prostate open lateral to the colliculus seminalis.

The prostate of the cat (471/*14*) is also bilobed. It has an uneven surface and covers the urethra only dorsally and laterally. Its **disseminate part** consists of small lobules which extend in the urethral wall to the bulbourethral glands. The two small **BULBO-URETHRAL GLANDS** (*15*), about 5 mm. in diameter, lie dorsolaterally on the urethra at the level of the ischiatic arch.

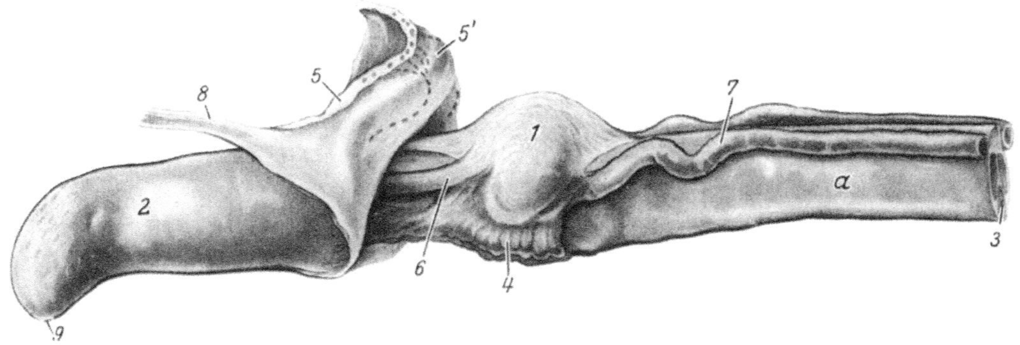

Fig. 468. Penis of the dog. Left lateral aspect. (After Vaerst 1938.)

a Body of penis; *1* Bulbus glandis; *2* Pars longa glandis; *3* Corpus spongiosum; *4* Anastomoses between corpus spongiosum and bulbus glandis; *5* Proximal part of prepuce pulled forward, the dotted lines represent veins draining the glans; *5'* Caudal extent of prepuce, pulled away from penis; *6* Vein connecting cavernous tissue of the pars longa with that of the bulbus glandis; *7* Dorsal veins of penis; *8* Insertion of retractor penis, turned forward; *9* Ext. urethral orifice

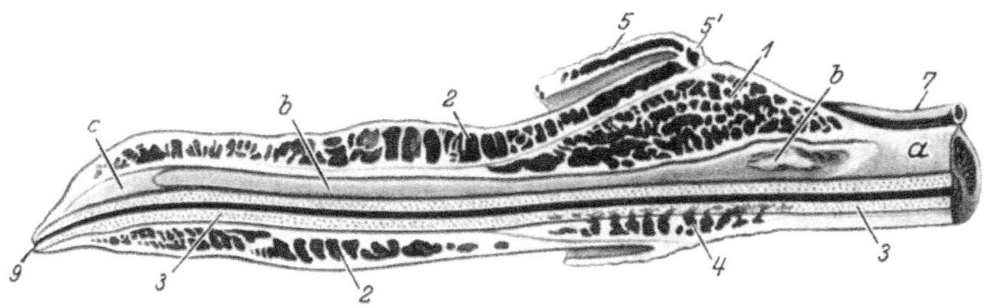

Fig. 469. Median section of distal part of canine penis. Cavernous spaces slightly schematic. (After Vaerst 1938.)

a Body of penis; *b* Os penis; *c* Fibrocartilaginous extension of os penis; *1* Bulbus glandis; *2* Pars longa glandis; *3* Corpus spongiosum penis; *4* Anastomoses between corpus spongiosum and bulbus glandis; *5* Int. lamina of prepuce; *5'* Caudal extent of prepuce; *7* Dorsal vein of penis; *9* Ext. urethral orifice

Penis

The penis of the dog is nearly cylindrical and extends from the ischiatic arch forward between the thighs (464). The short penis of the cat has retained its original embryonic orientation and is directed caudoventrally (471/*16, 17*).

The body of the canine penis, exclusive of the glans, can be divided into a proximal cavernous and a distal osseous part. The cavernous part (469/*a*) is formed by the two **corpora cavernosa.** These are united in the median plane, but their cavernous spaces remain separated by an intact median **septum.** Caudally, the corpora diverge and attach as the **crura of the**

penis (472/*e*) to the ischiatic arch. The osseous part (469/*b*) is formed by the **os penis** which is covered with tough periostium. The os penis develops after birth and is generally regarded to represent the ossified distal parts of the corpora cavernosa. In large breeds it attains a length of up to 11 cm. The proximal end of the bone is thick and firmly united with the corpora cavernosa. Distally, it tapers and at its tip has a tough fibrous extension (*c*). On the ventral surface of the os penis is a deep groove (467) which does not quite extend to the tip of the bone. The groove is continuous caudally with the **urethral groove** on the ventral surface of the corpora cavernosa and contains the urethra and the corpus spongiosum (458/*b*).

The **corpus spongiosum** begins at the level of the ischiatic arch. Its proximal expanded part, the **bulb of the penis,** is paired (472/*c*) and is covered by the **bulbospongiosus** (473/*a*). The bulb lies between the crura and extends to the proximal part of the body of the penis where the corpus spongiosum becomes tubular (469/*3*) and surrounds the urethra.

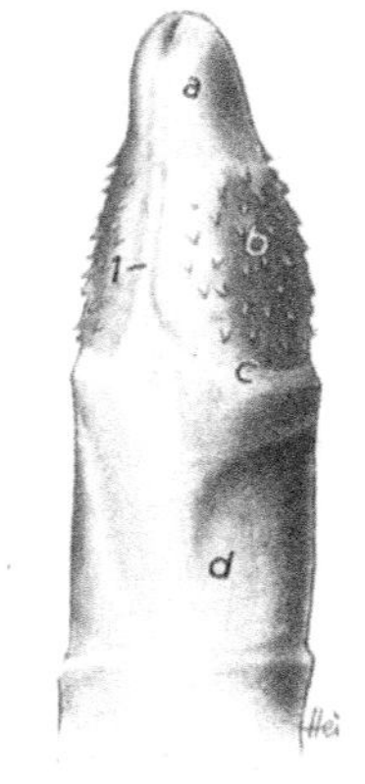

Fig. 470. Distal end of cat's penis, protruded from prepuce. Urethral surface

a Glans; *b* Free part of penis studded with cornified spines; *c* Caudal extent of prepuce; *d* Internal lamina of prepuce applied to shaft of penis
1 Frenulum preputii

The **GLANS PENIS** of the dog surrounds the os penis. Its erectile tissue (corpus spongiosum glandis) consists of elongated venous networks that communicate with the corpus spongiosum penis. The distal, long part of the glans, **pars longa glandis** (469/*2*), surrounds the slender portion of the os penis, while the very much shorter, bulbous part, **bulbus glandis** (*1*), surrounds the thick proximal portion of the bone, with the bulk of the bulbus lying dorsal to the bone. Because of the elasticity of its trabecular system, the glans of the dog, particularly the bulbus, is capable of considerable distension during coitus and contributes to the so-called

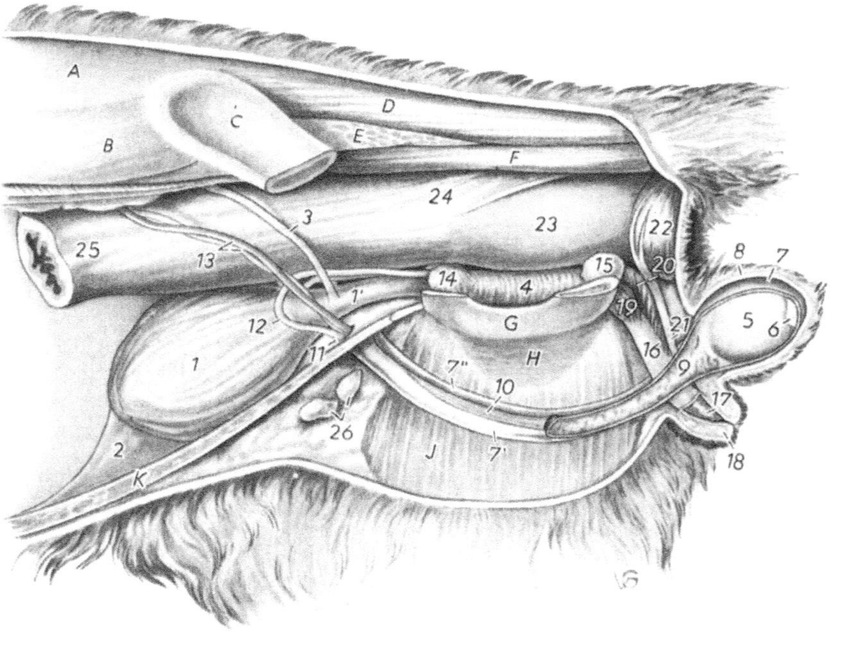

Fig. 471. Male genital organs of the cat in situ. Left lateral aspect.

A Longissimus; *B* Iliocostalis; *C* Wing of ilium; *D* Sacrocaudalis dorsalis lateralis; *E* Gluteus supf.; *F* Sacrocaudalis ventralis lateralis; *G* Floor of pelvis; *H* Symphysial tendon, origin of adductor and gracilis muscles; *J* Right gracilis; *K* Ventral abdominal wall

1 Urinary bladder; *1'* Neck of bladder; *2* Median lig. of bladder; *3* Left ureter; *4* Pelvic part of urethra, surrounded by urethralis; *5* Left testis; *6* Tail of epididymis; *7, 7'* Left tunica vaginalis parietalis, its distal part fenestrated; *7''* Proximal part of right tunica vaginalis parietalis; *8* Scrotum; *9* Spermatic cord; *10* Cremaster; *11* Vaginal ring; *12* Left ductus deferens; *13* Testicular artery and vein; *14* Prostate gland; *15* Bulbourethral gland; *16* Penis; *17* Free part of penis; *18* Prepuce; *19* Ischiocavernosus; *20* Bulbospongiosus; *21* Retractor penis; *22* Ext. anal sphincter; *23* Rectum; *24* Rectococcygeus; *25* Descending colon; *26* Supf. inguinal lymph nodes

"tie" between the partners which may last up to thirty minutes. (Hart and Kitchell [1965] have given a detailed description of the erected glans of the dog.)

The **PREPUCE** of the dog is fairly well separated from the abdominal wall, but remains attached to it by a median fold of skin. Only the distal part of the prepuce is free and forms a complete cylindrical fold that is tapered toward the small **preputial orifice** (464). The preputial cavity is relatively long and extends to the level of the bulbus glandis (469). Both the internal lamina of the prepuce and the skin covering the glans contain numerous lymph nodules which are visible without magnification.

Because of the peculiar orientation of the **CAT'S PENIS** (471/*16*), its urethral surface faces caudodorsally, while its dorsum faces cranioventrally. The cat's penis, like that of the dog, consists also of a proximal cavernous and a distal osseous part. The **os penis** is about .5 cm. long and is pointed distally. The **corpus spongiosum penis** is present and begins at the level of the ischiatic arch with the **bulb of the penis.** The **prepuce** (*18*) is a thick circular fold surrounding the caudally directed **preputial orifice.** The free part of the penis is conical and is capped by a thin indistinct **glans** (470/*a*). In the sexually mature cat, the free part is studded with small cornified papillae, **penile spines** (*b*), which are considered a secondary sex characteristic. The spines develop between two to six months of age, and regress following castration (Aronson and Cooper 1967).

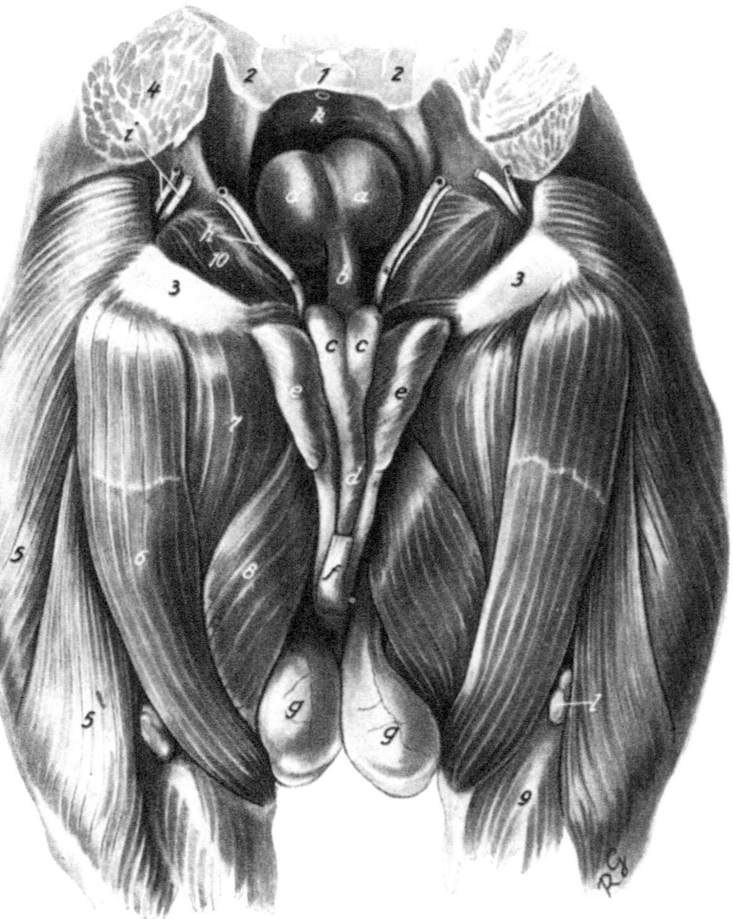

Fig. 472. Male genital organs of a dog in situ. Deep dissection. Caudal aspect. (After Grabowski 1937.)

a Right and left lobes of prostate gland; *b* Pelvic part of urethra; *c* Bulb of penis, exposed by removal of bulbospongiosus; *d* Urethra, enclosed by corpus spongiosum; *e* Crus penis covered by ischiocavernosus; *f* Distal stump of retractor penis; *g* Testis inside tunica vaginalis; *h* Int. pudendal artery and pudendal nerve; *i* Caudal gluteal artery and ischiatic nerve; *k* Aorta in cross section; *l* Popliteal lymph nodes

1 Sacrum; *2* Ilium; *3* Tuber ischiadicum; *4* Gluteal muscles; *5* Biceps femoris; *6* Semitendinosus; *7* Semimembranosus; *8* Gracilis; *9* Gastrocnemius

The **MUSCLES OF THE MALE GENITAL ORGANS,** except for differences in size, are similar in dog and cat. The **bulbospongiosus** (465/*7*; 473/*a*) is divided by a septum into right and left halves. It is a pear-shaped, striated muscle, which originates in the pelvic cavity as the continuation of the urethralis. Lying between the crura, it covers the bulb of the penis and extends to the junction of the crura. In the dog, a few bundles of the muscle (*a'*) may enter the scrotal septum. The bulbospongiosus of the cat (471/*20*) covers also the bulbourethral glands. The striated **ischiocavernosus** (471/*19*; 473/*b*) encloses the crus penis. It originates on the tuber ischiadicum and extends like the bulbospongiosus to the junction of the crura. The small **ischiourethralis** (465/*8*; 473/*d*) originates between the obturator internus and ischiocavernosus from the dorsal surface of the tuber ischiadicum and ends on the transverse perineal ligament. It is thought to influence the blood-flow in the dorsal vein of the penis. The **retractor penis** (464/*g*; 471/*21*) is a smooth muscle that originates on the sacrum or first two caudal vertebrae. Right and left muscles pass ventrocaudally across the lateral surface of the rectum, contributing many fibers (pars analis) to the wall of the rectum and anal canal and occasionally forming a subanal loop. The thinner penile part of the retractor (187/*r*) continues over the bulbospongiosus and along the urethra and ends in the skin covering the bulbus glandis.

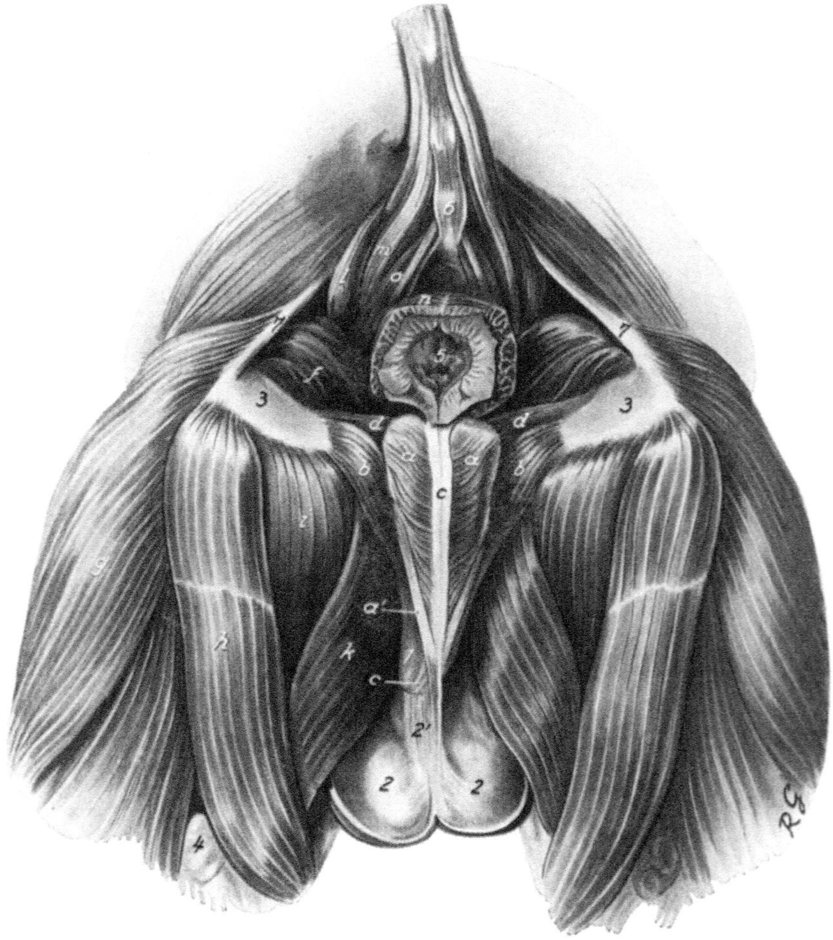

Fig. 473. Muscles associated with the male genital organs of a dog. Caudal aspect. (After Heinemann 1937.)

a Bulbospongiosus, covering bulb of penis; *a'* Fibers of bulbospongiosus to scrotal septum; *b* Ischiocavernosus; *c* Retractor penis; *d* Ischiourethralis; *f* Obturator internus; *g* Biceps femoris; *h* Semitendinosus; *i* Semimembranosus; *k* Gracilis; *l* Intertransversarius dorsalis caudae; *m* Coccygeus; *n* Ext. anal sphincter; *o* Levator ani

1 Body of penis; *2* Scrotum and testes; *2'* Scrotal septum; *3* Tuber ischiadicum; *4* Popliteal lymph nodes; *5* Anus; *6* Caudal vertebra; *7* Sacrotuberous ligament

Male Genital Organs of the Pig

(450, 459, 474—478)

Testis, Spermatic Cord, and Coverings

The large **SCROTUM** of the boar is divided by a deep groove into right and left halves, and lies flat against the caudal surface of the thighs a short distance ventral to the anus (474). Because of its broad base, it is not set off very much from the surface of the body and bulges only ventrally from the weight of the large testes. The scrotal skin has a few hairs and is much wrinkled in older animals; its **tunica dartos** is thin. The **cremaster muscle** (*e*) is a thin, long band which extends to the lateral surface of the tough **tunica vaginalis parietalis.** Because of the peculiar position of the testes in the pig, the cremasters cannot "raise" the testes as in other animals.

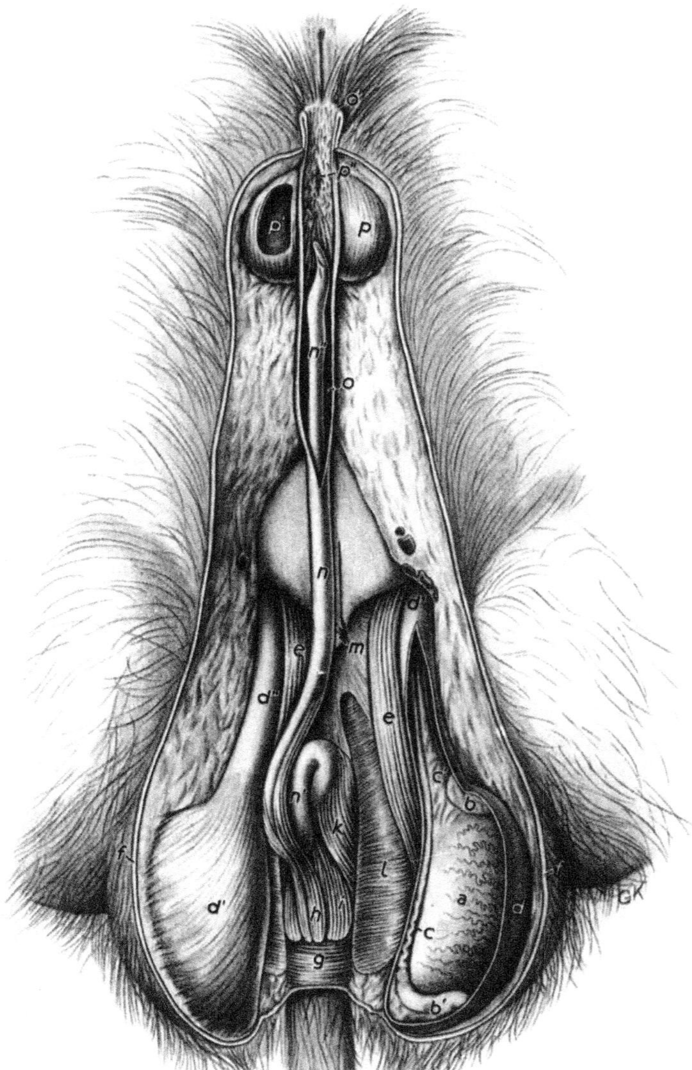

Fig. 474. Male genital organs of the pig in situ. Scrotum and prepuce have been split to expose the penis. Ventral aspect.

a Medial surface of left testis; *b, b'* Head and tail of left epididymis; *c* Ductus deferens; *c'* Spermatic cord; *d* Left tunica vaginalis parietalis, fenestrated; *d', d''* Right testis and spermatic cord, covered by tunica vaginalis parietalis; *e* Cremaster; *f* Scrotum; *g* Ext. anal sphincter; *h* Retractor penis; *i* Bulbospongiosus; *k* Ischiocavernosus; *l* Gracilis; *m* Stump of caudal preputial muscle; *n, n', n''* Penis; *n'* Sigmoid flexure; *n''* Free part of penis; *o* Prepuce; *o'* Preputial orifice; *p, p'* Preputial diverticulum; *p''* Opening into preputial diverticulum

The **TESTES** of the larger, more highly bred pigs are of considerable size, weighing on the average 400 gm. each in fully grown boars. (In one five-year-old purebred boar each testicle weighed 800 gm.) The testes (474/*a*) are long and elliptical, and when cut open longitudinally, the distinct **septula** are seen to combine centrally, forming an axially placed **mediastinum testis.** In the live animal, the extremitas caudata is close to the anus and the epididymal border lies against the caudal thigh muscles. The **EPIDIDYMIS** is also very large. Its **tail** (*b'*) forms a blunt projection at the extremitas caudata to which it is attached by the **proper ligament of the testis.** The **ligament of the tail of the epididymis** connects it to the tunica vaginalis parietalis. The part of the **mesorchium** attaching along the epididymal border of the testis is wide in young boars.

The **DUCTUS DEFERENS** (*c*) passes along the medial surface of the epididymis to the extremitas capitata of the testis, where it joins the testicular artery and the pampiniform plexus of veins to form the long **SPERMATIC CORD.** Enclosed by the tubelike proximal part of the tunica vaginalis (*d''*), the cord passes cranially between the thighs and, after crossing the lateral surface of the penis, passes through the inguinal canal to end at the **vaginal ring.** From there the testicular vessels continue craniodorsally to the aorta and vena cava, while the ductus deferens makes a sharp caudal turn into the pelvic cavity, crosses the ventral surface of the ureter, and disappears ventral to the extensive vesicular gland (475). The duct then enters the genital fold without forming an ampulla, passes through the prostate, and ends independently on the small **colliculus seminalis** (450). A few small glands are present in the terminal portion of the duct.

Accessory Genital Glands

The vesicular and bulbourethral glands of the pig are exceedingly large, while the prostate gland is relatively small. The pelvic urethra into which they open, is surrounded by a stratum cavernosum. The **VESICULAR GLAND** (475/*5*) consists of two pyramidal masses that are attached to each other medially by connective tissue. Each of these masses is 12—17 cm. long, 6—8 cm. wide, and 3—5 cm. thick, and is oriented so that its base is cranial and its apex caudal. The vesicular gland extends into the abdominal cavity between the layers of the genital fold, and is related ventrally to the caudal part of the bladder, neck of the bladder, prostate gland, and cranial part of the pelvic urethra. It is pinkish red in the fresh state, hard, and distinctly lobulated. On the cut surface are seen large collecting spaces which can store considerable amounts of the secretion. The collecting ducts of each side of the gland unite to form an **excretory duct,** which opens on the colliculus seminalis close to the ipsilateral ductus deferens (450).

The body of the **PROSTATE GLAND** (475/*6*) is relatively small. It is 3—4 cm. long, 2—3 cm. wide, and about 1 cm. thick, and lies like a small plaque on the dorsal surface of the urethra. The **disseminate part** (450) constitutes the bulk of the gland, surrounds the urethra with many lobules, and is covered by the urethralis muscle (475/*7*). Numerous excretory ducts carry the prostatic secretion of both parts of the gland directly into the lumen of the urethra.

The **BULBOURETHRAL GLANDS** (475/*8, 8'*) are very large. They consist of two dense cylindrical masses, 17—18 cm. long and about 5 cm. in diameter, which cover the greater part of the pelvic urethra and extend caudally to the pelvic outlet. They are invested by a thin striated muscle, known as the **bulboglandularis.** Like the vesicular gland, the bulbourethral glands contain large, cystlike collecting spaces filled with a clear mucoid secretion. An **excretory duct** leaves the caudal part of each gland, and opens into the urethra deep to the bulb of the penis. The bulbourethral glands can be palpated rectally.

The **vesicular** and **bulbourethral glands** in boars used for breeding are so large that they occupy a considerable part of the pelvic cavity. This is shown in Figure 475. Figure 476 shows the same but very much smaller glands in an animal that was castrated early in life. The accessory genital glands of boars castrated late in life and reaching slaughter after only a brief fattening period, do not differ appreciably in size from those of the intact animal.

Penis

The penis of the pig belongs to the fibroelastic type and when erect resembles a thin, tapering stick about 60 cm. long. In cross section it is dorsoventrally flattened proximally, round at the middle (459), and laterally flattened at the tip. The two **crura penis** originate laterally on the ischiatic arch. Distal to their union, the penis enters the space between the thighs and describes a tight **sigmoid flexure** (474/*n'*). The proximal curve of the flexure is open caudally, while the distal curve is open cranially. The tapering free part of the penis (*n''*) constitutes roughly one-third of the total length and is twisted counterclockwise like a corkscrew (477). The connective tissue structures of the penis such as the tick **tunica albuginea** and **trabeculae** (459) predominate over the erectile tissue, so that the penis is hard even in the quiescent state. Erection consists essentially of straightening the sigmoid flexure, which is accomplished by the blood distending the erectile tissue of the corpus cavernosum.

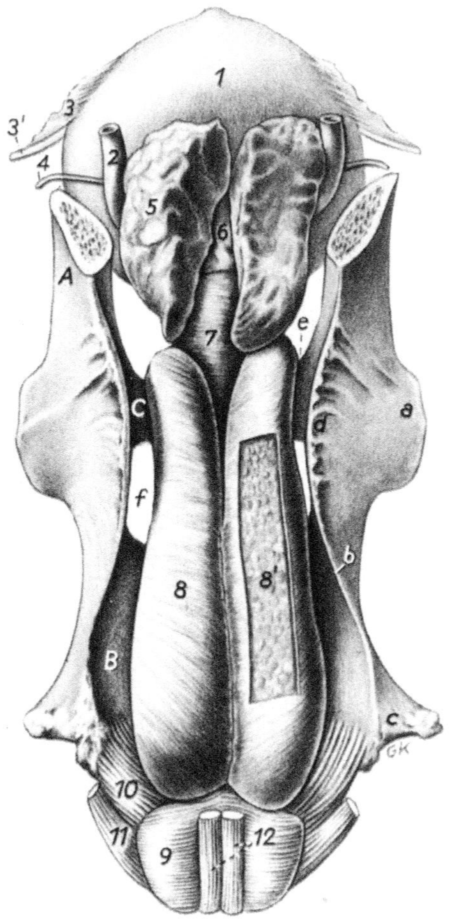

Fig. 475 (Intact boar)

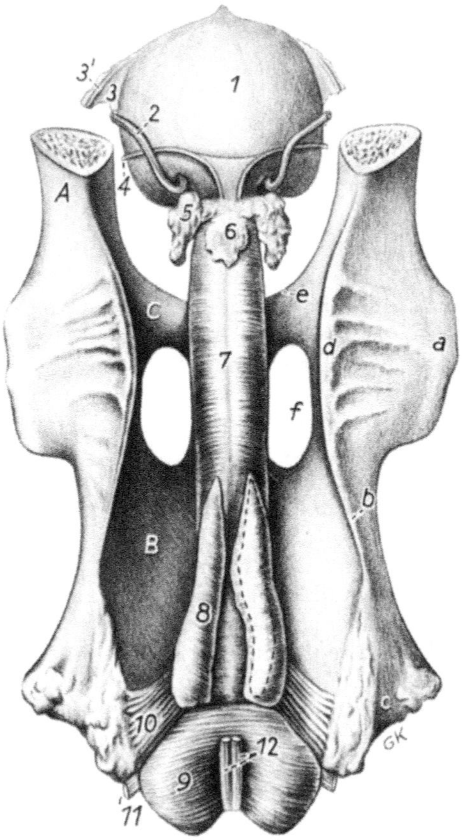

Fig. 476 (Castrated early in life)

Fig. 475 and 476. Urogenital organs and pelves of two male pigs of similar size. Dorsal aspect.

A Ilium, the wing has been removed; *B* Ischium; *C* Pubis; *a* Acetabulum; *b* Lesser ischiatic notch; *c* Tuber ischiadicum; *d* Ischiatic spine; *e* Brim of pelvis; *f* Obturator foramen

1 Urinary bladder; *2* Ureter; *3* Lateral ligament of bladder; *3'* Round ligament of bladder; *4* Ductus deferens; *5* Vesicular gland; *6* Body of prostate gland; *7* Pelvic part of urethra, enclosed by urethralis; *8* Bulbourethral gland, enclosed by bulboglandularis; *8'* Bulboglandularis, fenestrated to expose bulbourethral gland; *9* Bulb of penis and bulbospongiosus; *10* Ischiocavernosus; *11* Urethrocavernosus; *12* Retractor penis

The spongy part of the **urethra** begins at the ischiatic arch with an ampulliform enlargement of the lumen. Lying against the caudal surface of this part is the prominent **bulb of the penis** (475/*9*), which is divided by a median septum and is covered with a dense connective tissue capsule. Distally, the bulb is continued by the tubular part of the **corpus spongiosum** (459/*d*) which surrounds the urethra and gradually diminishes in diameter toward the glans. Urethra and corpus spongiosum occupy the **urethral groove,** which is shallow proximally, but becomes so deep toward the tip of the penis that it almost surrounds the urethra.

The pointed free end of the penis is capped by a small **GLANS** (477/*a*). The glans consists of a flat plexus of veins that is embedded in the skin covering the tip of the penis and forms vascular communications with the corpus spongiosum.

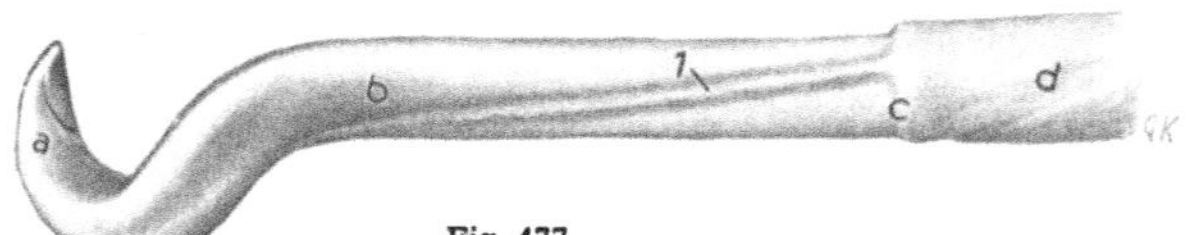

Fig. 477

Fig. 477. Distal end of boar's penis, protruded from prepuce. Ventral aspect.

a Glans; *b* Free part of penis; *c* Caudal extent of prepuce; *d* Int. lamina of prepuce applied to shaft of penis; *1* Raphe penis

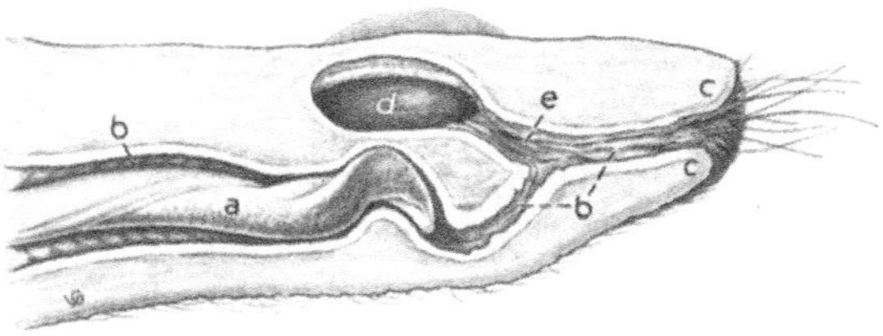

Fig. 478

Fif. 478. Sagittal section of the prepuce of a boar.

a Free part of penis, note the lymph nodules in its skin; *b* Int. lamina of prepuce; *c* Preputial orifice with bristles; *d* Left half of preputial diverticulum; *e* Opening between diverticulum and prepuce

The **PREPUCE** is considerably longer than the free part of the penis that occupies it (474). The **preputial orifice** (*o'*) is formed by a thick annular fold of skin covered with a tuft of stiff hairs. The preputial cavity is 20—25 cm. long, and may be divided at a transverse fold into proximal and distal parts. The proximal part contains the free part of the penis; it is narrow, so that its lightly folded lamina interna is in close contact with the penis. The distal part of the prepuce is wider and its lamina interna forms larger folds. The lining of both parts as well as the skin covering the free part of the penis contain lymph nodules. In the dorsal wall of the prepuce is an opening (*p''*) leading into the **preputial diverticulum** (*p*, *p'*). The opening is large enough to admit two fingers, but is usually closed by overlapping folds of mucous membrane. The preputial diverticulum is a dorsoventrally flattened pouch with a capacity of about 135 ml. It is partly divided into two oval halves by a median fold on its dorsal wall. Right and left halves lie against the dorsolateral surfaces of the prepuce and extend caudally over the wide part of the prepuce. Like the prepuce, the diverticulum is lined with a slightly folded mucous membrane that is covered with stratified squamous epithelium. Flakes of epithelial debris combined with decomposing urine always present in the pouch are the source of the characteristic and unpleasant odor surrounding a boar. The **cranial preputial muscles** form a loop around the caudal aspect of the diverticulum and can express its contents into the prepuce. In pigs castrated early in life, the preputial diverticulum remains rudimentary. In addition, the internal lamina of the prepuce fails to separate from the penis.

MUSCLES OF THE MALE GENITAL ORGANS. The **bulbospongiosus** (475/*9*) is a strong striated muscle divided into right and left parts by a median septum. It extends from the caudal end of the bulbourethral glands to the junction of the crura, and overlies the bulb of the penis to which it is closely attached. The **ischiocavernosus** (*10*) is a paired, striated muscle lying craniolateral to the bulbospongiosus. It originates on the medial surface of the tuber ischiadicum, encloses the crus penis, and is inserted broadly on the lateral surface of the penis proximal to the sigmoid flexure (474/*k*). The **urethrocavernosus** (475/*11*) is a thin band that arises on the ischiocavernosus and bulbospongiosus and after a short distal course is inserted into the distal end of the bulbospongiosus. The penile part of the unstriped **retractor penis** (474/*h*; 475/*12*) originates ventrally on the sacrum and passes the lateral surface of the rectum, then over the caudal surface of the bulbospongiosus and joins the penis at the distal curve of the sigmoid flexure, ending a short distance distal to the flexure. The rectal part of the retractor is independent, arises from the caudal vertebrae, and forms a thin loop around the ventral surface of the rectum (p. 145).

Male Genital Organs of the Ruminants

(443, 444, 451, 454, 460, 479—490)

Testis, Spermatic Cord, and Coverings

The **SCROTUM,** especially that of the small ruminants, is of considerable size. It is a long, pendulous pouch suspended between the thighs from the inguinal region, and is easily visible from behind the animal (485). It has a well-marked neck above the testes and a median groove dividing it into right and left halves (479). The hair on the bull's scrotum is sparse, but the scrotum of the small ruminants, particularly that of the ram, is well covered with hair. The long, pendulous scrotum in the ruminants results from the peculiar vertical orientation of the testes. The extremitas capitata is dorsal and from it the spermatic cord passes through the neck of the scrotum to the inguinal canal. Loose connective tissue between the thick tu-

Figs. 479—481. Testis and epididymis of the ruminants in situ. Caudal aspect. Scrotum and left tunica vaginalis parietalis fenestrated. Right tunica vaginalis intact.

a Left testis, testicular vessels shining through tunica albuginea; *b* Head, *b'* body, and *b''* tail of left epididymis; *c* Spermatic cord; *d* Cremaster

1 Scrotum; *1'* Scrotal raphe; *2* Tunica dartos; *2'* Scrotal septum; *3* Cut edge of tunica vaginalis parietalis; *4* Fat

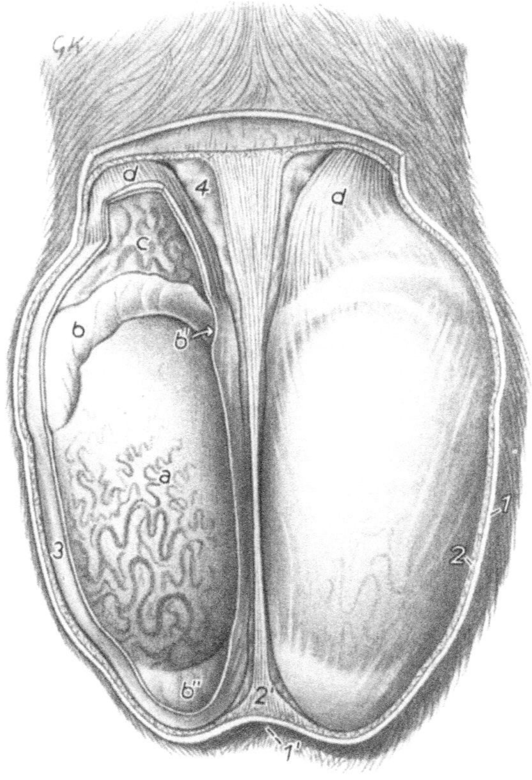

Fig. 479 (Bull)

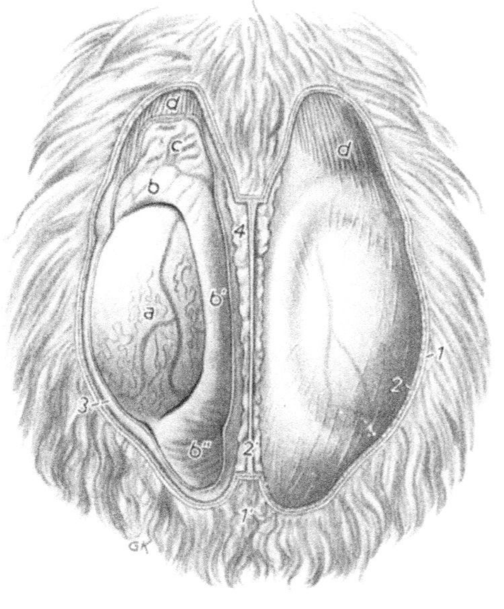

Fig. 480 (Ram)

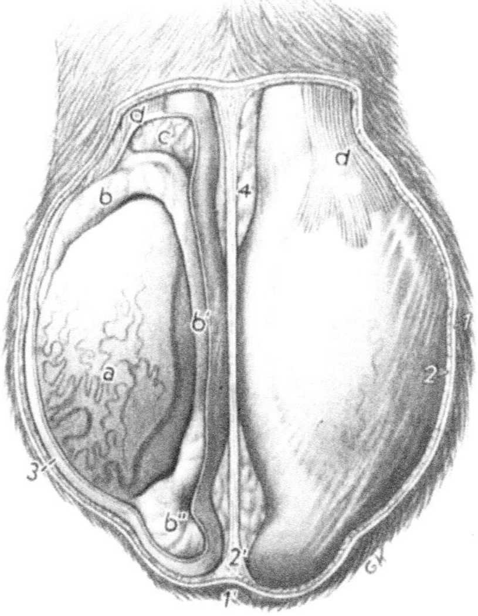

Fig. 481 (Billy goat)

nica dartos and the external spermatic fascia, and the fascia itself, facilitate the noticeable dorsoventral movements of the testes in the scrotum, brought about by the strong cremaster muscle (*d*). As the testes slide dorsally, the tunica dartos wrinkles the skin to accomodate the scrotum to the new position of the testes.

Right and left **TESTES** of the bull usually differ in size and position, one being slightly more dorsal than the other. They have the shape of a long oval, as compared to the more spherical shape in the small ruminants. The testes of the bull weigh 250—300 gm. each, those of the ram 200—300 gm., and those of the goat 145—150 gm. The epididymal border (443/*a''*) in the bull is medial, while in the small ruminants it is slightly more caudomedial (482). The head of the **EPIDIDYMIS** in the bull (479/*b*) curves over the extremitas capitata and is continued by the body along the medial surface of the testis. The tail of the epididymis (*b''*) is large and rounded and projects distally from the extremitas caudata. In the small ruminants (480, 481), the head of the epididymis extends onto the craniolateral surface of the testis. The body of the epididymis descends caudomedially along the testis and ends in the rounded tail which is directed slightly caudally and is visible and easily palpable through the scrotum. The **proper ligament of the testis** (454/*11*) and **ligament of the tail of the epididymis** (*11'*) are well developed in the ruminants.

The **DUCTUS DEFERENS** (443/*c*) ascends in the mesoductus on the medial surface of the testis to join the vessels of the **SPERMATIC CORD.** The latter is long and thick and ascends dorsally to the inguinal canal (454). The ease with which the spermatic cords can be palpated through the scrotum has led to a method of castration in which the cords are crushed with a special instrument without injury to the skin. The testes are left to degenerate in the scrotum.

In its abdominal course, the ductus deferens, before reaching the genital fold, enlarges to form the **ampulla** (483/*4*) which in the bull is 13—15 cm. long and 12—15 mm. in diameter. In sheep and goat (484/*5*), the ampulla is 6—8 cm. long and 4—8 mm. in diameter. Distal to the ampulla, the duct narrows, dips under the body of the prostate, and opens with the excretory duct of the vesicular gland on the **colliculus seminalis** (451).

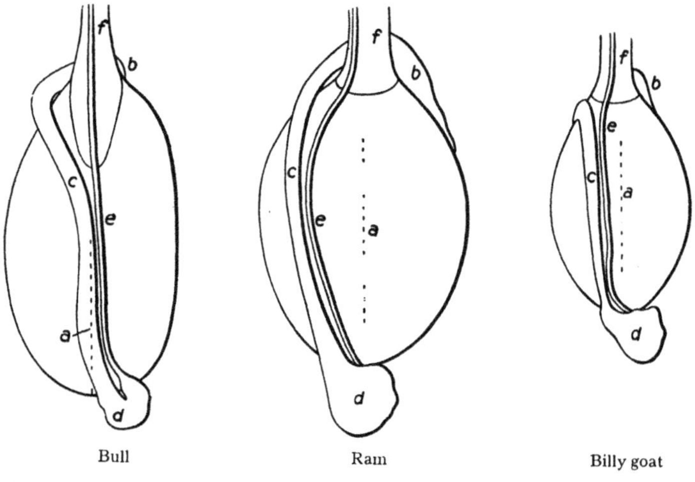

Fig. 482. Left testis and epididymis of bull, ram, and billy goat to show relation of epididymis to testis. Medial aspect. Schematic. (After v. Schlumperger 1954.)

a Dotted line dividing medial surface of testis into caudal and cranial halves; *b* Head, *c* body, and *d* tail of epididymis; *e* Ductus deferens; *f* Spermatic cord

Accessory Genital Glands

The paired **VESICULAR GLAND** is the largest of the accessory genital glands in the ruminants. It is hard and compact, has an uneven surface, and in the bull an irregular, elongated form, often bent on itself or S-shaped (483/*5*). The vesicular gland of the small ruminants is shorter and more rounded (484/*4*). In the bull, the gland is 10—12 cm. long and 1.5—3.5 cm. thick, having a total weight of 45—80 gm. In the small ruminants, the vesicular gland is 2.5—4 cm. long, 2—2.5 cm. wide, and 1—1.5 cm. thick. The free ends of the gland are directed cranially and slightly laterally, and lie in the **genital fold** dorsal to the bladder. Between right and left glands are the terminal segments of the ureters and deferent ducts

(483/*2, 4*). The vesicular glands of the bull are easily palpated rectally. The **excretory ducts,** one for each gland, pass through the prostate gland before opening on the colliculus seminalis (451).

The **PROSTATE GLAND** of the bull (483/*6*) is bilobed and lies transversely on the dorsal surface of the urethra, just caudal to the vesicular gland. It is 1—1.5 cm. thick, about 1 cm. wide (from cranial to caudal), and about 3 cm. long (from right to left). The **disseminate part** (451) surrounds a 12 cm. section of the urethra and is itself enclosed by the **urethralis muscle,** the rhythmical contractions of which can be felt during rectal palpation. The small ruminants have only a disseminate part, which in the goat surrounds the urethra completely and in the ram only the dorsal and lateral surfaces. Numerous small **secretory ducts** open directly into the urethra.

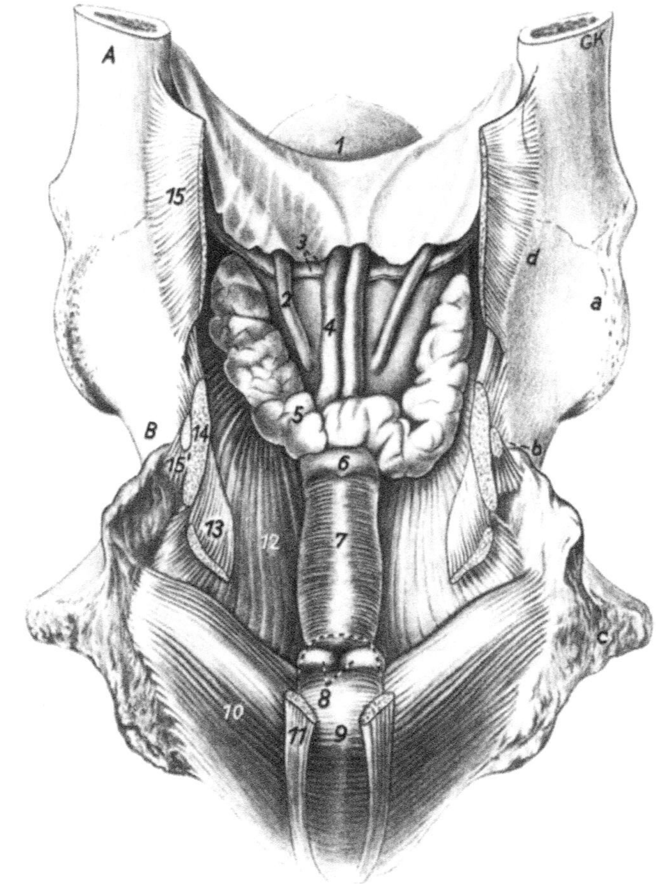

Fig. 483. Urogenital organs associated with the pelvis of the bull. Dorsal aspect.

A Ilium, the wing has been removed; *B* Ischium; *a* Acetabulum; *b* Lesser ischiatic notch; *c* Tuber ischiadicum; *d* Ischiatic spine

1 Urinary bladder, moderately full; *2* Ureter; *3* Dorsal and ventral layers of genital fold, caudal part removed; *4* Ampulla of ductus deferens; *5* Vesicular gland; *6* Body of prostate gland; *7* Pelvic part of urethra, enclosed by urethralis muscle; *8* Bulbourethral glands, exposed by removal of cranial portion of bulbospongiosus (dotted line); *9* Bulb of penis and bulbospongiosus; *10* Ischiocavernosus; *11* Retractor penis; *12* Obturator internus; *13* Levator ani; *14* Coccygeus; *15* Sacroischiatic ligament; *15'* Sacrotuberous part of sacroischiatic ligament

The **BULBOURETHRAL GLAND** (483/*8*) consists of right and left club-shaped, independent lobes which in the bull are about 2.8 cm. long craniocaudally and 1.8 cm. thick dorsoventrally. They lie on the dorsal surface of the urethra opposite the ischiatic arch and are covered by the proximal part of the thick **bulbospongiosus muscle** (490/*8*), so that under normal circumstances the gland cannot be felt rectally. Each lobe has an **excretory duct** that opens on the dorsal wall of the urethra under cover of a caudoventrally-directed fold of mucous membrane (Ellenberger and Baum 1943). This fold would prevent a catheter from entering the pelvic part of the urethra, and is likely the reason why it is not possible to catheterize a bull even after straightening the sigmoid flexure. The bulbourethral glands of the small ruminants (484/*3*) are about 1 cm. in diameter, but otherwise similar to those of the bull. (McMillan and Hafs [1969], studying the reproductive development in Holstein bulls, have published measurements of testis, epididymis, ductus deferens, vesicular gland, and penis of 65 one to twelve-month-old bulls.)

Penis

The penis of the ruminants belongs to the fibroelastic type. It is firm and nearly cylindrical, and when extended is not unlike a round, elastic stick with a pointed end. The penis of the bull when erect is 90—100 cm. long; that of the small ruminants is about 30—50 cm. long. In the quiescent state, the penis forms a **sigmoid flexure,** the proximal curve of which is open caudally and the distal curve is open cranially (485). The sigmoid flexure is between the thighs and related laterally to the spermatic cords.

The **tunica albuginea** of the bull's penis (460/*b*) is thick for the relatively small diameter of the penis. It forms a closed tube which is filled with a dense system of **trabeculae.** The trabeculae traverse the lumen of the tube in all transverse directions, forming a central axial column of fibrous tissue, from which the trabeculae seem to radiate toward the tunica albuginea. Between the many trabeculae are small cavernous blood spaces which extend the entire length of the penis and collectively represent the **corpus cavernosum penis.** Engorgement of the corpus cavernosum with blood produces erection, which manifests itself essentially by the straightening of the sigmoid flexure. Since the penis is anchored caudally to the pelvis by the crura and the tendinous suspensory ligaments, straightening of the sigmoid flexure causes the penis to protrude from the prepuce. Although the tunica albuginea and trabeculae consist mainly of fibrous tissue and only small amounts of elastic tissue, a slight absolute increase in length of the penis occurs at erection. Retraction of the penis does not depend on the action of the retractor penis muscle. It is brought about largely by the return of elastic elements in the proximal curve of the flexure which have been tensed during erection.

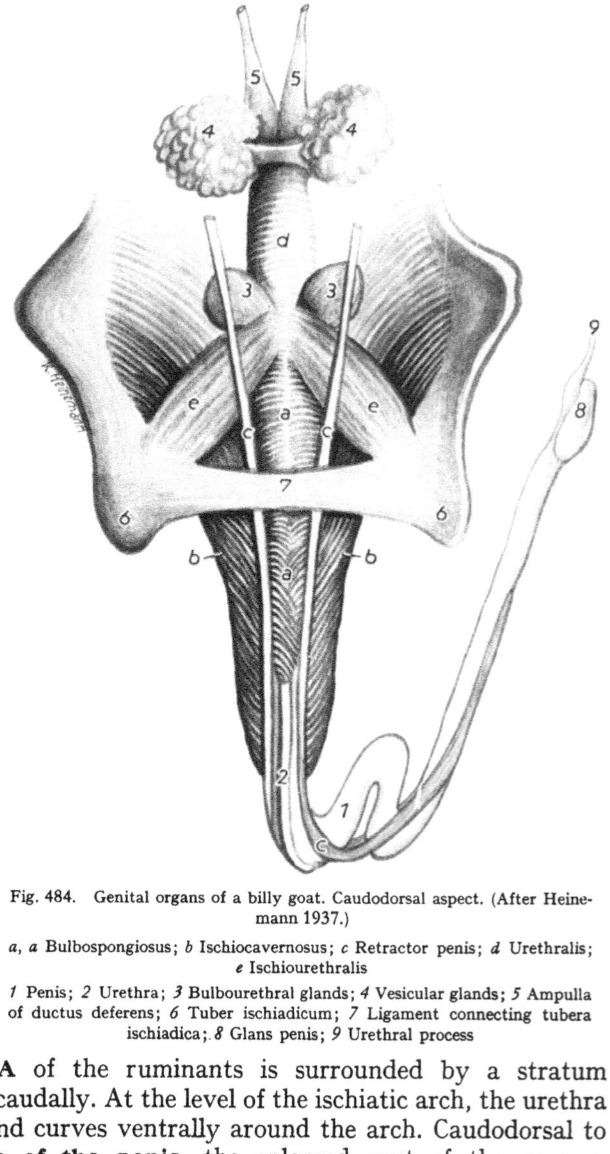

Fig. 484. Genital organs of a billy goat. Caudodorsal aspect. (After Heinemann 1937.)

a, a Bulbospongiosus; *b* Ischiocavernosus; *c* Retractor penis; *d* Urethralis; *e* Ischiourethralis

1 Penis; *2* Urethra; *3* Bulbourethral glands; *4* Vesicular glands; *5* Ampulla of ductus deferens; *6* Tuber ischiadicum; *7* Ligament connecting tubera ischiadica; *8* Glans penis; *9* Urethral process

The pelvic part of the **URETHRA** of the ruminants is surrounded by a stratum cavernosum that decreases in thickness caudally. At the level of the ischiatic arch, the urethra becomes narrower (isthmus urethrae) and curves ventrally around the arch. Caudodorsal to the urethra at this point is the **bulb of the penis,** the enlarged part of the **corpus spongiosum,** which is covered by the **bulbospongiosus** (490/*8*). The bulb extends from the bulbourethral gland to the proximal end of the body of the penis, where it is continued by the tubular part of the corpus spongiosum. This part encloses the urethra and lies with it in the shallow urethral groove. It is very wide proximally, but decreases gradually toward the glans. Its well-developed tunica albuginea (460/*e*) bridges the urethral groove, and with the wall of the groove forms a tube around urethra and corpus spongiosum.

The **FREE PART OF THE PENIS** of the bull is twisted slightly counterclockwise when viewed from behind. This causes the urethral groove and **raphe penis** (486/*1*), both ventral

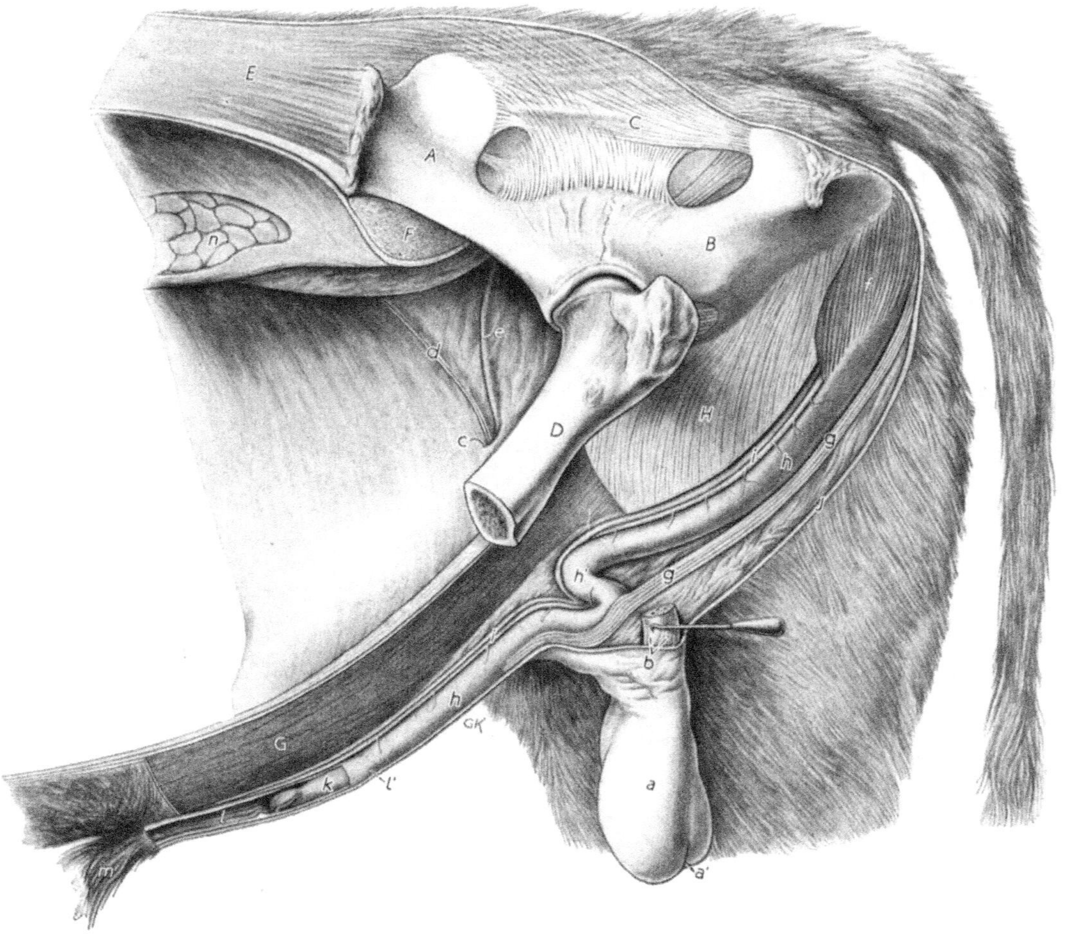

Fig. 485. Genital organs of a bull, fixed in situ. Left lateral aspect.

A Ilium; *B* Ischium; *C* Sacroischiatic ligament; *D* Femur; *E* Epaxial muscles; *F* Hypaxial muscles; *G* Ventral abdominal wall, tunica flava exposed; *H* Symphysial tendon; *J* Skin of perineal region

a Scrotum; *a'* Scrotal raphe; *b* Stump of spermatic cord; *c* Vaginal ring; *d* Testicular vessels; *e* Ductus deferens; *f* Ischiocavernosus; *g* Retractor penis; *h* Penis; *h'* Sigmoid flexure of penis related to spermatic cords; *i* Dorsal artery and vein of penis; *k* Free part of penis, abnormally twisted so that urethral process is visible on left side; *l* Prepuce, fenestrated; *l'* Caudal extent of prepuce; *m* Hair at preputial orifice; *n* Lateral surface of left kidney

structures originally, to lie on the right surface of the penis. (Figure 485 shows the end of the penis twisted abnormally so that the glans is dorsal and the urethral process on the left rather than on the right.) The **glans** (486/*a*) is poorly developed and consists of a thin layer of plexiform veins embedded in a cushion of loose connective tissue. The veins communicate with the corpus spongiosum, but do not form proper erectile tissue. The terminal 2—3 cm. of the urethra form the **urethral process** (*2*) which is visible in a groove on the right side of the tip of the penis. The **external urethral orifice** is a narrow slit on the papilliform end of the process. Because the orifice is small, urination is slow and takes place in squirts rather than in a continuous flow.

Habel (1970), referring to Seidel and Foote (1967) and Ashdown (1968b, 1969b), states that "just before ejaculation in the bull, a variable degree of counterclockwise spiraling of the free part of the penis occurs. This is caused by the fiber pattern of the subcutaneous tissue and tunica albuginea, and by the apical ligament of the tunica albuginea."

The **PREPUCE** of the bull is a narrow but easily distended tube, 25—40 cm. long, of which only the caudal third is occupied by the penis (489). The **preputial orifice** is surrounded by a thick, annular fold (*d*) from which hangs a tuft of long hairs that hides the

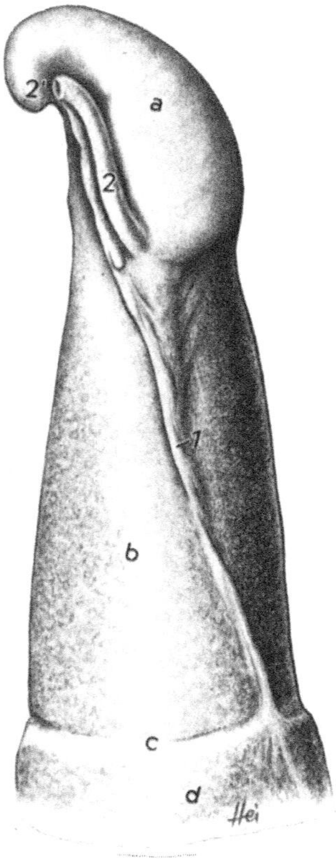

Fig. 486. Distal end of bull's penis, protruded from prepuce. Seen from the right and slightly ventral.

a Glans; *b* Free part; *c* Caudal extent of prepuce in non-erect penis; *d* Int. lamina of prepuce applied to shaft of penis,

1 Raphe penis, more proximally, frenulum of prepuce; *2* Urethral process; *2'* Ext. urethral orifice

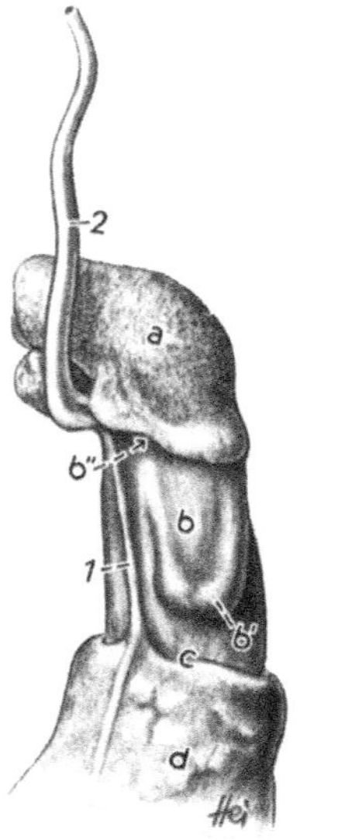

Fig. 487. Distal end of ram's penis. Seen from the left and ventral.

a Glans; *b* Free part; *b'* Tuberculum spongiosum; *b''* Recess between edge of glans and free part; *c* Caudal extent of prepuce in non-erect penis; *d* Int. lamina of prepuce applied to shaft of penis

1 Raphe penis, more proximally, frenulum of prepuce; *2* Urethral process

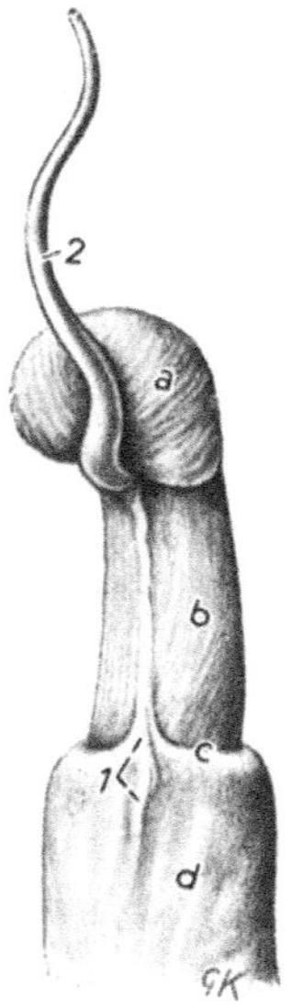

Fig. 488. Distal end of goat's penis. Ventral aspect.

a Glans; *b* Free part; *c* Caudal extent of prepuce in non-erect penis; *d* Int. lamina of prepuce applied to shaft of penis

1 Frenulum of prepuce; *2* Urethral process

orifice. Some of the hairs originate from the internal lamina and pass through the orifice. The mucous membrane lining the prepuce is covered with stratified squamous epithelium and contains lymph nodules. It forms longitudinal and, in places, transverse folds. The cranial two-thirds are dark gray, while the caudal third, which is in contact with the penis, is lighter in color. The **frenulum** of the prepuce is continued by the already mentioned raphe penis (486/*1*) that passes obliquely across the right side of the free end of the penis.

With the exception of the free part and glans, the **PENIS OF THE SMALL RUMINANTS** is similar to that of the bull. The **glans** of the billy goat (488/*a*) is an elongated, round cushion, well set off from the body of the penis. The venous networks it contains shine through its thin covering so that the glans appears red in the fresh state. Peculiar to the small ruminants is the **urethral process** (*2*) which projects beyond the glans by about 2.5 cm. in the goat. At the end of this slightly tapered process is the narrow **external urethral orifice.** The free part of the ram's penis is similar in appearance (487). The rounded **glans** is well set off from the penis, and with its overhanging edge forms a small recess (*b''*) on the left side. Characteristic for the ram is a small, erectile protuberance (tuberculum spongiosum, *b'*) on the left surface of the free part of the penis. It is a unilateral outgrowth of the corpus spongiosum. The **urethral process** of the ram is about 4 cm. long and is slightly twisted proximally, while distally it is straight. The urethral process of the small ruminants is composed of erectile tissue. It stiffens during erection and perhaps is introduced into the cervical canal of the female. The **prepuce** of the small ruminants is short, and its internal lamina contains lymph nodules.

The **MUSCLES OF THE MALE GENITAL ORGANS,** apart from size, are similar in the three ruminant species. The **bulbospongiosus** is thick and is divided by a median connective tissue septum. Covering the bulb of the penis, it extends from the bulbourethral glands to the junction of the crura. In the bull (483/*9*;490/*8*), it is about 17 cm. long, while in the small ruminants (484/*a*) about 9 cm. The small **ischiourethralis** of the bull originates from the medial part of the ischiatic arch and ends on the ventral surface of the urethralis (Habel, 1970). In the goat (484/*e*), it comes from the medial surface of the tuber ischiadicum and passes craniomedially to end opposite the bulbourethral glands on the dorsal surface of the pelvic urethra, blending here with the septum of the bulbospongiosus. The **ischiocavernosus** is a wide, paired, striated muscle (483/*10*, 484/*b*; 485/*f*; 490/*9*). It originates on the medial surface of the tuber ischiadicum, completely covering the crura of the penis, and has a wide insertion on the shaft of the penis. The **retractor penis** (490/*10'*, *10*) is a long, paired, smooth muscle which originates on the ventral surface of the first two caudal vertebrae. Right and left muscles pass the lateral surface of the rectum and, in the bull, receive contributions from the internal anal sphincter. Joining in the midline ventral to the anus, the retractor, enclosed in a fibrous sheath, continues ventrally to the caudal surface of the bulbospongiosus. It can be palpated here through the skin of the perineum, and is (as is the urethra) also accessible surgically. The retractor penis does not follow the sigmoid flexure but runs along the lateral surfaces of the distal curve of the flexure to be inserted farther distally on the tunica albuginea of the free part of the penis.

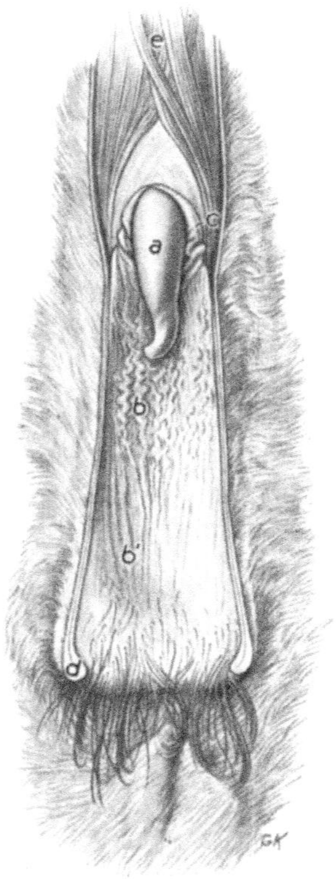

Fig. 489. Prepuce (opened ventrally) and free part of penis of the bull. Ventral aspect.

a Free part of penis; *b* Caudal, lighter zone of int. lamina of prepuce; *b'* Cranial, darker zone of int. lamina of prepuce; *c* Caudal extent of prepuce; *d* Thick annular fold and hair surrounding preputial orifice; *e* Caudal preputial muscles (the cranial preputial muscles have not been exposed)

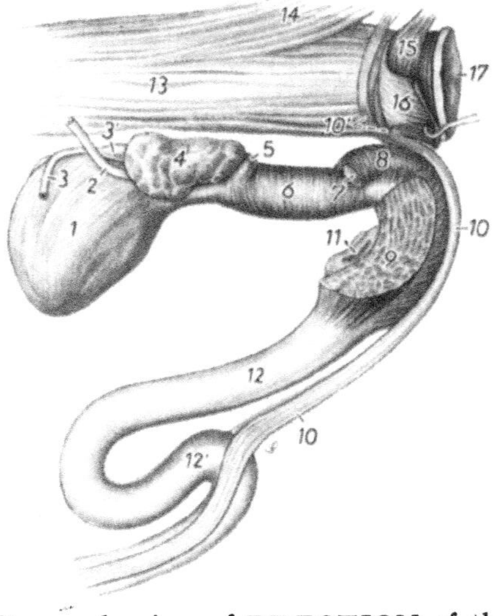

Fig. 490. Genital organs of a bull. Left lateral aspect.

1 Urinary bladder; *2* Left ureter; *3* Left ductus deferens; *3'* Ampulla of left ductus deferens; *4* Vesicular gland; *5* Body of prostate gland; *6* Pelvic part of urethra and urethralis; *7* Bulbourethral gland; *8* Bulbospongiosus; *9* Stump of left ischiocavernosus; *10, 10'* Retractor penis; *11* Left crus penis; *12* Body of penis; *12'* Sigmoid flexure; *13* Rectum; *14* Rectococcygeus; *15* Ext. anal sphincter; *16* Levator ani, reflected; *17* Anus

The mechanism of **ERECTION** of the fibroelastic penis differs from that of the musculocavernous type described on p. 320 by virtue of the corpus cavernosum penis of the former being a closed system (Watson 1964, Ashdown 1973). Blood enters the crura of the corpus cavernosum in the deep arteries of the penis and leaves at the same site in the deep veins. There are normally no veins passing through the tunica albuginea cranial to the crura. During sexual stimulation the ischiocavernosus muscles repeatedly compress the crura and pump blood forward through dorsal and ventrolateral blood channels inside the corpus cavernosum penis. Venous outflow is prevented by the same pumping action. From the venous channels the blood rapidly fills the spaces of the corpus, and the penis becomes erect straightening the sigmoid flexure. Pressures up to 60 times normal systolic blood pressure have been measured in the corpus cavernosum (Beckett et al. 1974). When the pumping of the ischiocavernosus ceases after ejaculation blood once again flows from the deep veins and erection subsides.

Male Genital Organs of the Horse

(15, 416, 452, 457, 461—463, 491—500)

Testis, Spermatic Cord, and Coverings

The **SCROTUM** is in the inguinal region and is placed high between the thighs (494). It is globular in form and set off from the ventral abdominal wall by a slight constriction. It is indented ventrally by a median groove along which runs the **scrotal raphe** (492/*n*).

The skin of the horse's scrotum is thin, mostly pigmented and, due to the secretions of its sebaceous and sweat glands, shiny and oily to the touch. A few fine hairs are present. The **tunica dartos** forms the distinct **scrotal septum** (*f'*) and upon contraction wrinkles the skin and reduces the size of the scrotum. This occurs when the testes are raised toward superficial inguinal ring (in young stallions often through the ring) by the strong **cremaster.** The loose external spermatic fascia between the tunica dartos and the tunica vaginalis parietalis facilitates the movements of the testes in the scrotum.

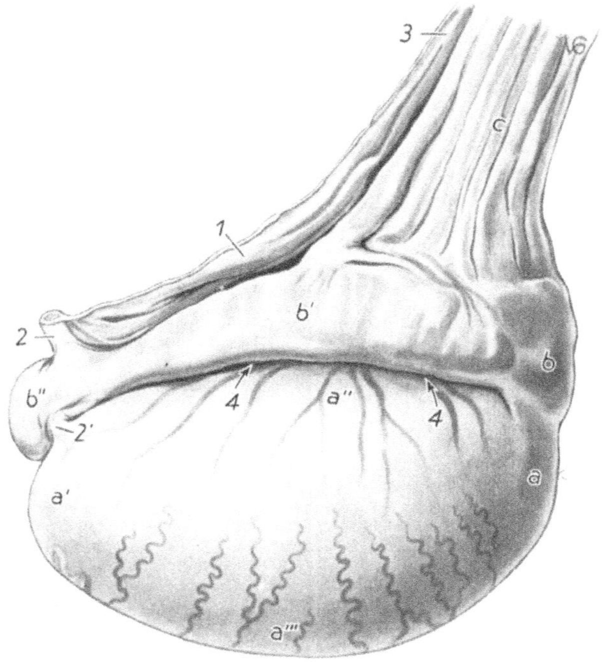

Fig. 491. Right testis and epididymis of a stallion. Lateral aspect.

a Extremitas capitata; *a'* Extremitas caudata; *a''* Epididymal border of testis; *a'''* Free border; *b* Head, *b'* body, *b''* tail of epididymis; *c* Spermatic cord

1 Mesorchium; *2* Ligament of tail of epididymis, cut; *2'* Proper ligament of testis; *3* Mesofuniculum; *4* Entrance to testicular bursa

The **TESTES** of the stallion (491) are oval and weigh, depending on age and breed, between 150—300 gm. each. The **mediastinum testis** is not, as in the other domestic mammals, an axial strand, but is confined to the extremitas capitata where it penetrates the tunica albuginea and is succeeded by usually 14—17 efferent ductules. The testes are oriented with the extremitas capitata cranial and with the long axis nearly parallel to the longitudinal axis of the animal. However, they are raised in front so that the free border points slightly cranioventrally.

Head and body of the **EPIDIDYMIS** are attached to the dorsolateral surface of the testis. On the lateral side, between the body of the epididymis and testis, is the deep **testicular bursa** (491/*4*). The large rounded tail of the epididymis (*b''*) lies on the extremitas caudata and bends around to the medial side of the testis where it is continued by the ductus deferens. The tail of the epididymis is connected to the extremitas caudata by the **proper ligament of the testis** (*2'*) and to the tunica vaginalis parietalis by the **ligament of the tail of the epididymis** (*2*).

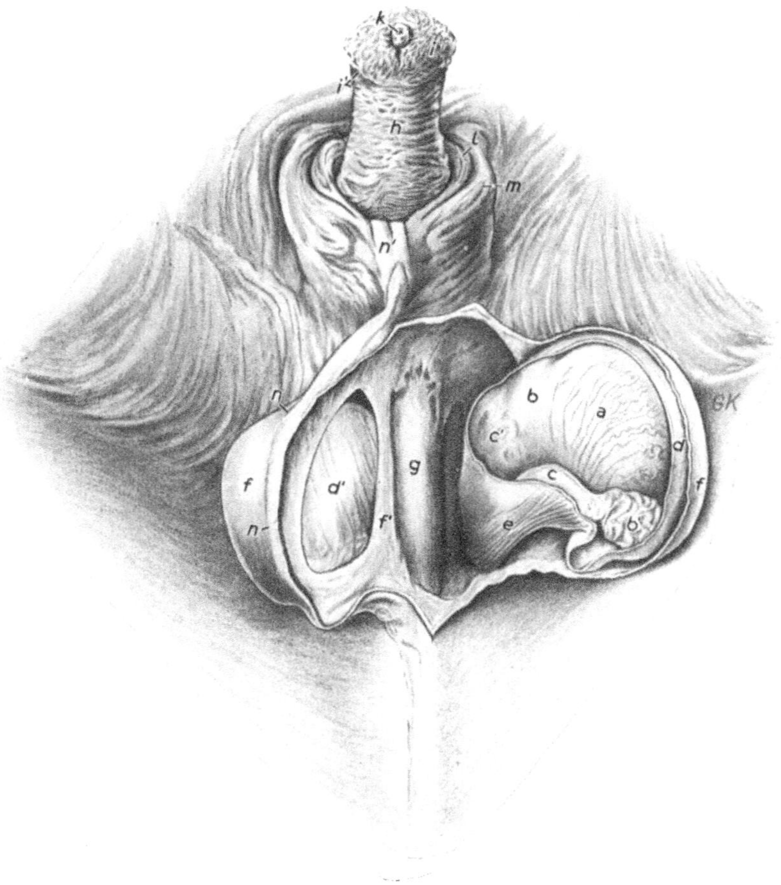

Fig. 492. External genitals of the stallion in situ. Ventral aspect.

a Medial surface of left testis; *b* Head and *b'* tail of left epididymis; *c* Ductus deferens; *c'* Vascular part of spermatic cord; *d* Tunica vaginalis parietalis, fenestrated; *d'* Tunica vaginalis parietalis, intact; *e* Cremaster; *f* Scrotum, split ventrally; *f'* Scrotal septum, fenestrated; *g* Penis; *h* Free part of penis, partly protruded from prepuce; *i, i'* Glans penis; *k* Urethral process; *l* Preputial fold, forming preputial ring; *m* Ext. fold of prepuce, free edge surrounding preputial orifice; *n* Scrotal raphe; *n'* Preputial raphe

The **DUCTUS DEFERENS** (457/*b*), somewhat flexuous at first, passes along the medial side of the epididymal border of the testis and joins the mass of testicular vessels (*c*) to form the **SPERMATIC CORD.** The latter, enclosed by the tunica vaginalis, crosses the lateral surface of the penis, and ascends almost vertically to the superficial inguinal ring (493/*c*), passes through the inguinal canal, and ends at the vaginal ring. The **vaginal ring** of the horse (*d*) is a slit up to 4.5 cm. long. Through it the testicular artery and vein (*e'*) enter the abdominal cavity and take a straight course to the aorta and caudal vena cava in the lumbar area. The ductus deferens (*f*) leaves the vessels and curves toward the pelvis. Before entering the pelvic cavity dorsal to the bladder, it increases in diameter and forms the prominent **ampulla** (494/*i'*) which is 20—25 cm. long and about 2—2.5 cm. in diameter. Right and left ampullae, together with the seminal vesicles and ureters, lie between the layers of the sizable **genital fold** (15/*5*; 495/*3*). The ampullae converge over the neck of the bladder and dip under the seminal vesicles (*5*), taking the **uterus masculinus** (*12*) between them. Beyond the ampullae, the deferent ducts narrow and open with the ducts of the seminal vesicles at the **ejaculatory orifices** (496/*6*) on the colliculus seminalis.

Accessory Genital Glands

The horse has all accessory genital glands (452). They open into the pelvic part of the urethra, which is enclosed by an erectile stratum cavernosum and the urethralis muscle. The latter covers part of the prostate gland.

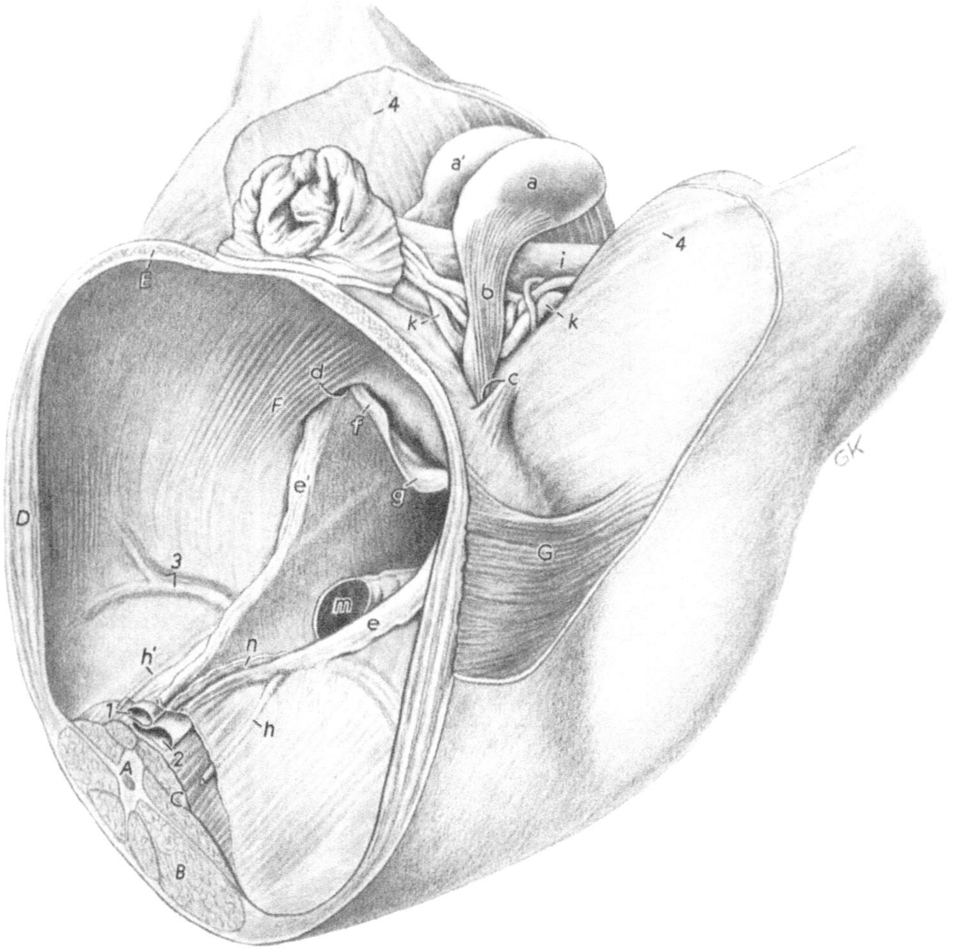

Fig. 493. Male genital organs of the horse in situ; the caudal abdominal and inguinal areas are also exposed.

A 3rd lumbar vertebra; *B* Epaxial muscles; *C* Psoas musculature; *D* Lateral abdominal wall; *E* Ventral abdominal wall with rectus abdominis in cross section; *F* Caudal part of int. abdominal oblique, visible through peritoneum and fasciae; *G* Cutaneus trunci in the fold of the flank

a, a' Right and left testes inside tunica vaginalis; *b* Cremaster on the tunica vaginalis parietalis; *c* Supf. inguinal ring; *d* Vaginal ring; *e, e'* Right and left testicular vessels in mesorchium; *f* Ductus deferens and mesoductus; *g* Genital fold; *h, h'* Right and left ureters; *i* Penis, exposed between thighs; *k* Network of veins alongside penis, mainly from the ext. pudendal vein; *l* Prepuce; *m* Rectum; *n* Stump of descending mesocolon

1 Aorta; *2* Caudal vena cava; *3* Deep circumflex iliac vessels; *4* Medial saphenous vein

The vesicular glands are hollow, pear-shaped sacs (494/*m*) and because of this are known as the **SEMINAL VESICLES** in the horse. The wall of the seminal vesicles is muscular and is lined with a thick mucous membrane. In the stallion, each vesicle is 10—15 cm. long and has a diameter of 3—6 cm. They are partly inside the genital fold and lie lateral to the terminal parts of the ureters and to the ampullae of the deferent ducts. As they converge caudally, each seminal vesicle decreases in diameter and is continued by a short **excretory duct** which dips under the prostate and ends as described above with the ipsilateral ductus deferens on the colliculus seminalis (416).

The **PROSTATE GLAND** is retroperitoneal and consists of two firm, nodular **lobes** connected across the midline by an **isthmus** (495/*6*) that is about 3 cm. long. Each lobe is 5—9 cm. long, 3—6 cm. wide, and about 1 cm. thick. A disseminate part is absent. Numerous ductules carry the prostatic secretion to small slitlike openings (496/*7*) on the side of the colliculus seminalis.

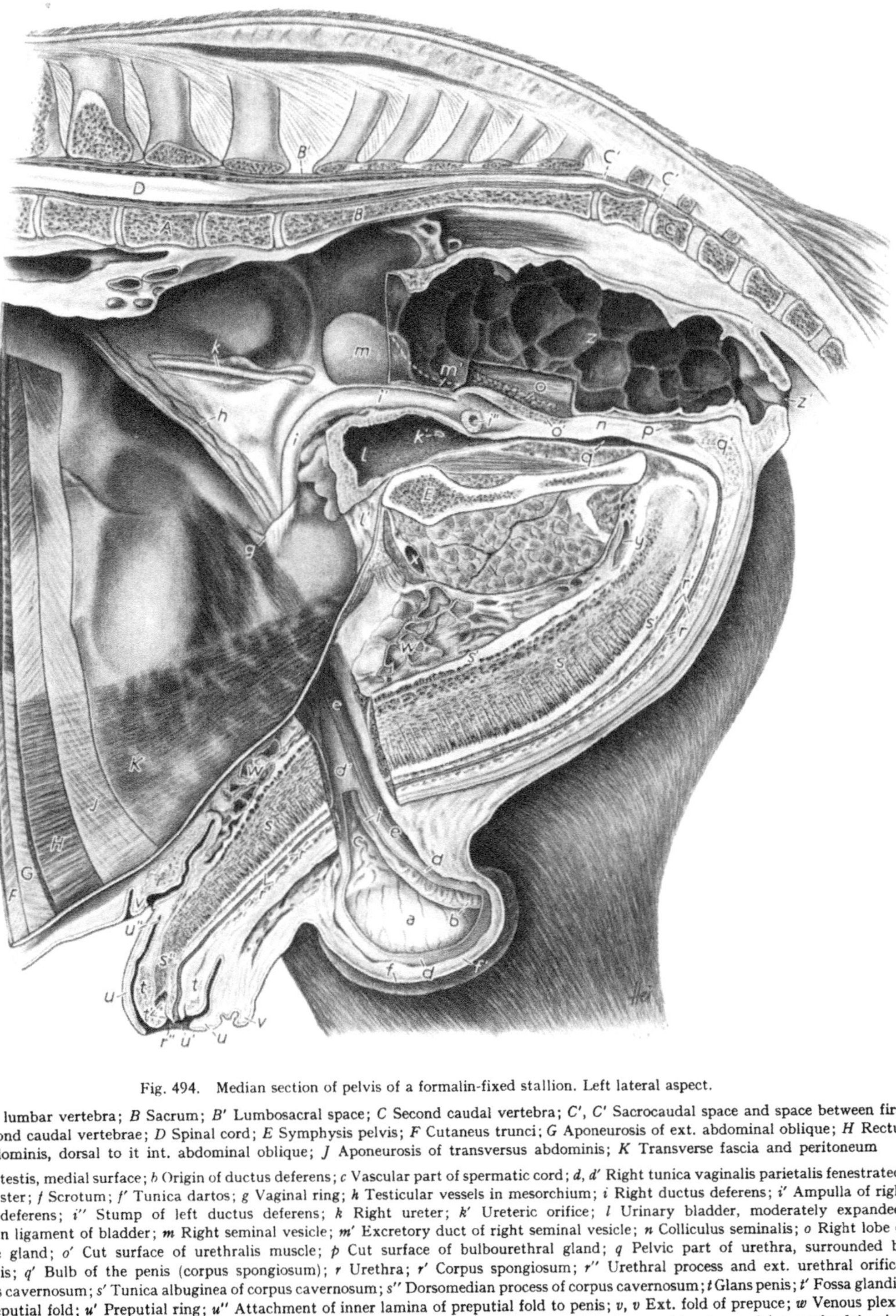

Fig. 494. Median section of pelvis of a formalin-fixed stallion. Left lateral aspect.

A Fifth lumbar vertebra; *B* Sacrum; *B'* Lumbosacral space; *C* Second caudal vertebra; *C'*, *C'* Sacrocaudal space and space between first and second caudal vertebrae; *D* Spinal cord; *E* Symphysis pelvis; *F* Cutaneus trunci; *G* Aponeurosis of ext. abdominal oblique; *H* Rectus abdominis, dorsal to it int. abdominal oblique; *J* Aponeurosis of transversus abdominis; *K* Transverse fascia and peritoneum

a Right testis, medial surface; *b* Origin of ductus deferens; *c* Vascular part of spermatic cord; *d, d'* Right tunica vaginalis parietalis fenestrated; *e* Cremaster; *f* Scrotum; *f'* Tunica dartos; *g* Vaginal ring; *h* Testicular vessels in mesorchium; *i* Right ductus deferens; *i'* Ampulla of right ductus deferens; *i''* Stump of left ductus deferens; *k* Right ureter; *k'* Ureteric orifice; *l* Urinary bladder, moderately expanded; *l'* Median ligament of bladder; *m* Right seminal vesicle; *m'* Excretory duct of right seminal vesicle; *n* Colliculus seminalis; *o* Right lobe of prostate gland; *o'* Cut surface of urethralis muscle; *p* Cut surface of bulbourethral gland; *q* Pelvic part of urethra, surrounded by urethralis; *q'* Bulb of the penis (corpus spongiosum); *r* Urethra; *r'* Corpus spongiosum; *r''* Urethral process and ext. urethral orifice; *s* Corpus cavernosum; *s'* Tunica albuginea of corpus cavernosum; *s''* Dorsomedian process of corpus cavernosum; *t* Glans penis; *t'* Fossa glandis; *u, u* Preputial fold; *u'* Preputial ring; *u''* Attachment of inner lamina of preputial fold to penis; *v, v* Ext. fold of prepuce; *w* Venous plexus dorsal to penis; *x* Anastomosis between right and left ext. pudendal veins; *y* Anastomosis between obturator and int. pudendal veins; *z* Interior of rectum; *z'* Anus

Since the prostate and the seminal vesicles are united by connective tissue, it is difficult to distinguish them on rectal palpation. Only the cranial ends of the seminal vesicles, which are covered with peritoneum and protrude from the cranial edge of the genital fold, can be identified with certainty.

The two **BULBOURETHRAL GLANDS** of the stallion (495/*8*) are club-shaped and are attached to the dorsolateral surface of the urethra at the pelvic outlet. They are about 4—5 cm. long and 2.5 cm. wide. Each gland has several small excretory ducts which open in two longitudinal rows on the dorsal surface of the urethra (496/*8*).

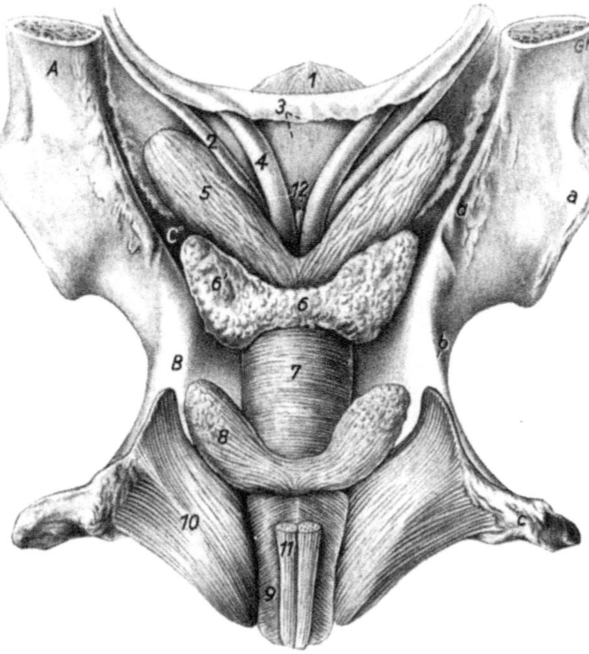

Fig. 495. Male genital organs associated with the pelvis in the stallion. Dorsal aspect.

A Ilium, the wing has been removed; *B* Ischium; *C* Pubis

a Acetabulum; *b* Lesser ischiatic notch; *c* Tuber ischiadicum; *d* Ischiatic spine

1 Urinary bladder; *2* Ureter; *3* Genital fold, caudal part of dorsal layer removed; *4* Ampulla of ductus deferens; *5* Seminal vesicle; *6* Isthmus of prostate gland; *6'* Left lobe of prostate gland; *7* Pelvic part of urethra, enclosed by urethralis; *8* Bulbourethral gland, covered by bulboglandularis; *9* Spongy part of urethra and bulb of penis, covered by bulbospongiosus; *10* Ischiocavernosus; *11* Retractor penis; *12* Uterus masculinus

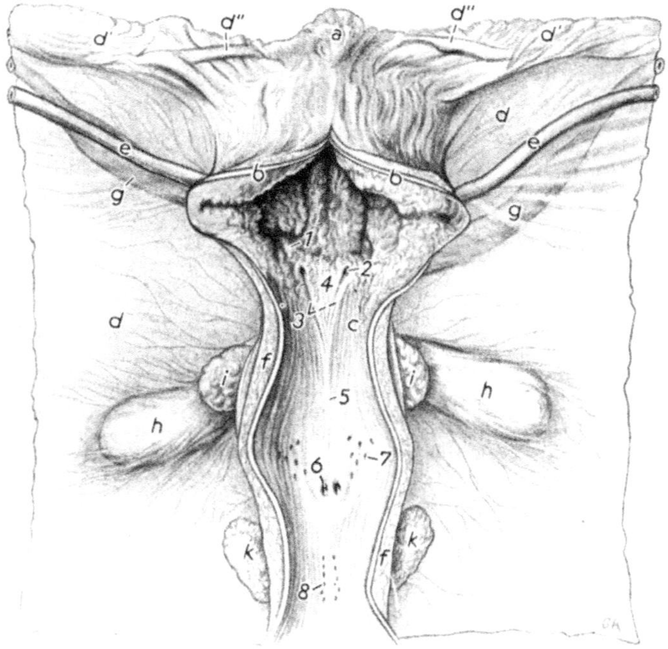

Fig. 496. Urinary bladder, pelvic urethra, and genital fold of a young stallion. Ventral aspect.

a—*c* Urinary bladder, contracted and partly opened ventrally: *a* apex, *b* ventral wall, *c* neck; *d* Genital fold; *d'* Lateral ligament of bladder; *d''* Round ligament of bladder (remnant of umbilical artery); *e* Ureter; *f* Wall of ventrally opened urethra; *g* Ampulla of ductus deferens; *h* Seminal vesicle; *i* Lobe of prostate gland; *k* Bulbourethral gland

1 Ureteric column; *2* Ureteric orifice; *3* Ureteric folds; *4* Trigonum vesicae; *5* Urethral crest; *6* Ejaculatory orifice on the colliculus seminalis (openings of ductus deferens and excretory duct of seminal vesicle); *7* Openings of prostatic ductules on the side of the colliculus seminalis; *8* Openings of the ducts of the bulbourethral gland

Penis

The penis of the horse (494) is an organ of considerable size, which because of the preponderance of cavernous spaces over connective tissue trabeculae belongs to the musculocavernous type. In the quiescent state it is soft and compressible, about 50 cm. long, and has a maximum circumference of 16 cm. The part of the penis becoming visible at erection is about 30—50 cm. long and has a diameter of 5—6 cm. (497). The **crura** (499/*1*, *2*) are thick and unite ventral to the ischiatic arch to form the laterally compressed body of the penis (461), which is attached caudally to the ventral surface of the pelvis by two short but very strong

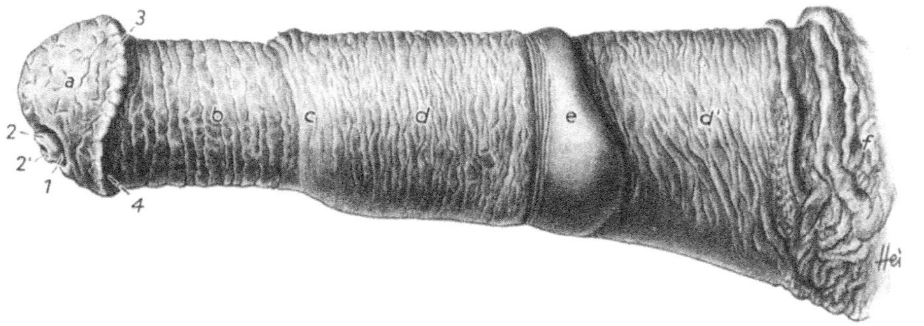

Fig. 497. Distal end of penis of a stallion (protruded from the prepuce). Left lateral aspect.

a Glans penis; *b* Free part of penis; *c* Attachment of inner lamina of preputial fold to penis; *d* Inner lamina of preputial fold; *d'* Outer lamina of preputial fold; *e* Preputial ring; *f* Int. lamina of ext. fold of prepuce

1 Fossa glandis; *2* Urethral process; *2'* Ext. urethral orifice; *3* Corona glandis; *4* Collum glandis

suspensory ligaments. The **CORPUS CAVERNOSUM** ends cranially in a long median and two blunt ventrolateral projections (500/*a''*, *a'''*). The former gives support to the glans, while the latter flank the urethra. Along the ventral surface of the corpus is the groove for the urethra (a^{IV}). The **tunica albuginea** of the corpus cavernosum is very thick and on closer inspection can be seen to consist of three not very distinct layers in which the fibers run in different directions (463). The **trabeculae** traverse the interior of the corpus from right to left, obliquely, and dorsoventrally (461; 462). Those in the median plane represent the incomplete septum penis (463/*b*). Associated with the trabecular network are thick smooth muscle bundles with predominantly longitudinal orientation. They are attached to the trabeculae, blood vessels, and walls of the many large cavernous spaces, and by their tonus keep the unerect penis short and in the prepuce. When the muscle tonus lessens, the penis, although flaccid, protrudes from the prepuce. This occurs fairly regularly at micturition. During erection, the smooth muscle of the corpus cavernosum also relaxes, but blood from the helicine arteries is allowed to fill the cavernous spaces. This engorgement, which is controlled and finally stopped by the unyielding tunica albuginea and trabeculae, lengthens and stiffens the organ.

The spongy part of the **URETHRA** begins with a narrow part (isthmus urethrae) at the ischiatic arch. As it turns ventrally and cranially around the arch, it is related caudodorsally to the club-shaped **bulb of the penis** (494/*q'*). The bulb, the proximal enlarged part of the **CORPUS SPONGIOSUM,** is about 7.5 cm. long and is divided by a median septum. It extends proximally to the bulbourethral glands (*p*) and is continued distally by the thick tubular part of the corpus spongiosum (*r'*) which encloses the urethra (*r*). The latter lies in the urethral groove on

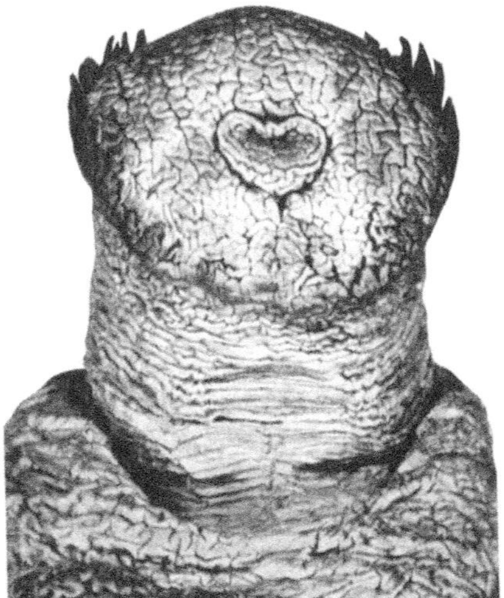

Fig. 498. Glans penis of an older stallion. Note the papillae on the corona, and the urethral process inside the fossa glandis on the cranial surface

the ventral surface of the corpus cavernosum, and ends with a free cylindrical **urethral process** (r'') that is 1.5—3 cm. long and located inside a spacious fossa in the center of the glans (t').

The **GLANS PENIS** of the horse (497/*a*) is well developed and as regards shape and internal structure unlike that of the other domestic mammals. It consists of soft erectile tissuy (corpus spongiosum glandis) covered by a thin elastic skin. The glans caps the free end of the corpus cavernosum (500) and with a long and flat **dorsal process** (c''') extends far caudalle on the dorsal surface. At about the middle of the glans is a constriction, **collum glandis** (c''), which is succeeded cranially by a circular edge, **corona glandis** (c') with conical papillae (498). The cranial surface of the glans is rounded and is idented by the **fossa glandis** (500/*c*), through the center of which passes the urethral process as has been mentioned. The fossa consists of a dorsal recess (sinus urethralis) and two ventrolateral ones. These recesses usually contain a caseous or sometimes granular material known as **smegma.** The glans and the free part of the penis are covered with a nonglandular skin that is often pigmented. The **corpus spongiosum glandis** consists of wide-meshed venous plexuses, which are directly connected to the corpus spongiosum penis. Both stroma and skin of the glans are highly elastic, allowing it to enlarge two or three times its initial size. This usually happens toward the end of coitus, and the glans then has a diameter of 13—16 cm. at the corona.

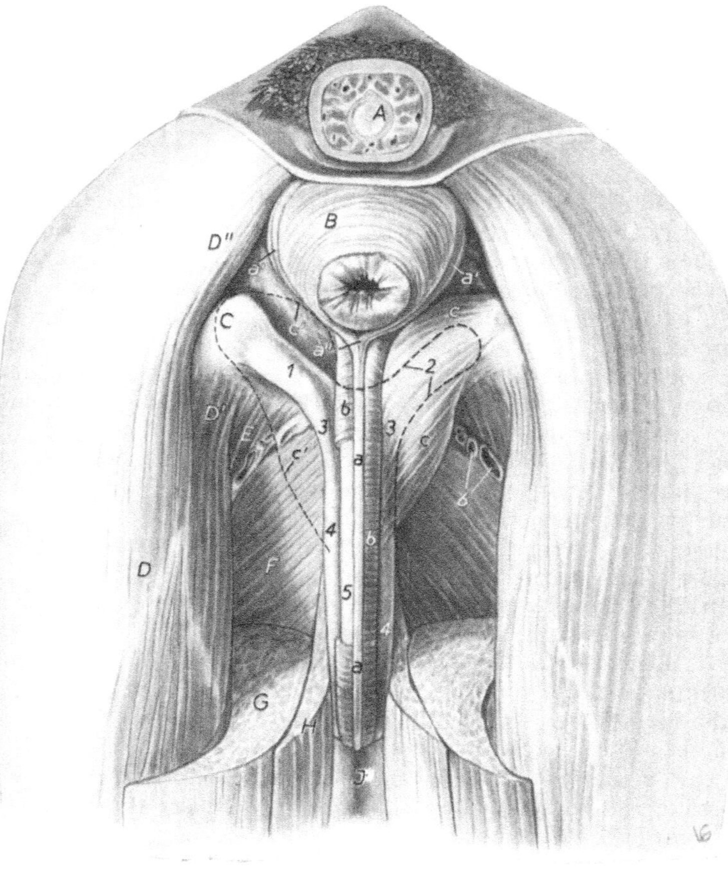

Fig. 499. Perineum of stallion, deep dissection. Caudal aspect.

A Cross section through root of tail; *B* Ext. anal sphincter; *C* Tuber ischiadicum; *D, D', D''* Semitendinosus, *D'* short head from tuber ischiadicum, *D''* vertebral head; *E* Obturator externus; *F* Adductor; *G* Ventral stump of semimembranosus, dorsal part of the muscle has been removed; *H* Gracilis; *J* Caudal wall of scrotum

a Penile part of retractor penis; *a', a''* Rectal part of retractor penis; *b* Bulbospongiosus, partly removed on the left side to expose the urethra; *c* Right ischiocavernosus, covering right crus penis (broken line); *c'* Outline of left ischiocavernosus, which has been removed to expose left crus penis

1 Left crus penis; *2* Outline of right crus penis under cover of ischiocavernosus; *3* Union of crura penis; *4* Corpus cavernosum penis; *5* Urethra, surrounded by corpus spongiosum; *6* Muscular branches of obturator vessels

The **PREPUCE**, or sheath, of the horse differs from the prepuce of the other species in that it is formed by two folds of preputial skin, one inside the other (492). The external fold corresponds to the prepuce of the other animals and consists of external and internal laminae which meet cranially at the **preputial orifice** (*m*). The internal fold (plica preputialis, preputial fold) arises from the internal lamina of the external fold and its inner and outer laminae meet at the **preputial ring** (*l*). (Fig. 494/*u, u, v, v* shows the arrangement of the two folds in section; the penis and preputial fold have slightly protruded from the preputial orifice.) The preputial cavity of the horse is thus divided into internal and external parts, of which the external is the more spacious. Likewise, the preputial orifice is a larger opening than that formed by the preputial ring. The preputial fold is an elastic reserve fold of skin which is straightened during erection and applied to the penis (497/*d, e, d'*).

Except for the inner lamina of the preputial fold the preputial skin resembles in structure the common integument: hairs, and ample sweat and sebaceous glands are present. The secretion of these glands combine swith epithelial debris to form the dirty, gray **smegma** which is often found in the prepuce, especially in geldings. The inner lamina of the preputial fold is nonglandular and resembles the skin covering the glans and free part of the penis.

MUSCLES OF THE MALE GENITAL ORGANS. The **bulboglandularis** of the horse represents part of the ischiourethralis. It arises from the ischiatic arch, covers the bulbourethral glands, and unites cranially with the urethralis. The **ischiourethralis** extends from the center of the ventral surface of the ischiatic arch, passes around the arch into the pelvic cavity, and ends at the ventral surface of the urethra, also uniting with the urethralis. The **bulbospongiosus** of the horse (499/*b*) differs from that of the other domestic mammals in that it extends the entire length of the penis (462). It originates at the bulbourethral glands as a direct continuation of the urethralis and, hemisected by a median septum, covers the bulb of the penis, and more distally, the urethra. Its transversely directed fibers arise on the edges of the urethral groove, pass ventral to urethra and corpus spongiosum

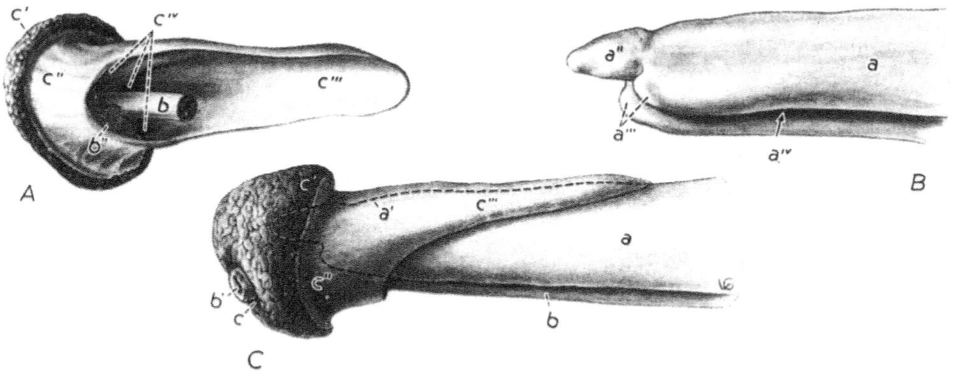

Fig. 500. Distal end of the penis of the horse.

A Caudoventral aspect of the glans, and of the terminal part of the urethra with corpus spongiosum; *B* Ventrolateral aspect of corpus cavernosum; *C* Lateral aspect of tip of penis, the skin of the penis has been removed proximal to the corona glandis

a, a' Corpus cavernosum; *a''* Dorsomedian process of corpus cavernosum; *a'''* Ventrolateral processes of corpus cavernosum; *a*iv Urethral groove; *b* Urethra, surrounded by corpus spongiosum; *b'* Urethral process and ext. urethral orifice; *b''* Stump of bulbospongiosus; *c* Fossa glandis; *c'* Corona glandis; *c''* Collum glandis; *c'''* Dorsal process of glans; *c*iv Recesses on the interior of the glans for the three processes (*a'', a'''*) of the corpus cavernosum

to end on a median raphe (461). Becoming thinner distally, the muscle ends under the skin of the free part of the penis near the glans. The powerful, paired **ischiocavernosus** (495/*10*) has a wide origin on the medial surface of the tuber ischiadicum and sacrotuberous part of the sacroischiatic ligament. It encloses and attaches on the crus penis and ends at about the union of the crura. The smooth **retractor penis** (499/*a, a'*) arises from the ventral surface of the first two caudal vertebrae, and passes ventrally on each side of the rectum. The greater part of the muscle (pars rectalis) passes under the terminal part of the rectum and with its fellow from the other side forms a loop, formerly known as the suspensory ligament of the anus. From the loop, the two narrow bands of the retractor (pars penina, 495/*11*) continue along the midline and lie first on the caudal and then on the ventral surface of the bulbospongiosus. At about the middle of the penis, the retractor passes obliquely through the substance of the bulbospongiosus (462), continues distally, and ends on the glans.

BIBLIOGRAPHY

Male Genital Organs

Almeida, M. de, O. S. Garcia, and J. Biondini: Anatomical study of the terminal parts of the excretory ducts of the vesicula seminalis in *Sus domesticus*. Arq. Esc. Vet. **17,** 1965.

Ammann, R. P., L. Johnson and B. W. Pickett: Connection between the seminiferous tubules and the efferent ducts in the stallion. Am. J. Vet. Res. **38** (1977).

Archibald, J., and E. J. Bishop: Radiographic visualization of the canine prostate gland. J. A. V. M. A. **128,** 1956.

Armingaud, A.: L' os pénien des carnivores domestiques. Rev. Méd. Vét. **84,** 1932.

Aronson, L. R., and M. L. Cooper: Penile spines of the domestic cat: Their endocrine-behavior relations. Anat. Rec. **157,** 1967.

Ashdown, R. R.: Persistence of the penile frenulum in young bulls. Vet. Rec. **74,** 1962.

Ashdown, R. R.: Angioarchitecture of the sigmoid flexure of the bovine corpus cavernosum penis and its significance in erection. Proc. Anat. Soc. Gr. Brit., Nov. 1969 a.

Ashdown, R. R.: The anatomy of the corpus cavernosum penis of the bull and its relationship to spiral deviation of the penis. J. Anat. **104,** 1969 b.

Ashdown, R. R.: Development of penis and sheath in the bull calf. J. Agr. Sci. **54,** 348—352 (1960).

Ashdown, R. R.: Functional anatomy of the penis in ruminants. Vet. Ann. **14** (1973).

Ashdown, R. R., and M. A. Coombs: Experimental studies on spiral deviation of the bovine penis. Vet. Rec. **82,** 1968 a.

Ashdown, R. R., S. W. Ricketts, and R. C. Wardley: The fibrous architecture of the integumentary covering of the bovine penis. J. Anat. **103,** 1968 b.

Backhouse, K. M., and H. Butler: The gubernaculum testis of the pig (*Sus scropha*). J. Anat. **94,** 1960.

Bascom, K. F., and H. L. Osterud: Quantitative studies of the testicle. II. Pattern and total tubal length in the testicles of certain common mammals. Anat. Rec. **31,** 1925.

Bassett, E. G.: Observations on the retractor clitoridis and retractor penis muscles of mammals, with special reference to the ewe. J. Anat. **95,** 1961.

Bastrop, H.: Form, Gewicht, Maße, Mediastinum testis und die oberflächliche Gefäßzeichnung des Hodens von Schwein und Pferd. Berlin (Humboldt-Univ.) Diss., 1958.

Beckett, S. D., D. F. Walker, R. S. Hudson, T. M. Reynolds and R. I. Vachon: Corpus cavernosum penis pressure and penile muscle activity in the bull during coitus. Am. J. Vet. Res. **35** (1974).

Bergin, W. C., H. T. Gier, G. B. Marion, and J. R. Coffmann: A developmental concept of equine cryptorchidism. Biol. Reprod. **3,** 1970.

Bharadwaj, M. B., and M. L. Calhoun: Mode of formation of the preputial cavity in domestic animals. Am. J. Vet. Res. **22,** 1961.

Bharadwaj, M. B., M. L. Calhoun: Histology of the bulbourethral gland of the domestic animals. Anat. Rec. **142,** 216 (1962).

Biborski, J.: Morphology of ram epididymis with special regard to the terminal part of the Cauda epididymal duct. Acta biol. cracoviensia, Zool. **10,** 195—203 (1967).

Blechschmidt, E.: Wachstumsfaktoren des Descensus testis. Zschr. Anat. Entw. gesch. **118,** 175—182, (1955).

Buchholz, O.: Zum Bau des Peniskörpers beim Pferde und seine funktionelle Bedeutung. Hannover, Diss. med. vet., 1951.

Böhm, A.: Zur Innervation der Glans penis beim Rind. Diss. med. vet. München, 1969.

Budras, K.-D.: Leistenband, Leistenkanal und M. cremaster ext. der Katze. Anat. Anz. **121,** 148—165 (1967).

Budras, K.-D., F. Preuss, W. Traeder, E. Henschel: Der Leistenspalt und die Leistenringe unserer Haussäugetiere in neuer Sicht. Berl. Münch. Tierärztl. Wschr. **85,** 427—431 (1972).

Celiščev, L. I.: Das Präputium des Bullen und Schafbockes. Anatomisch-physiologische Angaben. Veterinarija, Moskva **44,** 79—80 (1968).

Christensen, G. C.: Angioarchitecture of the canine penis and the process of erection. Am. J. Anat. **95,** 1954.

Dorst, J. und H. Sajonski: Zur Quantität der Gewebskomponenten des Schweinehodens während der postnatalen Entwicklung. Mh. Vet. Med. **29** (1974).

Dorst, J. und H. Sajonski: Morphometrische Untersuchungen am Tubulussystem des Schweinehodens während der postnatalen Entwicklung. Mh. Vet. Med. **29** (1974).

Egli, A.: Zur funktionellen Anatomie d. Bläschendrüse (Glandula vesiculosa) d. Rindes. Acta Anat. **28,** 1956.

Ellenberger, W., and H. Baum: Handbuch der vergleichenden Anatomie der Haustiere. 18th ed. Berlin, Springer, 1943.

Esser, P. H.: Über Funktion und Bau des Scrotums. Z. mikrosk. anat. Forsch. **31,** 108—174 (1932).

Fehér, G., and A. Haraszti: Beiträge zur Morphologie und zu den altersbedingten Veränderungen der akzessorischen Geschlechtsdrüsen von Stieren. Acta Vet. Budapest **14,** 1964.

Fitzgerald, T. C.: A study of the deviated penis of the bull. Vet. Med. **58,** 1963.

Ford, L.: Testicular maturation in dogs. Am. J. Vet. Res. **30,** 1969.

Garcia, O. S., M. de Almeida, and J. Biondini: Anatomical study of the terminal parts of the excretory ducts of the vesicula seminalis and the ductus deferens in cattle. Arq. Esc. Vet. **17,** 1965.

Geiger, G.: Die Hodenhüllen des Pferdes, ein ergänzender Beitrag zum Prinzip des Schichtenaufbaues der Skrotalwand. Berl. Münch. Tierärztl. Wschr. **69,** 1956.
Gerber, H.: Zur funktionellen Anatomie der Prostata des Hundes unter Berücksichtigung verschiedener Altersstufen. Schweiz. Arch. Tierheilk. **103,** 1961.
Gerstenberger, F.: Die Analbeutel des Hundes und ihre Beziehungen zum Geschlechtsapparat. Leipzig, Diss. med. vet., 1919.
Ghetie, V.: Präparation und Länge des Ductus epididymidis beim Pferd und Schwein. Anat. Anz. **87,** 1939.
Gordon, N.: Surgical anatomy of the bladder, prostate gland, and urethra in the male dog. J. A. V. M. A. **136,** 1960.
Gordon, N.: The position of the canine prostate gland. Am. J. Vet. Res. **22,** 1961.
Grabowski, K.: Über die Schwellräume der Harnröhre der männlichen Haussäugetiere unter besonderer Berücksichtigung ihres Bulbusstückes. Diss. med. vet. Hannover, 1937.
Grau, H., A. Karpf: Das innere Lymphgefäßsystem des Hodens. Zbl. Vet. Med. A **10,** 553—558 (1963).
Gutzschebauch, A.: Der Hoden der Haussäugetiere und seine Hüllen in biologischer und artdiagnostischer Hinsicht. (Pferd, Rind, Schaf, Ziege, Schwein, Hund, Katze, Kaninchen, Meerschweinchen). Zschr. Anat. Entw. gesch. **115,** 1936.
Habel, R. E.: Guide to the dissection of domestic ruminants. 3rd. ed. Ithaca, New York, Author, 1977.
Harrison, R. G.: The comparative anatomy of the blood-supply of the mammalian testis. Proc. Zool. Soc. London **119,** 1949.
Hart, B. L., and R. L. Kitchell: External morphology of the erect glans penis of the dog. Anat. Rec. **152,** 1965.
Hartig, F.: Ein Beitrag zur Anatomie der Hodenhüllen. Zentralbl. Vet. Med. **2,** 1955.
Hartig, F.: Das Stratum perivaginale im Bereich des Scrotum und der Inguinalgegend und seine chirurgische Bedeutung. Zbl. Vet. Med. A **12,** 881—887 (1965).
Heinemann, K.: Einige Muskeln des männlichen Geschlechtsapparates der Haussäugetiere (M. bulbocavernosus, M. ischiocavernosus, M. retractor penis). Hannover, Diss. med. vet., 1937.
Heinze, W., W. Lange: Beitrag zum artefiziellen Penisprolaps unter besonderer Berücksichtigung der anatomischen Verhältnisse beim Bullen. Mh. Vet. Med. **20,** 402—412 (1965).
Hemeida, N. A., W. O. Sack and K. McEntee: Ductuli efferentes of boar, goat, ram, bull and stallion. Am. J. Vet. Res. **39,** 1978.
Henning, Christa: Zur Kenntnis des M. retractor ani et penis s. clitoridis et constrictor recti (M. retractor cloacae) beim Hund. Diss. med. vet. Berlin, 1964.
Holý, L.: The relation of the size of the cauda epididymis to the production and quality of the ejaculate in bulls. Acta. vet. Brno **40,** 405—413 (1971).
Holý, L., F. Barba: Entwicklung und Größe der Hoden von Bullen der Rassen Schweizer Braunvieh und Holstein unter subtropischem Klima. Rev. Cubana Cienc. Vet. La Habana **3,** 31—43 (1972).
Ippensen, E., Ch. Klug-Simon, E. Klug: Der Verlauf der Blutgefäße vom Hoden des Pferdes im Hinblick auf eine Biopsiemöglichkeit. Zuchthygiene, Berlin **7,** 35—45 (1972).
Jackson, C. M.: On the structure of the corpora cavernosa in the domestic cat. Am. J. Anat. **2,** 1902.
Johnson, A. D., W. R. Gomes, and N. L. Vandemark, editors: The Testis. Vol. I: Development, Anatomy, and Physiology. London, Academic Press, 1970.
Kainer, R. A., L. C. Faulkner and M. Abdel-Raouf: Glands associated with the urethra of the bull. Am. J. Vet. Res. **30,** 1969.
Kirchner, A.: Zur Struktur der männl. Genitalorgane von Pferd und Rind. Z. Säugetierk. **4,** 90—121 (1929).
Lanz, T. v.: Über die Biologie des Säugetiernebenhodens. Klin. Wschr. **6,** 1927.
Larson, L. L., and R. L. Kitchell: Neural mechanisms in sexual behavior. II. Gross neuroanatomical and correlative neurophysiological studies of the external genitalia of the bull and the ram. Am. J. Vet. Res. **19,** 1958.
Long, S. E.: Comparison of some anatomical features of bulls which evert preputial epithelium and those which do not. Intern. Kongr. tier. Fortpflanz. u. Haustierbes., München, Kongr. Ber. **7,** 757—760 (1972).
MacMillan, K. L., and H. D. Hafs: Reproductive tract of Holstein bulls from birth through puberty. J. Animal Sci. **28,** 1969.
Magilton, J. H., R. Getty: Blood supply to the genitalia and accessory genital organs of the goat. Iowa State J. Sci. **43,** 285—305 (1969).
Marschner, F.: Größe und Wachstum der Hoden beim Ziegenbock. Leipzig, Diss., 1923.
McKenzie, F. F., J. C. Miller and L. C. Bauguess: The reproductive organs and semen of the boar. Missouri Agr. Exp. Sta. Bull. No. 279, 1938.
Metzdorff, H.: Untersuchungen an Hoden v. Wild- u. Hausschweinen. Zschr. Anat. Entw. gesch. **110,** 1940.
Meyen, I.: Neue Untersuchungen z. Funktion d. Präputialbeutels des Schweines. Zentralbl. Vet. Med. **5,** 1958.
Meyer, P.: Palpatorische Befunde zum Descensus testis beim Deutsch Kurzhaar. Dtsch. Tierärztl. Wschr. **79,** 590, 595—597 (1972).
Mobilio, C. and A. Campus: Osservazioni sull 'epididimo dei nostri animali domestici. Arch. Ital. Anat. Embriol. **11,** 1912—1913.
Mollerus, F. W.: Zur funktionellen Anatomie des Eberpenis. Diss. med. vet. Berlin, 1967.
Nickel, R.: Zur Topographie der akzessorischen Geschlechtsdrüsen bei Schwein, Rind und Pferd. Tierärztl. Umsch. **9,** 1954.
Nitschke, T. und F. Preuss: Zur Homologie der Mm. rectococcygeus und coccygeorectalis. Anat. Anz. **130,** 1972.
Oehmke, P.: Anatomisch-physiologische Untersuchungen über den Nabelbeutel des Schweines. Arch. wiss. Tierheilk. **23,** 146—191 (1897).

Okólski, A.: Über Bau und Funktion des Penis beim Schafbock. Med. Weteryn. **26,** 48—50 (1970).
Osman, A. M., K. Zaki: Die Wachstumsrate der Fortpflanzungsorgane schwarzbunter Bullen. Dtsch. Tierärztl. Wschr. **72,** 34—38 (1965).
Osman, A. M., K. Zaki: Clinical and anatomical studies on the scrotum and its contents in Buffaloes. Fortpflanz., Besam., Aufzucht Haustiere **7,** 57—81 (1971).
Perk, K.: Über den Bau und das Sekret der Glandula bulbo-urethralis (Cowperi) von Rind und Katze. Diss. med. vet. Bern, 1957.
Podaný, J.: Vergleichende testimetrische Untersuchungen bei Bullen verschiedener Rassen. Vet. Med. **14** (42), 561—574 (1969).
Podaný, J., V. Kral: Testimetrische Untersuchungen bei Ebern verschiedener Rassen. Vet. Med. **14** (42), 511—521 (1969).
Podaný, J., P. Sztwiertna: Testikuläre Biometrie bei Schafböcken. Veterínarni Med. **14** (42), 505—510 (1969).
Preuss, F.: Die Tunica albuginea penis und ihre Trabekel bei Pferd und Rind. Anat. Anz. **101,** 1954.
Preuss, F.: Seröse Hodenhüllen und Leistenkanal. Berl. Münch. Tierärztl. Wschr. **70,** 1957.
Preuss, F., K. D. Budras: Die Mm. supramammarius und praeputialis der Katze. Anat. Anz. **122,** 315—323 (1968).
Redenz, E.: Versuch einer biologischen Morphologie des Nebenhodens. Arch. mikr. Anat. Entw. gesch. **103,** 1924; ref. Anat. Ber. **5,** 1926.
Redlich, G.: Das Corpus penis des Katers und seine Erektionsveränderung, eine funktionell-anatomische Studie. Gegenbaurs Morph. Jb. **104,** 1963.
Roth, E., D. Smid: Untersuchungen zur Keimdrüsenentwicklung bei männlichen Veredelten Landschweinen. Züchtungsk. **42,** 144—160 (1970).
Schenk, A.: Topographie, makroskopische Anatomie, Maße, Gewichte, Gefäße und Nerven der Prostata des Hundes. Berlin (Humboldt-Univ.) Diss. med. vet., 1960.
Schenker, J.: Zur funktionellen Anatomie der Prostata des Rindes. Acta Anat. **9,** 1949.
Schlumperger, O.-R. v.: Der Nebenhoden und seine Lage zum Hoden bei Rind, Schaf und Ziege. Hannover, Diss. med. vet., 1954.
Schwarze, E.: III. Beitrag zur „Anatomie für den Tierarzt". Hoden und Nebenhoden. Dtsch. Tierärztl. Wschr. **47,** 1939.
Seidel, G. E., and R. H. Foote: Motion picture analysis of bovine ejaculation. J. Dairy Sc. **50,** 1967.
Seiferle, E.: Über die Leistengegend der Haussäugetiere. Schweiz. Arch. Tierheilk. 1933.
Setchell, B. P.: The mammalian testis. Ithaca, New York, Cornell University Press, 1978.
Shioda, T., K. Mochizuki, S. Nishida: Nerve terminations in the Vas deferens of large domestic animals. Jap. J. Vet. Sci. **30,** 323—330 (1968).
Sieg, E.: Hodenmessungen an lebenden Schafböcken. Diss. med. vet. Hannover, 1966.
Slijper, E. J.: Vergleichend anatomische Untersuchungen über den Penis der Säugetiere. Acta Neerl. Morph. norm. path. **1,** 1938.
Starflinger, F.: Zum Bau des Begattungsorganes beim Ziegenbock mit besonderer Berücksichtigung der Angioarchitektonik. Zbl. Vet. Med. C, **1,** 1972.
Steger, G.: Penisknochen bei einigen Tierarten. Tierärztl. Umsch. **14,** 1959.
Stolla, R., W. Leidl: Quantitative, histologische Untersuchungen des Hodenwachstums bei Bullen nach der Pubertät. Zbl. Vet. Med. A **18,** 563—574 (1971).
Stoss, A. O.: Die Begriffsbestimmung „Samenstrang". Berl. Münch. Tierärztl. Wschr. 1939.
Takahata, K., N. Kudo, K. Furuhata, M. Sugimura, and T. Tamura: Fine angioarchitectures in the penis of the dog. Jap. J. Vet. Res. **10,** 1962.
Thon, H.: Zur Struktur der Hodensackwand des Rindes unter besonderer Berücksichtigung der Tunica dartos. München, Diss. med. vet., 1954.
Tonutti, E., O. Weller, E. Schuchardt und E. Heinke: Die männliche Keimdrüse. Struktur — Funktion — Klinik. Grundzüge der Andrologie. Stuttgart, Thieme, 1960.
Traeder, W.: Zur Anatomie der Leistengegend des Rindes. Diss. med. vet. Berlin, 1968.
Vaerst, L.: Über die Blutversorgung des Hundepenis. Gegenbaurs Morph. Jb. **81,** 1938.
Wagner, R.: Die männlichen Geschlechtsorgane von Felis domestica. Leipzig, Diss. med. vet., 1909.
Watson, J. W.: Mechanism of erection and ejaculation in the bull and ram. Nature **204,** 1964.
Wensing, C. J. G.: Testicular descent in some domestic mammals. I. Anatomical aspect of testicular descent. Proc. Koninkl. Nederl. Akad. Wetensch. **71** C, 1968.
Wensing, C. J. G.: Testicular descent in some domestic mammals. II. The nature of the gubernacular change during the process of testicular descent in the pig. Proc. Koninkl. Nederl. Akad. Wetensch. **76** C (1973).
Wensing, C. J. G.: Testicular descent in some domestic mammals. III. Search for the factors that regulate the gubernacular reaction. Proc. Koninkl. Nederl. Akad. Wetensch. **76** C, (1973).
Widenmayer, H.: Über die Dermadartos bei Rind und Schwein. Diss. med. vet. München, 1958.
Wrobel, K.-H.: Morphologische Untersuchungen an der Glandula bulbourethralis der Katze. Z. Zellforsch. **101,** 607—620 (1969).
Wrobel, K.-H.: Über die Samenleiterampulle der Ziege. Zbl. Vet. Med., A **18,** 250—263 (1971).
Wrobel, K. H.: Zur Morphologie der Ductuli efferentes des Bullen. Zschr. Zellf. **135** (1972): 129—148.
Ziegler, H.: Comparative Morphology of the Prostate. Urol. internation. **3,** 1956.
Zietzschmann, O.: Über die Hodenhüllen im weiteren Sinne, mit Vorschlägen zur Vereinheitlichung der Namen. Arch. wiss. prakt. Tierheilk. **73,** 1938.
Zimmermann, A.: Zur Anatomie der Glans penis des Pferdes. Anat. Anz. **74,** 1932.

Female Genital Organs

General and Comparative

The method of reproduction in viviparous mammalian species—those giving birth to living young—is related to the morphology and internal structure of the female generative organs. These organs not only produce the female germinal cells, the ova, but they provide protection and nourishment for the embryo that develops from the fertilization of the ovum. This "incubation" or gestation period allows the embryo to reach a stage of development at which it is mature enough that it can be expelled by the mother and continue development and growth in the same environment in which its parents live.

Before describing the morphology of the female genital organs, the general pattern of reproduction should be considered. Shortly before puberty, the reproductive organs in both male and female undergo morphological and physiological changes, as a result of which they become functional. Following puberty, the male organs become fully mature and, except for a gradual involution in old age, they do not undergo any further morphological changes. In the female, however, even though sexual maturity is attained at puberty—that is, the female is capable of bearing young—full reproductive capacity comes only after the first few pregnancies. The uterus must actually have carried one or several fetuses to term, and the birth canal must actually have been dilated by passage of the fetus, before the reproductive organs attain their fullest degree of development.

Animals that have not been pregnant are said to be nulliparous. Those that have been pregnant are called parous animals. Uniparous animals are those that usually give birth to only one young per pregnancy; e. g., the ruminants and horse. Multiparous animals give birth to several offspring at a time; e. g., the pig and carnivores.

Puberty in the female signals the beginning of noticeable rhythmic changes, which are controlled by hormones and attended by reversible but also by irreversible modifications of the genital organs. These periodic phenomena constitute the **ESTROUS** or **SEXUAL CYCLE,** in which one can distinguish an **ovarian** and a **uterine cycle.**

Heat (estrus) is the principal external manifestation of the sexual cycle in animals. During heat, the temperament and behavior of the female changes; i. e., she will accept the male, and the genital tract is in optimal condition for conception to take place. In polyestrous animals, such as the pig, ruminants, and horse, estrus occurs several times a year. In diestrous animals, such as the carnivores, heat occurs about twice yearly, with a long anestrous period intervening. In monestrous animals, such as some of the wild-living mammals, heat occurs only once yearly, and they have only one breeding season per year.

The length of gestation, or pregnancy, differs from species to species as does the degree of independence of the young at the time of birth. Among the domesticated mammals, for instance, the ruminants, pig, and horse give birth to young that are relatively independent of the care and protection of their parents except for nourishment. The young of carnivores, on the other hand, are born at a relatively more immature stage (e. g., with closed eyes) and are confined to a nest where they receive intense parental attention for a relatively long period.

General Organization of the Female Genital Organs

The female genital organs can be divided functionally and morphologically into the ovaries, which are located in the abdominal cavity, and the tubular organs, which lead from the ovaries through the pelvis to the outside. The tubular organs consist of the uterine tubes, the uterus, and the copulatory organ. The latter is subdivided into vagina, vestibule, and vulva.

The **OVARIES** (538/*a*) are the reproductive glands. They produce the female germ cells, the ova, and are homologous to the testes of the male. After the ova are released from the ovaries, they enter the **UTERINE TUBES** (*b*) in which they mature and, under favorable circumstances, are fertilized. After fertilization has taken place, the conceptus is carried in the tubes to the **UTERUS** (*c, c'*) which protects and nourishes it for a period of time characteristic of the species. The uterus is highly adapted to changing with the changes that occur in it, and is thus well suited to becoming a temporary "incubator" for the developing embryo. The mucous membrane that lines the uterus is especially important and is capable of considerable structural differentiation. It forms the maternal part of the **placenta** which, with the fetal membranes, becomes the medium of contact and physiologic exchange between the embryo and the mother. Through the placenta, nutrient-rich substances flow to the fetus, and nutrient-poor and waste substances are carried away from it. The **cervix** (*d*), tightly closed during pregnancy, connects the uterus to the **VAGINA** (*e*), which, through the **VESTIBULE** (*f*) and vulva, forms a direct passage from the uterus to the outside. The urethra opens into the vestibule. The two labia of the **VULVA** (*i*) form the caudal end of the female genital tract. The ovaries, uterus, and cervix are attached to the abdominal and pelvic walls by the **broad ligaments** (*1, 2*).

The detailed description of the reproductive organs that follows is based on organs of parous animals—those that have been pregnant. At the end of the chapter (p. 366) the important differences in the organs of juvenile and sexually mature but nulliparous animals will be mentioned.

Ovaries

(519—522, 535—537, 542, 546, 557)

The ovaries of domestic mammals are oval or round, firm in consistency, and often nodular or tuberculate as a result of protruding follicles. In carnivores (535/*a*), the ovaries are oval and their surface is rough and nodular. In a medium-size bitch, they are about 2 cm. long and 1.5 cm. in diameter; in the cat, they are 8—9 mm. long. The ovaries of the pig (542/*1*) have a tuberculate surface, but are more cylindrical; they are about 5 cm. long, and weigh 8—14 gm. each in sows approximately 150 kg. in weight. In the cow (546/*1*), the ovaries are relatively small. Each weighs 15—19 gm., and has an average length of 3 cm., a width of 2 cm., and is a little less than 2 cm. thick. The ovaries of the small ruminants are spherical but slightly flattened, weigh 1—2 gm. each, and are 1.5 cm. long. The mare has the largest ovaries (558/*a*). They are bean-shaped, 5—8 cm. long, 2—4 cm. in diameter, weight 40—80 gm. each, and differ in several respects from those of the other species (see p. 385). These data are only averages because the ovaries of sexually mature animals vary in shape, size, and weight with the stages of the estrous cycle.

The ovaries of domestic mammals are usually located in the sublumbar area, caudal to the kidneys—roughly where they originated in the embryo. In the pig and ruminants, however, a caudal migration takes place, comparable to the descent of the testes in the male, with the result that the ovaries of these species lie at the pelvic inlet.

The cranial part of the **broad ligament,** the large peritoneal sheet suspending the genital tract, attaches to the ovary and is known as the **mesovarium** (538/*1*). In it, blood vessels, lymphatics, and nerves pass to the ovary and enter it at the **hilus.** The border along which the mesovarium attaches to the ovary is the **attached** or **mesovarian border** (margo mesovaricus); opposite is the **free border** (margo liber). The end of the ovary associated with the beginning of the uterine tube is the **tubal end** (extremitas tubaria, 542). The opposite **uterine end** (extremitas uterina) is connected by the **proper ligament of the ovary** (*4*) with the ipsilateral uterine horn. The surfaces of the ovary between the attached and free borders are referred to as **medial** and **lateral surfaces,** although they do not always face exactly in these directions in the live animal.

Remnants of the mesonephric duct and of the cranial excretory tubules of the mesonephros may be found in the mesovarium as small vesicles known as **epoophoron.** Remnants of the more caudally located excretory tubules are known as the **paroophoron.**

STRUCTURE AND FUNCTION OF THE OVARY OF SEXUALLY MATURE ANIMALS. Except in the horse, the ovaries consist of an outer dense **zona parenchymatosa** (501/*a*) and a central, less dense, vascular **zona vasculosa** (*b*). The **epithelium** (502/*a*) that covers the ovary is derived from the lining of the celom and in younger animals consists of a single layer of cuboidal or cylindrical cells that become flattened with age. Deep to the epithelium is the **tunica albuginea** (*b*), a layer of dense connective tissue. The zona paren-

PLATE XIII

Fig. 505 (Day 1)

Fig. 506a (Day 3)

Fig. 506b (Day 3)

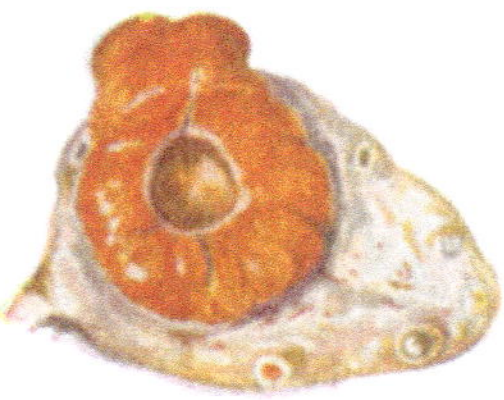

Fig. 507 (Day 7)

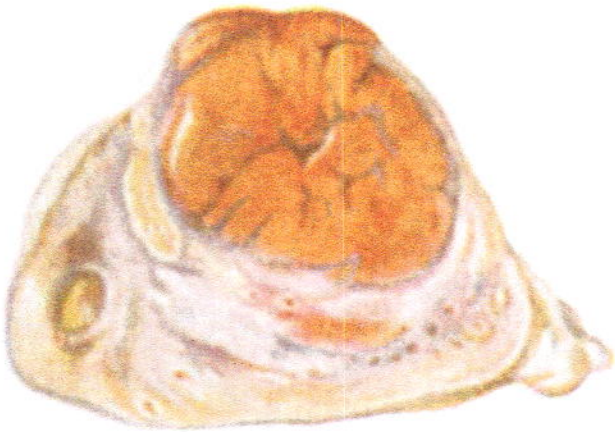

Fig. 508 (Day 9)

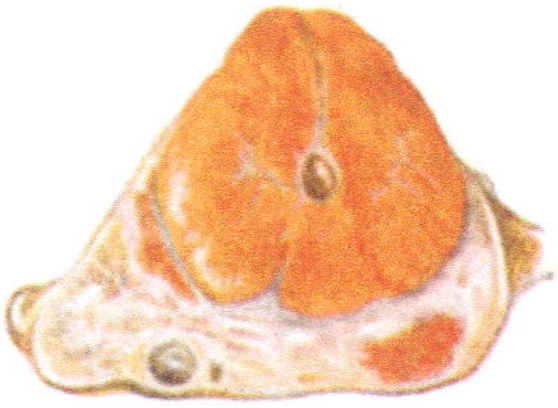

Fig. 509 (Day 11)

Fig. 510a (Day 13)

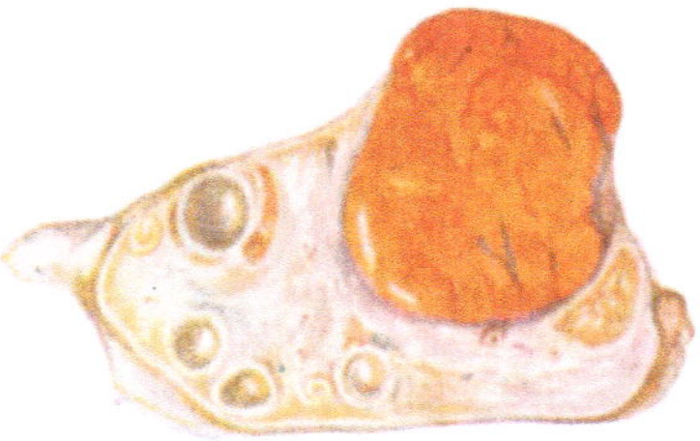

Fig. 510b (Day 13)

Fig. 511a (Day 16)

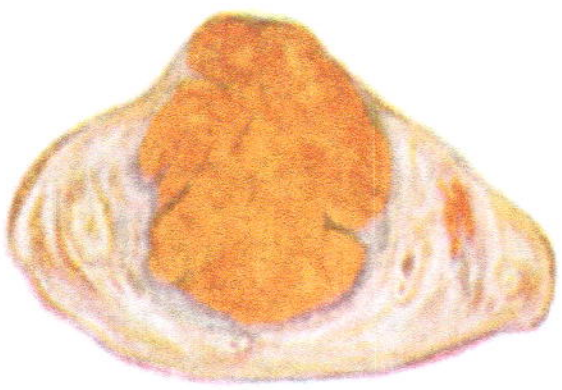

Fig. 511b (Day 16)

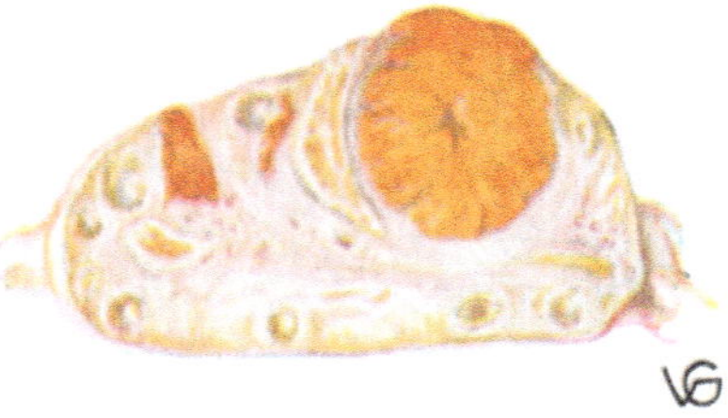

Fig. 512 (Day 19)

Figs. 505—512. Ovaries of the cow in different stages of the ovarian cycle.

505: Follicle about to rupture, day 1; 506a: Freshly ruptured follicle, day 3; 506b: Freshly ruptured follicle in section, day 3; 507: Growing corpus luteum in section, day 7; 508: Mature corpus luteum in section, day 9; 509: Mature corpus luteum and two regressing corpora lutea of previous cycles in section, day 11; 510a: Mature corpus luteum, regressing corpora lutea of previous cycles, and growing follicles, day 13; 510b: Same as 510a, but sectioned; 511a: Regressing corpus luteum, several small and one large growing follicle, day 16; 511b: Same as 511a, but sectioned; 512: Regressing corpus luteum, also older corpora lutea and growing follicles, in section, day 19

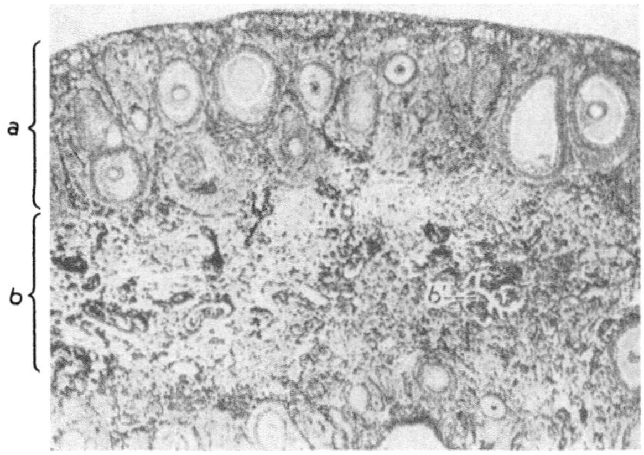

Fig. 501. Section of an ovary of a cat. Microphotograph.

a Zona parenchymatosa containing primary and vesicular follicles, corpora lutea, and interstitial cells; *b* Zona vasculosa with vessels and nerves; *b'* Remnants of the rete ovarii

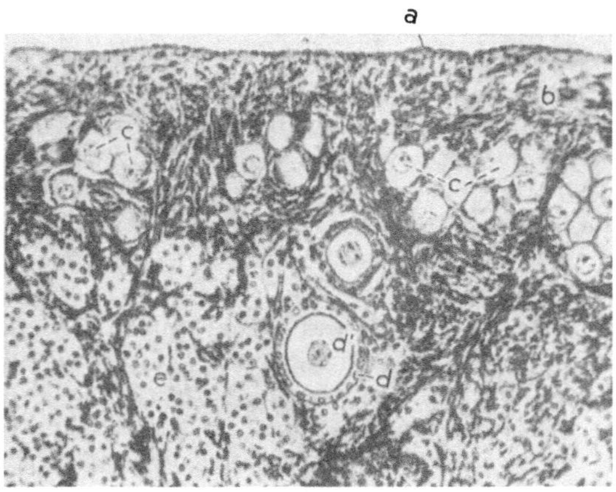

Fig. 502. Zona parenchymatosa of the feline ovary with primary follicles and interstitial cells. Microphotograph.

a Supf. epithelium; *b* Tunica albuginea; *c* Oocytes; *d, d'* Primary follicle, *d* follicular cells, *d'* oocyte; *e* Insterstitial cells

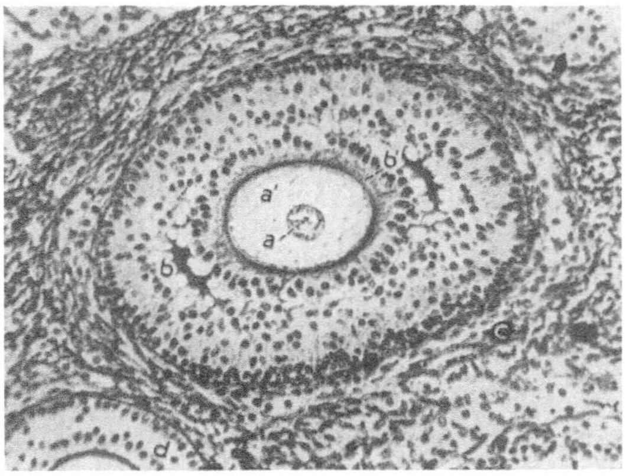

Fig. 503. Ovarian follicle of the cat. Microphotograph.

a Nucleus of oocyte; *a'* Ooplasm; *b* Follicular cells, surrounding oocyte; *b'* Zona pellucida; *c* Theca folliculi; *d* Neighboring follicle

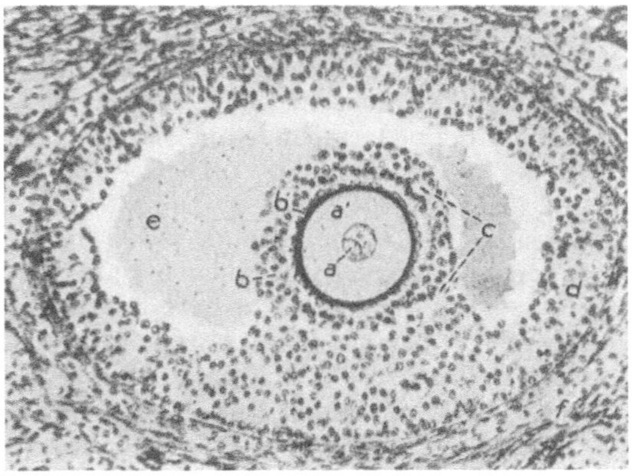

Fig. 504. Vesicular follicle from the ovary of a cat. Microphotograph.

a Nucleus of oocyte; *a'* Ooplasm; *b* Corona radiata; *b'* Zona pellucida; *c* Cumulus oophorus; *d* Stratum granulosum; *e* Follicular cavity; *f* Theca folliculi

chymatosa consists of a **stroma,** in which are embedded the oocytes (*c*). Most of the cells of the stroma are the cortical or stroma fibrocytes. These are pluripotent cells capable of various functions and of further differentiation. They are also thought to be involved in the development of the **interstitial cells** (502/*e*), which are particularly numerous in the ovary of the cat and pig. The interstitial cells have been linked with hormonal activity (progestins) and thus can be compared to the glandular cells of the testis. The **oocytes** (*c*) present in the zona parenchymatosa originate from the oogonia by mitotic division mostly before, but also shortly after birth. It has been estimated that one ovary of the pig contains about 60,000 oocytes, and that the two human ovaries contain about 400,000 oocytes. Most of the oocytes degenerate at this stage. The others grow and become surrounded by a single layer of follicular cells (*d*), forming a **primary follicle** (*d, d'*). Many of the oocytes that go on to form follicles later also regress (follicular atresia).

MATURATION OF THE CVARIAN FOLLICLE. The process of maturation of the primary follicles does not begin until puberty and is governed by the gonadotropic hormones of the adenohypophysis. It is not a continuous process, but occurs at the regular intervals of the ovarian cycle. Maturation begins with growth of the oocyte and with the proliferation of the follicular cells surrounding it. The follicular cells produce a translucent envelope around the oocyte, which, when it later thickens, is known as the **zona pellucida*** (503/*b'*). As these changes take place, the follicle is pushed into the deeper layers of the zona parenchymatosa. With continued growth and particularly with the appearance of a cavity, the primary follicle becomes a **vesicular follicle** (504), which is large enough to be seen with the naked eye. The vesicular follicle is surrounded by a capsule of modified stroma known as the **theca folliculi** (*f*). The theca consists of two layers: the **tunica externa,** made up of concentrically orientated, spindle-shaped stroma cells; and the **tunica interna,** made up of epitheloid cells containing lipoid droplets, and numerous blood capillaries. On the inside of the tunica interna and surrounding the follicular cavity (*e*) are several layers of follicular cells, known as the **stratum granulosum** (*d*). The follicular cavity is filled with fluid containing proteins and estrogenic hormones. Estrogen secretion is responsible for the external signs of estrus and for preparing the uterus to receive the fertilized egg. The expansion of the follicular cavity gradually displaces the oocyte, which is at first centrally located, toward the wall of the vesicle where it remains until ovulation on a small mound of follicular cells, known as the **cumulus oophorus** (*c*). The cells of the cumulus adjacent to the zona pellucida (*b'*) are tall and orientated with their long axes toward the center of the oocyte. Because they thus seem to radiate from the oocyte, they are known as the **corona radiata** (*b*). Through fine pores in the zona pellucida, the cells of the corona radiata send delicate processes into the space between the cell membrane of the oocyte and the zona pellucida, and are thought to be involved in the nutrition of the oocyte. The latter, in the meantime, has grown to a considerable size (120—140 microns in diameter). Its cytoplasm (ooplasm) contains yolk granules, and the sharply defined vesicular nucleus has a distinct nucleolus.

MATURATION OF THE OVUM. While the **primary oocyte** is in the vesicular follicle, and just prior to ovulation, the first of the two maturation divisions takes place.** During this division, the cytoplasm of the primary oocyte divides very unevenly so that a large **secondary oocyte** and a small **polar body** results, both sharing the space inside the zona pellucida of the original cell. The second maturation division occurs after ovulation, when the secondary oocyte is in the uterine tube, and is dependent on the penetration of a spermatozoon that fertilizes the cell. Again, the cytoplasm divides unevenly, resulting in a large mature **ovum** and a small polar body. During the two maturation divisions (meiosis) the chromosomal material is distributed in such a way that the ovum and the polar bodies end up with only half the number of chromosomes typical for the species. Union of the ovum with a spermatozoon at fertilization results in the **zygote,** the first cell of the new organism. The polar bodies are not viable and usually degenerate.

OVULATION is the rupture of the mature ovarian follicle on the surface of the ovary and the attendant discharge of ovum, adhering corona radiata cells, and follicular fluid. It is the most significant event during estrus, but usually does not occur until the external signs of heat have passed their peak and are beginning to subside.

Prior to rupture, the vesicular follicle increases considerably in size and in the cow, for example, reaches a diameter of 1—1.7 cm. As a result of the growth, the follicle pushes toward the periphery of the ovary and becomes visible as a highly distended, translucent vesicle, bulging from the surface (505). Follicle rupture is dependent on a combination of factors: an increase in the internal pressure of the follicle, enzymatic changes at the point of rupture, acute hyperemia at the point of rupture, and contraction of the hilus musculature, which brings about tension of the ovarian tissue and an additional increase in follicular pressure. Following rupture, the external part of the follicle collapses and the follicular cavity becomes filled with clotted blood or serous fluid (506).

After the discharge of the ovum, the granulosa cells and the cells of the theca interna, under the influence of the luteinizing hormone of the adenohypophysis, begin to proliferate

---

* L., transparent zone.

** In the bitch and mare, the first division occurs after ovulation.

exuberantly and initiate the formation of the **corpus luteum***, the name being derived from the yellow pigment associated with the lipids produced by the luteal** cells. Blood vessels and connective tissue grow into the corpus luteum from the periphery. The mature corpus luteum is larger than the follicle it has replaced and attains, in the cow for instance, a length of up to 3 cm. Except in the mare, it projects markedly from the surface of the ovary. The phases in the life span of the corpus luteum are: Proliferation (506, 507); Vascularization (508, 509); Maturity (510); and Regression (511, 512).

The corpus luteum is a temporary endocrine gland secreting hormones known as progestins. The progestins cause changes in the uterine lining that facilitate implantation of a fertilized ovum; at the same time they prevent new ovarian follicles from maturing. If fertilization fails to occur, the phase of the mature corpus luteum—and therefore the period of greatest hormone production—is short, and the corpus luteum regresses. This allows a new set of follicles to grow, and causes regressive changes in the endometrium. Such a short-lived corpus luteum is known as a cyclic corpus luteum. In polyestrous animals, i. e., those with recurring short sexual cycles, the ovaries usually contain several cyclic corpora lutea (from previous ovarian cycles) in various stages of regression. The fibrous remnant of a corpus luteum is a **corpus albicans.**

If, as a result of fertilization, pregnancy ensues, early luteal regression does not take place. The corpus luteum (of pregnancy) persists and becomes an indispensable factor in the maintenance of pregnancy. In some domestic mammals, the placenta begins production of progestins later in pregnancy, thus partially releasing the corpus luteum from this function. In the mare, accessory corpora lutea are formed during pregnancy after the original corpus luteum has regressed. Following parturition, the corpus luteum of pregnancy degenerates, and a new sexual cycle can occur.

Cyclic corpora lutea may persist longer than usual, particularly in nonpregnant cows. This disturbs and prolongs the sexual cycle of the animals and prevents them from coming into heat. Since animals that do not breed regularly become an economic liability, hormone treatment, but also manual removal (enucleation) of the persistent corpus luteum per rectum, are employed with varying degrees of success to bring on heat, so that the animals will breed again.

The **BLOOD SUPPLY OF THE OVARY** is from the ovarian artery, a branch of the aorta. The ovarian vein returns the blood into the caudal vena cava or the ipsilateral renal vein. Branches of the uterine artery may also supply the ovary. The **LYMPHATICS** go to the medial iliac lymph nodes or, depending on species and on the position of the ovaries, to the lumbar aortic nodes. **INNERVATION** reaches the ovaries with the blood vessels via the plexuses of the sympathetic and parasympathetic parts of the autonomic nervous system.

Tubular Genital Organs

The tubular genital organs consist of the uterine tubes, uterus, vagina, vestibule, and vulva. With the exception of the vestibule (which develops from the ventral part of the cloaca), they originate from the paired paramesonephric ducts †. These ducts are derived from mesenchyme, and in the sexually undifferentiated state of development run medial and parallel to the mesonephric ducts†† (which form the tubular genital organs in the male). The paramesonephric ducts extend from the gonads in the lumbar area to the urogenital sinus at the caudal end of the embryo, their caudal segments fusing to a greater or lesser degree. In primitive mammals such as the monotremes and marsupials, this fusion does not take place so that vagina and uterus, which arise from the caudal segments, remain paired (**vagina duplex** and **uterus duplex,** 513). In more highly evolved mammals, fusion of the

* L., yellow body.

** Pertaining to the corpus luteum.

† Formerly Müllerian ducts.

†† Formerly Wolffian ducts.

caudal segments of the genital ducts and differentiation have occurred. The rabbit, for example, has a common vagina and a paired uterus (**vagina simplex** and **uterus duplex,** 514). Further fusion of the genital ducts results in a common cervix and uterus, from which the genital ducts extend like "horns" to the vicinity of the ovaries (**uterus bicornis*,** 515—518), as in the domestic mammals. But even in this group there is considerable species variation in the length of the uterine body. Fusion of the genital ducts except for their most cranial segments from which the uterine tubes arise, results in the **uterus simplex** found in man and the other primates.

Uterine Tube

(519—522, 533—538, 542, 543, 546, 547, 557, 558)

The uterine tube (tuba uterina), or oviduct, is a narrow, muscular tube, which conveys the oocytes released from the ovary to the uterus. The second maturation division of the oocyte

Fig. 513

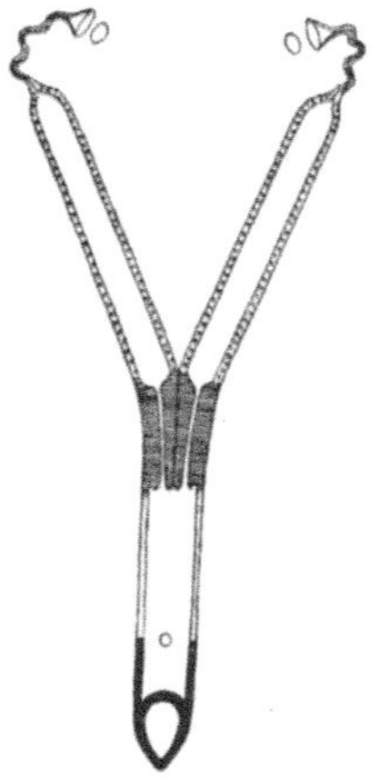

Fig. 514 (Rabbit)

Fig. 515 (Dog)

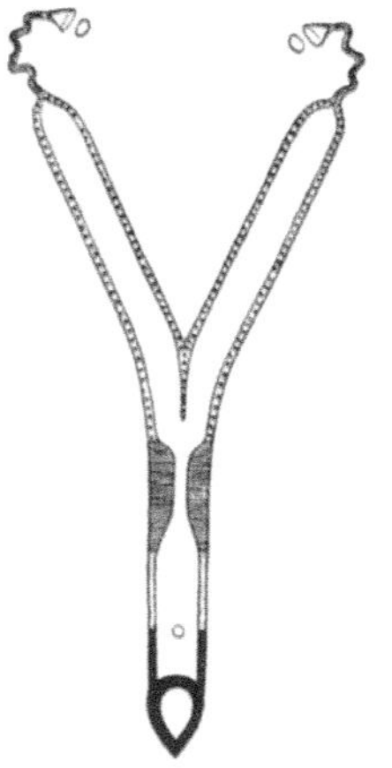

Fig. 516 (Pig)

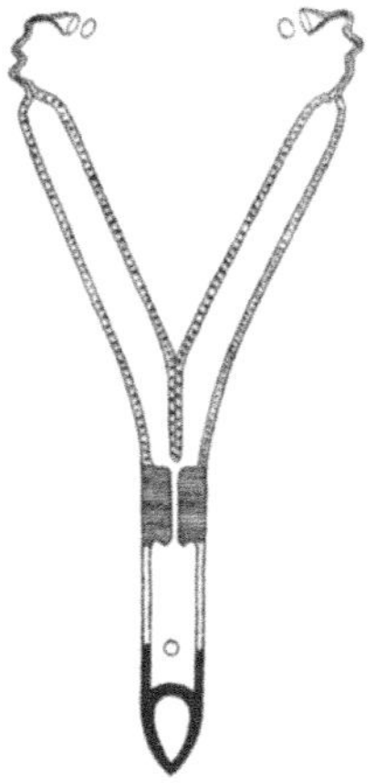

Fig. 517 (Cow)

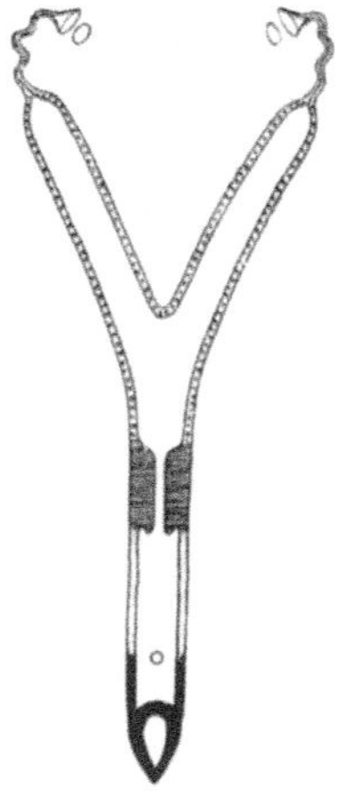

Fig. 518 (Horse)

Figs. 513—518. Comparative anatomy of the uterus in domestic mammals. Schematic. (After Seiferle 1933.)

Fig. 513: Uterus duplex — Vagina duplex (Original pattern of female genital organs, no union of the paramesonephric ducts has occurred.); Fig. 514: Uterus duplex — Vagina simplex (Rabbit); Figs. 515—518: Uterus bicornis (Dog, pig, cow, horse)

Black: Vulva and vestibule, the small circle is the external urethral orifice; *Gray:* Vagina; *Crosshatched:* Cervix; *Dotted:* Body and horns of uterus. At the top of each figure: uterine tube, infundibulum, and ovary

---

* L., having two horns.

takes place, as was shown, in the uterine tube, and since the ovum has a life span of only a few hours during which it can be fertilized, fertilization also takes place here. The spermatozoa are transported into the vicinity by contractions of the uterus and uterine tube and ascend in the tube until they meet an ovum and penetrate it. The fertilized ovum takes several days (cat, pig, and sheep about 3; cow 3—5; dog and horse about 8) to reach the uterine horn and during that time passes through the first few cleavage divisions. If the uterine tube for some reason becomes impassable, the fertilized ovum may continue to develop in the tube and a so-called tubal pregnancy results.

Fig. 519. Section through ovary and ovarian bursa of the bitch. Schematic. (After von Bönninghausen 1936.)

a Mesovarium; *b'* Mesosalpinx; *x* Entrance to ovarian bursa; *y* Ovarian bursa; *1* Ovary; *2'* Infundibulum and abdominal opening of uterine tube

Fig. 520. Left ovary, uterine tube, uterine horn, and associated ligaments of the sow. Medial aspect. Schematic. (After von Bönninghausen 1936.)

a, a', c Broad ligament; *a* Mesovarium; *b* Mesosalpinx; *c* Mesometrium; *x* Entrance to ovarian bursa; *1* Ovary; *2* Uterine tube; *2'* Infundibulum of uterine tube; *3, 3'* Uterine horn; *4* Proper ligament of ovary

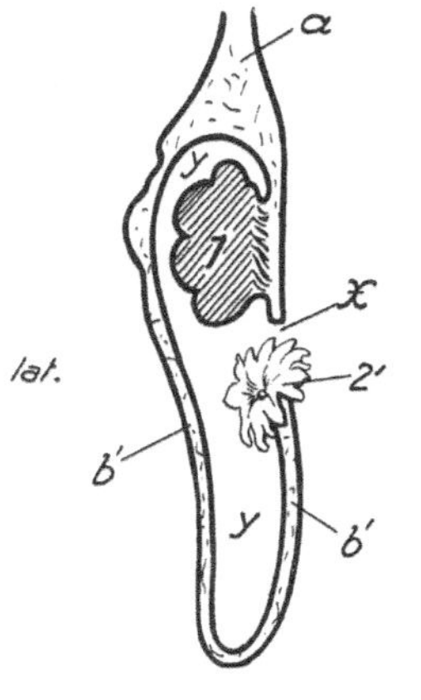

Fig. 519

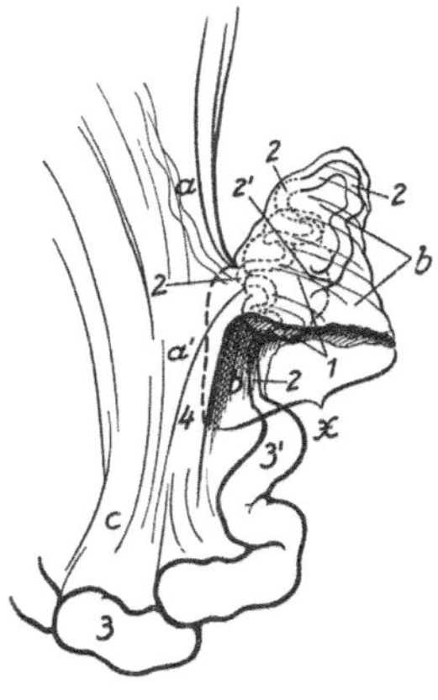

Fig. 520

The ovarian end of the uterine tube is formed by the funnel-shaped **infundibulum** (519/*2'*) near the center of which is the small **abdominal opening** (ostium abdominale) of the tube. This opening leads into the relatively wide, initial segment (ampulla) of the uterine tube. The remainder of the tube (isthmus) is narrow and just wide enough for the ovum to pass through. The uterine tube is very tortuous and consequently much longer than the short distance between ovary and uterine horn would indicate. It ends at the uterine horn with the **uterine opening,** which in dog and horse is on a small papilla (558/*b''*). The length of the uterine tube is 5—9 cm. in the dog, 15—30 cm. in the pig, 15—16 cm. in the small ruminants, 25—28 cm. in the cow, and 25—30 cm. in the horse.

The free margin of the infundibulum is indented to form irregular processes, **fimbriae,** of which some are permanently attached to the ovary (fimbriae ovaricae).

The uterine tube is enclosed in a peritoneal fold, the **mesosalpinx** (521/*b*), which arises from the lateral surface of the mesovarium (*a*) and differs in shape and length with the species. Regardless of these differences, a pouch, the **ovarian bursa** (*x*), is formed between the mesosalpinx laterally and the proper ligament of the ovary (*4*), mesovarium, and ovary (*1*) medially. In the mare, the bursa (522/*x*) is shallow and is too small to contain the large ovary. In the ruminants and pig, the mesosalpinx is very thin and translucent and is large enough to envelope the ovary from the lateral side so that the ovary appears to lie in the depth of the ovarian bursa (520, 521). In the carnivores, the mesosalpinx encloses the laterally flattened, very deep bursa, into which the ovary projects from the medial wall, and the fimbriae of the uterine tube can be seen through the narrow entrance of the bursa (519). The mesosalpinx of the well-nourished bitch contains much fat, which hides the ovary almost completely (533—535). Because of the fat, the mesosalpinx is large and pendulous, hanging from the roof of the abdominal cavity at the cranial and of the uterine horn (539/*8'*, *9'*).

STRUCTURE. The wall of the uterine tube consists of a serous coat, a muscular coat, and a mucous membrane which lines it. The latter is covered with simple columnar epithelium that may be pseudostratified and have cilia in some parts. During estrus, the epithelium assumes secretory character. The tunica propria mucosae, which, especially in the ampulla of the tube, forms high longitudinal folds, shows cyclic hyperemic* changes. The mucous membrane, lining the infundibulum, is arranged in high folds, which converge toward the abdominal opening of the tube. The muscular coat of the uterine tube consists of smooth muscle fibers arranged primarily in circular, but also in longitudinal and oblique directions. It produces peristaltic waves which, with the ciliary activity of the epithelium, propel the ovum toward the uterus. Between the muscular and serous coats are many fine blood vessels.

Although the mechanism by which the released oocyte is guided into the uterine tube is not completely understood, observations on laboratory animals and in man have shown that the infundibulum plays a major role. During estrus, and more particularly when ovulation is about to take place, the infundibulum becomes highly hyperemic and motile, invests the ovary more closely, and by peristaltic contractions draws the oocyte into the abdominal ostium. Oocytes that are not immediately caught up in the infundibulum usually enter the ovarian bursa, but are eventually brought into the uterine tube by the fimbriae at the entrance to the bursa before they can escape into the general peritoneal cavity.

Uterus

(515—518, 523—525, 538—541, 543—545, 547—555, 558, 559)

The uterus of domestic mammals varies both in shape and internal organization with the species. Controlled by hormones, it receives the fertilized ovum (several ova in multipara), facilitates its implantation and the nourishment of the embryo until birth, and at the end of gestation expels the fetus through the birth canal by muscular contraction.

The uterus consists of a cervix, a body (corpus), and of two uterine horns (cornua) which extend to the ovaries and connect with the uterine tubes (518).

The **CERVIX** (558/*d*) is the most caudal part of the uterus and connects it to the vagina. It is a cylindrical structure with a thick, firm wall consisting of smooth muscle and dense fibrous tissue, and functions as the sphincter of the uterus. The lumen of the cervix, the **cervical canal,** is narrow and extends from the **internal uterine orifice** (*d'*) to the **external uterine orifice** (*d''*), connecting the lumen of the uterine body with that of the vagina. The cervical canal varies markedly from species to species in both morphology and length (see below). The caudal portion of the cervix (portio vaginalis), except in the sow, projects for a short distance into the vagina. The musculature of the cervix, the firm mucosal folds and prominences present in some species, as well as the presence of a clear mucoid secretion, render the cervical canal impassable in its normal closed state (549). The canal is open only during estrus, during birth, and for a short pericd thereafter. The extreme dilatation of the normally tightly closed canal at the time of parturition is a complex neuro-hormonally controlled process due only in part to its expansion by the fetus.

The **BODY OF THE UTERUS** (538/*c'*) is a simple muscular tube of varying length cranial to the cervix.

The two **HORNS OF THE UTERUS** (*c, c*), like the body, are muscular tubes that diverge from the cranial end of the uterine body. They are fairly straight in the carnivores and the mare, but coiled in the sow and ruminants (548). The peritoneal fold(s) present in some species between the caudal ends of the horns is known as the **intercornual ligament** (*7, 7'*).

In **CARNIVORES,** the body of the uterus is about 1—3 cm. long in the bitch and about 1.5 cm. long in the cat. The slender uterine horns (3—8 mm. thick) are very long, however, and diverge from the level of the sixth or seventh lumbar vertebra to the vicinity of the kidneys (539). Caudally, the uterine horns are united by peritoneum for a short distance and lie close together on each side of the median plane, though they open separately into the uterine body. The cervix of the carnivores is very short so that external and internal uterine orifices lie close together.

In the **SOW,** the body of the uterus (543/*c'*) is also short and measures only about 5 cm. in length. The uterine horns, however, are extremely long and flexuous, have thick walls, and in

* From Gr., having an excess of blood.

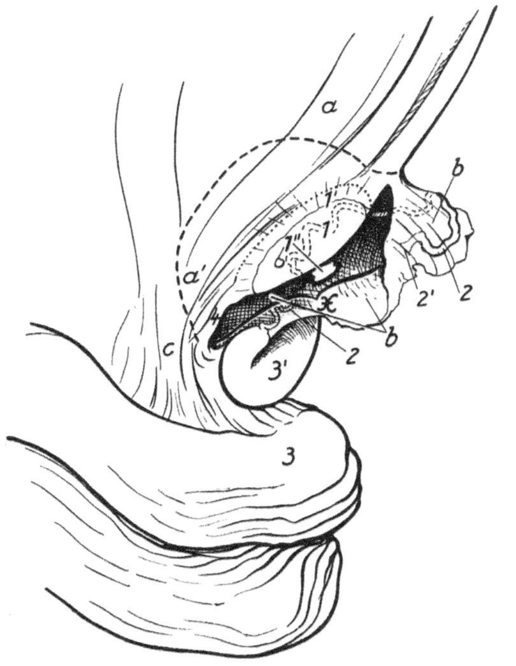

Fig. 521. Left ovary, uterine tube, uterine horn, and associated ligaments of the cow. Medial aspect. Schematic. The uterine horns have been pulled ventrally to expose ovary and bursa. (After von Bönninghausen 1936.)

a, a', c Broad ligament; *a* Mesovarium; *b* Mesosalpinx; *c* Mesometrium; *x* Entrance to ovarian bursa

1 Ovary; *1'* Border between peritoneum and epithelium (margo limitans peritonei); *1''* Corpus luteum; *2* Uterine tube; *2'* Infundibulum and abdominal opening of uterine tube; *3* Left uterine horn; *3'* Tip of left uterine horn; *4* Proper ligament of ovary

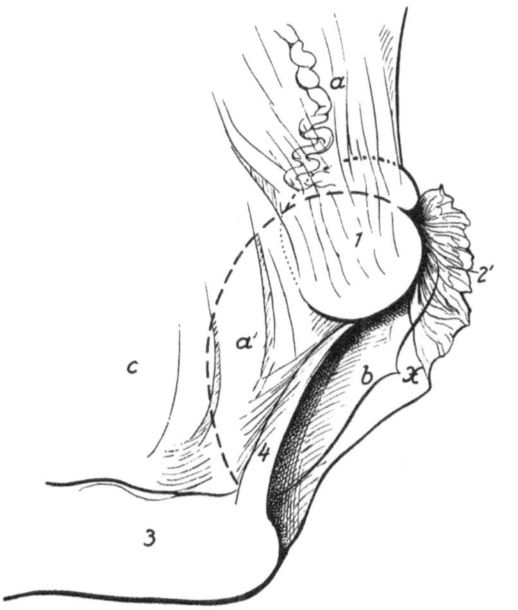

Fig. 522. Left ovary and tip of uterine horn with associated ligaments of the mare. Medial aspect. Schematic. (After von Bönninghausen 1936.)

a, a', c Broad ligament; *a* Mesovarium and ovarian artery running in mesovarium; *b* Mesosalpinx; *c* Mesometrium; *x* Entrance to ovarian bursa

1 Ovary; *2'* Infundibulum of uterine tube with fimbriae; *3* Tip of uterine horn; *4* Proper ligament of ovary

the live animal resemble coils of intestine. The cervix is also long (15—20 cm.) and has a wall up to 8 mm thick. It opens gradually into uterus and vagina without noticeable demarcation. Rows of rounded prominences (pulvini cervicales, *d*) on low longitudinal folds are peculiar to the cervical canal of the sow. When the cervix is closed, these prominences interdigitate and occlude the canal (544).

The uterus of the **RUMINANTS** has a short body and long horns that form two tightly wound spirals. The horns (547 A/*c, c''*) taper and lie parallel to each other as they leave the body of the uterus (*c'*), and then diverge, spiraling away from each other. They curve at first ventrally, then caudally, and finally dorsally. Caudally, the uterine horns are united by peritoneum, connective tissue, and parts of the outer longitudinal muscle layer, and give the impression of a long uterine body (547 B). That both horns, though united, are independent, can be seen when they are transected. With the exception of the peritoneum and the outer longitudinal muscle layer, they have all the layers of separate horns. The horns open independently into the uterine body (*c'*) which is only 1—3 cm. long. The cervix of the cow is about 10 cm. long and has numerous low longitudinal folds internally, and usually four thick transverse folds (plicae circulares, 549/*d*) which have a muscular base. The cervix of the goat (553) has up to eight transverse folds, while that of the ewe (552) in addition to two transverse folds has five to six hard prominences arranged longitudinally.

The uterus of the **MARE** (558) has a wide lumen. Its body is relatively large and about as long as the horns (22—25 cm.). The latter are suspended from the roof of the abdominal cavity by the broad ligaments and hang with a cranioventral convexity (which is more marked when the intestines are removed, 559/*12, 13*) between the body of the uterus and the ovaries. The cervix has a thick wall, is 6—7 cm. long, and has a straight canal with longitudinal mucosal folds on the inner surface. There are no mucosal and muscular elevations such as found inside the ruminant or porcine cervix. The internal uterine orifice is funnel-shaped: the external orifice is in the center of a prominent intravaginal projection (558/*d''*).

POSITION AND ATTACHMENT OF THE UTERUS. Most of the uterus of the domestic mammals is in the abdominal cavity, only the cervix is in the pelvic acvity. The uterus usually rests on the coils of the intestines, but may, because of its loose suspension, intermingle with them. In the pig it may reach the ventral abdominal wall (545). In the mare and cow, the entire uterus and the ovaries can be palpated rectally.

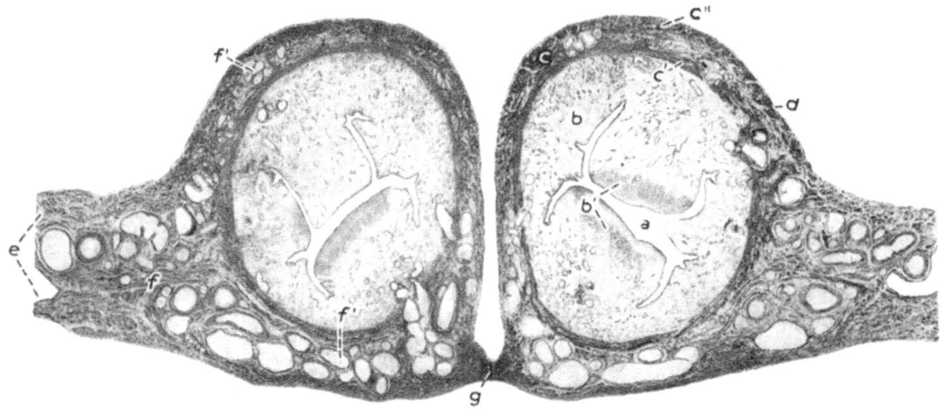

Fig. 523. Section through the uterine horns of the goat. Microphotograph.

a Lumen; *b* Endometrium with uterine glands; *b'* Caruncles; *c* Myometrium: *c'* inner circular layer, *c''* outer longitudinal layer, which is continuous with the smooth muscle in the mesometrium; *d* Serosal coat (perimetrium); *e* Mesometrium; *f* Parametrium, containing vessels which are continuous with the vessels in *f'*, the vascular layer between circular and longitudinal muscle layers; *g* Intercornual ligament

The body and horns of the uterus are suspended by the right and left **broad ligaments** (ligamenta lata uteri, 545/*11*) which are homologous to the mesoductus deferens and genital fold in the male. They arise from the dorsolateral wall of the pelvic cavity (16/*5*) and from the roof of the abdominal cavity and contain substantial amounts of smooth muscle in the form of flat strands or sheets. The part of the broad ligament ending on the horn and body of the uterus is the **mesometrium,** and the border of the horn along which the mesometrium attaches is the **mesometrial border** (margo mesometricus), in contrast to the **free border** (margo liber) on the opposite side of the horn. At the mesometrial border, the two serosal layers of the mesometrium are continuous with the visceral peritoneum of the uterus, which is known as the **perimetrium** (523/*d*). The smooth muscle fibers in the broad ligament are continuous with the longitudinal muscle layer of the uterus (*c''*). The presence of muscle in the mesometrium indicates that the uterus is not passively suspended—as is the gut, by the mesentery—but actively "carried" by the mesometrium. This may, at the time of parturition for instance, position the uterus in such a way as to facilitate the expulsion of the fetus. Unfortunately, this extensive suspension apparatus may also play a role in occasional displacements and torsions of the uterus. A secondary fold of varying width arises from the lateral surface of the mesometrium. In the free edge of this fold runs the **round ligament of the uterus** (lig. teres uteri, 540/*f*) which begins at the tip of the uterine horn or its general vicinity, and blends with the peritoneum over the deep inguinal ring. In the bitch, the round ligament enters the inguinal canal inside a narrow peritoneal evagination, known as the **vaginal process** (*h, h'*). This evagination is homologous to the tunica vaginalis enclosing the testis and extending into the scrotum in the male; and the round ligament of the uterus is homologous to the ligament of the tail of the epididymis. The caudal parts of the broad ligaments form part of the walls of the peritoneal excavations that extend caudally between the organs of the pelvic cavity (see p. 18).

Blood vessels, lymphatics, and nerves reach the uterus in the mesometrium and enter it along the mesometrial border. The accumulation of these vessels and nerves, and the loose connective tissue in which they are embedded along the mesometrial border, is known as the **parametrium** (523/*f*).

STRUCTURE OF THE UTERINE WALL. This description of the uterine wall is based on the resting, nongravid organ. The uterus is lined with a thick mucous membrane known as the **endometrium** (523/*b*). Covering the endometrium are one or more layers of cylindrical epithelium capable of secretion and said to be ciliated at times. Below the epithelium are numerous branched, tubular **uterine glands,** which become active following estrus and which secrete large amounts of fluid, the uterine milk, during pregnancy (530). The glands are embedded in a very vascular and cellular, reticular-like connective tissue. The contour of the endometrial surface varies in domestic mammals and is described in the special chapters.

The endometrium rests directly on the muscular coat, the **myometrium** (523/*c*), which consists of a thick, inner, circular layer and a thinner, outer, longitudinal layer of smooth muscle. The longitudinal layer is continuous with the smooth muscle in the mesometrium (*e*). Between the two layers of the myometrium is a prominent layer of blood vessels (*f'*) which is continuous with the vessels in the parametrium. External to the myometrium is the serous coat of the uterus, the **perimetrium** (*d*).

The **WALL OF THE CERVIX** consists of a mucous membrane lining the cavity, a thick circular and a thinner longitudinal muscle coat (524/*c*), and a serous coat externally. The **mucous membrane** (*b*) bears a single layer of columnar cells which, like the uterine epithelium, shows cyclic changes manifested by increased or decreased mucus production by its mucigenous cells. Tubular **cervical glands** present in the cat and goat pass through similar rhythmic secretory phases. In the cow, there are deep mucus-containing recesses between the tall cervical folds, and in the pig and sheep the epithelium forms conical invaginations in which mucus is stored. The tunica propria mucosae is rich in cells and collagenous fibers and has an almost tendonlike consistency. Large amounts of this firm "cervical tissue" are also found between the bundles of the well-developed circular muscle layer; and, because of its internal organization and ability to swell, it seems to play an important role in the dilation of the cervical canal. An often extensive vascular layer embedded in the parametrial connective tissue (*e*) is present along the wall of the cervix.

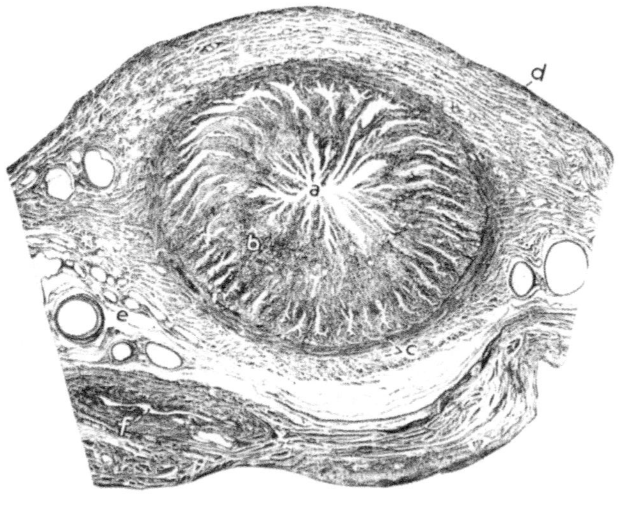

Fig. 524. Section through the cervix uteri of the goat. Microphotograph.
a Cervical canal; *b* Mucous membrane in longitudinal folds and containing cervical glands; *c* Muscular coat; *d* Serous coat; *e* Vascular layer; *f* Urethra

Vagina

(16, 525, 538, 541, 543, 545, 547, 554, 555, 558, 559)

The vagina has evolved from the fusion of the caudal parts of the paramesonephric ducts. Together with vestibule and vulva, it constitutes the copulatory organ of the female, receives the penis during copulation, and in ruminants is the receptacle for the seminal fluid. (In the dog, pig and horse, the semen is usually deposited in the cervical canal or directly into the uterus.) It is also an important segment of the birth canal.

The vagina (525/*e*) is a relatively thin-walled tube extending longitudinally inside the pelvic cavity, and related dorsally to the rectum and ventrally to the urinary bladder and urethra. Most of it is retroperitoneal; only a short cranial portion is covered with peritoneum (17/*b'*). The cranial end of the vagina is occupied by the intravaginal part of the cervix (525/*d'*), around which the vaginal lumen forms an annular recess known as the **fornix vaginae.** The vaginal wall consists of a coat of smooth muscle which is lined with a nonglandular mucous membrane arranged in longitudinal folds and loosely attached to the muscle coat. The mucous membrane is covered with stratified squamous epithelium, which changes in appearance and function with the sexual cycle. During proestrus and estrus, it increases in thickness while its superficial layers cornify. After estrus the cornified cells are shed. Externally, an abundant loose adventitia covers the muscle coat and contains numerous blood vessels, of which the veins may form extensive plexuses.

The junction between vagina and the vestibule (*h*), which connects the vagina with the outside, is at the level of the external urethral orifice (*f'*) in the floor of the genital tract. This would correspond in the embryo to the level at which the paramesonephric ducts joined the urogenital sinus. In women, this junction is marked by a well-developed transverse fold, the hymen, which separates the two parts. In domestic mammals, the **hymen** is poorly developed and is represented in the foal and juvenile pig by a low annular fold, and in the other species by a few small transverse folds at the vaginovestibular junction.

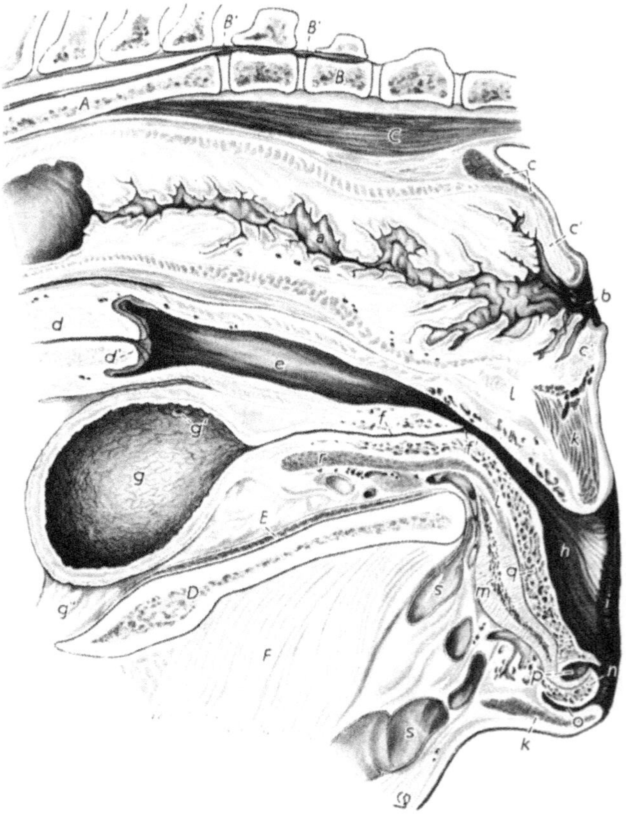

Fig. 525. Median section of the pelvic cavity of a formalin-fixed mare.

A Sacrum; *B* Second caudal vertebra; *B'*, *B'* Sacrocaudal space, and space between first and second caudal vertebrae; *C* Sacrocaudalis ventralis medialis; *D* Pelvic symphysis; *E* Obturator internus; *F* Symphysial tendon

a Rectum; *b* Anus; *c* Ext. anal sphincter; *c'* Int. anal sphincter; *d* Cervix uteri; *d'* Intravaginal part of cervix with ext. uterine orifice; *e* Vagina; *f* Urethra; *f'* Ext. urethral orifice; *g* Bladder; *g'* Ureteric orifice; *g''* Median ligament of bladder; *h* Vestibule of vagina; *i* Right labium of vulva; *k* Constrictor vulvae; *l* Constrictor vestibuli; *m* Body of clitoris; *n* Glans of clitoris; *o* Fossa clitoridis; *p* Fossa glandis; *q* Cavernous tissue in the floor of vestibule; *r* Urethralis; *s* Venous plexus in the vicinity of the clitoris, forming anastomoses between int. pudendal and obturator veins

Vestibule

Since the vestibule of the vagina develops from the ventral part of the cloaca, it belongs to both urinary and genital systems. It extends from the level of the external urethral orifice to the labia of the vulva. In the floor of the vestibule of the sow and cow is the entrance to the **suburethral diverticulum** (555/*10*), a short blind pouch ventral to the external urethral orifice.

The vestibule is lined with a more or less reddish mucous membrane, covered with stratified squamous epithelium, and contains mucus-producing glands of varying size and number. The **minor vestibular glands** are rows of single glands with separate ducts (558/*h*), present in the dog, pig, sheep, and horse in the lateral and/or ventral wall, and opening along the floor or sides of the vestibule. The **major vestibular glands** (547B/*h*), present in cow, cat, and occasionally in sheep, are two compact glandular masses, one in each lateral wall, with single excretory ducts. The mucus secreted by the vestibular glands lubricates the vestibule in preparation for copulation and for the passage of the fetus at birth. Large amounts of mucus are secreted during estrus. This mucus is thought to contain odorous substances which attract the male. In the wall of the vestibule are more or less extensive venous plexuses, which in the bitch and mare form a discrete patch of erectile tissue, the **vestibular bulb** (539, 559/*17'*), on each side. Remnants (ductus epoophori longitudinales)* of the mesonephric ducts may be found on each side of the external urethral orifice. These often break through the epithelium and open with small orifices into the vestibule (547B/*f'*). In the cat, pig, and sheep, short blind epithelial tubes (ductus paraurethrales) have been found next to the ductus epoophori. They are homologous to the primordium of the prostate gland of the male.

Vulva and Clitoris

(525, 526—529, 538, 539, 543, 547, 556, 558)

The **VULVA** is the terminal, and the only really external, part of the female genital tract. It consists of right and left **labia** (528/*2*, *3*) which meet dorsally in the rounded **dorsal commissure** (*1*) and ventrally in the more pointed **ventral commissure** (*4*). In the mare (529), the ventral commissure is also rounded. Between the labia is the **vulvar cleft** (rima vulvae) which leads into the vestibule. The labia of domestic mammals are homologous to the labia minora in women. The labia majora in women develop from the genital swellings which give rise to the scrotum in the male. Homologues of the labia majora are not usual in the female domestic mammal, except possibly for the two cutaneous elevations (526/*d*), derivatives of the genital swellings, that are occasionally seen lateral to the vulva of the cat and bitch.

The skin of the labia resembles the common integument. It has numerous sebaceous and sweat glands and hair follicles with soft fine hairs. Depending on the species, the labial skin is completely, partly, or not pigmented (black), but individual variations occur. Inside the vulvar cleft, the skin of the labia gradually merges into the mucous membrane lining the vestibule. The labia consist largely of adipose tissue, embedded in which are the bundles of the **constrictor vulvae** (559/*19'*).

Between dorsal commissure and anus is a cutaneous bridge (526—529/*b*) which may occasionally be torn (perineal laceration) during birth when the fetus is oversized or malpositioned. Deep to this cutaneous bridge is a mass of muscle and connective tissue without definable lateral boundaries, known as the **perineal body.** The **perineum** of the female is the body wall that surrounds anus and vulva and covers the pelvic outlet. Its deep boundaries are therefore roughly those of the pelvic outlet. (In the male, the deep boundaries of the perineum are the same as in the female.) The superficial boundaries of the perineum are similar to the deep boundaries dorsally and laterally, but extend ventrally to the base of the mammary gland (or the scrotum in the male).

The ventral commissure of the vulva is somewhat pendulous and always ventral and caudal to the ischiatic arch. It encloses the **CLITORIS** (529/*5*), the homologue of the penis, ventrally and laterally. The attachment and the structure of the clitoris are very similar to those of the penis; the main difference being that in the female the urethra is not part of the clitoris, whereas in the male the urethra is incorporated in the penis. Except for slight species differences, the clitoris consists of two crura, a body, and a glans. The **crura clitoridis** arise from the ischiatic arch and unite like those of the penis to form the **body** (525/*m*), which lies under the floor of the vestibule and is composed of erectile tissue (corpus cavernosum clitoridis). The **glans clitoridis** (*n*) is at the tip of the body and is the only exposed part of the organ. It protrudes from a depression known as the **fossa clitoridis** (*o*) and in some species is connected dorsally by a short **frenulum** to the dorsal part of the **prepuce of the clitoris**

* Canals of Gartner.

The lateral and ventral parts of the prepuce are formed by the ventral commissure of the labia. In the dog, the clitoris (539/*18'*, *18''*) has a well-developed body and ends in a glans of considerable size; in the cat, the glans is poorly developed. In the pig, the clitoris (543/*k*;

Fig. 526 (Bitch)

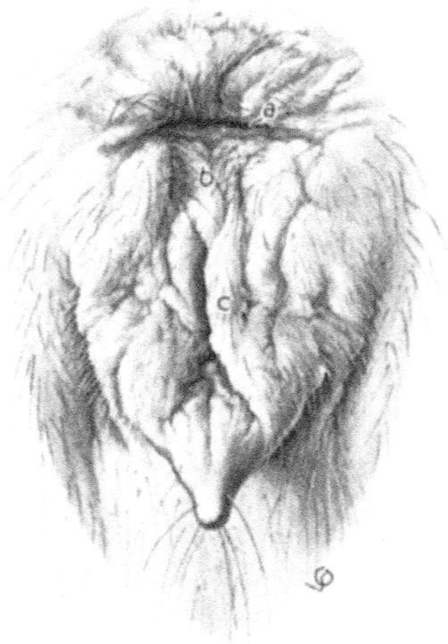

Fig. 527 (Sow)

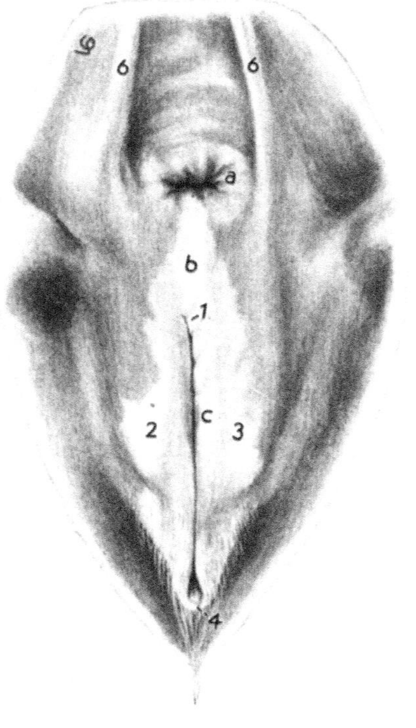

Fig. 528 (Cow)

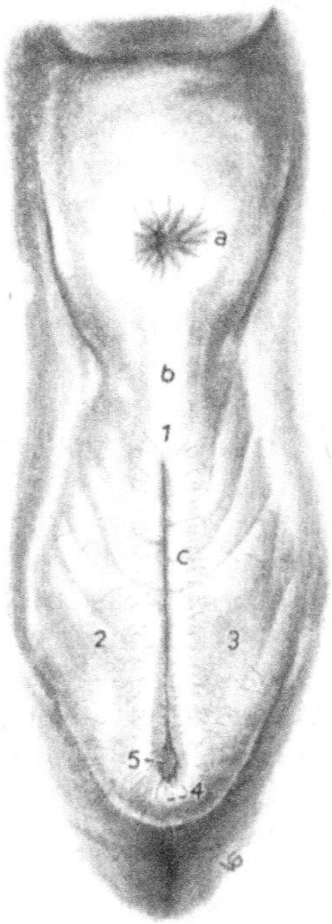

Fig. 529 (Mare)

Figs. 526—529. Anus and vulva of the domestic mammals. *a* Anus; *b* Cutaneous bridge between anus and vulva; *c* Vulvar cleft; *d* Cutaneous swellings lateral to the vulva of the bitch. *1* Dorsal commissure; *2* Left labium; *3* Right labium; *4* Ventral commissure; *5* Glans clitoridis of the mare; *6* Tail folds of the cow

545/*18'*) consists of a long tortuous body with a conical glans inside a narrow and shallow fossa. In the clitoris of the cow (547 B/*k*; 555/*14*), the body is also tortuous, but the glans lies in a slight mucosal depression, too shallow to be regarded a true fossa. In the small ruminants

(556/*5*) the fossa surrounding the glans is somewhat deeper. The clitoris of the mare (525) is relatively large and has a prominent glans which is often visible in the ventral end of the vulvar cleft (529).

Muscles Associated with the Female Genital Organs

The muscles described here are mostly striated and associated with the vestibule and the vulva. Being relatively superficial, they are part of the **perineum,** the part of the body wall that covers the pelvic outlet and has the anus and vulva at its center. The muscles of the female genital tract are homologues of muscles in the male perineum, and despite considerable structural differences, the same names are used for the same muscles in both sexes, wherever possible. These muscles do not vary much in females of domestic mammals and are therefore described only at this point. For detailed descriptions of all the muscles of the perineum in the small ruminants see the studies by Geiger (1956) and Bassett (1965); in the cow, Geiger (1954) and Habel (1966); and in the mare, Habel (1953).

The **ISCHIOCAVERNOSUS** is a small muscle that arises from the ischium and surrounds the crura of the clitoris. It can evert the ventral commissure of the labia and thus expose the clitoris. The **CONSTRICTOR VESTIBULI** (525/*l*; 539, 545, 554, 559/*19*) and the **CONSTRICTOR VULVAE** (525/*k*; 541/*19*; 545, 554, 559/*19'*) are homologous to the bulbospongiosus of the male. They are striated and circular, and constrict the vestibule and vulva, respectively. The constrictor vulvae is connected dorsally to the sphincter ani and forms the muscular base of the labia. (See also p. 133) A somewhat rudimentary **RETRACTOR CLITORIDIS**—the homologue of the retractor penis—has been described, although not always under that name, for the domestic mammals (except the dog). Like the retractor in the male, this smooth muscle arises from the caudal vertebrae (559/*26'*), passes the lateral surface of the rectum, and after decussating ventral to the anus, passes toward the clitoris, without reaching it, however, in some species. (See also pp. 111, 145, 194) For more detail consult the study of this muscle made by Bassett (1961). The **ISCHIOURETHRALIS,** which facilitates micturition, arises from the caudal end of the pelvic symphysis and passes to the vicinity of the external urethral orifice.

Blood Vessels, Lymphatics, and Innervation of the Female Genital Organs

(See also page 355)

The blood vessels supplying the female genital tract form numerous anastomoses. The uterine branch of the **ovarian artery** supplies the cranial part of the uterine horns. The **uterine artery***, the main artery to the uterus, arises from the umbilical artery, a branch of the internal iliac, except in the horse, where it comes from the external iliac artery. It is especially prominent in the pregnant cow and is palpated rectally in this species for the diagnosis of pregnancy (554/*i*). The uterine branch of the **urogenital artery** supplies the caudal part of the uterus, the cervix, and parts of the vagina. The urogenital artery of the carnivores and horse is a branch of the internal pudendal, while in the pig and ruminants it comes from the internal iliac. The remaining parts of the genital tract are supplied by other branches of the urogenital and internal pudendal arteries. The blood returns to the caudal vena cava in satellite veins.

The **lymphatics** of the uterus pass to the medial iliac and oartic lumbar nodes. Lymph from vagina and vulva is drained to the medial iliac, internal sacral (dog), anorectal (horse), and external sacral (ox and pig) nodes. Lymph from the ventral commissure of the vulva may also pass to the superficial inguinal nodes.

The **innervation** of the female genital tract is both autonomic and voluntary. Parasympathetic innervation comes from the sacral outflow and reaches the genital tract via the pelvic nerves. Sympathetic innervation comes from the caudal mesenteric ganglion and plexus and goes to the organs via the hypogastric nerves and pelvic plexus. The pudendal and caudal rectal nerves from the sacral plexus carry sensory and motor fibers, the latter for the voluntary muscles at the caudal end of the tract.

---

* Formerly the middle uterine artery.

Postnatal Changes in the Female Genital Organs

The female genital organs undergo certain morphological changes related either to their ability to function or to previous functioning. These changes are distinct enough to make it possible to decide whether a tract of unknown origin came from a juvenile or prepuberal animal, from a postpuberal but nulliparous animal, or from a parous animal.

The genital organs of **JUVENILE ANIMALS** display all the features characteristic of the species, but they are small and delicate and reveal immediately that they have not functioned in reproduction. The ovaries are small and have not begun to produce normal mature follicles. Vesicles of varying sizes are present, but the oocytes they contain are degenerating and appear to be similar to those found in atretic follicles of the adult ovary. The vesicles, moreover, do not rupture so that there are neither scars nor corpora lutea present. The mesometrium and mesovarium are delicate and translucent. The uterus, besides being small, is completely symmetrical. Its wall is smooth, thin and soft to the touch, and the layers are not very distinct. The vascular layer is thin and has narrow thin-walled vessels. The mucous membrane is without the pigmentation of the adult uterus that possibly results from repeated bleeding at estrus. Uterine glands are sparse. The most striking characteristic at this stage is the uterine vessels—especially the arteries—which have narrow lumina, thin walls, and are as yet not very tortuous. The cervix is soft and flaccid, very different from the tough and hard consistency it attains in the mature animal. Except for size and a pale yellowish mucosa, the vagina, vestibule, and vulva do not differ much from the adult organs.

POSTPUBERAL, NULLIPAROUS ANIMALS. When the female animal reaches puberty, the genital organs gradually increase in size and become functional, and the animal comes into heat at regular intervals. Depending on the species but also on the breed puberty in animals occurs at rather different ages.

The ovaries at this stage are of rubbery consistency, and follicles are present in various stages of development, as well as corpora lutea and their remnants. The wall of the entire tract is thicker and firmer. The mucosa is grayish red, yellowish brown, or even brownish red, and its surface is more sculptured. The arteries are larger but not very tortuous as yet. (For measurements and growth rates of the genital tracts of 65 calves and young nulliparous cows 1—12 months old see the study by Desjardins and Hafs [1969]).

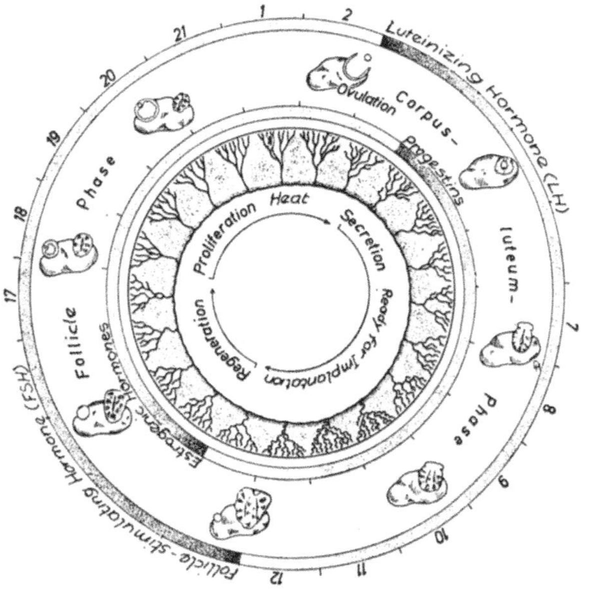

Fig. 530. Schematic representation of the sexual cycle in the cow. (After Vollmerhaus, unpublished.)
1—21 Days of sexual cycle; *Outer ring:* Level of gonadotropic hormones of the adenohypophysis; *Between outer and inner rings:* Changes in the ovary during ovarian cycle, caused by the gonadotropic hormones; *Inner ring:* Level of the ovarian hormones (estrogenic hormones and progestins); *Inside of inner ring:* Endometrium with uterine glands as influenced by the ovarian hormones

PAROUS ANIMALS. In animals that have been pregnant, the uterine wall is thick and firm and has a well-developed, grossly visible vascular layer. The horns of the ruminant uterus are often found to be of unequal size. In the cow, the uterus, including the mucous membrane, becomes yellow or yellowish brown. The cervix, especially that of the cow, is very dense and firm. The folds of the cervical mucosa become considerably larger, while the intravaginal part of the cervix takes on a peculiar foliate appearance. Abusineina (1969a, b) reports increases in size and weight of the cervix during pregnancy in the cow and ewe. The mesometrium shows a general thickening resulting from an increase in muscular and connective tissue fibers. The uterine blood vessels—particularly the uterine artery—are larger, thicker-walled, and much more tortuous, as are also the vessels in the vascular layer of the uterine wall. The lumen of the vagina is larger, and the vaginal wall thicker. The vulva is also larger.

These structural changes are of course the result of pregnancy. In the course of pregnancy, but especially close to term, the uterus increases to an astonishing size as a result of the marked hypertrophy* and hyperplasia** of all components of the uerine wall. Not only the uterus but also other parts of the tract enlarge: the broad ligaments, cervix, vagina, vestibule, vulva, and the entire associated vascular apparatus. After the fetus has been expelled, the genital tract decreases in size in a process known as **involution.** In involution, however, the genital organs never return to their original condition before pregnancy so that traces of pregnancy and parturition remain.

Placentation and the Gravid Uterus

With the exception of the egg-laying monotremes, mammalian species—and thus also the domestic mammals—are placental animals, placentalia; that is, the embryo develops membranes which unite intimately, though grossly in very species-specific ways, with the uterine mucosa. These membranes serve for implantation, and for an efficient exchange of nutrients and other metabolites between mother and embryo.

To prepare for **IMPLANTATION** of an embryo, the changes in the uterine mucosa begin before the external signs of heat and ovulation manifest themselves, and thus extend over the entire sexual cycle, regardless of whether the oocytes released at ovulation have been fertilized. Briefly, the progestational† changes occuring in the endometrium consist of a gradually increasing activation of its components, such as the surface epithelium, uterine glands, blood vessels, and the highly cellular connective tissue of the propria mucosae. The other parts of the genital tract are to some extent included in this process, but are not considered here. Except for differences in degree and intensity, these progestational changes are similar in all domestic mammals. If fertilization fails to occur, the endometrium regresses to its resting state. If fertilization takes place, the conceptus, now a blastocyst, is implanted on the endometrium, which, as the blastocyst grows into an embryo and fetal membranes develop, is transformed to form the **MATERNAL PLACENTA.**

Shortly after entering the uterus, the blastocyst contacts the endometrium with its outer cells, the trophoblast. The cells of the trophoblast are capable of selecting and absorbing nutrients present in the secretion (uterine milk) of the uterine glands, and in maternal cells that have been destroyed by the invasion of the trophoblastic cells; they also take up oxygen and other metabolites. These materials are used in the metabolism of the as yet very small embryonic cell mass. Metabolic wastes are returned to the maternal circulation in the endometrium in a similar way.

Later, when the trophoblast has been transformed, by the addition of cell layers and blood vessels to its interior surface, into the allantoic chorion (part of the fetal membranes), the nutrients are picked up and transported to the embryo by blood vessels. Certain areas of the trophoblastic surface of the chorion develop **villi,** which penetrate into the endometrium (maternal placenta), thus increasing the absorptive surface of the chorion, and providing the most intimate contact possible between the fetal and maternal tissues. The fetal membranes thus in contact with the maternal placenta are known as the **FETAL PLACENTA.**

In a **diffuse placenta** (horse, pig), the chorionic villi are evenly or nearly evenly distributed over the surface of the chorion. In a **cotyledonary placenta** (ruminants), the villi on the surface of the chorionic sac are grouped in numerous patches, **cotyledons,** which are separated by stretches of smooth chorion. In a **zonary placenta** (carnivores), the villi occupy a bandlike

---

* Growth due to increase in size of existing cells.

** Growth due to addition of new cells.

† Favoring gestation or pregnancy.

area around the middle of the chorionic sac, and in a **discoidal placenta** (primates, rodents and others), the villi are limited to a disc-shaped area on the chorion.

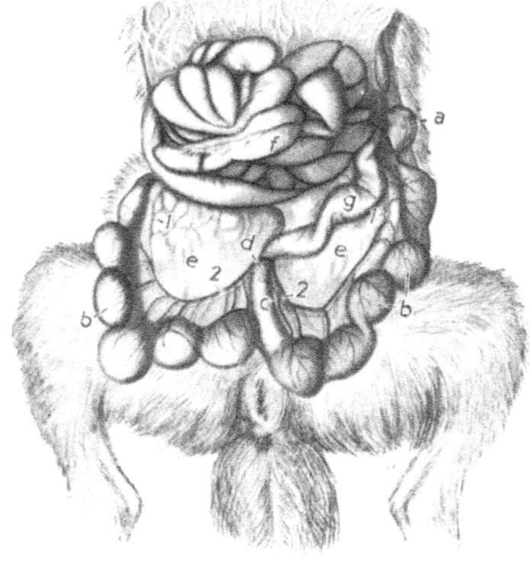

Fig. 531. Abdominal organs of a pregnant bitch. Ventral aspect. The greater omentum has been reflected cranially, and the pregnant uterus has been exteriorized.

a Left ovary; *b* Uterine horns with five loculi each; *c* Body of uterus; *d* Cervix; *e* Broad ligaments; *f* Jejunum; *g* Descending colon; *1* Uterine branch of ovarian artery; *2* Uterine artery and vein

With the outgrowth of villi from the surface of the chorion, corresponding cryptlike depressions develop in the maternal placenta into which the villi embed themselves. Furthermore, through the specific activity of the trophoblastic cells, the most superficial cell layers of the endometrium are destroyed and the nutrient substances thus freed (histotroph) conveyed to the embryo. This process of invasion and digestion by the trophoblastic cells occurs in horse and ruminants, but is particularly marked in the carnivores, and reduces the tissue barrier remaining between the maternal and fetal capillaries of the placenta. The thickness of the barrier remaining between the two circulations is being used for a histophysiological classification of placentas.

A further classification of placentas is based on the loss of endometrium (decidua) at the time of parturition. With the **deciduate* placentas,** present in the carnivores, there is loss of endometrium at birth because of the more intimate fusion of fetal and maternal tissues. In the simpler, **non-deciduate placentas,** such as those of the horse, pig, and ruminants, maternal and fetal parts are merely in apposition and little or no maternal tissue is lost.

Fig. 532. Placentome in advanced pregnancy of the cow, ewe, and goat. (After Andresen 1927)

a Endometrium with uterine glands (lamina propria mucosae); *a'* Epithelium of endometrium; *a''* Caruncular tissue; *b* Myometrium; c Allantoic chorion; *c'* Chorionic villi

Note the different shapes of the placentomes: convex in the cow, and concave in the ewe and goat

Because of the hormonal control mechanisms of pregnancy, the gestational processes taking place in the uterus and causing marked changes in its shape and size also influence the other parts of the genital tract, including the mammary glands. In fact, the entire animal is affected as evidenced by changes in behavior, all adapted to leading to uneventful pregnancy, birth, and nursing period.

Only the gross changes on the **GRAVID UTERUS** are mentioned here. In the multipara (carnivores and pig), the developing embryos and their membranes are implanted at equal distances along the uterine horns. The diameter of the uterine horn increases at the places of implantation, forming the **loculi.** Between loculi there are constrictions, and the pregnant uterus of these species takes on a beaded appearance (531/*b*). Closer to term the uterine horns as a whole enlarge markedly, and the constrictions between the loculi are less distinct. In the unipara (ruminants and horse), the embryo is implanted in one uterine horn. However, parts

* From L., characterized by shedding.

of the fetal membranes usually extend into the other horn. Because of the enlargement of the gravid horn, the uterus of these species becomes asymmetrical unless, of course, bicornual twins are carried, which is rare in the large species but frequent in sheep and goats.

The enlargement of the uterus during pregnancy is not due only to a passive stretching of its wall by the gradual increase in the size of the fetus, fetal membranes and fluid. The uterine wall seems to respond to the pressure from within by an enlargement (hypertrophy) of its muscle cells and connective tissue, and by production (hyperplasia) of new muscle and connective tissue cells. This results in an absolute increase in the weight of the uterus, which in the cow for instance is 12—16 times that of the nongravid uterus.

POSITION OF THE GRAVID UTERUS. As the uterus of the carnivores, the sow, and the mare increases in size and weight, it advances cranioventrally and by gradually displacing the intestines it sinks to the abdominal floor. It reaches the level of the xiphoid cartilage and makes contact here with the diaphragm, and in the carnivores it is covered ventrally by the extensive greater omentum. In the ruminants, the gravid uterus, as it advances cranially along the abdominal floor, passes the rumen on the right and occupies, together with the intestines, the supraomental recess. Occasionally, it is ventral to the greater omentum. As it increases in weight, it occupies an increasingly larger area of the ventral abdominal wall and displaces the rumen upward and to the left. Occasionally, the gravid uterus of the cow advances on the left side of the rumen and displaces it to the right.

The extent of the gross changes, particularly in the large domestic species in which fetus and fetal fluids attain considerable weight, is such that the uterus is gradually transformed from an initially delicate and soft organ—well suited to receive the extremely fragile blastocyst—into a tough-walled, supportive sack. So drastic are these changes, in fact, that upon involution the tissues do not regress to the virginal state, and thus traces of previous pregnancies and parturitions persist.

Female Genital Organs of the Carnivores

(17, 515, 519, 526, 531, 533—541)

In carnivores, the bitch is diestrous, i.e., she comes into heat twice a year, the heat period lasting for about three weeks. She is also multiparous, and after an average gestation of 63 days (range: 58—66), gives birth to 4—10 young, depending on the number typical of her breed. The female cat has two to three periods of heat per year, each of which lasts about one week, and the gestation period is 60—61 days, after which she gives birth to 3—5 young.

Ovaries

The ovaries (535—538/*a*) are elongated and flat, and in the bitch are on the average about 2 cm. long and 1.5 cm. thick, varying with the size of the breed. In the cat they are only 8—9 mm. long. They are suspended by the mesovaria caudal to the kidneys at about the level of the third to fourth lumbar vertebra. In the bitch they are completely within the ovarian bursa, but in the cat they are only partly enclosed by it. The surface of the ovary is tuberculate because of the many follicles and corpora lutea present in multipara.

The ovary is attached cranially to the diaphragm in the region of the last rib by the **suspensory ligament** (541/*11''*). This is a peritoneal fold with a thickened edge derived from the fold connecting the mesonephros with the diaphragm in the embryo. The uterine end of the ovary is attached by the very short **proper ligament** (536/*e*) to the tip of the uterine horn. The **mesovarium** (541/*11'*) is continuous cranially with the suspensory ligament and caudally with the mesometrium. The **mesosalpinx** of the bitch (519/*b'*) is an extensive lateral outgrowth of the mesovarium. It contains a large amount of fat, and is drawn out ventrally to form the laterally compressed **ovarian bursa,** which contains the ovary and hides it from view. The bursa measures 2.5—5 cm. dorsoventrally and .5—3.5 cm. craniocaudally. The entrance to the bursa is a small slit .2—1.8 cm. long on the medial side (533/*a*). Immediately dorsolateral to the entrance, the relatively small hard ovary can be palpated high in the ovarian bursa. On the lateral surface of the ovarian bursa is a round, fatfree area (534/*a*) through which the ovary can usually be seen. The mass of fat associated

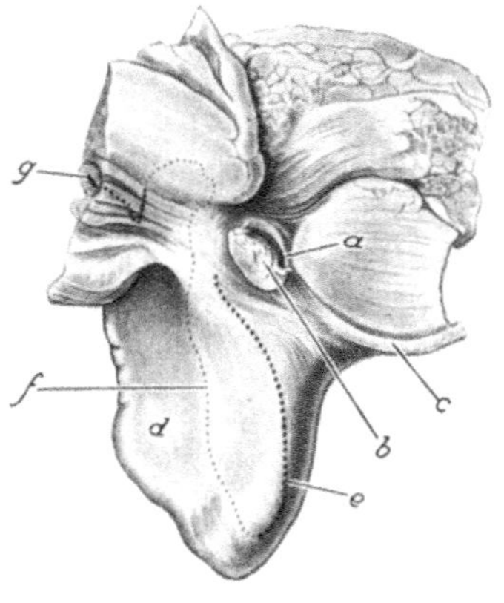

Fig. 533 (Medial aspect)

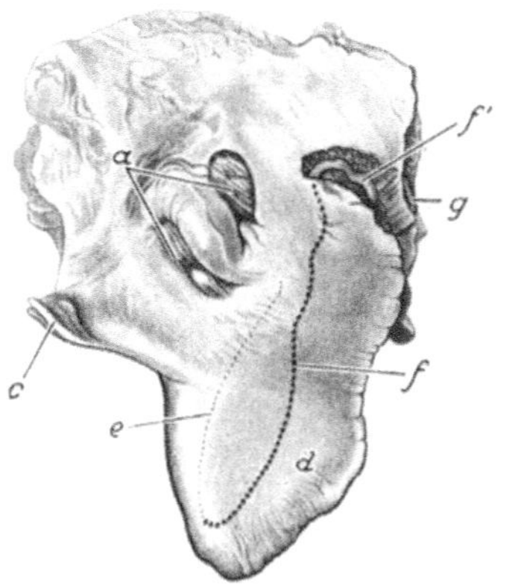

Fig. 534 (Lateral aspect)

Fig. 533. Left ovary, uterine tube, and associated ligaments of the bitch. Medial aspect. (After v. Bönninghausen 1936.)
a Entrance to ovarian bursa; *b* Part of infundibulum of uterine tube; *c* Suspensory ligament of ovary; *d* Fatfilled mesosalpinx; *e*, *f* Course of uterine tube in mesosalpinx indicated by dotted lines; *g* Tip of uterine horn

Fig. 534. Left ovary, uterine tube, and associated ligaments of the bitch. Lateral aspect. (After v. Bönninghausen 1936.)
a Ovary visible through fatfree area of mesosalpinx; *c* Suspensory ligament of ovary; *d* Fatfilled mesosalpinx; *e*, *f* Course of uterine tube in mesosalpinx indicated by dotted lines; *f'* Uterine tube, exposed as it joins the tip of the uterine horn; *g* Tip of uterine horn

Fig. 535. Left ovary, uterine tube, and associated ligaments of the bitch. Lateral aspect. The ovarian bursa has been opened sagittally, and the lateral half of the bursa reflected dorsally. (After v. Bönninghausen 1936.)
a Ovary; *b* Abdominal opening of uterine tube in center of infundibulum; *c* Suspensory ligament of ovary; *d* Mesosalpinx, medial wall of ovarian bursa; *d'* Mesosalpinx, lateral wall of ovarian bursa; *d''* Cut edge of ovarian bursa; *e*, *e'*, *f*, *f'*, *f''* Uterine tube: *e'*, *f''* cut surface, *e* in the medial wall of ovarian bursa, *f'* exposed as it joins the tip of the uterine horn; *g* Tip of uterine horn

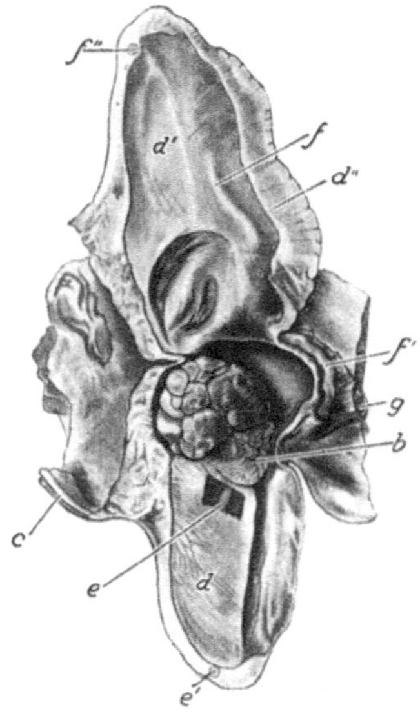

Fig. 535

with the bursa (539/*9'*) is related to the caudal pole of the kidney and is an important landmark for locating the ovary during surgery.

The ligaments associated with the ovary of the cat (536, 537) are similar to those of the bitch, but the mesosalpinx of the cat contains no fat, is much smaller, and covers only the lateral surface of the ovary. The medial surface of the ovary is covered ventrally and cranially by the infundibulum of the uterine tube. The narrow entrance to the ovarian bursa is bounded dorsally by the proper and suspensory ligaments of the ovary and ventrally by the fimbriae of the uterine tube.

Uterine Tube

The uterine tube of the bitch (533—535/*e*, *f*) is 6—10 cm. long and begins with the flattened, reddish **fimbriae** of its **infundibulum** ventral to the entrance of the ovarian bursa. Some of the fimbriae can usually be seen through the entrance. The infundibulum with the centrally located **abdominal opening of the uterine tube** is closely related to the ovary. The uterine tube runs first in the medial wall of the bursa and then in the lateral wall, the **uterine opening** being on a small papilla at the tip of the uterine horn.

The uterine tube of the cat (536, 537/*b*) is 4—5 cm. long and begins at the craniomedial surface of the ovary. From here it passes cranially, turns laterally around the ovary, and returns in the mesosalpinx along the lateral surface of the ovary, ending at the tip of the uterine horn.

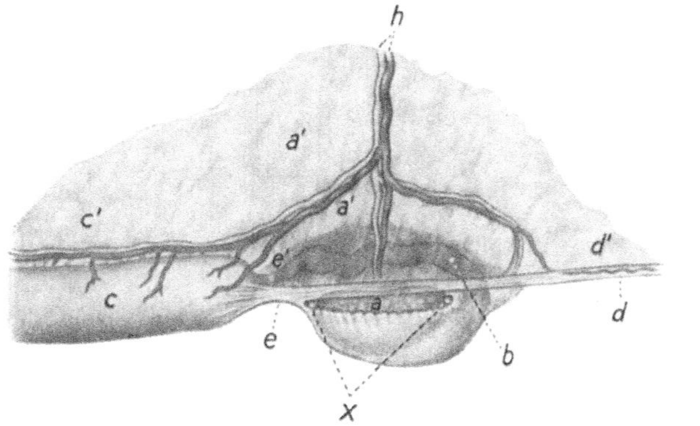

Fig. 536 (Medial aspect)

Fig. 536. Left ovary, uterine tube, uterine horn, and associated ligaments of the cat. Medial aspect. (After Merkt 1948.)

a Ovary; *a'*, *c'*, *d'*, *e'* Broad ligament; *a'* Mesovarium; *b* Uterine tube; *c* Uterine horn; *c'* Mesometrium; *d* Suspensory ligament of ovary; *e* Proper ligament of ovary; *h* Ovarian artery and vein; *x* Entrance to ovarian bursa

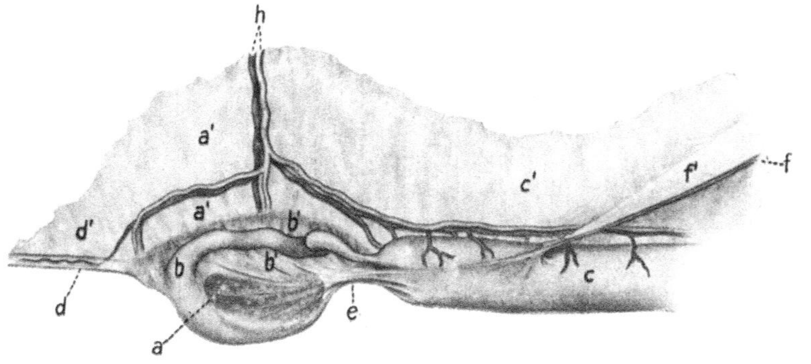

Fig. 537 (Lateral aspect)

Fig. 537. Left ovary, uterine tube, uterine horn, and associated ligaments of the cat. Lateral aspect. (After Merkt 1948.)

a Ovary; *a' c' d'* Broad ligament; *a'* Mesovarium; *b'* Uterine tube; *b'* Mesosalpinx; *c* Uterine horn; *c'* Mesometrium; *d* Suspensory ligament of ovary; *e* Proper ligament of ovary; *f* Round ligament of uterus; *f'* Fold supporting *f*; *h* Ovarian artery and vein

Uterus

As is common for multipara, the **horns** of the uterus of dog and cat are very long and of uniform diameter, about 8 mm. in the medium-size bitch. In the cat, the horns are 3—4 mm. thick and 9—10 cm. long. They extend with a slight ventral convexity from the uterine end of the ovaries to the body of the uterus and, before joining it, are united by peritoneum for a short distance (515). The **body** of the uterus is 2—3 cm. long in the bitch, and 2 cm. long in the cat, with the two openings into the horns at its cranial end. The **mucous membrane** of the uterus is a reddish gray, yellow, or brown, is generally smooth, and may be pigmented (black) in places. In the bitch it may be arranged in low, longitudinal and, less often, transverse folds, while in the cat it forms wide spirally arranged longitudinal folds.

The **CERVIX** of the bitch (538/*d*) is only about 1 cm. long. The internal and external uterine orifices lie thus close together. The intravaginal part of the cervix is semi-cylindrical and projects only ventrally into the vagina. Dorsally, it blends into a rounded longitudinal fold (*d''*) on the roof of the vagina. In the cat, the cervix feels like a hard oval knot at the utero-vaginal junction. Its small papilliform intravaginal part bears the **external uterine orifice.** The **internal uterine orifice** is not very distinct in the carnivores because the uterine lumen narrows gradually to the diameter of the cervical canal. The cervical mucosa presents a few oblique folds and in the cat contains **glands.**

The **broad ligament** (538/*2*) are wide serosal sheets containing much fat in the bitch, even if poorly nourished, but little fat in the cat. They attach on the mesometrial border of the uterine horns, which in the nongravid animal rest on the coils of the jejunum; and because the ligaments are wide, they allow the horns a fair amount of range. The **round ligament of the uterus** (537/*f*; 539/*11'''*; 540/*f*) is attached by a wide serosal fold to the lateral surface of the mesometrium and extends from the tip of the uterine horn to the deep inguinal ring. In the bitch, it usually leaves the abdominal cavity through the inguinal canal, and extends into the cutaneous swellings lateral to the vulva (526/*d*). The round ligament of the bitch, therefore, can be divided into abdominal and extra-abdominal parts. The extra-ab-

Fig. 538. Genital organs of the bitch. Dorsal aspect. Body of uterus, vagina, and vestibule have been opened dorsally.

a Right ovary with numerous vesicular follicles, in the opened ovarian bursa; *a'* Entrance to ovarian bursa; *b* Uterine tube; *b'* Infundibulum with fimbriae; *c* Uterine horn; *c'* Body of uterus; *d* Cervix; *d''* Mucosal fold connecting ext. uterine orifice with the roof of the vagina; *e* Vagina; *f* Vestibule; *f'* Mucosal folds (homologue of hymen); *g* Ext. urethral orifice; *h* Openings of minor vestibular glands; *i* Labia; *k* Fossa clitoridis; *k'* Mucosal fold, partly covering *k*; *l* Constrictor vulvae and constrictor vestibuli

1 Mesovarium; *2* Mesometrium; *3* Mesosalpinx, cut to expose ovary; *4* Left ovarian bursa, unopened; *5* Ovarian artery and vein; *5''* Uterine branch of ovarian artery; *6* Uterine artery

dominal part is usually enclosed in a strand of fat which in over half the animals examined (Zietzschmann 1928) is invaded by a fingerlike peritoneal evagination (processus vaginalis, 540/*h*) in such a way that the round ligament travels in the center of the processus. In extreme cases, the processus vaginalis extends also into the cutaneous swellings lateral to the vulva.

Vagina, Vestibule, and Vulva

The **VAGINA** of the bitch is very long and covered cranially with peritoneum (*17*). The long urethra (539/*7*) runs between the vagina and the floor of the pelvis and indents the ventral wall of the vagina. The **fornix vaginae** is a small recess between the floor of the vagina and the semi-cylindrical intravaginal part of the cervix. The mucous membrane of the vagina is covered with stratified squamous epithelium and forms longitudinal or circular folds (538).

The **VESTIBULE** contains a nodular area of erectile tissue on each lateral wall, the **vestibular bulb** (539/*17'*) which may be up to 1.5 cm. in diameter. The remainder of the wall contains venous plexuses with cavernous spaces. On the floor of the vestibule, forming two rows, are the **minor vestibular glands** (538/*h*). The mucous membrane is covered with stratified squamous epithelium which is bluish red, folded, and contains lymph nodules.

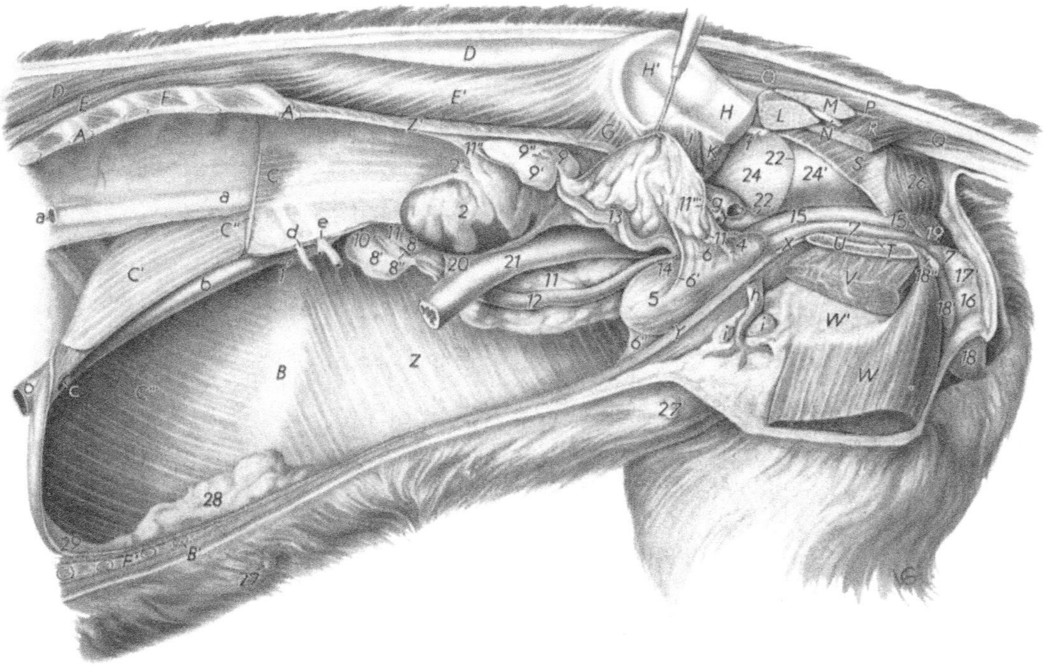

Fig. 539. Urinary and genital organs of a bitch in situ. Left lateral aspect.

A Tenth rib; *A'* Thirteenth rib; *B* Right costal arch; *B'* Left tenth costal cartilage; *C* Left crus of diaphragm; *C', C''* Right crus of diaphragm; *C'''* Costal part of diaphragm; *D* Longissimus; *E, E'* Iliocostalis; *F, F'* Ext. intercostal muscles; *G* Int. abdominal oblique; *H, H'* Ilium; *J* Caudal part of sartorius; *K* Iliopsoas; *L* Gluteus medius; *M* Gluteus supf.; *N* Sacrotuberous ligament; *O* Sacrocaudalis dorsalis lateralis; *P* Intertransversarius dorsalis; *Q* Sacrocaudalis ventralis lateralis; *R* Coccygeus; *S* Levator ani; *T* Obturator int.; *U* Floor of pelvis; *V* Adductor magnus et brevis; *W* Gracilis; *W'* Symphysial tendon; *X, Y* Rectus abdominis; *Z* Right transversus abdominis, covered with transverse fascia and peritoneum; *Z'* Origin of left transversus abdominis

a Aorta; *b* Caudal vena cava; *c* Stump of hepatic veins as they enter the caudal vena cava; *d* Celiac artery; *e* Cranial mesenteric artery; *f* Ischiatic nerve; *g* Femoral artery and vein; *h* Ext. pudendal vein; *i* Supf. inguinal lymph nodes

1 Right kidney; *2* Left kidney; *2'* Hilus of left kidney; *4* Left ureter, visible through lateral lig. of bladder; *5* Urinary bladder; *6* Left lateral lig. of bladder; *6'* Round lig. of bladder (degenerated umbilical artery); *6''* Median lig. of bladder; *7* Urethra; *7''* Level of ext. urethral opening; *8* Position of right ovary; *8'* Fatfilled mesosalpinx; *8''* Entrance to ovarian bursa; *9* Position of left ovary; *9'* Fatfilled mesosalpinx; *9''* Fatfree area in mesosalpinx; *10* Right uterine tube; *11* Mesometrium; *11'* Mesovarium; *11''* Suspensory lig. of ovary; *11'''* Round lig. of uterus; *11''''* Vaginal ring; *12* Right uterine horn; *13* Left uterine horn; *14* Body of uterus; *15* Vagina; *16* Vestibule; *17'* Vestibular bulb; *18* Vulva; *18'* Body of clitoris; *18''* Left crus clitoridis, transected; *19* Constrictor vulvae; *20* Mesocolon; *21* Descending colon; *22* Pelvic fascia and parietal peritoneum, fenestrated to expose rectum; *24* Rectum, covered with peritoneum; *24'* Retroperitoneal part of rectum; *26* Ext. anal sphincter; *27* Teat of mammary gland; *28* Fatfilled part of falciform ligament; *29* Falciform ligament

The labia (*i*) of the **VULVA** are round, often pigmented, and covered more or less densely with hair. The ventral commissure is pointed, while the dorsal commissure is usually covered by a transverse cutaneous fold (526). The crura and body of the **clitoris** (539/*18'*, *18''*) are relatively large and lie ventral to the floor of the vestibule. The body consists of dense adipose tissue and ends in a cavernous glans, which lies in the floor of the fossa clitoridis (538/*k*) without being visible. The lining of the fossa forms small intersecting mucosal folds between which are shallow depressions. The fossa itself is partly overhung by a mucosal fold (*k'*) and should not be mistaken, when catheterizing a bitch, for the external urethral orifice which is a short distance cranial to it.

In the cat (541), **vagina** and **vestibule** together are about 4 cm. long. The urethra (*7*) ends at the vagino-vestibular junction in a deep mucosal groove. A discrete vestibular bulb is lacking, but diffuse cavernous tissue is present. The cat has major vestibular glands, about 5 mm. in size, in the lateral walls of the vestibule, with small openings that can be detected in the vestibular floor. The **vulva** (*18*) is covered with dense hair. Its ventral commissure is rounded, and the dorsal is more pointed. The body of the clitoris is about 1 cm. long and ends in a poorly developed glans, which protrudes slightly from the shallow fossa clitoridis.

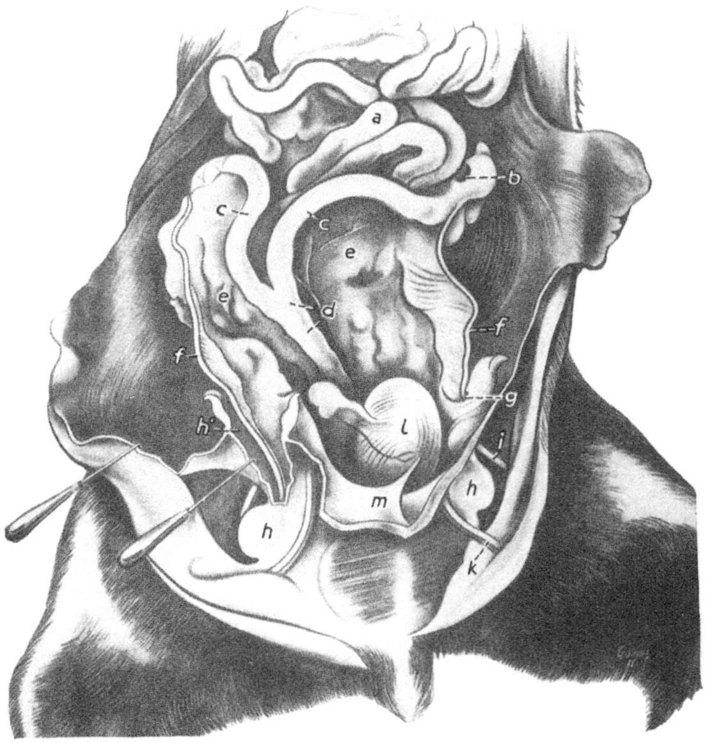

Fig. 540. Uterus, its ligaments, and exposed processus vaginalis of a bitch. Ventral aspect. (After Zietzschmann 1928.)

a Jejunum; *b* Ovary in ovarian bursa; *c* Uterine horns; *d* Body and cervix of uterus; *e* Broad ligament; *f* Round ligament of uterus; *g* Vaginal ring; *h* Processus vaginalis; *h'* Cavity of opened processus vaginalis; *i* Femoral artery; *k* Artery and vein of round ligament of uterus (drawn too large); *l* Urinary bladder; *m* Median ligament of bladder

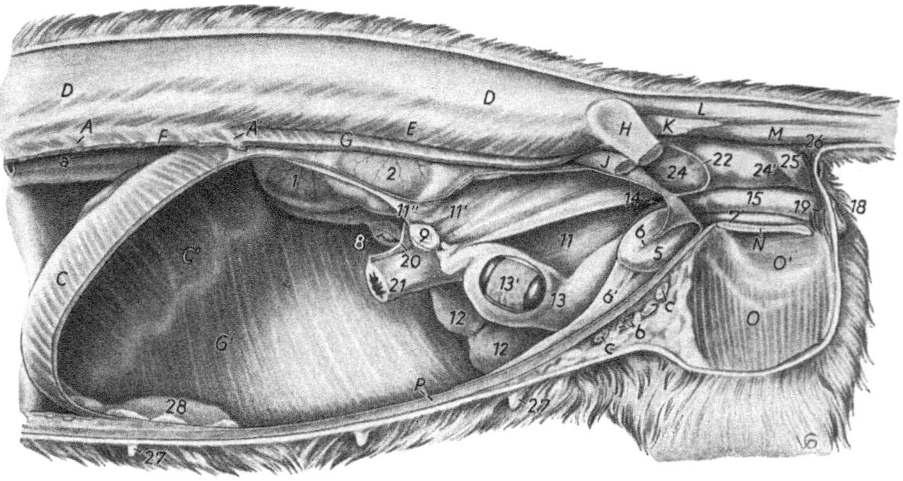

Fig. 541. Urinary and genital organs of a pregnant cat in situ. Left lateral aspect.

A Tenth rib; *A'* Thirteenth rib; *C, C'* Costal part of diaphragm; *D* Longissimus; *E* Iliocostalis; *F* Intercostal muscles; *G* Origin of left transversus abdominis; *G'* Right transversus abdominis, covered with transverse fascia and peritoneum; *H* Wing of ilium; *J* Iliopsoas; *K* Gluteus supf.; *L* Sacrocaudalis dorsalis lateralis; *M* Sacrocaudalis ventralis lateralis; *N* Floor of pelvis; *O* Gracilis; *O'* Symphysial tendon; *P* Rectus abdominis; *a* Aorta; *b* Ext. pudendal artery and vein; *c* Supf. inguinal lymph nodes

1 Right kidney; *2* Left kidney; *5* Urinary bladder; *6* Left lateral lig. of bladder; *6'* Median lig. of bladder; *7* Urethra; *8* Right ovary in ovarian bursa; *9* Left ovary in ovarian bursa; *11* Mesometrium; *11'* Mesovarium; *11''* Suspensory lig. of ovary; *12* Loculi on right uterine horn; *13* Left uterine horn with opened loculus; *13'* Exposed zonary placenta; *14* Body of uterus; *15* Vagina; *18* Vulva; *19* Constrictor vulvae; *20* Mesocolon; *21* Descending colon; *22* Pelvic fascia and parietal peritoneum, fenestrated to expose rectum; *24* Rectum, covered with visceral peritoneum; *24'* Retroperitoneal part of rectum; *25* Anal sac; *26* Ext. anal sphincter; *27* Teat of mammary gland; *28* Fatfilled falciform ligament

Female Genital Organs of the Pig

(516, 520, 527, 542—545)

The pig is a multiparous and polyestrous animal with a sexual cycle of 21 days. The sow is in heat for three to five days, and pregnant on the average for 115 days. She usually gives birth to 9—12, more rarely, up to 16 young.

Ovaries

The ovaries of the sow (542/*1*) are cylindrical, about 5 cm. long, and very uneven because of the many follicles and corpora lutea on the surface. The **corpora lutea** are either cherry or bright grayish red, or whitish. In sows that have borne young, the **mesovarium** is long, giving the ovaries considerable range and therefore an inconstant position between the kidneys and the pelvic inlet (545). The **proper ligament of the ovary** (542/*4*) is muscular, arises from the uterine end of the ovary, and radiates into the mesometrium. The **mesosalpinx** (520, 543/*b*), an extensive, gossamer, and very vascular serosal fold, arises from the lateral surface of the mesovarium and the adjacent part of the mesometrium. It contains the uterine tube and forms a large cone-shaped **ovarian bursa** which encloses the ovary. The entrance to the bursa is wide and directed ventrally. (Akins and Morisette [1968], using 79 nulliparous gilts, published color photographs showing the typical appearance of the ovary for each day of the estrous cycle.)

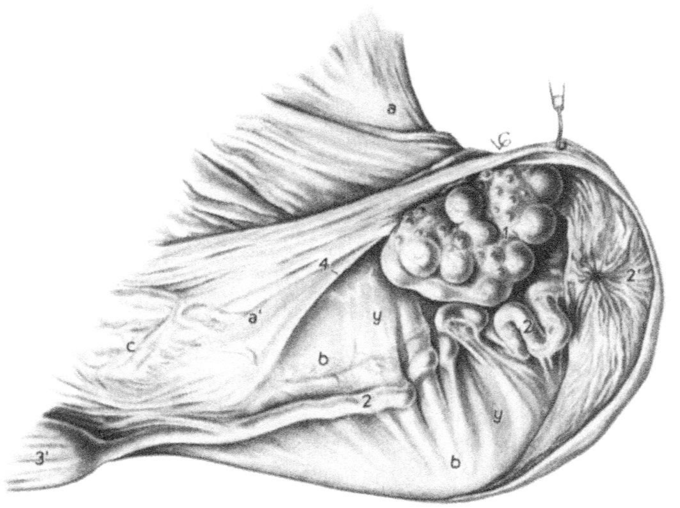

Fig. 542. Left ovary, uterine tube, and associated ligaments of the sow. Medial aspect.

a, a', c Broad ligament; *a* Mesovarium; *b* Mesosalpinx; *c* Mesometrium, *y* Ovarian bursa; *1* Ovary with several corpora lutea and small vesicular follicles; *2* Uterine tube; *2'* Infundibulum with abdominal opening of uterine tube; *3'* Tip of uterine horn; *4* Proper ligament of ovary

Uterine Tube

The thin-walled **infundibulum** (542/*2'*) with the abdominal opening of the uterine tube lies on the inner surface of the mesosalpinx and faces the ovary. The uterine tube, 19—22 cm. long, takes a tortuous course at first in the medial wall of the ovarian bursa, then runs over the apex of the cone-shaped bursa into the lateral wall, and ends at the uterine opening in the tip of the uterine horn, which is only 3 cm. away from the ovary (520, 542).

Uterus

The uterus if the pig is remarkable for its length. In mature sows, the **horns** are 120—140 cm. long and, in the nongravid state, arranged like intestinal coils. They are suspended by extensive muscular **broad ligaments** (545/*11*), which give the uterine horns enough range that they intermingle with the intestines and occasionally reach the abdominal floor. The uterine horns converge caudally and are united for a short distance before opening separately into the slightly expanded **body** of the uterus which is about 5 cm. long (543). The endometrium is gray or bluish red, very vascular, and arranged in folds of varying height, which may be particularly high in the body of the uterus.

The **CERVIX** of the sow is also very long. It is hard and cordlike, 15—25 cm. long, and lies with its cranial part in the abdominal cavity (545/*14'*). The **cervical canal** is lined with a light pink mucosa arranged in longitudinal folds. Projecting into the cervical canal from the wall are rows of firm prominences (pulvini cervicales, 543, 544/*d*), which interdigitate and, together with the circular muscle coat, effectively occlude the cervix. They are highest in the center and decrease in height toward both ends. The cervical canal gradually widens at the cervico-vaginal junction so that the **external uterine orifice** is not well marked and an intravaginal part of the cervix is not formed. The junction between cervix and vagina (*b''*) is at the level where the lowest of the cervical prominences fade into the longitudinal folds of the vagina.

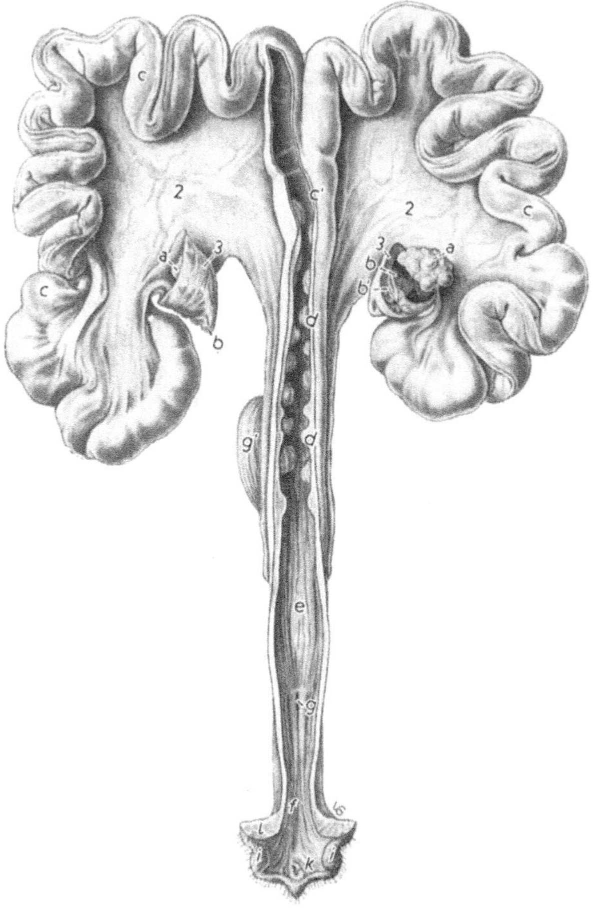

Fig. 543. Genital organs of the sow. Dorsal aspect. Vestibule, vagina, and part of uterus have been opened dorsally.

a Right ovary with corpora lutea and small vesicular follicles; *a'* Left ovary in ovarian bursa; *b* Uterine tube in mesosalpinx; *b'* Infundibulum with abdominal opening of uterine tube; *c* Uterine horns; *c'* Body of uterus; *d* Cervix, opened to show the rows of prominences (pulvini cervicales) in the cervical canal; *e* Vagina; *f* Vestibule; *g* Ext. urethral orifice; *g'* Urinary bladder; *i* Labia; *k* Fossa clitoridis with tip of glans clitoridis in its center; *l* Constrictor vulvae

2 Broad ligaments; *3* Mesosalpinx, forming most of the cone-shaped ovarian bursa, on the left enclosing the ovary, on the right reflected to expose the ovary

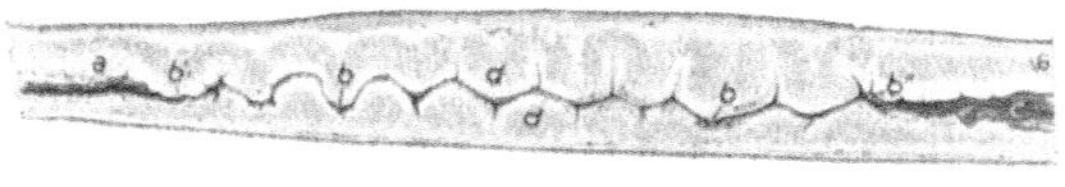

Fig. 544. Median section of cervix uteri of the sow.

a Lumen of uterine body; *b* Cervical canal; *b'* Int. uterine orifice; *b''* Junction of cervix and vagina (ext. uterine orifice); *c* Lumen of vagina; *d* Interdigitating prominences (pulvini cervicales), occluding cervical canal

Vagina, Vestibule, and Vulva

The muscular **VAGINA** is 10—12 cm. long. Its mucous membrane is covered with stratified squamous epithelium and is arranged in high longitudinal folds (543/*e*). At the junction with the vestibule, in the virginal animal, is an annular fold 1—3 mm. high, homologous to the hymen. Just caudal to it is the external orifice of the long urethra (545/7), which opens together with a small **suburethral diverticulum** on the floor of the vestibule.

The bluish red mucous membrane of the **VESTIBULE** forms longitudinal folds and contains solitary lymph nodules. Two rows of small round openings in its floor indicate the position of the **minor vestibular glands.**

The labia of the **VULVA** are round and wrinkled and bear only a few hairs, and the ventral commissure is drawn out to a point (527). The body of the **clitoris** is tortuous and up to 8 cm. long. It lies under the floor of the vestibule and ends with a poorly developed glans in the fossa clitoridis (543/*k*). The lining of the fossa adheres to the epithelial covering of the glans so that only the tip of the glans is visible.

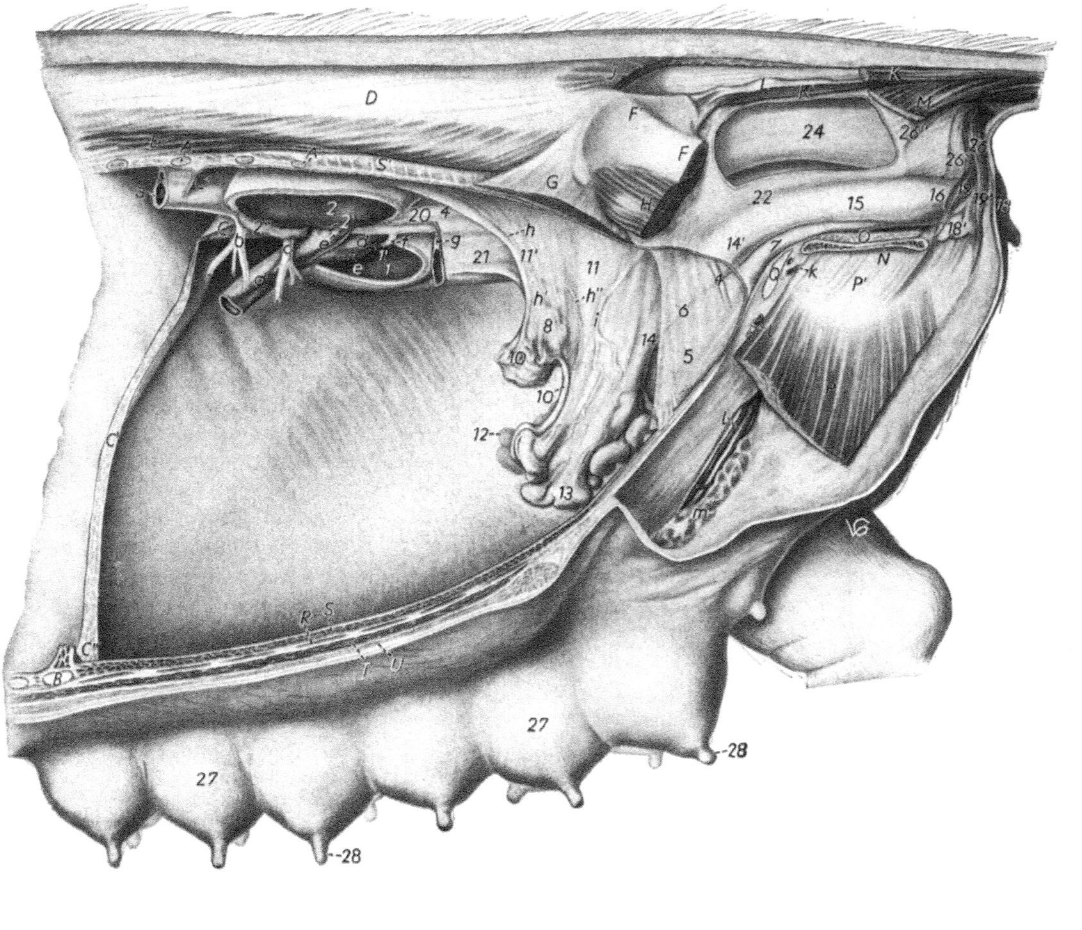

Fig. 545. Urinary and genital organs of a parous sow in situ. Left lateral aspect.

A Twelfth rib; *A'* Fourteenth rib; *B* Costal cartilage; *C* Right crus of diaphragm; *C'* Costal part of diaphragm; *C''* Sternal part of diaphragm; *D* Longissimus; *E* Iliocostalis; *F*, *F'* Ilium; *G* Int. abdominal oblique; *H* Iliopsoas; *J* Gluteus medius; *K* Sacrocaudales dorsales; *K'* Intertransversarius; *L* Sacrum; *M* Coccygeus; *N* Pelvic symphysis; *O* Obturator internus; *P* Gracilis; *P'* Symphysial tendon; *Q*, *R* Rectus abdominis; *S* Transversus abdominis; *S'* Origin of left transversus abdominis on the lumbar transverse processes; *T* Ext. abdominal oblique; *U* Cutaneus trunci

a Aorta; *b* Celiac artery; *c* Cranial mesenteric artery; *d* Caudal vena cava; *e*, *e* Renal veins; *f* Renal artery; *g* Left colic artery; *h*, *h'* Left ovarian artery; *h''* Uterine branch of left ovarian artery; *i* Uterine artery; *k* Deep femoral artery and vein; *l* Ext. pudendal artery and vein; *m* Supf. inguinal lymph nodes

1 Right kidney; *1'* Hilus; *2* Left kidney; *2'* Hilus; *2''* Left adrenal gland; *4* Left ureter; *5* Urinary bladder; *6* Lateral lig. of bladder; *7* Urethra; *8* Left ovary in ovarian bursa; *10* Uterine tube; *11* Mesometrium; *11'* Mesovarium; *12* Right uterine horn; *13* Left uterine horn; *14* Body of uterus; *14'* Cervix; *15* Vagina; *16* Vestibule; *18* Vulva; *18'* Crus clitoridis; *19* Constrictor vestibuli; *19'* Constrictor vulvae; *20* Descending mesocolon; *21* Descending colon; *22* Pelvic fascia and parietal peritoneum, fenestrated to expose rectum; *24* Rectum; *26* Ext. anal sphincter; *26'* Rectal part of retractor clitoridis; *26''* Rectococcygeus; *27* Mammary glands; *28* Teats of mammary glands

Female Genital Organs of the Ruminants

(505—512, 517, 521, 523, 524, 528, 532, 546—556)

The ruminants are unipara. The cow gives birth occasionally to twins; triplets and quadruplets have also been seen. In sheep, twin births are common; in some breeds over 50 per cent of the young are twins, but triplets are rare. In goats, up to 75 per cent of the young are twins, and triplets are not uncommon.

The domestic ruminants are polyestrous, with a sexual cycle of three weeks. Heat lasts 2—30 hrs. (average 14) in the cow, 24—36 hrs. in the sheep, and 48 hrs. in the goat. The length of pregnancy in the cow is nine months (281—290, with an average of 285 days), five months (149 days) in the sheep, and about the same length (154 days) in the goat.

The description that follows is based on the cow. The female genital organs of the small ruminants are similar to those of the cow, except for being smaller and for some other differences described on page 384.

Ovaries

The ovary of the cow is oval, laterally flattened, and remarkably small. On an average it is about 4 cm. long, 2 cm. wide, and 1—2 cm thick, though these measurements may vary widely. Near its attachment (margo mesovaricus) it is covered with peritoneum and is smooth, but by far the greatest part of its surface is covered with cuboidal or columnar epithelium, which is not as shiny as the peritoneum, and is uneven because of the presence of follicles and corpora lutea in various stages (546/*1*). Mature **follicles** with a diameter of up to 2 cm. (505), and **corpora lutea** measuring up to 3 cm. longitudinally (510b) protrude from the surface of the ovary enough that they can be identified with certainty on rectal palpation. Because of the short estrous cycle, corpora lutea in various stages of growth or regression are regularly present on the ovaries of healthy cows. Cyclic corpora lutea are an intense orange-yellow; those in regression, however, are more rust colored and later whitish.

The ovaries of mature cows are located near the lateral margin of the pelvic inlet slightly below its middle (548). They are suspended dorsal to the ends of their respective uterine horns by a tough muscular **mesovarium** (547B/*1*), and are fixed in addition by the short, strong **proper ligaments** (*4*) which extend from the uterine end of the ovaries to the mesometrium.

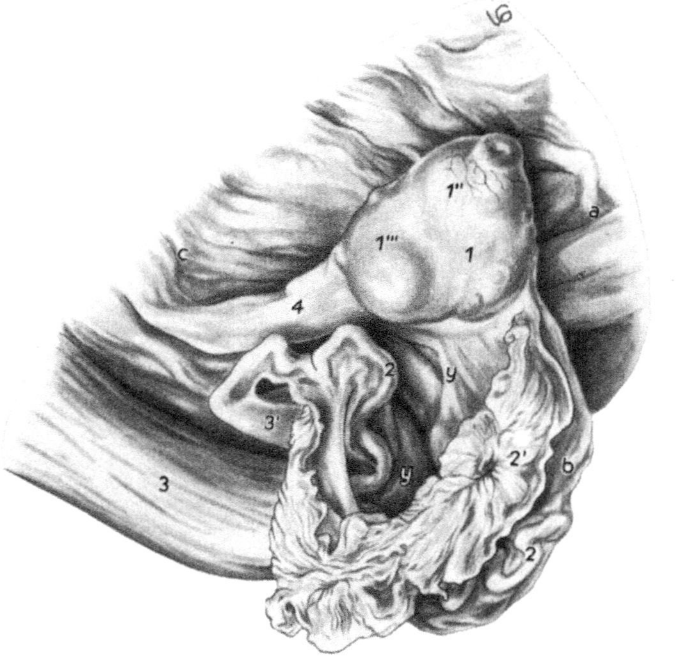

Fig. 546. Left ovary, uterine tube, and associated ligaments of the cow. Medial aspect. The uterine horn has been reflected ventrally, and the ovary dorsally.

a Mesovarium; *b* Mesosalpinx; *c* Mesometrium; *y* Interior of ovarian bursa

1 Ovary, lifted from ovarian bursa; *1''* Corpus luteum; *1'''* Vesicular follicle; *2* Uterine tube; *2'* Infundibulum with abdominal opening of uterine tube; *3* Uterine horn; *3'* Tip of uterine horn; *4* Proper ligament of ovary

The **mesosalpinx** (521/*b*) is delicate and translucent. It originates from the ventrolateral surface of the mesovarium and envelops the ovary cranially and laterally so that the ovary is usually inside the ovarian bursa. The entrance to the bursa is wide and directed cranioventromedially.

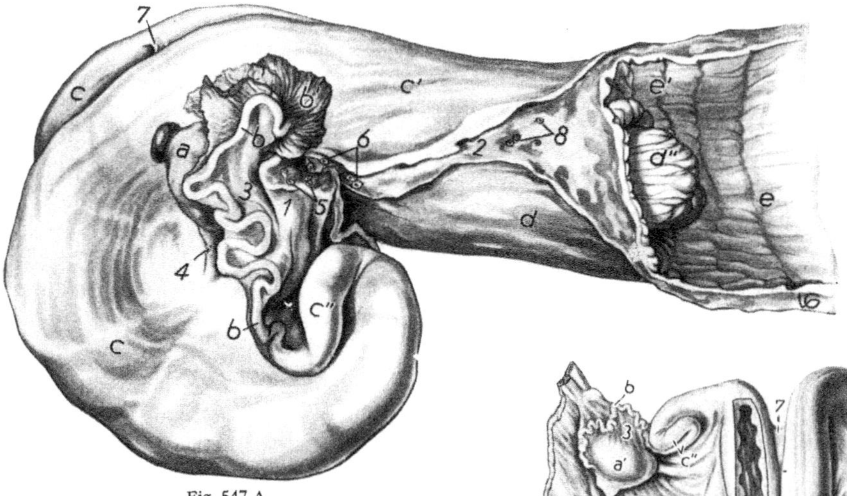

Fig. 547 A

Fig. 547. Genital organs of the cow.

A Left lateral aspect: the left wall of the vagina has been removed to expose the cervix; *B* Dorsal aspect: most of the genital tract has been opened dorsally. The left ovary is enclosed by the mesosalpinx and has been lifted dorsally and caudally to lie on the dorsal surface of the mesovarium (1). The right ovary has been similarly elevated but without the mesosalpinx which still hangs from the cranial free edge of the mesovarium.

a Ovary with growing corpus luteum and small follicles; *a'* Left ovary in ovarian bursa; *b* Uterine tube in mesosalpinx; *b'* Infundibulum with abdominal opening of uterine tube; *c* Uterine horn; *c'* Body of uterus; *c''* Tip of uterine horn; *d* Cervix; *d'* Int. uterine orifice; *d''* Intravaginal part of cervix with ext. uterine orifice; *e* Vagina; *e'* Fornix vaginae; *f* Vestibule; *f', f'* Openings of the ductus epoophori longitudinales (ducts of Gartner); *g* Entrance to suburethral diverticulum and ext. urethral orifice; *g'* Bladder; *g''* Ureter; *h, h* Openings of major vestibular glands, note outline of glands through the mucous membrane; *h'* Depression into which the ducts of the minor vestibular glands open; *i* Labium; *k* Glans and adhering prepuce of clitoris; *l* Constrictor vestibuli and constrictor vulvae

1, 2 Broad ligament; *1* Mesovarium; *2* Mesometrium; *3* Mesosalpinx, in Fig. B left, covering the ovary, and right, reflected to expose the ovary and the interior of the ovarian bursa; *4* Proper lig. of ovary; *5* Branches of ovarian artery and vein; *6* Branches of uterine artery and vein; *7* Dorsal intercornual lig.; *8* Branches of the ramus uterinus of the urogenital artery and vein

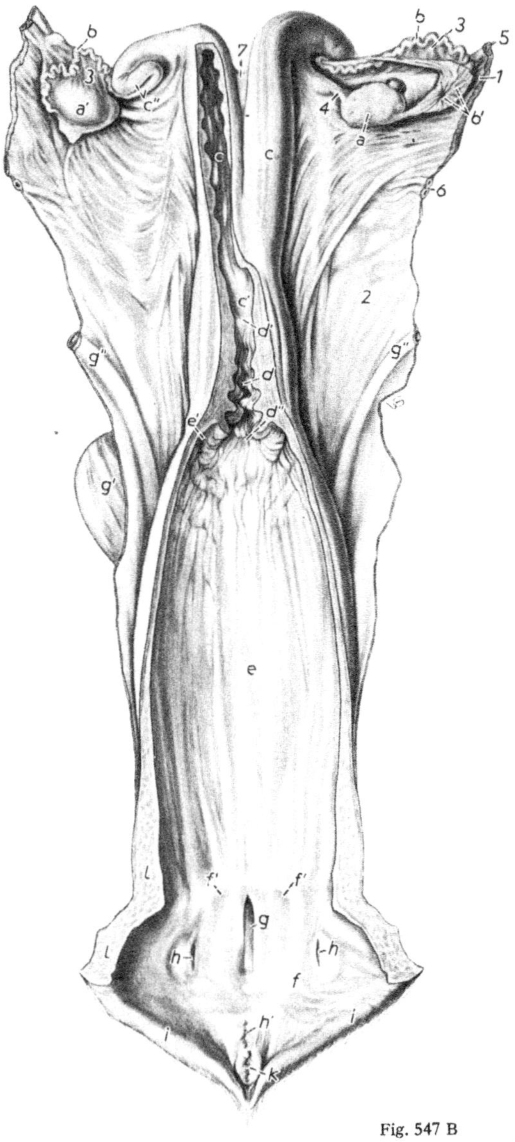

Fig. 547 B

Uterine Tube

The uterine tube of the cow (546/*2*) is 20—28 cm. long and only moderately tortuous. Its fimbriated **infundibulum** (*2'*), which is large and may envelop the ovary completely, is attached to the internal surface of the mesosalpinx near its free edge and, with the abdominal opening of the tube, faces the ovary. From the abdominal opening, the uterine tube passes cranially in the medial part of the mesosalpinx, and turns laterally into the lateral wall of the ovarian bursa in which it passes caudally to the uterus, gradually widening at its end to the diameter of the uterine horn.

Uterus

The **uterine horns** of the cow are 35—45 cm. long and gradually taper from the body of the uterus to their junction with the uterine tubes. The horns pass cranially from the body of the uterus and are at first closely united by a common peritoneal covering (perimetrium) and by connective and muscular tissue, giving the false impression that the body of the uterus of the cow is relatively long. The horns (547A/*c*) then form two diverging spirals, passing first ventrally, then caudally, and finally dorsally. At the point of divergence they are connected by dorsal and ventral **intercornual ligaments** (548/*7, 7'*), between which is a shallow recess open cranially. The tip of the uterus is S-shaped and blends with the uterine tube (547A). The **mesometrial border** of the uterine horn, i. e., the border to which the mesometrium

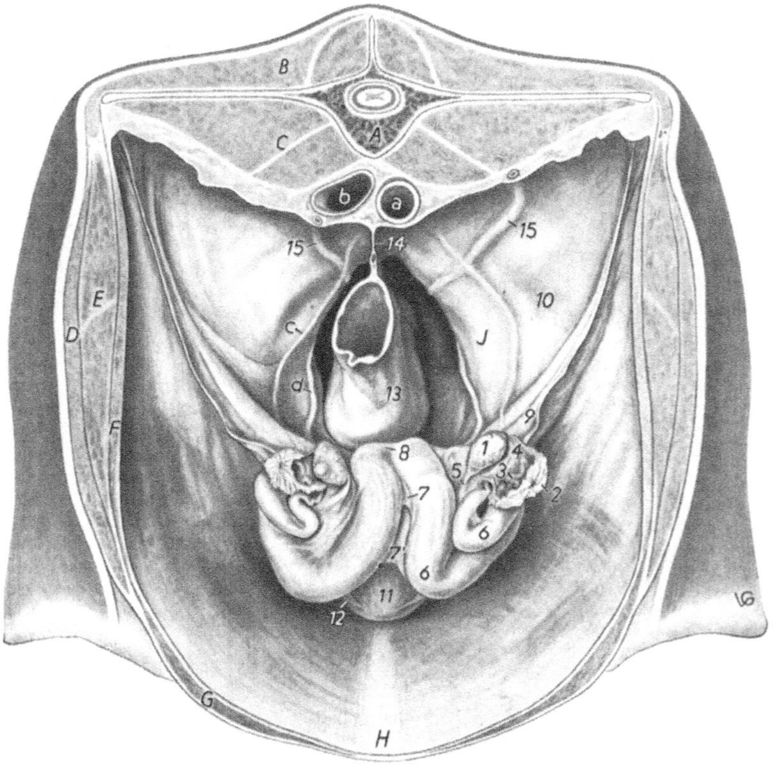

Fig. 548. Pelvic organs of a cow in situ. Cranial aspect. The right ovary of the cow has been lifted dorsally. The left mesosalpinx has been pulled ventrally away from its ovary.

A Fifth lumbar vertebra; *B* Epaxial muscles; *C* Hypaxial muscles; *D* Ext. abdominal oblique; *E* Int. abdominal oblique; *F* Transversus abdominis; *G* Rectus abdominis; *H* Linea alba; *J* Contour of the shaft of the ilium and the psoas minor and iliopsoas muscles; *a* Aorta; *b* Caudal vena cava; *c* Ovarian artery; *d* Uterine artery

1 Left ovary; *2* Infundibulum with abdominal opening of uterine tube; *3* Left uterine tube; *4* Mesosalpinx; *5* Proper lig. of ovary; *6* Left uterine horn; *7, 7'* Intercornual ligaments; *8* Body of uterus; *9, 10* Broad ligament; *9* Mesovarium; *10* Mesometrium; *11* Bladder; *12* Lateral lig. of bladder; *13* Descending colon; *14* Descending mesocolon with cranial rectal artery and vein; *15* Ureter

is attached, faces the inside of the spiral. In cows that have been pregnant repeatedly, the spirals formed by the uterine horns are somewhat flatter. The **body of the uterus** (*c'*) is only about 3 cm. long, and its lumen is continuous cranially with the lumina of the uterine horns through two separate openings.

The uterus, particularly of older cows, has a thick wall and is firm to the touch. Its **endometrium** is grayish or bluish red and forms longitudinal and transverse folds. It presents the characteristic **caruncles** (carunculae), which are mucosal prominences distributed over the internal surface of the uterus, usually in four irregular rows. Most of them are round or oval, some have a shallow central depression. Shape and size of the caruncles vary with the functional state of the uterus and the age of the cow. They increase tremendously in size during pregnancy, attaining a length of up to 10 cm., and becoming porous in appearance. The caruncles are the maternal part of the **cotyledonary bovine placenta** and, with the fetal

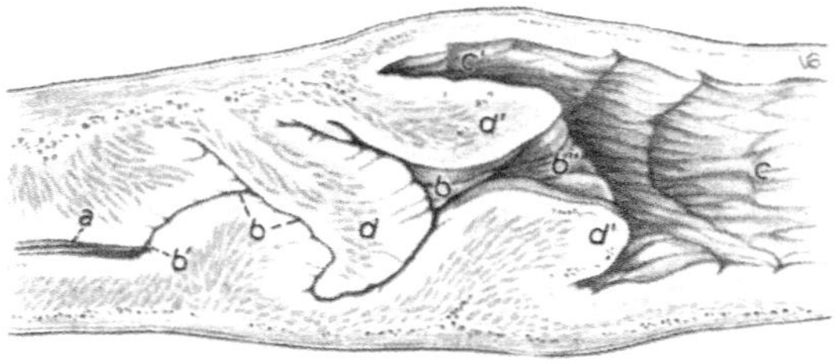

Fig. 549. Median section through the cervix of a cow.

a Lumen of body of uterus; *b* Cervical canal; *b'* Int. uterine orifice; *b''* Ext. uterine orifice; *c* Vagina; *c'* Fornix vaginae; *d, d'* Plicae circulares; *d'* Intravaginal part of cervix

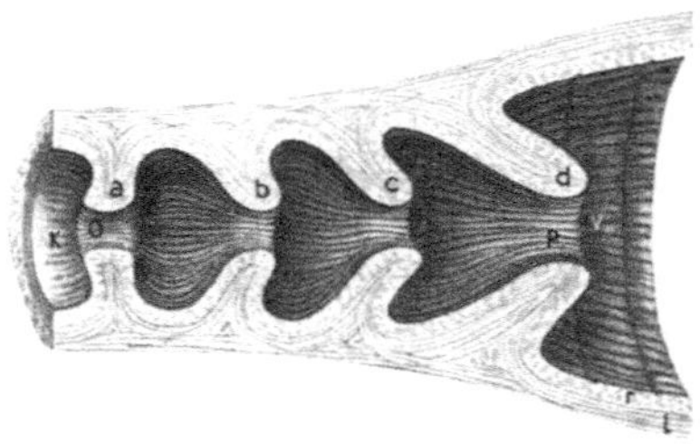

Fig. 550. Median section through the cervix of a calf. Semi-schematic. (After Preuss 1953.)

K Body of uterus; *O* Int. uterine orifice; *P* Ext. uterine orifice in the center of intravaginal part of cervix; *V* Vagina; *a—d* Plicae circulares; *a* Plica circularis surrounding int. uterine orifice; *d* Plica circularis surrounding ext. uterine orifice; Longitudinal muscle layer; *r* Circular muscle layer

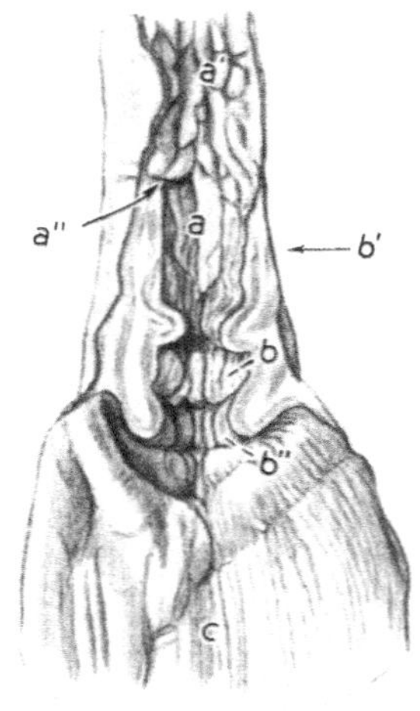

Fig. 551 (Calf)

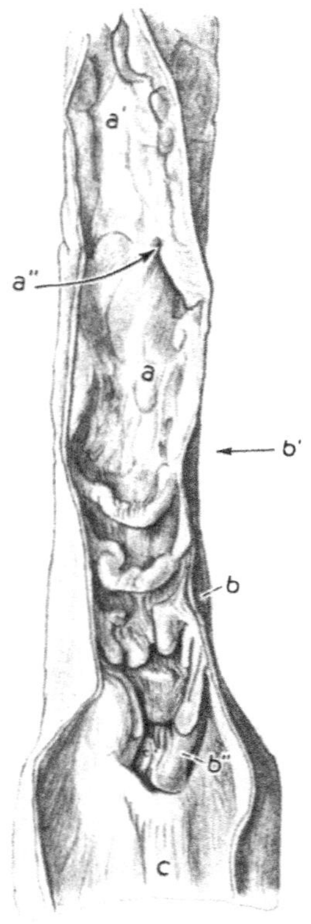

Fig. 552 (Sheep)

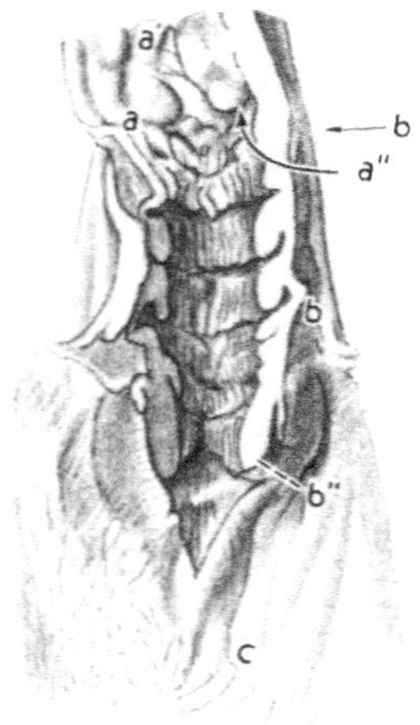

Fig. 553 (Goat)

Figs. 551—553. Cervix of calf, sheep, and goat, opened dorsally. (After Seiferle 1933.)

a Body of uterus; *a'* Opened left uterine horn (sheep and goat) and right uterine horn (calf); *a''* Entrance to right (sheep and goat) and left (calf) uterine horn, marked by arrow; *b* Opened cervix with plicae circulares; *b'* Int. uterine orifice; *b''* Ext. uterine orifice; *c* Vagina

part, the cotyledon, form the **placentome** (532A). Deep crypts on the surface of the caruncle contain the branching villi of the cotyledon, ensuring the most intimate contact between the material and fetal tissues.

The uterus is related to the dorsal sac of the rumen, to the intestines, and ventrally to the urinary bladder. It projects into the abdominal cavity (555), farther in mature parous cows than in heifers, and can be palpated rectally.

The **broad ligaments** (548/*9, 10*) are very muscular and extensive, especially in cows with several pregnancies. They originate from the lateral wall of the pelvic cavity and more cranially from the area ventral to the tuber coxae, and end along the mesometrial border of the uterine horns. The **round ligament of the uterus** arises from the lateral surface of the broad ligament. The smooth muscle fibers in the broad ligaments elevate the uterus toward the rectum, or in the more relaxed state let it sink ventrally.

The **CERVIX** of the cow's uterus (547B/*d*) is a thick-walled, firm tube that can readily be distinguished from the softer uterus in front and the vagina behind by palpation of the excised genital tract or rectally. It is 6—7 cm. long in young cows and about 10 cm. in older cows. The wall of the cervix consists of an external longitudinal muscle layer, a vascular layer, an internal circular muscle layer containing very dense, tough connective tissue, and a pale mucous membrane. Characteristic of the bovine cervix are three to four firm **plicae circulares** (549/*d, d'*) projecting into the lumen. These are of varying height and may be annular, spiral, or falciform in shape. They are formed by the mucous membrane, but also include portions of the inner circular muscle layer. The most cranial of these is more or less annular and surrounds the internal uterine orifice (*b'*). The most caudal is also annular, surrounds the external uterine orifice (*b''*), and projects into the vagina. In addition to the plicae circulares, the mucous membrane of the cervical canal forms secondary longitudinal folds of varying height, which pass over the plicae and give them a notched or foliated appearance. The secondary folds, after passing over the intravaginal part of the cervix, continue along the wall of the vagina (*c*). The plicae circulares make the cervical canal very tortuous and when tightly interdigitating in the normal closed state render the cervix impassable. The cervical seal is completed by the secretion of a mucous plug, very tenacious during pregnancy, which fills all the crevices and recesses of this irregular passage, including the external uterine orifice where the plug can be palpated.

Vagina, Vestibule, and Vulva

The **VAGINA** of the cow (547/*e*) is about 30 cm. long, has muscular thick walls, and is capable of great distension, although its lumen is collapsed in situ. Because of the caudoventral inclination of the intravaginal part of the cervix, the **fornix vaginae** (549/*c'*) is deeper dorsally than ventrally. The cranial part of the vagina and, of course, the entire cervix, are covered dorsally with peritoneum and help form the rectogenital pouch (554/*22*) making it possible, therefore, to enter the peritoneal cavity through the vagina and the dorsal wall of the fornix, as is done in one of the methods (vaginal approach) for the removal of the ovaries in mature cows. The mucous membrane of the vagina bears stratified squamous epithelium and forms longitudinal and transverse folds. At the junction with the vestibule just cranial to the external urethral orifice is a low transverse fold (rudimentary hymen).

The **VESTIBULE** in the cow (547B/*f*) is short and, with the vulva, hangs down from the deeply concave ischiatic arch. Because of this, the **external urethral orifice** is easily located, as it can often be seen when the labia are parted. The orifice is in the dorsal wall of the **suburethral diverticulum** (547B/*g*; 555/*10*), a small mucosal pouch about 2 cm. deep and 2 cm. wide, with the blind end cranial. The diverticulum must be avoided when catheterizing a cow. On each side of the external urethral orifice are the elongated openings of the ductus epoophori longitudinales (canals of Gartner, 547B/*f'*), and caudal to these are the **major vestibular glands** (*h*) which are 2—3 cm. in diameter and flat. Their ducts open into small mucosal recesses on the floor of the vestibule, a short distance caudal and lateral to the external urethral orifice. There are also many pale, somewhat translucent lymph nodules in the yellowish-brown mucosa of the vestibule.

The labia of the **VULVA** are rounded and slightly wrinkled, and meet in a pointed ventral commissure. They are covered with fine hair, but have a tuft of longer hairs ventrally (528).

The two crura of the **clitoris** originate from the ischiatic arch and unite to form a sinuous body (555/*14*), about 12 cm. long, which passes in the floor of the vestibule to the ventral commissure of the vulva. The **glans clitoridis** usually adheres to the internal lamina of the **preputium clitoridis** so that, at least dorsally, a fossa clitoridis is not present. Therefore,

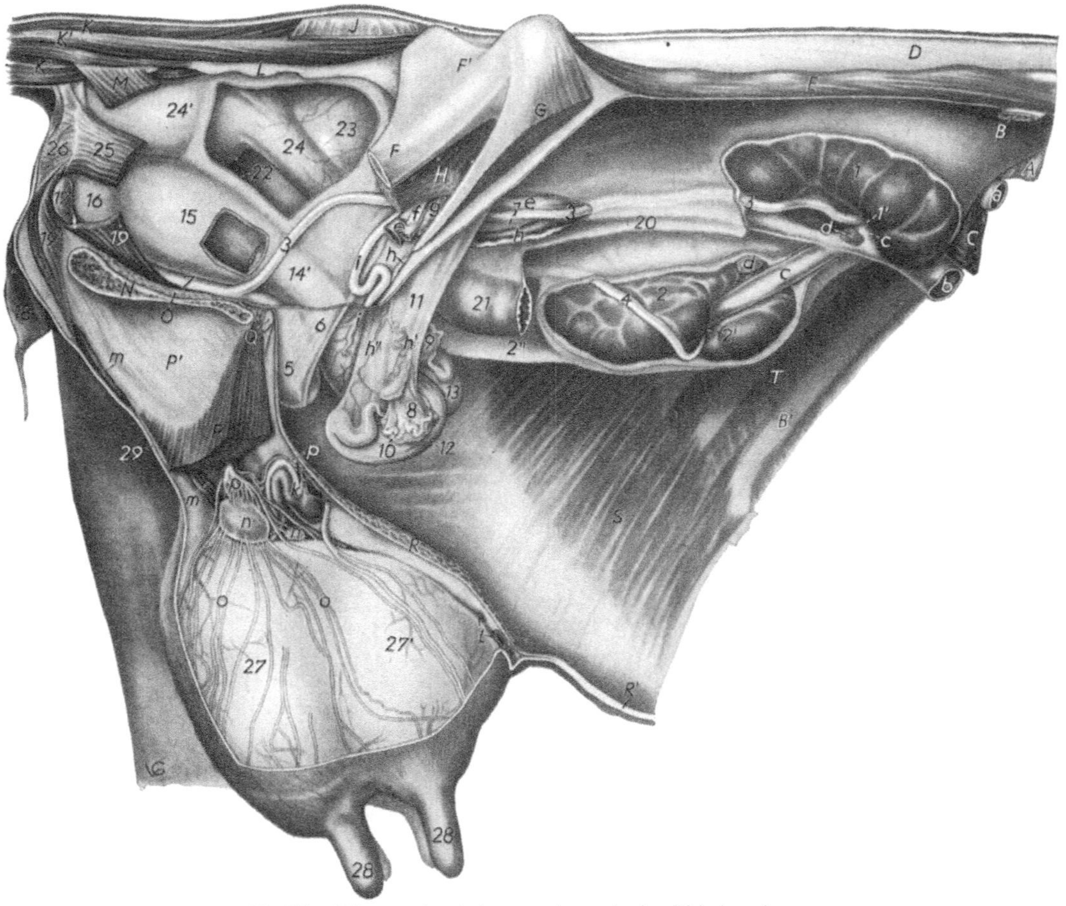

Fig. 554. Urinary and genital organs of a cow in situ. Right lateral aspect.

A Thirteenth thoracic vertebra; *B* Right thirteenth rib; *B'* Left eleventh rib; *C* Right crus of diaphragm; *D* Longissimus; *E* Iliocostalis and tips of lumbar transverse processes; *F, F'* Ilium; *G* Int. abdominal oblique; *H* Iliopsoas; *J* Biceps femoris; *K* Sacrocaudalis dorsalis; *K'* Intertransversarius; *K'''* Sacrocaudalis ventralis; *L* Lateral part of sacrum; *M* Coccygeus; *N* Symphysis pelvis; *O* Obturator internus; *P* Gracilis; *P'* Symphysial tendon; *Q* Cranial pubic ligament; *R* Rectus abdominis; *R'* Linea alba; *S* Transversus abdominis, covered by transverse fascia and peritoneum; *T* Int. intercostal muscles

a Aorta; *b* Vena cava; *c* Renal veins; *d* Renal lymph nodes; *e* Ext. iliac a. and v.; *f* Femoral nerve; *g* Iliofemoral (deep inguinal) l. n.; *h, h'* Ovarian artery; *h''* Uterine branch of ovarian artery; *i* Uterine artery; *k* Ext. pudendal artery and vein; *l* Branches of mammary vein; *m* Perineal vein; *n* Supf. inguinal (supramammary) lymph nodes; *o* Afferent lymphatics of *n*; *p* Efferent lymphatics of *n*

1, 2 Right and left kidneys, exposed by removal of perirenal fat; *1', 2'* Hilus of right and left kidneys; *2''* Fatfilled sagittal fold of peritoneum caudal to left kidney; *3* Right ureter; *4* Left ureter; *5* Bladder; *6* Right lateral lig. of bladder; *7* Urethra; *8* Right ovary in ovarian bursa; *9* Left ovarian bursa; *10* Right uterine tube; *11* Broad ligament; *12* Right uterine horn; *13* Left uterine horn; *14'* Cervix; *15* Vagina, fenestrated to expose intravaginal part of cervix; *16* Vestibule; *17* Major vestibular gland; *18* Vulva; *19* Constrictor vestibuli; *19'* Constrictor vulvae; *20* Mesocolon; *21* Descending colon; *22* Pelvic fascia and parietal peritoneum, fenestrated to expose rectum; *23* Mesorectum; *24* Rectum; *24'* Ampulla recti, part of it retroperitoneal; *25* Levator ani; *26* Ext. anal sphincter; *27, 27'* Right quarters of udder, exposed by removal of skin; *28* Teats; *29* Skin of urogenital region

the slight protuberance (547B/*k*), visible in the ventral end of the vulvar cleft when the labia are parted, consists of both glans and prepuce. Only the cone-shaped tip of the glans is free and projects from the center of the protuberance. Occasionally, however, the adhering mucous membranes separate in places, and the fossa between glans and prepuce becomes apparent.

Small Ruminants

The **OVARIES** of the sheep and goat are oval to round, about 1.5 cm. long and 1—1.8 cm. wide, and have an uneven surface. Because of the high incidence of twin births, two corpora lutea of the same cycle may be present either both on one ovary or one on each ovary. The corpora lutea are large and when fully developed make the ovary almost twice its usual size. On section, the corpora lutea are grayish red or, in regression, more whitish. The **mesovarium** holds the ovaries in position lateral to the coiled uterine horns at about the level of the fifth lumbar vertebra. The proper ligament of the ovary is muscular and extends from the uterine end of the ovary to the mesometrium. A suspensory ligament can be demonstrated in both species. The **mesosalpinx** is extensive and forms a deep conical **ovarian bursa** which has a wide entrance and covers the lateral surface of the ovary.

The **UTERINE TUBE** is remarkably long (14—15 cm.). Its **infundibulum** is attached to the tubal end of the ovary and to the mesosalpinx and, with the abdominal opening of the tube, faces the ovary. It is more tortuous than in the cow and passes in a large curve from medial to lateral around the cranial end of the ovary and then in the lateral wall of the bursa to the uterine horn.

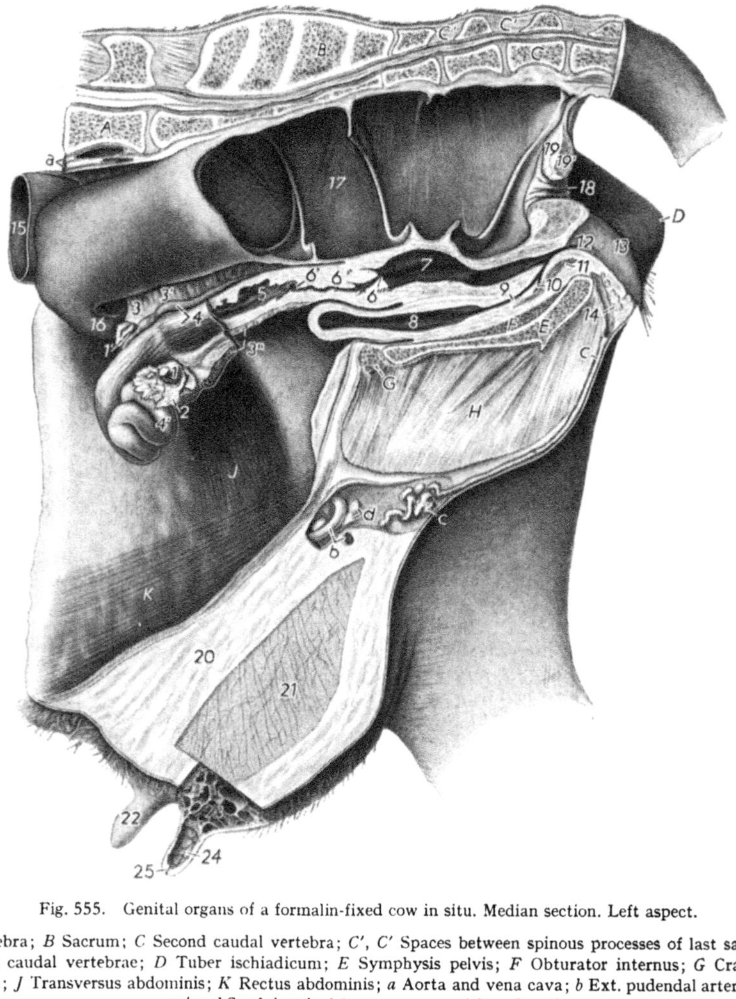

Fig. 555. Genital organs of a formalin-fixed cow in situ. Median section. Left aspect.

A Sixth lumbar vertebra; *B* Sacrum; *C* Second caudal vertebra; *C'*, *C'* Spaces between spinous processes of last sacral and first caudal, and first and second caudal vertebrae; *D* Tuber ischiadicum; *E* Symphysis pelvis; *F* Obturator internus; *G* Cranial pubic ligament; *H* Symphysial tendon; *J* Transversus abdominis; *K* Rectus abdominis; *a* Aorta and vena cava; *b* Ext. pudendal artery and vein; *c* Perineal vein; *d* Supf. inguinal (supramammary) lymph nodes

1 Left ovary with corpus luteum; *1'* Right ovary; *2* Infundibulum of uterine tube; *3* Right mesovarium; *3'* Right mesometrium; *3''* Distal stump of left mesometrium with cross sections of uterine vessels; *4* Uterine horns; *4'* Tip of left uterine horn; *5* Body of uterus; *6* Cervix; *6'* Int. uterine orifice; *6''* Ext. uterine orifice; *7* Vagina; *8* Bladder, empty; *9* Urethra; *10* Suburethral diverticulum; *11* Urethralis and constrictor vestibuli; *12* Vestibule; *13* Right labium; *14* Body of clitoris; *15* Descending colon; *16* Sigmoid colon; *17* Rectum with plicae transversales; *18* Anus; *19* Int. anal sphincter; *19'* Ext. anal sphincter; *20* Medial lamina of suspensory apparatus of udder; *21* Gland tissue of right quarters of udder with blood vessels and lymphatics; *22* Teat of right fore quarter; *23* Gland sinus; *24* Teat sinus; *25* Teat canal

Except for the interior of the cervix, the **UTERUS** of the small ruminants is similar to that of the cow. The uterine horns are relatively longer, coiled, and are united caudally by a common peritoneal covering. There is only one intercornual ligament at the bifurcation of the horns (523). The body of the uterus is about 3 cm. long. The endometrium is grayish pink, becoming brownish yellow in older animals, and forms folds on which are the **caruncles** (*b'*). These are usually convex, but in animals that have passed through several pregnancies they may have a more or less deep depression on the free surface (532BC/*a''*). In the ewe, there is typically a deposition of melanin confined mainly to the caruncles, but sometimes also present in the intercaruncular endometrium. This pigmentation ranges from light shades of gray to black. It is found in juvenile, puberal, and parous animals, but disappears during pregnancy. Such pigmentation is very rare in the goat.

CERVIX. The **internal uterine orifice** is not very distinct and several plicae circulares are present in the cervical canal. In the ewe (552), the intravaginal part of the cervix lies in a depression in the floor of the vagina. The **external uterine orifice** (*b''*) is thus level with the vaginal floor and is a transverse, oblique, or vertical slit surrounded by the very irregular folds and prominences that make up the intravaginal part of the cervix. Five to six hard, equally irregular plicae circulares project into the cervical canal from all sides, making the canal irregular, variable, and impassable in the closed state. In the goat (553), the plicae circulares are much more regular, concentric folds. They number from five to eight, and are grooved by small longitudinal folds. The most caudal of the plicae circulares projects into the vagina and, as in the ewe, lies in a depression in the floor of that organ.

The **VAGINA** and **VESTIBULE** (556) of the small ruminants are similar to those of the cow. Occasionally, major and minor **vestibular glands** are present in the vestibule of the ewe, but never in the goat. The labia of the **VULVA** are not very prominent, and at the ventral commissure taper to a point which is longer in the ewe. The body of the **clitoris** is short in both species, and the tip of the glans (*5*) projects only slightly from the fossa clitoridis.

For more detail on the genital organs of the goat see the studies by Lyngset (1968).

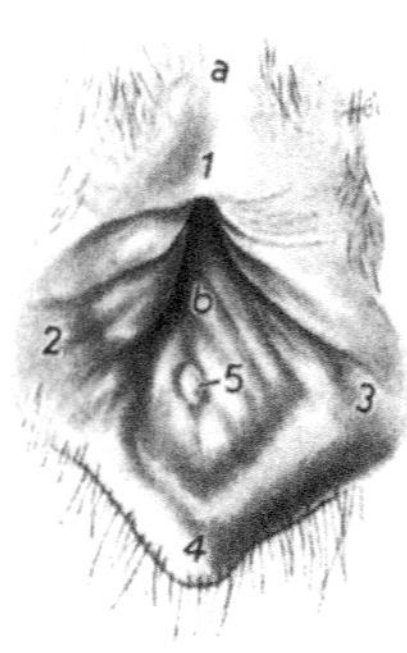

Fig. 556. Vulva of the goat. The labia have been parted to expose the vestibule.

a Cutaneous bridge between vulva and anus; *b* Vestibule

1 Dorsal commissure; *2* Left labium; *3* Right labium; *4* Ventral commissure; *5* Glans clitoridis

Female Genital Organs of the Horse

(16, 518, 522, 525, 529, 557—559)

The horse is uniparous and polyestrous. The sexual cycle varies in length: in about 40 per cent of mares it lasts 21 days, in the rest it ranges from 15—33 days; estrus lasts from 1—3 days. The length of gestation is on the average 336 days, but may range from 310—410 days. Twin births are very rare in the horse. Triplets or even quadruplets have on occasion been reported, but they were either dead at birth or did not survive.

Ovaries

The ovaries of the sexually mature mare are bean-shaped and relatively large, on the average 5—8 cm. long and 2—4 cm. thick (558). Most of their surface is covered with peritoneum, ovarian epithelium is present only in a notchlike depression, the **ovulation fossa** (fossa ovarii, *a'*) on the free border. This unusual, but for the horse typical, restriction of epithelium is not present at birth, but develops during the first 2—3 years of the mare's life. In the foal, the ovary is still oval, and the **zona vasculosa,** containing large numbers of yellowish pigment cells, occupies most of the gland. The **zona parenchymatosa,** which contains the follicles, is a platelike layer of tissue occupying the surface on the free border of the ovary. As the ovary matures, the zona vasculosa is gradually displaced from the center of the ovary and becomes more and more peripheral, while the zona parenchymatosa recedes toward the center. This reduces its exposed area on the surface of the ovary and gradually brings about an invagination on the free border and the formation of the funnel-shaped ovulation fossa. In the mature, 2—3 year old mare, the fossa is surrounded by the zona parenchymatosa, outside of which is the zona vasculosa, covered externally with peritoneum.

The **mesovarium** (559/*11'*) of the mare is about 15 cm. long and suspends the ovary roughly 10 cm. caudal to the kidney, where they are palpable rectally as two firm bodies. (The ovaries and uterine horns shown in Figure 559 are lower than they are in the live animal because of the absence of the intestines.) The **proper ligament of the ovary** (557/*4*) is a strong band of tissue connecting the uterine end of the ovary with the uterine horn. The **mesosalpinx** (522/*b*) has a curved origin from the lateral surface of the mesovarium and is relatively short. Together with the mesovarium, ovary, and proper ligament, it forms the shallow **ovarian bursa** (*x*). This has a large entrance, which is occupied by the ovary. The flat, fimbriated **infundibulum** (*2'*) is attached to the free border and the tubal end of the ovary, and with the abdominal opening of the uterine tube faces the ovulation fossa and the interior of the bursa.

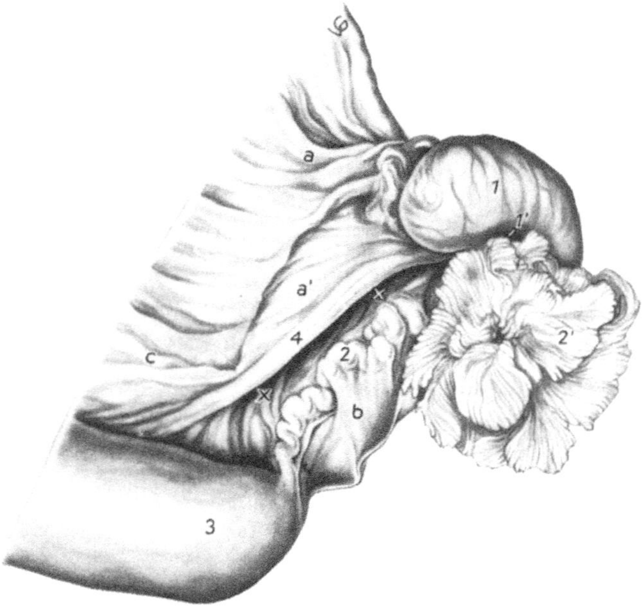

Fig. 557. Left ovary, uterine tube, and associated ligaments of the mare. Medial aspect.

a, a', c Broad ligament; *a* Mesovarium; *b* Mesosalpinx; *c* Mesometrium; *x, x* Entrance to ovarian bursa

1 Ovary, lifted from ovarian bursa; *1'* Ovulation fossa; *2* Uterine tube; *2'* Infundibulum with abdominal opening of uterine tube; *3* Tip of uterine horn; *4* Proper ligament of ovary

Uterine Tube

The uterine tube (558/*b*) is 20—30 cm. long and takes a tortuous course in the mesosalpinx, a short distance from the edge, toward the rounded cranial end of the uterine horn. The uterine opening of the tube is in the center of a small papilla (*b''*) in the wall of which is a small sphincter.

Uterus

The uterus of the mare has a large **body** and relatively short, 22—25 cm., tubular **horns** which are slightly longer than the body (558). They diverge from the cranial end of the uterine body (fundus uteri) and, with a slight ventral convexity, end bluntly immediately caudal to the ovaries. Both the horns and body have a large lumen. The **endometrium** is arranged in high permanent folds, and is yellowish or reddish brown in color. The horns are entirely within the abdominal cavity. The body of the uterus is partly in the abdominal and partly in the pelvic cavity (559). The **broad ligaments** (*11*) are long and arise from the sublumbar area and the lateral pelvic wall, extending from the level of the third to fourth lumbar vertebra to about the level of the fourth sacral vertebra. They contain large amounts of smooth muscle and attach to the mesometrial border of the uterine horns and to the body and cervix of the uterus. The **round ligament of the uterus** (*11''*) arises from the lateral surface of the broad ligament and extends from near the tip of the uterine horn to the deep inguinal ring, forming a round appendix at its cranial end. The nongravid uterus lies dorsal to, and intermingles with, the coils of jejunum and small colon, and is related also to the base of the cecum and to parts of the great colon, particularly the pelvic flexure.

The **CERVIX** (*d*) is firm and therefore easily identified on rectal palpation. The cervical canal is lined with pale mucous membrane, which forms many high longitudinal folds. It is straight and allows passage of suitable instruments when it is closed. The internal uterine orifice (*d'*) is slightly funnel-shaped, and the external orifice is in the center of the well-developed plicated intravaginal part of the cervix (*d''*).

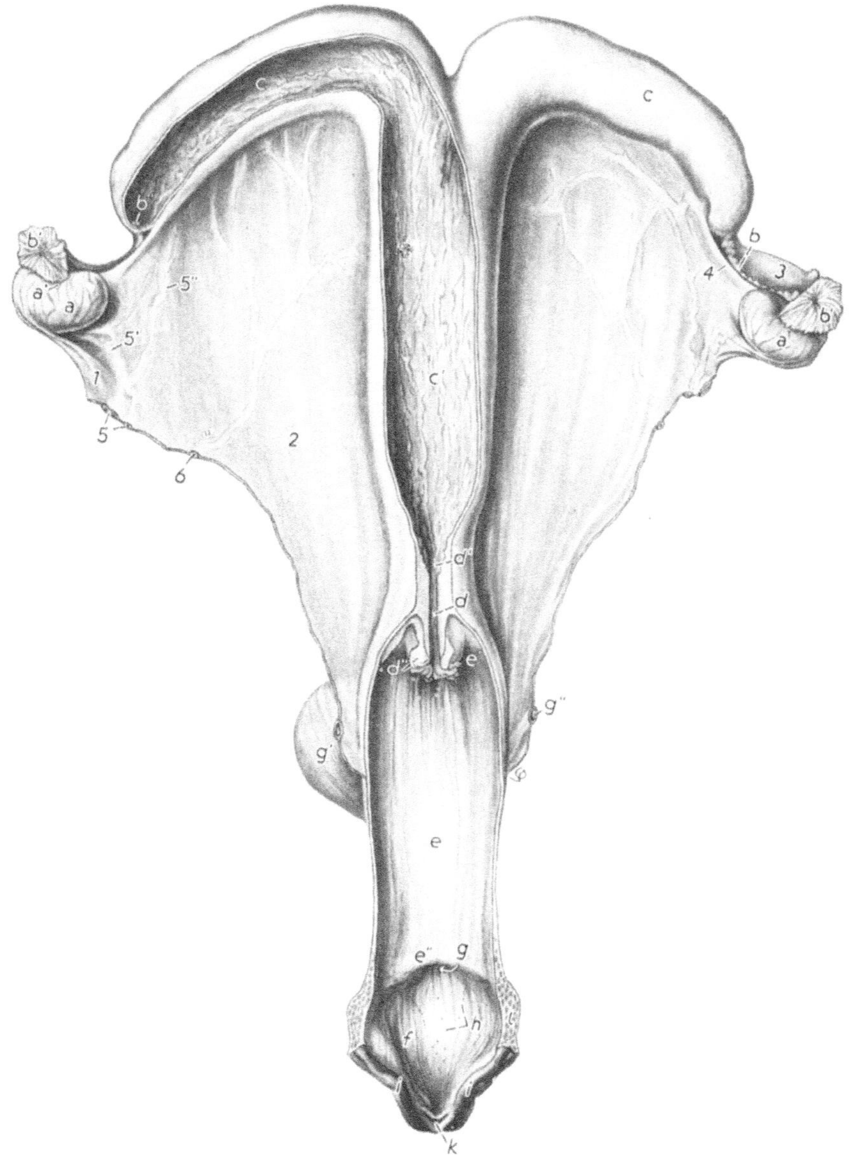

Fig. 558. Genital organs of the mare. Dorsal aspect. Partly opened dorsally.

a Ovary; *a'* Ovulation fossa; *b* Right uterine tube in mesosalpinx; *b'* Infundibulum with abdominal opening of uterine tube; *b''* Uterine opening of uterine tube on small papilla; *c* Uterine horn, on the left opened to expose folded mucous membrane; *c'* Body of uterus; *d* Cervix; *d'* Int. uterine orifice; *d''* Ext. uterine orifice in center of intravaginal part of cervix; *e* Vagina; *e'* Fornix vaginae; *e''* Transverse fold just cranial to ext. urethral orifice; *f* Vestibule; *g* Ext. urethral orifice; *g'* Bladder; *g''* Ureter; *h* Openings of minor vestibular glands; *i* Labia; *k* Fossa and glans clitoridis; *l* Constrictor vestibuli and constrictor vulvae

1, 2 Broad ligament, dorsomedial surface; *1* Mesovarium; *2* Mesometrium; *3* Mesosalpinx; *4* Proper lig. of ovary (between *3* and *4*: ovarian bursa); *5, 5'* Ovarian artery and vein; *5''* Uterine branch of ovarian artery; *6* Uterine artery

Vagina, Vestibule, and Vulva

The **VAGINA** of the mare is a long distensible tube, which projects cranially from the perineum into the peritoneal part of the pelvic cavity (559). Its cranial part, therefore, is covered with peritoneum and helps form the rectogenital and vesicogenital pouches (16/*c*, *5*, *2*). It is related dorsally to the rectum and ventrally to the bladder and urethra. Its lumen is usually collapsed dorsoventrally and surrounds, with a deep annular **fornix vaginae,** the intravaginal part of the cervix (525/*d'*). As in the cow, the fornix may be perforated, either dorsally or ventrally, to gain access to the peritoneal cavity, usually for the purpose of removing the ovaries.

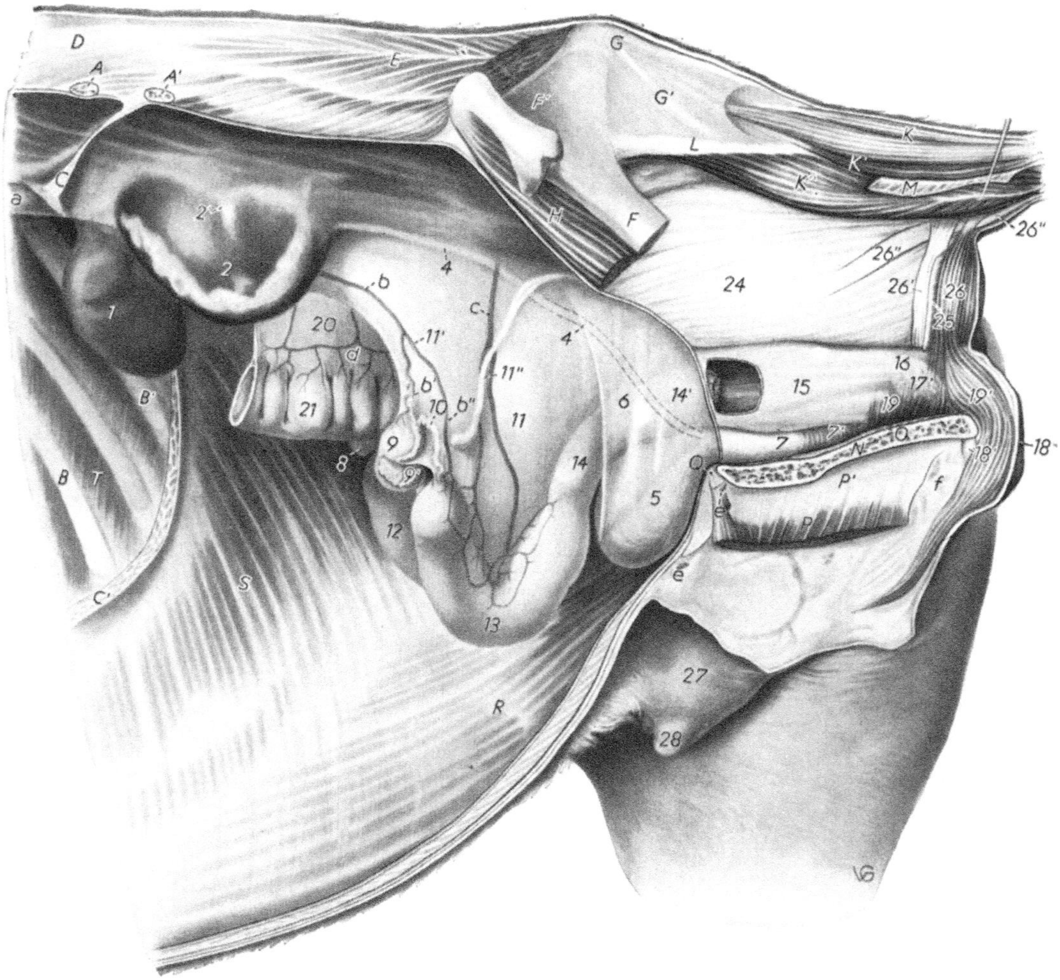

Fig. 559. Urinary and genital organs of a mare in situ. Left lateral aspect.

A, A' Left seventeenth and eighteenth ribs; *B, B'* Right fourteenth and sixteenth ribs; *C* Stump of left crus of diaphragm; *C'* Costal part of diaphragm; *D* Longissimus; *E* Lumbar part of gluteus medius; *F, F'* Ilium; *G, G'* Dorsal sacroiliac ligaments; *H* Iliacus; *K* Sacrocaudalis dorsalis; *K'* Intertransversarius; *K''* Sacrocaudalis ventralis; *L* Lateral part of sacrum; *M* Coccygeus; *N* Symphysis pelvis; *O* Obturator internus; *P* Gracilis; *P'* Symphysial tendon; *Q* Cranial pubic ligament; *R, S* Rectus abdominis and transversus abdominis, covered by transverse fascia and peritoneum; *T* Int. intercostal muscles

1, 2 Right and left kidneys; *2'* Hilus of left kidney; *4* Ureter; *5* Bladder; *6* Left lateral lig. of bladder; *7* Urethra; *7'* Urethralis; *8, 9* Right and left ovaries; *9'* Infundibulum of uterine tube; *10* Left uterine tube; *11* Mesometrium; *11'* Mesovarium; *11''* Round lig. of uterus; *12, 13* Right and left uterine horns; *14* Body of uterus; *14'* Cervix; *15* Vagina, fenestrated to expose intravaginal part of cervix; *16* Vestibule; *17'* Vestibular bulb; *18* Vulva; *18'* Clitoris; *19* Constrictor vestibuli; *19'* Constrictor vulvae; *20* Mesocolon; *21* Small colon; *24* Rectum; *25* Caudal stump of levator ani; *26* Ext. anal sphincter; *26'* Rectal part of retractor clitoridis; *26''* Rectococcygeus; *27* Mammary gland; *28* Left teat

a Aorta; *b, b'* Ovarian artery; *b''* Uterine branch of ovarian artery; *c* Uterine artery; *d* Branches of caudal mesenteric artery to small colon; *e* Branches of ext. pudendal vein, at symphysis pubis anastomosis between right and left ext. pudendal veins; *f* Veins in the region of the clitoris, anastomosis between int. pudendal and obturator veins

The vaginal mucous membrane is covered with stratified squamous epithelium, is arranged in longitudinal folds, and immediately cranial to the wide external urethral orifice forms an often prominent transverse fold (558/*e''*) which is homologous to the hymen.

The **VESTIBULE** continues the vagina caudally and hangs with the vulva over the ischiatic arch (525). Its mucous membrane is rusty brown and forms longitudinal folds. Submucosal venous plexuses can be seen through it. Only the **minor vestibular glands** are present. They can be detected by their craterlike small openings arranged in two ventral and two dorsolateral rows (558/*h*). The **vestibular bulbs** (559/*17'*) in the lateral wall of the vestibule are large.

The labia of the **VULVA** and the ventral commissure are rounded (529). The labial skin, usually pigmented, in rare cases mottled, is covered with a few soft hairs, and contains numerous sweat and sebaceous glands. The pigmentation also includes the parts of the labia facing the vulvar cleft. The **clitoris** (525/*m, n*) of the mare is well developed. It arises by two thick crura from the ischium and has a cavernous body 6—8 cm. long, which passes ventrally under the floor of the vestibule. The large glans occupies a deep fossa clitoridis and is usually visible in the ventral end of the vulvar cleft. It is covered with wrinkled, pigmented skin, and on its surface is a depression (fossa glandis, *p*) which is up to 1 cm. deep.

BIBLIOGRAPHY

Female Genital Organs

Abusineina, M. E.: Effect of parity and pregnancy on the dimensions and weight of the cervix uteri of cattle Brit. Vet. J. **125,** 1969a.

Abusineina, M. E.: Effect of pregnancy on the dimensions and weight of the cervix uteri of sheep. Brit. Vet. J. **125,** 1969b.

Afanašev, L., Ja. Uzulenš: Morphologie des Uterus und der Eierstöcke in verschiedenen Phasen des Sexualzyklus bei braunen lettischen Kühen. Trudy Latvijskoj sel'schochoz-jajstvenne akademii, Elgava **30,** 130—138 (1970).

Afiefy, M. M., W. Abul-Fadle, K. Zaki: The oviducts of the Egyptian cow in health and disease. Zbl. Vet. Med. B **20,** 256—264 (1973).

Akins, E. L., and M. C. Morrissette: Gross ovarian changes during estrous cycle of swine. Am. J. Vet. Res. **29,** 1968.

Ammann, K.: Histologie des Schweine-Eierstockes unter besonderer Berücksichtigung des Ovarialzyklus. Zürich, Diss., 1936.

Andersen, A. C. and M. E. Simpson: The ovary of the dog (beagle). Los Altos, California, Geron-X, Inc. 1973.

Andresen, A.: Die Placentome der Wiederkäuer. Gegenbaurs Morph. Jb. **57,** 1927.

Barone, R.: L'ovaire de la jument. Rev. Méd. Vét. **106,** 1955.

Barone, R.: La topographie des viscères abdominaux, chez la jument gravide. Rev. Méd. Vét. **113,** 1962.

Barone, R., et H. Massot: Le muscle utérin, chez la jument. Rev. Méd. Vét. **107,** 1956.

Barone, R., Cl. Pavaux, P. Frapart: Les vaisseaux sanguins de l'appareil génital chez la truie. Bull. Soc. Sci. Vét. Méd. comp. Lyon **64,** 337—346 (1962).

Bassett, E. G.: Observations on the retractor clitoridis and retractor penis muscles of mammals, with special reference to the ewe. J. Anat. **95,** 1961.

Bassett, E. G.: The anatomy of the pelvic and perineal region in the ewe. Aust. J. Zool. **13,** 1965.

Becher, H.: Über ein standortmäßiges, reaktives Wachstum im Bindegewebe des Ovariums. Ein Beitrag zur kausalen Genese und Histomechanik des Follikelepithels und der Theca interna. Zschr. wiss. Mikrosk. mikrosk. Techn. **51,** 1934.

Beck, W.: Anatomische und histologische Untersuchungen des Eierstockes und Eileiters der Ziege. Berlin, Diss. med. vet., 1912.

Bede, St.: Die weiblichen Geschlechtsorgane der Hauskatze (Felis domestica Briss.). German Summary. Budapest, Diss. med. vet., 1935.

Bergin, W. C., and W. D. Shipley: Observations concerning the ovulation fossa. Vet. Med. **63,** 1968.

Bönninghausen, H. v.: Die Bänder des weiblichen Geschlechtsapparates des Hundes. Mit einer vergleichenden Betrachtung der Eierstocksbänder von Pferd, Rind, Schwein und Hund. Hannover, Diss. med. vet., 1936.

Borell, H.: Untersuchungen über die Bildung des Corpus luteum und der Follikelatresie bei Tieren mit Hilfe der vitalen Färbungen. Beitr. path. Anat. **65,** 1919.

Brendecke, W.: Beiträge zur Kenntnis der Morphologie des Pferdeovars mit besonderer Berücksichtigung der Ovulationsgrube. Hannover, Diss. med. vet., 1926.

Bruyn-Ouboter, E. de: Über die Strukturverhältnisse des juvenilen und gravid gewesenen Uterus der Karnivoren, Canis familiaris und Felis domestica und von Lepus cunniculus mit spezieller Berücksichtigung der bleibenden, für den Nachweis einer bereits vorhanden gewesenen Trächtigkeit wichtigen anatomischen Merkmale. Bern, Diss. med. vet., 1911.

Corner, G. W.: Cyclic changes in the ovaries and uterus of the sow, and their relation to the mechanism of implantation. Contrib. Embryol. **13,** 1921.

Cowan, F. T., J. W. MacPherson: The reproductive tract of the porcine female. (A biometrical study.) Canad. J. Comp. Med. Vet. Sci. **30,** 107—108 (1966).

Davydenko, V. M.: Besonderheiten des Baues der Cervix uteri bei Askanischen und Zigajaschafen. Ref. Ž. Moskva, Životnovodsto i Veter. 1971.

Dellmann, H. D., and R. W. Carithers: Glands in the cervix uteri of the domestic goat (*Capra hircus* L.). Am. J. Vet. Res. **29,** 1968.

Desjardins, C., and H. D. Hafs: Maturation of bovine female genitalia from birth through puberty. J. Animal Sci. **28,** 1969.

Dobrowolski, W., E. S. E. Hafez: Ovariouterine vasculature in sheep. Amer. J. Vet. Res. **31,** 2121—2126 (1970).

Dohm, H.: Anatomische Unterschiede an den Geschlechtsorganen von Kalb und Kuh. Leipzig, Diss. med. vet., 1936.

Edgar, D. G., and S. A. Asdell: The valve-like action of the uterotubal junction of the ewe. J. Endocrin. **21,** 1960.

Edwards. M. J.: Observations on the anatomy of the reproductive organs of cows. With special reference to those features sought during examination per rectum. New Zealand Vet. J. **13,** 1965.

Erickson, B. H.: Development and senescence of the postnatal bovine ovary. J. Ani. Sci. **25,** 1966.

Flössner, O.: Untersuchungen über Ovarien. Zschr. Biol. **86,** 1927.

Forbes, J. M.: The physical relationships of the abdominal organs in the pregnant ewe. J. Agric. Sci. **70,** 1968.

Freni, S. C., and J. H. G. Wilson: Do bovine caruncles contain muscle cells? Vet. Res. **77,** 1965.

Fricke, E.: Topographische Anatomie der Beckenorgane bei Haussäugetieren (Pferd, Rind, Schaf, Ziege, Schwein, Hund, Katze). Diss. med. vet. Berlin, 1968.

Friemann, F. Kl.: Zur klinisch-anatomischen Unterscheidung juveniler und gravid gewesener Schweineuteri. Jber. Vet. med. **66,** 1940.

Gadev, Christa: Untersuchungen über die Pigmente in der Gebärmutter und der Plazenta beim Schaf. Zbl. Vet. Med. A **18,** 521—529 (1971).

Gapp, R.: Die Muskelschichtung und der Muskelfaserverlauf im Pferdeuterus. Wien. Tierärztl. Mschr. **25,** 1938.

Geiger, G.: Die anatomischen Grundlagen des „Hymenalringes" beim Rinde. Tierärztl. Umsch. **9,** 1954.

Geiger, G.: Die anatomische Struktur des Beckenausganges der kleinen Wiederkäuer. Anat. Anz. **103,** 1956.

Giménez, R. L.: Die Arteria ovarica im Eierstock mit Corpus luteum graviditatis beim Rind. Gac. vet., Buenos Aires **33,** 293—302 (1971).

Ginther, O. J., M. C. Garcia, E. L. Squireys, W. P. Steffenhagen: Anatomy of vasculature of uterus and ovaries in the mare. Amer. J. Vet. Res., Chicago, **33,** 1561—1568 (1972).

Ginther, O. J., C. H. del Campo, C. A. Rawlings: Vascular anatomy of the uterus and ovaries and the unilateral luteolytic effect of the uterus: a local venoarterial pathway between uterus and ovaries in sheep. Amer. J. Vet. Res., Chicago **34,** 723—728 (1973).

Grau, H., und P. Walter: Zu Feinbau und Schleimsekretion der Cervix uteri der Wiederkäuer. Berl. Münch. Tierärztl. Wschr. **71,** 1958.

Greenhoff, G. R. and R. M. Kenney: Evaluation of reproductive status ot nonpregant mares. (Rectal palpation of uterus and ovaries). J. Am. Vet. Med. Assoc. **167,** 1975.

Habel, R. E.: The perineum of the mare. Cornell Vet. **43,** 1953.

Habel, R. E.: The topographic anatomy of the muscles, nerves, and arteries of the bovine female perineum. Am. J. Anat. **119,** 1966.

Hadek, R., and R. Getty: The changing morphology in the uterus of the growing pig. Am. J. Vet. Res. **20,** 1959.

Hadek, R., and R. Getty: Age change studies of the ovray of the domesticated pig. Am. J. Vet. Res. **20,** 1959.

Haefried, O.: Der Eierstock der Stute in den verschiedenen Altersstadien. Leipzig, Diss. med. vet., 1923.

Hafez, E. S. E.: Mammalian oviduct: International symposium. Science **158,** 1967.

Hafez, E. S. E.: Functional anatomy of the Cervix uteri in domestic animals and primates. Intern. Kongr. tier. Fortpflanz. Haustierbesam., München, Kongr. Ber. **7,** 2303—2307 (1972).

Hafez, E. S. E., and R. J. Blandau, eds.: The mammalian oviduct. Comparative biology and methodology. Chicago, University of Chicago Press, 1969.

Hancock, J. L.: The clinical features of the reproductive organs of pregnant and nonpregnant cattle. Vet. Rec. **74,** 1962.

Hartmann, W.: The pelvic outlet in female goats. A morphological and functional study. Proefschrift, Rijksuniversiteit Utrecht, 1973.

Helm, F.-Chr.: Über das Corpus luteum des Hundes. Anat. Anz. **109,** 1960/61.

Henneberg, B.: Anatomie und Entwicklung der äußeren Genitalorgane des Schweines und vergleichend-anatomische Bemerkungen. Erster Teil: weibliches Schwein. Zschr. ges. Anat. **63,** 1922.

Henneberg, B.: Beitrag zur ontogenetischen Entwicklung des Skrotums und der Labia maiora. Zschr. Anat. Entw. gesch. **81,** 1926.

Hett, J.: Morphologische und experimentelle Untersuchungen am Eierstock. Handb. biol. Arb. method., 1928.

Höfliger, H.: Die Follikelatresie im Ovar des Kalbes. Festschr. Bürgi, Zürich, 1943.

Höfliger, H.: Das Ovar des Rindes in den verschiedenen Lebensperioden unter besonderer Berücksichtigung seiner funktionellen Feinstruktur. Acta Anat., 1948, Suppl. 5.
Hook, S. J., and E. S. E. Hafez: A comparative anatomical study of the mammalian uterotubal junction. J. Morph. **125**, 1968.
Hölscher, F. C. A.: Anatomische und histologische Untersuchungen der Uterusschleimhaut des Rindes in ihren Phasen. Berlin, Diss. med. vet., 1921.

Kanagawa, H., E. S. E. Hafez: Morphology of Cervix uteri of Rodentia, Carnivora, and Artiodactyla. Acta anat. **84**, 118—128 (1973).
Keller, L.: Das Bindegewebsgerüst des Eierstockes und seine funktionelle Bedeutung (1. Der Säugereierstock). Gegenbaurs Morph. Jb. **88**, 1943.
Kieschke, S.: Anatomische und histologische Untersuchungen über die Cervix uteri von Bos taurus. Leipzig, Diss. med. vet., 1919.
Kind, H. S.: Über die Pigmentation der Gebärmutter-Schleimhaut beim Schaf. Zürich, Diss. med. vet., 1943.
Klinge, A.: Zum zyklischen Verhalten vornehmlich der Höhe des Endometriums beim Rind. Zbl. Vet. med. **6**, 1959.
Koch, F.: Vergleichende anatomische und histologische Untersuchungen über den Bau der Vulva und Clitoris der Haustiere. Bern, Diss. med. vet., 1909.
Koch, T.: Über das Ovarium des Hundes. Zschr. Anat. Entw. gesch. **108**, 1938.
Kolewe, H.: Anatomische und histologische Untersuchungen der Gebärmutter und des weiblichen Begattungsorganes der Ziege. Berlin, Diss. med. vet., 1913.
Krafft, H.: Histologische Untersuchungen über die Involution des normalen Uterus des Rindes mit besonderer Berücksichtigung des elastischen Gewebes. Leipzig, Diss., 1923.
Küpfer, M.: Beiträge zur Morphologie der weiblichen Geschlechtsorgane bei den Säugetieren. Vj.schr. Naturforsch. Ges. Zürich **68**, 1923.
Kvačadze, I. S.: Zur Frage der Innervation der Geschlechtsorgane der Kuh. Ref. Ž. Moskva, Živótnovodstvo i Vet. 17—18 (1971).

Lanz, A.: Beiträge zur Kenntnis über die Entwicklung des Epoophorons und Paroophorons bei Schweine- Hunde- und Katzenföten. Berlin, Diss. med. vet., 1938.
Lyngset, O.: Studies on reproduction in the goat. I. The normal genital organs of the nonpregnant goat. Acta Vet. Scand. **9**, 1968.
Lyngset, O.: Studies on reproduction in the goat. II. The genital organs of the pregnant goat. Acta, Vet. Scand. **9**, 1968.

Merkt, H.: Die Bursa ovarica der Katze. Mit einer vergleichenden Betrachtung der Bursa ovarica des Hundes, Schweines, Rindes und Pferdes sowie des Menschen. Hannover, Diss. med. vet., 1948.
Müller, F.: Schwangerschaftsveränderungen am Uterus des Rindes. Leipzig, Diss., 1933.

Nagler, M.: Untersuchungen über Struktur und Funktion des Schweineuterus. München, Diss. med. vet., 1956.
Nitschke, Th.: Diaphragma pelvis, Clitoris und Vestibulum vaginae der Hündin. 1. Teil. Das Diaphragma pelvis. 2. Teil. Clitoris und Vestibulum vaginae. Anat. Anz. **127**, 76—125 (1970).

Osborne, V. E.: An analysis of the pattern of ovulation at it occurs in the annual reproductive cycle of the mare in Australia. Aust. Vet. J. **42**, 1966.
Otto, A.: Beiträge zur Anatomie der Cervix uteri des trächtigen Rindes. Hannover, Diss. med. vet., 1930.

Petry, G.: Das elastisch-muskulöse System der Plica lata uteri und seine Bedeutung für den Lymphabfluß des Uterus. Gegenbaurs Morph. Jb. **87**, 1942.
Pineda, M. H., R. A. Kainer and L. C. Faulkner: Dorsal median postcervical fold in the canine vagina. Am. J. Vet. Res. **34**, 1973.
Pivko, J., P. Majerčiak, D. Smidt: Histomorphologische Untersuchungen der Eileiterstruktur bei Jungsauen verschiedenen Alters. Pol'nohospodarstvo, Bratislava **18**, 392—399 (1972).
Popp, E.: Beitrag zur Kenntnis des juvenilen und des gravid gewesenen Uterus des Pferdes. Leipzig, Diss. med. vet., 1940 (Abstr. Jber. Vet. med. **68**, 1941.)
Porthan, L.: Morphologische Untersuchungen über die Cervix uteri des Rindes mit besonderer Berücksichtigung der Querfaltenbildung und des Kanalverlaufs. Leipzig, Diss., 1928.
Preuss, F.: Beschreibung und Einteilung des Rinderuterus nach funktionellen Gesichtspunkten. Anat. Anz. **100**, 1953.
Preuss, F.: Untersuchungen zu einer funktionellen Betrachtung des Myometriums vom Rind. Gegenbaurs Morph. Jb. **93**, 1953.
Preuss, F.: Geschlechtszyklus und Brunst beim Hunde. Schweizer Hundesport **12**, 1959.

Rautmann, H.: Zur Anatomie und Morphologie der Glandula vestibularis maior (Bartholini) bei den Säugetieren. Arch. mikrosk. Anat. Entw. gesch. **63**, 1903.
Richter, R.: Beitrag zur Kenntnis des juvenilen und des gravid gewesenen Uterus des Schweines. Leipzig, Diss. med. vet., 1936 (Abstr. Jber. Vet. med. **61**, 1937).
Rigby, J. P.: The structure of the uterotubal junction of the sow. J. Anat. **99**, 1965, 416.

Schulz, L. Cl., u. E. Grunert: Kritische Betrachtungen unserer Kenntnisse über den uterinen und ovariellen Geschlechtszyklus des Rindes. Zbl. Vet. med. **11**, 1964.

Seiferle, E.: Über Art- und Altersmerkmale der weiblichen Geschlechtsorgane unserer Haussäugetiere· Pferd, Rind, Kalb, Schaf, Ziege, Kaninchen, Meerschweinchen, Schwein, Hund und Katze. Zschr. ges· Anat. **101**, 1933.
Seiferle, E.: Die sog. interstitiellen Zellen des Eierstockes und ihre Beziehungen zu Stroma und Ovarialzyklus, im besonderen beim Schwein. Zschr. Zellforsch. mikros. Anat. **25**, 1936.
Seiferle, E.: Ovarialstroma und Ovarialzyklus. Schweiz. Arch. Tierheilk. **80**, 1938.
Seiferle, E.: Bauplan und Arbeitsweise des Säuger-Eierstockes. Dtsch. Tierärztl. Wschr. **50**, 201—205 (1942).
Shehata, R.: Mesonephric remnants in the female genital system of the domestic cat. Acta Anat. **87**, 1974.
Simić, V., H. Gadev: Röntgenanatomische Untersuchungen über die Arterien und deren funktionelle Rolle bei der Vaskularisation des Ovars, Eileiters und der kranialen Abschnitte der Cornua uteri bei einigen Hausequiden (Equus caballus, Equus asinus, Equus mulus und Equus hinnus). Acta Vet., Beograd **18**, 101—118 (1968).
Simić, V., H. Gadev: Anatomisch-radiologische Untersuchungen über die Arterien und ihre funktionelle Rolle in der Eierstock-, Eileiter- und Cornua uteri-Vaskularisation bei einigen Hausequiden (Equus caballus, Equus asinus, Equus mulus und Equus hinnus). Rev. Méd. Vét. N. S. **32** (**120**), 45—61 (1969).
Sokowlowski, J. H., R. G. Zimbleman, and L. S. Goyings: Canine reproduction: reproductive organs and related structures of the nonparous, parous, and postpartum bitch. Am. J. Vet. Res. **34**, 1973.
Stegmann, F.: Messungen und Wägungen am Uterus des Schweines. Berlin, Diss., 1923.

Tornow, U.: Zur „Septumfrage" und Nomenklatur des Uterus. Anat. Anz. **106**, 1959.

Überschär, S.: Zur makroskopischen und mikroskopischen Altersbestimmung am Corpus luteum des Rindes. Hannover, Diss. med. vet., 1961.

Vollmerhaus, B.: Untersuchungen über die normalen zyklischen Veränderungen der Uterusschleimhaut des Rindes. Zbl. Vet. Med. **4**, 1957.
Vollmerhaus, B.: Die Arteria und Vena ovarica des Hausrindes als Beispiel einer funktionellen Koppelung viszeraler Gefäße. Anat. Anz. **112**, Erg. H. 258—264 (1963).
Vollmerhaus, B.: Gefäßarchitektonische Untersuchungen am Geschlechtsapparat des weiblichen Hausrindes (Bos primigenius f. taurus, L., 1758). Teil I. Zbl. Vet. Med., A **11**, 538—596 (1964).
Vollmerhaus, B.: Gefäßarchitektonische Untersuchungen am Geschlechtsapparat des weiblichen Hausrindes (Bos primigenius f. taurus, L., 1758). Teil II. Zbl. Vet. Med., A **11**, 597—646 (1964).

Wachsmuth, U.: Längenveränderungen der Uterindrüsen des Rindes auf Grund graphischer Rekonstruktionen. Zbl. Vet. Med. A **15**, 185—203 (1968).
Wall, G. v. d.: Bursa ovarica des Schafes, der Ziege und des Kaninchens. Hannover, Diss. med. vet., 1951.
Wetli, W.: Die Entwicklung des Schweine-Eierstockes von der Geburt bis zur Geschlechtsreife. Zürich, Diss. med. vet., 1942.

Yamauchi, S.: A histological study on ovaries of aged cows. Jap. J. Vet. Sci. **25**, 1963.
Yamauchi, S.: A histological study on oviducts and uteri of aged cows. Jap. J. Vet. Sci. **26**, 1964.
Yamauchi, S., F. Sasaki: Studies on the vascular supply to the uterus of the cow. I. Morphological studies of arteries in a broad ligament. Bull. Univ. Osaka Prefect., Ser. B **20**, 30—47 (1968).
Yamauchi, S., F. Sasaki: Studies on the vascular supply of the uterus of a cow. II. Morphological investigation of arteries on the uterine wall with special reference to those in the caruncular region. Jap. J. Vet. Sci. **30**, 207—217 (1968).
Yamauchi, S., F. Sasaki: Studies on the vascular supply to the uterus of the cow. III. Morphological studies of veins in the broad ligament. Jap. J. Vet. Sci. **31**, 9—22 (1969).
Yamauchi, S., F. Sasaki: Studies on the vascular supply of the uterus of a cow. IV. Morphology of veins in the uterine wall, especially in the caruncular region. Jap. J. Vet. Sci. **31**, 253—264 (1969).

Zaki, S., M. A. Hadi, S. L. Manjrekar: A study on certain biometrical norms of porcine uteri. Indian Vet. J. **44**, 500—504 (1967).
Zietzschmann, O.: Über Funktionen des weiblichen Genitale bei Säugetier und Mensch. Vergleichendes über die cyclischen Prozesse der Brunst und Menstruation. Arch. Gynäk. **115**, 1921.
Zietzschmann, O.: Über den Processus vaginalis der Hündin. Deutsche Tierärztl. Wschr. **36**, 1928, Festschr.
Zinnbauer, M.: Untersuchungen über den Hymen bei Pferd, Rind, Ziege, Schwein und Hund. Wien, Diss., 1927.

Index

Items in this index are listed only under the nouns. Numerals refer to page numbers. Page numbers in italics refer to illustrations and may be followed, in parentheses, by the number of the illustration. Letters preceding the page numbers refer to the various animals: D (Dog), C (Cat), Ca (Carnivores), P (Pig), O (Ox), S (Sheep), G (Goat), Ru (Ruminants), H (Horse).

Zentralblatt für Veterinärmedizin, Series C

Anatomia, Histologia, Embryologia

Journal of the World Association of Veterinary Anatomists

Edited by: Prof. Dr. Robert Barone (Lyon, France), Prof. Julian J. Baumel, Ph. D. (Omaha, Nebraska, USA), Prof. Dr. James Breazile (Stillwater, Oklahoma, USA), Prof. Dr. Horst Dieter Dellmann (Ames, Iowa, USA), Prof. Carl Gans (Ann Arbor, Michigan, USA), Prof. Dr. Ekkehard Kleiss (Mérida, Venezuela), Prof. Dr. Bernd Vollmerhaus (Munich, Germany)
Editors-in-Chief: Prof. Dr. James Breazile (Stillwater, Oklahoma, USA), Prof. Dr. Bernd Vollmerhaus (Munich, Germany)
Scientific Advisory Board: Prof. Dr. H. A. Bern (Berkeley, California, USA), Prof. Dr. Nils H. Björkman (Copenhagen, Denmark), Prof. Dr. Gunnar D. Bloom (Umeå, Sweden), Univ.-Prof. Dr. György Fehér (Budapest, Hungary), Prof. Dr. Giovanni Godina (Torino, Italy), Prof. Dr. H. Kobayashi (Misaki, Kanagawa-Ken, Japan), Prof. Dr. Willy Mosimann (Bern, Switzerland), Prof. Narciso L. Murillo-Ferrol (Zaragoza, Spain), Prof. R. O'Rahilly (Davis, California, USA), Prof. Dr. Fritz Preuß (Berlin, Germany), M.A., Oh.D.B.V.Sc., M.R.C.V.S. Janis Priedkalns (Adelaide, Australia), Prof. Dr. Dr. Oskar Schaller (Vienna, Austria), Dr. Brian Weatherhead (Birmingham, Great Britain), Dr. Mikio Yasuda (Nagoya, Japan)
Mode of Publication: Four issues per volume. Each issue consists of about 100 pages and appears quarterly
Publication languages: German, English, French or Spanish. Summaries in German, English, French and Spanish

---

Zentralblatt für Veterinärmedizin

Edited by: Prof. Dr. Dr. h. c. Martin Lerche (Berlin, Germany), Prof. Dr. Dr. h. c. Anton Mayr (Munich, Germany), Prof. Dr. Dr. h. c. Heinrich Spörri (Zurich, Switzerland), Prof. Dr. E. G. White (Merseyside, Great Britain). In collaboration with a great number of leading and international authorities
Mode of publication: Ten issues per volume. Each issue consists of about 90 pages
Publication languages: German, English, French or Spanish. Summaries in German, English, French and Spanish

Series A

Physiology, Endocrinology, Biochemistry, Pharmacology, Internal Medicine, Surgery, Genetics, Animal Breeding, Obstetrics, Gynaecology, Andrology, Animal Nutrition and Feeding, General and Special Pathology (except Infectious and Parasitic Diseases)

Series B

Infectious and Parasitic Diseases, Microbiology (Bacteriology, Virology, Mycology), Immunology, Parasitology, Animal Hygiene, Food Hygiene, Pathology of Infectious and Parasitic Diseases

---

A supplement series "Fortschritte der Veterinärmedizin – Advances in Veterinary Medicine" is being published in irregular sequence

---

Published by Spitalerstraße 12, D-2000 Hamburg 1, Germany

Prices to be inquired from the publisher

VERLAG PAUL PAREY · BERLIN AND HAMBURG · GERMANY

GPSR Compliance

The European Union's (EU) General Product Safety Regulation (GPSR) is a set of rules that requires consumer products to be safe and our obligations to ensure this.

If you have any concerns about our products, you can contact us on ProductSafety@springernature.com

In case Publisher is established outside the EU, the EU authorized representative is:

Springer Nature Customer Service Center GmbH
Europaplatz 3
69115 Heidelberg, Germany

Batch number: 10755813

Printed by Printforce, the Netherlands